Computational Methods in Organometallic Catalysis

Computational Methods in Organometallic Catalysis

From Elementary Reactions to Mechanisms

Yu Lan

WILEY-VCH

Author

Prof. Yu Lan
Zhengzhou University
Green Catalysis Center, and College of Chemistry
450001 Zhengzhou
China

Cover Image: © Vikks/Shutterstock

All books published by **Wiley-VCH** are carefully produced. Nevertheless, authors, editors, and publisher do not warrant the information contained in these books, including this book, to be free of errors. Readers are advised to keep in mind that statements, data, illustrations, procedural details or other items may inadvertently be inaccurate.

Library of Congress Card No.:
applied for

British Library Cataloguing-in-Publication Data
A catalogue record for this book is available from the British Library.

Bibliographic information published by the Deutsche Nationalbibliothek
The Deutsche Nationalbibliothek lists this publication in the Deutsche Nationalbibliografie; detailed bibliographic data are available on the Internet at <http://dnb.d-nb.de>.

© 2021 WILEY-VCH GmbH, Boschstr. 12, 69469 Weinheim, Germany

All rights reserved (including those of translation into other languages). No part of this book may be reproduced in any form – by photoprinting, microfilm, or any other means – nor transmitted or translated into a machine language without written permission from the publishers. Registered names, trademarks, etc. used in this book, even when not specifically marked as such, are not to be considered unprotected by law.

Print ISBN: 978-3-527-34601-1
ePDF ISBN: 978-3-527-34605-9
ePub ISBN: 978-3-527-34603-5
oBook ISBN: 978-3-527-34602-8

Typesetting   SPi Global, Chennai, India
Printing and Binding   CPI Group (UK) Ltd, Croydon, CR0 4YY

Printed on acid-free paper

C092557_250321

Contents

Foreword *xv*
Preface *xvii*

Part I Theoretical View of Organometallic Catalysis *1*

1   Introduction of Computational Organometallic Chemistry *3*
1.1   Overview of Organometallic Chemistry *3*
1.1.1   General View of Organometallic Chemistry *3*
1.1.2   A Brief History of Organometallic Chemistry *6*
1.2   Using Computational Tool to Study the Organometallic Chemistry Mechanism *8*
1.2.1   Mechanism of Transition Metal Catalysis *8*
1.2.2   Mechanistic Study of Transition Metal Catalysis by Theoretical Methods *10*
        References *13*

2   Computational Methods in Organometallic Chemistry *19*
2.1   Introduction of Computational Methods *19*
2.1.1   The History of Quantum Chemistry Computational Methods *19*
2.1.2   Post-HF Methods *21*
2.2   Density Functional Theory (DFT) Methods *23*
2.2.1   Overview of Density Functional Theory Methods *23*
2.2.2   Jacob's Ladder of Density Functionals *25*
2.2.3   The Second Rung in "Jacob's Ladder" of Density Functionals *25*
2.2.4   The Third Rung in "Jacob's Ladder" of Density Functionals *26*
2.2.5   The Fourth Rung in "Jacob's Ladder" of Density Functionals *26*
2.2.6   The Fifth Rung in "Jacob's Ladder" of Density Functionals *26*
2.2.7   Correction of Dispersion Interaction in Organic Systems *27*
2.3   Basis Set and Its Application in Mechanism Studies *29*
2.3.1   General View of Basis Set *29*
2.3.2   Pople's Basis Sets *30*
2.3.3   Polarization Functions *31*

2.3.4	Diffuse Functions	*31*
2.3.5	Correlation-Consistent Basis Sets	*31*
2.3.6	Pseudo Potential Basis Sets	*32*
2.4	Solvent Effect	*33*
2.5	How to Choose a Method in Computational Organometallic Chemistry	*34*
2.5.1	Why DFT Method Is Chosen	*34*
2.5.2	How to Choose a Density Functional	*34*
2.5.3	How to Choose a Basis Set	*36*
2.6	Revealing a Mechanism for An Organometallic Reaction by Theoretical Calculations	*37*
2.7	Overview of Popular Computational Programs	*37*
2.8	The Limitation of Current Computational Methods	*40*
2.8.1	The Accuracy of DFT Methods	*40*
2.8.2	Exact Solvation Effect	*41*
2.8.3	Evaluation of Entropy Effect	*41*
2.8.4	The Computation of Excited State and High Spin State	*41*
2.8.5	Speculation on the Reaction Mechanism	*41*
	References	*42*
3	**Elementary Reactions in Organometallic Chemistry**	*51*
3.1	General View of Elementary Reactions in Organometallic Chemistry	*51*
3.2	Coordination and Dissociation	*52*
3.2.1	Coordination Bond and Coordination	*52*
3.2.2	Dissociation	*55*
3.2.3	Ligand Exchange	*57*
3.3	Oxidative Addition	*59*
3.3.1	Concerted Oxidative Addition	*60*
3.3.2	Substitution-type Oxidative Addition	*62*
3.3.3	Radical-type Addition	*67*
3.3.4	Oxidative Cyclization	*68*
3.4	Reductive Elimination	*70*
3.4.1	Concerted Reductive Elimination	*71*
3.4.2	Substitution-type Reductive Elimination	*73*
3.4.3	Radical-Substitution-type Reductive Elimination	*74*
3.4.4	Bimetallic Reductive Elimination	*75*
3.4.5	Eliminative Reduction	*78*
3.5	Insertion	*78*
3.5.1	1,2-Insertion	*79*
3.5.2	1,1-Insertion	*80*
3.5.3	Conjugative Insertion	*83*
3.5.4	Outer-Sphere Insertion	*84*
3.6	Elimination	*86*
3.6.1	β-Elimination	*86*
3.6.2	α-Elimination	*90*

3.7	Transmetallation	*92*
3.7.1	Concerted Ring-Type Transmetallation	*92*
3.7.2	Transmetallation Through Electrophilic Substitution	*98*
3.7.3	Stepwise Transmetallation	*99*
3.8	Metathesis	*100*
3.8.1	σ-Bond Metathesis	*100*
3.8.2	Olefin Metathesis	*102*
3.8.3	Alkyne Metathesis	*106*
	References	*106*

Part II On the Mechanism of Transition-metal-assisted Reactions *125*

4	**Theoretical Study of Ni-Catalysis**	***127***
4.1	Ni-Mediated C—H Bond Activation	*128*
4.1.1	Ni-Mediated Arene C—H Activation	*128*
4.1.2	Ni-Mediated Aldehyde C—H Activation	*132*
4.2	Ni-Mediated C—Halogen Bond Cleavage	*133*
4.2.1	Concerted Oxidative Addition of C—Halogen Bond	*133*
4.2.2	Radical-Type Substitution of C—Halogen Bond	*135*
4.2.3	C—Halogen Bond Cleavage by β-Halide Elimination	*137*
4.2.4	Nucleophilic Substitution of C—Halogen Bond	*139*
4.3	Ni-Mediated C—O Bond Activation	*140*
4.3.1	Ether C—O Bond Activation	*140*
4.3.2	Ester C—O Bond Activation	*142*
4.4	Ni-Mediated C—N Bond Cleavage	*148*
4.5	Ni-Mediated C—C Bond Cleavage	*151*
4.5.1	C—C Single Bond Activation	*151*
4.5.2	C=C Double Bond Activation	*152*
4.6	Ni-Mediated Unsaturated Bond Activation	*153*
4.6.1	Oxidative Cyclization with Unsaturated Bonds	*153*
4.6.2	Electrophilic Addition of Unsaturated Bonds	*156*
4.6.3	Unsaturated Compounds Insertion	*157*
4.6.4	Nucleophilic Addition of Unsaturated Bonds	*159*
4.7	Ni-Mediated Cyclization	*160*
4.7.1	Ni-Mediated Cycloadditions	*161*
4.7.2	Ni-Mediated Ring Substitutions	*163*
4.7.3	Ni-Mediated Ring Extensions	*166*
	References	*168*
5	**Theoretical Study of Pd-Catalysis**	***181***
5.1	Pd-Catalyzed Cross-coupling Reactions	*182*
5.1.1	Suzuki–Miyaura Coupling	*183*
5.1.2	Negishi Coupling	*186*

5.1.3	Stille Coupling 189
5.1.4	Hiyama Coupling 190
5.1.5	Heck–Mizoroki Reaction 192
5.2	Pd-Mediated C—Hetero Bond Formation 196
5.2.1	C—B Bond Formation 196
5.2.2	C—S Bond Formation 197
5.2.3	C—I Bond Formation 200
5.2.4	C—Si Bond Formation 201
5.3	Pd-Mediated C—H Activation Reactions 204
5.3.1	Chelation-Free C(sp^3)—H Activation 206
5.3.2	Chelation-Free C(sp^2)—H Activation 208
5.3.3	Coordinative Chelation-Assisted ortho- C(aryl)—H Activation 210
5.3.4	Covalent Chelation-Assisted ortho- C(aryl)—H Activation 212
5.3.5	Chelation-Assisted meta- C(aryl)—H Activation 214
5.3.6	Coordinative Chelation-Assisted C(sp^3)—H Activation 216
5.3.7	Covalent Chelation-Assisted C(sp^3)—H Activation 216
5.3.8	C—H Bond Activation Through Electrophilic Deprotonation 219
5.3.9	C—H Bond Activation Through σ-Complex-Assisted Metathesis 221
5.3.10	C—H Bond Activation Through Oxidative Addition 223
5.4	Pd-Mediated Activation of Unsaturated Molecules 224
5.4.1	Alkene Activation 225
5.4.2	Alkyne Activation 225
5.4.3	Enyne Activation 226
5.4.4	Imine Activation 229
5.4.5	CO Activation 229
5.4.6	Isocyanide Activation 231
5.4.7	Carbene Activation 231
5.5	Allylic Pd Complex 234
5.5.1	Formation from Allylic Oxidative Addition 235
5.5.2	Formation from Allylic Nucleophilic Substitution 236
5.5.3	Formation from the Nucleophilic Attack onto Allene 237
5.5.4	Formation from Allylic C—H Activation 237
5.5.5	Formation from Allene Insertion 238
	References 239
6	**Theoretical Study of Pt-Catalysis** 257
6.1	Mechanism of Pt-Catalyzed C—H Activation 258
6.1.1	Oxidative Addition of C—H Bond 259
6.1.2	Electrophilic Dehydrogenation 259
6.1.3	Carbene Insertion into C—H Bonds 261
6.2	Mechanism of Pt-Catalyzed Alkyne Activation 264
6.2.1	Nucleophilic Additions 264
6.2.2	Cyclopropanation 266
6.2.3	Oxidative Cycloaddition 268
6.3	Mechanism of Pt-Catalyzed Alkene Activation 270

Contents | ix

6.3.1	Hydroamination of Alkenes 270
6.3.2	Hydroformylation of Alkenes 272
6.3.3	Isomerization of Cyclopropenes 273
	References 274

7	**Theoretical Study of Co-Catalysis** 289
7.1	Co-Mediated C—H Bond Activation 289
7.1.1	Hydroarylation of Alkenes 291
7.1.2	Hydroarylation of Allenes 293
7.1.3	Hydroarylation of Alkynes 294
7.1.4	Hydroarylation of Nitrenoid 296
7.1.5	Oxidative C—H Alkoxylation 297
7.2	Co-Mediated Cycloadditions 297
7.2.1	Co-Mediated Pauson–Khand Reaction 298
7.2.2	Co-Catalyzed [4+2] Cyclizations 299
7.2.3	Co-Catalyzed [2+2+2] Cyclizations 299
7.2.4	Co-Catalyzed [2+2] Cyclizations 300
7.3	Co-Catalyzed Hydrogenation 301
7.3.1	Hydrogenation of Carbon Dioxide 301
7.3.2	Hydrogenation of Alkenes 304
7.3.3	Hydrogenation of Alkynes 306
7.4	Co-Catalyzed Hydroformylation 307
7.4.1	Direct Hydroformylation by H_2 and CO 308
7.4.2	Transfer Hydroformylation 309
7.5	Co-Mediated Carbene Activation 310
7.5.1	Arylation of Carbene 310
7.5.2	Carboxylation of Carbene 312
7.6	Co-Mediated Nitrene Activation 313
7.6.1	Aziridination of Olefins 314
7.6.2	Amination of Isonitriles 314
7.6.3	Amination of C—H Bonds 315
	References 317

8	**Theoretical Study of Rh-Catalysis** 329
8.1	Rh-Mediated C—H Activation Reactions 330
8.1.1	Rh-Catalyzed Arylation of C—H Bond 330
8.1.2	Rh-Catalyzed Alkylation of C—H Bond 332
8.1.3	Rh-Catalyzed Alkenylation of C—H Bond 335
8.1.4	Rh-Catalyzed Amination of C—H Bond 338
8.1.5	Rh-Catalyzed Halogenation of C—H Bond 340
8.2	Rh-Catalyzed C—C Bond Activations and Transformations 341
8.2.1	Strain-driven Oxidative Addition 341
8.2.2	The Carbon—Cyano Bond Activation 342
8.2.3	β-Carbon Elimination 343
8.3	Rh-Mediated C—Hetero Bond Activations 343

8.3.1	C—N Bond Activation 344
8.3.2	C—O Bond Activation 345
8.4	Rh-Catalyzed Alkene Functionalizations 345
8.4.1	Hydrogenation of Alkene 346
8.4.2	Diboration of Alkene 347
8.5	Rh-Catalyzed Alkyne Functionalizations 349
8.5.1	Hydroacylation of Alkynes 349
8.5.2	Hydroamination of Alkynes 349
8.5.3	Hydrothiolation of Alkynes 351
8.5.4	Hydroacetoxylation of Alkynes 351
8.6	Rh-Catalyzed Addition Reactions of Carbonyl Compounds 352
8.6.1	Hydrogenation of Ketones 352
8.6.2	Hydrogenation of Carbon Dioxide 353
8.6.3	Hydroacylation of Ketones 353
8.7	Rh-Catalyzed Carbene Transformations 354
8.7.1	Carbene Insertion into C—H Bonds 354
8.7.2	Arylation of Carbenes 357
8.7.3	Cyclopropanation of Carbenes 358
8.7.4	Cyclopropenation of Carbenes 359
8.8	Rh-Catalyzed Nitrene Transformations 359
8.8.1	Nitrene Insertion into C—H Bonds 360
8.8.2	Aziridination of Nitrenes 361
8.9	Rh-Catalyzed Cycloadditions 362
8.9.1	(3+2) Cycloadditions 365
8.9.2	Pauson–Khand-type (2+2+1) Cycloadditions 367
8.9.3	(5+2) Cycloadditions 367
8.9.4	(5+2+1) Cycloadditions 369
	References 369
9	**Theoretical Study of Ir-Catalysis** *387*
9.1	Ir-Catalyzed Hydrogenations 387
9.1.1	Hydrogenation of Alkenes 388
9.1.2	Hydrogenation of Carbonyl Compounds 391
9.1.3	Hydrogenation of Imines 393
9.1.4	Hydrogenation of Quinolines 396
9.2	Ir-Catalyzed Hydrofunctionalizations 397
9.2.1	Ir-Catalyzed Hydroaminations 397
9.2.2	Ir-Catalyzed Hydroarylations 397
9.2.3	Ir-Catalyzed Hydrosilylations 399
9.3	Ir-Catalyzed Borylations 401
9.3.1	Borylation of Alkanes 402
9.3.2	Borylation of Arenes 403
9.4	Ir-Catalyzed Aminations 405
9.4.1	Amination of Alcohols 405
9.4.2	Amination of Arenes 407

9.5	Ir-Catalyzed C—C Bond Coupling Reactions	*407*
	References *409*	
10	**Theoretical Study of Fe-Catalysis** *419*	
10.1	Fe-Mediated Oxidations *420*	
10.1.1	Alkane Oxidations *420*	
10.1.2	Arene Oxidations *422*	
10.1.3	Alkene Oxidations *423*	
10.1.4	Oxidative Catechol Ring Cleavage *425*	
10.2	Fe-Mediated Hydrogenations *426*	
10.2.1	Hydrogenation of Alkenes *427*	
10.2.2	Hydrogenation of Carbonyls *429*	
10.2.3	Hydrogenation of Imines *430*	
10.2.4	Hydrogenation of Carbon Dioxide *430*	
10.3	Fe-Mediated Hydrofunctionalizations *431*	
10.3.1	Hydrosilylation of Ketones *431*	
10.3.2	Hydroamination of Allenes *432*	
10.4	Fe-Mediated Dehydrogenations *434*	
10.4.1	Dehydrogenation of Alcohols *434*	
10.4.2	Dehydrogenation of Formaldehyde *435*	
10.4.3	Dehydrogenation of Formic Acid *435*	
10.4.4	Dehydrogenation of Ammonia-Borane *437*	
10.5	Fe-Catalyzed Coupling Reactions *438*	
10.5.1	C—C Cross-Couplings with Aryl Halide *438*	
10.5.2	C—N Cross-Couplings with Aryl Halide *438*	
10.5.3	C—C Cross-Couplings with Alkyl Halide *441*	
10.5.4	Iron-Mediated Oxidative Coupling *441*	
	References *441*	
11	**Theoretical Study of Ru-Catalysis** *451*	
11.1	Ru-Mediated C—H Bond Activation *452*	
11.1.1	Mechanism of the Ru-Mediated C—H Bond Cleavage *452*	
11.1.2	Ru-Catalyzed C—H Bond Arylation *456*	
11.1.3	Ru-Catalyzed *ortho*-Alkylation of Arenes *460*	
11.1.4	Ru-Catalyzed *ortho*-Alkenylation of Arenes *461*	
11.2	Ru-Catalyzed Hydrogenations *464*	
11.2.1	Hydrogenation of Alkenes *464*	
11.2.2	Hydrogenation of Carbonyls *465*	
11.2.3	Hydrogenation of Esters *466*	
11.2.4	Hydrodefluorination of Fluoroarenes *467*	
11.3	Ru-Catalyzed Hydrofunctionalizations *468*	
11.3.1	Hydroacylations *469*	
11.3.2	Hydrocarboxylations *470*	
11.3.3	Hydroborations *471*	
11.4	Ru-Mediated Dehydrogenations *472*	

11.4.1	Dehydrogenation of Alcohols	473
11.4.2	Dehydrogenation of Formaldehyde	473
11.4.3	Dehydrogenation of Formic Acid	474
11.5	Ru-Catalyzed Cycloadditions	475
11.5.1	Ru-Mediated (2+2+2) Cycloadditions	475
11.5.2	Ru-Mediated Pauson–Khand Type (2+2+1) Cycloadditions	478
11.5.3	Ru-Mediated Click Reactions	478
11.6	Ru-Mediated Metathesis	480
11.6.1	Ru-Mediated Intermolecular Olefin Metathesis	482
11.6.2	Ru-Mediated Intramolecular Diene Metathesis	484
11.6.3	Ru-Mediated Alkyne Metathesis	484
	References	485
12	**Theoretical Study of Mn-Catalysis**	**499**
12.1	Mn-Mediated Oxidation of Alkanes	500
12.1.1	C—H Hydroxylations	500
12.1.2	C—H Halogenations	501
12.1.3	C—H Azidations	501
12.1.4	C—H Isocyanations	501
12.2	Mn-Mediated C—H Activations	502
12.2.1	Electrophilic Deprotonation	503
12.2.2	σ-Complex-Assisted Metathesis	504
12.2.3	Concerted Metalation–Deprotonation	505
12.3	Mn-Mediated Hydrogenations	507
12.3.1	Hydrogenation of Carbon Dioxide	507
12.3.2	Hydrogenation of Carbonates	508
12.4	Mn-Mediated Dehydrogenations	510
12.4.1	Dehydrogenation of Alcohols	510
12.4.2	Dehydrogenative Couplings	511
	References	512
13	**Theoretical Study of Cu-Catalysis**	**517**
13.1	Cu-Mediated Ullmann Condensations	518
13.1.1	C—N Bond Couplings	520
13.1.2	C—O Bond Couplings	522
13.1.3	C—F Bond Couplings	522
13.2	Cu-Mediated Trifluoromethylations	524
13.2.1	Trifluoromethylations Through Cross-Coupling	524
13.2.2	Trifluoromethylations Through Oxidative Coupling	524
13.2.3	Radical-Type Trifluoromethylations	525
13.3	Cu-Mediated C—H Activations	527
13.3.1	C—H Arylations	527
13.3.2	C—H Aminations	529
13.3.3	C—H Hydroxylation	531
13.3.4	C—H Etherifications	532

13.4	Cu-Mediated Alkyne Activations	533
13.4.1	Azide–Alkyne Cycloadditions	533
13.4.2	Nucleophilic Attack onto Alkynes	535
13.4.3	Alkynyl Cu Transformations	536
13.5	Cu-Mediated Carbene Transformations	539
13.5.1	[2+1] Cycloadditions with Alkenes	539
13.5.2	Carbene Insertions	541
13.5.3	Rearrangement of Carbenes	542
13.6	Cu-Mediated Nitrene Transformations	542
13.6.1	[2+1] Cycloadditions with Alkenes	543
13.6.2	Amination of Nitrenes	543
13.6.3	Nitrene Insertions	544
13.7	Cu-Catalyzed Hydrofunctionalizations	545
13.7.1	Hydroborylations	547
13.7.2	Hydrosilylation	547
13.7.3	Hydrocarboxylations	548
13.8	Cu-Catalyzed Borylations	549
13.8.1	Borylation of Alkenes	551
13.8.2	Borylation of Alkynes	552
13.8.3	Borylation of Carbonyls	553
	References	554
14	**Theoretical Study of Ag-Catalysis**	**567**
14.1	Ag-Mediated Carbene Complex Transformations	568
14.1.1	Silver–Carbene Formation	568
14.1.2	Carbene Insertion into C—Cl Bond	572
14.1.3	Carbene Insertion into O—H Bond	573
14.1.4	Nucleophilic Attack by Carbonyl Groups	573
14.1.5	Carbene Insertion into C—H Bond	574
14.2	Ag-Mediated Nitrene Transformations	576
14.2.1	Silver–Nitrene Complex Formation	577
14.2.2	Nucleophilic Attack by Unsaturated Bonds	580
14.2.3	Nucleophilic Attack by Amines	582
14.3	Ag-Mediated Silylene Transformations	583
14.4	Ag-Mediated Alkyne Activations	585
14.4.1	π-Activation of Alkynes	586
14.4.2	C—H Activation of Alkynes	587
	References	588
15	**Theoretical Study of Au-Catalysis**	**597**
15.1	Au-Mediated Alkyne Activations	598
15.1.1	Isomerization of Alkynes	599
15.1.2	Nucleophilic Attack by Oxygen-Involved Nucleophiles	599
15.1.3	Nucleophilic Attack by Nitrogen-Involved Nucleophiles	602
15.1.4	Nucleophilic Attack by Arenes	606

15.2	Au-mediated Alkene Activations	*607*
15.2.1	Nucleophilic Addition of Alkenes	*607*
15.2.2	Allylic Substitutions	*608*
15.3	Au-mediated Allene Activations	*609*
15.3.1	Hydroamination of Allenes	*610*
15.3.2	Hydroalkoxylation of Allenes	*610*
15.3.3	Cycloisomerization of Allenyl Ketones	*613*
15.4	Au-mediated Enyne Transformations	*613*
15.4.1	1,5-Enyne Cycloisomerizations	*614*
15.4.2	1,6-Enyne Cycloisomerizations	*615*
15.4.3	Allenyne Cycloisomerizations	*617*
15.4.4	Conjugative Enyne Cycloisomerizations	*617*
	References *618*	

Index *629*

Foreword

Computational chemistry began in the 1940s with the earliest electronic computers and drastic approximations to the Schrödinger equation, such as Hückel molecular orbital theory. Since the 1960s, the possibilities of doing quantum mechanics calculations on large systems containing metals, indeed to study the heart of organometallic chemistry, began with the Mulliken–Wolfsberg–Helmholtz approach and then Roald Hoffmann's Extended Hückel Theory (EHT) in the 1950s and 1960s. While an amazingly useful method, EHT is really only of qualitative value. But actually what we need is a conceptual framework that is useful even today.

The 1960s saw remarkable advances in methods, approximations, and the beginnings of the flowering of computers for chemical calculations. The dawn of the modern hybrid density functional theory in the mid-1990s, borrowing exact exchange from wavefunction theory, made it possible to begin the quantitative calculations of structure, mechanics, and mechanisms, including the incredibly useful organometallic reactions. Both Ru and Mo catalysts for olefin metathesis, and Pd catalysis for cross-coupling reactions, have led to Nobel prizes for their discoverers.

Yu Lan has now written an introduction, a guide, a masterwork about quantum mechanical studies of organometallic reaction mechanisms. He is ideally equipped to write such a book, already doing outstanding work in the field as a student and postdoc, and becoming a leader in the field in his independent career. He has also trained many young experts in the field, and his influence will spread further from their achievements and through this book.

The book includes a historical introduction to organometallic chemistry, a survey of mechanisms, and an extensive introduction to quantum mechanical computational methods, especially density functional theory, as well as programs for quantum chemical calculations.

The description of organometallic structures and mechanisms is peppered with numerous calculations from the Lan group with relatively accurate density functionals. Part 2, the bulk of this book, is organized with a chapter for each of the most important metals used in organometallic chemistry: Ni, Pd, Pt, Co, Rh, Ir, Fe, Ru, Mn, Cu, Ag, and Au. The computational studies of the reactions of complexes of each of these metals are reviewed with great insights into mechanisms using computations.

The book will be a boon to organometallic chemists and computational chemists involved in the study of organometallic reactions. While a number of books on organometallic chemistry mechanisms are available, this is THE book describing the methods for computation and analysis of organometallic reactions using modern quantum mechanical methods.

29 August 2020　　　　　　　　　　　　　　　　　　　　　　　*Kendall N. Houk*

Preface

A long time ago, when I first came into contact with science, I was fascinated by the unique charm of organic chemistry. The tetravalent carbon atoms and their tetrahedral structures impressed me with elegant simplicity. Through the broken and formation of covalent bonds, various new molecules that possess unique properties could be generated. By manipulating the reaction conditions, catalysts, and ligands in organic reactions, chemists can effectively synthesize plenty of complex natural product molecules and pharmaceutical molecules in high regioselectivity and enantioselectivity. When I was a teenager, I took every organic chemical reaction as a puzzle, and the reaction mechanism is like the answer to the puzzle. In this way, I found great pleasure in thinking about the mechanism of organic reactions. The development of organic chemistry also requires a comprehensive mechanistic understanding. Generally, a significant amount of information about reaction mechanisms could be obtained by experimental techniques. However, the experimental mechanistic study mainly focuses on the macroscopically observed experimental phenomena. Therefore, in many cases, pure experimental observation is not sufficient for revealing the complete reaction pathway and clarifying the origin of selectivity. During my doctoral study at Peking University, I was fortunate to work with my supervisor Professor Yun-Dong Wu, from whom I learned how to use computational chemistry to investigate the organic reaction mechanisms. Theoretical calculations based on quantum mechanics, especially the density functional theory calculation, have constituted the most powerful tool for mechanistic study due to the development of supercomputer, computational theory, and corresponding software.

In the past several decades, one of the most important advances in organic chemistry has been the introduction of transition metal catalysts to organic synthesis. Transition metal species can react with organic compounds to generate intermediates that contain carbon–metal bonds. Subsequent conversion of these organometallic intermediates could enrich the synthetic approach toward new molecules. Different from organocatalysis, the organometallic catalysis usually goes through multiple steps and complicated catalytic cycles, which originated from the complex bonding pattern of organometallic catalysts and the variation of valance state of the central metal species. Thus, the utilization of theoretical calculation for understanding the reaction mechanism is imperative for the development of organic chemistry. My post-doctoral research with Professor K. N. Houk was working with experimental chemists to explore the mechanism of organometallic reactions through the collaboration of theoretical calculation with experimental study. The promotion of theoretical study on experimental development could be summarized into "3D," i.e. description, design, and direction. Based on the data obtained from experimental study, detailed descriptions of the organometallic reaction mechanisms could be fulfilled using theoretical calculation. The mechanistic study then provides further theoretical guidance for the rational design of new reactions, which points out the direction of experimental development.

Over recent decades, massive experimental and theoretical investigations on organometallic catalysis have been reported. In those works, theoretical studies have been proved to be an indispensable technique for modern organic chemistry. Consequently, this book is written to summarize and generalize the theoretical advances in the mechanistic study of organometallic catalysis. This book comprises two parts, which are the general overview of organometallic catalysis and the computational studies of reaction mechanisms classified by transition metals. I hope this book could inspire the mechanistic studies of complex reactions for theoretical chemists, and enable a better understanding of reaction mechanisms for experimental chemists.

29 July 2020

Yu Lan
Zhengzhou, P. R. China

Part I

Theoretical View of Organometallic Catalysis

It is time to write a book on computational organometallic chemistry.

The first part of this book can be considered as the introduction to computational organometallic chemistry. It is a long history since organometallic catalysis has been applied in organic synthesis; however, the mechanism of those reactions is too complicated to understand. Indeed, computational chemistry provided a powerful tool to reveal the mechanism of organometallic reactions. During recent two decades, the combination of computational chemistry and organometallic chemistry has made a series of progress in mechanistic studies, which has led to a new discipline, computational organometallic chemistry.

The first part would be composed of three chapters. In Chapter 1, a brief history of organometallics is given to reveal the significance of this chemistry. Computational chemistry, especially computational methods, is discussed in Chapter 2, which would be used in mechanism study of organometallic catalysis. Detailed processes for the familiar elementary reactions in organometallic catalysis discovered by theoretical calculations are summarized in Chapter 3.

1

Introduction of Computational Organometallic Chemistry

This chapter provides a brief introduction of computational organometallic chemistry, which usually focuses on the reaction mechanism of homogeneous organometallic catalysis.

1.1 Overview of Organometallic Chemistry

In this section, the historical footprint of organometallic chemistry is concisely given, which would help the readers better understand the role of computation in the mechanistic study of organometallic chemistry.

1.1.1 General View of Organometallic Chemistry

Creating new material is always entrusted with the important responsibility for the development of human civilization [1–3]. In particular, synthetic chemistry becomes a powerful tool for chemists, as it exhibits great value for the selective construction of new compounds [4–8]. Various useful molecules could be prepared by the strategies of synthetic chemistry, which provides material foundation, technological support, and drive force for science [9–20]. Synthetic chemistry is also the motivating force for the progress of material science, pharmaceutical science, energy engineering, agriculture, and electronics industry [21–41]. In this area, organic synthesis reveals broad interests from a series of research fields, which could target supply to multifarious functional molecules.

The synthetic organic chemistry usually focuses on "carbon" to widen related research, which could afford various strategies for the building of molecular framework, functional group transformations, and controlling stereochemistry in more sophisticated molecules [9, 22, 42–50]. Therefore, selective formation of new covalent bond between carbon atom and some other atom involving nitrogen, oxygen, sulfur, halogen, boron, and phosphorus becomes one of the most important aims for synthetic organic chemistry. In particular, nucleophiles and electrophiles are important for the construction of new covalent bonds.

Computational Methods in Organometallic Catalysis: From Elementary Reactions to Mechanisms,
First Edition. Yu Lan.
© 2021 WILEY-VCH GmbH. Published 2021 by WILEY-VCH GmbH.

1 Introduction of Computational Organometallic Chemistry

A nucleophile, which is a molecule with formal lone-pair electrophiles, can donate two electrons to its reaction partner for the formation of new covalent bond. Alternatively, an electrophile, which is a molecule with formal unoccupied orbitals, can accept two electrons from its partner for the formation of new covalent bond. Thereinto, coupling reactions could be categorized as redox-neutral cross-coupling with an electrophile and a nucleophile, oxidative coupling with two nucleophiles, and reductive coupling with two electrophiles (Scheme 1.1).

Cross-coupling:

Nu: + :E $\xrightarrow{\text{Cat.}}$ Nu–E

Oxidative cross-coupling:

Nu: + :Nu $\xrightarrow[\text{Cat.}]{-2e}$ Nu–Nu

Reductive cross-coupling:

E + :E $\xrightarrow[\text{Cat.}]{+2e}$ E–E

Scheme 1.1 Cross-coupling reactions with nucleophiles and electrophiles.

In organic chemistry, the nucleophile is an electron-rich molecule that contains a lone pair of electrons or a polarized bond, the heterolysis of which also could yield a lone pair of electrons (Scheme 1.2). According to this concept, organometallic compounds, alcohols, halides, amines, and phosphines with a lone pair of electrons are nucleophiles. Some nonpolar π bonds, including olefins and acetylenes, which could donate the π-bonding electrons, are often considered to be nucleophiles. Moreover, the C—H bonds of hydrocarbons can be considered to be nucleophiles because the electronegativity of carbon is higher than that of hydrogen, which could deliver a proton to form a formal carbon anion. Correspondingly, the electrophile is an electron-deficient molecule that contains unoccupied orbitals or

Lone-pair of electrons:

Unsaturated π Bonds:

Organometallic complexes:

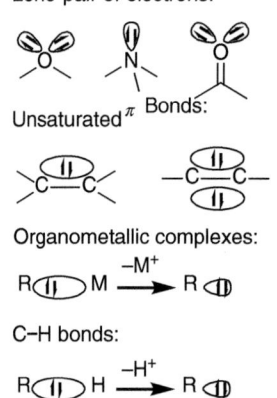

R–M $\xrightarrow{-M^+}$ R:

C–H bonds:

R–H $\xrightarrow{-H^+}$ R:

Scheme 1.2 Some selected examples of nucleophiles.

low-energy antibonding molecular orbital, which could accept the electrons from nucleophiles. In this chemistry, cationic carbons, which usually come from the heterolysis of carbon—halogen bonds, are electrophile. Polar π bonds, including carbonyl compounds and imines, also could be considered to be electrophile, which involve a low-energy π antibond. Interestingly, Fisher-type singlet carbene has an electron pair filling one sp^2 hybrid orbital and an unoccupied p orbital, which could be considered to be either nucleophile or electrophile in coupling reactions.

Superficially, at least, the reaction between nucleophile and electrophile could construct a covalent bond undoubtedly. However, the familiar nucleophiles and electrophiles, used in cross-coupling reactions, are usually inactive, which could not react with each other rapidly. Moreover, when more active nucleophiles and electrophiles are used in coupling reactions, it would become out of control, which would not selectively afford target products. In effect, introducing transition metal catalysis can perfectly solve this problem. The appropriate transition metal can be employed to selectively activate the nucleophiles and electrophiles and stabilize some others, which led to a specially appointed cross-coupling reaction.

High-valence transition metal can obtain electrons from nucleophile, which led to the transformation of nucleophile into electrophile. The newly generated electrophile can couple with other nucleophiles to form covalent bond, which is named oxidative coupling reaction [51–53]. Meanwhile, the reduced transition metal can be oxidized by exogenous oxidant for regeneration. Correspondingly, low-valence transition metal can donate electrons to electrophile leading to the transformation of electrophile into nucleophile, which can react with another electrophile to form covalent bond. Accordingly, it is named reductive coupling reaction. The oxidized transition metal also can be reduced by exogenous reductant.

The d orbital of some transition metals could be filled by unpaired electrons, which led to a unique catalytic activity in radical-involved reactions. The homolytic cleavage of transition metal–carbon (or some other atoms) bond is an efficient way for the generation of a radical species, which can promote further transformations. On the other hand, free radical can react with some transition metal leading to the stabilization of radical, which can cause further radical transformations [54–57]. Moreover, nucleophiles and electrophiles, activated by transition metals, also can react with radical to form new covalent bonds.

Although there is no electron barrier due to the appropriate symmetry of frontier molecular orbitals, a great deal of uncatalyzed pericyclic reactions would occur under harsh reaction conditions, which could be often attributed to the low-energy level of highest occupied molecular orbitals (HOMOs) and high-energy level of lowest unoccupied molecular orbitals (LUMOs) in reacting partners. Transition metals can play as a Lewis acid, which could significantly reduce the LUMO of coordinated organic moiety. Therefore, it has been widely adopted to catalyze pericyclic reactions, which leads to moderate reaction conditions and adjustable selectivity [58–62]. Moreover, the node of d orbital can change the symmetry of a conjugative compound, which involves a transition metal. Therefore, transition metal itself also could participate in a pericyclic reaction to reveal unique catalytic activity.

As an overview of organometallic chemistry, the core is the formation of a metal–carbon bond and its further transformation. Different from organocatalysis, organometallic catalysis process usually goes through multiple steps as well as complicated catalytic cycles, which originated from the complex bonding pattern of metallic catalyst and the variation of valence state for the central metal element. Consequently, improving the reaction efficiency and yield for organometallic catalysis encountered more difficulty than conventional organocatalysis. Moreover, the design of catalysis and ligand for transition metal-catalyzed reaction is still facing both opportunities and challenges. To solve the above-mentioned issues, the understanding of reaction mechanism is imperative, which could give more information for the detailed reaction process, and help to improve the reaction efficiency and yield.

1.1.2 A Brief History of Organometallic Chemistry

As an interdiscipline of organic and inorganic chemistry, organometallic chemistry has a history of almost 200 years since the first complex K[(C$_2$H$_4$)PtCl$_3$]·H$_2$O was reported by Zeise when he heated ethanol solution of PtCl$_4$/KCl [62]. The history of organometallic chemistry can be roughly divided into four stages. The chemists majorly focused on main group organometallic compounds in the nineteenth century. Later in the first half of the twentieth century, chemists paid more attention to understanding the structures of organometallic compounds involving transition metals. Then in the latter half of the twentieth century, various transition metal-catalyzed reactions had been widely reported. Since this century, chemists have been keen on using transition metal catalysis to selectively construct more complex organic compounds. The outlined history of organometallic chemistry could be concluded in Scheme 1.3 [63].

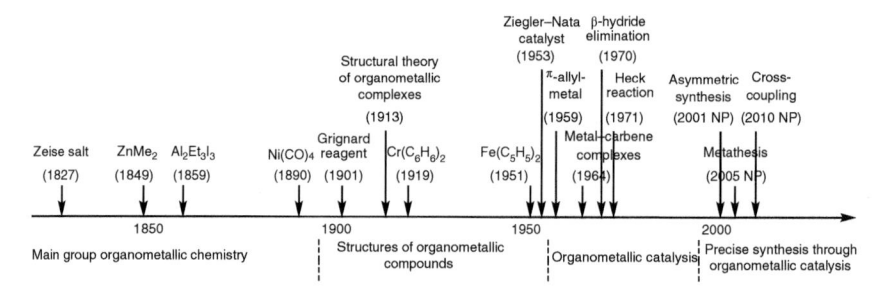

Scheme 1.3 A brief history of organometallic chemistry. Source: Based on Didier [63].

The nineteenth century could be considered as the enlightenment era of organometallic chemistry. Frankland first systemically investigated organometallic chemistry and prepared a series of alkyl metal compounds in 1850s. In the late nineteenth century, ZnMe$_2$ (in 1849 by E. Frankland), Sn(C$_2$H$_5$)$_4$ (in 1859 by E. Frankland), PbEt$_4$ (in 1853 by C. Löwig), Al$_2$Et$_3$I$_3$ (in 1859 by W. Hallwachs and

A. Schafarik), and RMgX (in 1900 by V. Grignard) had been prepared, and the chemical property of those compounds also had been studied [64–68].

In 1890, Ni(CO)$_4$ was found as the first metal carbonyl complex by L. Mond et al. in the study of the corrosion of stainless steel valves by CO [69]. Next year, Fe(CO)$_5$ was also found by the same group [70]. It could be considered as the beginning of the structural study of organometallic complexes. Two years later, Werner proposed structural theory of organometallic complexes involving the tetrahedral, octahedral, square planar, etc. which won him the Nobel prize in chemistry in 1913 [71]. In 1919, Cr(C$_6$H$_6$)$_2$ was prepared by Hein using MgPhBr to react with CrCl$_3$ [72]. However, the sandwich-like structure of this complex was proved by Fischer 36 years later. In 1951, Fe(C$_5$H$_5$)$_2$ had been synthesized by Kealy and Pauson individually [73]. The sandwich-like structure of that complex was confirmed by G. Wilkinson the following year, which aroused chemists' enthusiasm for the study of transition metal organic compounds. In 1964, tungsten carbene complex was reported by Fischer, who shared 1973 Nobel prize in chemistry with G. Wilkinson [74]. By the 1950s, with the appearance of representational methods, involving X-ray crystallography, infrared spectrum, and nuclear magnetic resonance spectrum, means of characterizing transition metal compounds were becoming more and more mature. Therefore, organometallic chemistry became an independent discipline.

From the middle of the twentieth century, organometallic compounds were gradually considered as a catalyst in organic reactions. In 1953, Ziegler and Natta found that TiCl$_4$/AlEt$_3$ could promote atmospheric polymerization of olefins, which helped them share 1963 Nobel prize in chemistry [75, 76]. In 1959, allylic palladium was prepared by Smidt and Hafner, which was the beginning of π-allyl metal chemistry [77]. The same year, Shaw and Ruddick reported an elementary reaction of oxidative addition [78]. In 1974, Wilkinson reported another elementary reaction of β-hydride elimination [79]. Those works led to a series of following mechanistic studies for organometallic reactions. In 1972, Heck and Nolley reported a palladium-catalyzed coupling reaction between aryl halides and olefins, which was named Heck reaction [80]. Meanwhile, a series of palladium-catalyzed cross-coupling reaction, including Kumada coupling with Grignard reagent [81], Suzuki coupling with aryl borane [82], Negishi coupling with organo zinc [83], Stille coupling with aryl tin [84], and Sonogashira coupling with alkynyl copper [85], were reported. Those reactions made transition metal-catalyzed cross-coupling reactions one of the most important ways to construct new C—C covalent bonds in synthetic chemistry. Therefore, R. F. Heck, E. Negishi, and A. Suzuki won the 2010 Nobel prize in chemistry. Also in 1971, W. S. Knowles applied chiral bisphosphine ligands as ligand in rhodium-catalyzed hydrogenation reactions, which had opened up a whole new field of asymmetric catalysis with transition metals [86]. W. S. Knowles shared 2001 Nobel prize in chemistry with K. B. Sharples and R. Noyori, who promoted the research upsurge of asymmetric catalysis. Moreover, Chauvin, Grubbs, and Schrock won the 2005 Nobel prize in recognition of their outstanding contributions in transition metal-mediated metathesis of olefins.

Based on the advances of methodology study and ligand design, transition metal catalysis has become one of the important means for synthetic chemists to construct more complex new substances in this century. The current pursuit is to selectively construct multiple covalent bonds in one reaction synchronously by transition metal catalysis. To achieve this goal, transition metal catalyst has been employed to selectively activate some inert covalent bonds. The most famous example – transition metal-mediated C—H bond activation – became the focus of chemists. This process could afford a carbon–metal bond directly, which could be used as a powerful nucleophile in further transformations. In modern organometallic chemistry, multistep elementary reactions in series have been extensively studied, which could afford a battery of new covalent bonds through one catalytic cycle. Synthetic efficiency in organometallic chemistry has become the focus of attention. Hereon, transition metal catalysis with higher turnover numbers was pursued to further improve the economy and environmental protection. Current research on transition metal catalysis is also devoted to improving the accuracy of synthesis, aiming at achieving specific functional group transformation in the exact location. To achieve these goals, the design of transition metal catalysis becomes more complex, and the requirements for suitable ligands are higher. It is necessary to design the corresponding ligands manually according to aspects of structure, electronic properties, steric effect, and coordination ability. These auxiliary designs also make the catalytic cycle with transition metal lengthier; meanwhile, the possibility of side reactions increases. Therefore, mechanistic studies for transition metal catalysis became more and more important, which were helpful for design of new catalysis, enhanced efficiency, increased selectivity, improved turnover number, and accurate synthesis.

1.2 Using Computational Tool to Study the Organometallic Chemistry Mechanism

Transition metal catalysis is one of the most powerful tools for the construction of new organic materials, whose development trend is more efficient as well as more complex. Therefore, studying the mechanism of organometallic catalysis has become even more essential, and has proved to be the basis for the design of new ligands, catalysts, and reactions.

1.2.1 Mechanism of Transition Metal Catalysis

Generally, reaction mechanism could be considered to be all elementary reactions used to describe a chemical change passing in a reaction. It is to decompose a complex reaction into several elementary reactions and then combine them according to certain rules, so as to expound the internal relations of complex reactions and the internal relations between total reactions and elementary reactions. The rate of chemical reaction is closely related to the specific pathways through which the reaction takes place.

To study the law of chemical reaction rate and find out the intrinsic causes of various chemical reaction rates, synthetic chemists must explore the reaction mechanism and find out the key to determine the reaction rate, so as to control the chemical reaction rate more effectively. As shown in Scheme 1.4, traditional research methods for reaction mechanism include: (i) determining the important intermediate or decisive step of a reaction by isotope tracing, (ii) determining the effect of different factors (e.g. reaction temperature, solvent, substituent effect, etc.) on reaction rate and selectivity by competitive test, (iii) studying the relationship between the reaction rate and the concentration of reactants and catalysts obtaining by kinetic experiments, and (iv) characterizing and tracking intermediates by instrumental analysis. However, these methods are often macroscopic observation of the average state of many molecules, which cannot watch a process of the transformation for one molecule from a micro-perspective. Fortunately, theoretical calculations based on first principles have become one of the important means to study the reaction mechanism with the development of software and the improvement of hardware computing capability in recent several decades. Through theoretical calculation and simulation, the transformation of one molecule in reaction process can be "watched" more clearly from the microscopic point of view. Actually, theoretical calculation can be considered to be a special kind of microscope, which can see the geometrical structure, electronic structure, spectrum, and dynamic process at atomic level, and is helpful for chemists to understand the real reaction mechanism.

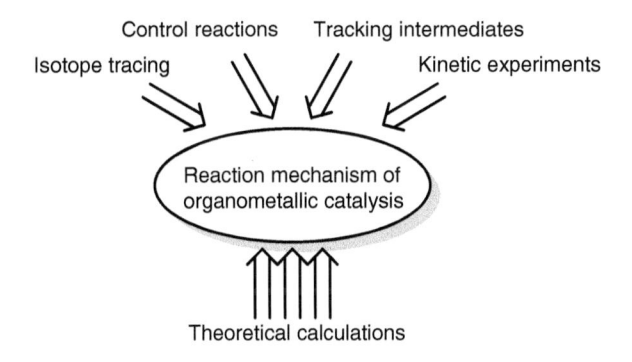

Scheme 1.4 Revealing the reaction mechanism of organometallic catalysis.

The combination of theoretical and experimental techniques could not only greatly improve the efficiency of reaction and yield of product, but also uncover the factors that control the selectivity of product more clearly. The promotion of theoretical study to experimental investigation could be summarized into "3D," i.e. description, design, and direction. Based on the data obtained from experimental technique, detailed description for the mechanism of organometallic catalysis could be fulfilled using theoretical calculations. Based on the results of computations, the mechanisms could be verified by the designed experiment. To put in a nutshell, theoretical calculations could play a critical role in the direction of transition-metal-organic synthesis.

1.2.2 Mechanistic Study of Transition Metal Catalysis by Theoretical Methods

Quantum chemical computation based on first principle provides a powerful tool for the mechanistic study of transition metal catalysis. Since the whole content of this book is to discuss the theoretical calculation-based study for the mechanism of transition metal catalysis, we will give only a few examples to show how to study the reaction mechanism by theoretical calculations.

Generally, mechanism research of transition metal catalysis initially faces a series of studies involving the molecular structure and electronic states. As an example, (Xantphos)Pd(CH$_2$NBn$_2$)$^+$ is an important precursor for aminomethylation reactions, the geometric structure of which has been confirmed by X-ray analysis [87]. However, why this complex could be formed and the electronic properties of this complex still remained unclear. As shown in Scheme 1.5, in resonance structure **1-1**, the Pd—C bond is a normal single bond, and Pd—N is a coordination bond. The formal positive charge is localized on palladium, and the formal oxidation state of palladium is +2. Alternatively, the iminium moiety acts as a monodentate ligand coordinated with Pd(0) in resonance structure **1-2**, and the formal positive charge is mainly localized on the iminium moiety. The real structure of this complex would be a mixture of resonance structures **1-1** and **1-2**. On the other hand, the bond orders of Pd—C, Pd—N, and C—N are determined to be 0.322, 0.135, and 0.965, respectively, which indicate that the Pd—C and Pd—N bonds are very weak. More importantly, these data support that the C—N is a double bond and **1-2** is most likely to be the main structure of this complex. Further frontier molecular orbital studies also supported this point.

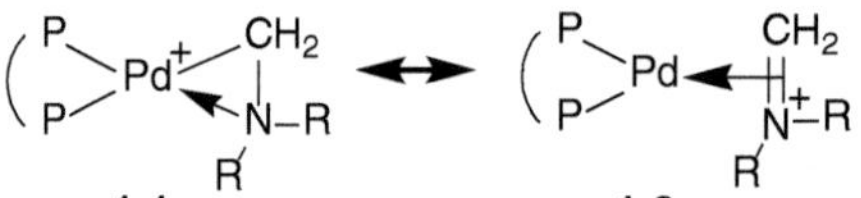

Scheme 1.5 The resonance structures of (Xantphos)Pd(CH$_2$NBn$_2$)$^+$.

To summarize, computational organometallic chemistry focuses on some of the stationary points on potential energy surface for the corresponding reactions, which could be used to compare the possible elementary reactions. The lowest energy reaction pathway could be found by theoretical calculations, which is helpful for chemists to understand the reaction mechanism and design new reactions.

As an example, computational method was used to study the mechanism of rhodium-catalyzed coupling reaction of quinoline *N*-oxide and acetylenes. As shown in Scheme 1.6, four possible pathways were taken into account (i.e. paths A–D) [88]. All these four pathways begin with the coordination of quinoline *N*-oxide to Rh(III) precursor **1-6**, which is followed by an *N*-oxide-directed electrophilic deprotonation by acetate and coordination of acetylene substrate to give rhodacycle **1-7**. The subsequent insertion of the acetylene substrate into the Rh—C bond of intermediate **1-7** gives intermediate **1-8**, which is a common intermediate for

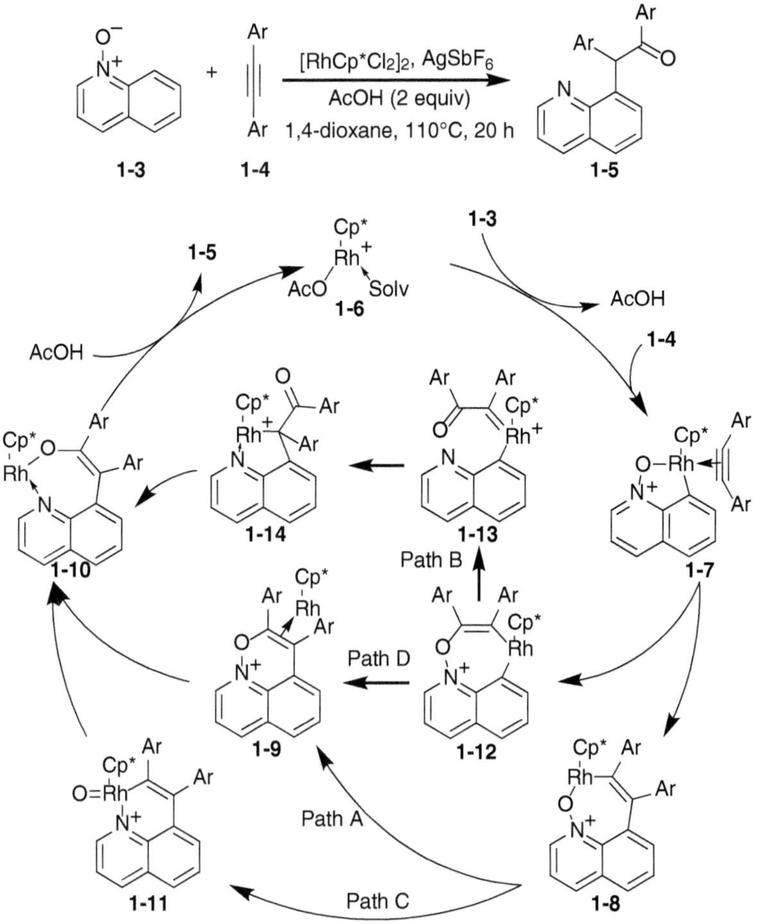

Scheme 1.6 Mechanism of rhodium-catalyzed coupling reaction of quinoline N-oxide and acetylenes. Source: From Li et al. [88].

both paths A and C. In path A, complex **1-8** is proposed to undergo a reductive elimination reaction to form a new C—O bond in π-coordinated complex **1-9**. The N—O bond in complex **1-9** would then oxidatively add back to Rh(I) to generate Rh(III) enolate intermediate **1-10**. Protonolysis of intermediate **1-10** by acetic acid would release one molecule of product **1-5** and regenerate complex **1-6**. Product **1-5** could also be generated by the protonation of intermediate **1-14**, which is an isomer of intermediate **1-10**. In path C, the migration of the oxygen atom from complex **1-8** would lead to the formation of Rh(V)–oxo intermediate **1-11**. A subsequent oxygen insertion into Rh—C(vinyl) bond may give rise to intermediate **1-10**, which is the common intermediate in path A. Alternatively, the insertion of the acetylene substrate into the Rh—O bond of intermediate **1-7** would give rise to a common intermediate **1-12**, which could proceed along path B or path D. For path B, cleavage of the N—O bond would lead to the formation of α-oxo carbenoid species **1-13**, which could undergo carbene insertion to give intermediate **1-14**.

This intermediate is either protonated directly or isomerizes to intermediate **1-10**, followed by protonolysis to afford product **1-5** together with the regeneration of active catalyst **1-6**. Intermediate **1-12** could also undergo reductive elimination through path D to give complex **1-10**. All four of these different pathways have been evaluated in the current study by using theoretical calculations. The computational results revealed that path A is the most favorable process for this reaction.

In fact, quantum chemical calculations have become one of the conventional methods to study the mechanism of transition metal catalysis (Scheme 1.7). The basis function of computational organometallic chemistry is to calculate the geometries and energies of stationary points, including local minimums and transition states, so as to construct potential energy surface for the catalytic cycle and explore the reaction mechanism. Information of molecular orbital, charge distribution, dipole moment, and so on can be achieved by computational population analysis, which is propitious to study factors of reactivity and selectivity. More useful information, including infrared, nuclear magnetic resonance, circular dichroism spectra, dispersion interactions, nucleophilicity/electrophilicity, and aromaticity, also can be achieved by theoretical calculations for organometallic chemistry.

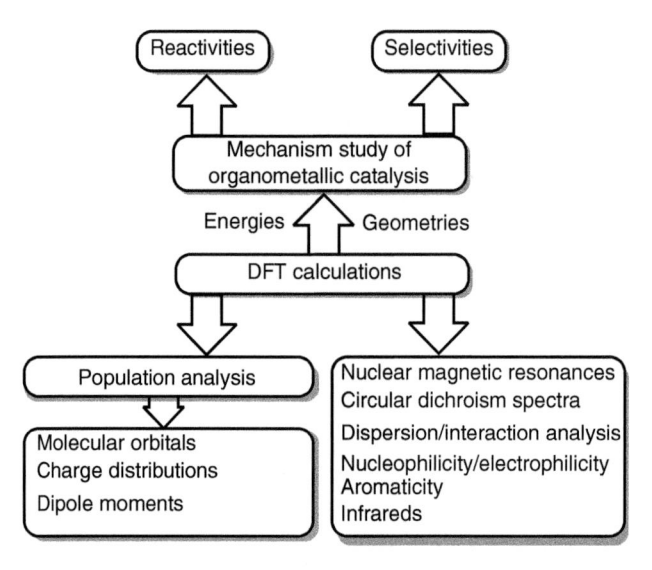

Scheme 1.7 Mechanism study of organometallic catalysis by density functional theory (DFT) calculations.

Over recent several decades, massive experimental and theoretical investigations were reported about organometallic catalysis. In those works, theoretical studies have proved to be indispensable routine technique for modern synthetic organic chemistry. Consequently, we herein provide a book to summarize and generalize the theoretical advances regarding organometallic catalysis. In Chapter 2, we summarize the popular computational methods, which are of benefit for the mechanism study of organometallic catalysis. We discuss the theoretical studies of elementary reactions in Chapter 3. Detailed processes for all the familiar elementary reactions in organometallic catalysis discovered by theoretical calculations are summarized in this chapter. Based on those two chapters, the second part of this book is organized by the type of transition metals, which are used as catalyst in organic reactions.

References

1 Olson, G.B. (2000). Designing a new material world. *Science* 288 (5468): 993–998.
2 Collins, T. (2001). Toward sustainable chemistry. *Science* 291 (5501): 48–49.
3 Feng, S. and Chien, S. (2003). Chemotherapeutic engineering: application and further development of chemical engineering principles for chemotherapy of cancer and other diseases. *Chemical Engineering Science* 58: 4087–4114.
4 Thuronyi, B.W. and Chang, M.C.Y. (2015). Synthetic biology approaches to fluorinated polyketides. *Accounts of Chemical Research* 48: 584–592.
5 Xu, L., Zhang, S., Li, P. et al. (2015). Boron-selective reactions as powerful tools for modular synthesis of diverse complex molecules. *Chemical Society Reviews* 44: 8848–8858.
6 Barboiu, M. and Gilles, A. (2013). From natural to bioassisted and biomimetic artificial water channel systems. *Accounts of Chemical Research* 46: 2814–2823.
7 Patel, M., Saunthwal, R.K., Verma, A.K. et al. (2017). Base-mediated hydroamination of alkynes. *Accounts of Chemical Research* 50: 240–254.
8 Golder, M.R. and Jasti, R. (2015). Syntheses of the smallest carbon nanohoops and the emergence of unique physical phenomena. *Accounts of Chemical Research* 48: 557–566.
9 Kolb, H.C., Finn, M.G., Sharpless, K.B. et al. (2001). Click chemistry: diverse chemical function from a few good reactions. *Angewandte Chemie International Edition* 40: 2004–2021.
10 Mosesa, J.E. and Moorhousea, A.D. (2007). The growing applications of click chemistry. *Chemical Society Reviews* 36: 1249–1262.
11 Devaraj, N.D. and Weissleder, R. (2011). Biomedical applications of tetrazine cycloadditions. *Accounts of Chemical Research* 44: 816–827.
12 Hu, R., Leung, N.L.C., Zhong, B. et al. (2014). AIE macromolecules: syntheses, structures and functionalities. *Chemical Society Reviews* 43: 4494–4562.
13 Luo, Z., Yuan, X., Yu, Y. et al. (2012). From aggregation-induced emission of Au(I)–thiolate complexes to ultrabright Au(0)@Au(I)–thiolate core–shell nanoclusters. *Journal of the American Chemistry Society* 134: 16662–16670.
14 Dreyer, D.R., Park, S., Bielawski, C.W. et al. (2010). The chemistry of graphene oxide. *Chemical Society Reviews* 1: 228–240.
15 Zou, X. and Zhang, Y. (2015). Noble metal-free hydrogen evolution catalysts for water splitting. *Chemical Society Reviews* 44: 5148–5158.
16 Lee, Y.H., Zhang, X.Q., Zhang, W. et al. (2012). Synthesis of large-area MoS_2 atomic layers with chemical vapor deposition. *Advanced Materials* 24: 2320–2325.

17 König, H.M. and Kilbinger, A.F. (2007). Learning from nature: β-sheet-mimicking copolymers get organized. *Angewandte Chemie International Edition* 46: 8334–8340.

18 Biernacki, J.J., Bullard, J.W., Sant, G. et al. (2017). Cements in the 21st century: challenges, perspectives, and opportunities. *Journal of the American Chemistry Society* 100 (7): 2746–2773.

19 Long, N.J. and Williams, C.K. (2003). Metal alkynyl sigma complexes: synthesis and materials. *Angewandte Chemie International Edition* 42: 2586–2617.

20 Palacci, J., Sacanna, S., and Steinberg, A.P. (2013). Living crystals of light-activated colloidal surfers. *Science* 339: 6122.

21 Lutolf, M.P. and Hubbell, J.A. (2005). Synthetic biomaterials as instructive extra-cellular microenvironments for morphogenesis in tissue engineering. *Nature Biotechnology* 23: 47–55.

22 James, S.L. (2003). Metal–organic frameworks. *Chemical Society Reviews* 32: 276–288.

23 Hollister, S.J. (2005). Porous scaffold design for tissue engineering. *Nature Materials* 4: 518–524.

24 Hoyle, C.E. and Bowman, C.N. (2010). Thiol–ene click chemistry. *Angewandte Chemie International Edition* 49: 1540–1573.

25 Tan, C., Cao, X., and Wu, X. (2017). Recent advances in ultrathin two-dimensional nanomaterials. *Chemical Reviews* 117: 6225–6331.

26 Larcher, D. and Tarascon, J.M. (2015). Towards greener and more sustainable batteries for electrical energy storage. *Nature Chemistry* 7: 19–29.

27 Ghidiu, M., Lukatskaya, M.R., Zhao, M.Q. et al. (2014). Conductive two-dimensional titanium carbide 'clay' with high volumetric capacitance. *Nature* 516: 78.

28 Cheng, F., Shen, J., Peng, B. et al. (2011). Rapid room-temperature synthesis of nanocrystalline spinels as oxygen reduction and evolution electrocatalysts. *Nature Chemistry* 3: 79–84.

29 Duncan, R. (2003). The dawning era of polymer therapeutics. *Nature Reviews Drug Discovery* 2: 347–360.

30 Cho, S.H., Kim, J.Y., Kwak, J. et al. (2011). Recent advances in the transition metal-catalyzed twofold oxidative C—H bond activation strategy for C—C and C—N bond formation. *Chemical Society Reviews* 40: 5038–5083.

31 Seregina, I.V. and Gevorgyan, V. (2007). Direct transition metal-catalyzed functionalization of heteroaromatic compounds. *Chemical Society Reviews* 36: 1173–1193.

32 Fischbach, M.A. and Walsh, C.T. (2009). Antibiotics for emerging pathogens. *Science* 325: 1089–1093.

33 Shimizu, M. and Hiyama, T. (2005). Modern synthetic methods for fluorine-substituted target molecules. *Angewandte Chemie International Edition* 44: 214–231.

34 Sun, J.Y., Zhao, X., Illeperuma, W.R.K. et al. (2012). Highly stretchable and tough hydrogels. *Nature* 489: 133–136.

35 Nugent, P., Belmabkhout, Y., Burd, S.D. et al. (2013). Porous materials with optimal adsorption thermodynamics and kinetics for CO_2 separation. *Nature* 495: 80–84.
36 Cortright, R.D., Davda, R.R., Dumesic, J.A. et al. (2002). Hydrogen from catalytic reforming of biomass-derived hydrocarbons in liquid water. *Nature* 418: 964–967.
37 Hu, B., Wang, K., and Wu, L. (2010). Engineering carbon materials from the hydrothermal carbonization process of biomass. *Advanced Materials* 22: 813–828.
38 Hirst, A.R., Escuder, B., Miravet, J.F. et al. (2008). High-tech applications of self-assembling supramolecular nanostructured gel-phase materials: from regenerative medicine to electronic devices. *Angewandte Chemie International Edition* 47: 8002–8018.
39 Li, L., Wu, G., Yang, G. et al. (2013). Focusing on luminescent graphene quantum dots: current status and future perspectives. *Nanoscale* 5: 4015–4039.
40 Katz, H.E., Bao, Z., Gilat, S. et al. (2001). Synthetic chemistry for ultrapure, processable, and high-mobility organic transistor semiconductors. *Accounts of Chemical Research* 34: 359–369.
41 Pron, A. and Rannou, P. (2002). Processable conjugated polymers: from organic semiconductors to organic metals and superconductors. *Progress in Polymer Science* 27: 135–190.
42 Furukawa, H., Cordova, K.E., O'Keeffe, M. et al. (2013). The chemistry and applications of metal–organic frameworks. *Science* 341: 974.
43 Burke, M.D. and Schreiber, S.L. (2004). A planning strategy for diversity-oriented synthesis. *Angewandte Chemie International Edition* 43: 46–58.
44 Jiang, H., Taranekar, P., Reynolds, J.R. et al. (2009). Conjugated polyelectrolytes: synthesis, photophysics, and applications. *Angewandte Chemie International Edition* 48: 4300–4316.
45 Salonen, M.S.L.M., Ellermann, M., Diederich, F. et al. (2011). Aromatic rings in chemical and biological recognition: energetics and structures. *Angewandte Chemie International Edition* 50: 4808–4842.
46 Khodagholy, D., Rivnay, J., Sessolo, M. et al. (2013). High transconductance organic electrochemical transistors. *Nature Communication* 2013 (4): 2133.
47 Yu, J., Shi, F., Gong, L.Z. et al. (2011). Bronsted-acid-catalyzed asymmetric multicomponent reactions for the facile synthesis of highly enantioenriched structurally diverse nitrogenous heterocycles. *Accounts of Chemical Research* 2011 (44): 1156–1171.
48 Moonen, K., Laureyn, I., Stevens, C.V. et al. (2014). Synthetic methods for aza-heterocyclic phosphonates and their biological activity. *Chemical Reviews* 104: 6177–6215.
49 Bartoli, G., Bencivennia, G., Dalpozzob, R. et al. (2010). Organocatalytic strategies for the asymmetric functionalization of indoles. *Chemical Society Reviews* 39: 4449–4465.
50 Qin, Y., Zhu, L., Luo, S. et al. (2017). Organocatalysis in inert C—H bond functionalization. *Chemical Reviews* 107: 9433–9520.

51 Shi, W., Liu, C., Lei, A. et al. (2011). Transition-metal catalyzed oxidative cross-coupling reactions to form C—C bonds involving organometallic reagents as nucleophiles. *Chemical Society Reviews* 2011 (40): 2761.

52 Liu, C., Zhang, H., Shi, W. et al. (2011). Bond formations between two nucleophiles: transition metal catalyzed oxidative cross-coupling reactions. *Chemical Reviews* 2011 (111): 1780.

53 Lei, A., Shi, W., Liu, C. et al. (2016). *Oxidative Cross-Coupling Reactions*. Wiley-VCH.

54 Twilton, J., Le, C., Zhang, P. et al. (2017). The merger of transition metal and photocatalysis. *Nature Reviews Chemistry* 1: 0052.

55 Mayer, J.M. (2011). Understanding hydrogen atom transfer: from bond strengths to Marcus theory. *Accounts of Chemical Research* 44: 36–46.

56 Soleilhavoup, M. and Bertrand, G. (2015). Cyclic (alkyl)(amino)carbenes (CAACs): stable carbenes on the rise. *Accounts of Chemical Research* 48: 256–266.

57 Tellis, J.C., Kelly, C.B., and Primer, D.N. (2016). Single-electron transmetalation via photoredox/nickel dual catalysis: unlocking a new paradigm for sp^3–sp^2 cross-coupling. *Accounts of Chemical Research* 49 (7): 1429–1439.

58 Houk, K.N. (1975). Frontier molecular-orbital theory of cycloaddition reactions. *Accounts of Chemical Research* 8: 361–369.

59 Mcleod, D., Thogersen, M.K., Jessen, N.I. et al. (2019). Expanding the frontiers of higher-order cycloadditions. *Accounts of Chemical Research* 52: 3488–3501.

60 Li, J., Liu, T., Chen, Y. et al. (2012). Aminocatalytic asymmetric Diels–Alder reactions via HOMO activation. *Accounts of Chemical Research* 45 (9): 1491–1500.

61 Afewerki, S. and Cordova, A. (2016). Combinations of aminocatalysts and metal catalysts: a powerful cooperative approach in selective organic synthesis. *Chemical Reviews* 116 (22): 13512–13570.

62 Zeise, W.C. (1831). Von der Wirkung zwischen Platinchlorid und Alkohol, und von den dabei entstehenden neuen Substanzen. *Annual Physical Chemistry (in German)* 97 (4): 497–541.

63 Didier, A. (2007). History of organometallic chemistry. In: *Organometallic Chemistry and Catalysis*, 5–20. Springer-Verlag Berlin Heidelberg.

64 Frankland, E. (1849). Notiz über eine neue Reihe organischer Körper, welche Metalle, Phosphor u. s. w. enthalten. *Liebigs Annalen der Chemie und Pharmacie* 71 (2): 213–216.

65 Löwig, C. (1853). Ueber Methplumbäthyl. *Annalen der Chemie und Pharmacie* 88: 318–322.

66 Hallwachs, W. and Schafarik, A. (1859). Ueber die Verbindungen der Erdmetalle mit organischen Radicalen. *Liebigs Annalen der Chemie und Pharmacie* 109: 207.

67 Grignard, V. (1900). Sur quelques nouvelles combinaisons organométalliques du magnèsium et leur application à des synthèses d'alcools et d'hydrocarbures. *Comptes Rendus* 130: 1322.

68 Frankland, E. (1859). Researches on Organo-metallic Bodies. *Philosophical Transactions of the Royal Society of London* 149: 401–415.

References

69 Mond, L., Langer, C., and Quincke, F. (1890). Action of carbon monoxide on nickel. *Journal of the Chemical Society, Faraday Transactions* 57: 749–753.

70 Mond, L. and Langer, C. (1891). On iron carbonyls. *Journal of the Chemical Society, Faraday Transactions* 59: 1090–1093.

71 Werner, A. (1893). Beitrag zur Konstitution anorganischer Verbindungen. *Anorganic Chemistry* 3: 267.

72 Hein, F. (1919). Notiz über Chromorganoverbindungen. *Chemische Berichte* 52: 195.

73 Kealy, T.J. and Pauson, P.L. (1951). A new type of organo-iron compound. *Nature* 168 (4285): 1039–1040.

74 Fischer, E.O. and Maasböl, A. (1964). On the existence of a tungsten carbonyl carbene complex. *Angewandte Chemie (International Edition in English)* 3 (8): 580–581.

75 Ziegler, K., Holzkamp, E., Breil, H. et al. (1955). Das Mülheimer Normaldruck-Polyäthylen-Verfahren. *Angewandte Chemie (International Edition in English)* 67: 541–547.

76 Natta, G. (1955). Une nouvelle classe de polymeres d'α-olefines ayant une régularité de structure exceptionnelle. *Journal of Polymer Science* 16 (82): 143–154.

77 Smidt, J. and Hafner, W. (1959). Eine Reaktion von Palladiumchlorid mit Allylalkohol. *Angewandte Chemie (International Edition in English)* 71: 284.

78 Ruddick, J.D. and Shaw, B.L. (1969). Transition metal–carbon bonds. Part XXI. Methyl derivatives of platinum(II) and platinum(IV) containing dimethylphenylarsine as ligand. *Journal of the American Chemical Society* 123 (13): 2964–2969.

79 Wilkinson, G. (1974). The long search for stable transition metal alkyls. *Science* 185: 109–112.

80 Heck, R.F. and Nolley, J.P. (1972). Palladium-catalyzed vinylic hydrogen substitution reactions with aryl, benzyl, and styryl halides. *The Journal of Organic Chemistry* 37 (14): 2320–2322.

81 Tamao, K., Sumitani, K., and Kumada, M. (1972). Selective carbon–carbon bond formation by cross-coupling of Grignard reagents with organic halides. Catalysis by nickel–phosphine complexes. *Journal of the American Chemical Society* 94 (12): 4374–4376.

82 Miyaura, N., Yamada, K., and Suzuki, A. (1979). A new stereospecific cross-coupling by the palladium-catalyzed reaction of 1-alkenylboranes with 1-alkenyl or 1-alkynyl halides. *Tetrahedron Letters* 20 (36): 3437–3440.

83 King, A.O., Okukado, N., and Negishi, E.-i. (1977). Highly general stereo-, regio-, and chemo-selective synthesis of terminal and internal conjugated enynes by the Pd-catalysed reaction of alkynylzinc reagents with alkenyl halides. *Journal of the Chemical Society, Chemical Communications* 19: 683.

84 Milstein, D. and Stille, J.K. (1978). A general, selective, and facile method for ketone synthesis from acid chlorides and organotin compounds catalyzed by palladium. *Journal of the American Chemical Society* 100: 3636–3638.

85 Sonogashira, K., Tohda, Y., and Hagihara, N. (1975). A convenient synthesis of acetylenes: catalytic substitutions of acetylenic hydrogen with bromoalkenes, iodoarenes and bromopyridines. *Tetrahedron Letters* 16 (50): 4467–4470.

86 Vineyard, B.D., Knowles, W.S., Sabacky, M.J. et al. (1977). Asymmetric hydrogenation. Rhodium chiral bisphosphine catalyst. *Journal of the American Chemical Society* 99 (18): 5946–5952.

87 Qi, X., Liu, S., and Lan, Y. (2016). Computational studies on an aminomethylation precursor: (Xantphos)Pd(CH$_2$NBn$_2$)$^+$. *Organometallics* 35: 1582–1585.

88 Li, Y., Liu, S., Qi, Z. et al. (2015). The mechanism of N—O bond cleavage in rhodium-catalyzed C—H bond functionalization of quinoline N-oxides with alkynes: a computational study. *Chemistry – A European Journal* 21: 10131–10137.

2

Computational Methods in Organometallic Chemistry

2.1 Introduction of Computational Methods

2.1.1 The History of Quantum Chemistry Computational Methods

The core requirement of quantum chemistry is the solution of the time-independent Schrödinger equation

$$\hat{H}\Psi = E\Psi$$

where $\hat{H}$ is the Hamiltonian operator, Ψ is the wavefunction for all of the nuclei and electrons, and E is the energy associated with this wavefunction. The Hamiltonian contains all operators that describe the kinetic and potential energy of the molecule at hand. Schrödinger equation is the basis of quantum mechanics, which was proposed by E. Schrödinger, an Austrian theoretical physicist, in 1926 [1–3]. It describes the law of the state of microparticles changing with time. The state of microsystem can be described by wavefunctions, whose differential equation is Schrödinger equation. It means that the wavefunctions can be solved by the equation, when the initial conditions and boundary conditions are given.

The Hamiltonian operator $\hat{H}$ can be broken into two operators

$$\hat{H} = T + V$$

Those two operators represent kinetic energy (T) and potential energy (V), respectively. It could be further broken according to the nuclei and electron parts as

$$H = -\frac{\hbar^2}{2}\sum_A^N M_A^{-1}\nabla_A^2 + \sum_{A<B} e^2 Z_A Z_B r_{AB}^{-1} - \frac{\hbar}{2m}\sum_i^n \nabla_i^2 - \sum_A\sum_i e^2 Z_A r_{Ai}^{-1} + \sum_{i<j} e^2 r_{ij}^{-1}$$

$$\quad\quad\quad\text{i}\quad\quad\quad\quad\text{ii}\quad\quad\quad\quad\text{iii}\quad\quad\quad\quad\text{iv}\quad\quad\quad\text{v}$$

which represent (i) kinetic energy of nuclei, (ii) nuclear–nuclear repulsions, (iii) kinetic energy of electrons, (iv) nuclear–electron attraction, and (v) electron–electron repulsion. In most cases, it is impossible and unnecessary to find an analytic solution for the existing Schrödinger equation.

For organic system, several assumptions are indispensable to arrive at a solution due to the excessive degree of freedom. One of the most important assumptions is the Born Oppenheimer approximation proposed by M. Born and R. Oppenheimer in

Computational Methods in Organometallic Catalysis: From Elementary Reactions to Mechanisms, First Edition. Yu Lan.
© 2021 WILEY-VCH GmbH. Published 2021 by WILEY-VCH GmbH.

1927 [4]. Considering that the nucleus is much larger than the electron generally by 3–4 orders, the moving of the electron should be much more rapid than that of the nucleus under the same interaction. The result of this difference in velocity is that the electron moves at every moment as if it were in a potential field composed of stationary nucleus, while the nucleus cannot observe the specific position of electrons and can only be averaged interactions. Thus, the variables of nuclear coordinates and electronic coordinates can be approximately separated. The complex process of solving the whole-wave function can be decomposed into two relatively simple processes, solving the electronic wavefunction and solving the nuclear wavefunction

$$\Psi(R, r) = \Phi(R) \phi(r)$$

This separation of the total wavefunction into an electronic wavefunction $\phi(r)$ and a nuclear wavefunction $\Phi(R)$ means that the positions of the nuclei can be fixed and then one only has to solve the Schrödinger equation for the electronic part. Usually, we only focus on the electron energy of the potential energy surface, which is determined by the electronic wavefunction $\phi(r)$. Therefore, the Hamiltonian obtained after applying the Born–Oppenheimer approximation and neglecting relativity is

$$\hat{H} = -\frac{1}{2}\sum_i^n \nabla_i^2 - \sum_i^n \sum_I^N \frac{Z_I}{r_{Ii}} + \sum_{i<j}^n \frac{1}{r_{Ii}} + V^{\text{nuc}}$$

where V^{nuc} is the nuclear–nuclear repulsion energy.

In 1930, Hartree–Fock (HF) theory was formulated by V. Fock and D. R. Hartree, which is the basis of any other methods for solving Schrödinger equation [5, 6]. The core of HF theory is to simplify the problem of solving the multiparticle system in the external field into a problem of solving the wavefunction of a single particle. The electronic wavefunction can be separated into a product of functions that depend only on one electron

$$\psi(r_1, r_2, \ldots, r_n) = \varphi_1(r_1)\varphi_2(r_2)\cdots\varphi_n(r_n)$$

Unfortunately, the effect of electron–electron repulsion cannot be solved; therefore, the Schrödinger equation still cannot be solved exactly. Alternatively, the exact electron–electron repulsion is replaced with an effective field V_i^{eff} produced by the average positions of the remaining electrons. With this assumption, the separable functions φ_i satisfy the Hartree equations

$$\left(-\frac{1}{2}\nabla_i^2 - \sum_I^N \frac{Z_I}{r_{Ii}} + V_i^{\text{eff}}\right)\varphi_i = E_i\varphi_i$$

In the above-mentioned equation, solving for a set of functions φ_i is still problematic because the effective field V_i^{eff} is dependent on the wavefunctions. To solve this problem, an iterative procedure, named self-consisted field (SCF), was proposed by D. R. Hartree in 1927. First, a set of functions $(\varphi_1, \varphi_2, \ldots, \varphi_n)$ is assumed, which can be used to produce the set of effective potential operators V_i^{eff} and the Hartree equations are solved to produce a set of improved functions φ_i. These new functions produce an updated effective potential, which, in turn, yields a new set of functions φ_i. This procedure is repeated until the functions φ_i no longer change (converge) and produce an SCF. SCF convergence is a necessity for energy calculations.

To further simplify approximation of wavefunctions, linear combination of atomic orbitals (LCAO) theory was proposed by Roothaan in 1951 [7, 8]. A LCAO is a quantum superposition of atomic orbitals and a technique for calculating molecular orbitals in quantum chemistry. In a mathematical sense, these wavefunctions are the basis set of functions, the basis functions, which describe the electrons of a given atom. In chemical reactions, orbital wavefunctions are modified, i.e. the electron cloud shape is changed, according to the type of atoms participating in the chemical bond.

An initial assumption is that the number of molecular orbitals is equal to the number of atomic orbitals included in the linear expansion. In a sense, n atomic orbitals combine to form n molecular orbitals, which can be numbered $i = 1$ to n and which may not all be the same. The linear expansion for the ith molecular orbital would be

$$\varphi_i = \sum_r c_{ri} \chi_r$$

where ϕ_i is a molecular orbital represented as the sum of n atomic orbitals χ_r, each multiplied by a corresponding coefficient c_{ri}, and r (numbered 1 to n) represents which atomic orbital is combined in the term. The coefficients are the weights of the contributions of the n atomic orbitals to the molecular orbital. The Hartree–Fock procedure is used to obtain the coefficients of the expansion. The orbitals are thus expressed as linear combinations of basis functions, and the basis functions are one-electron functions, which may or may not be centered on the nuclei of the component atoms of the molecule. In either case, the basis functions are usually also referred to as atomic orbitals, which are typically those of hydrogen-like atoms since these are known analytically, i.e. Slater-type orbitals (STOs), but other choices are possible such as the Gaussian-type functions from standard basis sets or the pseudo-atomic orbitals from plane-wave pseudopotentials.

2.1.2 Post-HF Methods

Because HF theory uses an effective electron–electron repulsion term, HF energy, E_{HF}, will always be greater than the exact energy E. The instantaneous electron–electron repulsion is referred to as electron correlation, which is the best-case error of HF theory.

$$E_{corr} = E - E_{HF}$$

This equation ignores relativistic effects, which are very small for typical organometallic molecules; however, it can be significant for heavier elements. In computational chemistry, post-HF methods are the set of methods developed to improve on the HF method. They add electron correlation, which is a more accurate way of including the repulsions between electrons than in the HF method where repulsions are only averaged.

Methods that use more than one determinant are not strictly post-HF methods, as they use a single determinant as reference, but they often use similar perturbation, or configuration interaction (CI) methods to improve the description of electron correlation. The usually used post-HF methods include configuration interaction (CI) [9], coupled cluster (CC) [10], Møller–Plesset (MP) perturbation theory (MP2, MP3, MP4, etc.) [11–15], quadratic configuration interaction (QCI) [16], and quantum chemistry composite methods [G2, G3, complete basis set (CBS), W1, etc.] [17–22].

CI is a post-HF linear variational method for solving the nonrelativistic Schrödinger equation within the Born–Oppenheimer approximation for a quantum chemical multielectron system. Mathematically, configuration simply describes the linear combination of Slater determinants or configuration state functions (CSFs) used for the wavefunction. In terms of a specification of orbital occupation, interaction means the mixing (interaction) of different electronic configurations (states). Due to the long CPU time and large memory required for full CI calculations, the method is limited to relatively small systems. Truncating the CI space is important to save computational time. For example, the method configuration interaction with doubles (CID) is limited to double excitations only [13]. The method configuration interaction with singles and doubles (CISD) is limited to single and double excitations. Single excitations on their own do not mix with the HF determinant [23]. The Davidson correction can be used to estimate a correction to the CISD energy to account for higher excitations. It was used to solve the important problem of truncated CI methods is their size inconsistency, which means the energy of two infinitely separated particles is not double that of the single particle.

QCI is an extension of CI that corrects for size-consistency errors in single- and double-excitation CI methods (CISD). Size consistency means that the energy of two noninteracting (i.e. at large distance apart) molecules calculated directly will be the sum of the energies of the two molecules calculated separately. This method called quadratic configuration interaction with singles and doubles (QCISD) was developed in the group of J. Pople in 1989 [24, 25]. It gives results that are comparable to the CC method, coupled cluster single double (CCSD) [26, 27]. QCISD can be improved by the same perturbative inclusion of unlinked triples to give QCISD(T) [16]. This gives similar results to CCSD(T) [28].

CC is a numerical technique used for describing many-body systems. CC essentially takes the basic HF molecular orbital method and constructs multielectron wavefunctions using the exponential cluster operator to account for electron correlation. Some of the most accurate calculations for small-to-medium-sized molecules use this method. The choice of the exponential ansatz is opportune because it guarantees the size extensive of the solution. Size consistency in CC theory, unlike other theories, does not depend on that of the reference wavefunction. This is easily seen, for example, in the single-bond breaking of F_2 when using a restricted HF reference, which is not size consistent, at the coupled-cluster single double triple (CCSDT) [29, 30] level of theory, which provides an almost exact, full CI quality, potential energy surface and does not dissociate the molecule into F^- and F^+ ions, like the restricted Hartree–Fock (RHF) wavefunction, but rather into two neutral F atoms. If one were to use, for example, the CCSD, or CCSD(T) levels of theory, they would not provide reasonable results for the bond breaking of F_2, with the latter one approaches unphysical potential energy surfaces, though this is for reasons other than just size consistency.

MP perturbation theory improves on the HF method by adding electron correlation effects by means of Rayleigh–Schrödinger perturbation theory (RS-PT) [31], usually to second (MP2), third (MP3), or fourth (MP4) order. Its main idea was published as early as 1934 by Møller and Plesset [11].

Quantum chemistry composite methods are computational chemistry methods that aim for high accuracy by combining the results of several calculations. They combine methods with a high level of theory and a small basis set with methods that employ lower levels of theory with larger basis sets. They are commonly used to calculate thermodynamic quantities such as enthalpies of formation, atomization energies, ionization energies, and electron affinities. They aim for chemical accuracy, which is usually defined as within 1 kcal/mol of the experimental value.

The first systematic model chemistry of this type with broad applicability was called Gaussian-1 (G1) introduced by Pople et al. [32]. This was quickly replaced by the Gaussian-2 (G2), which has been used extensively [33]. The Gaussian-3 (G3) was introduced later [17]. The current version of this series of methods is Gaussian-4 (G4) [34]. G4 theory is an improved modification of the earlier approach G3 theory.

The CBS methods are a family of composite methods, the members of which are: CBS-4M, CBS-QB3, and CBS-APNO, in increasing order of accuracy [35]. The CBS methods were developed by G. Petersson and coworkers, and they extrapolate several single-point energies to the "exact" energy. In comparison, the Gaussian-n methods perform their approximation using additive corrections. Similar to the modified G2(+) method, CBS-QB3 has been modified by the inclusion of diffuse functions in the geometry optimization step to give CBS-QB3(+).

As a summary, although the above-mentioned post-HF methods could be often obtained with higher accuracy than HF method, they are still unable to be directly applied to the mechanistic study of transition metal catalysis because of the rather expensive time consumption. In fact, those methods are often used as a reference of energy to test the accuracy of some other more efficient computational methods.

2.2 Density Functional Theory (DFT) Methods

2.2.1 Overview of Density Functional Theory Methods

The inefficiency of HF and post-HF methods is the high computational effort that is required for the treatment of relatively large molecular systems. Therefore, it is rather complicated to solve. Fortunately, the total electron density of a molecule is only dependent on three variables in space, which is simpler than the electronic wavefunction and is also observable. It would offer a more direct way to obtain the molecular properties by the calculation of electron density.

The Hohenberg–Kohn existence theorem proves just that the ground state of any interacting many particle system with a given fixed inter-particle interaction is a unique functional of the electron density [36]. There exists a unique functional such that

$$E[\rho(r)] = E_{\text{elec}}$$

where E_{elec} is the exact electronic energy. It could be considered to be a functional, in which the function $\rho(r)$ depends on the spatial coordinates, and the energy depends on the values (is a functional) of $\rho(r)$.

To solve for the energy via the density functional theory (DFT) method, Kohn and Sham proposed that the functional has the form

$$E[\rho(r)] = T_{e'}[\rho(r)] + V_{\text{ne}}[\rho(r)] + V_{\text{ee}}[\rho(r)] + E_{\text{xc}}[\rho(r)]$$

where $T_{e'}$ is the kinetic energy of noninteracting electrons whose density is the same as the density of the real electrons, the true interacting electrons. V_{ne} is the nuclear–electron attraction term. V_{ee} is the classical electron–electron repulsion [37]. The last term is called the exchange–correlation functional, and is a catch-all term to account for all other aspects of the true system. However, it offers no guidance as to the form of that functional.

The exchange–correlation functional is generally written as a sum of two components, an exchange part and a correlation part. This is an assumption, an assumption that we have no way of knowing is true or not. These component functionals are usually written in terms of an energy density ε

$$E_{\text{XC}}[\rho(r)] = E_{\text{X}}[\rho(r)] + E_{\text{C}}[\rho(r)] = \int \rho(r)\, \varepsilon_{\text{X}}[\rho(r)] \mathrm{d}r + \int \rho(r)\, \varepsilon_{\text{C}}[\rho(r)] \mathrm{d}r$$

The major problem with DFT is that the exact functionals for exchange and correlation are not known except for the free electron gas. However, approximations exist, which permit the calculation of certain physical quantities quite accurately. One of the initial simple approximations of exchange–correlation functional is the local-density approximation (LDA), in which the exchange–correlation functional of uniform electron gas with same density is used as the approximation of the corresponding nonuniform system [38]. Unexpectedly, such a simple approximation often yields good results, which directly led to the widespread application of DFT currently. If the electron densities of different spin components are further considered, the local spin density approximation (LSDA) can be obtained. Despite the great success of L(S)DA, there are many shortcomings, such as systematic overestimation of binding energies.

To make improvements over the L(S)DA, one has to assume that the density is not uniform. The approach that has been taken is to develop functionals that are dependent on not just the electron density but also derivatives of the density. This constitutes the generalized gradient approximation (GGA). It is at this point that the form of the functionals begins to cause the eyes to glaze over and the acronyms to appear to be random samplings from an alphabet soup. The method of constructing GGA exchange–correlation functional can be divided into two ways. One is the group headed by Becke, which believes that "everything is allowed." Any formation of exchange–correlation functionals for any reason can be chosen, while the quality

2.2 Density Functional Theory (DFT) Methods

of this formation only depends on the comparison with real system. Another group, led by Perdew, believes that the development of exchange–correlation functionals should be based on certain physical laws, such as scaling relations and progressive behavior.

2.2.2 Jacob's Ladder of Density Functionals

Indeed, there is no unified standard for the classification of density functionals in the physical chemistry field. In 2001, J. P. Perdew et al. proposed using "Jacob's ladder" to classify the level of density functionals [39, 40]. As shown in Scheme 2.1, the ground in the "Jacob's ladder" is HF theory, which is an imprecise method with neither exchange energy nor correlation energy. In fact, HF calculation is rarely used in theoretical and computational chemistry nowadays.

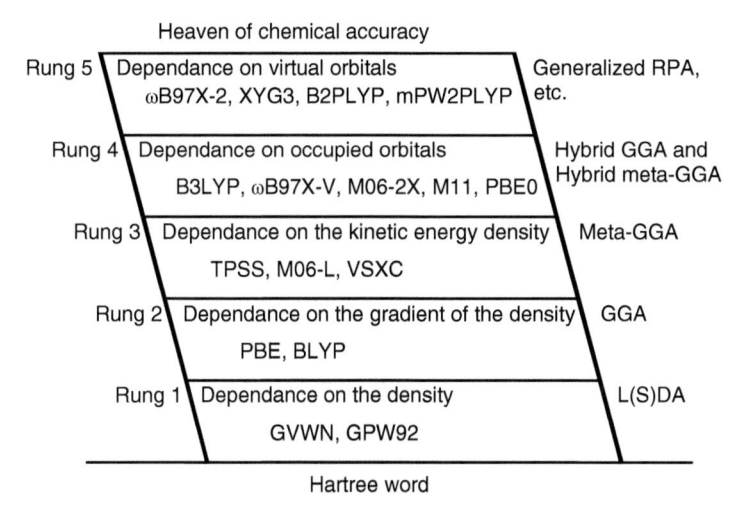

```
            Heaven of chemical accuracy
Rung 5  Dependance on virtual orbitals              Generalized RPA,
        ωB97X-2, XYG3, B2PLYP, mPW2PLYP             etc.

   Rung 4  Dependance on occupied orbitals          Hybrid GGA and
           B3LYP, ωB97X-V, M06-2X, M11, PBE0        Hybrid meta-GGA

      Rung 3  Dependance on the kinetic energy density   Meta-GGA
              TPSS, M06-L, VSXC

         Rung 2  Dependance on the gradient of the density   GGA
                 PBE, BLYP

            Rung 1  Dependance on the density                L(S)DA
                    GVWN, GPW92

                      Hartree word
```

Scheme 2.1 Jacob's ladder of density functionals.

The first rung in "Jacob's ladder" is the density functional based on L(S)DA, the variable in which kind of functionals is the local spin density. The exchange functional of L(S)DA can be written as analytic expressions, which is often called Slater or Dirac exchange functional. However, the correlation functional of L(S)DA has no analytic expression, and can only be fitted by a functional with parameters from the results of high-level calculations on some uniform electron gases. L(S)DA has achieved surprising success in the early works on the computational study of solid-state physics. However, it is failure in computational chemistry because L(S)DA usually overestimates the bonding energy.

2.2.3 The Second Rung in "Jacob's Ladder" of Density Functionals

The second rung in "Jacob's ladder" of density functionals is the GGA. The variables in this kind of functionals are local spin density and its gradient. Therefore, there are no analytic expressions for both exchange functionals and correlation functionals of GGA density functionals. The first successful GGA density functional for chemical calculation was Becke–Lee–Yang–Parr hybrid DFT functional (BLYP) [41, 42]. Thereinto, B was Becke88 exchange functional, while LYP was Lee–Yang–Parr correlation functional. Around 2000, the commonly used GGA functionals in computational organometallic chemistry are the Perdew–Burke–Ernzerh (PBE) and BP86 functionals [43, 44]. Although these functionals are seldom used at present, many of popular functionals are developed on the basis of these functionals.

2.2.4 The Third Rung in "Jacob's Ladder" of Density Functionals

The third rung in "Jacob's ladder" of density functionals is meta-GGA functionals. The variables with more functionals than GGA are the kinetic energy density or the second derivative of the local spin density. The most common meta-GGA involved are M06-L, TPSS, and VSXC [45–47], which are often used in computational organometallic chemistry currently.

2.2.5 The Fourth Rung in "Jacob's Ladder" of Density Functionals

The fourth rung in "Jacob's ladder" of density functionals is hybrid-GGA and hybrid-meta-GGA. This kind of functionals are the most popular functional in computational chemistry currently, into which HF exchange is introduced. In the field of computational organometallic chemistry, the commonly used hybrid-GGA functionals involves B3LYP [42, 48], B97 [49], O3LYP [50], PBE0 [51], mPW1PW [52], X3LYP [53], etc.; the commonly used hybrid-meta-GGA functionals involves M05, M05-2X, M06, M06-HF, M06-2X, TPSSh, MPW1K, etc. [46, 54–58].

Undoubtedly, B3LYP is the most widely used functional in computational chemistry, which combines exact HF exchange with Becke's gradient-corrected exchange, the LYP correlation functional, and VWN for the local correlation terms, as the following function:

$$E_{XC}^{B3LYP} = (1-a_0)E_X^{LSDA} + a_0 E_X^{HF} + a_x \nabla E_X^{B88} + a_c E_C^{LYP} + (1-a_c)E_C^{VMN}$$

According to Becke's parametrization against the G1 database, a_0, a_x, and a_c are taken as 0.20, 0.72, and 0.81, respectively.

2.2.6 The Fifth Rung in "Jacob's Ladder" of Density Functionals

In the fifth rung of "Jacob's ladder," the information of virtual orbital is used to build double-hybrid functional. As examples, in this type, functionals B2PLYP and mPW2PLYP, Kohn–Sham unoccupied orbitals are used to calculate MP2-type correlation functional [59, 60].

2.2.7 Correction of Dispersion Interaction in Organic Systems

Dispersion interactions are generally described as the interaction between instantaneous dipole moments within the electron distributions of two atoms or molecules. The simplest model of dispersion is the interaction between two Drude oscillators, where the instantaneous dipole moments of the oscillators cause a stabilizing interaction between them.

The dispersion energy, E_{disp}, between two atoms or molecules at large separation takes the form of a series expansion

$$E_{disp} = -\frac{C_6}{R^6} - \frac{C_8}{R^8} - \frac{C_{10}}{R^{10}} - \cdots$$

The dispersion coefficients can be determined experimentally or theoretically. Also, there are several approximate methods to evaluate the dispersion coefficients such as the London, Slater–Kirkwood, and Salem models. In most of the cases, dispersion attraction is modeled by the first term in the series only. For example, in molecular mechanics (MM) force fields, commonly used in biochemistry, a Lennard–Jones 6-12 potential is used to describe nonbonded interactions with the R^6 term accounting for dispersion attraction and the R^{12} term corresponding to Pauli repulsion.

Most of the classical density functionals (e.g. B3LYP) cannot describe the dispersion interaction because the long-range behavior of the correlation functional is not correct. Therefore, the results used to study dispersion-dominated problems are poor, such as physical adsorption, molecular conformations, ligand coordination, ligand–substrate interaction, and π-stacking; however, those systems are important in organometallic chemistry. Indeed, classical density functionals reveal poor dispersion interaction is due to the incorrect behavior of the exchange–correlation functional in the medium and long range, especially in the long range, which results in the failure of van der Waals C_6/R^6 behavior.

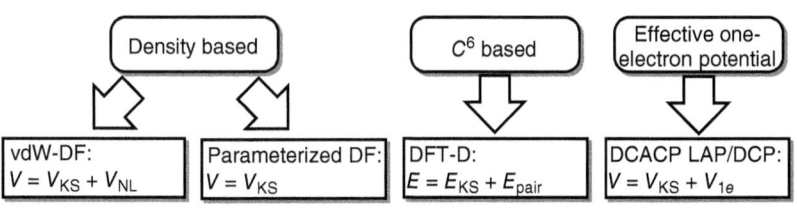

The frequently used approaches for the dispersion correction of density functionals are grouped in Scheme 2.2, which includes nonlocal van der Waals density functionals (vdW-DFs), semilocal density functionals parameterizing from the forms of some standard meta-hybrid-GGA functionals, DFT-D methods, and dispersion-correcting atom-centered one-electron potentials (DACACP) [59, 61–65]. The properties of those groups of methods are accounted for in Table 2.1.

Dispersion corrections:

- Density based
  - vdW-DF: $V = V_{KS} + V_{NL}$
  - Parameterized DF: $V = V_{KS}$
- C^6 based
  - DFT-D: $E = E_{KS} + E_{pair}$
- Effective one-electron potential
  - DCACP LAP/DCP: $V = V_{KS} + V_{1e}$

Scheme 2.2 The frequently used approaches for the dispersion correction of density functionals.

Table 2.1 Properties of frequently used approaches for the dispersion correction.

Properties	vdW-DF	DF	DFT-D	DCACP LAP/DCP
Correct R^{-6}	Yes	No	Yes	No
Thermochemistry	—	Yes	Yes	—
Numerical complexity	High	Medium	Low	Low
Simple forces	No	Yes	Yes	Yes
System dependency	Yes	Yes	Yes	No
Electronic effect	Yes	Yes	No	Yes
Empiricism	Low	Medium	Medium	High
Analysis insight	—	No	Good	—

In computational chemistry, some highly parameterized meta-GGA functionals incorporating kinetic energy density have been assessed to quantitatively account for dispersion effect. One of the most popular parameterized functionals, which can accurately describe dispersion interaction, is Minnesota series of functionals proposed by Truhlar group. Currently, the usually used Minnesota functionals in computational organometallic chemistry include M06, M06-2X, M06-L, M11, M11-L, MN12, MN12-L, MN15, etc. One serious problem of this and related highly parameterized functionals with many terms in a power series expansion is numerical instability that can lead to artificial van der Waals minima and "noisy" potential energy curves.

DFT-D methods with semiclassical corrections are alternative choice for computational organometallic chemistry, which treat the difficult dispersion interactions semiclassically and combine the resulting potential with a quantum chemical approach. Currently, the most widely used DFT-D method, named DFT-D3 series, was proposed by S. Grimme group in 2006, which represents an update of DFT-D1. DFT-D3 has recently been refined regarding higher accuracy, broader range of applicability, and less empiricism [65]. The main new ingredients are atom pairwise-specific dispersion coefficients and a new set of cutoff radii, both computed from first principles. The coefficients for eighth-order dispersion terms are computed using established recursion relations. System-dependent information is used for the first time in a DFT-D type approach by employing the new concept of fractional coordination numbers. This allows one to distinguish the different hybridization states of atoms in molecules in a differentiable way, which in particular for the first two rows of the periodic table have quite different dispersion coefficients. The method only requires adjustment of two global parameters for each density functional is asymptotically same for a gas of weakly interacting neutral atoms and easily allows the computation of atomic forces. Accurate dispersion coefficients and cutoff radii are available for all elements up to $Z = 94$. The revised DFT-D3 method can be used as a general tool for the computation of the dispersion

2.3 Basis Set and Its Application in Mechanism Studies

2.3.1 General View of Basis Set

In quantum chemistry calculations, a common approach is to represent a molecular orbital as a LCAO

$$\Psi_i = \sum_k c_{ik} \varphi_k$$

in which each atomic orbital is represented by a single mathematical function. The atomic orbitals used in this procedure are represented by what is known as the basis set. Following this idea, the mathematical form of atomic orbitals should be considered, when it is used to be linear combinations for the construction of molecular orbitals. One choice would be to simply use the hydrogenic wavefunctions adapted for other atoms, which is called a Slater-type orbital (STO). These wavefunctions have radial forms possessing terms such as $r^{n-1}e^{-\zeta r}$ ($\zeta = Z/n$), whose function form has clear physical meaning (Scheme 2.3a). However, it is very difficult to evaluate the complex two-electron integrals.

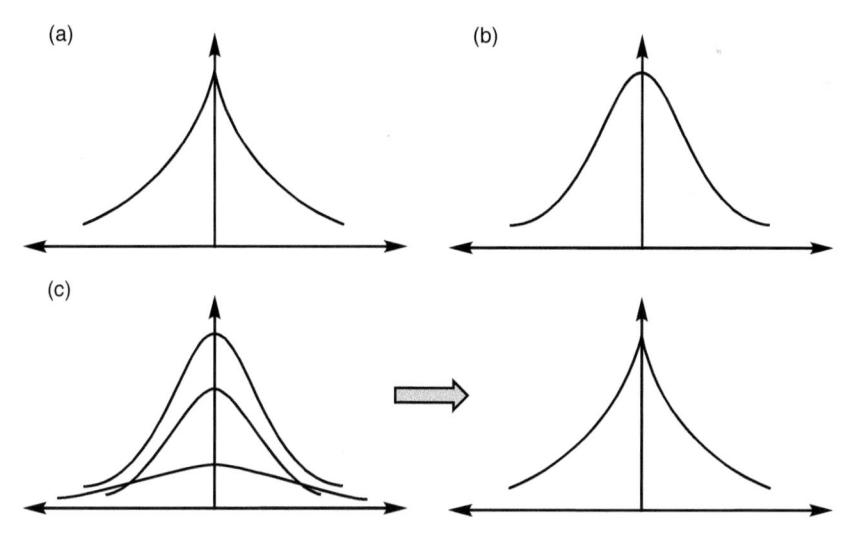

Scheme 2.3 The combination of Gaussian-type orbitals (GTOs) for the construction of Slater-type orbital (STO). (a) STO. (b) GTO. (c) The combination of GTOs.

Following a suggestion of Boys, Pople decided to use a combination of Gaussian-type functions to mimic the STO, which was named Gaussian-type orbital (GTO). These orbitals have a different spatial function, $X'Y'''Z'e^{-\zeta r^2}$; therefore, the integrals required to build the Fock matrix can be evaluated exactly

(Scheme 2.3b). The tradeoff is that GTOs do differ in shape from the STOs, particularly at the nucleus where the STO has a cusp, but the GTO is continually differentiable. The computational advantage is so substantial that it is more efficient to represent a single atomic orbital as a combination of several GTOs rather than a single STO (Scheme 2.3c). When a few GTOs with differing shapes are added, the result is a function that resembles an STO. Now, the basis set is not just the atomic orbitals, but is instead all the GTOs that are used to make up the atomic orbitals. The minimum Gaussian-type basis set is STO-3G, in which "STO" is the abbreviation of STO, and "3G" means that each STO is obtained by a linear combination of three GTOs [66, 67].

The minimum basis set has one basis function for every formally occupied or partially occupied orbital in the atom, which is referred to as a single-zeta (SZ) basis set. The use of the term zeta here reflects that each basis function mimics a single STO, which is defined by its exponent, zeta (ζ). The minimum basis set is usually inadequate, failing to allow the core electrons to get close enough to the nucleus and the valence electrons to delocalize. An obvious solution is to double the size of the basis set, creating a double-zeta (DZ) basis. Further improvement can be had by choosing a triple-zeta (TZ) or even larger basis set [68].

2.3.2 Pople's Basis Sets

As most of chemistry focuses on the action of the valence electrons, Pople developed the split-valence basis sets, single zeta in the core and double zeta in the valence region, which won him the 1998 Nobel prize in chemistry. A double-zeta split-valence basis set for carbon has three s basis functions and two p basis functions for a total of nine functions, a triple-zeta split valence basis set has four s basis functions and three p functions for a total of 13 functions, and so on.

For the vast majority of basis sets, including the split-valence sets, the basis functions are not made up of a single Gaussian-type function. Rather, a group of Gaussian-type functions are contracted together to form a single basis function. An example is split-valence basis set 6-31G, which is popular in computational organic chemistry. In this basis set, the left value means that each core basis function comprises six Gaussian functions. Meanwhile, the valence space is split into two basis functions, which referred to the inner and outer parts of valence space. The inner basis function is composed of three contracted Gaussian-type functions, and each outer basis function is a single Gaussian-type function. Thus, for carbon, the core region is a single s basis function made up of six s-GTOs. The carbon valence space has two s and two p basis functions. The inner basis functions are made up of three Gaussians, and the outer basis functions are each composed of a single Gaussian-type function. Therefore, the carbon 6-31G basis set has nine basis functions made up of 22 Gaussian-type functions. This type of split-valence basis sets involves 3-21G, 4-31G, 6-31G, 6-311G, etc. [69–72]. The accuracy of those basis sets depends mainly on the number of basis functions, and secondly on the number of Gaussians. However, the time consumed in calculation increases accordingly with the improvement of accuracy.

2.3.3 Polarization Functions

A critical problem with a simple split-valence basis set, such as 6-31G, is that the flexibility of wavefunctions is insufficient to distort to the actual shape. Extending the basis set by including a set of functions that mimic the atomic orbitals with angular momentum one greater than in the valence space greatly improves the basis flexibility. These added basis functions are called polarization functions. For second and third periodic elements, adding polarization functions means adding a set of d GTOs; therefore, a basis set involving polarization functions for heavy atoms can be written as 6-31G(d) or 6-31G*. For hydrogen, polarization functions are a set of p functions. Therefore, a basis set involving polarization functions for all atoms can be written as 6-31G(d,p) or 6-31G**. As adding multiple sets of polarization functions has become broadly implemented, the use of asterisks has been abandoned in favor of explicit indication of the number of polarization functions within parentheses, that is, 6-311G(2df,2p) means that two sets of d functions and a set of f functions are added to heavy atoms and two sets of p functions are added to the hydrogen atoms. The polarization functions are simply mathematical tools that allow to give the basis set more flexibility, and thus produce a better calculation.

2.3.4 Diffuse Functions

For anions or molecules with many adjacent lone pairs, the basis set must be augmented with diffuse functions to allow the electron density to expand into a larger volume. For split-valence basis sets, this is designated by "+" as in 6-31+G(d). The diffuse functions added are a full set of additional functions of the same type as are present in the valence space. So, for carbon, the diffuse functions would be an added s basis function and a set of p basis functions. If a molecule involving hydride, diffuse functions for hydrogen atom is necessary, which would be an added s basis function, the corresponding split-valence basis set can be written as 6-31++G(d).

2.3.5 Correlation-Consistent Basis Sets

The split-valence basis sets developed by Pople are widely used. The correlation-consistent basis sets developed by Dunning are popular alternatives. The split-valence basis sets were constructed by minimizing the energy of the atom at the HF level with respect to the contraction coefficients and exponents. The correlation-consistent basis sets were constructed to extract the maximum electron correlation energy for each atom. The correlation-consistent basis sets are designated as "cc-pVNZ," to be read as correlation-consistent polarized split valence N-zeta, where N designates the degree to which the valence space is split [73–76]. The "cc-p," stands for "correlation consistent polarized" and the "V" indicates they are valence only basis sets. As N increases, the number of polarization functions also increases. So, for example, the cc-pVDZ basis set for carbon is double-zeta in the valence space and includes a single set of d functions, and the cc-pVTZ basis set is triple-zeta in the valence space and has two sets of d functions and a set of f functions. The addition of diffuse functions to the correlation-consistent basis sets is designated with the prefix aug-, as in aug-cc-pVTZ. Usually, the correlation-consistent basis sets are used in post-HF level calculations, which revealed a significantly better accuracy.

2.3.6 Pseudo Potential Basis Sets

Pseudo potential basis sets are widely used in computational organometallic chemistry, which shows a rather high efficiency for transition metals. Pseudo potential basis sets are not to calculate the inner electrons, but to describe the contribution of the inner electrons by a potential, which is placed in Hamiltonian. Pseudo potential basis sets actually include two parts, pseudo potential and the basis sets. The inner electrons adopt effective core potential (ECP), while the outer valence electrons adopt general basis sets. When pseudo potential basis sets are used, inner electrons do not need to be considered directly, so that the consumption of calculation is significantly reduced, which is more obvious for more heavy elements. In transition metal chemistry, relativistic effects mainly come from inner electrons. If the relativistic effect is taken into account in fitting the pseudo potential, that effect can be reflected equivalently in the calculation based on the pseudo potential.

Generally, the pseudo potential basis sets used in computational organometallic chemistry include three categories: Stuttgart pseudo potentials, Los Alamos National Laboratory (LANL) pseudo potentials, and def series pseudo potentials [66, 77–82]. SDD is the most popular Stuttgart pseudo potential basis set, which often are used to describe the atomic orbitals of transition metals in the fifth and sixth periods. The accuracy of SDD is roughly equivalent to a triple-zeta basis set.

LANL series pseudo potentials were proposed by Hay and Wadt in 1980s, which involve relativistic effect for the elements in the fifth and sixth periods. The frequently used double-zeta LANL series basis set is LANL2DZ [81]. LANL2TZ is restricted from LANL2DZ, which is a quasi-triple-zeta basis set [82]. In LANL2TZ+ or LANL2TZ(f) basis set, diffuse function or polarization function is involved, respectively. LANL08 is a completely derestricted basis set based on LANL2DZ. Diffuse function or polarization function also can be involved in LANL08+ or LANL08(f), respectively. Even now, LANL2DZ basis set has been widely used; however, it is not recommended because of its obviously low accuracy. For transition metals, LANL2TZ(f) and LANL08(f) are recommended because of great improvement in accuracy though it is slightly expensive [83].

The accuracy of def2 series basis sets increases from def2-SV(P), def2-SVP, def2-TZVP, def2-TZVPP, def2-QZVP, and def2-QZVPP [83–87]. For the first four periods, def2 series basis sets are full electronic basis set. From the fifth period, they become pseudo potential basis sets, which are combined with Stuttgart small core pseudo potential. The main advantage of the def2 series basis set is that it covers all elements in the first six periods; therefore, def2 series basis set can be used solidly in most systems instead of a mixed basis set. In computational organometallic chemistry, def2-TZVP basis set is suggested in energy calculation.

2.4 Solvent Effect

Solvent effect is very important in organometallic chemistry; therefore, in theoretical calculations, it should be considered in energy calculations. Usually, the solvent effect in homogeneous catalysis calculations is considered by implicit solvent model.

In implicit solvent model, the solvent environment is simply considered as a polarizable continuous medium; meanwhile, the structure and distribution of solvent molecules close to the solute are not specifically described. The advantage of implicit solvent model is that it can represent the average effect of solvents without the consideration of various possible molecular arrangements of solvent layer as explicit solvent model does, and it does not increase the computational time. Therefore, it is widely used in the field of computational organic and organometallic chemistry. The weakness of the implicit solvent model is that the strong interaction between solvent and solute cannot be represented, such as hydrogen bond. Moreover, the accuracy of solvation energy for ionic solute case is significantly lower than that of neutral solute case.

In implicit solvent model, the solvent effect can be divided into polar and non-polar parts. The polar part, which is the main body of the implicit solvent model, reflects the electrostatic interaction between solvent and solute molecule, and also includes the polarization of solute electron distribution contributing by solvent. The nonpolar part reflects various nonelectrostatic interactions between solute molecule and solvent, which includes the solvent exclusion energy, the influence of solute molecule on the entropy effect of solvent, and the dispersion between solute molecule and solvent.

The frequently used implicit solvent model is a polarizable continuum model (PCM) [88], in which the solvation free energy can be decomposed into three terms

$$G_{sol} = G_{es} + G_{dr} + G_{cav}$$

in which G_{es} represents the electrostatic energy, G_{dr} represents the dispersion–repulsion energy, and G_{cav} represents the hole energy. All those three terms are calculated by a hole defined by a chain of van der Waals spheres centered on the solute atoms. Dielectric formulation (D-PCM) is the early member of PCM family, which only includes the charge density of the solute wavefunction within the solute surface into the solute–solvent interaction. The integral equation formalism polarizable continuum model (IEF-PCM), developed by Cances and Mennucci, also includes the charge density of the wavefunction beyond the solute surface into the solute–solvent interaction [89]. The conductor-like polarizable continuum model (C-PCM), developed by Barone and Cossi, is the implementation of conductor-like screening model in the PCM framework, which works well for solvents with a high dielectric constant such as water solvent [90, 91]. In isodensity polarizable continuum model (I-PCM) and self-consistent IPCM (SCI-PCM), the solute cavity can be defined as a surface with a constant electron density (isodensity surface) [92].

In computational organometallic chemistry, SMD (solvation model based on density), proposed by Cramer and Truhlar in 2009, is almost the best implicit solvent model for DFT calculations at present [93]. In the SMD model, the bulk electrostatic contribution is calculated on the basis of the IEF-PCM protocol.

2.5 How to Choose a Method in Computational Organometallic Chemistry

2.5.1 Why DFT Method Is Chosen

The current computational studies of organic and organometallic systems are frequently subjected to a compromise of accuracy and time consumption: while it is always desired to reach the highest possible accuracy in the computational assessment, the necessity to complete the calculations within a reasonable time scale does not always allow for that.

The HF method ignores instantaneous electron correlation; therefore, it can be excluded firstly. Post-HF methods attempt to treat electron correlation through several methods, which usually can provide a good accuracy in theoretical calculations. However, rather high scaling behaviors of those methods (Table 2.2) restrict the application in computational organometallic chemistry. In fact, some of the post-HF methods are often used to calculate some small systems to obtain accurate results as a benchmark reference for other computational methods.

In DFT, electron density functional is used instead of wavefunction, which provides a better accuracy than HF method. In Kohn–Sham DFT framework, the most difficult multielectron interaction is simplified to a problem of the motion of electrons in an effective potential field without interaction, which includes the influence of external potential field and Coulomb interaction between electrons, such as exchange–correlation interaction. Therefore, DFT method provides an excellent computational efficiency.

In computational organometallic chemistry, the two results we are most concerned about are the activation energy for one pathway and the difference of activation energy between two pathways, which represent the reaction rate and selectivity, respectively. The precision orders required by those two results are 1 and 0.1 kcal/mol, respectively. Although the accuracy of DFT method seems to be insufficient to meet the requirements, the error cancelation makes the accuracy sufficient for the DFT calculations of reaction rate and selectivity.

2.5.2 How to Choose a Density Functional

Strictly, this paragraph may be the least serious part in this book, because the selection of a functional in computational organometallic chemistry has many other reasons, even might be that I like this functional. If there must be a way, the selection of density functional can be based on the targeted benchmark results because of the lack of clear physical meaning leading to the inestimable error. However, there

Table 2.2 Scaling behaviors of computational methods.

Methods	Behaviors
HF	N^4
MP2	N^5
MP3, CISD, CCSD	N^5
MP4, CCSD(T)	N^7
MP5, CISDT	N^8
MP6	N^9
MP7	N^{10}
DFT	N^3–N^4

are often insufficient data for that in a specific organometallic system. Therefore, it is rather hard to choose a density functional for computational organometallic chemistry.

Until now, B3LYP functional has always been the preferred functional for theoretical study of reaction mechanism, although it has been proposed for more than 20 years. There are hundreds of density functionals, many of which would perform better than B3LYP in their own field of expertise. However, very few of them have more comprehensive performance than B3LYP, which leads to popularity of that functional. There are two main weaknesses in B3LYP: (i) disregard of dispersion interaction and (ii) the bad performance on charge transfer and Rydberg excitations. The first problem can be completely modified by DFT-D3 correction without additional computing time. The second one can be solved by the variant CAM-B3LYP functional. These amendments continue to extend the service life of B3LYP functional.

From a large number of calculation results, M06-2X, which is a Minnesota series functional proposed by Truhlar in 2007, proved to be one of the best alternative functionals of B3LYP. In this functional, dispersion effect is introduced in its fitting parameters, which reveals good weak interactions. The 54% HF exchange component leads to a better performance on the calculation of charge transfer and Rydberg excitation. The shortcoming is that Minnesota series functionals require much higher accuracy of DFT integration grid than B3LYP. It could be solved by improvement of that; however, it will obviously be more time consuming. Moreover, M06-2X functional is parameterized for the main group elements; therefore, it is unsuited for the calculation of transition metal involved system. As an alternative, M06-L functional can be used in this case. It is noteworthy that MN15, also proposed by Truhlar in 2016, achieves a good balance between main group and transition metal elements in computational study.

The ωB97XD functional is another alternative of B3LYP, which is proposed by Head-Gordon group in 2008 [94, 95]. This functional included empirical dispersion at DFT-D2 level, which gives a good accuracy of weak interaction. Moreover, the introduction of long-range correction into ωB97XD gives a good result in the calculation of charge transfer and Rydberg excitation. The time consumed for ωB97XD is also significantly higher than that of B3LYP.

Based on the author's experience, it is better to use hybrid-GGA or hybrid-meta-GGA functionals involving dispersion correction on the geometry and thermodynamics calculation of organometallic system.

2.5.3 How to Choose a Basis Set

Generally, the time taken for DFT calculations is mostly used in the calculation of double-election integral, which is positively correlated with N^4, in which N is the total Gaussian-type functions for a specific molecule. Therefore, the calculation of time taken for a specific molecule is dependent on the selectivity of basis set for all atoms in this molecule. In fact, the selection of the basis set is also arbitrary, which even could be based on experience and preferences. Fortunately, the accuracy of most DFT methods is not too dependent on the size of selected basis set, while most of large enough basis sets can yield a same result in DFT calculations. Additionally, when the number of base functions for using a basis set is the same, the time spent can be saved by using a basis set with less Gaussian-type function under the same precision. Therefore, segmented contraction basis sets, such as Pople's basis sets and def2 series of basis sets, are better choice in DFT calculations.

For DFT calculations, regular polarization functions (d,p) are necessary, which can improve the accuracy to a great extent. However, the larger angular momentum functions are unnecessary. A triple-zeta basis set is usually slightly better than a double-zeta one. As an example, the basis set of 6-311G(d,p) for DFT calculations is better than that of 6-31G(d,p); however, the basis set of 6-311(3df,2pd) has not shown improvement over that of 6-311G(d,p). Following this idea, 6-31G(d) is the smallest acceptable basis set in modern computational organometallic chemistry. The basis set of 6-311G(d,p) is a better choice for both accuracy and efficiency. When def2 series of basis set is used, def2-SVP is acceptable, while def2-TZVP is a better one. In an anionic molecule, defuse function is necessary; therefore, 6-31+G(d) is the acceptable basis set, while 6-311++G(d,p) would provide a good accuracy.

For the DFT calculation onto transition metals, employing a pseudo potential basis set is strongly recommended for the consideration of both accuracy and efficiency. In this area, LANL2DZ is the smallest basis set for transition metals, which provides only acceptable accuracy for some geometry optimization. The larger ones, such as LANL2TZ, LANL08, LANL08(d), and LANL08(f), are recommended for 4–6th periodic elements involving transition metals.

Notably, because the accuracy is not obviously dependent on the basis set in geometry optimizations and harmonic vibrational frequency calculations, while the time consumption is large, a smaller basis set usually can be chosen, such as 6-31G(d), 6-31+G(d), def2-SVP, or LANL2DZ. By contrast, larger basis sets are often more suitable in the single-point calculations with high accuracy, such as 6-311G(d,p), 6-311+G(d,p), def2-TZVP, LANL08 series. Additionally, for some highly parameterized functionals, a basis set consistent with that used in parameterizations is recommended in both geometry optimization and single-point energy calculation. As an example, a basis set of 6-311+G(d,p) is recommended for the DFT calculation with M06-2X functional by that reason.

2.6 Revealing a Mechanism for An Organometallic Reaction by Theoretical Calculations

It has been more than 20 years since the computational tool became a powerful and popular way to reveal the mechanism for organic and organometallic reactions, which starts with the development of density functionals that gave excellent geometries and reasonable energetics for organic and organometallic compounds. Generally, the approach to the exploration of reaction mechanism and catalytic cycle involves the use of calculations to test hypotheses.

Collaboration with experimental chemists is helpful for the theoretical study of reaction mechanism. The proposed reaction mechanism can be assumed based on the known reaction conditions, phenomena obtained from experiments, and previous experience. Several possible mechanisms can be presumed, but each of them should be reasonable and not contradictory to the present experimental observations. In computational organometallic chemistry, the proposed reaction mechanism needs to involve all stationary points on the whole reaction pathway. It means each of the two local minimums should be connected by only one transition state. One can get the geometries of stationary points from geometry optimization of each structure usually by DFT calculations. The thermodynamic (Gibbs free energy and enthalpy) corrections also can be got from the same level of theory. Solvation effect is usually considered by the higher accuracy calculation with implicit solvent models. The total Gibbs free energy for each stationary points can be observed by

$$\Delta G_{real} = \Delta E_{solv} + \Delta G_{corr}$$

which is used to construct free-energy profiles. The calculated free-energy profiles for all of the possible reaction pathways are considered by energy criterion. The most energetically feasible pathway usually is considered to be the best reaction mechanism for a specific reaction. Herein, initial geometries are acquired by conformational searches with force fields. Input geometries of the transition structures are identified either from experience or through scanning along the reaction coordinate.

If results are in agreement with experiment, we analyze the transition states and the structural factors responsible for asymmetric induction are proposed. Qualitative models are developed to understand and to make new predictions for catalyst and reaction design.

2.7 Overview of Popular Computational Programs

Relevant calculation programs are often needed to implement various calculation methods. In computation organic and organometallic chemistry, various programs are provided for theoretical calculations involving Gaussian, ADF, ORCA, GAMESS, Molpro, MOLCAS, NWchem, Q-Chem, etc. [96–103].

Undoubtedly, Gaussian series is the most frequently used program for computational chemistry, which is also the most popular one in computational organic and organometallic chemistry [96]. It is capable of predicting many properties of molecules and reactions, including molecular energies and structures, energies and structures of transition states, bond and reaction energies, molecular orbitals, multipole moments, atomic charges and electrostatic potential, vibrational frequencies, NMR properties, and reaction pathways. Computation can be carried out on systems in the gas phase or in solutions, and in their ground state or in an excited state. It is very flexible in the choice of basis sets, ECPs, and density functionals, and it has excellent geometry optimizers and initial guesses for the SCF iterations. It supports analytic Hessians for all functionals, and has classical dynamics capabilities and a very useful interface to external programs. Gaussian supports CPU parallelization with TCP Linda in the distributed memory multiprocessor environments, and GPU acceleration in the latest version Gaussian 16. Gaussian is commercially available with single, site, and Lina licenses. Website: https://gaussian.com/.

Q-Chem is a comprehensive ab initio quantum chemistry software for accurate predictions of molecular structures, reactivities; and vibrational, electronic, and NMR spectra [102]. The package is available either as a standalone code or integrated with the Spartan package. Fully integrated graphic interface IQmol includes molecular builder, input generator, contextual help, and visualization toolkit. The new release of Q-Chem 5 represents the state of the art of methodology from the highest-performance DFT/HF calculations to high-level post-HF correlation methods. Q-Chem supports ONIOM model and an interface of CHARMM for QM/MM calculation. Q-Chem is commercially available with academic, government, and industry licenses. Website: http://www.q-chem.com/.

ORCA is an ab initio and semiempirical quantum chemistry program and has become one of the most widely used programs developed in the research group of Professor Frank Neese [98]. Various modern electronic structure methods have been implemented, including DFT, single-reference correlation methods, multireference correlation methods, local correlation methods, excited state methods, solvation, solids and surfaces, and semiempirical quantum chemistry methods. ORCA has special features of robustness, efficiency, and focus on transition metals, and spectroscopic properties. QM/MM calculations could be performed using the interfaces of external programs, such as ASE, Gromas, NAMD, and Chemshell. Academic use version at academic institutions is freely available, and the commercial version is also available. Website: https://orcaforum.kofo.mpg.de/app.php/portal.

NWChem is a high-performance computational chemistry code with ab initio, band-structure, molecular mechanics, and molecular dynamics methods [103, 104]. This program achieves an efficient computational chemistry code with high parallel scalability of running on tens of thousands of processors. The ab initio methods available include HF, DFT, MPn, multiconfiguration self-consistent field (MCSCF), CI, and CC. There are a large number of basis sets available, including ECP sets. Ab initio dynamic calculations can be performed at respective levels of theory. QM/MM

2.7 Overview of Popular Computational Programs

minimization and dynamics calculations are also possible. This computational chemistry code is fully open source from Pacific Northwest National Laboratory. Website: http://www.nwchem-sw.org/.

Turbomole is a software package for large-scale quantum chemical simulations of molecules, clusters, and periodic solids [105]. Turbomole has electronic structure methods with excellent performance, such as time-dependent DFT, second-order Møller–Plesset theory, and explicitly correlated CC methods. These methods are implemented by efficient algorithms such as integral-direct and Laplace transform methods, resolution-of-the-identity, pair natural orbitals, fast multipole, and low-order scaling techniques. The properties of optics, electrical density, and magnetics are accessible for ground and excited states. Recently, new methods were released such as post-Kohn–Sham calculations within the random-phase approximation, periodic calculations, spin–orbit couplings, explicitly correlated CC singles doubles and perturbative triples methods, CC singles doubles excitation energies, and nonadiabatic molecular dynamics simulations using time-dependent density functional theory (TDDFT). This code is commercially available. Website: https://www.turbomole.org/.

The general atomic and molecular electronic structure system (GAMESS (United States)) package has many electronic structure capabilities [99]. The electronic structure methods have been implemented, including Hartree–Fock method, DFT, generalized valence bond, and multi-configurational SCF. Post-SCF correlation corrections can be performed by configuration interaction, second-order Møller–Plesset perturbation theory, and CC theory. Solvent effect can be considered explicitly by QM/MM calculations or inexplicitly by PCM (polarized continuum models). The third-order Douglas–Kroll scalar terms were used to calculated relativistic corrections. For large systems, using fragment molecular orbital method is highly recommended. Particularly, it has the interfaces to the valence bond programs, like VB2000 and Xiamen valence bond (XMVB). This computational chemistry code is fully open sourced available. Website: https://www.msg.chem.iastate.edu/gamess.

The Amsterdam Density Functional (ADF) is a program for first-principles electronic structure calculations in understanding and predicting structure, reactivity, and spectra of molecules [97]. The program can run various types of calculation, including optimization of geometry, transition structure optimization, frequency analysis, and intrinsic reaction coordinate (IRC) calculation and electronic excited states. Solvation effects can be considered using the conductor-like screening models (COSMO) method. External electric fields, point charges, and relativistic correction calculations can be also performed. Molecular properties such as ESR and NMR simulations, Raman intensities, and hyperpolarizabilities can be computed. The "BAND" module is used for periodic materials to perform a single-point calculation. This code is commercially available. Website: https://www.scm.com/product/adf/.

Molpro is an ab initio program package for advanced molecular electronic structure calculations [100, 106]. The package has state-of-the art methods with efficient parallelized algorithms, such as DFT, high-level CC and multireference wavefunction methods, such as MCSCF/CASSCF (complete active space self-consistent field), CASPT2, multireference configuration interaction (MRCI), or full configuration interaction (FCI) methods, or response methods such as TDDFT, CC2, and EOM-CCSD. Computing modules include molecular properties, geometry optimization, vibrational frequencies, and further wavefunction analysis. Density-fitting approximations can speed up DFT and MP2 calculations for large systems. F12 explicitly correlated methods with triple-zeta basis sets can yield results near CBS quality. Methods with local approximations can be applied to large molecules accurately. This code is commercially available. Website: http://www.molpro.net.

MOLCAS is an ab initio computational chemistry program package that is specialized in multiconfigurational wavefunction theory [101, 107]. The package has the CASSCF and restricted MCSCF wavefunctions restricted active space self-consistent-field (RASSCF) modules. The program can perform the geometries optimization for equilibrium and transition states and vibrational frequency calculations. Second-order perturbation theory codes CASPT2 and RASPT2 were also implemented. A unique technique named Cholesky decomposition is implemented robustly and efficiently. This code is commercially available. Website: https://www.molcas.org/. It also has an open source version, OpenMolcas. Website: https://gitlab.com/Molcas/OpenMolcas.

PySCF is a quantum chemistry program and a lightweight and efficient platform for calculations and code development [108]. It uses the Python language to equilibrate the convenience in development and computational efficiency. The SCF module includes implementations of HF and DFT for restricted, unrestricted, closed-shell, and open-shell Slater determinant references. Post-HF methods are also available, including Møller–Plesset second-order perturbation theory, configuration interaction, and CC theory. Multiconfigurational calculations can be performed by an interface to external solvers. Relativistic correction can be added by relativistic effective basis set and other complexed methods. This computational chemistry code is fully open source available. Website: https://github.com/pyscf/pyscf.

2.8 The Limitation of Current Computational Methods

Although the application of DFT methods has greatly flourished computational organic and organometallic chemistry, there are still a series of shortcomings and challenges in this field, including the accuracy of DFT methods, exact solvation effect, evaluation of entropy effect, the computation of excited state and high spin state, speculation on the reaction mechanism, mechanism study in biological system, etc.

2.8.1 The Accuracy of DFT Methods

Over the past several decades, hundreds of DFT methods have been developed. Generally speaking, the accuracy has been improved according to the developed date of density functionals; however, their application scope and field are often restricted by their own profiles. In the field of computational organic and organometallic chemistry, the choice of density functional and basis set is often uncertain, which is more often based on past experience of experimenters. This results in computational data depending on the selection of density functional and basis, which may lead to inaccurate and incomparable computational results. The challenge now is to find a density functional that can be adapted to as many systems as possible and with sufficient precision.

2.8.2 Exact Solvation Effect

In computational organometallic chemistry, most of the computing systems are homogeneous catalysis in solution. However, the existing computational capability can only support the calculation method based on the implicit solvent model level. This model has a good effect in dealing with weak polar solvents, but still has a large error for polar solvent. Especially for protonic solvents, it is difficult to get satisfactory results from the implicit solvent model.

2.8.3 Evaluation of Entropy Effect

Entropy effect is very important for multimolecular reactions in organometallic chemistry. In the current computational studies, the entropy correction data often come from the calculation results in gas phase or in the implicit solvent model, which is not enough to reflect the influence of entropy in homogeneous solvent system. The influence of solvents on the entropy majorly restricted translational and rotational entropy and would make the actual free energy of a molecule between the calculated free energy and enthalpy. Sometimes, the inevitable calculation error of the entropy even affects the judgment of the reaction mechanism.

2.8.4 The Computation of Excited State and High Spin State

The d orbitals of transition metals are often occupied by unpaired electrons. Due to the limitation of principle, DFT method often exhibited a critical error in dealing with high spin states. On the other hand, photocatalytic processes involving transition metals currently became the focus of chemists, whose core steps would include the formation, transformation, and quenching of excited states. For this system, DFT calculation also cannot obtain reliable spectral results. In fact, some completely active space-based methods can obtain high-precision results for those systems, but due to the huge amount of calculation requirement, it is still difficult for the exploration of reaction mechanism in photocatalysis with transition metals and large ligands.

2.8.5 Speculation on the Reaction Mechanism

As mentioned in Section 2.6, the initial step of studying the reaction mechanism by computational chemistry is the hypotheses of possible pathways. The critical disadvantage of this approach is that the correctness of mechanism depends on the experience of researchers. If the correct pathway of one reaction is ignored, the true one cannot be obtained by theoretical calculation. Although some computational methods based on the full potential energy surface analysis are being applied to the study of reaction mechanism, the computational efficiency limits their application. Maybe artificial intelligence would be the way out for mechanism exploration in the near future. The possible mechanism of an unknown reaction can be obtained directly by analyzing and screening a large number of previous mechanism research information.

References

1 Cramer, C.J. (2002). *Essential of Computational Chemistry: Theories and Models, 2nd Edition*. New York: John Wiley & Sons.
2 Schrödinger, E. (1926). Quantisierung als Eigenwertproblem. *Annals of Physics* 384: 361–376.
3 Szabo, A. and Ostlund, N.S. (1996). *Modern Quantum Chemistry, Introduction to Advanced Electronic Structure Theory*. New York: Dover Publications.
4 Born, M. and Oppenheimer, R. (1927). Zur Quantentheorie der Molekeln. *Annals of Physics* 389: 457–484.
5 Hartree, D.R. (1928). The wave mechanics of an atom with a non-Coulomb central field. Part I. Theory and methods. *Mathematical Proceedings of the Cambridge Philosophical Society* 24: 89.
6 Fock, V. (1930). Näherungsmethode zur Lösung des quantenmechanischen Mehrkörperproblems. *Zeitschrift für Physik* 61: 126.
7 Clark, T. and Koch, R. (1999). Linear combination of atomic orbitals. In: *The Chemist's Electronic Book of Orbitals*. Berlin, Heidelberg: Springer.
8 Roothaan, C.C.J. (1951). New developments in molecular orbital theory. *Reviews of Modern Physics* 23: 69.
9 Foresman, J.B., Head-Gordon, M., Pople, J.A. et al. (1992). Toward a systematic molecular orbital theory for excited states. *Journal of Chemical Physics* 96 (1): 135–149.
10 Bartlett, R.J. and Purvis, G.D. (1978). Many-body perturbation-theory, coupled-pair many-electron theory, and importance of quadruple excitations for correlation problem. *International Journal of Quantum Chemistry* 14: 561–581.
11 Møller, C. and Plesset, M.S. (1934). Note on an approximation treatment for many-electron systems. *Physical Review* 46: 618–622.
12 Pople, J.A., Binkley, J.S., Seeger, R. et al. (1976). Theoretical models incorporating electron correlation. *International Journal of Quantum ChemistrySupply* 10: 1–19.
13 Pople, J.A., Seeger, R., Krishnan, R. et al. (1977). Variational configuration interaction methods and comparison with perturbation theory. *International Journal of Quantum Chemistry* 12 (Suppl 11): 149–163.

14 Head-Gordon, M., Pople, J.A., Frisch, M.J. et al. (1988). MP2 energy evaluation by direct methods. *Chemical Physics Letters* 153: 503–506.

15 Raghavachari, K. and Pople, J.A. (1978). Approximate 4th-order perturbation-theory of electron correlation energy. *International Journal of Quantum Chemistry* 14: 91–100.

16 Pople, J.A., Head-Gordon, M., and Raghavachari, K. (1987). Quadratic configuration interaction – a general technique for determining electron correlation energies. *Journal of Chemical Physics* 87: 5968–5975.

17 Curtiss, L.A., Raghavachari, K., Redfern, P.C.V. et al. (1998). Gaussian-3 (G3) theory for molecules containing first and second-row atoms. *Journal of Chemical Physics* 109: 7764–7776.

18 Nyden, M.R. and Petersson, G.A. (1981). Complete basis set correlation energies. I. The asymptotic convergence of pair natural orbital expansions. *Journal of Chemical Physics* 75: 1843–1862.

19 Petersson, G.A., Bennett, A., Tensfeldt, T.G. et al. (1988). A complete basis set model chemistry. I. The total energies of closed-shell atoms and hydrides of the first-row atoms. *Journal of Chemical Physics* 89: 2193–2218.

20 Petersson, G.A. and Al-Laham, M.A. (1991). A complete basis set model chemistry. II. Open-shell systems and the total energies of the first-row atoms. *Journal of Chemical Physics* 94: 6081–6090.

21 Martin, J.M.L. and Oliveira, G.de. (1999). Towards standard methods for benchmark quality ab initio thermochemistry – W1 and W2 theory. *Journal of Chemical Physics* 111: 1843–1856.

22 Parthiban, S. and Martin, J.M.L. (2001). Assessment of W1 and W2 theories for the computation of electron affinities, ionization potentials, heats of formation, and proton affinities. *Journal of Chemical Physics* 114: 6014–6029.

23 Raghavachari, K. and Pople, J.A. (1981). Calculation of one-electron properties using limited configuration-interaction techniques. *International Journal of Quantum Chemistry* 20: 67–71.

24 Gauss, J. and Cremer, D. (1988). Analytical evaluation of energy gradients in quadratic configuration-interaction theory. *Chemical Physics Letters* 150: 280–286.

25 Salter, E.A., Trucks, G.W., Bartlett, R.J. et al. (1989). Analytic energy derivatives in many-body methods. I. First derivatives. *Journal of Chemical Physics* 90: 1752–1766.

26 Čížek, J. (1969). Correlation and effects in atoms molecules. *Advances in Chemical Physics* 14: 35.

27 Scuseria, G.E. and Schaefer, H.F. III, (1989). Is coupled cluster singles and doubles (CCSD) more computationally intensive than quadratic configuration-interaction (QCISD)? *Journal of Chemical Physics* 90: 3700–3703.

28 Purvis, G.D. III, and Bartlett, R.J. (1982). A full coupled-cluster singles and doubles model – the inclusion of disconnected triples. *Journal of Chemical Physics* 76: 1910–1918.

29 Scuseria, G.E., Janssen, C.L., and Schaefer, H.F. III, (1988). An efficient reformulation of the closed-shell coupled cluster single and double excitation (CCSD) equations. *Journal of Chemical Physics* 89: 7382–7387.

30 Watts, J.D., Gauss, J., and Bartlett, R.J. (1993). Coupled-cluster methods with noniterative triple excitations for restricted open-shell Hartree–Fock and other general single determinant reference functions. Energies and analytical gradients. *Journal of Chemical Physics* 98: 8718.

31 Strutt, J.W. and Rayleigh, L. (1894). *Theory of Sound*, 2e. London: Macmillan.

32 Pople, J.A., Head-Gordon, M.D., Fox, J.K. et al. (1989). Gaussian-1 theory: a general procedure for prediction of molecular energies. *Journal of Chemical Physics* 90: 5622–5629.

33 Curtiss, L.A., Raghavachari, K.G., Trucks, W. et al. (1991). Gaussian-2 theory for molecular energies of first- and second-row compounds. *Journal of Chemical Physics* 94: 7221–7230.

34 Curtiss, L.A., Redfern, P.C., Raghavachari, K. et al. (2007). Gaussian-4 theory. *Journal of Chemical Physics* 126: 084108.

35 Ochterski, J.W., Petersson, G.A., Montgomery, J.A. et al. (1996). A complete basis set model chemistry. V. Extensions to six or more heavy atoms. *Journal of Chemical Physics* 104: 2598–2619.

36 Hohenberg, P. and Kohn, W. (1964). Inhomogeneous electron gas. *Physical Review* 136: B864–B871.

37 Kohn, W. and Sham, L.J. (1965). Self-consistent equations including exchange and correlation effects. *Physical Review* 140: A1133–A1138.

38 Slater, J.C. (1974). *The Self-Consistent Field for Molecular and Solids, Quantum Theory of Molecular and Solids*, vol. 4. New York: McGraw-Hill.

39 Perdew, J.P. and Schmidt, K. (2001). Jacob's ladder of density functional approximations for the exchange-correlation energy. *AIP Conference Proceedings* 577: 1–20.

40 Perdew, J.P., Ruzsinszky, A., Constantin, L.A. et al. (2009). Some fundamental issues in ground-state density functional theory: a guide for the perplexed. *Journal of Chemical Theory and Computation* 5 (4): 902–908.

41 Becke, A.D. (1988). Density-functional exchange-energy approximation with correct asymptotic-behavior. *Physical Review A* 38: 3098–3100.

42 Lee, C., Yang, W., Parr, R.G. et al. (1988). Development of the Colle–Salvetti correlation-energy formula into a functional of the electron density. *Physical Review B* 37: 785–789.

43 Perdew, J.P., Burke, K., Ernzerhof, M. et al. (1996). Generalized gradient approximation made simple. *Physical Review Letters* 77: 3865–3868.

44 Perdew, J.P. (1986). Density-functional approximation for the correlation energy of the inhomogeneous electron gas. *Physical Review B* 33: 8822–8824.

45 Zhao, Y. and Truhlar, D.G. (2006). A new local density functional for main-group thermochemistry, transition metal bonding, thermochemical kinetics, and noncovalent interactions. *Journal of Chemical Physics* 125: 194101–194118.

46 Tao, J.M., Perdew, J.P., Staroverov, V.N. et al. (2003). Climbing the density functional ladder: nonempirical meta-generalized gradient approximation designed for molecules and solids. *Physical Review Letters* 91: 146401.
47 Van Voorhis, T. and Scuseria, G.E. (1998). A never form for the exchange-correlation energy functional. *Journal of Chemical Physics* 109: 400–410.
48 Becke, A.D. (1993). Density-functional thermochemistry. III. The role of exact exchange. *Journal of Chemical Physics* 98: 5648–5652.
49 Becke, A.D. (1997). Density-functional thermochemistry. V. Systematic optimization of exchange-correlation functionals. *Journal of Chemical Physics* 107: 8554–8560.
50 Cohen, A.J. and Handy, N.C. (2001). Dynamic correlation. *Molecular Physics* 99: 607–615.
51 Adamo, C. and Barone, V. (1999). Toward reliable density functional methods without adjustable parameters: the PBE0 model. *Journal of Chemical Physics* 110: 6158–6169.
52 Adamo, C. and Barone, V. (1998). Exchange functionals with improved long-range behavior and adiabatic connection methods without adjustable parameters: the *m*PW and *m*PW1PW models. *Journal of Chemical Physics* 108: 664–675.
53 Xu, X. and Goddard, W.A. (2004). The X3LYP extended density functional for accurate descriptions of nonbond interactions, spin states, and thermochemical properties. *Proceedings of the National Academy of Sciences of the United States of America* 101: 2673–2677.
54 Zhao, Y., Schultz, N.E., Truhlar, D.G. et al. (2005). Exchange-correlation functional with broad accuracy for metallic and nonmetallic compounds, kinetics, and noncovalent interactions. *Journal of Chemical Physics* 123: 161103.
55 Zhao, Y. and Truhlar, D.G. (2008). The M06 suite of density functionals for main group thermochemistry, thermochemical kinetics, noncovalent interactions, excited states, and transition elements: two new functionals and systematic testing of four M06-class functionals and 12 other functionals. *Theoretical Chemistry Accounts* 120: 215–241.
56 Zhao, Y. and Truhlar, D.G. (2006). Comparative DFT study of van der Waals complexes: rare-gas dimers, alkaline-earth dimers, zinc dimer, and zinc-rare-gas dimers. *Journal of Physical Chemistry* 110: 5121–5129.
57 Zhao, Y. and Truhlar, D.G. (2006). Density functional for spectroscopy: no long-range self-interaction error, good performance for Rydberg and charge-transfer states, and better performance on average than B3LYP for ground states. *Journal of Physical Chemistry A* 110: 13126–13130.
58 Henderson, T.M., Izmaylov, A.F., Scalmani, G. et al. (2009). Can short-range hybrids describe long-range-dependent properties? *Journal of Chemical Physics* 131: 044108.
59 Grimme, S. (2006). Semiempirical hybrid density functional with perturbative second-order correlation. *Journal of Chemical Physics* 124: 034108.

60 Schwabe, T. and Grimme, S. (2006). Towards chemical accuracy for the thermodynamics of large molecules: new hybrid density functionals including non-local correlation effects. *Physical Chemistry Chemical Physics* 8: 4398.

61 Manna, D., Kesharwani, M.K., Sylvetsky, N. et al. (2017). Conventional and explicitly correlated ab initio benchmark study on water clusters: revision of the BEGDB and WATER27 data sets. *Journal of Chemical Theory and Computation* 13: 3136–3152.

62 Bühl, M., Reimann, C., Pantazis, D.A. et al. (2008). Geometries of third-row transition-metal complexes from density-functional theory. *Journal of Chemical Theory and Computation* 4: 1449–1459.

63 Kesharwani, M.K., Karton, A., Martin, J.M.L. et al. (2016). Benchmark ab initio conformational energies for the proteinogenic amino acids through explicitly correlated methods. Assessment of density functional methods. *Journal of Chemical Theory and Computation* 12: 444–454.

64 Theresa, S., Sanhueza, I.A., Kalvet, I. et al. (2015). Computational studies of synthetically relevant homogeneous organometallic catalysis involving Ni, Pd, Ir, and Rh: an overview of commonly employed DFT methods and mechanistic insights. *Chemical Reviews* 115: 9532–9586.

65 Grimme, S., Antony, J., Ehrlich, S. et al. (2010). A consistent and accurate ab initio parameterization of density functional dispersion correction (DFT-D) for the 94 elements H-Pu. *Journal of Chemical Physics* 132: 154104.

66 Hehre, W.J., Stewart, R.F., Pople, J.A. et al. (1969). Self-consistent molecular orbital methods. 1. Use of Gaussian expansions of Slater-type atomic orbitals. *Journal of Chemical Physics* 51: 2657–2664.

67 Collins, J.B., von Schleyer, P.R., Binkley, J.S. et al. (1976). Self-consistent molecular orbital methods. 17. Geometries and binding energies of second-row molecules. A comparison of three basis sets. *Journal of Chemical Physics* 64: 5142–5151.

68 Ditchfield, R., Hehre, W.J., Pople, J.A. et al. (1971). Self-consistent molecular orbital methods. 9. Extended Gaussian-type basis for molecular-orbital studies of organic molecules. *Journal of Chemical Physics* 54: 724.

69 Rassolov, V.A., Ratner, M.A., Pople, J.A. et al. (2001). 6-31G* basis set for third-row atoms. *Journal of Computational Chemistry* 22: 976–984.

70 Binkley, J.S., Pople, J.A., Hehre, W.J. et al. (1980). Self-consistent molecular orbital methods. 21. Small split-valence basis sets for first-row elements. *Journal of the American Chemical Society* 102: 939–947.

71 Wachters, A.J.H. (1970). Gaussian basis set for molecular wavefunctions containing third-row atoms. *Journal of Chemical Physics* 52: 1033.

72 McLean, A.D. and Chandler, G.S. (1980). Contracted Gaussian-basis sets for molecular calculations. 1. 2nd row atoms, $Z = 11$–18. *Journal of Chemical Physics* 72: 5639–5648.

73 Dunning, T.H. (1989). Gaussian basis sets for use in correlated molecular calculations. I. The atoms boron through neon and hydrogen. *Journal of Chemical Physics* 90: 1007–1023.

References

74 Kendall, R.A., Dunning, T.H., Harrison, R.J. et al. (1992). Electron affinities of the first-row atoms revisited. Systematic basis sets and wave functions. *Journal of Chemical Physics* 96: 6796–6806.

75 Woon, D.E. and Dunning, T.H. (1993). Gaussian-basis sets for use in correlated molecular calculations. 3. The atoms aluminum through argon. *Journal of Chemical Physics* 98: 1358–1371.

76 Davidson, E.R. (1996). Comment on 'comment on Dunning's correlation-consistent basis sets. *Chemical Physics Letters* 260: 514–518.

77 Fuentealba, P., Preuss, H., Stoll, H. et al. (1982). A proper account of core-polarization with pseudopotentials – single valence-electron alkali compounds. *Chemical Physics Letters* 89: 418–422.

78 Wadt, W.R. and Hay, P.J. (1985). Ab initio effective core potentials for molecular calculations – potentials for main group elements Na to Bi. *Journal of Chemical Physics* 82: 284–298.

79 Weigend, F. and Ahlrichs, R. (2005). Balanced basis sets of split valence, triple zeta valence and quadruple zeta valence quality for H to Rn: design and assessment of accuracy. *Physical Chemistry Chemical Physics* 7: 3297–3305.

80 Hay, P.J. and Wadt, W.R. (1985). Ab initio effective core potentials for molecular calculations – potentials for the transition-metal atoms Sc to Hg. *Journal of Chemical Physics* 82: 270–283.

81 Schwerdtfeger, P., Dolg, M., Schwarz, W.H.E. et al. (1989). Relativistic effects in gold chemistry. 1. Diatomic gold compounds. *Journal of Chemical Physics* 91: 1762–1774.

82 Stevens, W.J., Basch, H., and Krauss, M. (1984). Compact effective potentials and efficient shared-exponent basis-sets for the 1st-row and 2nd-row atoms. *Journal of Chemical Physics* 81: 6026–6033.

83 Roy, L.E., Hay, P.J., Martin, R.L. et al. (2008). Revised basis sets for the LANL effective core potentials. *Journal of Chemical Theory and Computation* 4: 1029–1031.

84 Schäfer, A., Horn, H., Ahlrichs, R. et al. (1992). Fully optimized contracted Gaussian basis sets for atoms Li to Kr. *Journal of Chemical Physics* 97: 2571.

85 Schäfer, A., Horn, H., Ahlrichs, R. et al. (1994). Fully optimized contracted Gaussian basis sets of triple zeta valence. *Journal of Chemical Physics* 100: 5829.

86 Hättig, C. (2005). Optimization of auxiliary basis sets for RI-MP2 and RI-CC2 calculations: core–valence and quintuple-ζ basis sets for H to Ar and QZVPP basis sets for Li to Kr. *Physical Chemistry Chemical Physics* 7: 59–66.

87 Hellweg, A., Hättig, C., Höfener, S. et al. (2007). Optimized accurate auxiliary basis sets for RI-MP2 and RI-CC2 calculations for the atoms Rb to Rn. *Theoretical Chemistry Accounts* 117: 587–597.

88 Tomasi, J., Mennucci, B., Cammi, R. et al. (2005). Quantum mechanical continuum solvation models. *Chemical Reviews* 105: 2999–3093.

89 Tomasi, J., Mennucci, B., Cancès, E. et al. (1999). The IEF version of the PCM solvation method: an overview of a new method addressed to study molecular

solutes at the QM ab initio level. *Journal of Molecular Structure THEOCHEM* 464: 211–226.
90 Cossi, M., Rega, N., Scalmani, G. et al. (2003). Energies, structures, and electronic properties of molecules in solution with the C-PCM solvation model. *Journal of Computational Chemistry* 24: 669–681.
91 Barone, V. and Cossi, M. (1998). Quantum calculation of molecular energies and energy gradients in solution by a conductor solvent model. *Journal of Physical Chemistry A* 102: 1995–2001.
92 Foresman, J.B., Keith, T.A., Wiberg, K.B. et al. (1996). Solvent effects 5. The influence of cavity shape, truncation of electrostatics, and electron correlation on ab initio reaction field calculations. *Journal of Physical Chemistry* 100: 16098–16104.
93 Marenich, A.V., Cramer, C.J., Truhlar, D.G. et al. (2009). Universal solvation model based on solute electron density and a continuum model of the solvent defined by the bulk dielectric constant and atomic surface tensions. *Journal of Physical Chemistry B* 113: 6378–6396.
94 Chai, J.D. and Head-Gordon, M. (2008). Long-range corrected hybrid density functionals with damped atom–atom dispersion corrections. *Physical Chemistry Chemical Physics* 10: 6615–6620.
95 Chai, J.D. and Head-Gordon, M. (2008). Systematic optimization of long-range corrected hybrid density functionals. *Journal of Chemical Physics* 128: 084106.
96 Frisch, M.J., Trucks, G.W., Schlegel, H.B. et al. (2010). *Gaussian 09*. Wallingford, CT: Gaussian.
97 Velde, G., Bickelhaupt, F.M., Baerends, E.J. et al. (2001). Chemistry with ADF. *Journal of Computational Chemistry* 22: 931.
98 Neese, F. (2012). The ORCA program system. *Wiley Interdisciplinary Reviews: Computational Molecular Science* 2: 73–78.
99 Barca, G.M.J., Bertoni, C., Carrington, L. et al. (2020). Recent developments in the general atomic and molecular electronic structure system. *Journal of Chemical Physics* 152: 154102.
100 Werner, H.-J., Knowles, P.J., Knizia, G. et al. (2012). Molpro: a general-purpose quantum chemistry program package. *Wiley Interdisciplinary Reviews: Computational Molecular Science* 2: 242–253.
101 Aquilante, F., Autschbach, J., Carlson, R.K. et al. (2016). MOLCAS 8: new capabilities for multiconfigurational quantum chemical calculations across the periodic table. *Journal of Computational Chemistry* 37: 506–541.
102 Shao, Y., Gan, Z., and Epifanovsky, E. (2015). Advances in molecular quantum chemistry contained in the Q-Chem 4 program package. *Molecular Physics* 113: 184–215.
103 Valiev, M., Bylaska, E.J., and Govind, N. (2010). NWChem: a comprehensive and scalable open-source solution for large scale molecular simulations. *Computer Physics Communications* 181: 1477.
104 Dam, H.J.J., Jong, W.A., and Bylaska, E. (2011). NWChem: scalable parallel computational chemistry. *Wiley Interdisciplinary Reviews: Computational Molecular Science* 1: 888–894.

105 Balasubramani, S.G., Chen, G.P., and Coriani, S. (2020). TURBOMOLE: modular program suite for ab initio quantum-chemical and condensed-matter simulations. *Journal of Chemical Physics* 152: 184107.

106 Ma, Q. and Werner, H.-J. (2018). Explicitly correlated local coupled-cluster methods using pair natural orbitals. *Wiley Interdisciplinary Reviews: Computational Molecular Science* 8: e1371.

107 Aquilante, F., Pedersen, T.B., and Veryazov, V. (2012). MOLCAS – a software for multiconfigurational quantum chemistry calculations. *Wiley Interdisciplinary Reviews: Computational Molecular Science* 3: 143–149.

108 Sun, Q., Berkelbash, T.C., and Blunt, N.S. (2018). PYSCF: the Python-based simulations of chemistry framework. *Wiley Interdisciplinary Reviews: Computational Molecular Science* 8: e1340.

3

Elementary Reactions in Organometallic Chemistry

Classically, a reaction can be considered as the reconstruction of interatomic connections from reactants to products. In organometallic chemistry, metals are involved in this process, which could play as reactant and/or catalyst. In this area, the redistribution of connectivity of atoms often involves many steps of formation and cleavage of covalent bonds; therefore, this process would reveal great complexity. Due to the diversity of bonding with metal, the introduction of metal into organic reactions further increases the complexity of corresponding structures and reactions. To have a clearer understanding of how a reaction takes place, the whole process can be usually divided into many individual steps with specific sequence, which is called reaction mechanism. Understanding reaction mechanism is one of the fundamental goals of chemists. The factors affecting the reactivity and selectivity can be revealed by studying the reaction mechanism, which is helpful to further optimization of the reaction conditions and to selection of reaction conditions. A mechanism describes the changes of molecular structure along time coordinate, which can be likened to a motion picture of the chemical transformation of molecular structure. Each step of the reaction can be observed in a sequential manner by analyzing each frame of the motion picture individually.

3.1 General View of Elementary Reactions in Organometallic Chemistry

In an organic reaction, the reaction mechanism, i.e. the reassembly of atomic connections, can be considered as the change of the bonded electrons sharing way by the movement of nuclei. For the main group elements, because their valence electronic structure is relatively simple, there are three major ways to change the way of electron sharing: the polar process with a pair of electrons transfer from nucleophile to electrophile, the radical process with unpaired electron transfer, and the pericyclic process with a series of electrons cooperative transfer. In contrast, the d-orbitals leads to more complex valence electronic structure of transition metals, in which both occupied and unoccupied orbitals are presented. In other words, the presence of transition metal atoms can act as both nucleophile and electrophile, which greatly increases the complexity of an organometallic reaction. Moreover, the

Computational Methods in Organometallic Catalysis: From Elementary Reactions to Mechanisms,
First Edition. Yu Lan.
© 2021 WILEY-VCH GmbH. Published 2021 by WILEY-VCH GmbH.

core of transition metal catalysis lies in the formation and is followed by transformation of carbon–metal bonds, which can undergo a series of bonding and breaking processes in the reaction course. To describe the mechanism of organometallic reaction, a reaction is often divided into several subprocesses, each of which undergoes a regularity of change; therefore, it is defined as elementary reaction. To be exact, the elementary reactions are not a reaction mechanism, but a part of reaction mechanism. There are many possible mechanisms for each elementary reaction itself, while the mechanism of each elementary reaction also can be distinguished into polar process, radical process, and pericyclic process. Therefore, we will elaborate on the related mechanism of the common elementary reactions in organometallic chemistry with the combination of theoretical computational chemistry in this chapter. Understanding the mechanism of each elementary reaction should help us explore the mechanism of the whole reaction.

3.2 Coordination and Dissociation

3.2.1 Coordination Bond and Coordination

Coordination bond can be regarded as a special covalent bond, which is formed when the shared pair of electrons in the covalent bond is provided by one of those two atoms as well as the empty orbital is supplied by another one [1–3]. The formation of coordination bonds is similar to that of covalent bonds. The two electrons shared between bonded atoms are not provided by each one, but from one atom only. In organometallic chemistry, transition metal atom usually has empty d orbitals, which can accept the electrons from other atoms or groups (Scheme 3.1) [4, 5]. Those atoms or groups, which can share electrons with metal, are usually named *ligand* in organometallic chemistry. Coordination can be considered as a polar reaction, in which a metal with unoccupied d orbital plays as an electrophile as well as a ligand with electrons plays as a nucleophile. Therefore, coordination is the combination of an electrophile (metal) and a nucleophile (ligand).

The most common ligand is to use lone-pair electrons to coordinate with metals to form coordination bonds. Due to the weak binding ability of nuclei to the electrons in π bonding orbital, a π bond can share the electrons in bonding orbital with the empty d orbital of transition metals to form a weak coordination bond (Scheme 3.2a). Ethylenes and acetylenes are the most common π ligands [6–10]. In some cases, electrons in σ bond can also be shared with empty d orbital of transition metals, thus forming σ coordination. For instance [11–16], dihydrogen can coordinate onto some transition metals by the σ bond between its two hydrogen

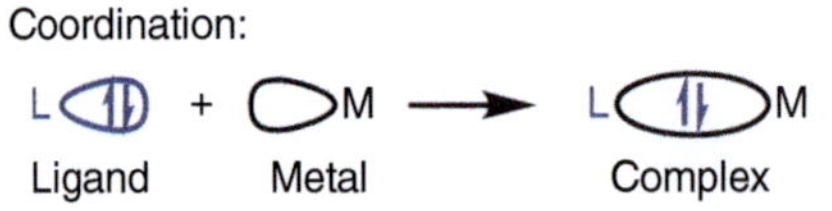

Coordination:

L⬡ + ◯M ⟶ L⬡M
Ligand   Metal   Complex

Scheme 3.1 The coordination of ligand onto metal.

atoms (Scheme 3.2b). Moreover, the occupied d orbital in transition metals also can donate electrons to the unoccupied orbital of ligand, which leads to the formation of back-donation π bond. In most cases, the acceptor of back-donation π bond is molecules with low-energy level anti π* bond orbital, such as ethylene, acetylene, and carbon monoxide. The current understanding on the nature of back donation bond is based on the Dewar–Chatt–Duncanson model [17]. Because the radius of hydrogen atom is very small, σ* antibonding orbital of dihydrogen molecule can also form back-donation bond with some transition metal [18, 19]. In particular, metal-(η^2-silane) interactions also can be considered as the coordination of σ-bonding orbital of Si—H onto unoccupied d orbital of metal and the back-donation of occupied d orbital of metal onto unoccupied σ* antibonding orbital of Si—H (Scheme 3.2c) [13, 20–22].

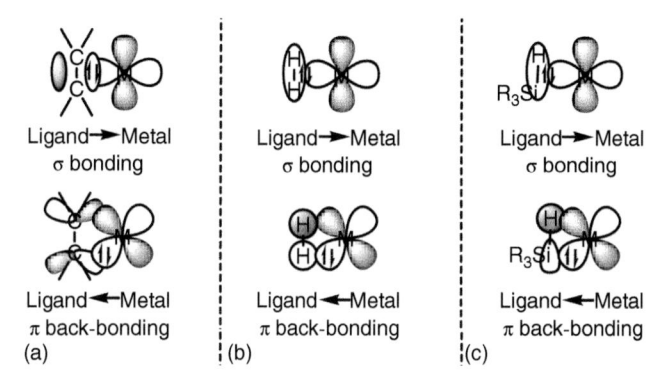

Scheme 3.2 Back-donation bonds in metal-η^2–alkene (a), metal-η^2–dihydrogen (b), and metal-η^2–silane (c) complexes.

The formation of coordination bond involves sharing electrons between ligand and metal; therefore, this process is often accompanied by the reduction of enthalpy, which is favorable for the total free energy. On the other hand, due to the formation of coordination bond, the degree of freedoms and/or number of molecules decreases also lead to a lower entropy, which is unfavorable for the total free energy. Therefore, the change of total free energy is dependent on the effects of both enthalpy and entropy. When a strong ligand coordinates with metal, reduction of enthalpy is enough to offset the disadvantage of entropy; therefore, it is a thermodynamically favorable process. In contrast, when a weak ligand is used, entropy loss is dominant, and the coordination is an unfavorable process. The change of free energy caused by coordination can be accurately calculated by density functional theory (DFT) methods.

Generally, the overlap of orbitals in forming coordination bond between ligand and metal increases when a ligand approaches a metal; therefore, the enthalpy decreases gradually in the process of coordination. It means that there is no energy barrier in this process according to the approach of the ligand and the metal. As shown in Figure 3.1, DFT calculations clearly revealed that the relative enthalpy is decreased when the phosphine ligand is close to palladium. However, the ligand

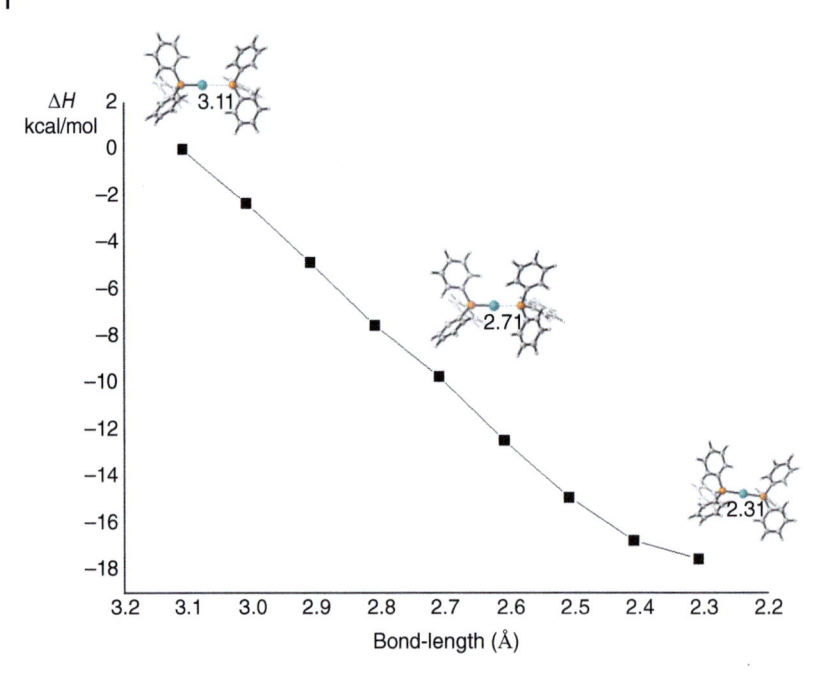

Figure 3.1 The free-energy profiles for the coordination of phosphine onto palladium. The energies were calculated at M06L/6-311+G(d,P) (SDD for Pd)/SMD(Tetrahydrofuran)// B3LYP/6-31G(d) (SDD for Pd) level of theory.

coordination may result in a change in the geometry of the metal center in some cases. Because the change of geometry will lead to the increase followed by decrease of relative enthalpy, the coordination process will undergo a transition state with an energy barrier.

In another case (Figure 3.2), Lin and coworkers [23] found that the coordination of acetylene onto osmium (Os) will change the geometry of Os from trigonal bipyramidal geometry **3-1** to octahedral **3-3**; therefore, a transition state of **3-2ts** is located with a free-energy barrier of 14.6 kcal/mol. Interestingly, this free-energy barrier would be majorly attributed to the entropy loss because the relative activation enthalpy is only 2.3 kcal/mol. Usually, the barrier of coordination is relatively low. Coordination will not be the rate-determining step; therefore, the detailed process of that is often ignored in theoretical studies.

In fact, the change of coordination modes between ligand and metal can be a complicated process. $Cr(CO)_3$ can coordinate with polycyclic arenes [24]. Figure 3.3 depicts the free-energy profiles for moving $Cr(CO)_3$ from a center of six-membered ring to the center of its neighbor. η^6-oligoacene-coordinated $Cr(CO)_3$ **3-4** is the stable species in this case, which can isomerize to a η^1-oligoacene-coordinated intermediate **3-6** via a η^3-oligoacene-coordinated transition state **3-5ts** with an energy barrier of 12.3 kcal/mol. The second η^3-oligoacene-coordinated transition state **3-7ts** accomplishes the migration of $Cr(CO)_3$ from a center of six-membered ring to its neighbor.

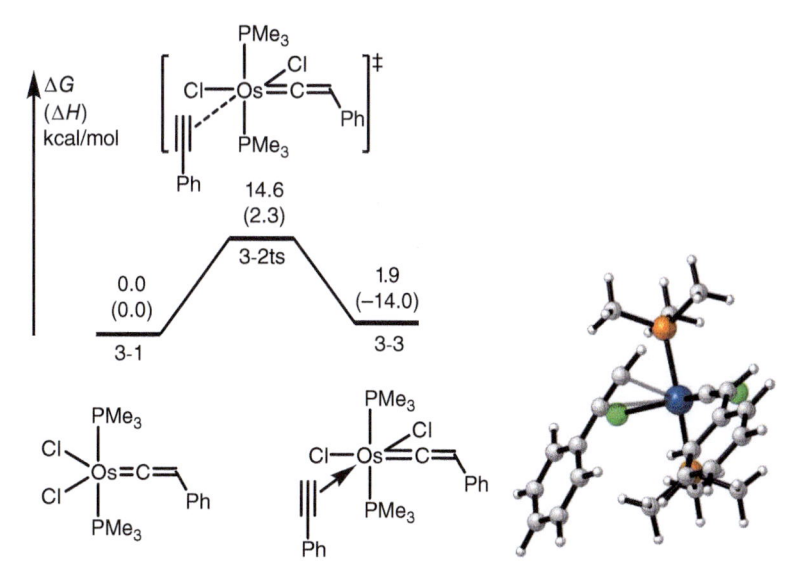

Figure 3.2 The free-energy profiles for the coordination of acetylene onto Os. The energies were calculated at B3LYP-D3/6-31G(d) (Lanl2dz for Os, P, and Cl)/SMD(DCM)// B3LYP-D3/6-31G(d) (Lanl2dz for Os, P, and Cl) level of theory. Values given in parentheses are relative enthalpies. Source: Based on Wen, T. B., Lee, K.-H., Chen, J. et al. (2016). Preparation of Osmium η3-Allenylcarbene Complexes and Their Uses for the Syntheses of Osmabenzyne Complexes. Organometallics 35 (10): 1514-1525.

Consideration for the coordination of various ligands is significant for the analysis of potential energy surface of organometallic catalysis. As shown in Figure 3.4, in theoretical study on the mechanism of Ni-catalyzed pyridine alkenylation [25], Ni(0)–carbene complex **3-9** can be further coordinated by pyridine, pyridine–AlMe$_3$ complex, or acetylene. DFT calculation showed that the relative free energy of last one (**3-14**) is lowest. Therefore, it is chosen as the relative zero for the potential energy surface. Maybe the chosen relative zero does not affect the theoretical predicted selectivity; however, improper selection may affect the theoretical estimation of the overall activation energy for the whole catalytic cycle.

3.2.2 Dissociation

Dissociation is often considered as the reverse reaction of coordination, in which a heterolytic cleavage of coordination bond takes place with the ligand leaving with its own electrons [26–28]. From the point of view of Gibbs free energy, the dissociation process is favorable for entropy, but unfavorable for enthalpy [29]. Because dissociation is an inverse reaction of coordination, in many cases, dissociation is an enthalpy-increasing process; therefore, transition state is often not required in this process. Following this idea, only the initial and final states of dissociation process are studied in general mechanistic studies of organometallic catalysis.

In fact, the activation energy of this process is often very low even if the geometry changes lead to the dissociation undergoing a transition state. For instance [30], in

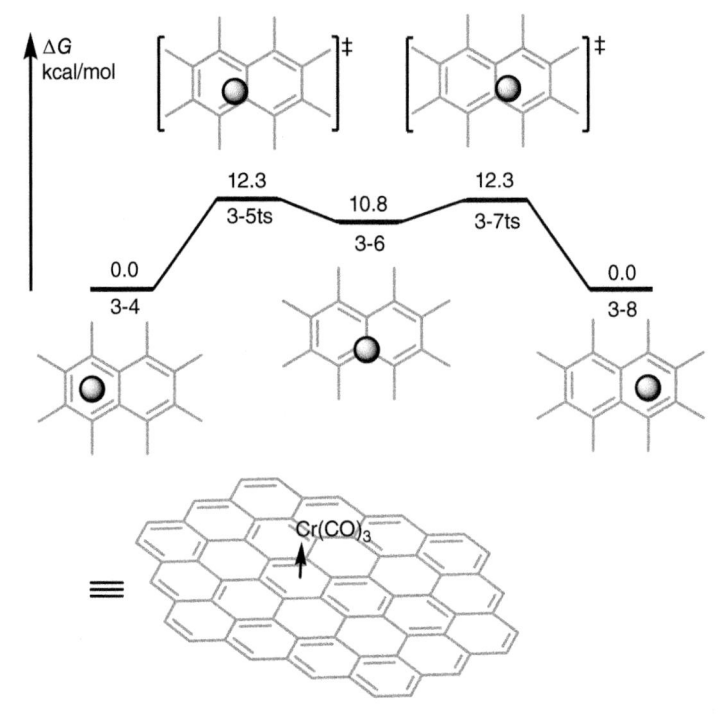

Figure 3.3 The free-energy profiles for the moving of Cr(CO)$_3$ on a polycyclic arene. The free energies were calculated at M11-L/6-311+G(d)(SDD for Cr)//B3LYP/6-31G(d) (SDD for Cr) level of theory.

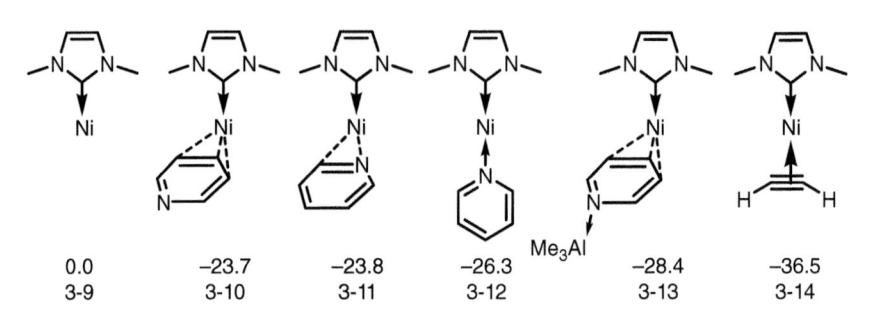

Figure 3.4 The relative free energies of Ni–carbene complexes. The free energies were calculated at B3LYP-D3/6-311+G(d)(SDD for Ni)/CPCM(toluene)//B3PW91/6-31G(d)(Lanl2dz for Ni) level of theory.

a tetrahedral Ni(0) species **3-15**, the dissociation of one phosphine ligand can take place via transition state **3-16ts** with a free-energy barrier of only 8.7 kcal/mol to afford a triangular Ni(0) species **3-17**. This process does not affect either mechanism or selectivity of the whole catalytic cycle. Merely, the calculation for the details of dissociation is beneficial for understanding the detail of an organometallic reaction (Figure 3.5).

Figure 3.5 The free-energy profiles for the dissociation of phosphine ligand from Ni-complex. The energies were calculated at B3LYP-D3/6-311G(d,p) (SDD for Ni)/IEF-PCM(toluene)//B3LYP-D3/6-31 (SDD for Ni) level of theory. Values given in parentheses are relative enthalpies.

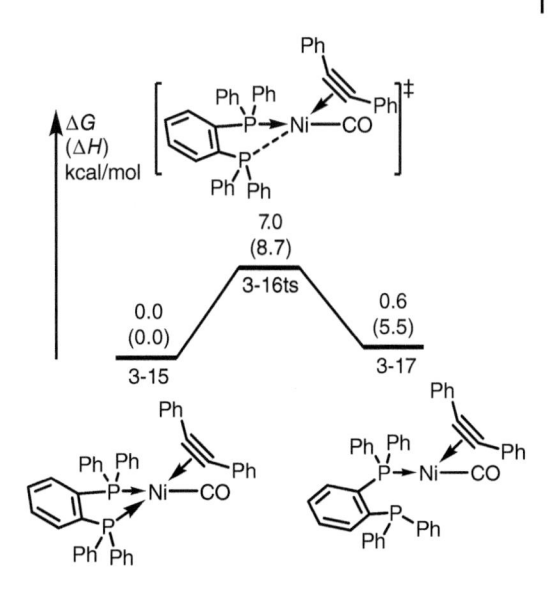

3.2.3 Ligand Exchange

Ligand exchange can be regarded as the result of the nucleophilic substitution of one ligand in the complex by another ligand. As shown in Scheme 3.3, there are three possible pathways for the ligand exchange process, including dissociative, associative replacement, and associative course [31–34].

Dissociative pathway:

L$_1$ + M—L$_2$ ⟶ L$_1$ + M + L$_2$ ⟶ L$_1$—M + L$_2$

Associative pathway:

L$_1$ + M—L$_2$ ⟶ L$_1$—M—L$_2$ ⟶ L$_1$—M + L$_2$

Coincerted associative replacement pathway:

L$_1$ + M—L$_2$ ⟶ [L$_1$---M---L$_2$]‡ ⟶ L$_1$—M + L$_2$

Scheme 3.3 Possible pathways for the ligand exchange.

An example given by Houk and coworkers revealed the mechanism of product release step in Pd-catalyzed alkyl halide formation reaction [35]. The dissociative mechanism (path A) involves dissociation of the alkyl halide from complex **3-18** to form a monoligated Pd(0) complex **3-19**, which associates with one P(t-Bu)$_3$ ligand to regenerate the catalyst **3-20**. The associative replacement mechanism (path C) involves a concerted replacement of the alkyl halide with P(t-Bu)$_3$ via a three-centered transition state **3-22ts**. It is called associative replacement by McMullin et al. [36]. The associative mechanism (path B) involves a stepwise

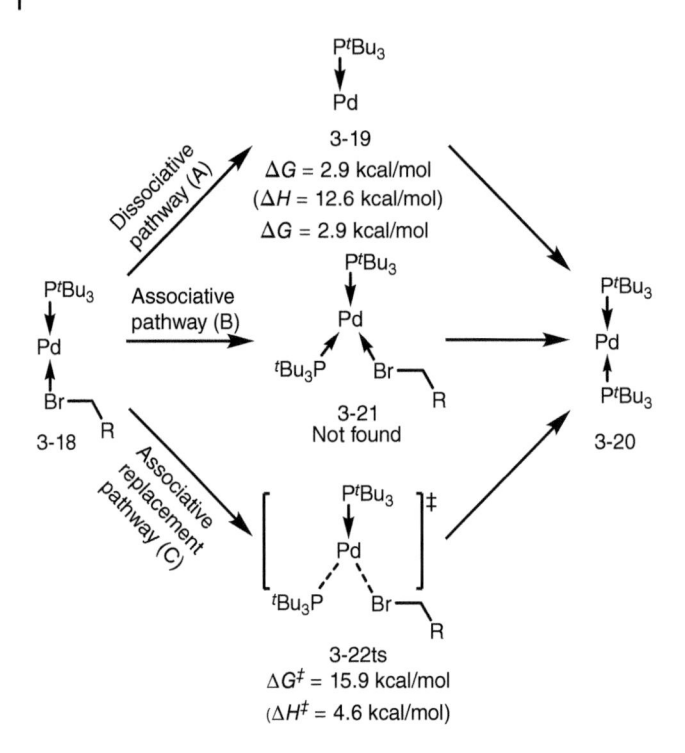

Figure 3.6 The mechanism of ligand exchange in the regeneration of Pd-catalyst. The energies were calculated at BP86/6-31+G(d) (SDD for Pd and Br)/CPCM(toluene)//BP86/6-31+G(d) (SDD for Pd and Br) level of theory. Values given in parentheses are relative enthalpies.

displacement of the alkyl halide via a tricoordinated Pd(0) intermediate **3-21**; however, it is not found by DFT calculations. The intermediates and transition states involved in the different pathways of reductive elimination of alkyl iodide were calculated, and the results are summarized in Figure 3.6.

Interestingly, the configuration isomerization of some organometallic complexes can take place through a ligand exchange–assisted process. As shown in Figure 3.7, the $Cl_2Pd(PH_3)_2$ complex can exist as two isomers [37]. The trans isomer is 6.4 kcal/mol more stable than cis one by DFT calculation. A trans-to-cis isomerization becomes possible when a third phosphine ligand is involved. The new PH_3 enters the metal coordination sphere and determines the expulsion of one of the two PH_3 groups attached to the Pd atom by overcoming a barrier of 38.3 kcal/mol via transition state **3-24ts**. In the geometry of this transition state, the lengths of forming and breaking P—Pd bond are 3.441 and 3.393 Å, which are longer than a typical P—Pd coordination bond.

In some cases, a ligand can coordinate with a metal by using different ligand sites, resulting in isomerization between the yielded complexes, which also can be considered as a unique ligand exchange. As shown in Figure 3.8 [38], both of the carbonyl and ethylene moieties of ketene can be used as a ligand site to coordinate with Ni, resulting in complexes **3-26** and **3-28**, respectively. The isomerization

Figure 3.7 The free-energy profiles for the cis-/trans-isomerization of Cl$_2$Pd(PH$_3$)$_2$. The energies were calculated at B3LYP/6-31G(d) (SDD for Pd)/PCM(CH$_2$Cl$_2$)//B3LYP/6-31G(d) (SDD for Pd) level of theory.

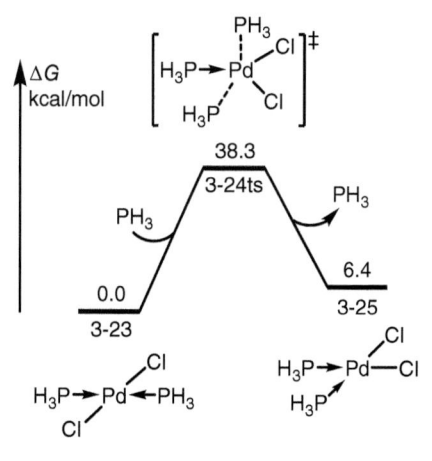

Figure 3.8 The free-energy profiles for the Ni-shift on the ketene. The energies were calculated at BP86/TZVP-SVP/PCM(benzene)//BLYP/6-31G(d) (SDD for Pd) level of theory.

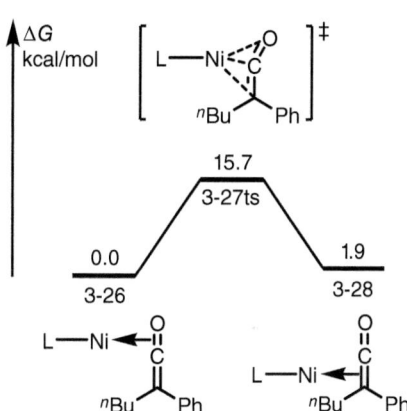

from **3-26** to **3-28** can take place via transition state **3-27ts** with a free-energy barrier of 15.7 kcal/mol, which can be considered as an associative replacement mechanism.

3.3 Oxidative Addition

A typical oxidative addition is the addition of covalently linked A–B molecule onto low oxidative state metal (Scheme 3.4). The result of oxidative addition is that metal complex with A and B as ligands is obtained, in which both the oxidative state and coordination number of metal increase. From mechanistic point of view, oxidative addition can take place through concerted addition [39–49], substitution-type addition [50–54], or radical-type addition [55–62]. Moreover, oxidative cycloaddition is

$$L_n-M^n + A-B \longrightarrow L_n-M^{a+2}{\underset{A}{\overset{B}{|}}}$$

Scheme 3.4 Typical oxidative addition.

also considered as a particular oxidative addition [63–70]. The significance of oxidative addition is to use a metal to activate inert covalent bonds in organic compounds and obtain active metal–carbon covalent bonds for further conversions.

3.3.1 Concerted Oxidative Addition

As a double-electron transformation, concerted oxidative addition is the most common form of this type of elementary reactions [71–73]. As shown in Scheme 3.5, when a covalent bond A—B in organic molecule reacts with metal, metal inserts into A—B covalent bond to cleave this bond and lead to the formation of new M—A and M—B covalent bonds synchronously in formed A—M—B complex. In this type of elementary reactions, A—B covalent bond is often a polar bond formally formed by nucleophile A and electrophile B. After oxidative addition process, both of A and B moieties are nucleophiles because M—A and M—B covalent bonds are formed. It means that the oxidative addition leads to the conversion of electrophile to nucleophile, which would change the reactivity of that species. The oxidative addition leads to the breaking of an A—B bond and formation of an A—M bond and a B—M bond; therefore, it is often an exergonic process [35, 74, 75].

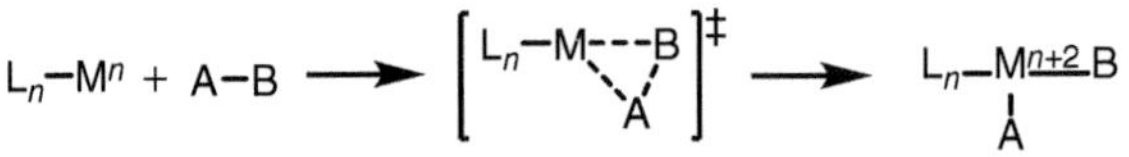

$$L_n\text{—}M^n + A\text{—}B \longrightarrow \left[L_n\text{—}M\text{---}B \atop A \right]^{\ddagger} \longrightarrow L_n\text{—}M^{n+2}\text{—}B \atop A$$

Scheme 3.5 Concerted oxidative addition.

The oxidative addition of aryl halides to Pd is a very common process undergoing a concerted three-membered ring-type transition state, which is the initial step in a series of Pd-catalyzed cross-coupling reactions (Scheme 3.6a) [76–80]. In this transition state, the C(Ar)—X bond is partial cleavage and the C(Ar)—Pd and X—Pd bonds are still not formed. Therefore, an energy barrier is necessary for this process. Analysis of the frontier molecular orbitals (FMOs) of the oxidative addition transition state is shown in Scheme 3.6b [81]. σ_{C-X} orbital donates electrons to the empty p_y orbital of Pd, while a back-donation of occupied d_{xy} orbital in Pd to σ^*_{C-X} orbital is also present. Interestingly, a second-order orbital iteration of occupied d_{xy} orbital in Pd with empty π^*C—X bond in aryl halides promotes this process. The FMO analysis for concerted oxidative addition revealed that the metal provides one pair of electrons to form two covalent bonds; however, the electrons of those two covalent bonds are often formally placed to other groups due to the low electronegativity of metal. Therefore, the formal oxidative state of the metal is increased by +2.

The DFT calculated free-energy profiles for a typical oxidative addition process are shown in Figure 3.9 [82]. The reaction starts to form a diphosphine coordinated Pd(0) species **3-29**, which could undergo a ligand exchange with aryl chloride to form an arene-coordinated Pd(0) complex **3-30** with 18.1 kcal/mol endergonic. The oxidative addition takes place via a three-membered ring-type transition state **3-31ts** with a free-energy barrier of 8.8 kcal/mol to form an aryl Pd(II) complex **3-32**. The geometry of **3-31ts** is also given in Figure 3.9. The lengths of breaking C—Cl and

3.3 Oxidative Addition

Scheme 3.6 (a) The oxidative addition of aryl halide onto Pd(0). (b) The secondary orbital interaction in transition state.

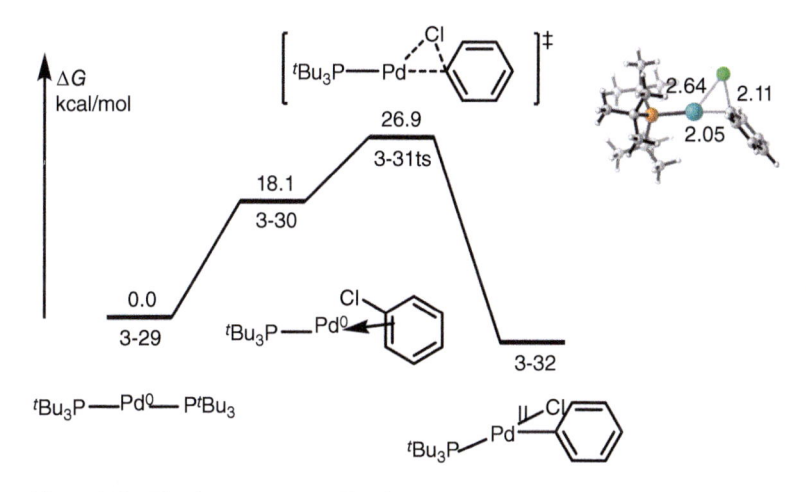

Figure 3.9 The free-energy profiles for the oxidative addition of aryl chloride onto Pd(0). The energies were calculated at B3LYP/LACVP*/PB-SCRF(DMF)//B3LYP/LACVP* level of theory.

forming Pd—C and Pd—Cl bonds are 2.11, 2.05, and 2.64 Å, respectively, which are significantly longer than the corresponding typical single bonds.

Sometimes, the oxidative addition even can occur with a double bond. As an example (Figure 3.10) [38], when ketene coordinates with Ni(0) species by its C=C double bond in complex **3-28**, the oxidative addition can take place via transition state **3-33ts** with a free-energy barrier of 26.8 kcal/mol to afford a carbonyl-coordinated Ni–carbene complex **3-34**. The geometry of transition state **3-33ts** is shown in Figure 3.10, the lengths of breaking C=C bond, forming C(carbonyl)=Ni and C(carbene)=Ni bonds are 2.04, 1.75, and 1.96 Å, respectively, which are longer than corresponding typical double bond; however, they are

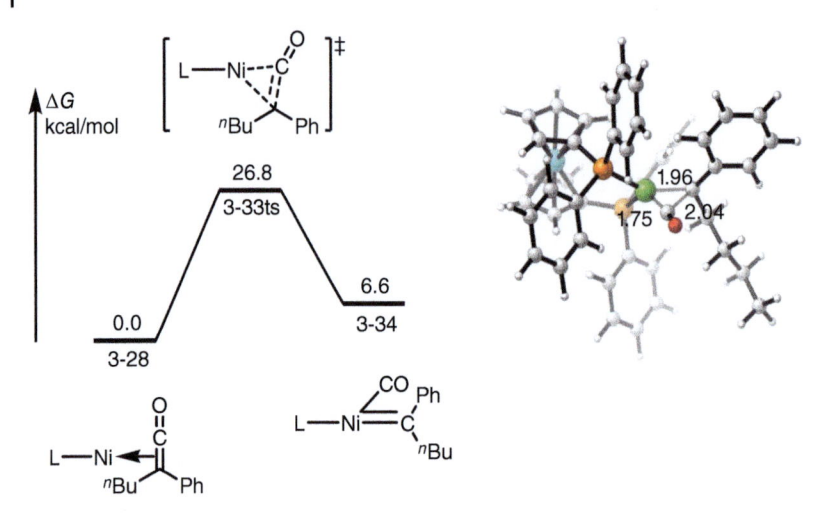

Figure 3.10 The free-energy profiles for the oxidative addition of ketene onto Ni(0). The energies were calculated at BP86/TZVP-SVP/PCM(benzene)//BLYP/6-31G(d) (SDD for Pd) level of theory. Source: Based on Staudaher et al. [38].

significantly shorter than the corresponding single bond. This process can be considered as a rearrangement of a three-membered metal cycle.

As a variant, if there are conjugated groups in the oxidant, the corresponding oxidative addition might undergo via a five-membered ring-type transition state instead of three-membered ring [31, 83–85]. As shown in Figure 3.11 [86], when N-bromosuccinimide (NBS) is used as an oxidant to react with Rh(III) species **3-35**, the oxidative addition can occur via a three-membered ring-type transition state **3-36ts** with an energy barrier of 16.6 kcal/mol. The geometry of transition state **3-36ts** reveals that the Rh center is steric crowded. Alternatively, oxidative addition also could take place via a five-membered ring-type transition state **3-38ts**, which involves a carbonyl moiety in the ring. The relative free energy of **3-38ts** is 2.7 kcal/mol lower than that of **3-36ts**. Although 2.7 kcal/mol free-energy difference by DFT calculation cannot exclude three-membered ring-type concerted oxidative addition, the computational results at least show that the five-membered ring-type transition state is an alternative choice.

3.3.2 Substitution-type Oxidative Addition

Substitution-type oxidative addition is another double-electron transformation. The d orbital with a pair of occupied electrons in transition metal plays as a nucleophile, which can react with a polar A(nucleophile)–B(electrophile) bond leading to the heterolytic cleavage of A—B covalent bond with the formation of new M—B bond in cationic B–M$^+$ complex and release of anionic A$^-$. The polarity of electrophile is also reversed after the substitution-type oxidative addition. The substitution-type oxidation addition can be divided into the following categories (Scheme 3.7): (i) S_N2-type substitution with neutral electrophiles, (ii) S_N2-type

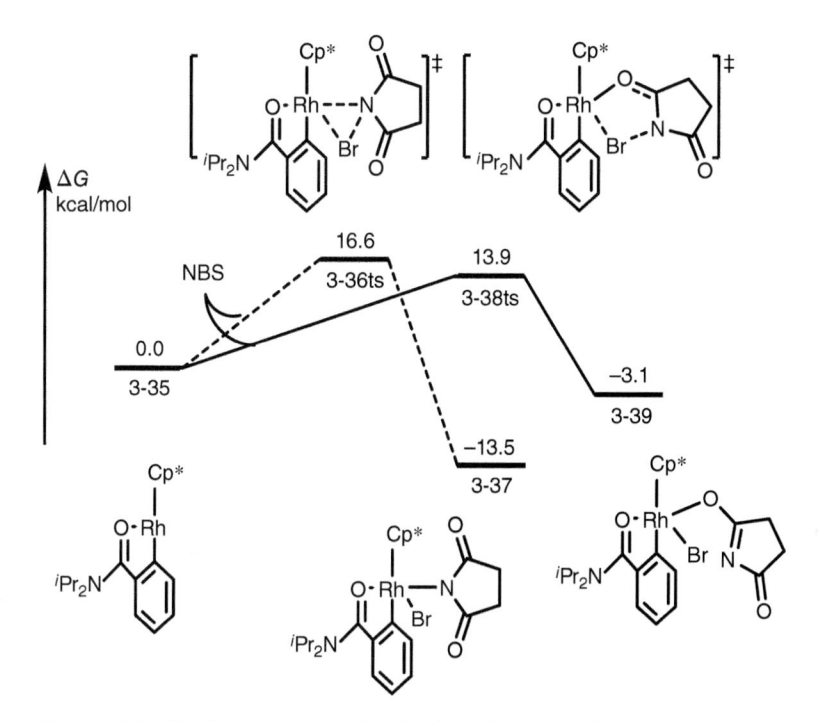

Figure 3.11 The free-energy profiles for the oxidative addition of NBS onto Rh(III). The energies were calculated at M11-L/6-311+G(d) (LANL08-f for Rh)/SMD(acetic acid)//B3LYP/6-31+G(d) (SDD for Rh) level of theory. Source :Based on Zhang, T., Qi, X., Liu, S. et al. (2017). Computational Investigation of the Role Played by Rhodium(V) in the Rhodium(III)-Catalyzed ortho-Bromination of Arenes. Chemistry--A European Journal 23 (11): 2690-2699.

substitution with cationic electrophiles, (iii) arene nucleophilic substitution, (iv) radical-type substitution, (v) substitution using ligand as nucleophiles, and (vi) bis-substitution [71, 87].

In organic chemistry, primary alkyl iodide is prone to S_N2-type substitution because of the small steric effect of primary alkyl group and the good leaving group of iodide. Therefore, when primary alkyl iodide reacts with palladate (Figure 3.12) [88], an S_N2-type substitution process was found to be the lowest activation free-energy pathway with an energy barrier of 21.4 kcal/mol via transition state **3-41ts**. The geometry information of this transition state is also given in this figure. The lengths of forming Pd—C bond and breaking C—I bond are 2.53 and 2.76 Å, respectively. The bond angle of Pd—C—I is 148.1°, which clearly revealed an S_N2-type transition state. The negative charge of palladate species increases the nucleophilicity, which is beneficial for the nucleophilic substitution with primary alkyl iodide.

When a cationic oxidant is used in substitution-type oxidative addition, it can react with weaker nucleophile to release a neutral-leaving group. As shown in Figure 3.13 [89], when Selectfluor is used as an oxidant to oxidize dirhodium(II) **3-43** to dirhodium(III) **3-45**, the oxidative addition can take place

64 | *3 Elementary Reactions in Organometallic Chemistry*

(1) $L_n\text{-}M^- + A\text{-}B \longrightarrow [L_n\text{-}M\text{---}A\text{---}B]^{\ddagger} \longrightarrow L_n\text{-}M\text{-}A + B^-$

(2) $L_n\text{-}M + A\text{-}B^+ \longrightarrow [L_n\text{-}M\text{---}A\text{---}B]^{\ddagger+} \longrightarrow L_n\text{-}M^+\text{-}A + B$

(3) $L_n\text{-}M\text{-}X=Y + A\text{-}B \longrightarrow [L_n\text{-}M\overset{Y}{\underset{A\text{---}B}{\text{-}X}}]^{\ddagger} \longrightarrow L_n\text{-}M\text{-}\overset{Y}{\underset{A}{X}} + B^-$

(4) $L_n\text{-}M^- + \text{Ar-R with X} \longrightarrow L_n\text{-}M\text{-Ar-R} \longrightarrow [L_n\text{-}M\text{-Ar-R}]^{\ddagger} \longrightarrow L_n\text{-}M\text{-Ar-R} + X^-$

(5) $L_n\text{-}M + A\text{-}B \longrightarrow [L_n\text{-}M\text{---}A\text{---}B]^{\ddagger} \longrightarrow L_n\text{-}M\text{-}A + B$

(6) $L_n\text{-}M + A{\overset{X}{\diagdown}}B \longrightarrow [L_n\text{-}M{\overset{A}{\underset{B}{\diagup\diagdown}}}X]^{\ddagger} \longrightarrow L_n\text{-}M{\overset{A}{\diagdown}}_B + X$

Scheme 3.7 Oxidative addition through substitutions.

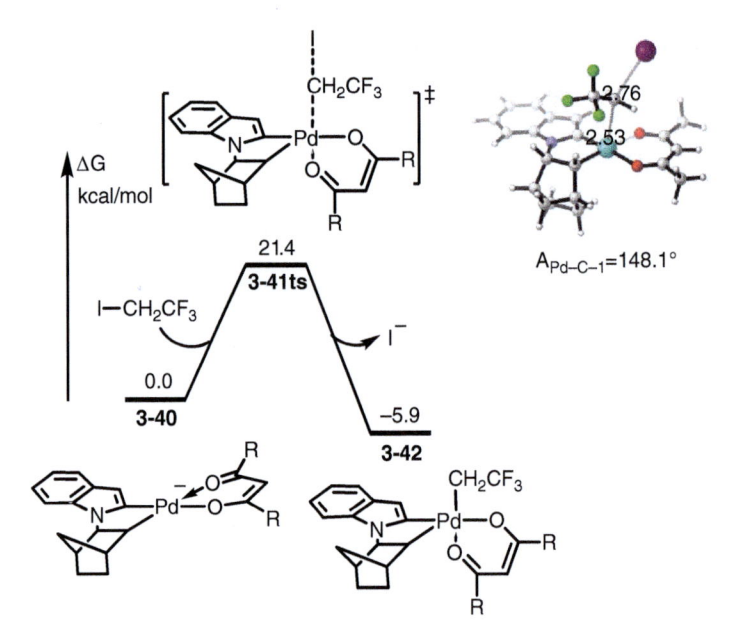

Figure 3.12 The free-energy profiles for the S_N2-substitution-type oxidative addition of Pd(II). The energies were calculated at M11-L/6-311+G(d,p) (SDD for Pd)/SMD(DMF)// B3LYP/6-31+G(d) (SDD for Pd) level of theory. Source: Based on Zhang, H., Wang, H.-Y., Luo, Y. et al. (2018). Regioselective Palladium-Catalyzed C-H Bond Trifluoroethylation of Indoles: Exploration and Mechanistic Insight. ACS Catalysis 8 (3): 2173-2180.

via an S_N2-substitution-type transition state **3-44ts** to form intermediate **3-45**. The release of amine leads to exergonic. In geometry information of **3-44ts**, the Rh—F—N bond angle is 124.7°, which reveals a substitution-type transition state.

In oxidative addition of metalate cases, a nucleophilic addition of metalate **3-46** with reacting aryl halide led to the formation of a π-complex **3-47** (Figure 3.14).

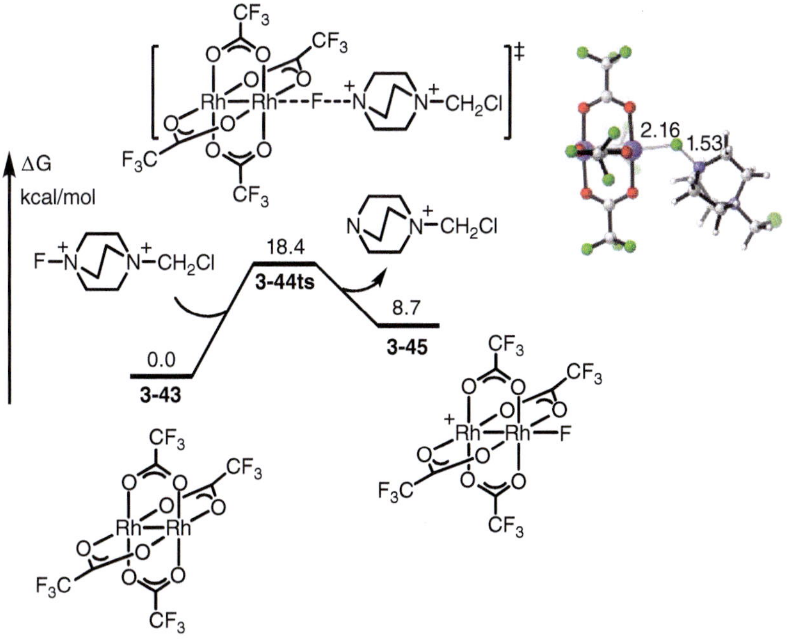

Figure 3.13 The free-energy profiles for the S$_N$2-substitution-type oxidative addition of dirhodium. The energies were calculated at N12/6-311+G(d) (SDD for Rh)/SMD(acetic acid)//M11-L/6-31G(d) (SDD for Rh) level of theory. Source: Based on Lin, Y., Zhu, L., Lan, Y. et al. (2015). Development of a Rhodium(II)-Catalyzed Chemoselective C(sp(3))-H Oxygenation. Chemistry--A European Journal 21 (42): 14937-14942.

Figure 3.14 The free-energy profiles for the S$_N$Ar-substitution-type oxidative addition of Pd. The energies were calculated at BP86/LANL2DZ/CPCM(THF)//BP86/LANL2DZ level of theory.

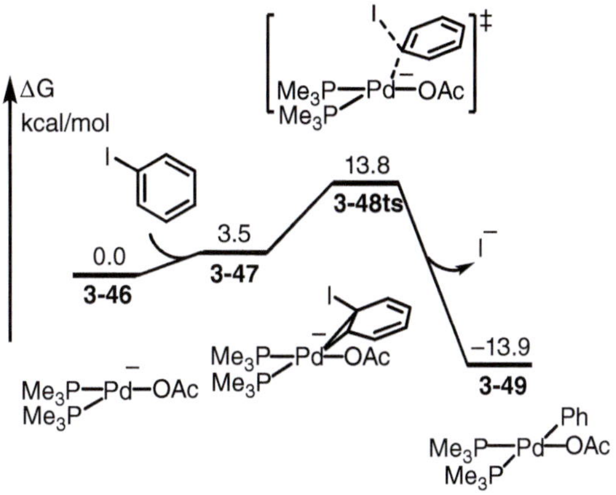

The back-donation effect increases the electron density of arene, which leads to a C—I bond cleavage via transition state **3-48ts** to afford an aryl metal complex **3-49**, which releases anionic iodide. The mechanism of this course can be considered as an S$_N$Ar process, which is different with concerted oxidative addition, because the Pd—I bond is not formed in this course [90, 91].

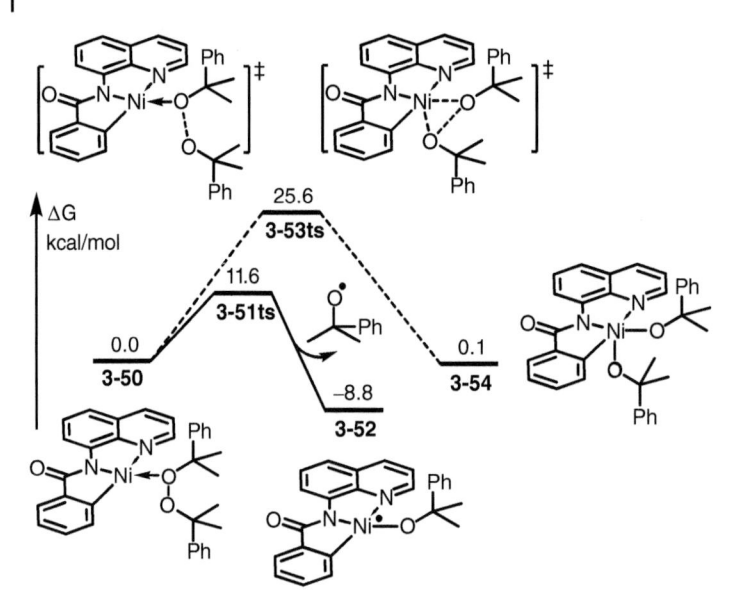

Figure 3.15 The free-energy profiles for the oxidative addition of Ni(II) species by using peroxide. The energies were calculated at M06/6-311+G(d,p) (SDD for Ni)/SMD(DMF)// B3LYP/6-31G(d) (LANL2DZ for Ni) level of theory. Source: Based on Wang, F., Chen, P., Liu, G. (2018). Copper-Catalyzed Radical Relay for Asymmetric Radical Transformations. Accounts of Chemical Research 51 (9): 2036-2046.

When the oxidative addition takes place with some fourth periodic late-transition metals, such as Fe, Co, Ni, Cu, etc., a radical-type substitution would afford a free radical and the oxidative state of metal is increased by +1 only [59, 92, 93]. As shown in Figure 3.15 [94], when a Ni(II) species **3-50** reacts with peroxide, a radical-type substitution can occur via transition state **3-51ts** to afford alkoxy free radical and alkoxy Ni(III) complex **3-52**. The calculated activation free energy of this step is only 11.6 kcal/mol. This transformation increases the oxidative state and coordination number of Ni by +1. However, the barrier of concerted oxidative addition to afford Ni(IV) species **3-54** is as high as 25.6 kcal/mol via transition state **3-53ts**.

In some cases, the coordination onto metal with unpaired electron would lead to a radical character of ligand. As shown in Figure 3.16 [95], the methylation of carbonyl coordinated Ni(I) complex **3-55** by methyl iodide can occur on either Ni atom or carbonyl group by a radical substitution process to afford a iodine free radical. The computational results show that the radical substitution by Ni can take place via transition state **3-56ts** with a free-energy barrier of 16.6 kcal/mol, to yield a methyl Ni(II) species **3-57** with 4.3 kcal/mol exergonic. In this process, both the oxidative state and the coordination number of Ni are increased by +1. In an alternative process, the radical-type substitution also can occur on coordinated carbonyl group via transition state **3-58ts** with a free-energy barrier of 16.0 kcal/mol to form an acetyl Ni(II) species **3-59**. In this process, only the oxidative state of Ni is increased by +1. The second process is 35.3 kcal/mol exergonic because new C—C covalent bond is

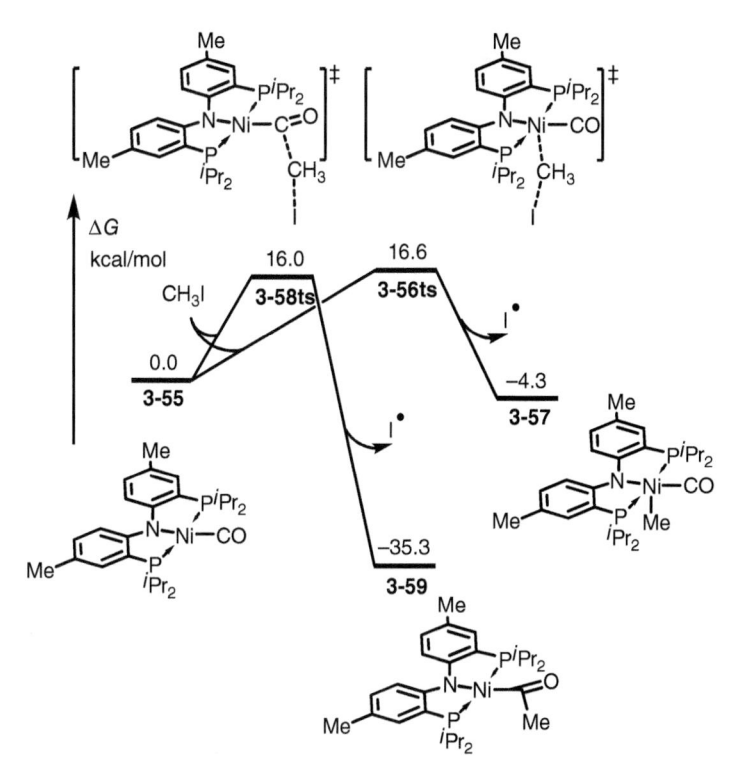

Figure 3.16 The free-energy profiles for the radical substitution by Ni(I)–carbonyl complex. The energies were calculated at B3LYP-D3/6-311+G(d,p) (6-311G(d) for I, and LANL2DZ for Ni)/IEF-PCM(THF)//B3LYP/6-31G(d) (6-311G(d) for I, and LANL2DZ for Ni) level of theory. Source: Based on Yoo, C., Ajitha, M. J., Jung, Y. et al. (2015). Mechanistic Study on C–C Bond Formation of a Nickel(I) Monocarbonyl Species with Alkyl Iodides: Experimental and Computational Investigations. Organometallics 34 (17): 4305-4311.

formed here. Indeed, only a 0.6 kcal/mol free-energy difference reveals that both of those two pathways are possible.

Oxidative addition may be accompanied by multi-covalent bond transformation. As shown in Figure 3.17 [96], Cu(I) species **3-60** can be oxidized by Togni's reagent **3-61** involving a hypervalent iodine. Computational result shows that the oxidative addition can take place via transition state **3-62ts** with a free-energy barrier of 16.0 kcal/mol to afford a Cu(III) specie **3-63**. In transition state **3-62ts**, both O—I and C—I bonds break with the formation of O—Cu and Cu—C bonds. It can be considered as Cu atom is used to substitute iodine atom on both O and C atoms. After this process, the oxidative state and the coordination number of Cu are increased by +2.

3.3.3 Radical-type Addition

Free radical can provide an unpaired electron to combine another unpaired election from metal for the construction of a new covalent bond. After this process, the oxidation state and the coordination number of metal are increased by +1 [97, 98].

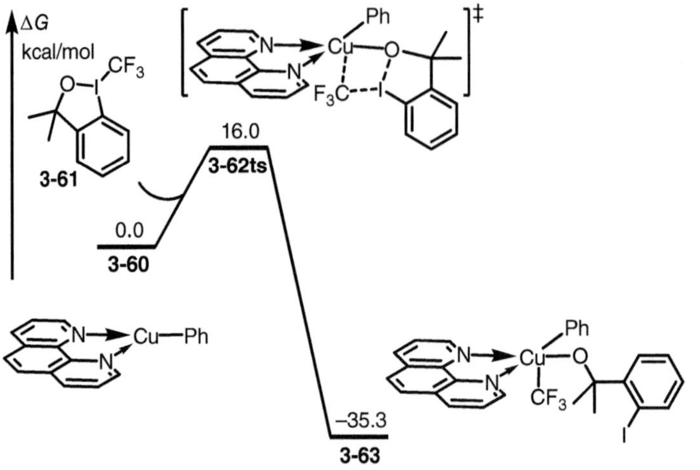

Figure 3.17 The free-energy profiles for the oxidative addition of Cu(I) with Togni's reagent through multi-covalent bond transformation. The energies were calculated at M11-L/6-311+G(d) (SDD for Cu and I)/SMD(dibutyl ether)//B3LYP/6-31G(d) (SDD for Cu and I) level of theory. Source: Based on He, X., Lan, Y., Shan, C. et al. (2017). Mechanistic insights into copper-catalyzed trifluoromethylation of aryl boronic acids: a theoretical study. SCIENTIA SINICA Chimica 47 (7): 859-864.

For instance, alkyl radical can coordinate with doublet Cu(II) species **3-64d** to afford a high spin complex **3-65t** reversibly (Figure 3.18) [99]. In geometry of **3-65t**, the length of Cu—C is 3.03 Å, which reveals a weak coordination bond. A spin cross-over can take place via minima energy cross-point (MECP) **3-66MECP** to afford a low-spin alkyl Cu(III) complex **3-67s** with an energy barrier of only 2.2 kcal/mol. In the geometry of complex **3-67s**, the bond length of Cu—C is 2.02 Å, which reveals a typical single covalent bond.

3.3.4 Oxidative Cyclization

As shown in Scheme 3.8, in oxidative cyclization, a metal reacts with two molecules or fragments containing unsaturated bonds to form two covalent bonds involving a metal and a covalent bond between the two reacting moieties. These covalent bonds together form a ring structure; therefore, it is named oxidative cyclization. During this process, both the oxidative state and the coordination number of metal are increased by +2. Nevertheless, the oxidative cyclization also can be considered as the coordination of unsaturated bond in one molecule to form a metal-involved three-membered ring, then, another molecule containing unsaturated bond inserts into this formed metal-involved three-membered ring to achieve ring enlargement [63, 64, 69].

Oxidative cyclization is usually considered as the initial step of Co-mediated Pauson–Khand reaction [100, 101]. As shown in Figure 3.19 [102], $Co_2(CO)_8$ **3-68** can react with acetylene to form a Co_2/acetylene complex **3-69** with loss of gaseous CO. The further ligand exchange with ethylene forms intermediate **3-70** with a

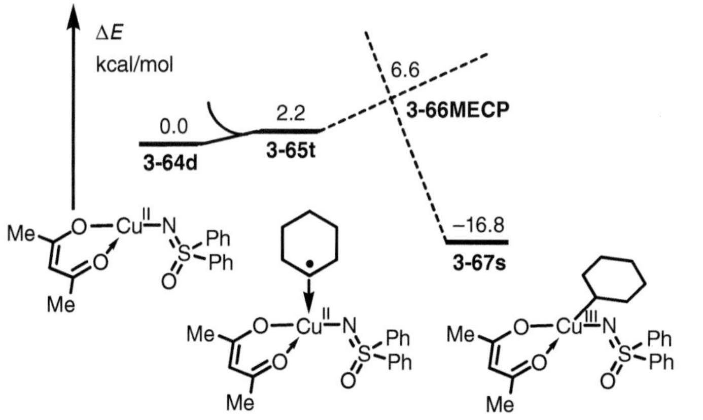

Figure 3.18 The energy profiles for the radical-type oxidation of Cu(II) by cyclohexyl radical. The energies were calculated at M06/6-311+G(d,p) (def2-TZVP for Cu)/SMD(cyclohexane)//B3LYP/6-31G(d) (SDD for Cu) level of theory. Source: Based on Qi, X., Zhu, L., Bai, R. et al. (2017). Stabilization of Two Radicals with One Metal: A Stepwise Coupling Model for Copper-Catalyzed Radical-Radical Cross-Coupling. Scientific Reports 7: 43579.

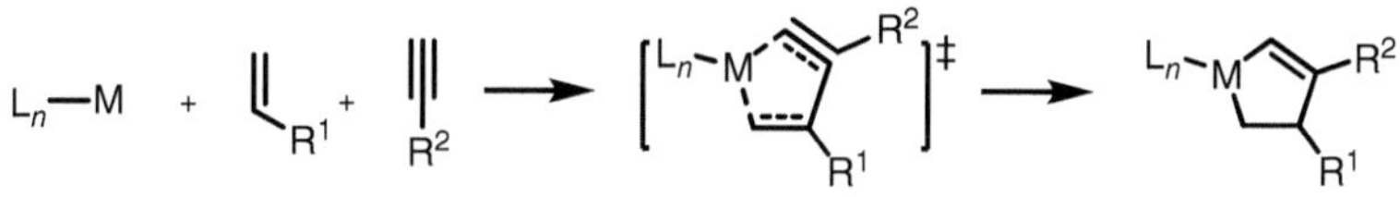

Scheme 3.8 Oxidative cyclization.

total 21.6 kcal/mol endergonic. The oxidative cyclization takes place via transition state **3-71ts** with an overall activation free energy of 36.2 kcal/mol to afford a Co(0)–Co(II) species **3-72**. The geometry of transition state **3-71ts** is shown in Figure 3.19. The lengths of Co–C1, Co–C2, Co–C3, and Co–C4 are 1.89, 2.10, 2.26, and 2.00 Å, respectively, which reveal an insertion-type oxidative cyclization due to the short distance between Co and C2.

As a contrast, the DFT-calculated free-energy profiles of the oxidative cyclization onto Ni(0) species **3-73** by aldehyde and acetylene are shown in Figure 3.20 [103]. The oxidative cyclization can occur via transition state **3-75ts** with an energy barrier of 20.8 kcal/mol. In geometry information of **3-75ts**, the lengths of Ni–O, Ni–C1, Ni–C2, and Ni–C3 are 1.90, 2.29, 2.22, and 1.81 Å, respectively, which reveal an oxidative addition-type cyclization due to the long distance between Ni and C2/C3.

Interestingly, the oxidative addition of vinyl aziridine onto Rh(I) is also can be considered as oxidative cyclization (Figure 3.21) [104]. The coordination of vinyl aziridine leads to the formation of intermediate **3-78**. The oxidative addition can occur via transition state **3-79ts**, in which the cleavage of aziridine ring leads to the formation of allylic Rh(III) complex **3-80**, which contains a formal five-membered rhodacycle.

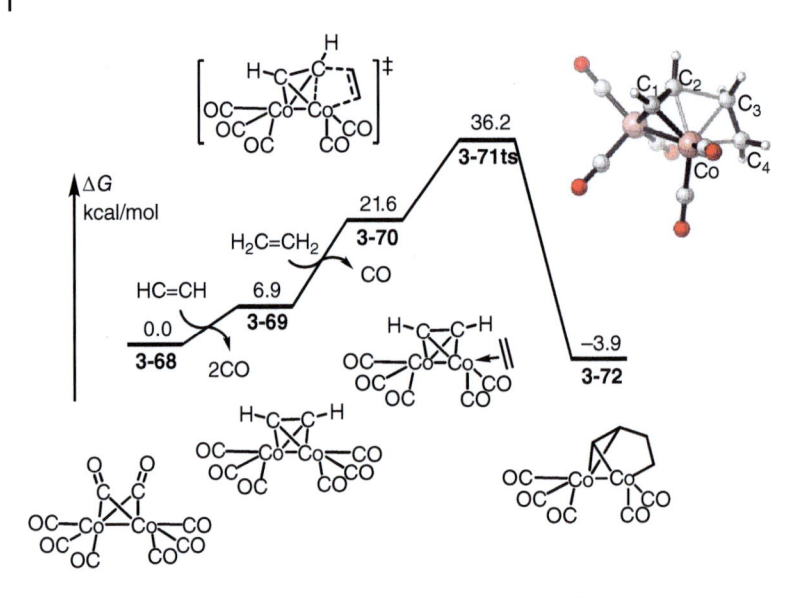

Figure 3.19 The free-energy profiles for the oxidative cyclization in Pauson–Khand reaction. The energies were calculated at B3LYP/6-311+G(d) (SDD for Co)//B3LYP/6-31G(d) (LANL2DZ for Co) level of theory. Source: Based on Yamanaka, M., Nakamura, E. (2001). Density functional studies on the Pauson--Khand reaction. Journal of the American Chemical Society 123 (8): 1703-1708.

In more specialized cases, the oxidative cyclization even could occur between allene and Pd(0) species. As shown in Scheme 3.9 [105], allene can react with Pd(0) to afford complex **3-81**, in which one of the π bonds in allene coordinates with Pd(0). Complex **3-81** can isomerize to a four-membered palladacycle(II) **3-82**, in which the internal carbon of allene moiety plays carbene character.

Scheme 3.9 The isomerization of allene-coordinated Pd(0). The energies were calculated at M06/6-311+G(d,p) (SDD for Pd)/CPCM(THF)//M06/6-311+G(d,p) (SDD for Pd)/CPCM(THF) level of theory.

3.4 Reductive Elimination

Reductive elimination is the inverse reaction of oxidative addition. The results of reductive elimination are that both the oxidative state and the coordination number of metal are decreased. From mechanistic point of view, this elementary reaction can occur through concerted elimination [106–109], substitution-type elimination [110, 111], bimetallic reductive elimination [112–114], and eliminative reduction [115–117]. The significance of reductive elimination in organometallic chemistry

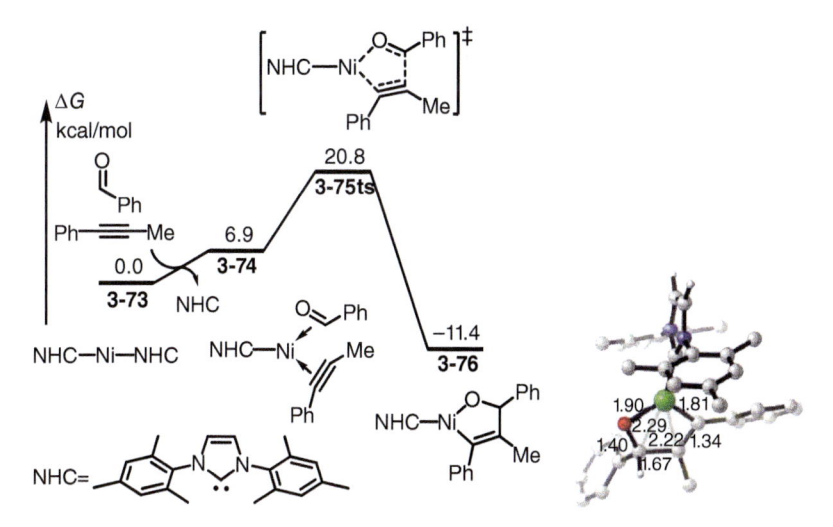

Figure 3.20 The free-energy profiles for the oxidative cyclization of Ni(0). The energies were calculated at M11-L/6-311++G(d,p) (SDD for Ni)/SMD(THF)//B3LYP/6-31G(d) (SDD for Ni) level of theory. Source: Based on Luo, X., Qi, X., Chen, C. et al. (2017). Ligand effect on nickle-catalyzed alkyne-aldehyde coupling reactions: a computational study. SCIENTIA SINICA Chimica 47 (3): 341-349.

is that the conversion of newly generated carbon–metal by this process leads to the formation of new C—X covenant bonds.

3.4.1 Concerted Reductive Elimination

As a double-electron transformation, concerted reductive elimination can be considered as the inverse reaction of concerted oxidative addition, which is also the most

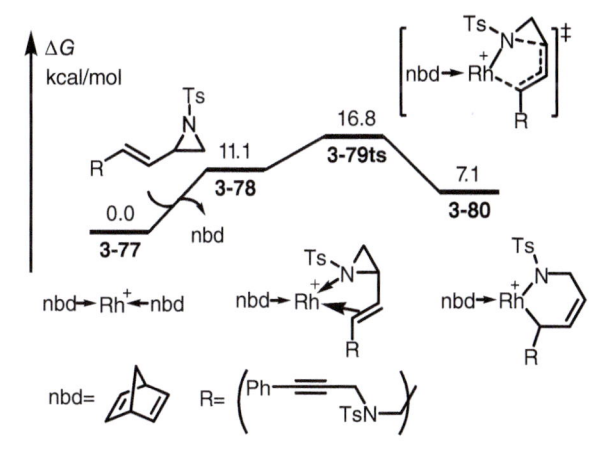

Figure 3.21 The free-energy profiles for the oxidative cyclization of Rh(I). The energies were calculated at M11-L/6-311+G(d) (SDD for Rh)/SMD(DCE)//B3LYP/6-31G(d) (SDD for Rh) level of theory. Source: Based on Zhu, L., Qi, X., Lan, Y. (2016). Rhodium-Catalyzed Hetero-(5 + 2) Cycloaddition of Vinylaziridines and Alkynes: A Theoretical View of the Mechanism and Chirality Transfer. Organometallics 35 (5): 771-777.

common process in organometallic chemistry. As shown in Scheme 3.10, in complex A—M—B, A and B groups can eliminate from metal M to form an A—B covalent bond. In this process, two polar covalent bonds M—A and M—B are changed to a weak- or non-polar covalent bond A—B, while two nucleophilic moieties A and B are converted to a pair of nucleophile–electrophile.

Scheme 3.10 Concerted reductive elimination.

Concerted reductive elimination is the most popular way for the construction of new C—C and C—hetero covalent bonds [118–123]. Particularly, it is often involved in the catalytic cycle of Pd-catalyzed cross-coupling reactions [35, 124–129]. As an example, Maseras and coworkers reported mechanism study of Pd-catalyzed Suzuki–Miyaura cross-coupling reaction [130]. In the last step (Figure 3.22), a concerted reductive elimination occurs via a three-membered ring-type transition state **3-84ts** to form the new $C(sp^2)$—$C(sp^2)$ bond. The geometry information of transition state **3-84ts** is shown in Figure 3.22. The lengths of two breaking C—Pd bonds and forming C—C bond are 2.07 and 2.07 Å, respectively.

In some unique conditions, concerted reductive elimination might be involved in the rate-determining step. As shown in Figure 3.23 [131], the calculated barrier of aryl$_F$ reductive elimination from an Au(III) species **3-86** is 22.6 kcal/mol via a concerted three-membered ring-type transition state **3-87ts**. In this case, the dissociation of ligand would lead to a lower barrier of this process; however, the difference of activation energy cannot offset the loss of dissociation energy.

The concerted reductive elimination can take place through a larger ring-type transition state. As shown in Scheme 3.11 [132], in the Pd-catalyzed cross-coupling reaction of benzyl chloride **3-89** and allylic tin **3-90**, a dearomatized allyl

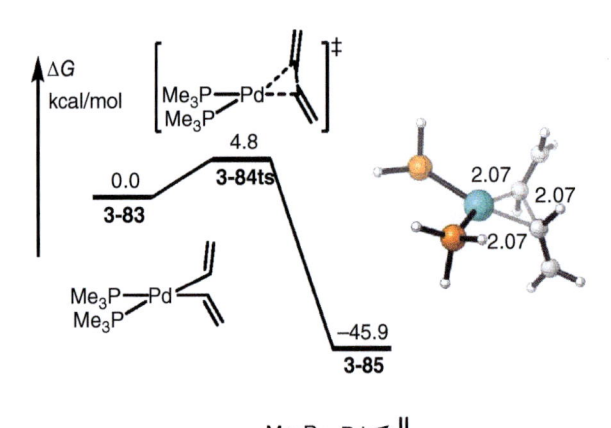

Figure 3.22 The free-energy profiles for the reductive elimination of Pd–divinyl complex. The energies were calculated at B3LYP/6-31+G(d) (LANL2DZ for Pd)/PCM(water)// B3LYP/6-31+G(d) (LANL2DZ for Pd) level of theory.

Figure 3.23 The free-energy profiles for the reductive elimination of Au–diaryl complex. The energies were calculated at B3LYP-D3/6-311+G(d) (SDD for Au)/SMD(CDCl$_3$)//B3LYP/6-31G(d) (SDD for Au) level of theory. Source: Based on Kang, K., Liu, S., Xu, T. et al. (2017). C(sp2)–C(sp2) Reductive Elimination from Well-Defined Diarylgold(III) Complexes. Organometallics 36 (24): 4727-4740.

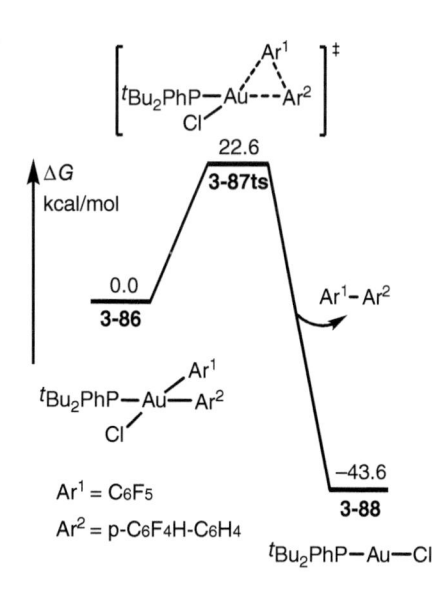

methylenecyclohexadiene **3-91** is found as major product experimentally. A DFT study found that a concerted reductive elimination can take place via a large ring-type transition state **3-92ts**, in which allylic Pd is an important intermediate leading to dearomatization.

Scheme 3.11 Pd-catalyzed cross-coupling reaction of benzyl chloride and allylic tin.

3.4.2 Substitution-type Reductive Elimination

Substitution-type reductive elimination can be considered to be the reverse reaction of substitution-type oxidative addition, which is also a double-electron transformation. In this process (Scheme 3.12), an extra nucleophile (Nu$^-$) attacks the bonding group (R) to form a new molecular R–Nu and release the metal. During this process, the oxidative state and coordination number of metal are reduced by 2 and 1, respectively [133, 134].

As shown in Figure 3.24 [135], when an extra base, such as phenolate, is added in an octahedral Ni(IV) complex **3-93**, there is non-vacancy to accept the coordination of phenolate. Therefore, a substitution-type reductive elimination takes place via

74 | *3 Elementary Reactions in Organometallic Chemistry*

$$M\overset{n}{-}R]^+ + Nu^- \longrightarrow [M\text{---}R\text{---}Nu]^{\ddagger} \longrightarrow M^{n-2} + R\text{—}Nu$$

Scheme 3.12 Reductive elimination through nucleophilic substitution.

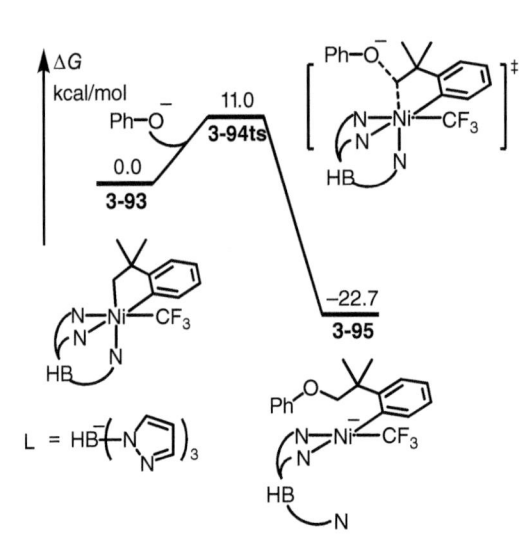

Figure 3.24 The free-energy profiles for the substitution-type reductive elimination of Ni(IV)-alkyl. The energies were calculated at B3LYP-D3/6-311+G(2d,p) (def2-QZVP for Ni)/IEF-PCM (acetonitrile)//B3LYP/6-31G(d) (SDD for Pd) level of theory. Source: Based on Camasso, N. M., Canty, A. J., Ariafard, A. et al. (2017). Experimental and Computational Studies of High-Valent Nickel and Palladium Complexes. Organometallics 36 (22): 4382–4393.

transition state **3-94ts** with an energy barrier of 11.0 kcal/mol. In the geometry of this transition state, the bond angle of Ni—C—O is 153.4°, which indicates a S_N2-type substitution. In intermediate **3-93**, the binding alkyl group is nucleophile, while after this process, it turns to electrophile in the formed phenyl ether.

In another case (Figure 3.25) [136], when a square planar bisphosphine-coordinated allylic Pd(II) complex **3-96** reacts with amine, an intermolecular nucleophilic attack of amine takes place via transition state **3-97ts** to afford a olefin-coordinated Pd(0) species **3-98**. In geometry, information of that transition state shows that the lengths of forming N—C bond and breaking C—Pd bond are 2.19 and 2.18 Å, respectively. The bond angle of N—C—Pd is 161.8°, which reveals a nucleophilic substitution.

3.4.3 Radical-Substitution-type Reductive Elimination

As shown in Scheme 3.13, in the reaction of spin-active metal complex M–R and a free radical X·, the interaction between free radical and a metal–organic covalent bond M—R leads to homolysis of this bond and forms new covalent bond in X—R compound. Both the oxidative state and coordination number of metal are reduced by 1 [137–140].

$$M\overset{n}{-}R + X^{\bullet} \longrightarrow [M\text{---}R\text{---}X]^{\ddagger} \longrightarrow M^{n-1}{}^{\bullet} + R\text{—}X$$

Scheme 3.13 Reductive elimination through radical substitution.

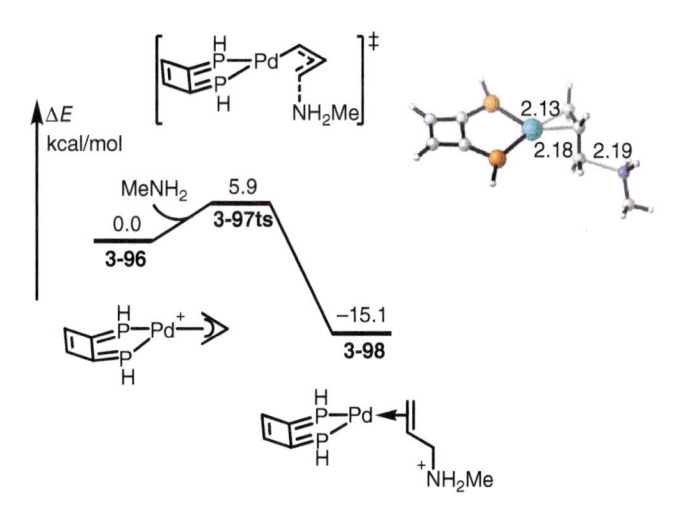

Figure 3.25 The energy profiles for the substitution-type reductive elimination of allylic Pd(II) cation. The energies were calculated at B3PW91/6-31+G(d) (LANL2DZ for Pd)/PCM(THF)//B3PW91/6-31+G(d) (LANL2DZ for Pd) level of theory. Source: Based on Piechaczyk, O., Thoumazet, C., Jean, Y. et al. (2006). DFT study on the palladium-catalyzed allylation of primary amines by allylic alcohol. Journal of the American Chemical Society 128 (44): 14306-14317.

A simple example of radical-substitution-type reductive elimination is shown in Figure 3.26 [99]. A sulfoximide Cu(II) complex **3-64d** with a single electron on Cu can react with cyclohexyl radical through a radical-substitution-type reductive elimination. In this process, cyclohexyl radical attacks the nitrogen atom in sulfoximide group leading to the Cu—N bond cleavage. One electron in Cu—N covalent bond is used to form a new covalent bond with cyclohexyl group, while another one can pair with the single electron on Cu atom. Therefore, a low-spin complex **3-100s** is formed, in which the oxidative state of Cu is reduced to +1. In this process, a high-spin state complex **3-64d** and cyclohexyl radical undergoes a radical–substitution to yield a low-spin complex **3-100s**. Therefore, a spin crossover should take place. In theoretical calculation, a MECP is found as **3-99MECP**, which is only 9.0 kcal/mol higher than the relative state.

In another case, an intramolecular radical-type reduction would occur on a carbamate Cu(II) species **3-101t** [97]. As shown in Figure 3.27, an intramolecular substitution can occur through an MECP **3-102MECP**, in which the carbon radical attacks the carboxylate group to afford oxazolidinone-coordinated Cu(I) complex **3-103s** by forming new C—O bond. The calculated relative energy of MECP **3-102MECP** is 5.7 kcal/mol higher than that of complex **3-101t**.

3.4.4 Bimetallic Reductive Elimination

Sometimes, reductive elimination may not be as simple as the textbook shows. It may undergo a more complex process, in which more than one metal participates [141–144]. As shown in Figure 3.28 [145], two molecules of alkynyl Cu(II) complex can

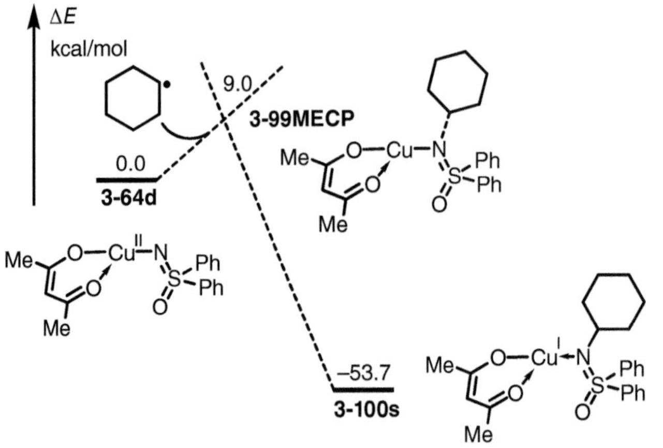

Figure 3.26 The energy profiles for the reduction of Cu(II) through radical substitution. The energies were calculated at M06/6-311+G(d,p) (def2-TZVP for Cu)/SMD(cyclohexane)//B3LYP/6-31G(d) (SDD for Cu) level of theory. Source: Based on Qi, X., Zhu, L., Bai, R. et al. (2017). Stabilization of Two Radicals with One Metal: A Stepwise Coupling Model for Copper-Catalyzed Radical-Radical Cross-Coupling. Scientific Reports 7: 43579.

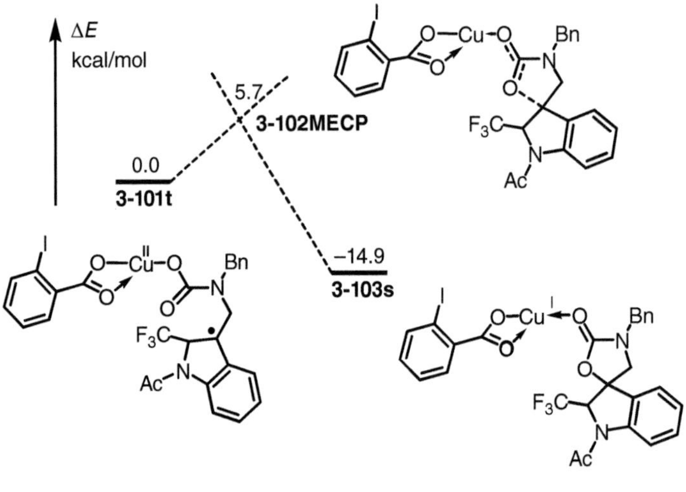

Figure 3.27 The energy profiles for the reduction of Cu(II) through intramolecular radical substitution. The energies were calculated at B3LYP/6-311+G(d,p) (SDD for Cu and I)/IEF-PCM(acetonitrile)//B3LYP/6-31G(d) (SDD for Cu and I) level of theory.

form a high-spin dimmer **3-104t**, in which the oxidative state of both Cu atoms is +2. The direct binuclear reductive elimination would occur via a triplet transition state **3-105ts** with an energy barrier of 22.1 kcal/mol. DFT calculation found that triplet **3-104t** undergoing an MECP **3-107MECP** can afford a singlet species **3-108s** only with a 5.6 kcal/mol endergonic. In low-spin species **3-108s**, the oxidative states of two Cu atoms can be considered as a combination of +1 and +3. When this singlet species **3-108s** is formed, the following reductive elimination could occur via a concerted

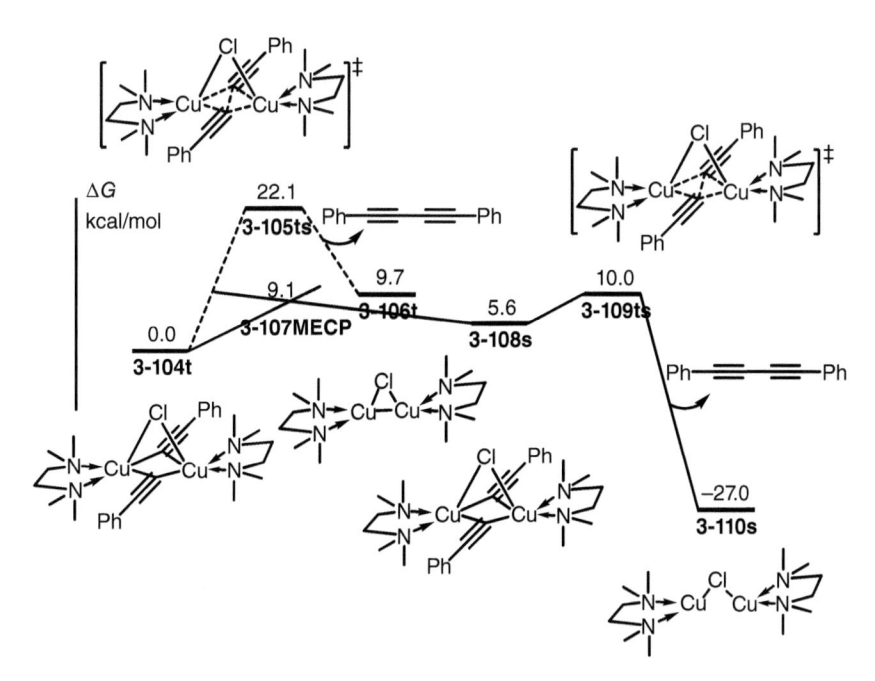

Figure 3.28 The free-energy profiles for the bimetallic reductive elimination of dicopper alkynyl. The energies were calculated at M11-L/6-311+G(d) (SDD for Cu)/SMD(DMF)//B3LYP/6-31G(d) (SDD for Cu) level of theory. Source: Based on Qi, X., Bai, R., Zhu, L. et al. (2016). Mechanism of Synergistic Cu(II)/Cu(I)-Mediated Alkyne Coupling: Dinuclear 1,2-Reductive Elimination after Minimum Energy Crossing Point. The Journal of Organic Chemistry 81 (4): 1654-1660.

Figure 3.29 The free-energy profiles for the eliminative reduction of Pd(IV) through either inner- or outer-sphere pathways. The energies were calculated at M11-L/6-311+G(d,p) (SDD for Pd)/SMD (acetonitrile)//B3LYP/6-31G(d) (SDD for Pd) level of theory. Source: Based on Chen, C., Luo, Y., Fu, L. et al. (2018). Palladium-Catalyzed Intermolecular Ditrifluoromethoxylation of Unactivated Alkenes: CF3O-Palladation Initiated by Pd(IV). Journal of the American Chemical Society 140 (4): 1207-1210.

transition state **3-109ts** with an energy barrier of only 4.4 kcal/mol. Therefore, the reductive elimination of alkynyl Cu(II) complex can be described in the following steps, including dimerization for a triplet dimer, spin crossover for a singlet dimer, and a concerted bimetallic reductive elimination.

3.4.5 Eliminative Reduction

In particular cases, elimination would lead to the reduction of metal center. As shown in Figure 3.29 [115], in a trifluoromethanolate alkyl Pd(IV) complex **3-111**, the bonding trifluoromethanolate group can play as a base, which leads to an inner-sphere alkyl β-hydrogen elimination to afford an olefin-coordinated Pd(II) complex **3-113** via transition state **3-112ts**. The calculated barrier of this step is only 6.1 kcal/mol. Moreover, an outer-sphere elimination of alkyl β-hydrogen was also considered to go through transition state **3-114ts** with an energy barrier of 8.9 kcal/mol. In this process, the elimination by base leads to the reduction of metal's oxidative state; therefore, it can be named eliminative reduction.

3.5 Insertion

The insertion reaction can be regarded as the process of an unsaturated molecule insertion into the metal–organic covalent bond to form two new covalent bonds with an unsaturated molecule. After insertion reaction, the coordination number and the oxidative state of metal remain unchanged [146–151]. In organometallic chemistry, insertion is often accompanied by the initial coordination of unsaturated molecule with metals, which can be considered as the migration of bonding nucleophile on metal to unsaturated molecule, so insertion is also named migratory insertion. Generally, olefins, acetylenes, carbonyl compounds, imines, arenes, etc., are often used as unsaturated molecule in metal-mediated insertion reaction [146, 152–155]. In those cases, π bond reacts with metal–organic covalent bond; therefore, they are named 1,2-insertion (Scheme 3.14a) [74, 156–161]. Alternatively, some compounds, such as carbene, carbon monoxide, nitrene, etc., contain an unsaturated atom, which also can insert into metal–organic covalent bond. This type of insertion is named 1,1-insertion (Scheme 3.14b) [162–165].

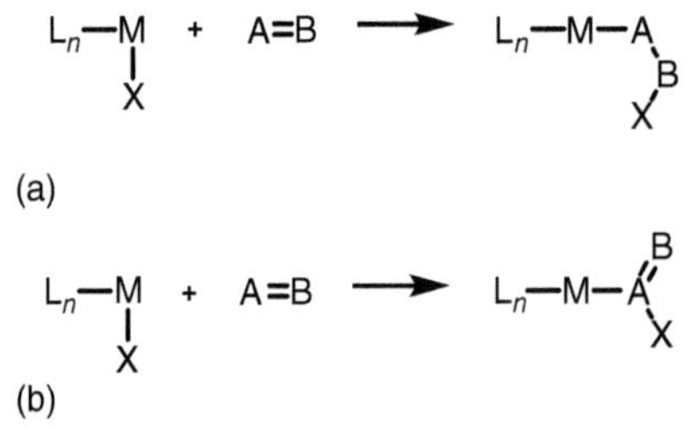

(a) L$_n$—M—X + A=B ⟶ L$_n$—M—A—B—X

(b) L$_n$—M—X + A=B ⟶ L$_n$—M—A(B)(X)

Scheme 3.14 Insertion. (a) 1,2-Insertion, (b) 1,1-insertion.

3.5.1 1,2-Insertion

1,2-Insertion is a common process, which breaks a π bond in unsaturated molecule and a metal–organic covalent bond with the formation of two covalent bonds. In 1,2-insertion process, the initial step is the dissociation of one ligand or counter iron to result in a vacancy. The coordination of π bond leads to the activation of unsaturated molecule, which leads to a migration of bonding nucleophile onto unsaturated molecule accompanied by the formation of new covalent bond with metal. Usually, this process takes place through a four-membered ring-type transition state.

As shown in Figure 3.30, 1,2-olefin insertion is one of the critical processes in Pd-catalyzed Mizoroki–Heck reaction [156]. The oxidative addition of aryl halide leads to the formation of a neutral Pd(II) species **3-115**. The dissociation of halide anion yields a vacancy in a tri-coordinated Pd(II) intermediate **3-116**, which can be coordinated by olefin to give a tetra-coordinated Pd(II) complex **3-117**. A migratory insertion of olefin into Pd–aryl bond can occur via a four-membered ring-type transition state **3-118ts** to afford a benzylic Pd(II) intermediate **3-119**. The overall result can be considered as a transformation of a π bond to a C—C σ bond; therefore, it is an exergonic process.

As shown in Figure 3.31 [161], the insertion of acetylene was also initiated by the coordination onto a 16-electron Co(III) intermediate **3-120** to afford a more stable 18-electron species **3-121**. The migratory insertion of coordinated acetylene into Co–vinyl bond occurs via a four-membered ring-type transition state **3-122ts** to afford a 16-electron vinyl Co(III) complex **3-123** with 15.8 kcal/mol exergonic.

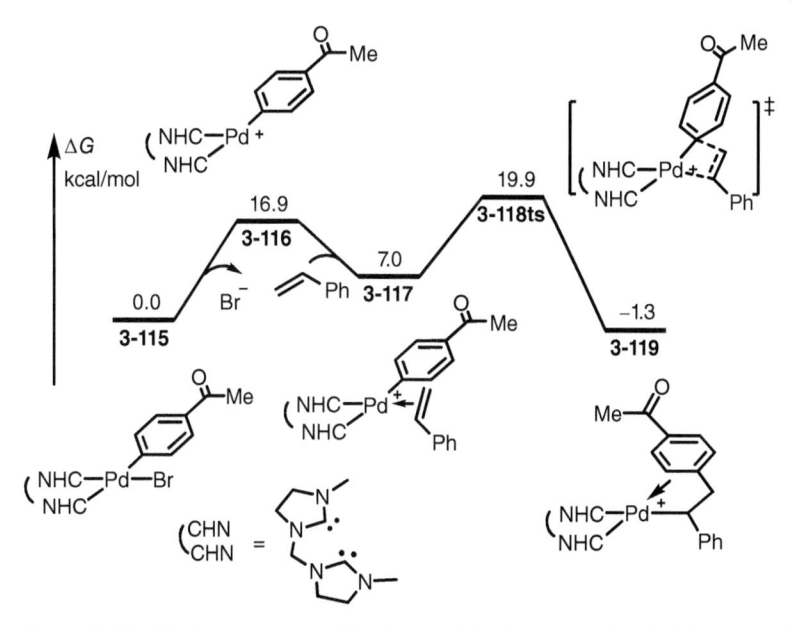

Figure 3.30 The free-energy profiles for the 1,2-alkene insertion in Pd-catalyzed Heck coupling reaction. The energies were calculated at M06/def2-TZVP (LANLTZ-f for Pd)/PCM(DMAc)//B3LYP/6-31G(d) (LANL2DZ for Pd) level of theory.

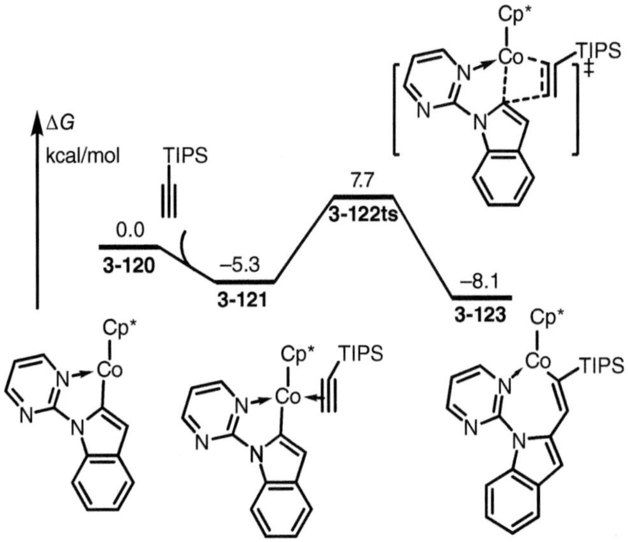

Figure 3.31 The free-energy profiles for the 1,2-alkyne insertion into Co—C(aryl) bond. The energies were calculated at M11-L/6-311+G(d,p) (SDD for Co)/SMD(TFE)//B3LYP/ 6-31+G(d) (SDD for Co) level of theory. Source: Based on Zhou, X., Luo, Y., Kong, L. et al. (2017). Cp*CoIII-Catalyzed Branch-Selective Hydroarylation of Alkynes via C—H Activation: Efficient Access to α-gem-Vinylindoles. ACS Catalysis 7 (10): 7296-7304.

Carbonyl compounds involve a polar C=O double bond, which can be activated by the coordination metal to increase the electrophilicity of carbon atom in it. Therefore, in a 1,2-acyl insertion reaction, oxygen atom often bonds with metal, while the carbon atom is attacked by the nucleophiles [158]. As shown in Figure 3.32 [160], the coordination of aldehyde forms complex **3-124** reversibly. Then, a 1,2-acyl insertion could occur via a four-membered ring-type transition state **3-125ts** with an energy barrier of 20.8 kcal/mol to afford an alkoxyl Zn(II) complex **3-126**. Alternatively, 1,2-acyl insertion could occur through a bimetallic process, in which one Zn is used to activate acyl group, while another one is used to stabilize alkynyl nucleophile. The bimetallic 1,2-acyl insertion can take place via a six-membered ring-type transition state **3-128ts** with an energy barrier of 19.2 kcal/mol, which is slightly lower than that of corresponding monometallic process.

3.5.2 1,1-Insertion

When a molecule owns an unsaturated atom, it can react with metal–organic compound to form two new covalent bonds with this atom. It can be considered as the unsaturated atom inserts into metal–organic covalent bond; therefore, it is named as 1,1-insertion. Usually, the unsaturated molecule would coordinate with metal before its insertion, while the insertion process can be thought of as the migration of organic moiety onto the coordinated unsaturated atom. Therefore, it also can be named as 1,1-migratory insertion. Mechanistically, a 1,1-insertion could occur through a concerted three-membered ring-type transition state, in which

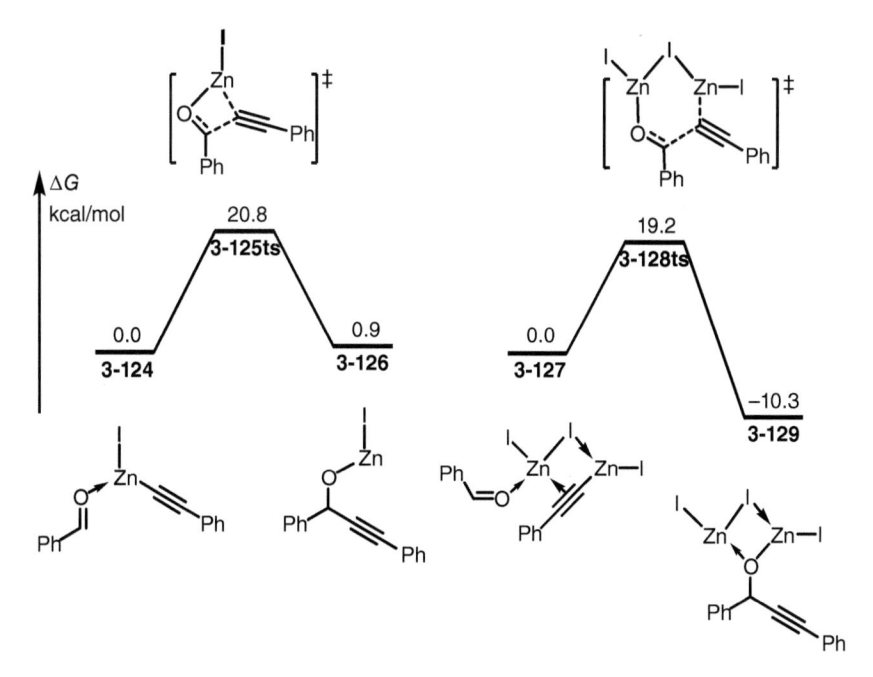

Figure 3.32 The free-energy profiles for the 1,2-acyl insertion through either four- or six-membered ring-type transition state. The energies were calculated at M11-L/6-311+G(d) (LANL08 for Zn and I)/SMD(toluene)//B3LYP/6-31G(d) (SDD for Zn and I) level of theory. Source: Based on Yue, X., Qi, X., Bai, R. et al. (2017). Mononuclear or Dinuclear? Mechanistic Study of the Zinc-Catalyzed Oxidative Coupling of Aldehydes and Acetylenes. Chemistry – A European Journal 23 (26): 6419-6425.

the metal–organic bond is partially broken with the partial formation of organic covalent bond.

Carbon monoxide, which involved a formal divalent carbon atom, can easily insert into a metal–organic covalent bond to form an acyl–metal complex [166]. As shown in Figure 3.33 [162], carbon monoxide can coordinate onto a 16-electron Rh(III) complex **3-130** to form an 18-electron carbonyl-coordinated complex **3-131** with 6.1 kcal/mol exergonic. The followed carbonyl migratory insertion into Rh—C(aryl) bond occurs via a three-membered ring-type transition state **3-132ts** with an energy barrier of 13.9 kcal/mol. In geometry of transition state **3-132ts**, the lengths of forming C(carbonyl)—C(aryl) bonds and breaking Rh—C(aryl) bond are 1.73 and 2.22 Å, respectively. The energetic information also reveals that the carbonyl insertion process is usually reversible, for which the reverse reaction is elimination.

In fact, except carbon monoxide, other molecules containing unsaturated atoms are often unstable, so they often need to be in situ generated with the coordination onto metal before insertion. As an example, carbene (R^1R^2C:) is an important organic intermediate that contains two unshared valence electrons and a neutral carbon atom with a valence of two, where R^1 and R^2 represent substituents or hydrogen atoms [163, 164, 167–170]. The center carbon atom in carbene has six valence

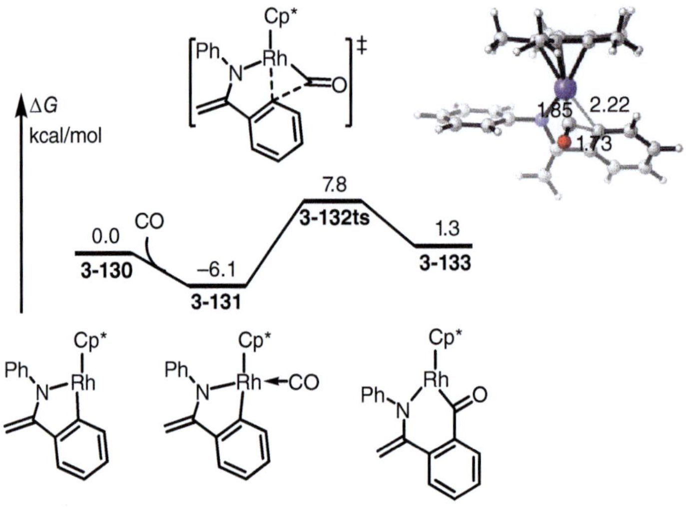

Figure 3.33 The free-energy profiles for the 1,1-carbonyl insertion. The energies were calculated at M11-L/6-311+G(d) (SDD for Rh)/SMD(toluene)//B3LYP/6-31G(d) (SDD for Rh) level of theory. Source: Based on Gao, B., Liu, S., Lan, Y. et al. (2016). Rhodium-Catalyzed Cyclocarbonylation of Ketimines via C—H Bond Activation. Organometallics 35 (10): 1480-1487.

electrons, which is two electrons less than allowed by the octet rule; therefore, carbenes are classified as either singlet or triplet depending upon their electronic behavior. Carbene can form metal–carbene complex to further stabilize it by two possible types. Schrock-type metallacarbene has a covalent double bond between the metal and carbene. Alternatively, carbene is considered as in the singlet state. While the σ lone pair of the carbene is donated to the vacant d orbital of the metal, and simultaneously π back-donation occurs from the metal d orbital to the unoccupied carbene p orbital, which collectively constitutes the metal–carbon bond in this Fischer-type metal–carbene complex.

In late-transition metal chemistry, metallacarbene complexes are also considered to be Fischer type, which usually can be generated by the metal-mediated denitrogenation of diazo compounds [168, 169, 171]. As shown in Figure 3.34 [164], diazo compound can coordinate with a 16-electron Rh(III) species **3-134** to form a carbon-coordinated complex **3-135** with 10.0 kcal/mol endergonic. The following denitrogenation takes place via a substitution transition state **3-136ts** to afford Rh(III)–carbene complex **3-137**. The release of gaseous dinitrogen leads to the irreversibility of this process. The coordinated carbene moiety can insert into either Rh—C(aryl) bond or Rh—C(alkyl) bond in complex **3-137**. The computational results showed that the carbene insertion into Rh–C(aryl) is more favorable, which occurs via a concerted three-membered ring-type transition state **3-138ts** with an energy barrier of 10.2 kcal/mol. The geometry information of that transition state is shown in Figure 3.34. The lengths of forming C(carbene)—C(aryl) bond and breaking Rh—C(aryl) bond are 2.18 and 2.17 Å, respectively.

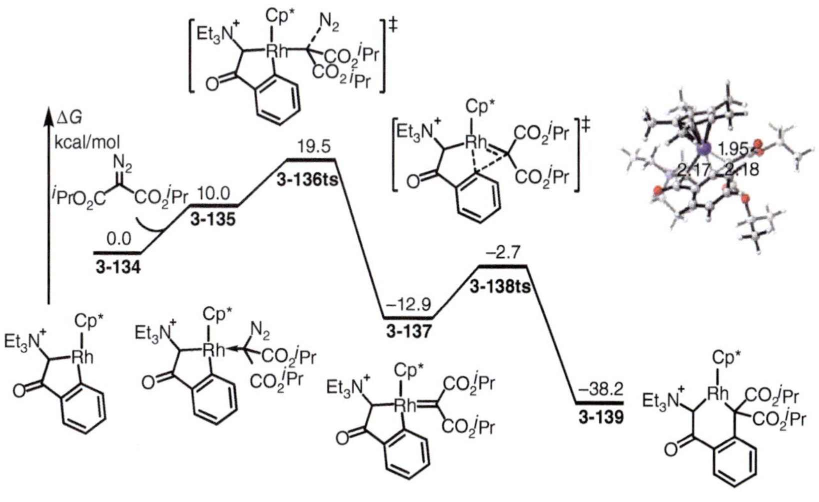

Figure 3.34 The free-energy profiles for the formation of Rh–carbene complex following by the 1,1-carbene insertion. The energies were calculated at M06/6-311+G(d) (SDD for Rh)/SMD(acetonitrile)//B3LYP/6-31+G(d) (SDD for Rh) level of theory. Source: Based on Yu, S., Liu, S., Lan, Y. et al. (2015). Rhodium-Catalyzed C−H Activation of Phenacyl Ammonium Salts Assisted by an Oxidizing C−N Bond: A Combination of Experimental and Theoretical Studies. Journal of the American Chemical Society 137 (4): 1623-1631.

Nitrene is a neutral monovalent species with six valence electrons that is the isoelectronic equivalent of carbene, sharing some similar characteristics. Correspondingly, nitrenes are also active intermediate, which can coordinate with metals to form metal–nitrene complexes. The in situ–generated nitrene coordinated with transition metal also can insert into other metal–organic bond to form an amino metal complex [172–176]. As shown in Figure 3.35 [165], the coordination of anthranil onto Rh forms a Rh(III) complex **3-141**. The N—O bond cleavage in anthranil generates a nitrene–Rh(III) complex B via transition state **3-142ts**. A migratory insertion of nitrene takes place via a three-membered ring-type transition state **3-144ts** with a free-energy barrier of 9.7 kcal/mol to afford amino-Rh(III) species **3-145**. The overall activation free energy of nitrene insertion is 21.8 kcal/mol given by DFT calculation, because the formation of Rh(III)–nitrene complex is endergonic.

3.5.3 Conjugative Insertion

When conjugative unsaturated bonds are involved in insertion precursors, a 1,4-insertion also could take place for a terminal functionalized product. As shown in Figure 3.36 [177], in butadiene-coordinated nickelacycle **3-146**, a 1,2-insertion of butadiene could take place via a four-membered ring-type transition state **3-147ts** to afford a seven-membered nickelacycle **3-148**. Alternatively, a 1,4-insertion occurs via a six-membered ring-type transition state **3-149ts** to afford a nine-membered nickelacycle **3-150**. The computational results suggested that the relative free energy of 1,4-insertion transition state **3-149ts** is 7.4 kcal/mol lower than that of

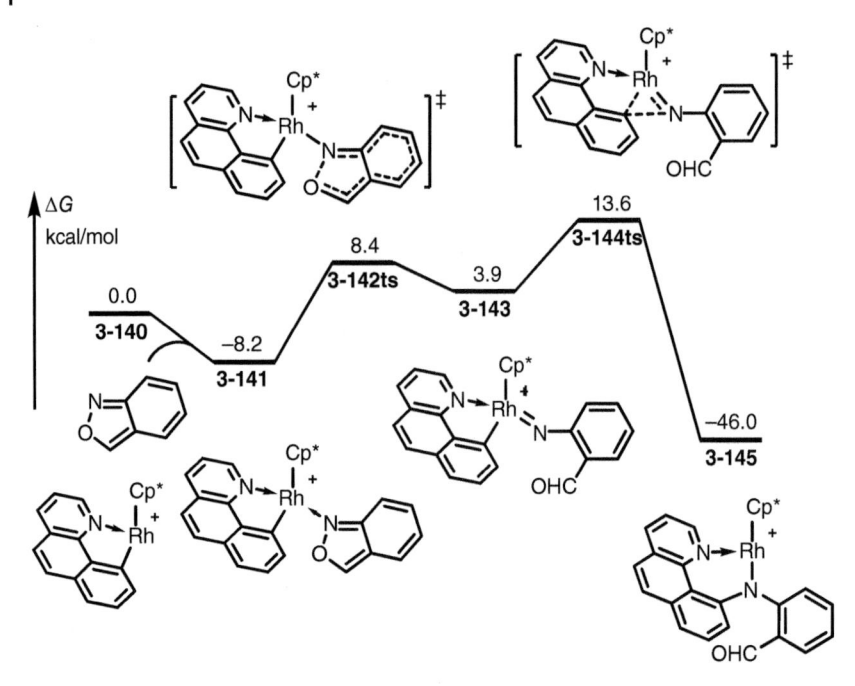

Figure 3.35 The free-energy profiles for the formation of Rh−nitrene complex following the 1,1-nitrene insertion. The energies were calculated at M06/6-311+G(d) (LANL08-f for Rh)/SMD(DCE)//B3LYP/6-31+G(d) (LANL08-f for Rh) level of theory. Source: Based on Yu, S., Tang, G., Li, Y. et al. (2016). Anthranil: An Aminating Reagent Leading to Bifunctionality for Both C(sp3)−H and C(sp2)−H under Rhodium(III) Catalysis. Angewandte Chemie International Edition 55 (30): 8696-8700.

1,2-insertion transition state **3-147ts**. Therefore, 1,4-insertion of butadiene is a favorable process.

In some unique examples, an atypical insertion would lead to a special reaction mechanism. As shown in Figure 3.37 [178], when an aryl rhodacycle(III) **3-151** is formed, what could have happened was reductive elimination of aryl and vinyl. However, complex **3-151** also can be considered as the resonance structure of cyclic rhodium−carbene, which can undergo a carbene insertion into Rh—C(aryl) bond via transition state **3-152ts** with a free-energy barrier of 17.7 kcal/mol. In this step, a nucleophilic addition onto vinyl imine moiety leads to π bond shift to afford an amino rhodacycle **3-153**, in which the oxidative state of Rh remains unchanged.

3.5.4 Outer-Sphere Insertion

In a regular insertion process, a metal−organic bond can be considered as one where metal and organic groups play the role of electrophile and nucleophile individually, while conjugative bond or atom inserts into the pair of nucleophile−electrophile. In some cases, the nucleophilic position of organic group is not bonding with metal; thus, an outer-sphere insertion can take place through a nucleophilic attack [179–183].

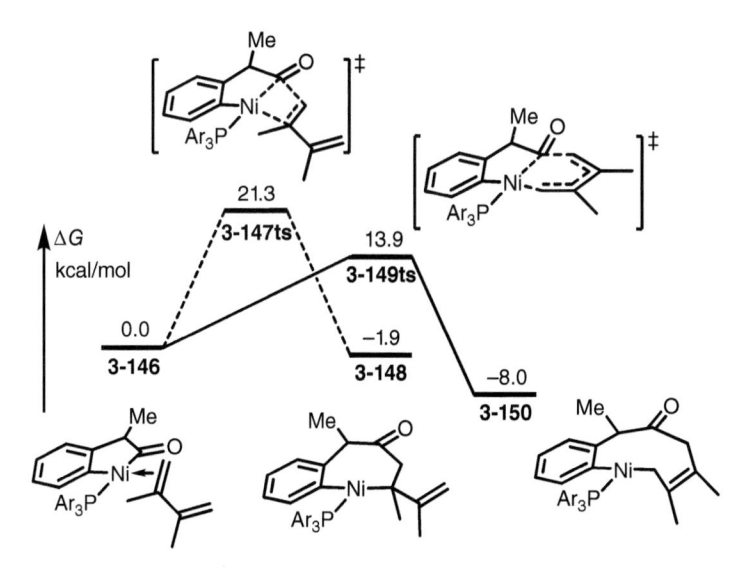

Figure 3.36 The free-energy profiles for the 1,4-conjugative insertion of diene. The energies were calculated at B3LYP/6-311+G(d,p) (SDD-f for Ni)/SMD(toluene)//B3LYP/6-31G(d) (LANL2DZ-f for Ni) level of theory.

Figure 3.37 The free-energy profiles for the carbene insertion. The energies were calculated at M11-L/6-311+G(d,p) (SDD-f for Rh)/SMD(DCE)//B3LYP/6-31G(d) (SDD for Rh) level of theory. Source: Based on Tan, G., Zhu, L., Liao, X. et al. (2017). Rhodium/Copper Cocatalyzed Highly trans-Selective 1,2-Diheteroarylation of Alkynes with Azoles via C-H Addition/Oxidative Cross-Coupling: A Combined Experimental and Theoretical Study. Journal of the American Chemical Society 139 (44): 15724-15737.

As shown in Figure 3.38 [184], when a four-coordinated cyanomethyl Ni(II) complex **3-154** reacts with benzaldehyde, the barrier of concerted 1,2-insertion is as high as 37.4 kcal/mol via a four-membered ring-type transition state **3-155ts**. Alternatively, cyanomethyl Ni(II) **3-154** can isomerize to ethenamide Ni(II) complex **3-158** with an energy barrier of 28.5 kcal/mol via transition state **3-157ts**. An outer-sphere nucleophilic attack of ethenamide onto benzaldehyde takes

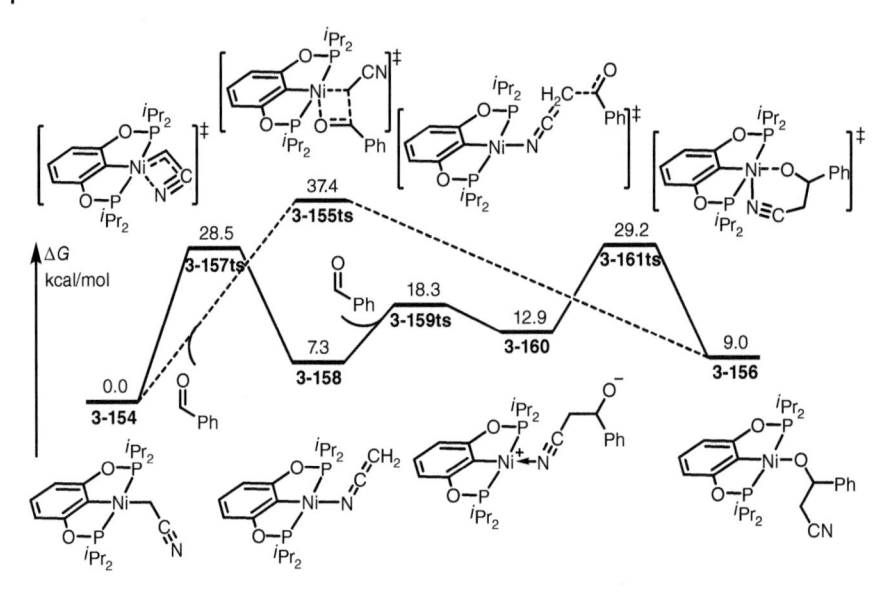

Figure 3.38 The free-energy profiles for the acyl insertion through an outer-sphere nucleophilic attack. The energies were calculated at B3LYP-D3/def2-TZVP/CPCM (acetonitrile)//M06/6-31G(d) (LANL2DZ-f for Ni)/CPCM(acetonitrile) level of theory. Source: Based on Ariafard, A., Ghari, H., Khaledi, Y. et al. (2016). Theoretical Investigation into the Mechanism of Cyanomethylation of Aldehydes Catalyzed by a Nickel Pincer Complex in the Absence of Base Additives. ACS Catalysis 6 (1): 60-68.

place via transition state **3-159ts** with an energy barrier of only 11.0 kcal/mol to afford a zwitterionic intermediate **3-160**, which can isomerize to common alkoxyl Ni(II) complex **3-156**. The DFT calculation revealed that carbonyl insertion in this case cannot occur via a concerted process; however, a more complex involving isomerization to ethenamide, outer-sphere nucleophilic attack, and coordination exchange resulted by DFT calculations.

3.6 Elimination

Elimination reaction is the reverse of insertion [185–190]. As shown in Scheme 3.15, in a metal involving compound M–A–X, the elimination yields an M—X covalent bond and releases an unsaturated molecule A. Consistent with insertion, elimination is usually a reversible process. When an unsaturated bond is formed in A=B, it can be named β-elimination [191–193], while α-elimination forms an A=B molecule with an unsaturated atom [194, 195]. After the elimination, the newly generated unsaturated molecule can either coordinate with metal or release out.

3.6.1 β-Elimination

When a σ covalent bond is formed between the metal and the α-position of an organic moiety, the metal can interact with the group or the atom at β-position of that moiety,

3.6 Elimination | 87

(a) β-elimination:

$L_n-M-A \atop B \atop X$ ⟶ $L_n-M \atop X$ + A=B

(b) α-elimination:

$L_n-M-A \atop X$ (with B double-bonded to A) ⟶ $L_n-M \atop X$ + A=B

Scheme 3.15 Elimination. (a) β-Elimination, (b) α-elimination.

which leads to an elimination of metal and that group to result in an α,β unsaturated bond.

The β-elimination is one of the critical steps involved in the catalytic cycle of Mizoroki–Heck reaction. As shown in Figure 3.39 [156], in an alkyl Pd(II) species **3-162**, an agostic bond is observed, in which the σ orbital of C—H bond donates electrons to the empty d orbital of Pd to afford a three-center two-electron bond. With the activation of agostic bond, a hydrogen transfer from β-position of alkyl to Pd occurs through a concerted four-membered ring-type transition state **3-163ts** with a barrier of 6.7 kcal/mol to form an olefin-coordinated hydride Pd(II) complex **3-164**. It can be considered as a β-hydrogen elimination. In that

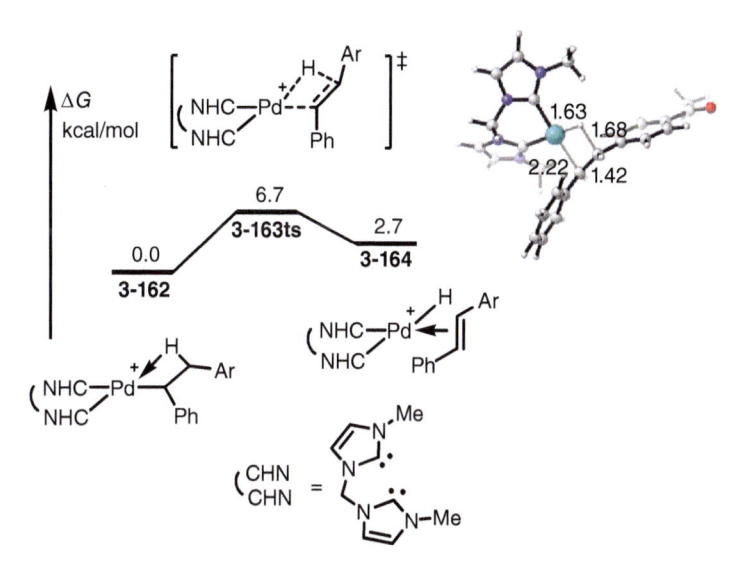

Figure 3.39 The free-energy profiles for the β-elimination in Pd-catalyzed Heck coupling reaction. The energies were calculated at M06/def2-TZVP (LANLTZ-f for Pd)/PCM(DMAc)// B3LYP/6-31G(d) (LANL2DZ for Pd) level of theory. Source: Based on Allolio, C., Strassner, T. (2014). Palladium Complexes with Chelating Bis-NHC Ligands in the Mizoroki–Heck Reaction—Mechanism and Electronic Effects, a DFT Study. The Journal of Organic Chemistry 79 (24): 12096-12105.

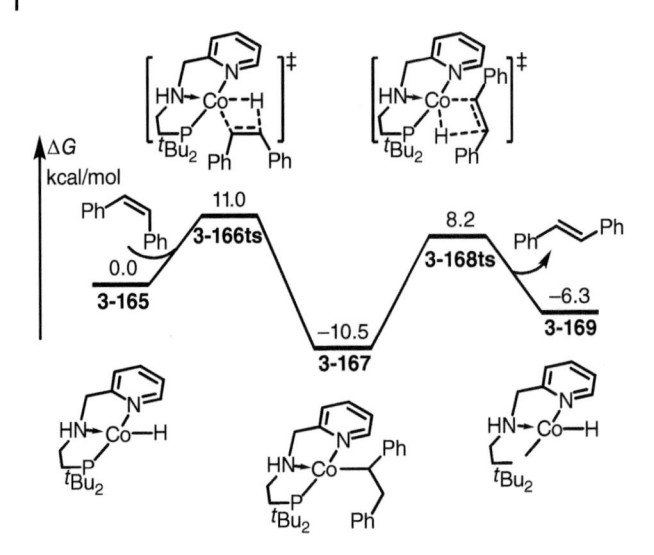

Figure 3.40 The free-energy profiles for the Co-hydride catalyzed Z-/E-isomerization of olefins through 1,2-insertion and β-hydride elimination. The energies were calculated at M06/6-311+G(d,p) (SDD for Co)/SMD(methanol)//B3LYP/6-31G(d) (SDD for Co) level of theory. Source: Based on Xue, L., Ng, K. C., Lin, Z. (2009). Theoretical studies on β-aryl elimination from Rh(i) complexes. Dalton Transactions (30): 5841-5850.

transition state, the distances of Pd—H and C—H are 1.63, and 1.68 Å, respectively. It reveals that the Pd—H bond is partially formed, while C—H bond is not fully broken.

The combination of insertion and elimination would lead to the Z-/E-isomerization of olefins. As shown in Figure 3.40 [196], when (Z)-diphenylethene coordinates with hydride Co(I) species **3-165**, an olefin insertion occurs via transition state **3-166ts** to afford an alkyl–Co(I) complex **3-167**. Then, a β-hydrogen elimination takes place via transition state **3-168ts** to yield a (E)-diphenylethene as product and regenerate hydride Co(I) species **3-169**.

The β-hydrogen elimination is also a common process in Rh(I) chemistry. As shown in Figure 3.41 [197], an agostic bond can present in a benzyloxy Rh(I) complex **3-170**. A β-hydrogen elimination can occur via a concerted four-membered ring-type transition state **3-171ts** with a barrier of only 4.5 kcal/mol to afford a benzaldehyde-coordinated hydride Rh(I) complex **172**.

In addition to β-hydrogen elimination, other β-heteroatom (C, F, N, etc.) elimination can also occur through a concerted four-membered ring-type transition state. As shown in Figure 3.42 [197], when a triarylmethoxy Rh(I) species **3-173** is formed, a β-aryl elimination can occur via a four-membered ring-type transition state **3-174ts** with an energy barrier of 19.8 kcal/mol to afford a benzaldehyde-coordinated aryl Rh(I) complex **3-175**. In contrast, corresponding β-aryl elimination of triarylethyl Rh(I) species **3-176** also could occur via transition state **3-177ts** to afford an olefin-coordinated aryl Rh(I) complex **3-178**. However, the barrier of the later one is 6.1 kcal/mol higher than that of the early one.

Figure 3.41 The free-energy profiles for the Rh(I) mediated β-hydride elimination. The energies were calculated at B3LYP/6-31G(d) (LANL2DZ-f for Rh)//B3LYP/6-31G(d) (LANL2DZ-f for Rh) level of theory. Source: Based on Qi, X., Liu, X., Qu, L.-B. et al. (2018). Mechanistic insight into cobalt-catalyzed stereodivergent semihydrogenation of alkynes: The story of selectivity control. Journal of Catalysis 362: 25-34.

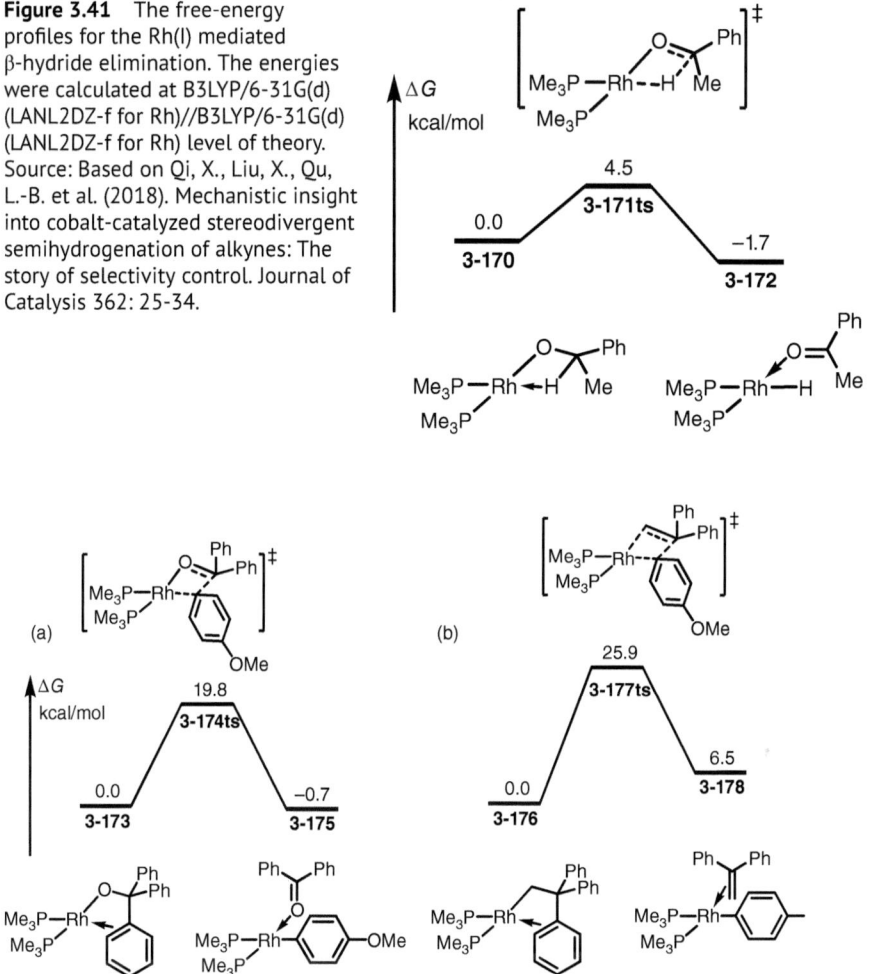

Figure 3.42 The free-energy profiles for the Rh(I)-mediated β-aryl elimination. The energies were calculated at B3LYP/6-31G(d) (LANL2DZ-f for Rh)//B3LYP/6-31G(d) (LANL2DZ-f for Rh) level of theory. Source: Based on Xue, L., Ng, K. C., Lin, Z. (2009). Theoretical studies on β-aryl elimination from Rh(i) complexes. Dalton Transactions (30): 5841-5850.

As shown in Figure 3.43 [197], a β-allyl elimination of complex **3-179** can take place via a chair-type six-membered ring-type transition state **3-180ts** with an energy barrier of 20.7 kcal/mol to afford an allylic Rh(I) complex **3-181** reversibly.

When the generated unsaturated bond is stable enough, β-carbon elimination can be an irreversible and rapid process. As shown in Figure 3.44 [198], when a Rh(II) cycle **3-182** is formed, a β-carbon elimination can occur without barrier via transition state **3-183ts** to afford a four-membered ring-type carboxyl Ru(II) complex **3-184** with the delivery of gaseous dinitrogen.

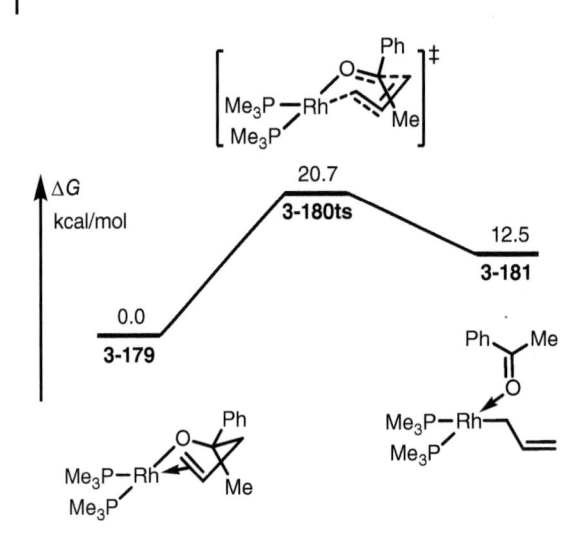

Figure 3.43 The free-energy profiles for the Rh(I)-mediated β-allyl elimination. The energies were calculated at B3LYP/6-31G(d) (LANL2DZ-f for Rh)//B3LYP/6-31G(d) (LANL2DZ-f for Rh) level of theory.

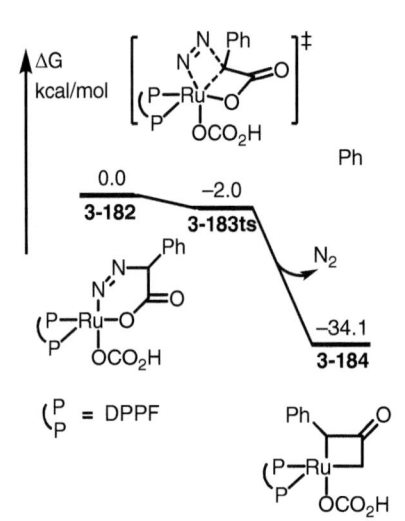

Figure 3.44 The free-energy profiles for the Ru-mediated β-carbon elimination. The energies were calculated at M06-L/6-311+G(d,p) (LANL2DZ for Ru)/SMD(DMF)//B3LYP/6-31G(d) (LANL2DZ for Ru) level of theory.

A gaseous leaving molecule with unsaturated bond can promote β-heteroatom elimination. As shown in Figure 3.45 [199], when a Ni(II)-cycle **3-185** is formed, isomerization leads to the coordination of nitrogen atom onto Ni in complex **3-187**. Then, a β-amino elimination occurs via a concerted four-membered ring-type transition state **3-188ts** with an energy barrier of only 6.7 kcal/mol to release one gaseous carbon dioxide.

3.6.2 α-Elimination

α-Elimination is often present in a decarbonylation reaction. The general process is shown in Figure 3.46 [194]. In an square planar acyl Ni(II) species **3-190**, the dissociation of phosphine affords a vacancy to the coming of aryl group in complex

Figure 3.45 The free-energy profiles for the Ni-mediated β-amino elimination. The energies were calculated at M06/6-311+G(2d,p) (SDD for Ni)/CPCM(toluene)//M06/6-31G(d) (LANL2DZ for Ni)/CPCM(toluene) level of theory.

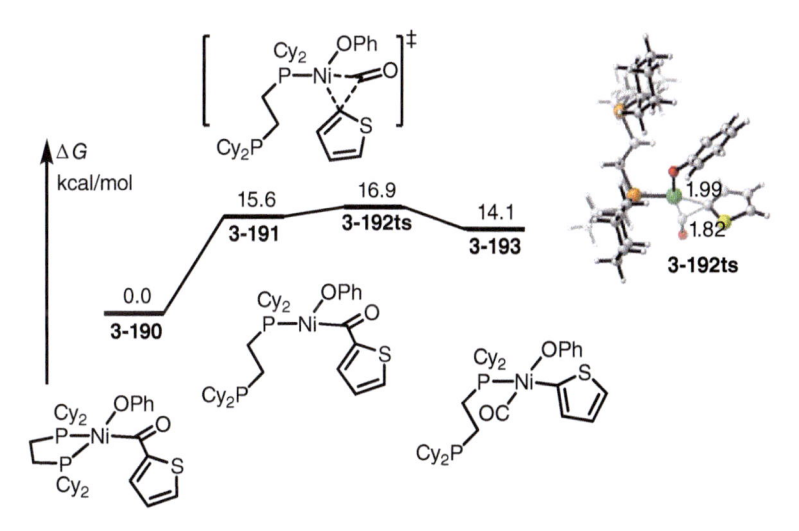

Figure 3.46 The free-energy profiles for the Ni-mediated α-aryl elimination. The energies were calculated at M06-L/6-311+G(2d,p) (SDD-f for Ni)/SMD(1,4-dioxane)//B3LYP/6-31G(d) (LANL2DZ for Ni) level of theory.

3-191. Then, an α-aryl elimination occurs via a three-membered ring-type transition state **3-192ts** to yield a carbonyl-coordinated aryl–Ni(II) complex **3-193** reversibly. In the geometry information of that transition state, the lengths forming Ni—C(aryl) bond and breaking C(carbonyl)—C(aryl) bond are 1.99 and 1.82, respectively, which reveal that the bond forming and breaking occur simultaneously.

3.7 Transmetallation

Transmetallation is a type of elementary reaction in organometallic chemistry that involves the transfer of anionic ligands from one metal to another. The general form of transmetallation is shown as Scheme 3.16a [200, 201]. When M–X reacts with M′–R, anionic ligand exchange leads to the formation of M–R and M′–X. In this type of reaction, some elements with metallic properties, such as boron and silicon, are also regarded as metals. Transferred groups can be, but are not limited to, an alkyl, aryl, alkynyl, allyl, halogen, or pseudo-halogen group. The reaction is usually an irreversible process due to thermodynamic and kinetic reasons. Thermodynamics will favor the reaction based on the electronegativities of the metals and kinetics will favor the reaction if there are empty orbitals on both metals. In a meaningful transmetallation process, anionic ligand transfer usually occurs between the catalytically activated metal and the metal reactant bonding with nucleophile (Scheme 3.16b) [202, 203]. As a result, the nucleophilic group in the reactant is introduced into catalytic cycle, which leads to further coupling reactions. Usually, there are majorly two possible pathways for transmetallation process, including concerted ring-type pathway and electrophilic pathway.

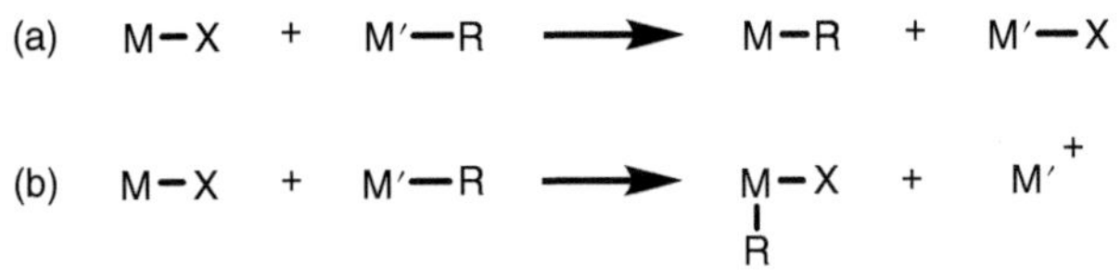

(a) M–X + M′–R ⟶ M–R + M′–X

(b) M–X + M′–R ⟶ M–X + M′⁺
                            |
                            R

Scheme 3.16 Transmetallation. (a) Metathesis type transmetallation, (b) Substitution type transmetallation.

3.7.1 Concerted Ring-Type Transmetallation

In concerted ring-type transmetallation pathway (Scheme 3.17) [204, 205], when M–X reacts with M′–R, the cleavage of M–X/M′–R covalent bond and the formation of M–R/M′–X covalent bond takes place synchronously to afford transmetallation product, though the coordination bonds are observed before transmetallation or remained after transmetallation. Usually, this type of transmetallation undergoes a four-membered ring-type transition state, which are composed of reacting M–X and M′–R moieties.

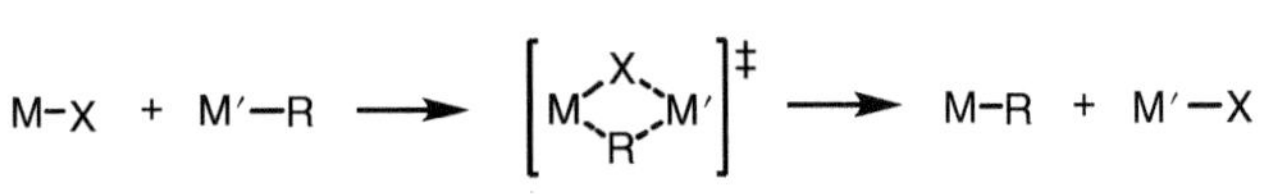

M–X + M′–R ⟶ [M⟨ˣ⟩M′ over R]‡ ⟶ M–R + M′–X

Scheme 3.17 Concerted transmetallation.

DFT calculations are used to reveal the detailed mechanism of transmetallation. The free-energy profiles for a typical concerted ring-type transmetallation are shown in Figure 3.47 [206]. (Bpin)$_2$ molecule can be considered as that one Bpin group

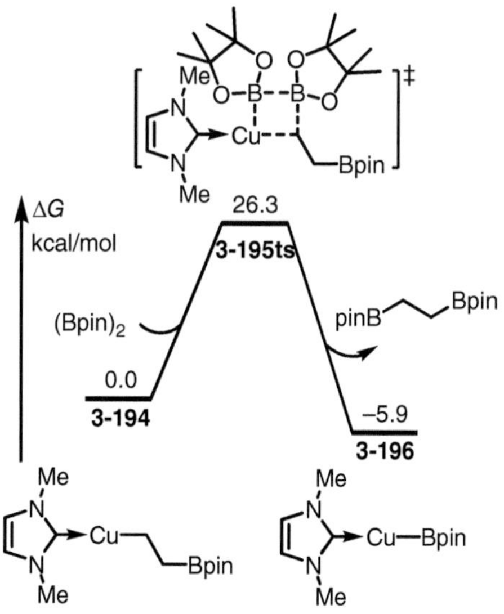

Figure 3.47 The free-energy profiles for the concerted transmetallation between alkyl-copper and diboryl. The energies were calculated at B3LYP/6-311G(d) (6-31G(d) for N and H)//B3LYP/6-311G(d) (6-31G(d) for N and H) level of theory.

revealing metallic character, while another one is transferred group. When an alkyl Cu(I) species **3-194** reacts with (Bpin)$_2$ compound, an intermolecular transmetallation takes place with those two molecules. A four-membered ring-type transition state **3-195ts** is located, in which the breaking B—B and Cu—C bond lengths are 2.50 and 2.08 Å, respectively, and the lengths of forming Cu—B and B—C bonds are 1.99 and 2.17 Å, respectively. It reveals a concerted process of the transfers of boryl and alkyl groups. After this transmetallation, a boryl Cu(I) complex **3-196** is formed with the release of alkylboronate.

When the transfer group, such as halide, hydroxyl, and alkoxyl, in one reactant for transmetallation contains lone-pair electrons, it could coordinate with the metal atom in another reactant to reversibly form an intermediate [143], while this coordination bond is transferred to covalent bond in further transmetallation. As shown in Figure 3.48 [207], when an alkoxyl Ni(II) **3-197** is involved in a transmetallation to react with triethylborane, the lone-pair electrons in alkoxyl group can coordinate with borane to form complex **3-198** with 4.6 kcal/mol exergonic. In that complex, the oxygen atom is shared by Pd and B atoms. Subsequently, an ethyl group shift from B to Pd takes place via transition state **3-199ts** with an energy barrier of 9.9 kcal/mol to afford ethyl Ni(II) complex **3-200**, in which an alkoxyborane coordinates with Ni. In this process, the Ni—O bond lengths of reactant **3-198**, transition state **3-199ts**, and product **3-200** are 1.87, 1.88, and 2.09 Å, respectively, which reveal a transformation of covalent bond to coordination bond instead of bond cleavage.

In another case (Figure 3.49) [208], when the transmetallation takes place between a square planar Ni(I) ethylate **3-201** and (Bpin)$_2$, the initial coordination of ethylate onto Bpin group forms complex **3-202** with 14.3 kcal/mol exergonic. The sequential transmetallation occurs via a four-membered ring-type transition state **3-203ts** with an energy barrier of only 3.3 kcal/mol to afford a boryl Ni(I)

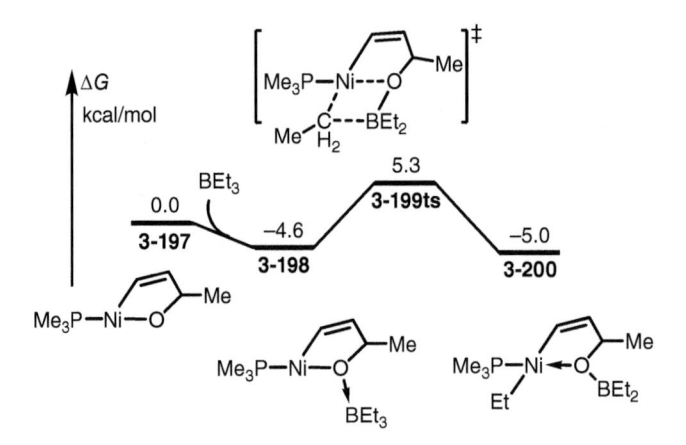

Figure 3.48 The free-energy profiles for the concerted transmetallation between alkoxyl-nickel and triethylborane. The energies were calculated at B3LYP/6-31G(d) (LANL2DZ for Ni)//B3LYP/6-31G(d) (LANL2DZ for Ni) level of theory.

Figure 3.49 The free-energy profiles for the concerted transmetallation between alkoxyl-nickel and diboryl. The energies were calculated at B3LYP-D3/6-31G(d)/CPCM (DMA)//B3LYP/6-31G(d) level of theory.

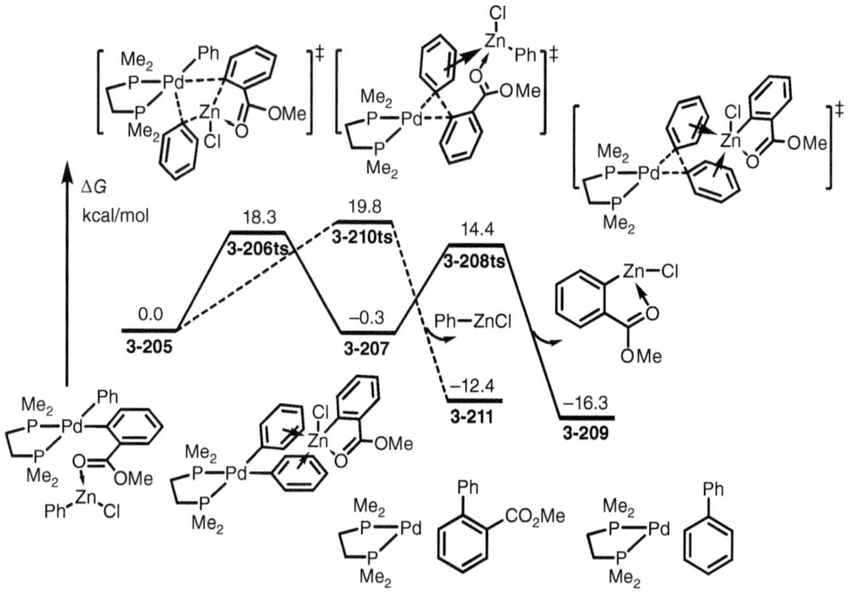

Figure 3.50 The free-energy profiles for the second transmetallation in Nigishi coupling reaction. The energies were calculated at B3LYP/6-31G(d) (SDD for Pd and Zn)/IEF-PCM(THF)//B3LYP/6-31G(d) (SDD for Pd and Zn) level of theory.

complex **3-204**. In that complex, Ni is square planar configuration with 17 electrons; therefore, the absence of vacancy in Ni leads to the direct release of borate without further coordination.

Usually, transmetallation is a rapid process without impact on reaction mechanism. However, transmetallation can affect the chemoselectivity of a coupling reaction. As shown in Figure 3.50 [209], in a Pd-catalyzed Negishi-type coupling, when a biaryl Pd(II) species **3-205** is formed from oxidative addition followed by the first transmetallation with aryl Zn reactant, the second transmetallation with aryl Zn can occur via transition state **3-206ts** with a barrier of 18.3 kcal/mol, which is 1.5 kcal/mol lower than that of regular reductive elimination process (via **3-210ts**). Therefore, the second transmetallation leads to the homo-coupling side product of biaryl.

In fact, the detailed mechanism of transmetallation can be very complex. Houk group provided an exhaustive theoretical study on the mechanism for an alkoxysilane-catalyzed transmetallation of aryl Li and chloride Pd(II) [210]. As shown in Figure 3.51, the first transmetallation occurs between aryl Li and alkoxysilane **3-212** via a four-membered ring-type transition state **3-213ts** to afford a phenyl silane **3-214** with alkoxyl Li moiety. The coordination of alkoxyl onto Pd forms a four-membered ring-type intermediate **3-215**, which can release LiCl to afford alkoxyl Pd species **3-216**. The second intramolecular transmetallation takes place via transition state **3-217ts** to afford phenyl Pd and regenerate active species **3-218**. The rate-limiting step is the second transmetallation with an energy barrier of 16.4 kcal/mol.

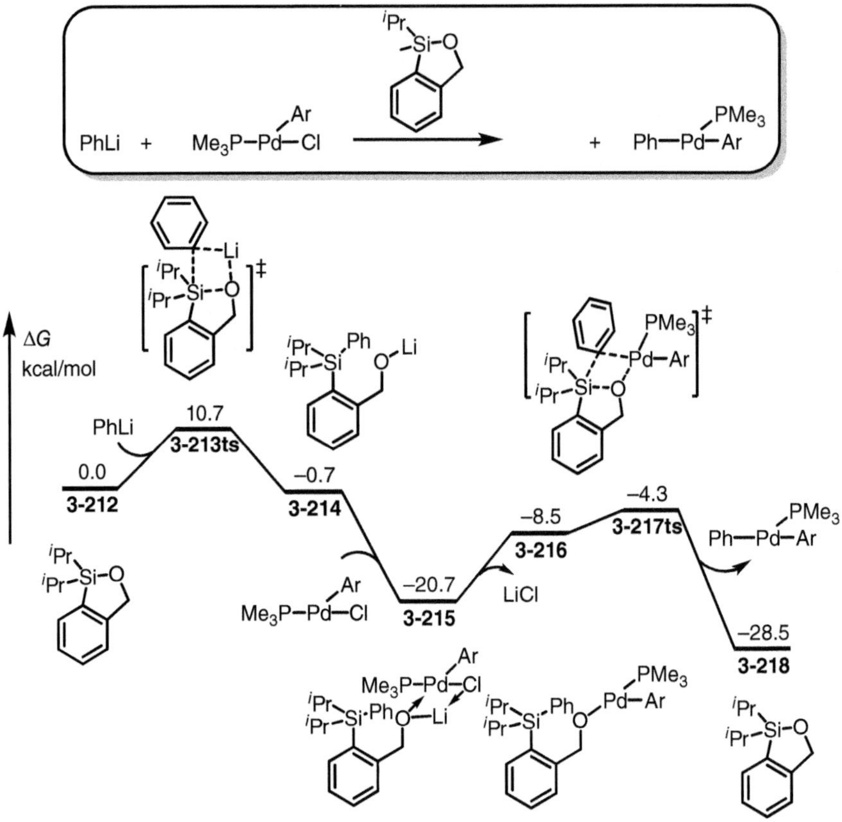

Figure 3.51 The free-energy profiles for the alkoxysilane-catalyzed transmetallation of aryl Li and chloride Pd(II). The energies were calculated at M06/6-311+G(d,p) (SDD for Pd)/CPCM(THF)//M06/6-31G(d) (SDD for Pd) level of theory.

In transmetallation process of Negishi coupling, a metal–metal bond is observed theoretically. As shown in Figure 3.52 [211], in transmetallation between chloride Pd(II) complex **3-219** and methyl Zn, a Cl-bridged complex **3-220** could be formed initially with 3.1 kcal/mol endergonic. The transmetallation can occur via a four-membered ring-type transition state **3-221ts** with an energy barrier of 7.4 kcal/mol. In the geometry information of this transition state, the distance between Zn and Pd is only 3.01 Å, which clearly revealed a bonding interaction. After transmetallation, a Zn—Pd bond is located in complex **3-222**, where the bond length of Zn—Pd is 3.01 Å. The reversible dissociation of Zn leads to the formation of methyl Pd complex **3-223**.

When the transferred group neighbored with a conjugative bond, a larger ring would be present in the transition state of transmetallation. As shown in Figure 3.53 [212], when bromide Ni(II) **3-224** reacts with lithium enolate, the initial coordination forms intermediate **3-226** with 13.0 kcal/mol exergonic. The following transmetallation takes place via a six-membered ring-type transition state **3-227ts** to afford an

3.7 Transmetallation | 97

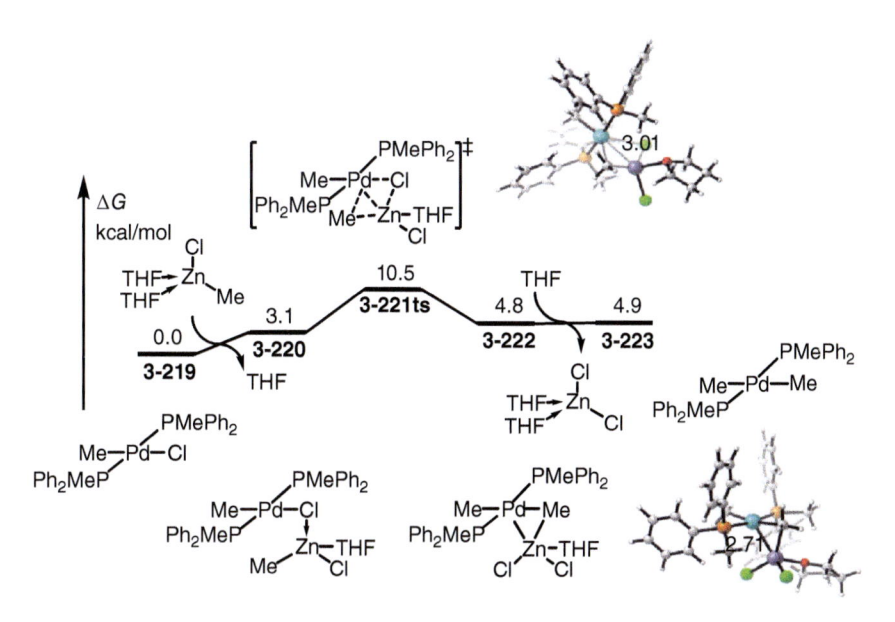

Figure 3.52 The free-energy profiles for the transmetallation in Negishi coupling involving a Pd–Zn complex. The energies were calculated at M06/DG-TZVP (DG-DZVP for Pd)/PCM(THF)//M06/6-31G(d) (LAN2DZ for Pd and Zn) level of theory.

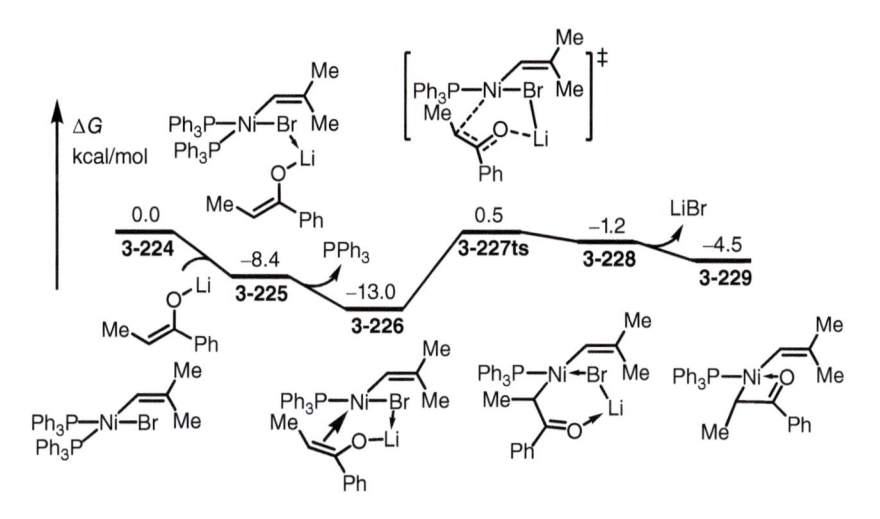

Figure 3.53 The free-energy profiles for the transmetallation of nickel bromide and lithium enolate via six-membered ring-type transition state. The energies were calculated at M06/6-311+G(d,p) (def2-TZVP for Ni)/SMD(THF)//M06/6-31G(d) (LAN2DZ for Ni) level of theory.

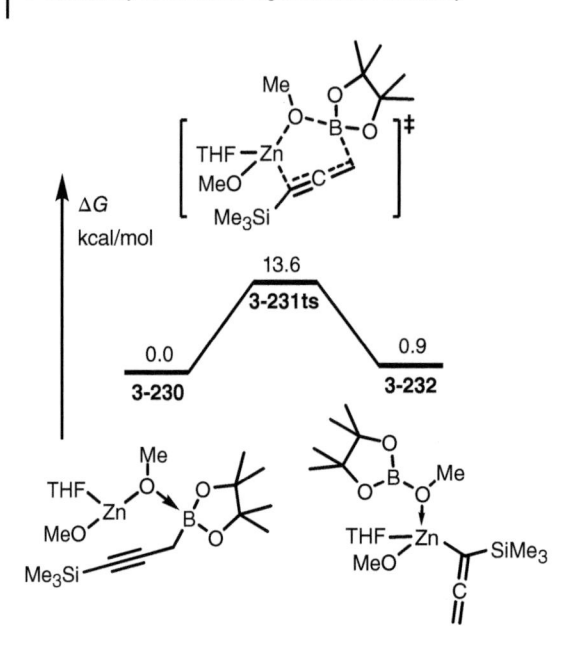

Figure 3.54 The free-energy profiles for the transmetallation of propargyl borane and methoxy zinc leading to the isomerization of propargyl. The energies were calculated at B3LYP/LANL2DZ//B3LYP/LANL2DZ level of theory.

alkyl Ni(II) complex **3-228**, which can further isomerize to its enolate isomer **3-229** exergonically.

Transmetallation also can lead to isomerization of transferred group. As shown in Figure 3.54 [213], the transmetallation of a propargyl borane and methoxy zinc **3-230** can occur via a six-membered ring-type transition state **3-231ts** with an energy barrier of 13.6 kcal/mol. However, after this process, propargyl is isomerized to propadienyl group in its Zn complex **3-232**. It can be considered that the boryl group increases the nucleophilicity of triple bond in propargyl borane, while the nucleophilic attack onto Zn leads to this isomerization.

3.7.2 Transmetallation Through Electrophilic Substitution

In a substitution-type transmetallation (Scheme 3.18), the metal can be considered as an electrophile, which can react with nucleophile coordinating on another metal to achieve transmetallation [31, 214].

$$M-X + M'^+ \longrightarrow [M---X---M']^{\ddagger} \longrightarrow M^+ + X-M'$$

Scheme 3.18 Transmetallation through electrophilic substitution.

As shown in Figure 3.55 [178], aryl Cu(I) species can be considered as nucleophile, which can react with cationic Rh(III) complex **3-233** via transition state **3-234ts** with an energy barrier of 21.8 kcal/mol to afford an aryl Rh(III) complex **3-235**. In this transition state, the bond lengths of Rh—C(aryl) and Cu—C(aryl) are 2.62 and 1.91 Å, respectively. The bond angle of Rh—C(aryl)—Cu is 95.2°, which reveals an arene-type electrophilic substitution.

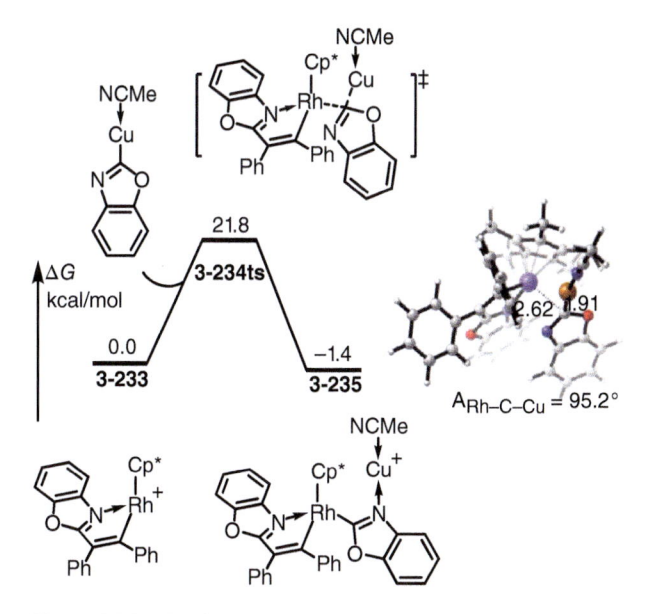

Figure 3.55 The free-energy profiles for the transmetallation of rhodium and aryl copper through electrophilic substitution. The energies were calculated at M11-L/6-311+G(d,p) (SDD-f for Rh)/SMD(DCE)//B3LYP/6-31G(d) (SDD for Rh) level of theory.

In another case [215], the transmetallation of vinyl stannane and cationic Pd(II) complex **3-236** also can carry out through a substitution process. The π-coordination of vinyl stannane onto Pd forms complex **3-238** with 5.5 kcal/mol endergonic. This coordination increases electrophilicity of stannane moiety; therefore, it can be nucleophilically attacked by bromide via transition state **3-240ts**. Then, a Br—Sn bond and a Pd—C(vinyl) bond are formed with the cleavage of Sn—C(vinyl) bond (Figure 3.56).

3.7.3 Stepwise Transmetallation

Strictly, transmetallation as an elementary reaction can even be decomposed into a combination of several other elementary reactions [216–219]. As shown in Figure 3.57 [220], in an intramolecular transmetallation between methyl silane and vinyl Rh(I) moieties in complex **3-242**, a direct transmetallation is considered to be processed via transition state **3-247ts** to afford a vinyl silane and a methyl Rh(I) **3-246**; however, the calculated activation free energy of this process is as high as 43.8 kcal/mol. Alternatively, the intramolecular oxidative addition of C(methyl)—Si bond onto Rh(I) can occur via a three-membered ring-type transition state **3-243ts** to afford a methyl silyl Rh(III) intermediate **3-244**. Subsequently, the reductive elimination via transition state **3-245ts** also can result in the common products, vinyl silane and a methyl Rh(I) **3-246**. The calculated overall activation free energy of this process is only 14.7 kcal/mol, which is much lower than that of direct transmetallation process. Therefore, transmetallation can be decomposed into a sequential oxidative addition and reductive elimination in some unique cases.

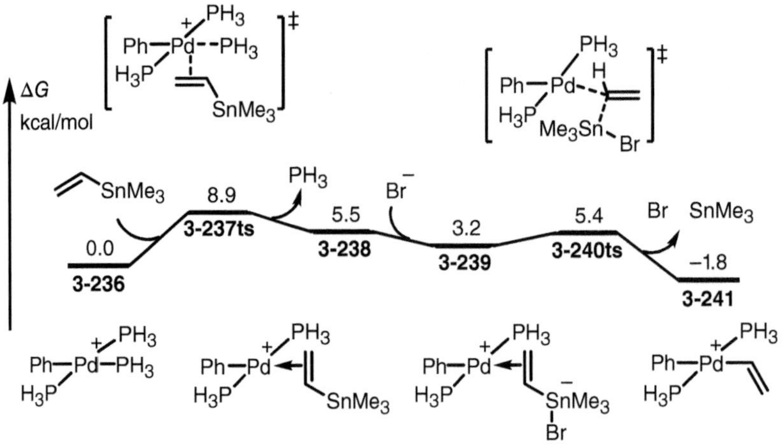

Figure 3.56 The free-energy profiles for the transmetallation of vinyl stannane and cationic Pd(II) through electrophilic substitution. The energies were calculated at B3LYP/6-31G(d) (LANL2DZ for Pd, P, and Sn; 6-31+G(d) for O)/CPCM(THF)//B3LYP/6-31G(d) (LANL2DZ for Pd, P, and Sn; 6-31+G(d) for O) level of theory.

3.8 Metathesis

In organometallic chemistry, metathesis is a metal-involved process that entails the redistribution of fragments of organic and organometallic compounds by the scission and regeneration of new metal–carbon bonds. In this kind of elementary reaction (Scheme 3.19), all of single, double, or triple bonds can be redistributed [221, 222].

$$M=X + R=Y \longrightarrow M=R + X=Y$$

Scheme 3.19 Metathesis.

3.8.1 σ-Bond Metathesis

A typical σ-bond metathesis, shown in Scheme 3.20, can be considered as one pair of electrophile–nucleophile (M–X) reacting with another one (H–Y). The σ-bond in H—Y can be activated by a metal with a vacant orbital at the metal center to achieve a highly organized four-membered ring-type transition state [223–226]. After σ-bond metathesis, the oxidative state of metal remained unchanged. Then, nucleophile exchange leads to the formation of M–Y and X–H. Frontier molecular orbital analysis can be used to explain why σ-bond metathesis can happen. As shown in Scheme 3.20 [227], in the σ and σ* bonds of M—X forming by metal's d orbital and nucleophile's p orbital, the involvement of d orbital leads to the presence of a node in those two molecular orbitals. Therefore, they can match the corresponding σ* and σ orbitals in H–Y. The orbital symmetry allows this type of σ-bond metathesis.

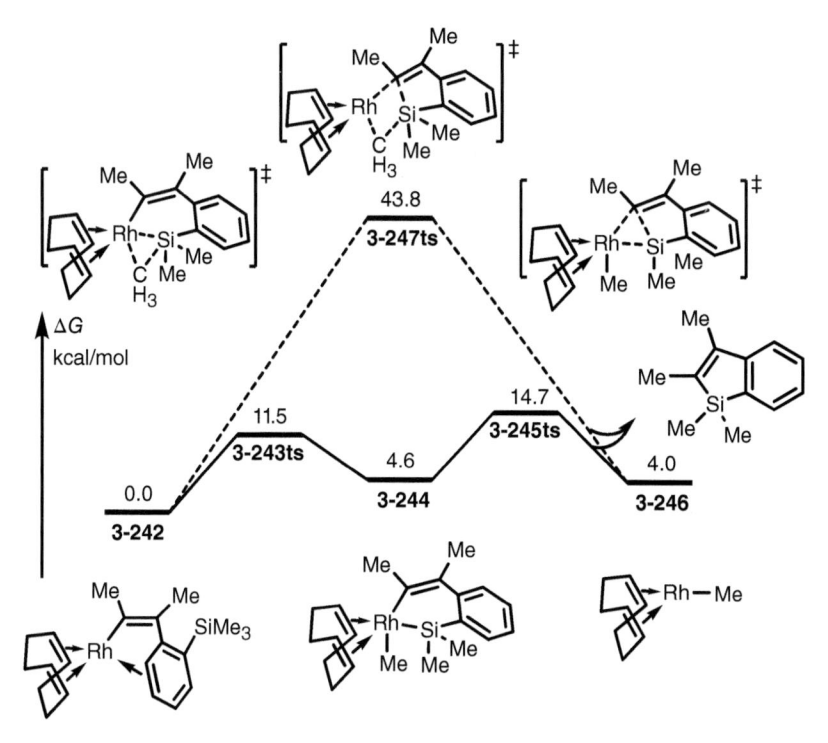

Figure 3.57 The free-energy profiles for the transmetallation through oxidative addition–reductive elimination. The energies were calculated at M11/6-311+G(d,p) (SDD for Rh)/IEF-PCM(1,4-dioxane)//B3LYP/6-31G(d) (LANL2DZ for Rh) level of theory.

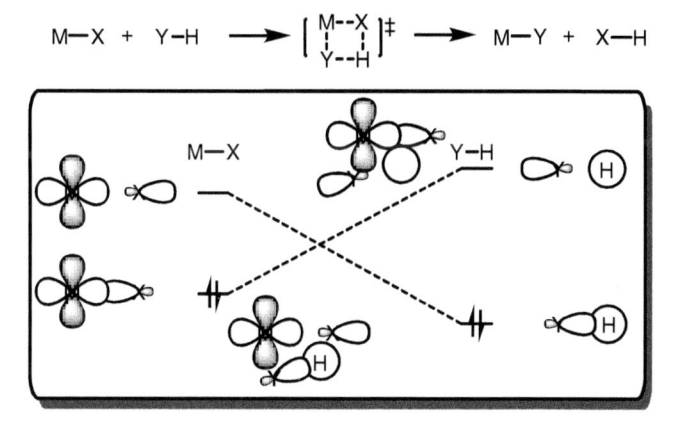

$$M-X + Y-H \longrightarrow \left[\begin{array}{c} M\cdots X \\ | \quad | \\ Y \cdots H \end{array}\right]^{\ddagger} \longrightarrow M-Y + X-H$$

Scheme 3.20 σ-bond metathesis and FMO interactions.

The σ-bond metathesis is a common process in early-transition-metal-mediated C—H bond activation [228–230]. As shown in Figure 3.58 [231], in the reaction of Zr(IV)(NMe$_2$)$_4$ complex **3-248** and thiophene of ligand, a four-membered ring-type σ-bond metathesis transition state can be found with an activation free energy of 19.2 kcal/mol. In geometry of this transition state **3-249ts**, the lengths of forming

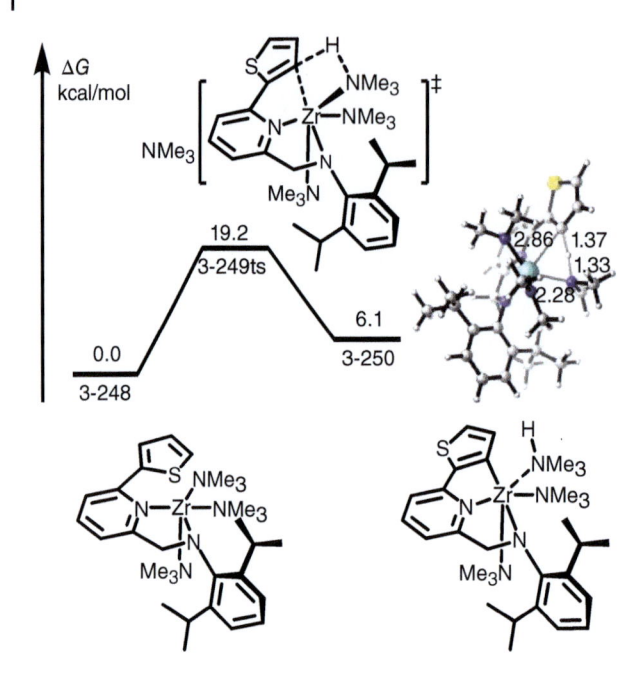

Figure 3.58 The free-energy profiles for the σ-bond metathesis of Zr(IV)(NMe$_2$)$_4$ and thiophene of ligand. The energies were calculated at M06/6-31G*/(LANL2DZ for Zr)/PCM (toluene) level of theory.

Zr—C and N—H bonds are 2.86 and 1.33 Å, while the lengths of breaking Zr—N and C—H bonds are 2.28, and 1.37 Å, respectively. Then a N$_2$ThZr-(NMe$_2$)$_2$(η^1-HNMe$_2$) species **3-250** can yield.

The σ-bond metathesis also can be mediated by late transition metals [232]. As shown in Figure 3.59 [233], when a Ni(II) *tert*-butoxide **3-251** is formed, it can be coordinated with gaseous dihydrogen to afford a 16-electron Ni(II) complex **3-252**. A σ-bond metathesis takes place via a four-membered ring-type transition state **3-253ts** with an overall activation free energy of 16.1 kcal/mol. Then, a *tert*-butyl alcohol coordinated hydride Ni(II) is afforded.

In a fluoride Rh(III)-mediated benzylic C—H bond activation reaction, the C—H bond cleavage takes place via a four-membered ring-type transition state **3-256ts** with an energy barrier of 18.4 kcal/mol. It is considered as a σ-bond metathesis process [234] (Figure 3.60).

3.8.2 Olefin Metathesis

Olefin metathesis entails the redistribution of fragments of olefins by the scission and regeneration of carbon–carbon double bonds. Mechanistically (Scheme 3.21), olefin metathesis starts from a direct [2+2] cycloaddition of metal–carbene and olefin to afford metallacyclobutane intermediate. Then, the metallacyclobutane

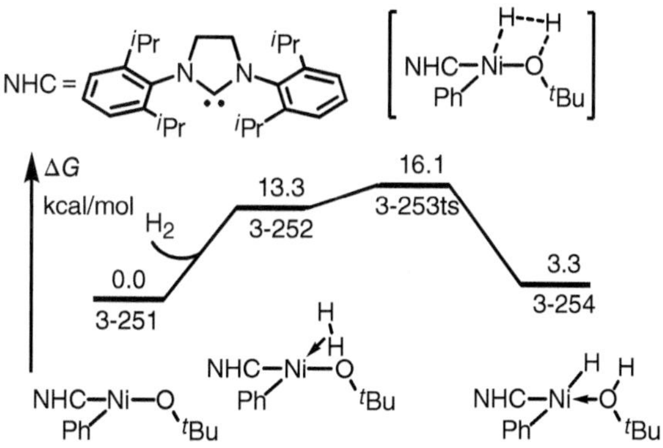

Figure 3.59 The free-energy profiles for the σ-bond metathesis of Ni(II) tert-butoxide and dihydrogen. The energies were calculated at M06/6-311++G(d,p) (SDD for Ni)/SMD(m-xylene)//M06/6-31G(d,p) (LANL2DZ for Ni) level of theory.

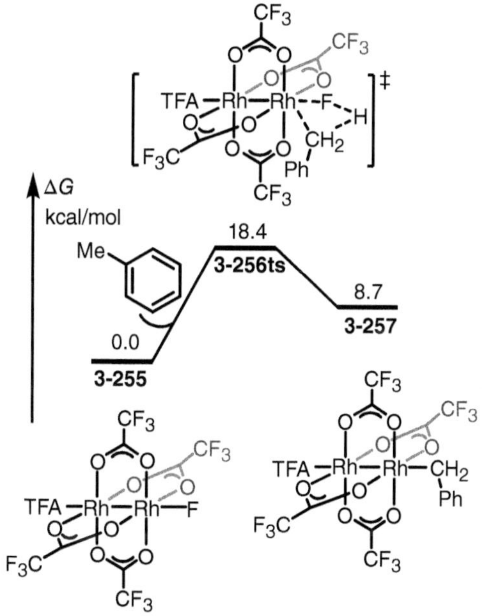

Figure 3.60 The free-energy profiles for the Rh-F-mediated C—H activation through σ-bond metathesis. The energies were calculated at N12/6-311+G(d) (SDD for Rh)/SMD(acetic acid)//M11-L/6-31G(d) (SDD for Rh) level of theory.

intermediate can reversibly cycloeliminate to yield a new olefin and metal–carbene [235, 236].

As shown in Scheme 3.22 [237], the [2+2] cycloaddition of metal–carbene and olefin is symmetry-allowed in frontier molecular orbitals. In the π and π* orbitals of metal–carbene, the node of d orbital in metal leads to the matching with the corresponding π* and π orbitals in olefin. Therefore, a concerted [2+2] cycloaddition can occur to afford a metallacyclobutane intermediate. Because of the relative simplicity of olefin metathesis, it often creates fewer undesired byproducts and hazardous wastes than alternative organic reactions.

The DFT calculated free-energy profiles for a typical olefin metathesis are shown in Figure 3.61 [237]. The reaction starts from a Ru-carbene species

Scheme 3.21 Olefin metathesis and its mechanism.

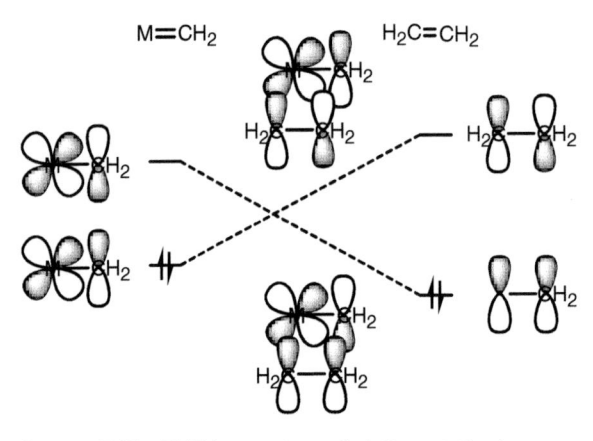

Scheme 3.22 FMO interactions of olefin metathesis.

3-258 with a vacancy generating from the dissociation of phosphine ligand from complex **3-259**. The coordination of olefin forms complex **3-260**. Subsequently, a [2+2] cycloaddition takes place via a four-membered ring-type transition state **3-261ts** with overall activation free energy of 17.2 kcal/mol, which is considered to be the rate-determining step. After this step, a ruthenacyclobutane **3-262** is formed reversibly. Then, a cycloelimination of another C—C bond breaks the four-membered ring to afford an olefin-coordinated Ru–carbene complex **3-264**, which can further react with another molecular olefin.

Interestingly, metathesis also can occur between olefin and an acetylene, which leads to the construction of diene product. As shown in Figure 3.62 [238], the

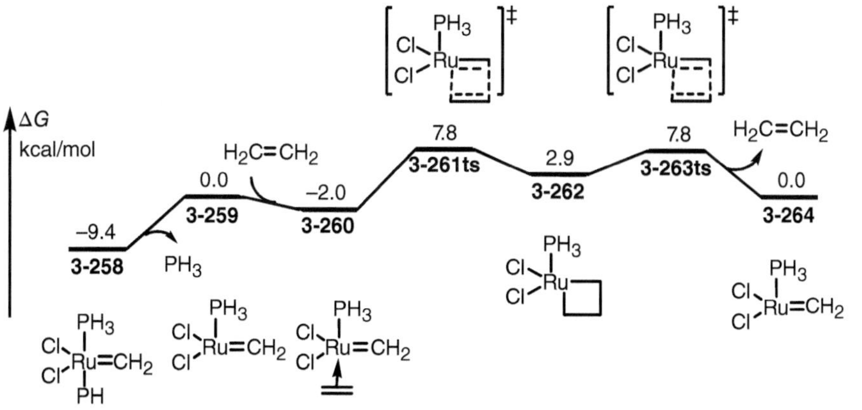

Figure 3.61 The free-energy profiles for the Ru-mediated olefin metathesis. The energies were calculated at PW91/DNP//PW91/DNP level of theory.

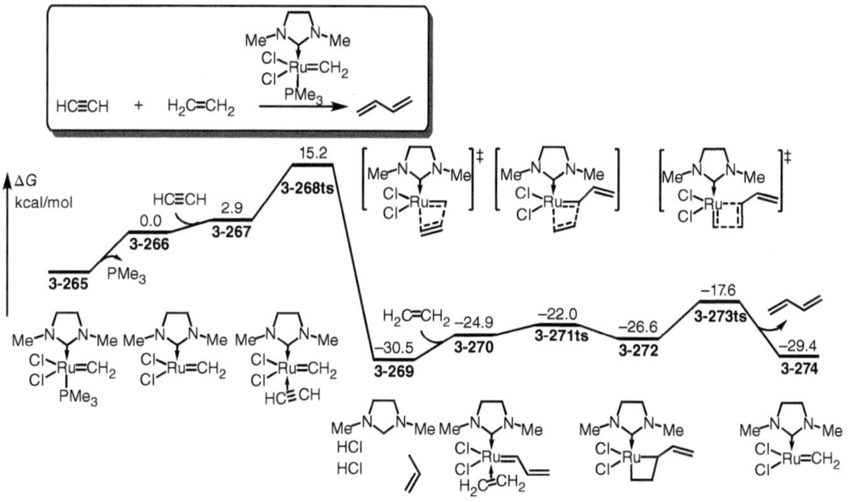

Figure 3.62 The free-energy profiles for the Ru-mediated olefin/acetylene metathesis. The energies were calculated at B3LYP/LACV3P**+//B3LYP/LACVP* level of theory.

ligand exchange with Ru-carbene species **3-265** by acetylene forms a π-coordinated Ru–carbene intermediate **3-267** with the release of phosphine ligand. The metathesis can occur via a four-membered ring-type transition state **3-268ts** to rearrange two π bonds. The result of this reaction is the formation of a vinylcarbene-coordinated Ru species **3-269** with 30.5 kcal/mol exergonic. Then, the coordination of another molecule olefin forms complex **3-270**, which undergoes a regular [2+2] cyclization via transition state **3-271ts** with an activation free energy of 8.5 kcal/mol to form a ruthenacycle **3-272**. Finally, the decomposition of four-membered ring via transition state **3-273ts** yields diene product and regenerates active catalyst **3-274**.

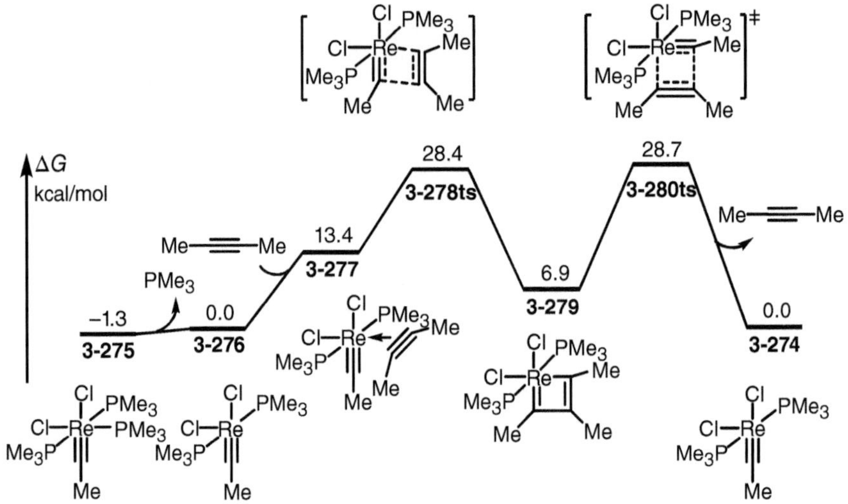

Figure 3.63 The free-energy profiles for the Re-mediated alkyne metathesis. The energies were calculated at B3LYP/6-311+G(d,p) (SDD for Re, Cl, and P)/SMD(toluene)//B3LYP/6-31G(d) (LANL2DZ for Re, Cl, and P) level of theory.

3.8.3 Alkyne Metathesis

As shown in Scheme 3.23, the alkyne metathesis is close to that for olefin, in which triple bonds are reorganized [239–242]. The calculated free-energy profiles for a typical alkyne metathesis are shown in Figure 3.63 [243]. The ligand exchange of Re-carbyne species **3-275** with acetylene forms a π-coordinated Re-carbyne **3-277** with 14.7 kcal/mol endergonic. A [2+2] cycloaddition takes place via a four-membered ring-type transition state **3-278ts** to afford a rhenacyclobutadiene **3-279**. In this intermediate, the lengths of both two Re—C bonds are 2.06 Å, indicating equivalent bonds. Then, another C—C bond can be broken by cycloelimination to afford another Re-carbyne and acetylene.

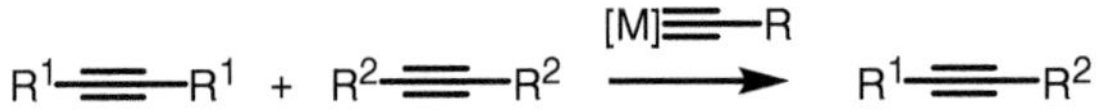

Scheme 3.23 Alkyne metathesis.

References

1. Bercaw, J.E. and Labinger, J.A. (2007). The coordination chemistry of saturated molecules. *Proceedings of the National Academy of Sciences of the United States of America* 104 (17): 6899–6900.
2. Qi, X., Liu, S., and Lan, Y. (2016). Computational studies on an aminomethylation precursor: (xantphos)Pd(CH$_2$NBn$_2$)$^+$. *Organometallics* 35 (10): 1582–1585.

3 van der Boom, M.E. and Milstein, D. (2003). Cyclometalated phosphine-based pincer complexes: mechanistic insight in catalysis, coordination, and bond activation. *Chemical Reviews* 103 (5): 1759–1792.

4 Bailey, P.J. and Pace, S. (2001). The coordination chemistry of guanidines and guanidinates. *Coordination Chemistry Reviews* 214 (1): 91–141.

5 Liu, S., Lei, Y., Yang, Z. et al. (2014). Theoretical study of the electron-donating effects of thiourea ligands in catalysis. *Journal of Molecular Structure* 1074: 527–533.

6 Bennett, M.A. (1962). Olefin and acetylene complexes of transition metals. *Chemical Reviews* 62 (6): 611–652.

7 Hartley, F.R. (1969). Olefin and acetylene complexes of platinum and palladium. *Chemical Reviews* 69 (6): 799–844.

8 Michalak, A. and Ziegler, T. (2001). DFT studies on the copolymerization of α-olefins with polar monomers: comonomer binding by nickel- and palladium-based catalysts with Brookhart and Grubbs ligands. *Organometallics* 20 (8): 1521–1532.

9 Tolman, C.A. (1974). Olefin complexes of nickel(0). III. Formation constants of (olefin)bis(tri-*o*-tolyl phosphite)nickel complexes. *Journal of the American Chemical Society* 96 (9): 2780–2789.

10 Wakatsuki, Y., Koga, N., Yamazaki, H. et al. (1994). Acetylene π-coordination, slippage to σ-coordination, and 1,2-hydrogen migration taking place on a transition metal. The case of a Ru(II) complex as studied by experiment and ab initio molecular orbital simulations. *Journal of the American Chemical Society* 116 (18): 8105–8111.

11 Braunschweig, H., Brenner, P., Dewhurst, R.D. et al. (2012). Unsupported boron–carbon σ-coordination to platinum as an isolable snapshot of σ-bond activation. *Nature Communications* 3: 872.

12 Kubas, G.J. (2002). Molecular hydrogen complexes: coordination of a σ-bond to transition metals. *Accounts of Chemical Research* 21 (3): 120–128.

13 Lin, Z. (2002). Structural and bonding characteristics in transition metal–silane complexes. *Chemical Society Reviews* 31 (4): 239–245.

14 Luo, X.-L., Kubas, G.J., Bryan, J.C. et al. (1994). η^2-Coordination of Si—H σ-bonds to transition-metal fragments that also bind η^2-dihydrogen ligands and agostic C—H bonds: synthesis and characterization of η^2-silane complexes cis-Mo(η^2-H-SiHR'$_2$)(CO)(R$_2$PC$_2$H$_4$PR$_2$)$_2$. *Journal of the American Chemical Society* 116 (22): 10312–10313.

15 Schubert, U. (1990). η^2 coordination of Si—H σ bonds to transition metals. *Advances in Organometallic Chemistry* 30: 151–187.

16 Tilley, T.D. (2002). The coordination polymerization of silanes to polysilanes by a "σ-bond metathesis" mechanism. Implications for linear chain growth. *Accounts of Chemical Research* 26 (1): 22–29.

17 Kubas, G.J. (2001). Metal–dihydrogen and σ-bond coordination: the consummate extension of the Dewar–Chatt–Duncanson model for metal–olefin π bonding. *Journal of Organometallic Chemistry* 635 (1–2): 37–68.

18 Crabtree, R.H. (1993). Transition metal complexation of σ bonds. *Angewandte Chemie International Edition in English* 32 (6): 789–805.

19 Heinekey, D.M. and Oldham, W.J. (1993). Coordination chemistry of dihydrogen. *Chemical Reviews* 93 (3): 913–926.

20 Fan, M.-F., Jia, G., and Lin, Z. (1996). Metal–silane interaction in the novel pseudooctahedral silane complex cis-Mo(CO)(PH$_3$)$_4$(H···SiH$_3$) and some related isomers: an ab initio study. *Journal of the American Chemical Society* 118 (41): 9915–9921.

21 Fan, M.-F. and Lin, Z. (1997). Competing metal–π-acetylene and metal–σ-(H–Si) interactions in the complex Ti(η^5-C$_5$H$_5$)$_2$(η^2-trans-RC⋮CSiHR$_2$). *Organometallics* 16 (3): 494–496.

22 Fan, M.-F. and Lin, Z. (1999). Stability of the *trans*-Bis(H···Si) structure in the complex RuH$_2$(PCy$_3$)$_2$(κ-η^2-H···SiMe$_2$-*o*-C$_6$H$_4$-SiMe$_2$···H), studied by density functional theory. *Organometallics* 18 (2): 286–289.

23 Wen, T.B., Lee, K.-H., Chen, J. et al. (2016). Preparation of osmium η^3-allenylcarbene complexes and their uses for the syntheses of osmabenzyne complexes. *Organometallics* 35 (10): 1514–1525.

24 Liu, S., Lei, Y., Li, Y. et al. (2014). Hexahapto–chromium complexes of graphene: a theoretical study. *RSC Advances* 4 (54): 28640–28644.

25 Singh, V., Nakao, Y., Sakaki, S. et al. (2017). Theoretical study of nickel-catalyzed selective alkenylation of pyridine: reaction mechanism and crucial roles of Lewis acid and ligands in determining the selectivity. *The Journal of Organic Chemistry* 82 (1): 289–301.

26 Brunner, H. and Tsuno, T. (2009). Ligand dissociation: planar or pyramidal intermediates? *Accounts of Chemical Research* 42 (10): 1501–1510.

27 Schmid, R., Herrmann, W.A., and Frenking, G. (1997). Coordination chemistry and mechanisms of metal-catalyzed CC-coupling reactions. 10.†Ligand dissociation in rhodium-catalyzed hydroformylation: a theoretical study‡. *Organometallics* 16 (4): 701–708.

28 Scott, J.D. and Puddephatt, R.J. (1983). Ligand dissociation as a preliminary step in methyl-for-halogen exchange reactions of platinum(II) complexes. *Organometallics* 2 (11): 1643–1648.

29 Yuan, R. and Lin, Z. (2014). Computational insight into the mechanism of nickel-catalyzed reductive carboxylation of styrenes using CO$_2$. *Organometallics* 33 (24): 7147–7156.

30 Jiang, J., Fu, M., Li, C. et al. (2017). Theoretical investigation on nickel-catalyzed hydrocarboxylation of alkynes employing formic acid. *Organometallics* 36 (15): 2818–2825.

31 Goossen, L.J., Koley, D., Hermann, H.L. et al. (2005). The palladium-catalyzed cross-coupling reaction of carboxylic anhydrides with arylboronic acids: a DFT study. *Journal of the American Chemical Society* 127 (31): 11102–11114.

32 Long, R., Huang, J., Shao, W. et al. (2014). Asymmetric total synthesis of (−)-lingzhiol via a Rh-catalysed [3+2] cycloaddition. *Nature Communications* 5: 5707.

References

33 Wang, M., Fan, T., and Lin, Z. (2012). DFT studies on the reaction of CO_2 with allyl-bridged dinuclear palladium(I) complexes. *Polyhedron* 32 (1): 35–40.

34 Yu, Z., Qi, X., Li, Y. et al. (2016). Mechanism, chemoselectivity and enantioselectivity for the rhodium-catalyzed desymmetric synthesis of hydrobenzofurans: a theoretical study. *Organic Chemistry Frontiers* 3 (2): 209–216.

35 Lan, Y., Liu, P., Newman, S.G. et al. (2012). Theoretical study of Pd(0)-catalyzed carbohalogenation of alkenes: mechanism and origins of reactivities and selectivities in alkyl halide reductive elimination from Pd(II) species. *Chemical Science* 3 (6): 1987.

36 McMullin, C.L., Jover, J., Harvey, J.N. et al. (2010). Accurate modelling of Pd(0) + PhX oxidative addition kinetics. *Dalton Transactions* 39 (45): 10833–10836.

37 Carvajal, M.A., Miscione, G.P., Novoa, J.J. et al. (2005). DFT computational study of the mechanism of allyl chloride carbonylation catalyzed by palladium complexes. *Organometallics* 24 (9): 2086–2096.

38 Staudaher, N.D., Arif, A.M., and Louie, J. (2016). Synergy between experimental and computational chemistry reveals the mechanism of decomposition of nickel–ketene complexes. *Journal of the American Chemical Society* 138 (42): 14083–14091.

39 Bishop, K.C. (1976). Transition metal catalyzed rearrangements of small ring organic molecules. *Chemical Reviews* 76 (4): 461–486.

40 Corey, J.Y. and Braddock-Wilking, J. (1999). Reactions of hydrosilanes with transition-metal complexes: formation of stable transition-metal silyl compounds. *Chemical Reviews* 99 (1): 175–292.

41 Deutsch, P.P. and Eisenberg, R. (1988). Stereochemistry of hydrogen oxidative addition and dihydride-transfer reactions involving iridium(I) complexes. *Chemical Reviews* 88 (7): 1147–1161.

42 Halpern, J. (2002). Oxidative-addition reactions of transition metal complexes. *Accounts of Chemical Research* 3 (11): 386–392.

43 Jones, W.D. and Feher, F.J. (2002). Comparative reactivities of hydrocarbon carbon–hydrogen bonds with a transition-metal complex. *Accounts of Chemical Research* 22 (3): 91–100.

44 Kaesz, H.D. and Saillant, R.B. (1972). Hydride complexes of the transition metals. *Chemical Reviews* 72 (3): 231–281.

45 Marciniec, B. (2005). Catalysis by transition metal complexes of alkene silylation–recent progress and mechanistic implications. *Coordination Chemistry Reviews* 249 (21–22): 2374–2390.

46 Rendina, L.M. and Puddephatt, R.J. (1997). Oxidative addition reactions of organoplatinum(II) complexes with nitrogen-donor ligands. *Chemical Reviews* 97 (6): 1735–1754.

47 Sharma, H.K. and Pannell, K.H. (1995). Activation of the Si—Si bond by transition metal complexes. *Chemical Reviews* 95 (5): 1351–1374.

48 Souillart, L. and Cramer, N. (2015). Catalytic C—C bond activations via oxidative addition to transition metals. *Chemical Reviews* 115 (17): 9410–9464.

49 Stille, J.K. and Lau, K.S.Y. (2002). Mechanisms of oxidative addition of organic halides to group 8 transition-metal complexes. *Accounts of Chemical Research* 10 (12): 434–442.

50 Fu, G.C. (2017). Transition-metal catalysis of nucleophilic substitution reactions: a radical alternative to S_N1 and S_N2 processes. *ACS Central Science* 3 (7): 692–700.

51 Griffin, T.R., Cook, D.B., Haynes, A. et al. (1996). Theoretical and experimental evidence for S_N2 transition states in oxidative addition of methyl iodide to cis-$[M(CO)_2I_2]^-$ (M = Rh, Ir). *Journal of the American Chemical Society* 118 (12): 3029–3030.

52 Huang, X., Zhang, K., Shao, Y. et al. (2019). Mechanism of Si—H bond activation for Lewis acid PBP-Ni-catalyzed hydrosilylation of CO_2: the role of the linear S_N2 type cooperation. *ACS Catalysis* 9 (6): 5279–5289.

53 Kameo, H., Baba, Y., Sakaki, S. et al. (2017). Iridium hydride mediated stannane–fluorine and –chlorine σ-bond activation: reversible switching between X-type stannyl and Z-type stannane ligands. *Organometallics* 36 (11): 2096–2106.

54 Senn, H.M. and Ziegler, T. (2004). Oxidative addition of aryl halides to palladium(0) complexes: a density-functional study including solvation. *Organometallics* 23 (12): 2980–2988.

55 Breitenfeld, J., Ruiz, J., Wodrich, M.D. et al. (2013). Bimetallic oxidative addition involving radical intermediates in nickel-catalyzed alkyl–alkyl Kumada coupling reactions. *Journal of the American Chemical Society* 135 (32): 12004–12012.

56 Gansauer, A., Fleckhaus, A., Lafont, M.A. et al. (2009). Catalysis via homolytic substitutions with C—O and Ti—O bonds: oxidative additions and reductive eliminations in single electron steps. *Journal of the American Chemical Society* 131 (46): 16989–16999.

57 Iqbal, J., Bhatia, B., and Nayyar, N.K. (1994). Transition metal-promoted free-radical reactions in organic synthesis: the formation of carbon–carbon bonds. *Chemical Reviews* 94 (2): 519–564.

58 Le, C., Chen, T.Q., Liang, T. et al. (2018). A radical approach to the copper oxidative addition problem: trifluoromethylation of bromoarenes. *Science* 360 (6392): 1010–1014.

59 Omer, H.M. and Liu, P. (2017). Computational study of Ni-catalyzed C–H functionalization: factors that control the competition of oxidative addition and radical pathways. *Journal of the American Chemical Society* 139 (29): 9909–9920.

60 Ouchi, M., Terashima, T., and Sawamoto, M. (2009). Transition metal-catalyzed living radical polymerization: toward perfection in catalysis and precision polymer synthesis. *Chemical Reviews* 109 (11): 4963–5050.

61 Seitz, L.C., Dickens, C.F., Nishio, K. et al. (2016). A highly active and stable $IrO_x/SrIrO_3$ catalyst for the oxygen evolution reaction. *Science* 353 (6303): 1011–1014.

References

62 Taniguchi, T., Sugiura, Y., Zaimoku, H. et al. (2010). Iron-catalyzed oxidative addition of alkoxycarbonyl radicals to alkenes with carbazates and air. *Angewandte Chemie International Edition* 49 (52): 10154–10157.

63 Heller, B. and Hapke, M. (2007). The fascinating construction of pyridine ring systems by transition metal-catalysed [2 + 2 + 2] cycloaddition reactions. *Chemical Society Reviews* 36 (7): 1085–1094.

64 Lautens, M., Klute, W., and Tam, W. (1996). Transition metal-mediated cycloaddition reactions. *Chemical Reviews* 96 (1): 49–92.

65 Lledo, A., Pla-Quintana, A., and Roglans, A. (2016). Allenes, versatile unsaturated motifs in transition-metal-catalysed [2+2+2] cycloaddition reactions. *Chemical Society Reviews* 45 (8): 2010–2023.

66 Louie, J., Gibby, J.E., Farnworth, M.V. et al. (2002). Efficient nickel-catalyzed [2 + 2 + 2] cycloaddition of CO_2 and diynes. *Journal of the American Chemical Society* 124 (51): 15188–15189.

67 Montgomery, J. (2000). Nickel-catalyzed cyclizations, couplings, and cycloadditions involving three reactive components. *Accounts of Chemical Research* 33 (7): 467–473.

68 Ogoshi, S., Nagata, M., and Kurosawa, H. (2006). Formation of nickeladihydropyran by oxidative addition of cyclopropyl ketone. Key intermediate in nickel-catalyzed cycloaddition. *Journal of the American Chemical Society* 128 (16): 5350–5351.

69 Schore, N.E. (1988). Transition metal-mediated cycloaddition reactions of alkynes in organic synthesis. *Chemical Reviews* 88 (7): 1081–1119.

70 Weding, N. and Hapke, M. (2011). Preparation and synthetic applications of alkene complexes of group 9 transition metals in [2+2+2] cycloaddition reactions. *Chemical Society Reviews* 40 (9): 4525–4538.

71 Crespo, M., Martínez, M., Nabavizadeh, S.M. et al. (2014). Kinetico-mechanistic studies on CX (X = H, F, Cl, Br, I) bond activation reactions on organoplatinum(II) complexes. *Coordination Chemistry Reviews* 279: 115–140.

72 Shan, C., Luo, X., Qi, X. et al. (2016). Mechanism of ruthenium-catalyzed direct arylation of C—H bonds in aromatic amides: a computational study. *Organometallics* 35 (10): 1440–1445.

73 Zhu, L., Qi, X., Li, Y. et al. (2017). Ir(III)/Ir(V) or Ir(I)/Ir(III) catalytic cycle? Steric-effect-controlled mechanism for the para-C–H borylation of arenes. *Organometallics* 36 (11): 2107–2115.

74 Lan, Y., Deng, L., Liu, J. et al. (2009). On the mechanism of the palladium catalyzed intramolecular Pauson–Khand-type reaction. *The Journal of Organic Chemistry* 74 (14): 5049–5058.

75 Xu, Z.-Y., Zhang, S.-Q., Liu, J.-R. et al. (2018). Mechanism and origins of chemo- and regioselectivities of Pd-catalyzed intermolecular σ-bond exchange between benzocyclobutenones and silacyclobutanes: a computational study. *Organometallics* 37 (4): 592–602.

76 Fihri, A., Meunier, P., and Hierso, J.-C. (2007). Performances of symmetrical achiral ferrocenylphosphine ligands in palladium-catalyzed cross-coupling

reactions: a review of syntheses, catalytic applications and structural properties. *Coordination Chemistry Reviews* 251 (15–16): 2017–2055.

77 Garcia-Melchor, M., Braga, A.A., Lledos, A. et al. (2013). Computational perspective on Pd-catalyzed C–C cross-coupling reaction mechanisms. *Accounts of Chemical Research* 46 (11): 2626–2634.

78 Martin, R. and Buchwald, S.L. (2008). Palladium-catalyzed Suzuki–Miyaura cross-coupling reactions employing dialkylbiaryl phosphine ligands. *Accounts of Chemical Research* 41 (11): 1461–1473.

79 Negishi, E. (2002). Palladium- or nickel-catalyzed cross coupling. A new selective method for carbon–carbon bond formation. *Accounts of Chemical Research* 15 (11): 340–348.

80 Terao, J. and Kambe, N. (2008). Cross-coupling reaction of alkyl halides with grignard reagents catalyzed by Ni, Pd, or Cu complexes with π-carbon ligand(s). *Accounts of Chemical Research* 41 (11): 1545–1554.

81 Legault, C.Y., Garcia, Y., Merlic, C.A. et al. (2007). Origin of regioselectivity in palladium-catalyzed cross-coupling reactions of polyhalogenated heterocycles. *Journal of the American Chemical Society* 129 (42): 12664–12665.

82 Ahlquist, M. and Norrby, P.-O. (2007). Oxidative addition of aryl chlorides to monoligated palladium(0): a DFT–SCRF study. *Organometallics* 26 (3): 550–553.

83 Melvin, P.R., Nova, A., Balcells, D. et al. (2017). DFT investigation of Suzuki–Miyaura reactions with aryl sulfamates using a dialkylbiarylphosphine-ligated palladium catalyst. *Organometallics* 36 (18): 3664–3675.

84 Schoenebeck, F. and Houk, K.N. (2010). Ligand-controlled regioselectivity in palladium-catalyzed cross coupling reactions. *Journal of the American Chemical Society* 132 (8): 2496–2497.

85 Xu, H., Muto, K., Yamaguchi, J. et al. (2014). Key mechanistic features of Ni-catalyzed C–H/C–O biaryl coupling of azoles and naphthalen-2-yl pivalates. *Journal of the American Chemical Society* 136 (42): 14834–14844.

86 Zhang, T., Qi, X., Liu, S. et al. (2017). Computational investigation of the role played by rhodium(V) in the rhodium(III)-catalyzed ortho-bromination of arenes. *Chemistry – A European Journal* 23 (11): 2690–2699.

87 Kameo, H., Baba, Y., Sakaki, S. et al. (2020). Experimental and theoretical investigation of an S_N2-type pathway for borate–fluorine bond cleavage by electron-rich late-transition metal complexes. *Inorganic Chemistry* 59 (7): 4282–4291.

88 Zhang, H., Wang, H.-Y., Luo, Y. et al. (2018). Regioselective palladium-catalyzed C—H bond trifluoroethylation of indoles: exploration and mechanistic insight. *ACS Catalysis* 8 (3): 2173–2180.

89 Lin, Y., Zhu, L., Lan, Y. et al. (2015). Development of a rhodium(II)-catalyzed chemoselective C(sp^3)–H oxygenation. *Chemistry – A European Journal* 21 (42): 14937–14942.

90 Goossen, L.J., Koley, D., Hermann, H. et al. (2004). The mechanism of the oxidative addition of aryl halides to Pd-catalysts: a DFT investigation. *Chemical Communications* 40 (19): 2141–2143.

91 Goossen, L.J., Koley, D., Hermann, H.L. et al. (2005). Mechanistic pathways for oxidative addition of aryl halides to palladium(0) complexes: a DFT study. *Organometallics* 24 (10): 2398–2410.

92 Allen, S.E., Walvoord, R.R., Padilla-Salinas, R. et al. (2013). Aerobic copper-catalyzed organic reactions. *Chemical Reviews* 113 (8): 6234–6458.

93 Tang, S., Liu, K., Liu, C. et al. (2015). Olefinic C–H functionalization through radical alkenylation. *Chemical Society Reviews* 44 (5): 1070–1082.

94 Wang, F., Chen, P., and Liu, G. (2018). Copper-catalyzed radical relay for asymmetric radical transformations. *Accounts of Chemical Research* 51 (9): 2036–2046.

95 Yoo, C., Ajitha, M.J., Jung, Y. et al. (2015). Mechanistic study on C—C bond formation of a nickel(I) monocarbonyl species with alkyl iodides: experimental and computational investigations. *Organometallics* 34 (17): 4305–4311.

96 He, X., Lan, Y., Shan, C. et al. (2017). Mechanistic insights into copper-catalyzed trifluoromethylation of aryl boronic acids: a theoretical study. *Scientia Sinica Chimica* 47 (7): 859–864.

97 Ye, J.-H., Zhu, L., Yan, S.-S. et al. (2017). Radical trifluoromethylative dearomatization of indoles and furans with CO_2. *ACS Catalysis* 7 (12): 8324–8330.

98 Zhu, L., Ye, J.-H., Duan, M. et al. (2018). The mechanism of copper-catalyzed oxytrifluoromethylation of allylamines with CO_2: a computational study. *Organic Chemistry Frontiers* 5 (4): 633–639.

99 Qi, X., Zhu, L., Bai, R. et al. (2017). Stabilization of two radicals with one metal: a stepwise coupling model for copper-catalyzed radical–radical cross-coupling. *Scientific Reports* 7: 43579.

100 Liu, S., Shen, H., Yu, Z. et al. (2014). What controls stereoselectivity and reactivity in the synthesis of a trans-decalin with a quaternary chiral center via the intramolecular Pauson–Khand reaction: a theoretical study. *Organometallics* 33 (22): 6282–6285.

101 Zhang, J., Wang, X., Li, S. et al. (2015). Diastereoselective synthesis of cyclopentanoids: applications to the construction of the ABCD tetracyclic core of retigeranic acid A. *Chemistry – A European Journal* 21 (36): 12596–12600.

102 Yamanaka, M. and Nakamura, E. (2001). Density functional studies on the Pauson–Khand reaction. *Journal of the American Chemical Society* 123 (8): 1703–1708.

103 Luo, X., Qi, X., Chen, C. et al. (2017). Ligand effect on nickle-catalyzed reductive alkyne–aldehyde coupling reactions: a computational study. *Scientia Sinica Chimica* 47 (3): 341–349.

104 Zhu, L., Qi, X., and Lan, Y. (2016). Rhodium-catalyzed hetero-(5 + 2) cycloaddition of vinylaziridines and alkynes: a theoretical view of the mechanism and chirality transfer. *Organometallics* 35 (5): 771–777.

3 Elementary Reactions in Organometallic Chemistry

105 Bernar, I., Fiser, B., Blanco-Ania, D. et al. (2017). Pd-catalyzed hydroamination of alkoxyallenes with azole heterocycles: examples and mechanistic proposal. *Organic Letters* 19 (16): 4211–4214.

106 Cavell, K.J. and McGuinness, D.S. (2004). Redox processes involving hydrocarbylmetal (N-heterocyclic carbene) complexes and associated imidazolium salts: ramifications for catalysis. *Coordination Chemistry Reviews* 248 (7–8): 671–681.

107 Grushin, V.V. (2002). Reductive elimination of hydrogen chloride from chloro hydrido transition metal complexes: an efficient and simple method for generation of electron-rich, coordinatively unsaturated, reactive intermediates. *Accounts of Chemical Research* 26 (5): 279–286.

108 Kühl, O. (2009). Sterically induced differences in N-heterocyclic carbene transition metal complexes. *Coordination Chemistry Reviews* 253 (21–22): 2481–2492.

109 Vigalok, A. (2015). Electrophilic halogenation-reductive elimination chemistry of organopalladium and -platinum complexes. *Accounts of Chemical Research* 48 (2): 238–247.

110 Rao, W.-H. and Shi, B.-F. (2016). Recent advances in copper-mediated chelation-assisted functionalization of unactivated C—H bonds. *Organic Chemistry Frontiers* 3 (8): 1028–1047.

111 Ye, L.W., Shu, C., and Gagosz, F. (2014). Recent progress towards transition metal-catalyzed synthesis of γ-lactams. *Organic & Biomolecular Chemistry* 12 (12): 1833–1845.

112 Inagaki, A. and Akita, M. (2010). Visible-light promoted bimetallic catalysis. *Coordination Chemistry Reviews* 254 (11–12): 1220–1239.

113 Murahashi, T. and Kurosawa, H. (2002). Organopalladium complexes containing palladium–palladium bonds. *Coordination Chemistry Reviews* 231 (1–2): 207–228.

114 Powers, I.G. and Uyeda, C. (2016). Metal–metal bonds in catalysis. *ACS Catalysis* 7 (2): 936–958.

115 Chen, C., Luo, Y., Fu, L. et al. (2018). Palladium-catalyzed intermolecular ditrifluoromethoxylation of unactivated alkenes: CF_3O-palladation initiated by Pd(IV). *Journal of the American Chemical Society* 140 (4): 1207–1210.

116 Hazari, N., Melvin, P.R., and Beromi, M.M. (2017). Well-defined nickel and palladium precatalysts for cross-coupling. *Nature Reviews Chemistry* 1 (3): 0025.

117 Lappert, M.F. and Lednor, P.W. (1976). Free radicals in organometallic chemistry. *Advances in Organometallic Chemistry* 14: 345–399.

118 Halpern, J. (2002). Formation of carbon–hydrogen bonds by reductive elimination. *Accounts of Chemical Research* 15 (10): 332–338.

119 Hartwig, J.F. (1998). Carbon–heteroatom bond-forming reductive eliminations of amines, ethers, and sulfides. *Accounts of Chemical Research* 31 (12): 852–860.

120 Hartwig, J.F. (2007). Electronic effects on reductive elimination to form carbon–carbon and carbon–heteroatom bonds from palladium(II) complexes. *Inorganic Chemistry* 46 (6): 1936–1947.

121 Phapale, V.B. and Cardenas, D.J. (2009). Nickel-catalysed Negishi cross-coupling reactions: scope and mechanisms. *Chemical Society Reviews* 38 (6): 1598–1607.

122 Wolf, W.J., Winston, M.S., and Toste, F.D. (2014). Exceptionally fast carbon–carbon bond reductive elimination from gold(III). *Nature Chemistry* 6 (2): 159–164.

123 Younesi, Y., Nasiri, B., BabaAhmadi, R. et al. (2016). Theoretical rationalisation for the mechanism of N-heterocyclic carbene–halide reductive elimination at Cu(III), Ag(III) and Au(III). *Chemical Communications* 52 (28): 5057–5060.

124 Iuliis, M.Z.D., Watson, I.D.G., Yudin, A.K. et al. (2009). A DFT investigation into the origin of regioselectivity in palladium-catalyzed allylic amination. *Canadian Journal of Chemistry* 87 (1): 54–62.

125 Jana, R., Pathak, T.P., and Sigman, M.S. (2011). Advances in transition metal (Pd, Ni, Fe)-catalyzed cross-coupling reactions using alkyl-organometallics as reaction partners. *Chemical Reviews* 111 (3): 1417–1492.

126 Kambe, N., Iwasaki, T., and Terao, J. (2011). Pd-catalyzed cross-coupling reactions of alkyl halides. *Chemical Society Reviews* 40 (10): 4937–4947.

127 Miyaura, N. and Suzuki, A. (1995). Palladium-catalyzed cross-coupling reactions of organoboron compounds. *Chemical Reviews* 95 (7): 2457–2483.

128 Mora, G., Piechaczyk, O., Le Goff, X.F. et al. (2008). Palladium-catalyzed deallylation of allyl ethers with a xanthene phosphole ligand. Experimental and DFT mechanistic studies. *Organometallics* 27 (11): 2565–2569.

129 Zeni, G. and Larock, R.C. (2006). Synthesis of heterocycles via palladium-catalyzed oxidative addition. *Chemical Reviews* 106 (11): 4644–4680.

130 Braga, A.A.C., Ujaque, G., and Maseras, F. (2006). A DFT study of the full catalytic cycle of the Suzuki–Miyaura cross-coupling on a model system. *Organometallics* 25 (15): 3647–3658.

131 Kang, K., Liu, S., Xu, T. et al. (2017). C(sp^2)–C(sp^2) reductive elimination from well-defined diarylgold(III) complexes. *Organometallics* 36 (24): 4727–4740.

132 Ariafard, A. and Lin, Z. (2006). DFT studies on the mechanism of allylative dearomatization catalyzed by palladium. *Journal of the American Chemical Society* 128 (39): 13010–13016.

133 Hopkinson, M.N., Gee, A.D., and Gouverneur, V. (2011). Au(I)/Au(III) catalysis: an alternative approach for C–C oxidative coupling. *Chemistry – A European Journal* 17 (30): 8248–8262.

134 Winston, M.S., Wolf, W.J., and Toste, F.D. (2014). Photoinitiated oxidative addition of CF_3I to gold(I) and facile aryl-CF_3 reductive elimination. *Journal of the American Chemical Society* 136 (21): 7777–7782.

135 Camasso, N.M., Canty, A.J., Ariafard, A. et al. (2017). Experimental and computational studies of high-valent nickel and palladium complexes. *Organometallics* 36 (22): 4382–4393.

136 Piechaczyk, O., Thoumazet, C., Jean, Y. et al. (2006). DFT study on the palladium-catalyzed allylation of primary amines by allylic alcohol. *Journal of the American Chemical Society* 128 (44): 14306–14317.

137 Chen, B., Fang, C., Liu, P. et al. (2017). Rhodium-catalyzed enantioselective radical addition of CX$_4$ reagents to olefins. *Angewandte Chemie International Edition* 56 (30): 8780–8784.

138 Lan, X.-W., Wang, N.-X., and Xing, Y. (2017). Recent advances in radical difunctionalization of simple alkenes. *European Journal of Organic Chemistry* 2017 (39): 5821–5851.

139 Matson, E.M., Franke, S.M., Anderson, N.H. et al. (2014). Radical reductive elimination from tetrabenzyluranium mediated by an iminoquinone ligand. *Organometallics* 33 (8): 1964–1971.

140 Sibi, M.P., Manyem, S., and Zimmerman, J. (2003). Enantioselective radical processes. *Chemical Reviews* 103 (8): 3263–3296.

141 Brenzovich, W.E. Jr.,, Benitez, D., Lackner, A.D. et al. (2010). Gold-catalyzed intramolecular aminoarylation of alkenes: C—C bond formation through bimolecular reductive elimination. *Angewandte Chemie International Edition* 49 (32): 5519–5522.

142 Norton, J.R. (2002). Organometallic elimination mechanisms: studies on osmium alkyls and hydrides. *Accounts of Chemical Research* 12 (4): 139–145.

143 Pye, D.R. and Mankad, N.P. (2017). Bimetallic catalysis for C–C and C–X coupling reactions. *Chemical Science* 8 (3): 1705–1718.

144 Xu, H., Diccianni, J.B., Katigbak, J. et al. (2016). Bimetallic C—C bond-forming reductive elimination from nickel. *Journal of the American Chemical Society* 138 (14): 4779–4786.

145 Qi, X., Bai, R., Zhu, L. et al. (2016). Mechanism of synergistic Cu(II)/Cu(I)-mediated alkyne coupling: dinuclear 1,2-reductive elimination after minimum energy crossing point. *The Journal of Organic Chemistry* 81 (4): 1654–1660.

146 Ansell, M.B., Navarro, O., and Spencer, J. (2017). Transition metal catalyzed element–element' additions to alkynes. *Coordination Chemistry Reviews* 336: 54–77.

147 Deubel, D.V. and Ziegler, T. (2002). DFT study of olefin versus nitrogen bonding in the coordination of nitrogen-containing polar monomers to diimine and salicylaldiminato nickel(II) and palladium(II) complexes. Implications for copolymerization of olefins with nitrogen-containing polar monomers. *Organometallics* 21 (8): 1603–1611.

148 Oestreich, M., Hartmann, E., and Mewald, M. (2013). Activation of the Si–B interelement bond: mechanism, catalysis, and synthesis. *Chemical Reviews* 113 (1): 402–441.

149 Qi, X., Li, Y., Bai, R. et al. (2017). Mechanism of rhodium-catalyzed C–H functionalization: advances in theoretical investigation. *Accounts of Chemical Research* 50 (11): 2799–2808.

150 Suginome, M. and Ito, Y. (2000). Transition-metal-catalyzed additions of silicon–silicon and silicon–heteroatom bonds to unsaturated organic molecules. *Chemical Reviews* 100 (8): 3221–3256.

151 Xiao, P., Gao, L., and Song, Z. (2019). Recent progress in the transition-metal-catalyzed activation of Si—Si bonds to form C—Si bonds. *Chemistry – A European Journal* 25 (10): 2407–2422.

152 Ahmad, T., Li, Q., Qiu, S.Q. et al. (2019). Copper-catalyzed regiodivergent 1,4- and 1,6-conjugate silyl addition to diendioates: access to functionalized allylsilanes. *Organic & Biomolecular Chemistry* 17 (25): 6122–6126.

153 Holmes, M., Schwartz, L.A., and Krische, M.J. (2018). Intermolecular metal-catalyzed reductive coupling of dienes, allenes, and enynes with carbonyl compounds and imines. *Chemical Reviews* 118 (12): 6026–6052.

154 Jakoobi, M., Tian, Y., Boulatov, R. et al. (2019). Reversible insertion of Ir into arene ring C—C bonds with improved regioselectivity at a higher reaction temperature. *Journal of the American Chemical Society* 141 (14): 6048–6053.

155 Li, W., Ma, X., Walawalkar, M.G. et al. (2017). Soluble aluminum hydrides function as catalysts in deprotonation, insertion, and activation reactions. *Coordination Chemistry Reviews* 350: 14–29.

156 Allolio, C. and Strassner, T. (2014). Palladium complexes with chelating bis-NHC ligands in the Mizoroki–Heck reaction – mechanism and electronic effects, a DFT study. *The Journal of Organic Chemistry* 79 (24): 12096–12105.

157 Li, Y., Shan, C., Yang, Y.-F. et al. (2017). Mechanism, regio-, and diastereoselectivity of Rh(III)-catalyzed cyclization reactions of *N*-arylnitrones with alkynes: a density functional theory study. *The Journal of Physical Chemistry A* 121 (23): 4496–4504.

158 Qi, X., Li, Y., Zhang, G. et al. (2015). Dinuclear versus mononuclear pathways in zinc mediated nucleophilic addition: a combined experimental and DFT study. *Dalton Transactions* 44 (24): 11165–11171.

159 Yu, S., Li, Y., Kong, L. et al. (2016). Mild acylation of C(sp^3)–H and C(sp^2)–H bonds under redox-neutral Rh(III) catalysis. *ACS Catalysis* 6 (11): 7744–7748.

160 Yue, X., Qi, X., Bai, R. et al. (2017). Mononuclear or dinuclear? Mechanistic study of the zinc-catalyzed oxidative coupling of aldehydes and acetylenes. *Chemistry – A European Journal* 23 (26): 6419–6425.

161 Zhou, X., Luo, Y., Kong, L. et al. (2017). Cp*Co^{III}-catalyzed branch-selective hydroarylation of alkynes via C–H activation: efficient access to α-*gem*-vinylindoles. *ACS Catalysis* 7 (10): 7296–7304.

162 Gao, B., Liu, S., Lan, Y. et al. (2016). Rhodium-catalyzed cyclocarbonylation of ketimines via C—H bond activation. *Organometallics* 35 (10): 1480–1487.

163 Herbert, M.B., Lan, Y., Keitz, B.K. et al. (2012). Decomposition pathways of Z-selective ruthenium metathesis catalysts. *Journal of the American Chemical Society* 134 (18): 7861–7866.

164 Yu, S., Liu, S., Lan, Y. et al. (2015). Rhodium-catalyzed C–H activation of phenacyl ammonium salts assisted by an oxidizing C—N bond: a combination of experimental and theoretical studies. *Journal of the American Chemical Society* 137 (4): 1623–1631.

165 Yu, S., Tang, G., Li, Y. et al. (2016). Anthranil: an aminating reagent leading to bifunctionality for both C(sp^3)–H and C(sp^2)–H under rhodium(III) catalysis. *Angewandte Chemie International Edition* 55 (30): 8696–8700.

166 Khumtaveeporn, K. and Alper, H. (1995). Transition metal mediated carbonylative ring expansion of heterocyclic compounds. *Accounts of Chemical Research* 28 (10): 414–422.

167 Gillingham, D. and Fei, N. (2013). Catalytic X–H insertion reactions based on carbenoids. *Chemical Society Reviews* 42 (12): 4918–4931.

168 Kirmse, W. (2003). Copper carbene complexes: advanced catalysts, new insights. *Angewandte Chemie International Edition* 42 (10): 1088–1093.

169 Xia, Y., Qiu, D., and Wang, J. (2017). Transition-metal-catalyzed cross-couplings through carbene migratory insertion. *Chemical Reviews* 117 (23): 13810–13889.

170 Zhao, X., Zhang, Y., and Wang, J. (2012). Recent developments in copper-catalyzed reactions of diazo compounds. *Chemical Communications* 48 (82): 10162–10173.

171 Davies, H.M.L. and Hedley, S.J. (2007). Intermolecular reactions of electron-rich heterocycles with copper and rhodium carbenoids. *Chemical Society Reviews* 36 (7): 1109–1119.

172 Alderson, J.M., Corbin, J.R., and Schomaker, J.M. (2017). Tunable, chemo- and site-selective nitrene transfer reactions through the rational design of silver(I) catalysts. *Accounts of Chemical Research* 50 (9): 2147–2158.

173 Braunstein, P. and Nobel, D. (1989). Transition-metal-mediated reactions of organic isocyanates. *Chemical Reviews* 89 (8): 1927–1945.

174 Gharpure, S.J., Naveen, S., Samala, G. et al. (2019). Transition-metal acetate-promoted intramolecular nitrene insertion to vinylogous carbonates for divergent synthesis of azirinobenzoxazoles and benzoxazines. *Chemistry – A European Journal* 25 (6): 1456–1460.

175 Li, Z. and He, C. (2006). Recent advances in silver-catalyzed nitrene, carbene, and silylene-transfer reactions. *European Journal of Organic Chemistry* 2006 (19): 4313–4322.

176 Zhang, X., Xu, H., Liu, X. et al. (2016). Mechanistic insight into the intramolecular benzylic C–H nitrene insertion catalyzed by bimetallic paddlewheel complexes: influence of the metal centers. *Chemistry – A European Journal* 22 (21): 7288–7297.

177 Yang, S., Xu, Y., and Li, J. (2016). Theoretical study of nickel-catalyzed proximal C–C cleavage in benzocyclobutenones with insertion of 1,3-diene: origin of selectivity and role of ligand. *Organic Letters* 18 (24): 6244–6247.

178 Tan, G., Zhu, L., Liao, X. et al. (2017). Rhodium/copper cocatalyzed highly trans-selective 1,2-diheteroarylation of alkynes with azoles via C–H addition/oxidative cross-coupling: a combined experimental and theoretical study. *Journal of the American Chemical Society* 139 (44): 15724–15737.

179 Abril, P., del Río, M.P., López, J.A. et al. (2019). Inner-sphere oxygen activation promoting outer-sphere nucleophilic attack on olefins. *Chemistry – A European Journal* 25 (64): 14546–14554.

180 Franzoni, I., Yoon, H., García-López, J.-A. et al. (2018). Exploring the mechanism of the Pd-catalyzed spirocyclization reaction: a combined DFT and experimental study. *Chemical Science* 9 (6): 1496–1509.

181 Guo, J., Pham, H.D., Wu, Y.-B. et al. (2020). Mechanism of cobalt-catalyzed direct aminocarbonylation of unactivated alkyl electrophiles: outer-sphere amine substitution to form amide bond. *ACS Catalysis* 10 (2): 1520–1527.

182 Iglesias, M., Fernández-Alvarez, F.J., and Oro, L.A. (2019). Non-classical hydrosilane mediated reductions promoted by transition metal complexes. *Coordination Chemistry Reviews* 386: 240–266.

183 Wu, Z., Zhang, M., Shi, Y. et al. (2020). Mechanism and origins of stereo- and enantioselectivities of palladium-catalyzed hydroamination of racemic internal allenes via dynamic kinetic resolution: a computational study. *Organic Chemistry Frontiers* 7 (12): 1502–1511.

184 Ariafard, A., Ghari, H., Khaledi, Y. et al. (2016). Theoretical investigation into the mechanism of cyanomethylation of aldehydes catalyzed by a nickel pincer complex in the absence of base additives. *ACS Catalysis* 6 (1): 60–68.

185 Burger, B.J., Thompson, M.E., Cotter, W.D. et al. (1990). Ethylene insertion and β-hydrogen elimination for permethylscandocene alkyl complexes. A study of the chain propagation and termination steps in Ziegler–Natta polymerization of ethylene. *Journal of the American Chemical Society* 112 (4): 1566–1577.

186 Forbes, J.G. and Gellman, A.J. (1993). The β-hydride elimination mechanism in adsorbed alkyl groups. *Journal of the American Chemical Society* 115 (14): 6277–6283.

187 Lemke, F.R. and Bullock, R.M. (1992). Insertion and β-hydride elimination reactions of ruthenium/zirconium complexes containing C_2 bridges with bond orders of 1, 2, and 3. *Organometallics* 11 (12): 4261–4267.

188 Peng, L., Li, Y., Li, Y. et al. (2018). Ligand-controlled nickel-catalyzed reductive relay cross-coupling of alkyl bromides and aryl bromides. *ACS Catalysis* 8 (1): 310–313.

189 Schneck, F., Ahrens, J., Finger, M. et al. (2018). The elusive abnormal CO_2 insertion enabled by metal-ligand cooperative photochemical selectivity inversion. *Nature Communications* 9 (1): 1161.

190 Zhang, J., Shan, C., Zhang, T. et al. (2019). Computational advances aiding mechanistic understanding of silver-catalyzed carbene/nitrene/silylene transfer reactions. *Coordination Chemistry Reviews* 382: 69–84.

191 Deeth, R.J., Smith, A., Hii, K.K. et al. (1998). The Heck olefination reaction; a DFT study of the elimination pathway. *Tetrahedron Letters* 39 (20): 3229–3232.

192 Wititsuwannakul, T., Tantirungrotechai, Y., and Surawatanawong, P. (2016). Density functional study of nickel N-heterocyclic carbene catalyzed C—O bond hydrogenolysis of methyl phenyl ether: the concerted β–H transfer mechanism. *ACS Catalysis* 6 (3): 1477–1486.

193 Zhang, X., Liu, Y., Chen, G. et al. (2017). Theoretical insight into C(sp^3)—F bond activations and origins of chemo- and regioselectivities of "tunable" nickel-mediated/-catalyzed couplings of 2-trifluoromethyl-1-alkenes with alkynes. *Organometallics* 36 (19): 3739–3749.

194 Lu, Q., Yu, H., and Fu, Y. (2014). Mechanistic study of chemoselectivity in Ni-catalyzed coupling reactions between azoles and aryl carboxylates. *Journal of the American Chemical Society* 136 (23): 8252–8260.

195 Luo, X., Bai, R., Liu, S. et al. (2016). Mechanism of rhodium-catalyzed formyl activation: a computational study. *The Journal of Organic Chemistry* 81 (6): 2320–2326.

196 Qi, X., Liu, X., Qu, L.-B. et al. (2018). Mechanistic insight into cobalt-catalyzed stereodivergent semihydrogenation of alkynes: the story of selectivity control. *Journal of Catalysis* 362: 25–34.

197 Xue, L., Ng, K.C., and Lin, Z. (2009). Theoretical studies on β-aryl elimination from Rh(I) complexes. *Dalton Transactions* 38 (30): 5841–5850.

198 Yan, S.-S., Zhu, L., Ye, J.-H. et al. (2018). Ruthenium-catalyzed umpolung carboxylation of hydrazones with CO_2. *Chemical Science* 9 (21): 4873–4878.

199 Guan, W., Sakaki, S., Kurahashi, T. et al. (2013). Theoretical mechanistic study of novel Ni(0)-catalyzed [6 − 2 + 2] cycloaddition reactions of isatoic anhydrides with alkynes: origin of facile decarboxylation. *Organometallics* 32 (24): 7564–7574.

200 Davies, G., El-Sayed, M.A., and El-Toukhy, A. (1992). Transmetallation and its applications. *Chemical Society Reviews* 21 (2): 101–104.

201 Guerrero, I. and Correa, A. (2018). Metal-catalyzed C–H functionalization processes with "click"-triazole assistance. *European Journal of Organic Chemistry* 2018 (44): 6034–6049.

202 Lu, H., Yu, T.-Y., Xu, P.-F. et al. (2020). Selective decarbonylation via transition-metal-catalyzed carbon–carbon bond cleavage. *Chemical Reviews* 120 https://doi.org/10.1021/acs.chemrev.0c00153.

203 MacNeil, C.S., Dickie, T.K.K., and Hayes, P.G. (2018). Chapter 7 – Actinide pincer chemistry: a new frontier. In: *Pincer Compounds* (ed. D. Morales-Morales), 133–172. Elsevier.

204 Osakada, K. and Yamamoto, T. (2000). Transmetallation of alkynyl and aryl complexes of group 10 transition metals. *Coordination Chemistry Reviews* 198 (1): 379–399.

205 Raubenheimer, H.G. and Cronje, S. (2001). Carbene complexes derived from lithiated heterocycles, mainly azoles, by transmetallation. *Journal of Organometallic Chemistry* 617–618: 170–181.

206 Dang, L., Zhao, H., Lin, Z. et al. (2008). Understanding the higher reactivity of B_2cat_2 versus B_2pin_2 in copper(I)-catalyzed alkene diboration reactions. *Organometallics* 27 (6): 1178–1186.

207 McCarren, P.R., Liu, P., Cheong, P.H.-Y. et al. (2009). Mechanism and transition-state structures for nickel-catalyzed reductive alkyne–aldehyde coupling reactions. *Journal of the American Chemical Society* 131 (19): 6654–6655.

208 Cheung, M.S., Sheong, F.K., Marder, T.B. et al. (2015). Computational insight into nickel-catalyzed carbon–carbon versus carbon–boron coupling reactions of primary, secondary, and tertiary alkyl bromides. *Chemistry – A European Journal* 21 (20): 7480–7488.

209 Liu, Q., Lan, Y., Liu, J. et al. (2009). Revealing a second transmetalation step in the Negishi coupling and its competition with reductive elimination: improvement in the interpretation of the mechanism of biaryl syntheses. *Journal of the American Chemical Society* 131 (29): 10201–10210.

References | 121

210 Martinez-Solorio, D., Melillo, B., Sanchez, L. et al. (2016). Design, synthesis, and validation of an effective, reusable silicon-based transfer agent for room-temperature Pd-catalyzed cross-coupling reactions of aryl and heteroaryl chlorides with readily available aryl lithium reagents. *Journal of the American Chemical Society* 138 (6): 1836–1839.

211 Fuentes, B., García-Melchor, M., Lledós, A. et al. (2010). Palladium round trip in the Negishi coupling of *trans*-[PdMeCl(PMePh$_2$)$_2$] with ZnMeCl: an experimental and DFT study of the transmetalation step. *Chemistry – A European Journal* 16 (29): 8596–8599.

212 Zhang, X., Tutkowski, B., Oliver, A. et al. (2018). Mechanistic study of the nickel-catalyzed α,β-coupling of saturated ketones. *ACS Catalysis* 8 (3): 1740–1747.

213 Fandrick, D.R., Reeves, J.T., Bakonyi, J.M. et al. (2013). Zinc catalyzed and mediated asymmetric propargylation of trifluoromethyl ketones with a propargyl boronate. *The Journal of Organic Chemistry* 78 (8): 3592–3615.

214 Álvarez, R., Pérez, M., Faza, O.N. et al. (2008). Associative transmetalation in the Stille cross-coupling reaction to form dienes: theoretical insights into the open pathway. *Organometallics* 27 (14): 3378–3389.

215 Nova, A., Ujaque, G., Maseras, F. et al. (2006). A critical analysis of the cyclic and open alternatives of the transmetalation step in the Stille cross-coupling reaction. *Journal of the American Chemical Society* 128 (45): 14571–14578.

216 Bluemke, T.D., Clegg, W., García-Alvarez, P. et al. (2014). Structural and reactivity insights in Mg–Zn hybrid chemistry: Zn–I exchange and Pd-catalysed cross-coupling applications of aromatic substrates. *Chemical Science* 5 (9): 3552–3562.

217 Zelinskii, G.E., Belov, A.S., Chuprin, A.S. et al. (2017). Clathrochelate iron(II) tris-nioximates with non-equivalent capping groups and their precursors: synthetic strategies, X-ray structure, and reactivity. *Journal of Coordination Chemistry* 70 (13): 2313–2333.

218 Zhang, C., Zhao, R., Dagnaw, W.M. et al. (2019). Density functional theory mechanistic insight into the base-free nickel-catalyzed Suzuki–Miyaura cross-coupling of acid fluoride: concerted versus stepwise transmetalation. *The Journal of Organic Chemistry* 84 (21): 13983–13991.

219 Zhao, S. and Mankad, N.P. (2019). Metal-catalysed radical carbonylation reactions. *Catalysis Science & Technology* 9 (14): 3603–3613.

220 Yu, Z. and Lan, Y. (2013). Mechanism of rhodium-catalyzed carbon–silicon bond cleavage for the synthesis of benzosilole derivatives: a computational study. *The Journal of Organic Chemistry* 78 (22): 11501–11507.

221 Basset, J.-M., Coperet, C., Soulivong, D. et al. (2010). Metathesis of alkanes and related reactions. *Accounts of Chemical Research* 43 (2): 323–334.

222 Grubbs, R.H. and Chang, S. (1998). Recent advances in olefin metathesis and its application in organic synthesis. *Tetrahedron* 54 (18): 4413–4450.

223 Alcaraz, G., Grellier, M., and Sabo-Etienne, S. (2009). Bis σ-bond dihydrogen and borane ruthenium complexes: bonding nature, catalytic applications, and reversible hydrogen release. *Accounts of Chemical Research* 42 (10): 1640–1649.

224 Bokka, A., Hua, Y., Berlin, A.S. et al. (2015). Mechanistic insights into Grubbs-type ruthenium-complex-catalyzed intramolecular alkene hydrosilylation: direct σ-bond metathesis in the initial stage of hydrosilylation. *ACS Catalysis* 5 (6): 3189–3195.

225 Esteruelas, M.A., López, A.M., and Oliván, M. (2016). Polyhydrides of platinum group metals: nonclassical interactions and σ-bond activation reactions. *Chemical Reviews* 116 (15): 8770–8847.

226 Perutz, R.N. and Sabo-Etienne, S. (2007). The σ-CAM mechanism: σ complexes as the basis of σ-bond metathesis at late-transition-metal centers. *Angewandte Chemie International Edition* 46 (15): 2578–2592.

227 Lin, Z. (2007). Current understanding of the σ-bond metathesis reactions of LnMR + R′–H → LnMR′ + R–H. *Coordination Chemistry Reviews* 251 (17): 2280–2291.

228 Diver, S.T. and Giessert, A.J. (2004). Enyne metathesis (enyne bond reorganization). *Chemical Reviews* 104 (3): 1317–1382.

229 Perrin, L., Maron, L., and Eisenstein, O. (2002). A DFT study of SiH$_4$ activation by Cp$_2$LnH. *Inorganic Chemistry* 41 (17): 4355–4362.

230 Vougioukalakis, G.C. and Grubbs, R.H. (2010). Ruthenium-based heterocyclic carbene-coordinated olefin metathesis catalysts. *Chemical Reviews* 110 (3): 1746–1787.

231 Luconi, L., Giambastiani, G., Rosin, A. et al. (2010). Intramolecular σ-bond metathesis/protonolysis on zirconium(IV) and hafnium(IV) pyridylamido olefin polymerization catalyst precursors: exploring unexpected reactivity paths. *Inorganic Chemistry* 49 (15): 6811–6813.

232 Sawatlon, B., Wititsuwannakul, T., Tantirungrotechai, Y. et al. (2014). Mechanism of Ni N-heterocyclic carbene catalyst for C—O bond hydrogenolysis of diphenyl ether: a density functional study. *Dalton Transactions* 43 (48): 18123–18133.

233 Xu, L., Chung, L.W., and Wu, Y.-D. (2016). Mechanism of Ni–NHC catalyzed hydrogenolysis of aryl ethers: roles of the excess base. *ACS Catalysis* 6 (1): 483–493.

234 Lin, Y., Zhu, L., Lan, Y. et al. (2015). Development of a rhodium(II)-catalyzed chemoselective C(sp^3)–H oxygenation. *Chemistry – A European Journal* 21 (42): 14937–14942.

235 Calderon, N. (1972). Olefin metathesis reaction. *Accounts of Chemical Research* 5 (4): 127–132.

236 Grubbs, R.H., Miller, S.J., and Fu, G.C. (1995). Ring-closing metathesis and related processes in organic synthesis. *Accounts of Chemical Research* 28 (11): 446–452.

237 Sabbagh, I.T. and Kaye, P.T. (2006). A computational study of Grubbs-type catalysts: structure and application in the degenerate metathesis of ethylene. *Journal of Molecular Structure: THEOCHEM* 763 (1): 37–42.

238 Lippstreu, J.J. and Straub, B.F. (2005). Mechanism of enyne metathesis catalyzed by grubbs ruthenium–carbene complexes: a DFT study. *Journal of the American Chemical Society* 127 (20): 7444–7457.

239 Coutelier, O. and Mortreux, A. (2006). Terminal alkyne metathesis: a further step towards selectivity. *Advanced Synthesis & Catalysis* 348 (15): 2038–2042.

240 Fürstner, A. (2013). Alkyne metathesis on the rise. *Angewandte Chemie International Edition* 52 (10): 2794–2819.

241 Fürstner, A. and Davies, P.W. (2005). Alkyne metathesis. *Chemical Communications* 41 (18): 2307–2320.

242 Zhang, W. and Moore, J.S. (2007). Alkyne metathesis: catalysts and synthetic applications. *Advanced Synthesis & Catalysis* 349 (1–2): 93–120.

243 Bai, W., Lee, K.-H., Sung, H.H.Y. et al. (2016). Alkyne metathesis reactions of rhenium(V) carbyne complexes. *Organometallics* 35 (22): 3808–3815.

Part II

On the Mechanism of Transition-metal-assisted Reactions

When transition metal atom is introduced into an organic molecule, it would reveal fantastic, amazing, and fascinating chemical properties. An organic molecule is usually composed of carbon, hydrogen, oxygen, sulfur, nitrogen, phosphorus, halogen, and so on. The covalent bonds in a molecule are usually formed by s and p orbitals of above-mentioned atoms. The structures of those covalent bonds are relatively simple. Therefore, the mechanism of a traditional organic reaction becomes easy for understanding. When a transition metal is used in organic reaction, particularly as a catalyst, the complexity of the reaction mechanism is greatly enhanced. Generally, d orbital is involved in a typical transition metal atom, which can further combine with valence s and p orbitals to provide d^5sp^3 valence shell. The complexity of orbital shape and filling mode leads to a variety of organometallic catalyses.

To be honest, it is very difficult to write a book on the mechanism of organometallic catalysis because these reactions are rather diverse that it is difficult to generalize and classify them. In this book, the author tries to categorize the reactions by the main transition metals involved in the catalysts, thus forming the Chapters 4–15 of the second part.

In organometallic catalysis, the key processes are the activation of inert covalent bond in organic molecules to afford an active metal–carbon bond and the further transformation of that metal–carbon bond. Therefore, in each chapter of the second part, the sections are organized majorly by the activation modes of inert covalent bond in organic molecules, such as C—H, C—X, C—O, C—N, and C—O bonds. As one important complement, some organometallic complexes have unique reactivity and are common intermediates in the catalytic cycle. Therefore, the formation and transformation of some special organometallic complexes will be discussed in some sections, such as metal–carbene, allylic–metal, metallocycle, metal-involved arene, etc.

Notably, the "reaction mechanism" discussed in Chapters 4–15 can be divided into abstract and concrete ones. In abstract mechanisms, a general catalytic cycle is drawn with labeling of a few key species. Alternatively, the concrete mechanisms

Computational Methods in Organometallic Catalysis: From Elementary Reactions to Mechanisms,
First Edition. Yu Lan.
© 2021 WILEY-VCH GmbH. Published 2021 by WILEY-VCH GmbH.

are often given by calculated free energy profiles, which start from the active species in the catalytic cycle and end at the same compound. For ease of reading, most of intermediates and transition states are given with relative free energies in free energy profiles. The materials in and out are represented by arrows on the free energy profiles. For comparison of some important processes, side pathways are usually given as dashed lines. As a contrast, the main pathways are expressed by solid lines.

4

Theoretical Study of Ni-Catalysis

58.6934
Ni 28
Nickle
[Ar]3d⁸4s²

As a VIII subgroup transition metal, nickel is abundant in the crust and easy to extract. Therefore, it has a long history since it was used as catalyst in organometallic chemistry [1–15]. As a homologous element of Pd and Pt, Ni usually exhibits similar chemical properties and catalytic activity to the other two ones, and even in some reactions, they can replace each other to provide same products. 0 and +2 are the two most common oxidation states of Ni, which are the same as that of Pd and Pt in same subgroup; however, the reducibility of Ni(0) is much stronger than that of the other two metals [16–26]. Therefore, in Ni chemistry, the catalytic cycle often starts from a Ni(0) species, which can be used to activate some covalent bonds by oxidative addition to afford corresponding Ni(II) species. In particular, some stable covalent bonds, such as C—H, C—O, C—F, and C—C, can be broken by Ni(0) catalyst, which can provide a powerful tool in organometallic synthesis for the transformation of those inert chemical bonds [27–39]. In this area, the mechanisms of bond activation and transformation have caught the attention of chemists; therefore, theoretical studies are usually used to reveal the mechanism of those reactions [6, 40–55].

As a transition metal in the fourth period, the chemical properties of Ni are similar to those of other transition metals in the same period. Ni can easily acquire or lose an electron and thus exists in the form of Ni free radicals. As result, +1 and +3 are also common oxidation states of Ni in some catalytic cycles, which undoubtedly further increases the mechanistic complexity of Ni-catalysis. In this chapter, sections are majorly divided by the corresponding Ni-activated substrates.

4.1 Ni-Mediated C−H Bond Activation

In Pd chemistry, a carboxylate-assisted C−H bond activation and functionalization through a concerted metalation–deprotonation mechanism is a very common process; however, as a congener, this type of reaction is less in Ni catalysis [56–59].

As an alternative, the oxidative addition of C−H bond onto Ni(0) species often occurs in Ni-catalyzed C−H bond activation reactions [60–64]. Moreover, Ni-mediated proton transfer is also a possible way for the C−H bond cleavage [65].

The general reaction mechanism of Ni-catalyzed C−H bond activation and functionalization can be summarized in Scheme 4.1. The catalytic cycle starts from a base-assisted C−H bond cleavage to afford a Ni−C bond. Then, further transformation of Ni−C bond provides new C−C covalent bond. The Ni-catalyst can be regenerated by the product release.

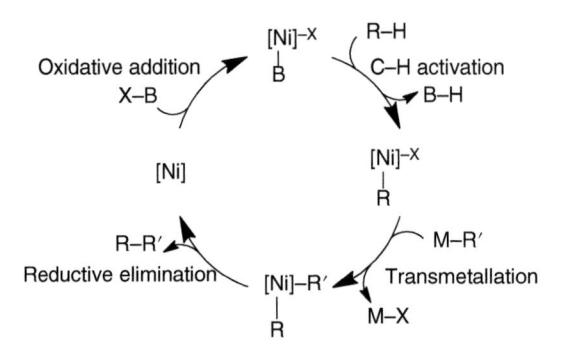

Scheme 4.1 A typical catalytic cycle for Ni-mediated C−H activation and functionalization.

4.1.1 Ni-Mediated Arene C−H Activation

In Ni-catalyzed arene iodination, when an amide oxizoline in compound **4-1** is used as directing group, a carboxylate-assisted concerted metalation–deprotonation process is proposed and revealed by density functional theory (DFT) calculation [66]. As shown in Figure 4.1, the coordination of oxizoline onto Ni(OAc)$_2$ forms intermediate **4-4** by 8.1 kcal/mol endergonic. Then a N−H bond cleavage by acetate forms an amino–Ni(II) complex **4-5**. The C−H bond cleavage takes place via a concerted metalation–deprotonation type transition state **4-6ts** with an overall activation free energy of 26.6 kcal/mol. The geometry information of transition state **4-6ts** is shown in Figure 4.1, which reveals a square planar configuration of Ni center. The lengths of forming Ni−C bond and breaking C−H bond are 2.03 and 1.40 Å, respectively, which show a concerted process. When the aryl Ni(II) complex **4.7** is formed, a nucleophilic substitution with polarized diiodine takes place via transition state **4-9ts** with an energy barrier of only 4.4 kcal/mol to afford iodinated arene product.

In the absence of directing group, arenes with acidic hydrogen are usually used in some Ni-mediated concerted metalation–deprotonation type C−H bond activation,

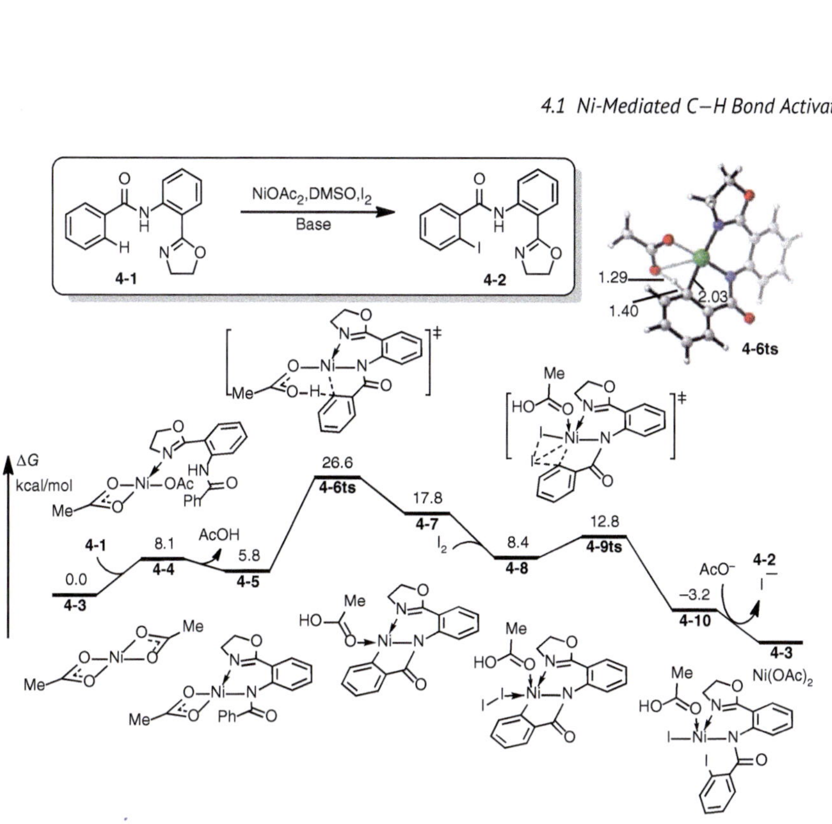

Figure 4.1 The free-energy profiles for the Ni catalyzed C—H activation and iodination. The energies were calculated at B3LYP-D3/6-31G(d,p) (LANL2DZ for I)/IEF-PCM(DMSO)//B3LYP-D3/6-31G(d,p) (LANL2DZ for I) level of theory. DMSO; dimethylsulfoxide. Source: Based on Haines et al. [66].

because Ni center is saturated by extra ligands. The mechanism of cross-coupling of aryl esters and oxazoles is studied by DFT calculations. When a bidentate phosphine ligand is used, the calculated free energy profiles for the catalytic cycle are shown in Figure 4.2 [43]. The aryl ester coordinated Ni(0) complex **4-11** is set to relative zero. A oxidative addition of C(aryl)—O bond onto Ni(0) takes place via transition state **4-12ts** with an energy barrier of 25.5 kcal/mol to afford a carboxylate Ni(II) species **4-13** irreversibly. A directing group free C—H activation occurs via transition state **4-14ts** with an energy barrier of 35.5 kcal/mol to afford an aryl Ni(II) complex 10 with 4.4 kcal/mol endergonic. Then a rapid reductive elimination takes place via transition state **4-16ts** to yield biarene coordinated Ni(0) complex **4-11**. The rate-determining step is considered as the C—H bond cleavage step via transition state **4-14ts**, whose geometry is shown in Figure 4.2. The lengths of forming Ni—C(aryl) bond and breaking C(aryl)—H bond are 2.17 and 1.31 Å, respectively, which reveals a concerted process.

In a Ni(0)-mediated C—H bond cleavage, oxidative addition of C—H bond onto Ni is a possible process. As shown in Figure 4.3, the mechanism of Ni-catalyzed olefin hydroarylation is studied by DFT calculation [67]. When a Ni(0)–carbene complex **4-17** is used as catalyst, the oxidative addition of C(aryl)—H bond can occur via a three-centered transition state **4-18ts** with an energy barrier of 14.9 kcal/mol. In the geometry of this transition state, the lengths of breaking C—H bond and forming

130 | *4 Theoretical Study of Ni-Catalysis*

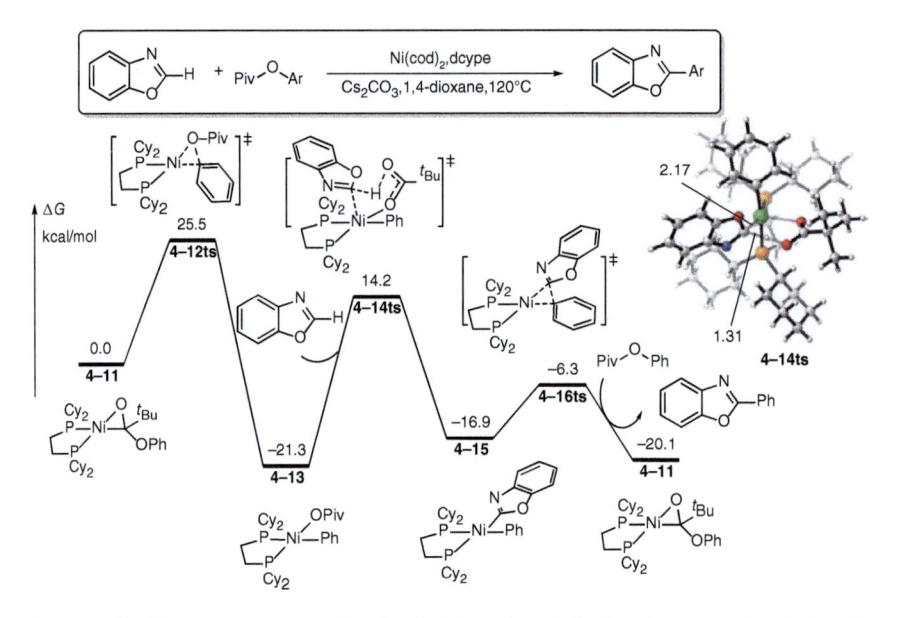

Figure 4.2 The free-energy profiles for the Ni catalyzed C−H activation and arylation. The energies were calculated at M06/6-311+G(d,p) (SDD for Ni)/SMD(1,4-dioxane)//B3LYP/6-31G(d) (SDD for Ni) level of theory. Source: Based on Hong et al. [43].

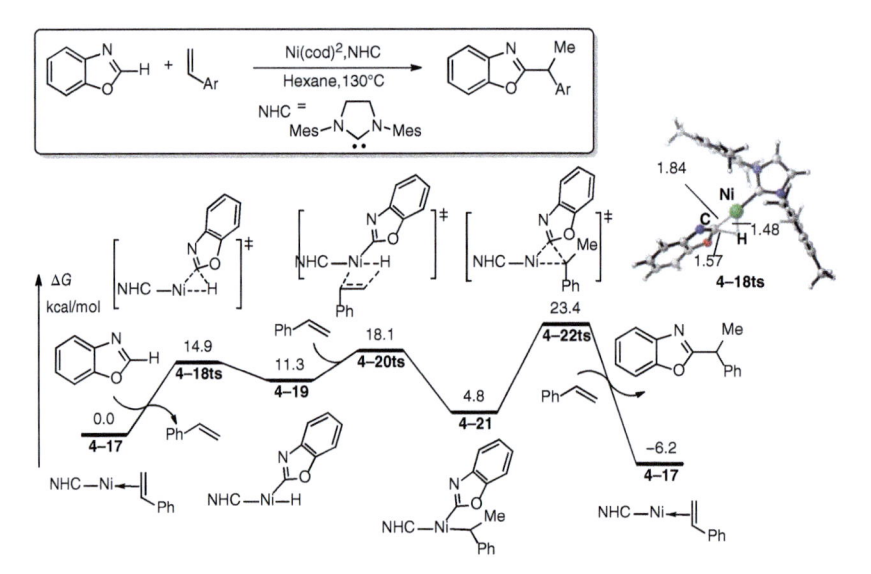

Figure 4.3 The free-energy profiles for the Ni catalyzed C−H activation and arylation through oxidative addition of C−H bond. The energies were calculated at B3LYP/6-311+G(2d,p) (SDD for Ni)/CPCM(hexane)//B3LYP/6-31G(d) (LANL2DZ for Ni) level of theory. Source: Based on Labinger [23].

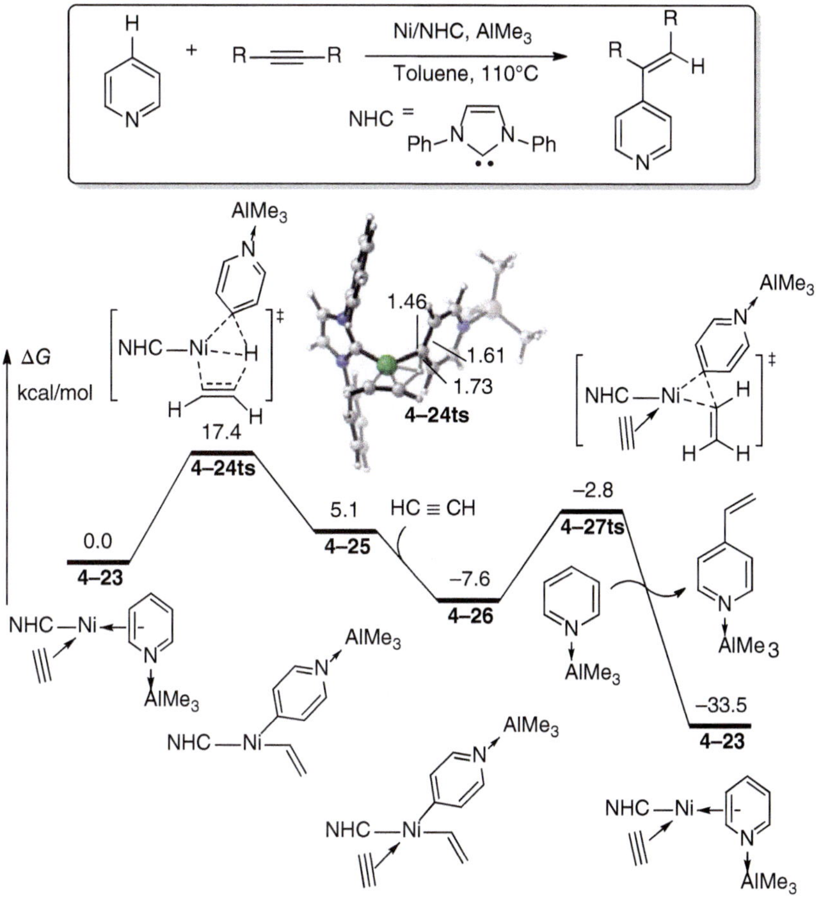

4.1 Ni-Mediated C−H Bond Activation | 131

Figure 4.4 The free-energy profiles for the acetylene-assisted Ni catalyzed C−H activation and vinylation. The energies were calculated at B3LYP-D3/6-311+G(d) (SDD for Ni)/CPCM(toluene)//B3PW91/6-31G(d) (LANL2DZ for Ni) level of theory.

Ni—H and Ni—C bonds are 1.57, 1.48, and 1.84 Å, respectively. Then, a hydride Ni(II) complex **4-19** is formed reversibly. The coordination of styrene onto Ni leads to a corresponding insertion of double bond into Ni—H bond via a four-membered ring type transition state **4-20ts** to afford a benzylic Ni(II). Subsequently, the reductive elimination of aryl and benzyl groups takes place via transition state **4-22ts** to afford hydroarylation product, which is considered to be the rate-determining step because of the generation of Ni(0) species.

The presence of acetylene can accelerate the oxidative addition in Ni(0)-mediated arene C—H bond activation. As shown in Figure 4.4, DFT calculations are performed to reveal the mechanism of Ni-catalyzed hydroarylation of acetylene [68]. An intermolecular oxidative addition of C—H bond in AlMe$_3$-activated pyridine with acetylene-coordinated Ni(0) species **4-23** can take place via transition state **4-24ts** with an free energy barrier of 17.4 kcal/mol. In this transition state, the cleavage of

C(pyridine)—H bond occurs simultaneously with the formation of C(acetylene)—H bond. **4-24ts** is not a usual transition state of concerted oxidative addition. In the geometry information of transition state **4-27ts**, the lengths of C(pyridine)—H, C(acetylene)—H, and Ni—H are 1.61, 1.73, and 1.46 Å, respectively. Therefore, it can be considered as a Ni-bridged hydrogen transfer from pyridine to acetylene, which can afford a vinyl pyridyl Ni(II) complex **4-25**. The followed reductive elimination of vinyl and pyridyl groups can be accelerated by the coordination of next molecular acetylene. The calculated barrier of reductive elimination is only 4.8 kcal/mol via transition state **4-27ts** to yield hydroarylation product.

4.1.2 Ni-Mediated Aldehyde C—H Activation

Due to the reducibility of Ni(0), the C—H bond in aldehydes can be activated by Ni(0) species. As shown in Figure 4.5, Fu and Yu revealed the mechanism of Ni-catalyzed Tishchenko reaction [69]. The catalytic cycle starts from an aldehyde-coordinated Ni(0) species **4-28**. The oxidative addition of the C—H bond in one of the coordinated aldehydes takes place via a three-membered ring type transition state **4-29ts** with an energy barrier of 19.0 kcal/mol to afford a hydride Ni(II) complex **4-30**. Then, a migratory insertion of acyl group in another coordinated aldehyde into Ni—H bond occurs via a four-membered ring-type transition state **4-31ts** to afford an alkoxyl Ni(II) complex **4-32**. The followed reductive elimination of alkoxyl and acyl groups

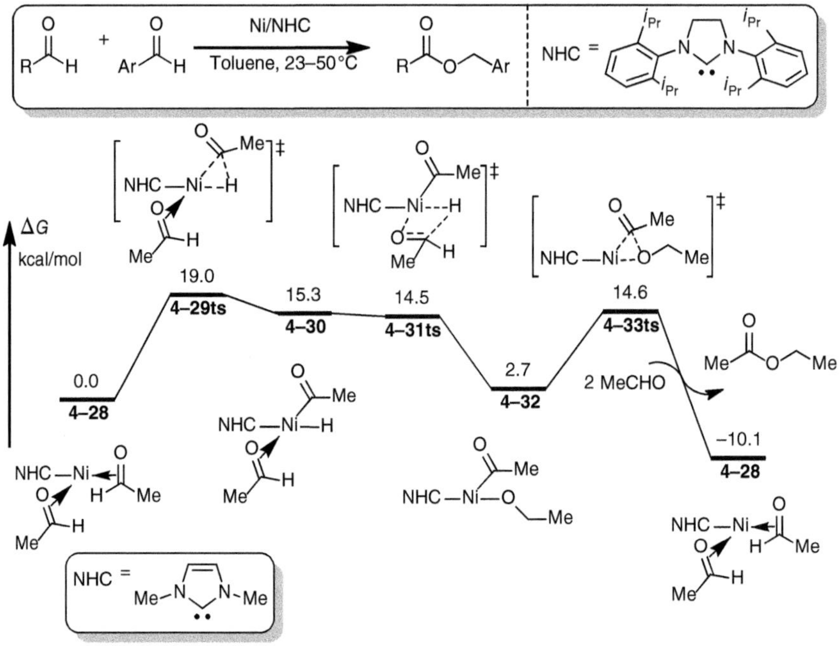

Figure 4.5 The free-energy profiles for the Ni-catalyzed Tishchenko reaction. The energies were calculated at M06-L/6-311+G(d,p) (SDD for Ni)/CPCM(toluene)//B3LYP/6-31G(d) (LANL2DZ for Ni) level of theory.

via transition state **4-33ts** leads to the formation of ester product with the regeneration of Ni(0) catalyst **4-28**.

4.2 Ni-Mediated C—Halogen Bond Cleavage

It is well known that Pd-catalyzed cross-coupling reactions of aryl halides is one of the most famous methods in the field of organometallic catalysis and has a wide range of applications [4, 25]. As a homologous element of Pd, Ni can also promote many cross-coupling reactions of aryl halides [70–82]. The mechanism of these reactions is very similar to that of Pd-catalysis. Therefore, less theoretical studies focus on the whole catalytic cycles of this type reaction. However, the mechanism of Ni-mediated C—X (X = F, Cl, Br, or I) bond functionalizations is more complicated due to the variety of the oxidation states for Ni. Different from Pd chemistry, Ni can activate C—halogen covalent bonds in a variety of ways. Therefore, many theoretical studies have focused on the mechanism of Ni-mediated C—halogen bond cleavage process, which will be discussed in this section.

4.2.1 Concerted Oxidative Addition of C—Halogen Bond

A general view of Ni-catalyzed C—halogen bond functionalization is shown in Scheme 4.2, which involved two possible pathways of coupling reactions and insertion reactions. Both of those two pathways start from a low oxidative state Ni species, which can undergo an oxidative addition of C—X(halogen) bond to afford C—Ni intermediate. In pathway (a), a ligand exchange or transmetallation can introduce another anion onto Ni. Then, a reductive elimination can afford cross-coupling product and regenerate low-valence Ni species. Alternatively, the insertion and reductive elimination according to pathway (b) can afford insertion product and also regenerate the same Ni species.

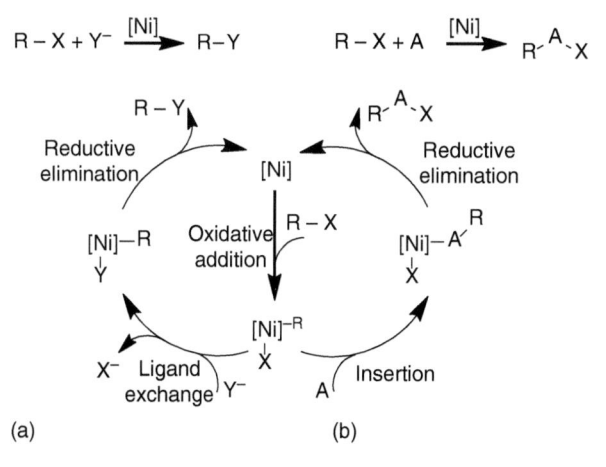

Scheme 4.2 General catalytic cycles for the Ni-mediated C—halogen bond activations. (a) Coupling reaction (b) insertion reaction.

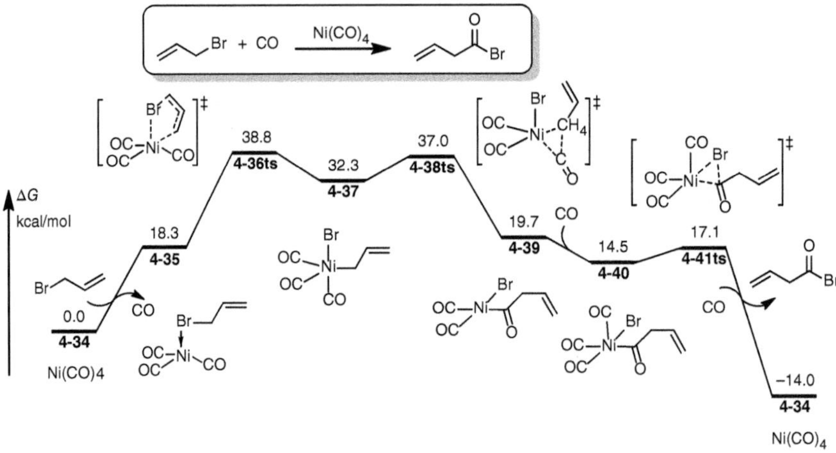

Figure 4.6 The free-energy profiles for the Ni-catalyzed carbonylation of allyl bromide. The energies were calculated at B3LYP/DZVP/PCM(DCM)//B3LYP/DZVP level of theory. DCM; dichloromethane.

The concerted oxidative addition of C—halogen bond on to Ni(0) species is one of the most common ways for the activation of C—halogen bond. The mechanism of a Ni-catalyzed carbonylation of allyl bromide involves this process, which has been proved by DFT calculation [83]. The calculated free energy profiles for the catalytic cycle are shown in Figure 4.6. The ligand exchange of Ni(CO)$_4$ with reactant allyl bromide forms a π-coordinated complex **4-35** with 18.3 kcal/mol endergonic. The concerted oxidative addition of coordinated allyl bromide undergoes a five-membered ring-type transition state **4-36ts** with an energy barrier of 20.5 kcal/mol. The high barrier of this step can be attributed to the fact that the Ni(0) is stabilized by the π-acidic CO ligands. After the oxidative addition, an allylic Ni(II) complex **4-37** is formed. Then, the coordinated CO inserts into Ni—C(allyl) bond via transition state **4-38ts** to afford an acyl Ni(II) complex **4-39**. The coordination of an extra CO leads to a reductive elimination of acyl and bromide to yield acyl bromide product with the regeneration of Ni(CO)$_4$ active species. The barrier for the reductive elimination via transition state **4-41ts** is only 2.6 kcal/mol because the coordinated π-acidic CO ligands can stabilize the formed Ni(0) species.

When a bidentate phosphine is used, the barrier of oxidative addition can be significantly reduced in Ni-mediated C—halogen bond activation. As shown in Figure 4.7, DFT studies were used to predict Ni-catalyzed trifluoromethylthiolation of aryl chlorides [84]. In calculated free-energy profiles, the relative zero is chosen as cyclooctadiene-coordinated Ni(0) species **4-42**. The overall activation free energy of the oxidative addition of phenyl chloride is considered to be 24.4 kcal/mol through a three-membered ring-type transition state **4-44ts**. It clearly revealed that the more stable C—Cl bond can be activated by Ni(0) in a relatively mild condition. A phenyl Ni(II) complex **4-45** is formed reversibly, which underwent a counterion exchange with NMe$_4$(SCF$_3$) to afford a sulfide Ni(II) complex **4-46**. The followed reductive elimination takes place via a three-membered ring-type transition state **4-47ts** to

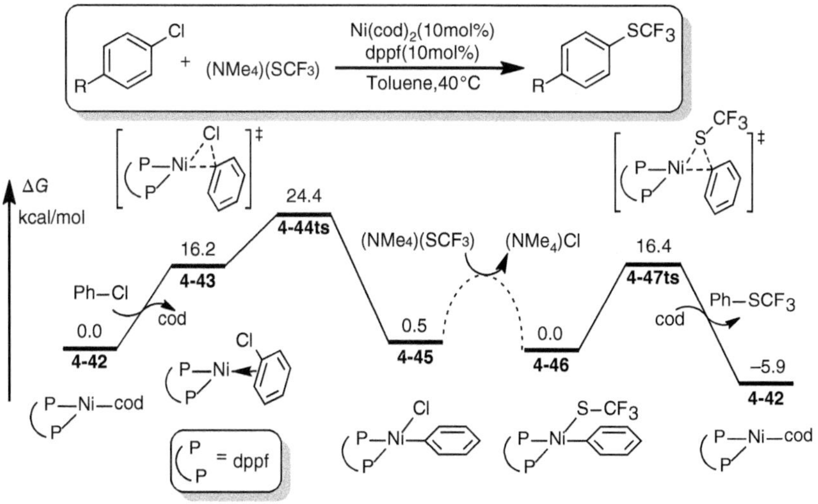

Figure 4.7 The free-energy profiles for the Ni-catalyzed trifluoromethylthiolation of aryl chlorides. The energies were calculated at M06-L/6-311++G(d,p) (LANL2DZ for Ni and Fe)/CPCM(toluene)//B3LYP/6-31G(d) (LANL2DZ for Ni and Fe) level of theory.

generate phenyl(trifluoromethyl)sulfane product and regenerate the Ni(0) catalyst with a barrier of 16.4 kcal/mol.

Due to the reducibility, C—F bond can be activated by Ni(0) species in a cross-coupling reaction between aryl fluoride and arenes [85] (Figure 4.8). DFT calculation found that the C(aryl)—F oxidative addition onto a nickelate(0) species **4-48** can occur via a concerted transitions state **4-49ts** with a barrier of 23.9 kcal/mol to afford a square planer aryl Ni(II) complex **4-50**. Then a base-assisted arene C—H bond activation takes place via an outer-sphere transition state **4-51ts** to afford an aryl Ni(II) complex **4-52**. Subsequently, reductive elimination yields the biarene product with the regeneration of active nickelate(0) species **4-48**.

Moreover, a Ni(I)–Ni(III) catalytic cycle is also acceptable in Ni catalysis. As shown in Figure 4.9, the mechanism of Ni-catalyzed Negishi-type coupling of aryl iodide and alkyl zinc is studied, in which a Ni(I)–Ni(III) catalytic cycle through oxidative addition/reductive elimination was proposed and proved by DFT calculations [86]. The catalytic cycle starts with an iodide Ni(I) species **4-54**, which can react with phenyl iodide to afford a phenyl Ni(III) complex **4-56** via a concerted oxidative addition transition state **4-55ts** with a barrier of only 10.1 kcal/mol. The followed transmetallation with methyl zinc reactant takes place via a concerted four-membered ring-type transition state **4-58ts** to afford a methyl phenyl Ni(III) complex **4-60**. The final reductive elimination yields a toluene and regenerates Ni(I) active species **4-54**.

4.2.2 Radical-Type Substitution of C—Halogen Bond

Instead of aryl bromide, the alkyl bromide also can be used in Ni-catalyzed Negishi type cross-coupling reactions. The mechanism of this type reaction has been proposed by Lin group [87], which is shown in Scheme 4.3. The reaction states

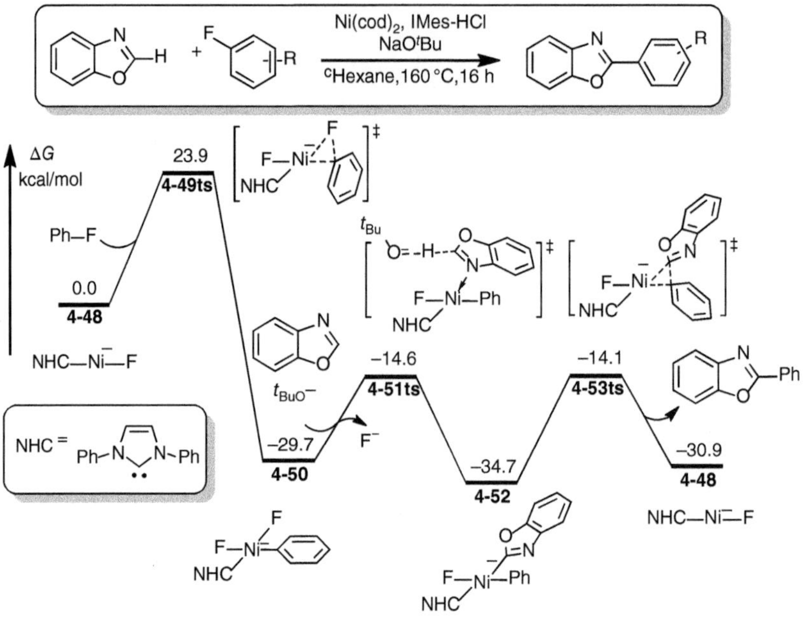

Figure 4.8 The free-energy profiles for the Ni-catalyzed cross-coupling reaction between aryl fluoride and arenes. The energies were calculated at M11-L/6-311++G(d,p) (SDD for Ni)/SMD(cyclohexane)//B3LYP/6-31G(d) (SDD for Ni) level of theory.

from a bromide Ni(I) complex **4-62**, which can undergo a transmetallation to afford an alkyl Ni(I) species **4-63**. A radical-type oxidation forms alkyl Ni(III) species **4-65** through a Ni(II) intermediate **4-64**. The reductive elimination formed the cross-coupling product.

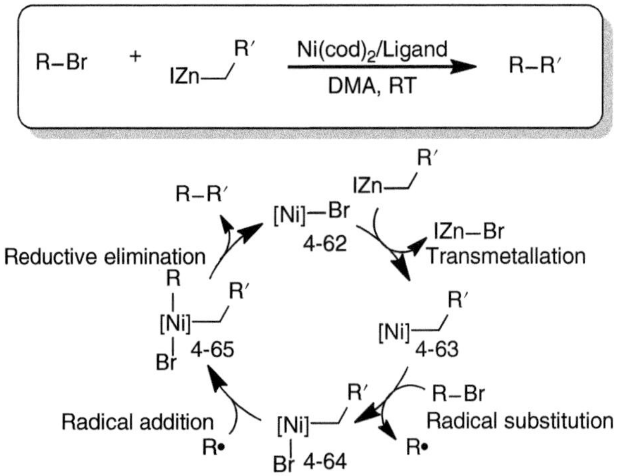

Scheme 4.3 Ni-catalyzed Negishi type cross-coupling reactions through a radical type pathway. Source: Based on Cheung et al. [87].

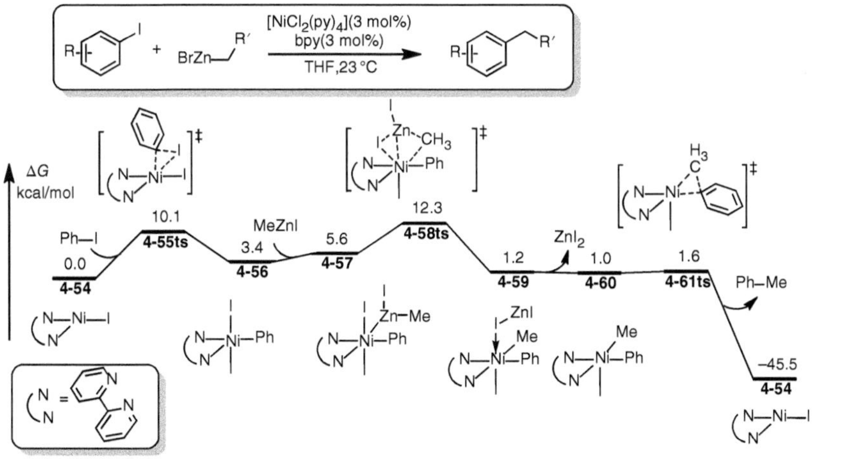

4.2 Ni-Mediated C—Halogen Bond Cleavage

Figure 4.9 The energy profiles for the Ni(I)-catalyzed Negishi cross-coupling reaction through oxidative addition/reductive elimination pathway. The energies were calculated at B3LYP/6-31G(d) (LANL2DZ for Ni, Zn, and I)/PCM(THF)//B3LYP/6-31G(d) (LANL2DZ for Ni, Zn, and I) level of theory.

A DFT study is taken to account this catalytic cycle, which is shown in Figure 4.10. When the catalytic cycle starts from a bromide Ni(I) species **4-66**, the transmetallation with ethyl zinc can take place via concerted transition state **4-67ts** with an energy barrier of 12.3 kcal/mol to form an ethyl Ni(I) complex **4-68**, which reveals a free radical character on metal center. Complex **4-68** can react with *tert*-butyl bromide through a radical substitution-type transition state **4-69ts** to afford a bromide Ni(II) complex **4-70** with the release of *tert*-butyl free radical. In the geometry of transition state **4-69ts**, the bond angle of Ni—Br—C is 175°, which clearly reveals a substitution process. Then the generated *tert*-butyl free radical can react with Ni(II) complex **4-70** via a radical addition transition state **4-71ts** to afford an alkyl Ni(III) complex **4-72**. Finally, the reductive elimination of two alkyl groups on Ni occurs via a concerted three-membered ring-type transition state **4-73ts** to yield alkane type cross-coupling product with the regeneration of bromide Ni(I) species **4-66**. The rate-determining step for the whole catalytic cycle is the reductive elimination with an energy barrier of 27.9 kcal/mol.

4.2.3 C—Halogen Bond Cleavage by β-Halide Elimination

Generally, C—F bond is more stable than the other C—halogen bonds. In Ni catalysis, it can be activated by the β-fluoride elimination. As shown in Figure 4.11, the mechanism of Ni-catalyzed reductive coupling between trifluoromethyl ethylene and acetylenes was studied by DFT calculations [88]. The computational results showed that the activation free energy of direct oxidative addition by C—F bond onto Ni(0) species **4-74** is as high as 32.5 kcal/mol via transition state **4-75ts**; therefore, this pathway can be excluded by the high barrier. Alternatively, the calculated activation free energy of the oxidative cycloaddition with trifluoromethyl ethylene

138 | *4 Theoretical Study of Ni-Catalysis*

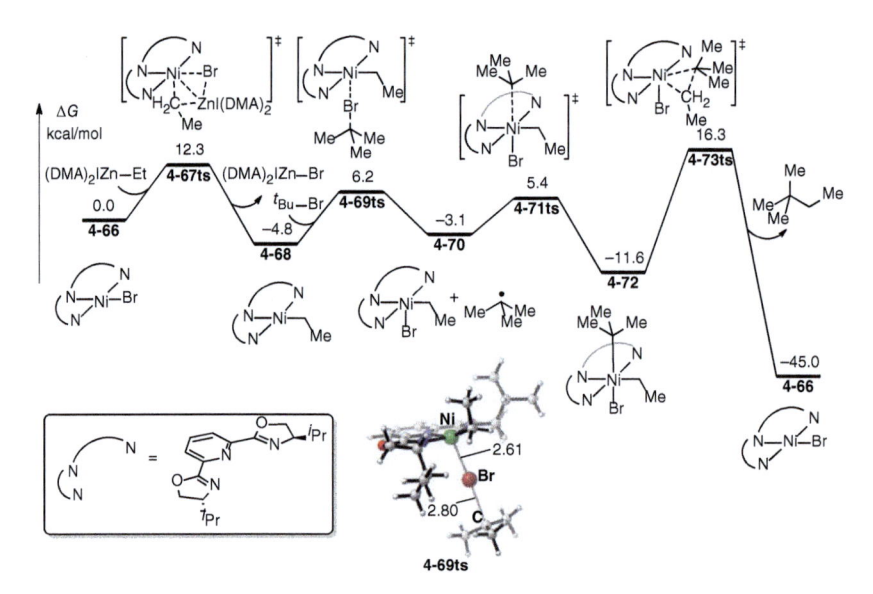

Figure 4.10 The free-energy profiles for Ni-catalyzed Negishi-type cross-coupling reactions through a radical-type pathway. The energies were calculated at B3LYP-D3/6-31G(d) (LANL2DZ for Br and I)/CPCM(DMA)//B3LYP/6-31G(d) (LANL2DZ for Br and I) level of theory. DMA; N,N-dimethylacetamide

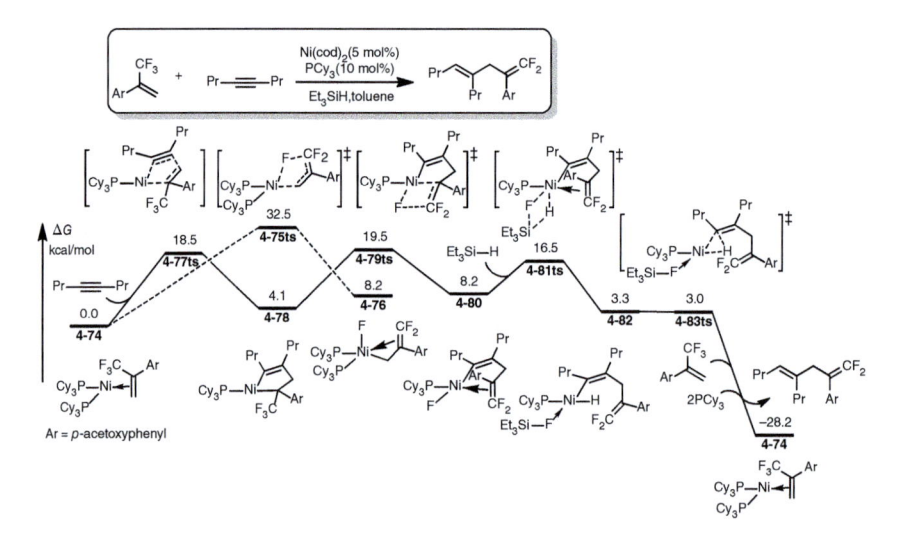

Figure 4.11 The free-energy profiles for Ni-catalyzed reductive coupling between trifluoromethyl ethylene and acetylenes. The energies were calculated at M06/6-311+G(d,p) (SDD for P, Si, and Ni)/SMD(toluene)//B3LYP/6-31G(d,p) (SDD for P, Si, and Ni) level of theory.

and acetylene onto Ni(0) is only 18.5 kcal/mol via transition state **4-77ts** to afford a five-membered nickelous ring **4-78**. Subsequently, a β-fluoride elimination takes place via a four-membered ring-type transitions state **4-79ts** to generate a fluoride Ni(II) complex **4-80**. Then, a transmetallation with silane occurs via transition state **4-81ts** to form a hydride Ni(II) intermediate **4-82**. A rapid reductive elimination yields coupling product with the regeneration of Ni(0) species **4-74**.

4.2.4 Nucleophilic Substitution of C–Halogen Bond

In a polar C–halogen bond, carbon can be considered as an electrophile. When a C–halogen bond reacts with a nucleophilic Ni species, a substitution-type C–halogen bond activation can be taken into account involving alkyl halide activation and alkyl halide activation.

Figure 4.12 shows the calculated free-energy profiles for the Ni-catalyzed cross-coupling reaction between aryl chloride and indoles [89]. When an electron-rich Ni(0)–carbene complex **4-84** is used as active catalyst, the aromatic nucleophilic substitution of coming chloropyridine can take place via transition state **4-85ts** with an energy barrier of only 21.8 kcal/mol. The geometry information of transition state **4-85ts** shows that the angle of Ni—C(aryl)—Cl is 93.9°, and the distance between Ni and Cl is 2.96 Å. Therefore, chloride anion would be released after this substitution transition state. After the substitution-type oxidation, the coordination of indolate provides aryl Ni(II) species **4-87**. The reductive elimination formed amination product with the regeneration of Ni(0) species **4-84**.

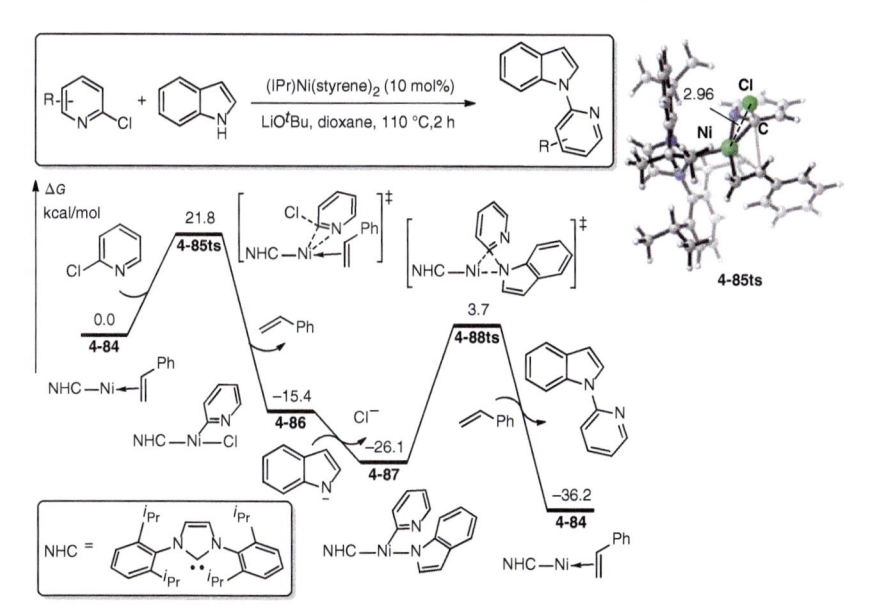

Figure 4.12 The free-energy profiles for Ni-catalyzed cross-coupling reaction between aryl chloride and indoles. The energies were calculated at ωB97X-D/6-311+G(d,p) (LANL2TZ(f) for Ni)/SMD((1,4-dioxane)//ωB97X-D/6-31G(d) (LANL2TZ(f) for Ni) level of theory. Source: Based on Rull et al. [89].

4.3 Ni-Mediated C—O Bond Activation

According to the reducibility of Ni(0) species, even stable C—O covalent bonds can be activated by Ni(0), while the formed alkyl or aryl Ni(II) intermediate can be obtained by oxidative addition to break the C—O bonds. In this type reaction, C—O bond in both of ethers and esters can react with Ni(0) to afford corresponding organonickel(II) species for further transformations [32, 90–100]. Generally, the reaction mechanism for this type of reactions is shown in Scheme 4.4. The catalytic cycle starts from a Ni(0) species, which can undergo an oxidative addition with C—O bond in ethers or esters to form an organonickel(II) species. The transmetallation introduces another alkyl or aryl group on Ni(II). The sequential reductive elimination constructs new C—C bond and regenerates Ni(0) species.

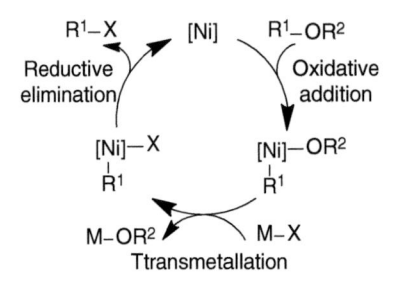

Scheme 4.4 Ni-mediated C—O bond activation.

4.3.1 Ether C—O Bond Activation

In ethers, C—O bonds are difficult for cleavage in organic reactions because of their relative high bond dissociation energies. When Ni(0) species is used, this transformation can be easily achieved. Many theoretical calculations have focused on this process, attempting to elucidate the reaction mechanism, activity, and selectivity.

In a Ni(0)–NHC (N-heterocyclic carbene) catalyzed hydrogenolysis of diphenyl ether, DFT calculation is used to reveal the mechanism of this reaction [101]. As shown in Figure 4.13, the coordination of diphenyl ether forms a Ni(0) species **4-89**. Then an oxidative addition carries out via a three-membered ring-type transition state **4-90ts** to afford an aryl Ni(II) intermediate **4-91**. The calculated barrier of this step is 24.0 kcal/mol, which is considered to be the rate-determining step for the whole catalytic cycle. The coordination of gaseous dihydrogen causes a σ-bond metathesis via transition state **4-92ts** to break the H—H single bond. After release a benzene molecule, a hydride Ni(II) intermediate **4-93** is formed. The reductive elimination yields a phenol and regenerates Ni(0) species **4-89** by the coordination of diphenyl ether reactant.

When an anisole is used as substrate, the oxidative addition would take place with either C(alkyl)—O or C(aryl)—O bond onto Ni(0) species. DFT calculation is taken to reveal the regioselectivity of this reaction [103]. Computational results showed that the relative free energy of transition state **4-95ts** is 4.7 kcal/mol lower than that

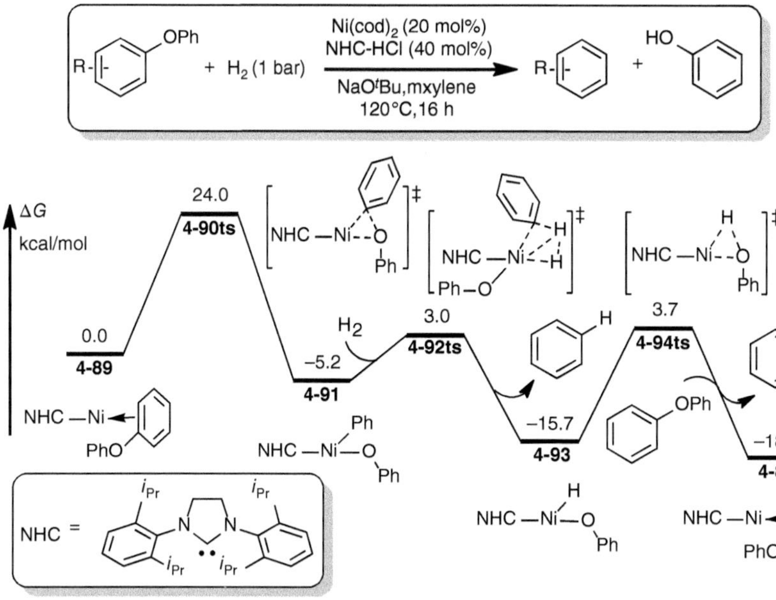

4.3 Ni-Mediated C—O Bond Activation | 141

Figure 4.13 The free-energy profiles for Ni-catalyzed catalyzed hydrogenolysis of diphenyl ether. The energies were calculated at M06/6-311++G(d,p)/6-31G(d) (SDD for Ni)/CPCM((m-xylene)//B3LYP/6-311++G(d,p)/6-31G (SDD for Ni) level of theory. Source: Based on Schwarzer et al. [102].

of **4-96ts** (Scheme 4.5). Therefore, the C(aryl)—O bond oxidative addition is easier than that by C(alkyl)—O bond. The further distortion-interaction analysis revealed that the difference of activation energy is primarily controlled by interaction energy, which can be attributed to the second-order orbital interaction of phenyl group with the d orbital of Ni.

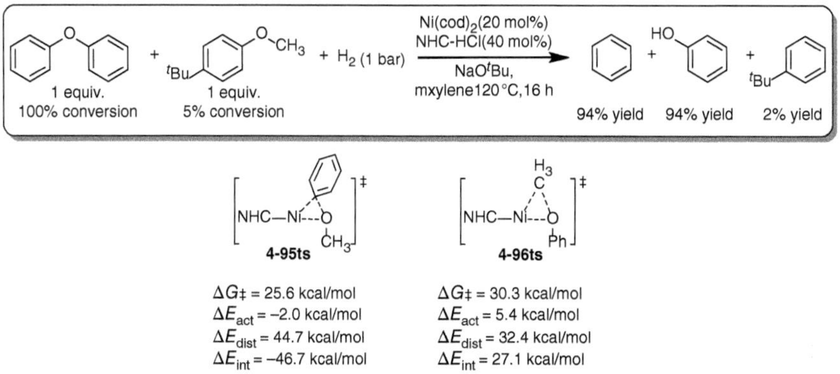

Scheme 4.5 The competition of the oxidative addition with C(aryl)—O bond or C(alkyl)—O bond in Ni-catalyzed hydrogenolysis.

The calculated activation free energy indicates that it is difficult for the oxidative addition of a C(aryl)—O bond to take place under low temperature. DFT

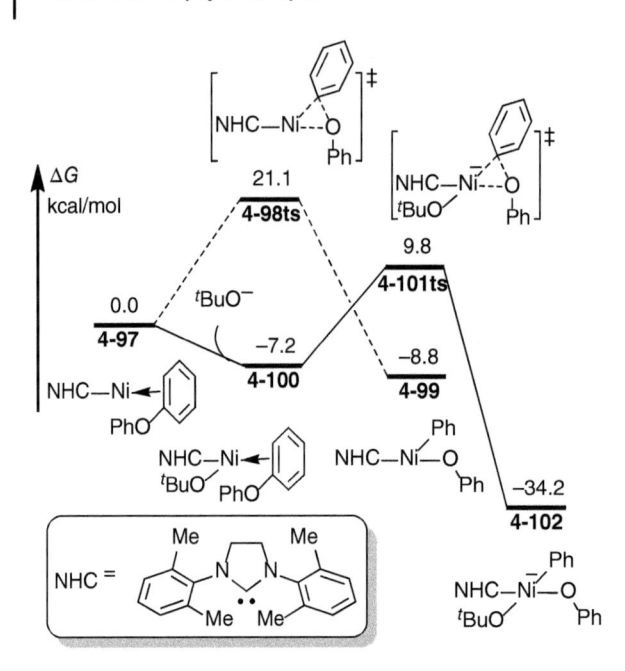

Figure 4.14 The free-energy profiles for base-assisted oxidative addition of C(aryl)–O bond onto Ni(0). The energies were calculated at M06/6-311++G(d,p) (SDD for Ni)/CPCM((m-xylene)//M06/6-31G(d) (LANL2DZ(f) for Ni) level of theory.

calculations predicted that the base additives can significantly reduce the activation energy for the oxidative addition step [53]. As shown in Figure 4.14, when *tert*-butoxide is added, it can coordinate with Ni(0) species to afford a nickelate intermediate **4-100** reversibly. The corresponding oxidative addition can take place via transition state **4-101ts** to afford aryl Ni(II) species **4-102**. The calculated activation free energy of this process is only 17.0 kcal/mol, which is 4.1 kcal/mol lower than that of the corresponding oxidative addition in the absence of base.

DFT calculations also predicted that the Lewis acid additives can activate the C—O bond in the oxidative addition process [102]. As shown in Figure 4.15, when PCy$_3$ is used as the ligand, an oxidative addition of naphthol methyl ether on to Ni(PCy$_3$)$_2$ complex could take place via a three-membered ring-type transition state **4-104ts** with an energy barrier of 33.0 kcal/mol. When CsF is used as Lewis acid additive, it can coordinate by the oxygen atom in naphthol methyl ether to afford intermediate **4-106** exergonically. The corresponding oxidative addition can occur via transition state **4-107ts** with a free energy barrier of only 25.3 kcal/mol, which is 7.7 kcal/mol lower than that for the absence of Lewis acid.

4.3.2 Ester C—O Bond Activation

In Ni(0)-mediated C—O bond activation reactions, there are three mechanistic models in Ni(0)-mediated C—O bond activation (Scheme 4.6). Type A is the classic three-membered ring transition state, in which the nickel catalyst directly inserts

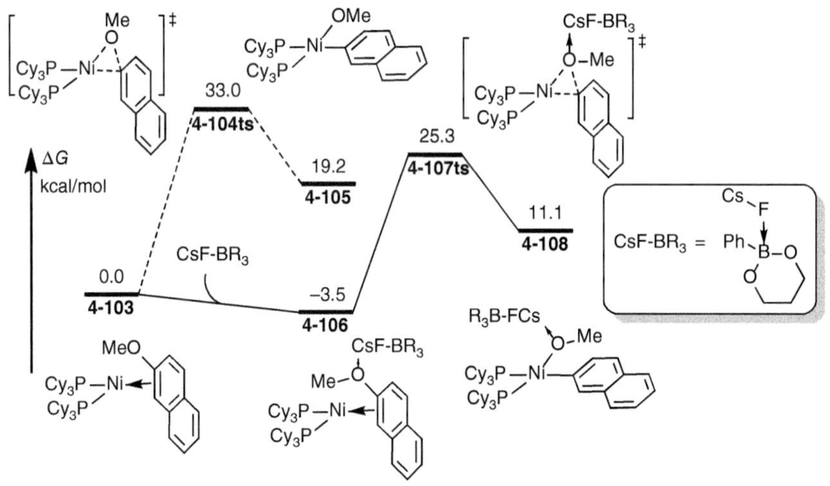

Figure 4.15 The free-energy profiles for Lewis acid-assisted oxidative addition of C(aryl)—O bond onto Ni(0). The energies were calculated at BP86-D3/def2-TZVP/PCM(toluene)//BP86-D3/6-31G(d) (def2-SVP for Ni, and SDD for Cs) level of theory. Source: Based on Schwarzer et al. [102].

into the C—O bond while interacting with the cleaving two fragments. Type B is the chelation-assisted transition state, in which the nickel catalyst interacts with the directing group of the substrate while cleaving the C—O bond. Type C is the S_N2-type transition state, in which the nickel catalyst attacks as a nucleophile to generate an ion pair during the C—O bond cleavage. Because carboxylate is a good leaving group, an S_N2 substitution can occur to cleave C—O bond, in which the Ni(0) species is used as nucleophile to generate a cationic Ni(II) intermediate.

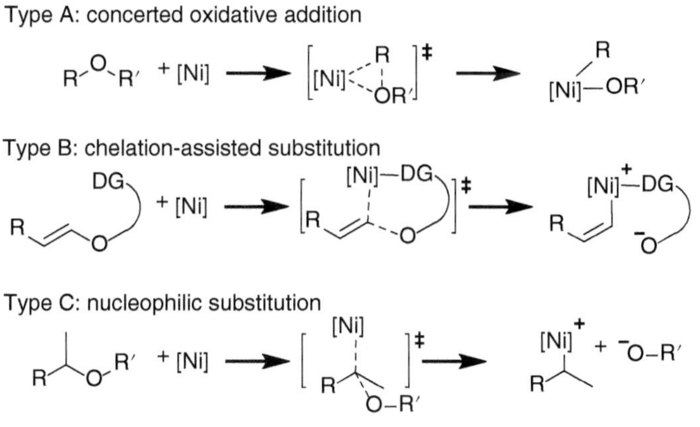

Type A: concerted oxidative addition

Type B: chelation-assisted substitution

Type C: nucleophilic substitution

Scheme 4.6 Possible models for Ni(0)-mediated C—O bond activation.

In a phenyl acetate (Scheme 4.7) [93], the bond dissociation energies of C(aryl)—O and C(acyl)—O bonds are 106 and 80 kcal/mol, respectively, which reveal that the C(acyl)—O bond cleavage is easier than the other one. However, in a Ni-catalyzed

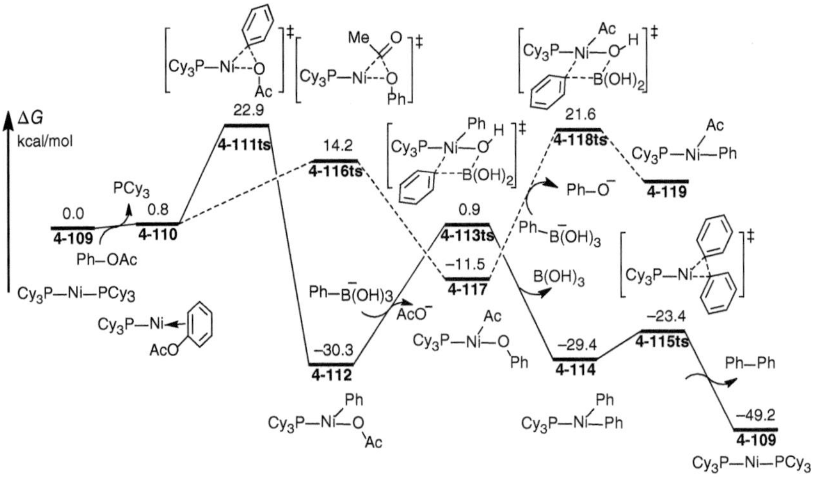

Figure 4.16 The free-energy profiles for the Ni-catalyzed decarboxylative arylation of phenyl acetate. The energies were calculated at B3PW91/6-311++G(2d,p) (SDD for Ni)/IEF-PCM(dioxane)//B3PW91/D95v(d) (LANL2DZ+p for P and Ni) level of theory.

decarboxylative arylation, the experimental result showed that the cleaved C—O bond is C(aryl)—O bond in phenyl acetate.

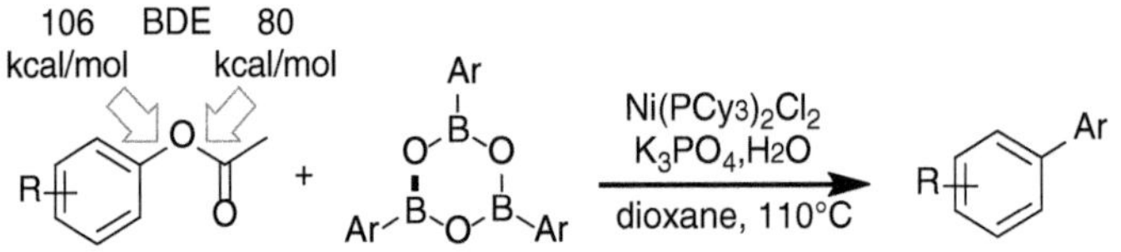

Scheme 4.7 The Ni-assisted C—O bond activation of phenyl acetate. Source: Based on Li et al. [93].

DFT calculations are taken to reveal the mechanism of Ni-catalyzed decarboxylative arylation of phenyl acetate (Figure 4.16). The computational result shows that the C(aryl)—O oxidative addition can occur via a three-membered ring-type transition state **4-111ts** with a free energy of 22.9 kcal/mol to afford a phenyl Ni(II) species **4-112** irreversibly. Then, the transmetallation with phenylboronic acid takes place via transition state **4-113ts** to afford a diphenyl Ni(II) species **4-114**. After a rapid reductive elimination, biphenyl product can be released with the regeneration of Ni(0) species **4-109**. The alternative oxidative addition with C(acyl)—O bond is also considered, which can take place via transition state **4-116ts**. Although the relative free energy of **4-116ts** is 8.7 kcal/mol lower than that of **4-111ts**, the relative free energy of generated acyl Ni(II) species is 18.8 kcal/mol higher than that of **4-117**. Therefore, in next transmetallation process, the calculated barrier is as high as 33.1 kcal/mol, which leads to a reversible process of oxidative addition. Therefore, only C(aryl)—O oxidative addition is effective. It means that the reacting

C—O bond is determined by the following processes in Ni(0)-mediated C—O bond activation.

Another example clearly shows the effect of subsequent processes on the regioselectivity of Ni-mediated C—O bond activation [94]. As shown in Scheme 4.8, in a Ni-catalyzed arylation reaction, when the R group is heteroaryl (entry a), the C(acyl)—O bond cleavage product is observed as decarboxylative arylation. In another case, when R group is *tert*-butyl (entry b), a C(aryl)—O bond activation is observed experimentally.

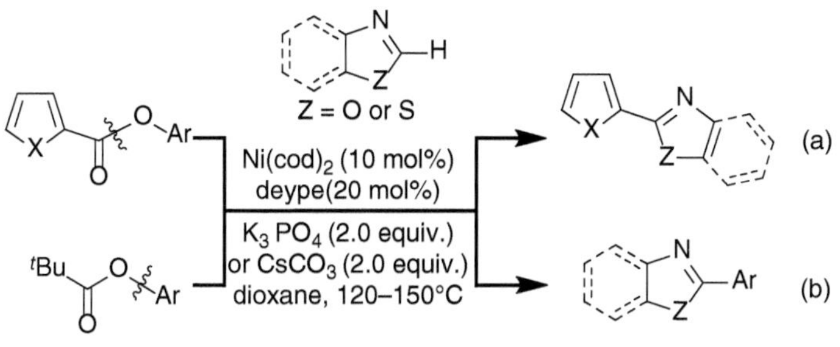

Scheme 4.8 Regioselectivity of Ni-catalyzed ester arylation. Source: Based on Lu et al. [94].

The calculated free energy profiles are shown in Figure 4.17. When R group is thiophenyl, the calculated energy barrier of C(acyl)—O bond oxidative addition onto Ni(0) is 19.8 kcal/mol via transition state **4-121ts(a)** to afford an acyl Ni(II) intermediate **4-122(a)**. The following phosphate-assisted intermolecular C—H bond activation occurs via transition state **4-124ts(a)** to afford an aryl acyl Ni(II). The decarboxylation generates a diaryl Ni(II) intermediate **4-127(a)** via transition state **4-126ts(a)**, then a reductive elimination via transition state **4-128ts(a)** yields biaryl product R-Ar and regenerates Ni(0) species **4-120(a)** by ligand exchange. The computational data showed that the relative free energy of transition states **4-124ts(a)**, **4-126ts(a)**, and **4-128ts(a)** are lower than that of **4-121ts(a)**. Therefore, the oxidative addition of C(acyl)—O bond is irreversible, when R group is thiophenyl. The C(aryl)—O bond oxidative addition of C(aryl)—O bond is also considered. The calculated activation free energy is 23.0 kcal/mol in a five-membered ring-type transition state **4-130ts(a)**, which is 3.2 kcal/mol higher than that of C(acyl)—O bond oxidative addition. Therefore, when the R group is thiophenyl, the C(acyl)—O bond cleavage is the major reaction pathway.

In another case, when R group is *tert*-butyl, the activation barrier of C(acyl)—O bond oxidative addition is 10.1 kcal/mol via transition state **4-121ts(b)** to afford the corresponding acyl Ni(II) intermediate **4-122(b)**. After an intermolecular C—H bond activation, an aryl acyl Ni(II) intermediate **4-125(b)** is formed; however, the relative free energy of the followed decarboxylation transition state **4-126ts(b)** is as high as 29.4 kcal/mol, which indicates a reversible process before decarboxylation. Alternatively, the calculated energy barrier of C(aryl)—O bond oxidative addition is 16.5 kcal/mol via transition state **4-130ts(b)** to afford a phenyl Ni(II) intermediate **4-131(b)**. The followed transmetallation and reductive elimination yield phenyl

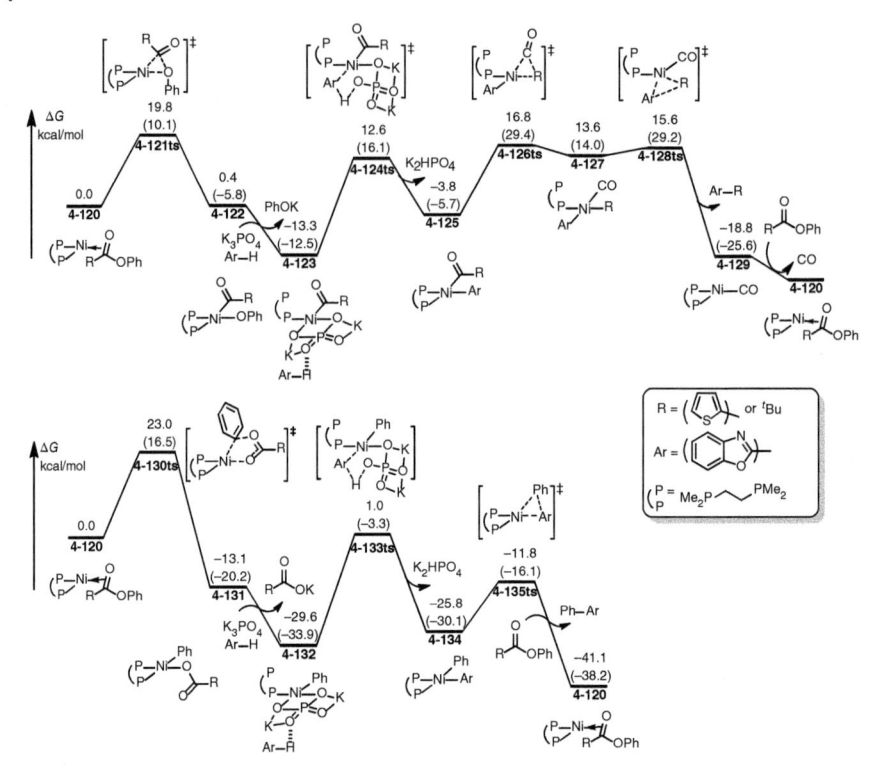

Figure 4.17 The free-energy profiles for the Ni-catalyzed regioselective ester arylation. The values given normally represent the case (a) that R is thiophenyl. The values in parentheses represent the case (b) that R is *tert*-butyl. The energies were calculated at M06-L/6-311++G(2d,p) (SDD for K and Ni)/SMD(1,4-dioxane)//B3LYP/6-31G(d) (LANL2DZ for K and Ni) level of theory.

arene product. The computational data show that the relative free energy of transition state **4-130ts(b)** is 12.9 kcal/mol lower than that of **4-126ts(b)**; therefore, the C(aryl)—O bond activation is the major path, when R group is *tert*-butyl group.

In the competition of oxidative addition onto Ni(0) with C(aryl)—O and C(acyl)—O bonds in esters, ligands usually play a decisive role. As shown in Scheme 4.9, when a monodentate phosphine ligand is used (case a), a C(aryl)—O bond activation is found experimentally [104]. Alternatively, when a bidentate phosphine ligand is used, C(acyl)—O bond might be cleaved by Ni(0) species (case b) [96].

A DFT calculation is taken to reveal the regioselectivity of Ni(0)-mediated C—O bond activation [43]. As shown in Figure 4.18, when a monodentate phosphine ligand is used (case a), three C—O bond activation models are calculated. The C(acyl)—O bond can be cleaved through a three-membered ring-type transition state **4-137ts** with an energy barrier of 23.5 kcal/mol, which is much lower than that of three-centered C(aryl)—O bond oxidative addition via transition state **4-138ts**. However, the C(aryl)—O bond activation can occur through a chelation-assisted

4.3 Ni-Mediated C—O Bond Activation | 147

Scheme 4.9 Ligand-controlled regioselectivity of Ni-catalyzed ester arylation.

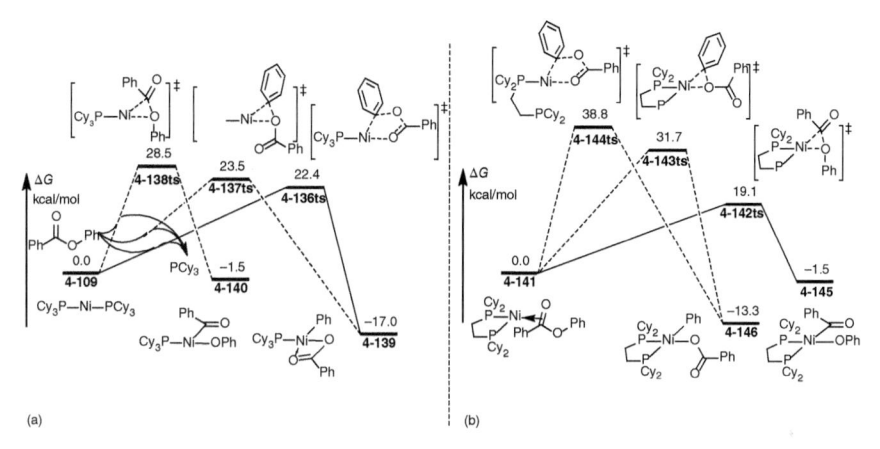

Figure 4.18 The free-energy profiles for the oxidative addition of C—O bond onto Ni(0) by using monodentate ligand (a) or bidentate ligand (b). The energies were calculated at M06/6-311+G(d,p) (SDD for Ni)/SMD(1,4-dioxane)//B3LYP/6-31G(d) (SDD for Ni) level of theory. Source: Based on Hong et al. [43].

process. The calculated activation free energy of that process via a five-centered transition state **4-136ts** is 1.1 kcal/mol lower than the corresponding step via **4-137ts**. Therefore, in monodentate ligand case, C(aryl)—O bond activation through a chelation-assisted process might be favorable.

In the presence of bidentate phosphine ligand (case b), the coordination of chelation group leads to the dissociation of one phosphine ligand. Accordingly, the calculated activation free energy in a chelation-assisted process is as high as 38.8 kcal/mol via a five-centered transition state **4-144ts**. Alternatively, the C—O bond activation can take place through three-centered processes. The calculated energy barrier of C(acyl)—O bond oxidative addition is 19.1 kcal/mol via transition state **4-142ts**, which is 12.6 kcal/mol lower than that of C(aryl)—O bond oxidative addition via transition state **4-143ts**. Therefore, C(acyl)—O bond oxidative addition might be favorable in the presence of bidentate ligand.

Because carboxylate is a good leaving group, a substitution type C—O bond activation would happen, if benzyl acetate is used as substrate [54]. As shown in Figure 4.19a, when monodentate phosphine is used as ligand, a chelation-assisted oxidative addition can take place via transition state **4-148ts** with an energy barrier

148 | 4 Theoretical Study of Ni-Catalysis

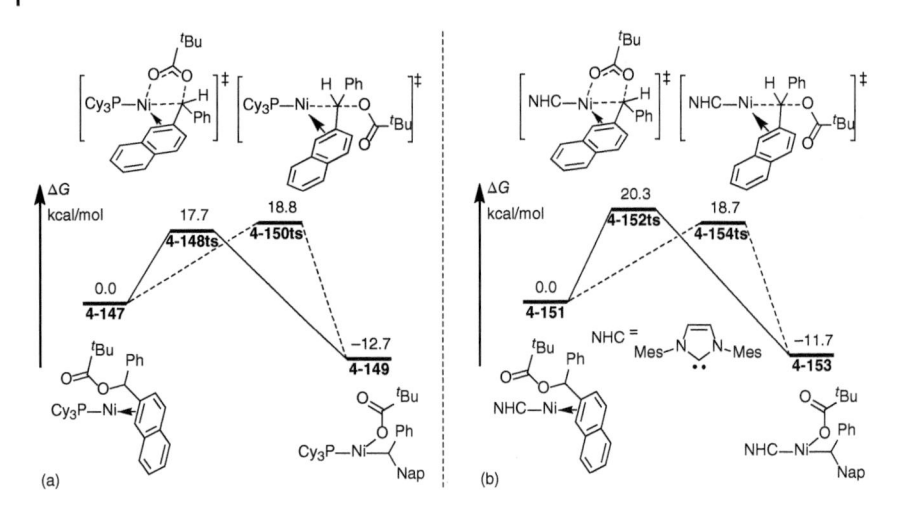

Figure 4.19 The free-energy profiles for the oxidative addition of C—O bond onto Ni(0) by using phosphine (a) or NHC ligand (b). The energies were calculated at M06/6-311+G(d,p) (SDD for Ni)/CPCM(THF)//B3LYP/6-31G(d) (LANL2DZ for Ni) level of theory.

of 17.7 kcal/mol to afford a carboxylate-coordinated benzylic Ni(II) intermediate **4-149**. Alternatively, assisted with the coordination of aryl group onto Ni(0), a substitution type C(benzyl)—O bond cleavage can occur via transition state **4-150ts** leading to the inversion of chiral carbon center. The calculated activation free energy of nucleophilic substitution by Ni(0) is 18.8 kcal/mol, which is 1.1 kcal/mol higher than that of chelation-assisted oxidative addition. Therefore, when monodentate phosphine is used, the chirality of benzylic carbon continues in further functionalizations.

In another case (Figure 4.19), when a N-heterocyclic carbene is used as ligand, the steric hindrance significantly increases the energy barrier of chelation-assisted oxidative addition. The calculated barrier is 20.3 kcal/mol via a five-centered transition state **4-152ts**. Alternatively, the barrier of substitution type C(benzyl)—O bond cleavage remains as 18.7 kcal/mol to afford a cationic Ni(II) intermediate, which can be coordinated with a carboxylate to form the corresponding benzylic Ni(II) intermediate **4-153**. Therefore, when N-heterocyclic carbene is used as ligand, the chiral benzylic carbon is inversed in further functionalizations.

4.4 Ni-Mediated C—N Bond Cleavage

The C—N bond can be activated by Ni(0) species through oxidative addition, while the insertion of Ni into C—N bond can be considered to form a Ni—C bond for further transformations. However, a C—N bond in regular amines is hard to be activated. In most cases, the reduced C—N bonds by Ni exist in amides [105–117].

As shown in Figure 4.20, when amide reacts with phenylboronate in the presence of Ni(0) catalyst, a deaminative Suzuki–Miyaura type cross-coupling product is observed experimentally [106]. In amide substrate, three types of potential C—N

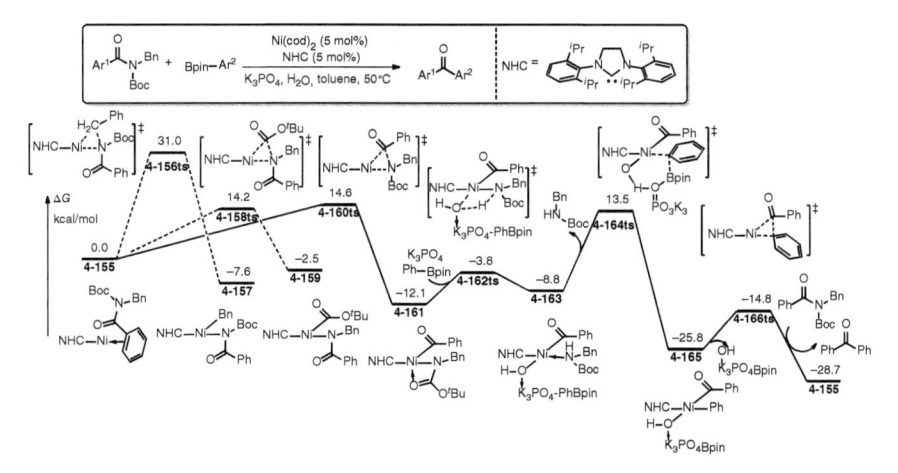

Figure 4.20 The free-energy profiles for the Ni-mediated deaminative Suzuki–Miyaura type cross-coupling of amides. The energies were calculated at M06/6-311++G(2d,p) (SDD for Ni, P, and K)/IEF-PCM(toluene)//M06/D95v(d) (LANL2DZ for Ni, P, and K) level of theory.

activation may occur, including C(benzyl)—N, C(acyl)—N, and C(carbonate)—N bonds. DFT calculation revealed that the activation free energy of C(benzyl)—N bond is as high as 31.0 kcal/mol via transition state **4-156ts**, which can be excluded. The activation free energy of C(acyl)—N bond and C(carbonate)—N bond cleavages is close via transition states **4-160ts** and **4-158ts**, respectively. Both of these two processes are proceeded via three-membered ring-type transition states. The C(acyl)—N bond cleavage is more exergonic, which leads to an irreversible process. The formed amino Ni(II) complex **4-161** can be reversibly hydrolyzed by water molecule in the presence of potassium phosphate additive via transition state **4-162ts** with an energy barrier of 8.3 kcal/mol to afford a hydroxide Ni(II) complex **4-163**. Further transmetallation takes place via transition state **4-164ts**, which is considered to be the turnover-determining step with an overall activation free energy of 25.6 kcal/mol. Then, a rapid reductive elimination of aryl and acyl group yields benzophenone product and regenerates Ni(0) active species **4-155**.

When a phosphine ligand is used in Ni-catalyzed Suzuki–Miyaura cross-coupling of amide, a decarbonylative product can be observed experimentally (Figure 4.21) [118]. This reaction also starts to form a Ni(0)-mediated C(acyl)—N bond activation. The C(acyl)—N bond cleavage takes place via a chelation-assisted five-membered ring-type transition state **4-168ts** with an energy barrier of only 9.6 kcal/mol to afford an amino Ni(II) complex **4-169**. A base-assisted transmetallation with naphthylboronic acid occurs via transition state **4-171ts** to afford a naphthyl acyl Ni(II) intermediate **4-172**. An intramolecular α-phenyl elimination takes place via transition state **4-173ts** to form a carbonyl-coordinated Ni(II) intermediate **4-174**. The reductive elimination forms biaryl product and regenerates Ni(0) species **4-167**.

Ni-mediated C—N bond activation could lead to the conversion of amides to ester. The mechanism of this reaction is revealed by DFT calculations (Figure 4.22)

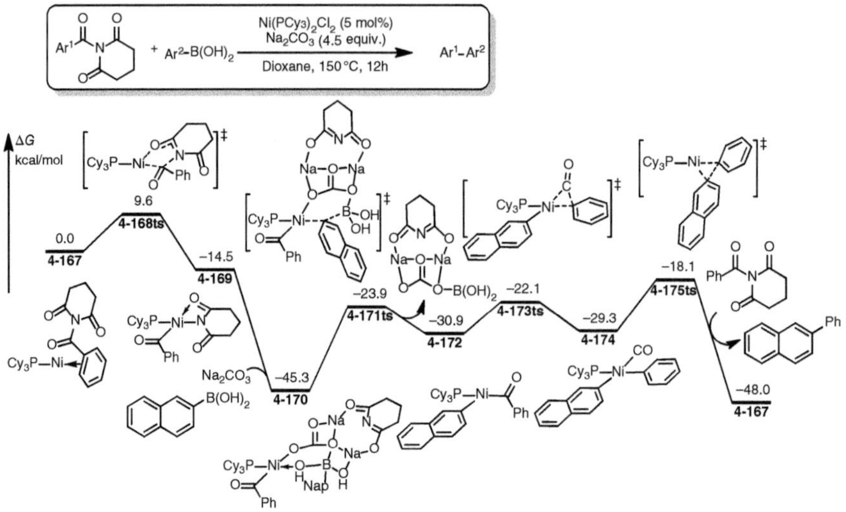

150 | *4 Theoretical Study of Ni-Catalysis*

Figure 4.21 The free-energy profiles for the Ni-catalyzed Suzuki–Miyaura cross-coupling of amide. The energies were calculated at M06/6-311+G(d,p) (SDD for Ni)/CPCM(dioxane)//B3LYP/6-31G(d) (LANL2DZ for Ni) level of theory. Source: Based on Ji and Hong [118].

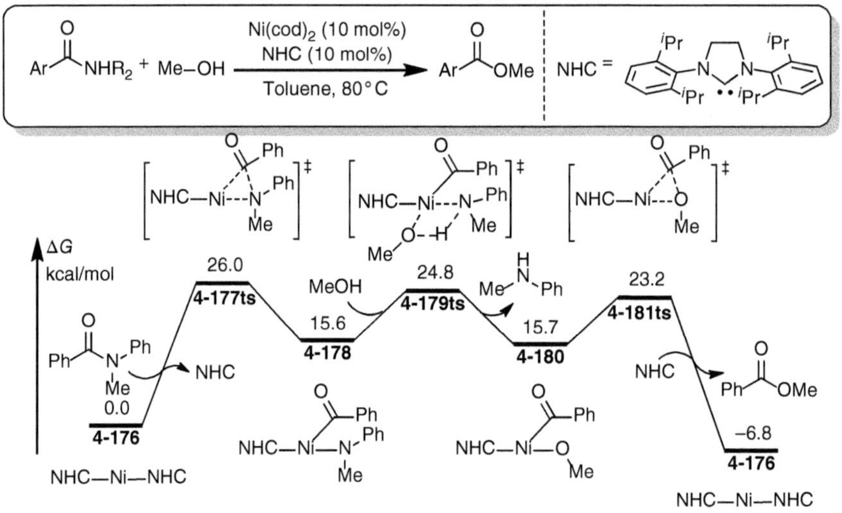

Figure 4.22 The free-energy profiles for the Ni-catalyzed esterification of amide. The energies were calculated at M06/6-311+G(d,p) (SDD for Ni)/SMD(toluene)//B3LYP/6-31G(d) (LANL2DZ for Ni) level of theory. Source: Based on Hie et al. [119].

[119]. The C—N bond activation occurs through an oxidative addition process via a three-membered ring-type transition state **4-177ts** with an energy barrier of 26.0 kcal/mol. Then, a metathesis-type methanolysis, occurring via a four-centered transition state **4-179ts**, forms methoxyl Ni(II) complex **4-180**. The sequential reductive elimination yields ester product.

4.5 Ni-Mediated C—C Bond Cleavage

4.5.1 C—C Single Bond Activation

In organometallic catalysis, C—C covalent bonds are generally difficult to activate and participate in coupling reactions due to their stability. However, some more active C—C bonds can be cleaved by Ni(0) species and take part in the followed transformations because of the reducibility of Ni(0) species [30, 34, 120–130]. The C—C bond in arylnitrile is one of the most active one, in which cyano group partially reveals halide character. Therefore, the oxidative addition of C(aryl)—C(cyano) bond onto Ni(0) species is a possible way to achieve the further transformation of this bond.

As shown in Figure 4.23, a Ni-catalyzed phenylcyanation of alkynes was reported experimentally [131, 132]. The reaction starts from a η^2-cyano-coordinated Ni(0) species **4-182**. The oxidative addition can occur via a three-membered ring-type transition state **4-183ts** to afford a cyano Ni(II) intermediate **4-184**. The activation free energy of this step is considered to be 25.1 kcal/mol, which is the rate-determining step of the whole catalytic cycle. The ligand exchange with acetylene forms intermediate **4-186** with 9.3 kcal/mol endergonic by releasing a phosphine ligand. Subsequently, an acetylene migratory insertion into C(aryl)—Ni bond forms a vinyl Ni(II) intermediate **4-188**. The coordination of phosphine ligand leads to a rapid reductive elimination of cyano and vinyl groups via transition state **4-190ts**. After releasing phenylcyanation product, Ni(0) species can be regenerated by the coordination of benzonitrile.

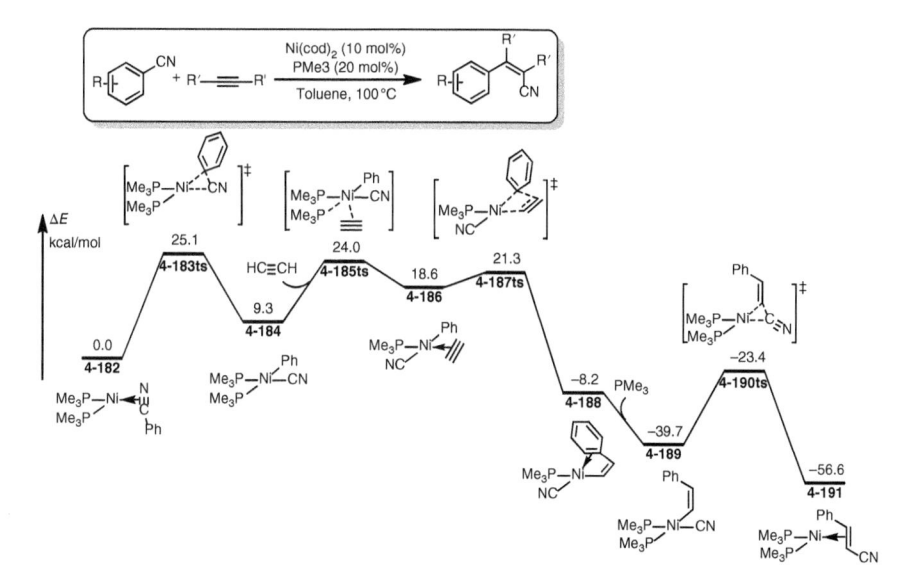

Figure 4.23 The energy profiles for the Ni-catalyzed phenylcyanation of alkyne. The energies were calculated at B3PW91/6-31G(d) (cc-pVDZ for Ni)/IEF-PCM(toluene)//B3PW91/6-31G(d) (SDD for Ni) level of theory.

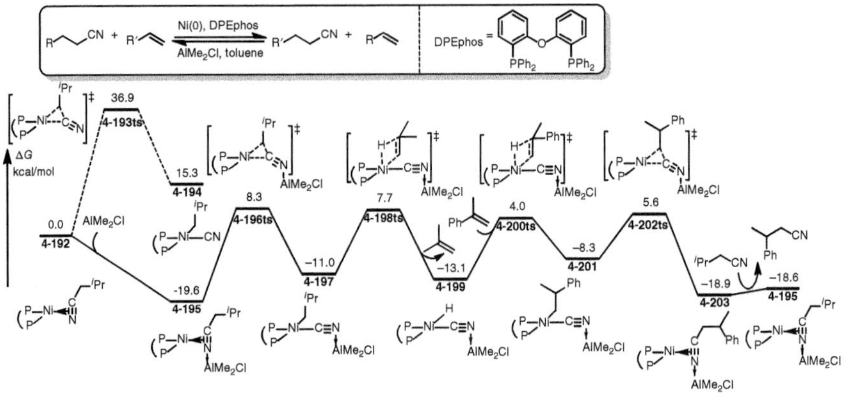

Figure 4.24 The free-energy profiles for the Ni-catalyzed transfer hydrocyanation in the presence of Lewis acid cooperatives. The energies were calculated at M11-L/6-311+G(d,p) (LANL2DZ for Ni, P, and Cl)/SMD(toluene)//wB97X-D/6-31G(d) (LANL2DZ for Ni, P, and Cl) level of theory. Source: Based on Ni et al. [133].

The activation of a C(alkyl)—C(cyano) bond is rather harder than that of a C(aryl)—C(cyano) bond. A DFT study is performed to reveal the mechanism of Ni(0)-catalyzed transfer hydrocyanation in the presence of Lewis acid cooperatives (Figure 4.24) [133]. DFT calculation found that the activation barrier for the oxidative addition of isopentanenitrile is as high as 36.9 kcal/mol via transition state **4-193ts**, which is hard to climb. When a Lewis acid AlMe$_2$Cl is added, it can be coordinated by cyano group to form intermediate **4-195** with 19.6 kcal/mol exergonic. The oxidative addition can take place via transition state **4-196ts** with an energy barrier of 27.9 kcal/mol, which is much lower than that in the absence of Lewis acid. When an alkyl Ni(II) intermediate **4-197** is formed, a β-hydride elimination occurs via transition state **4-198ts** to form a hydride Ni(II) species **4-199**. Then, an olefin insertion into Ni—H bond via transition state **4-200ts** forms a new C(alkyl)—Ni bond. Subsequently, a reductive elimination of alkyl and cyano groups yields transfer hydrocyanation product and regenerates Ni(0) species **4-195**.

4.5.2 C=C Double Bond Activation

In the presence of Ni(0), ketone can be decomposed to carbon monoxide and olefin, which involves a C=C double bond activation. The mechanism of this reaction is studied by DFT calculations [134]. As shown in Figure 4.25, both of C=O and C=C bond in ketene can coordinate with Ni(0) to form complexes **4-204** and **4-206** independently. The isomerization of those two complexes can undergo a low barrier transition state **4-205ts**. The Ni insertion into C=C double bond can occur via a three-centered transition state **4-207ts** with an energy barrier of 26.8 kcal/mol to afford a carbonyl-coordinated Ni–carbene complex **4-208**. The geometry of transition state **4-207ts** is shown in Figure 4.25. The lengths of breaking C—C and two forming C—Ni bonds are 2.04, 1.75, and 1.96 Å, respectively, which can be also considered

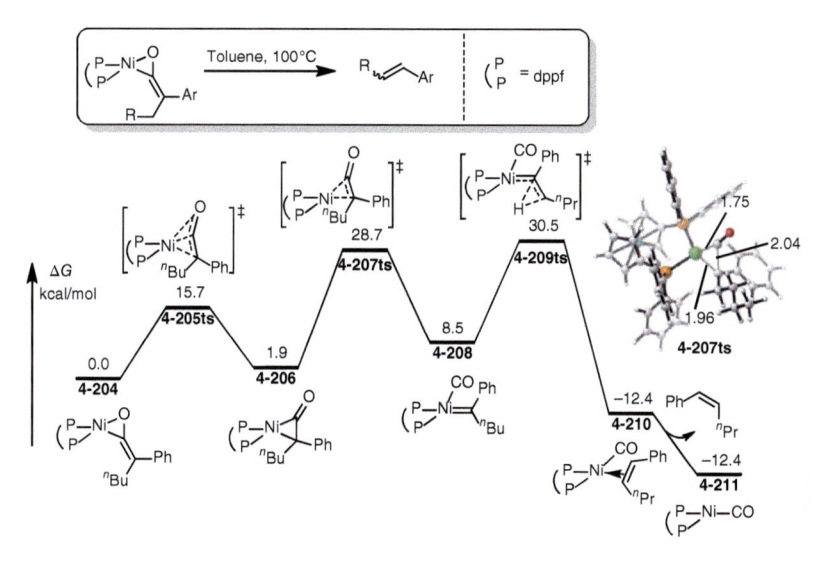

Figure 4.25 The free-energy profiles for the Ni-assisted decomposition of ketene to carbon monoxide and olefin. The energies were calculated at B3LYP/SVP/TZVP /PCM(benzene)//B3LYP/SVP/TZVP level of theory.

as a three-centered oxidative addition course. When Ni-carbene complex **4-208** is formed, a 1,2-hydrogen shift forms olefin product via transition state **4-209ts**.

4.6 Ni-Mediated Unsaturated Bond Activation

Mechanistically, the Ni-mediated unsaturated bond activation is rather complicated, which is generally determined by the oxidative state of active Ni species (Scheme 4.10). When a Ni(0) species is used, its reducibility can promote an oxidative cycloaddition of two unsaturated bonds to achieve a five-membered nickelacycle for further transformation (Mode A). In Mode B, the back-donation in the coordination of unsaturated bonds onto Ni(0) species would further increase the electron density of unsaturated bonds; therefore, an electrophilic attack would occur to perform an outer-sphere addition. In another case (Mode C), when Ni(II) species is used in unsaturated bond activation, the coordination of those compounds would lead to a migratory insertion of unsaturated bond. Moreover (Mode D), the coordination of unsaturated bond onto a cationic Ni(II) species would increase its electrophilicity; therefore, a nucleophilic attack is also possible.

4.6.1 Oxidative Cyclization with Unsaturated Bonds

A wildly accepted mechanism of Ni-mediated two component unsaturated compounds activation is shown in Scheme 4.11. The coordination of two unsaturated compounds would form a bis-π-Ni(0)-complex. The followed oxidative cyclization

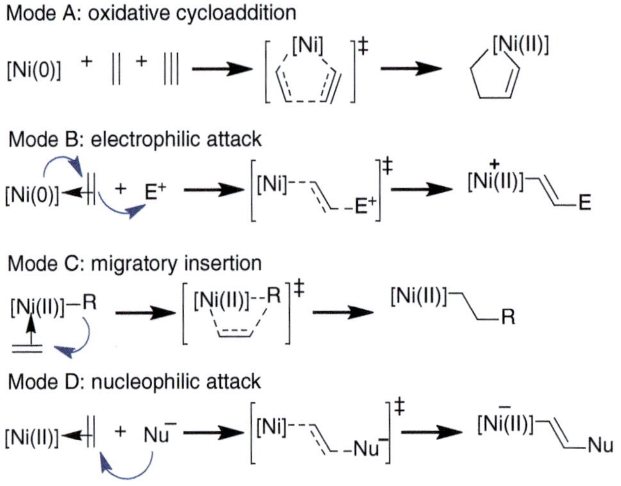

4 Theoretical Study of Ni-Catalysis

Mode A: oxidative cycloaddition

[Ni(0)] + || + ||| ⟶ [[Ni]]‡ ⟶ [Ni(II)]

Mode B: electrophilic attack

[Ni(0)] + E⁺ ⟶ [[Ni]---E⁺]‡ ⟶ [Ni(II)]—E

Mode C: migratory insertion

[Ni(II)]—R ⟶ [[Ni(II)]---R]‡ ⟶ [Ni(II)]—R

Mode D: nucleophilic attack

[Ni(II)] + Nu⁻ ⟶ [[Ni]---Nu⁻]‡ ⟶ [Ni(II)]—Nu

Scheme 4.10 Ni-mediated unsaturated bond activation.

would afford nickela(II)cycle. The further transformations could yield the coupling product of unsaturated compounds.

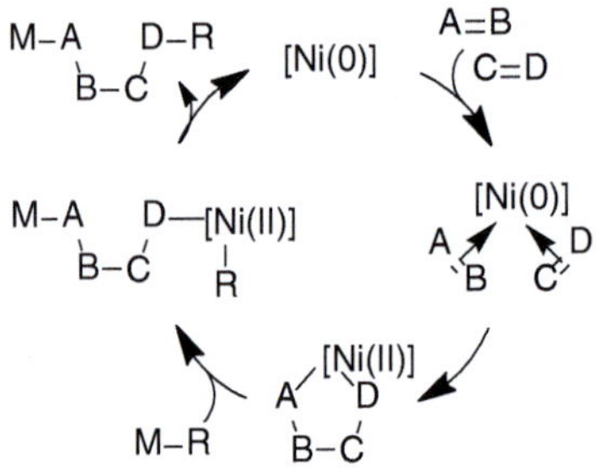

Scheme 4.11 Mechanism of Ni-mediated two-component unsaturated compounds activation.

DFT studies on a model reaction are taken to study the mechanism of Ni-catalyzed three-component addition reaction of enynes and dimethyl zinc [135]. As shown in Figure 4.26, when alkene and alkyne coordinate with Ni(0), the oxidative cyclization can take place via transition state **4-213ts** to form a five-membered nickelacycle **4-214**. The calculated activation free energy of this step is 30.8 kcal/mol. When a Lewis acid additive dimethyl zinc is added, the carbonyl group in alkene can be activated in intermediate **4-215**. The corresponding oxidative cyclization can take place via transition state **4-216ts**. The calculated barrier of this step is decreased by 4.5 kcal/mol.

A DFT calculation taken by Houk group studied the mechanism of Ni(0)-catalyzed alkyne–aldehyde reductive coupling reaction (Figure 4.27) [136]. DFT calculations found that the oxidative cyclization of acetylene and acetaldehyde can take place via transition state **4-219ts** with an energy barrier of 19.7 kcal/mol to afford a

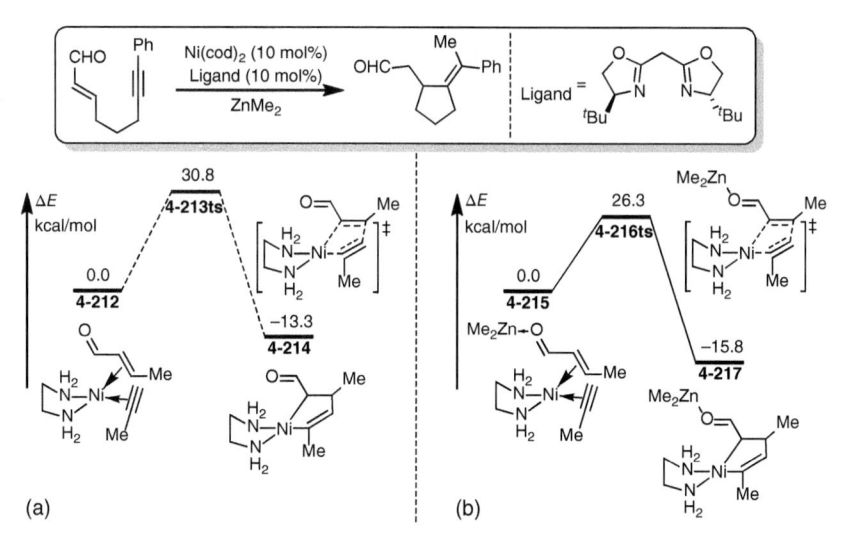

Figure 4.26 The energy profiles for the key step of Ni-mediated enyne cycloaddition in the absence (a) or presence (b) of Lewis acid. The energies were calculated at B3LYP/6-31G(d) level of theory.

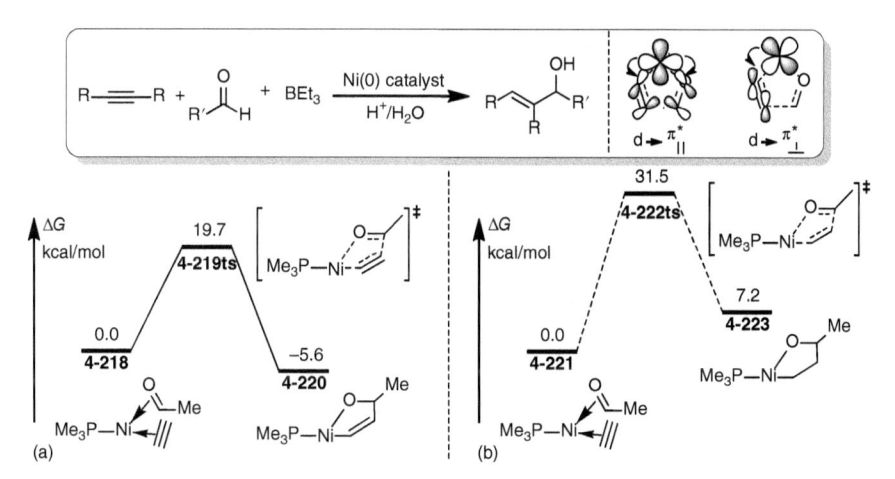

Figure 4.27 The free-energy profiles for the key step of Ni-mediated alkyne–aldehyde reductive coupling: (a) cycloaddition with alkyne, (b) cycloaddition with alkene. The energies were calculated at B3LYP/6-31G(d) (LANL2DZ for Ni) level of theory. Source :Based on McCarren et al. [136].

nickelacycle **4-220**. However, when ethylene is used to instead of acetylene, the barrier of corresponding oxidative cyclization is increased to 31.5 kcal/mol. The energy difference can be explained by the orbital analysis. As shown in the same figure (Figure 4.27), when the oxidative cyclization of acetylene and acetaldehyde occurs through transition state **4-219ts**, a second-order orbital interaction between d orbital in Ni and vertical π^* orbital in acetylene can significantly reduce the reacting barrier. However, this interaction is absent in the oxidative cyclization of ethylene and acetaldehyde.

4.6.2 Electrophilic Addition of Unsaturated Bonds

When an unsaturated compound coordinates with Ni(0) species, the back-donation bond would increase the electron density of that unsaturated bond, which would leads to an outer-sphere electrophilic attack. Proton is a frequently used electrophile in Ni(0)-mediated unsaturated bond activation process. The mechanism of Ni(0)-catalyzed hydroalkoxylation is studied by DFT calculation, whose results are summarized in Figure 4.28 [137]. When butadiene coordinates with Ni(0) in complex **4-224**, an intermolecular hydrogen transfer from solvated methanol to double bond can take place via transition state **4-225ts** with an energy barrier of 13.5 kcal/mol to give an cationic allylic Ni(II) complex **4-226** and a methoxide. The computed geometry of **4-225ts** shows the transferring hydrogen to be approximately midway between oxygen and carbon. Then, a nucleophilic attack of methoxide onto allyl group takes place via transition state **4-227ts** to yield hydroalkoxylation product.

In another example, DFT study is carried out to reveal the mechanism of Ni-catalyzed hydrocarboxylation of alkynes (Figure 4.29) [138]. The calculated free-energy profiles show that the out-sphere hydrogen transfer can occur via transition state **4-231ts** with an energy barrier of 17.4 kcal/mol to form a vinyl Ni(II) intermediate **4-232**. A sequential carbonyl migratory insertion takes place via transition state **4-233ts** to form an acyl Ni(II) intermediate **4-234**. The following reductive elimination of acyl and formate groups yields anhydride product and regenerates Ni(0) species **4-229**.

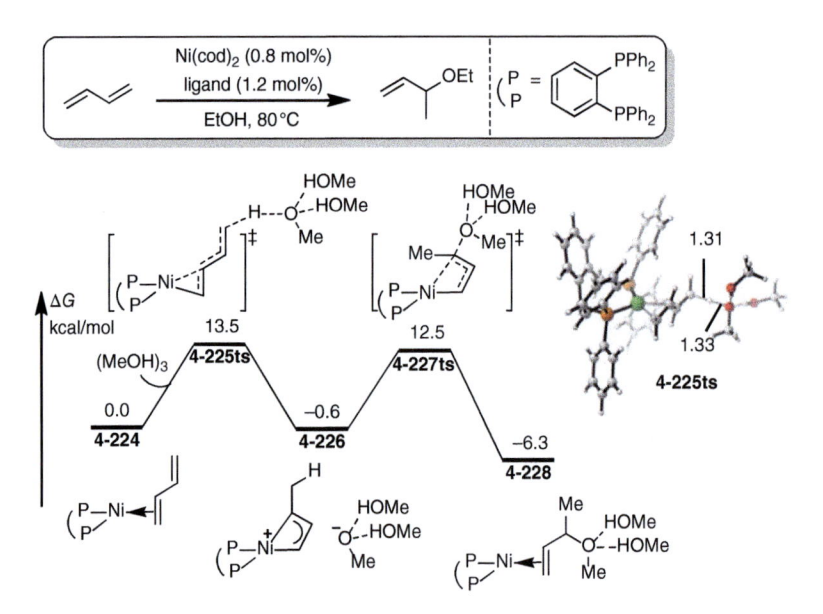

Figure 4.28 The free-energy profiles for the Ni(0)-catalyzed hydroalkoxylation. The energies were calculated at wB97X-D/def2-TZVP/PCM(methanol)//BP86/6-31G(d,p) (SDD for Ni and P) level of theory. Source: Based on Mifleur et al. [137].

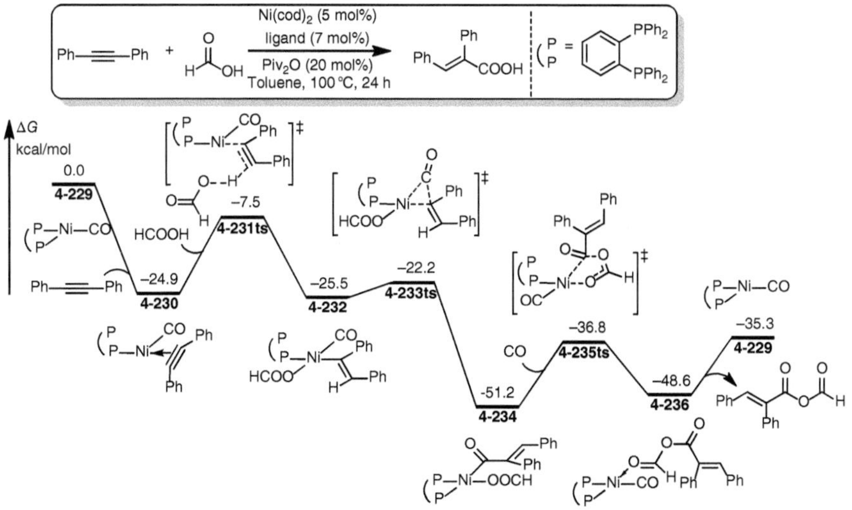

Figure 4.29 The free-energy profiles for the Ni-catalyzed hydrocarboxylation of alkynes. The energies were calculated at B3LYP/6-311G(d,p) (SDD for Ni)/IEF-PCM(toluene) level of theory. Source: Based on Jiang et al. [138].

Moreover, the electrophile can be a silyl group in a Ni(0)-mediated unsaturated compound activation process. As shown in Figure 4.30, DFT calculations are employed to reveal the mechanism of Ni-catalyzed hydrosilylation of allenes [139]. When both silane and allene are coordinated onto Ni(0) in complex **4-237**, the inner-sphere silyl group transfer onto allene can occur via transition state **4-238ts** with an energy barrier of 13.3 kcal/mol. The geometry information of that transition state is shown in Figure 4.30. The lengths of Si—H and Ni—H are 2.42 and 1.51 Å, respectively, which reveals that the Si—H bond is breaking in this transition state with the formation of Ni—H bond. Therefore, an allylic hydride Ni(II) complex **4-239** is formed irreversibly. Finally, a reductive elimination of allyl group and hydride yields hydrosilylation product.

4.6.3 Unsaturated Compounds Insertion

When a Ni(II) species is formed, the coordinated unsaturated compounds, such as ethylene and carbon monoxide, can insert into C—Ni or H—Ni bond leading to the addition of unsaturated bond or atom.

As shown in Figure 4.31, the mechanism of Ni(II)-catalyzed dihydrogenation of olefins is explored by DFT calculations [140]. The calculated free energy profiles show that a metathesis-type gaseous dihydrogen activation takes place via transition state **4-243ts** with an free energy barrier of 23.3 kcal/mol to afford a hydride Ni(II) intermediate **4-244**. The coordination of nitroethene forms complex **4-245** reversibly. Subsequently, a conjugative insertion of nitroethene into Ni—H bond takes place via transition state **4-246ts** with a barrier of only 1.3 kcal/mol. Then, a protonation by acetate acid yields dihydrogenation product and regenerates acetate Ni(II) species **4-242** to accomplish the catalytic cycle.

158 | *4 Theoretical Study of Ni-Catalysis*

Figure 4.30 The free-energy profiles for the Ni-catalyzed hydrosilylation of allenes. The energies were calculated at B3LYP/6-31G(d,p) (LANL2DZ(f) for Ni and Si)/CPCM(THF) //B3LYP/6-31G(d,p) (LANL2DZ(f) for Ni and Si) level of theory.

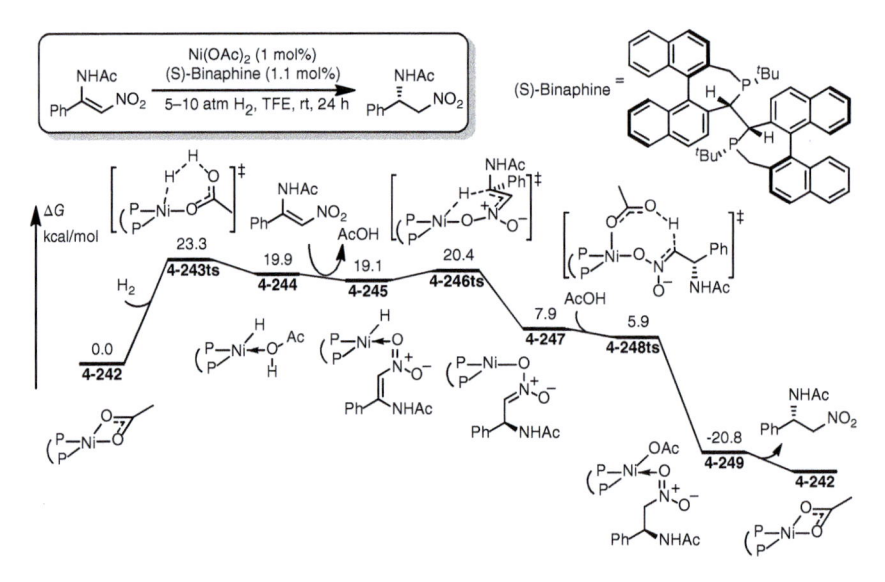

Figure 4.31 The free-energy profiles for the Ni(II)-catalyzed dihydrogenation of olefins. The energies were calculated at M06L-D3/6-31G(d)/SMD(TFE)//M06L-D3/6-31G(d) level of theory. TFE; trifluoroethanol.

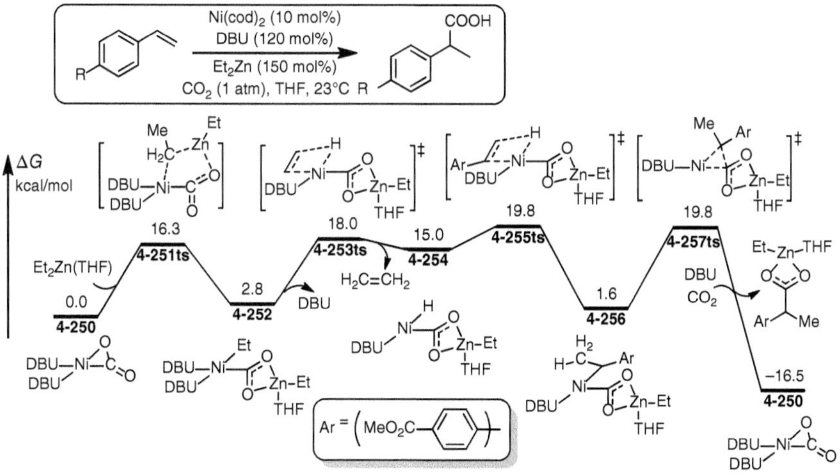

Figure 4.32 The free-energy profiles for the Ni-catalyzed reductive carboxylation of styrenes using CO_2. The energies were calculated at B3LYP/6-31G(d,p) (LANL2DZ for Ni, and 6-311+G(d) for O)/PCM(TFE) level of theory.

Insertion plays a key step in Ni-catalyzed reductive carboxylation of styrenes using CO_2. In this reaction (Figure 4.32), a nickel hydride pathway is proposed and supported by DFT calculations [141]. The first step is transmetallation between the CO_2-coordinated Ni(0) species **4-250** and $Et_2Zn(THF)$ to give a square-planar ethyl Ni(II) species **4-252** via transition state **4-251ts** with an energy barrier of 16.3 kcal/mol. In ethyl Ni(II) species **4-252**, β-hydride elimination occurs via transition state **4-253ts** to give a hydride Ni(II) intermediate **4-254**. Then, a styrene insertion into Ni—H bond occurs via transition state **4-255ts**, which is considered to be the rate-determining step. After this step, an alkyl Ni(II) intermediate **4-256** is formed. Finally, reductive elimination generates the C—C bond in carboxylic zinc. Ni(0) species **4-250** can be regenerated by the coordination of CO_2.

4.6.4 Nucleophilic Addition of Unsaturated Bonds

The coordination of unsaturated bond onto a Ni(II) species can increase its electrophilicity, which would lead to a nucleophilic attack onto coordinated unsaturated bonds. The mechanism of Ni-catalyzed hydroamination of olefins is studied by DFT calculations [65]. As shown in Figure 4.33, the coordination of acrylonitrile onto a cationic Ni(II) species **4-258** leads to the formation of complex **4-259**, in which the C=C double bond is activated by cationic Ni(II). A outer-sphere nucleophilic attack of amine takes place via transition state **4-260ts** with an energy barrier of only 8.8 kcal/mol. The geometry information of transition state **4-260ts** shows that the length of forming C—N bond is 2.50 Å. After this step, a vinylideneamide Ni(II) species **4-261** is formed. The deprotonation of ammonium group takes place via transition state **4-262ts** to form a neutral intermediate **4-263**. Then, a protonation yields hydroamination product. In the whole catalytic cycle of hydroamination, the oxidative state of Ni remains at +2.

160 | *4 Theoretical Study of Ni-Catalysis*

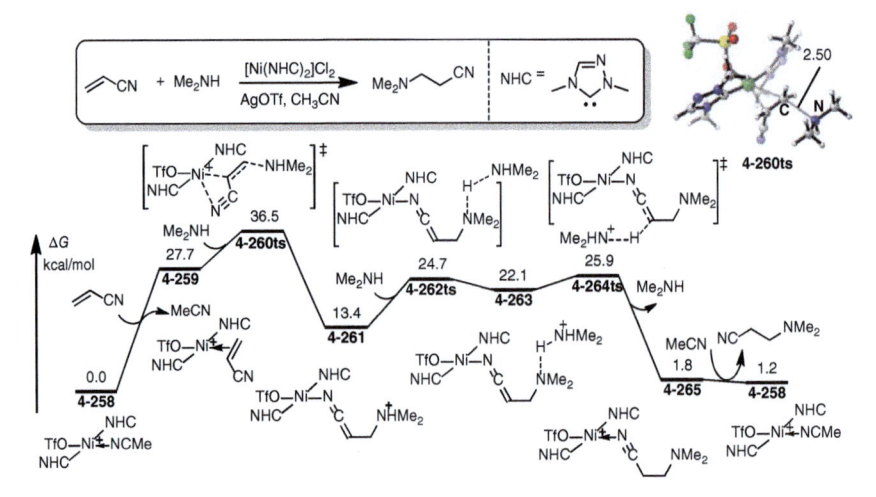

Figure 4.33 The free-energy profiles for the Ni-catalyzed hydroamination of olefins. The energies were calculated at B3LYP/TZVP/PCM(acetonitrile)//B3LYP/6-31G(d) (LANL2DZ for Ni) level of theory.

4.7 Ni-Mediated Cyclization

In Ni-catalysis, a Ni—C bond can be easily formed, which can be further transferred to other C—C bonds. Therefore, it provides a powerful tool for the construction of organic rings [142–153]. As shown in Scheme 4.12, three types of cyclizations in Ni-catalysis are studied, including cycloadditions, substitutions of organic ring, and ring extensions. When linear organic compounds are used to synthesize ring-type product through addition reactions, it is named cycloadditions. In a substitution of organic ring, a part of components in an organic ring is substituted by other organic compounds. When the forming organic ring is larger than that in substrate, it is a ring extension.

Scheme 4.12 Ni-mediated cyclizations. (a) Annulations, (b) ring substitutions, and (c) ring extensions.

4.7.1 Ni-Mediated Cycloadditions

As shown in Scheme 4.13, a common process of Ni-mediated annulations starts from a Ni(0) active species, which is coordinated by two unsaturated compounds, such as alkenes and/or alkynes. The oxidative cyclization of the coordinated unsaturated compounds with Ni(0) generates a nickela(II)cycle. After some further transformation of Ni—C bonds, the reductive elimination yields ring-type products with the regeneration of Ni(0) species.

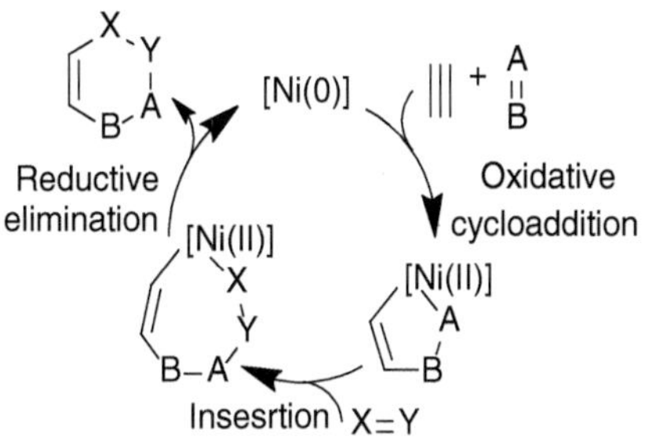

Scheme 4.13 The common mechanism of Ni-catalyzed cycloadditions.

DFT calculation is used to study the mechanism of a Ni-catalyzed tetramerization of acetylenes for the preparation of cyclooctatetraene (Figure 4.34) [154]. The reaction starts from a diacetrylene-coordinated Ni(0) species **4-266**, which can coordinate with another acetylene to afford a triacetrylene-coordinated Ni(0) intermediate **4-267** with 4.1 kcal/mol endergonic. The oxidative cyclization takes place via transition state **4-268ts** with an overall activation free energy of 20.0 kcal/mol to form a five-membered nickela(II)cycle **4-269**. Subsequently, the coordination of acetylene leads to an alkyne insertion into C(vinyl)—Ni bond to form a seven-membered nickelacycle **4-271**. The calculated barrier of this step is only 6.2 kcal/mol through transition state **4-270ts**. Then, another acetylene coordination followed by insertion forms a nickelacycle **4-273**. A sequential reductive elimination forms cyclooctatetraene coordinated Ni(0) species **4-274** without an intrinsic barrier.

In another Ni-catalysis to construct macrocyclic compounds, (1Z,4Z,7E)-cyclodeca-1,4,7-triene can be synthesized by the cycloaddition of butadienes and acetylenes (Figure 4.35) [155]. The DFT calculations found that when two molecules of butadiene are coordinated onto Ni(0) in species **4-275**, an oxidative cyclization of two terminal carbons of butadienes takes place via transition state **4-276ts** with an energy barrier of 13.8 kcal/mol to form an allylic Ni(II) intermediate **4-277**. Then, a ligand exchange with acetylene followed by an alkyne insertion occurs via transition state **4-278ts** to form an allylic vinyl–Ni(II) intermediate **4-279**. Then, a reductive elimination yields cyclodecatriene product and regenerates Ni(0) species **4-275**. The calculated rate-determining step of the whole catalytic cycle is the alkyne insertion step with an activation free energy of 25.5 kcal/mol.

In the mechanism study of Ni(0)-carbene catalyzed intramolecular (5+2) cycloadditions of enynes (Figure 4.36) [156], the computational results showed that the

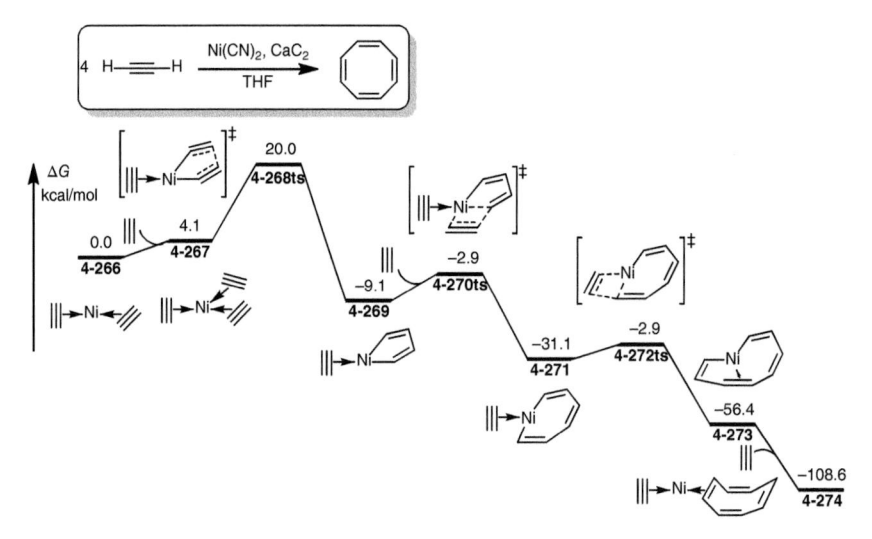

Figure 4.34 The free-energy profiles for the Ni-catalyzed tetramerization of acetylenes. The energies were calculated at B3LYP/6-311+G(2d,p) (SDD for Ni)/PCM(THF)//B3LYP/6-31G(d) (SDD for Ni) level of theory. Source: Based on Straub and Gollub [154].

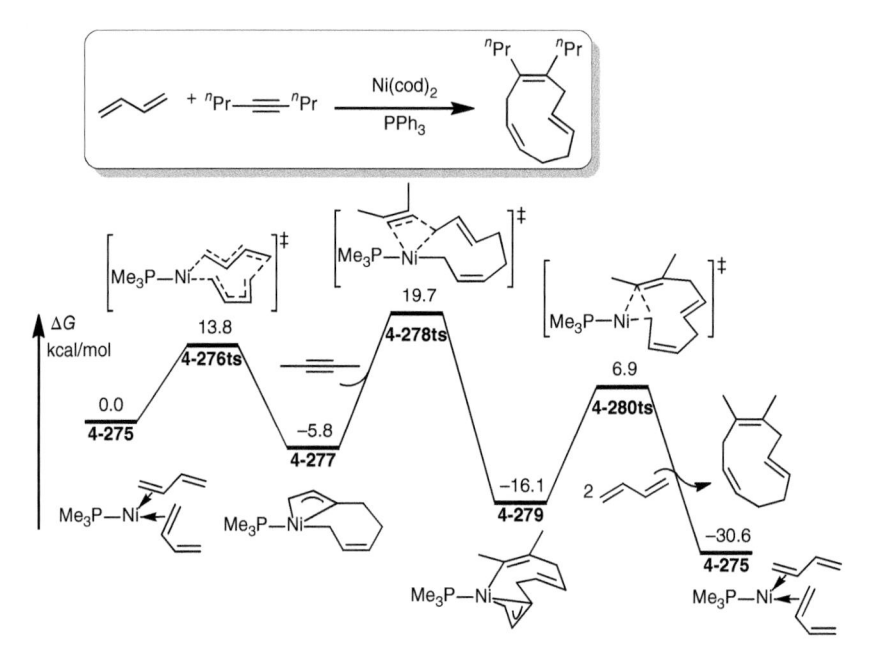

Figure 4.35 The free-energy profiles for the Ni-catalyzed cycloaddition of dienes and alkynes. The energies were calculated at M06/6-311+G(2d,p) (SDD for Ni)//B3LYP/6-31G(d) (SDD for Ni) level of theory. Source: Based on Hong et al. [155].

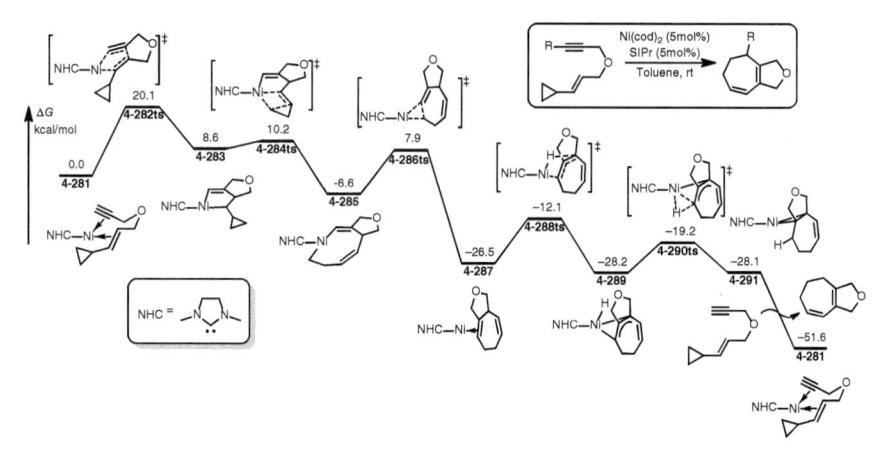

Figure 4.36 The free-energy profiles for the Ni(0)-carbene catalyzed intramolecular (5+2) cycloadditions of enynes. The energies were calculated at B3LYP-D3/6-311+G(2d,p) (SDD for Ni)/CPCM(toluene)//B3LYP/6-31G(d) (SDD for Ni) level of theory. Source: Based on Hong et al. [156].

reaction starts from an intramolecular oxidative cyclization of coordinated enyne in complex **4-281** to afford a nickela(II)cycle **4-283**. The calculated energy barrier of this step is 20.1 kcal/mol through transition state **4-282ts**. When nickela(II)cycle **4-283** is formed, a rapid ring open of cyclopropyl group takes place via β-carbon elimination transition state **4-284ts** with a barrier of only 1.6 kcal/mol to irreversibly afford an eight-centered nickelacycle **4-285**. Then, a reductive elimination leads to the formation of a cyclohepta-1,4-diene coordinated Ni(0) intermediate **4-287**, which can be isomerized to more stable isomer cyclohepta-1,3-diene **4-291** in the presence of Ni(0). The computational results suggested that when a larger N-heterocyclic carbene ligand is used, it is favorable for the formation of cycloaddition product.

In a Ni-carbine-catalyzed cycloaddition of diynes and tropone, an 8π insertion is predicted by DFT calculations (Figure 4.37) [157]. In free-energy profiles for the catalytic cycle of this reaction, the coordination of diyne leads to an intramolecular oxidative cyclization to form a nickelacyclopentadiene intermediate **4-294**. The calculated barrier of this step is 13.4 kcal/mol through transition state **4-293ts**. Then, an 8π insertion of tropone takes place via transition state **4-295ts** with an energy barrier of 9.8 kcal/mol to form an enolate Ni(II) intermediate **4-296**. The reductive elimination creates six-membered carbon ring, which coordinates with Ni(0) species in complex **4-298**. The further rearrangement generates the final product.

4.7.2 Ni-Mediated Ring Substitutions

As shown in Scheme 4.14, when an organic ring molecule involves an active covalent bond, the oxidative addition onto Ni(0) species leads to the cleavage of this bond. The sequential elimination releases an unsaturated molecule. After this process, the insertion of another unsaturated molecule followed by reductive elimination reconstructs a more stable organic ring. The process can be considered as a substitution reaction of an organic ring.

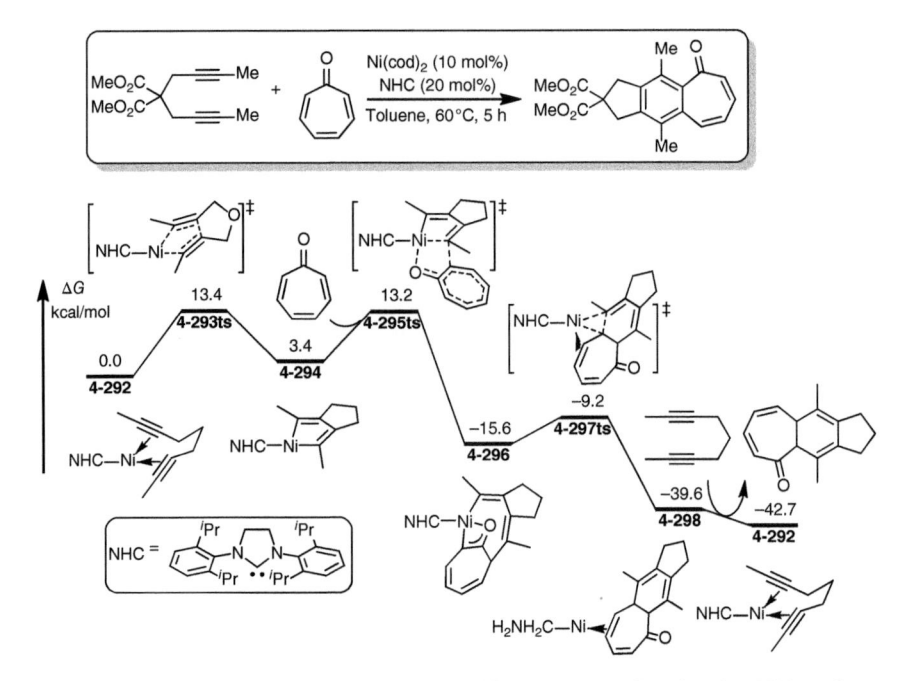

Figure 4.37 The free-energy profiles for the Ni(0)-carbene catalyzed cycloaddition of diynes and tropone. The energies were calculated at M06/6-311+G(2d,p) (SDD for Ni)/CPCM(toluene)//B3LYP/6-31G(d) (SDD for Ni) level of theory. Source: Based on Kumar et al. [157].

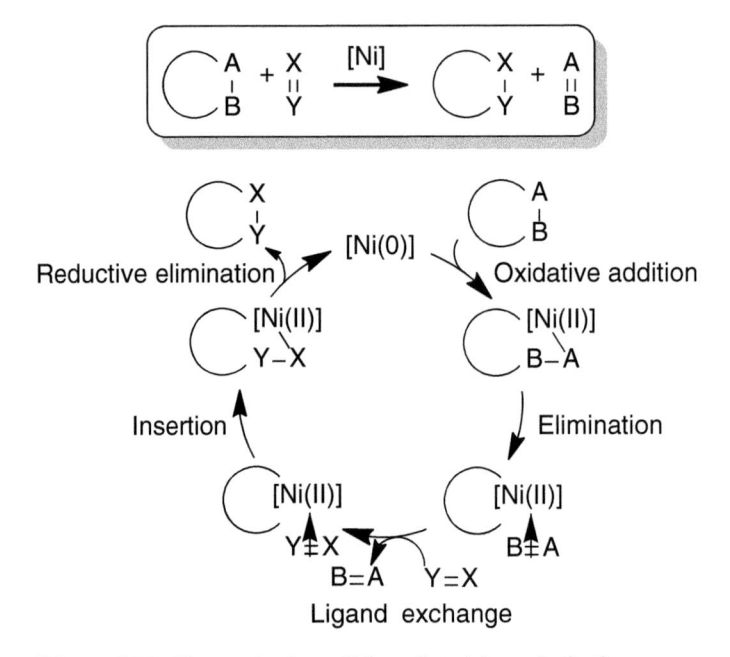

Scheme 4.14 The mechanism of Ni-mediated ring substitutions.

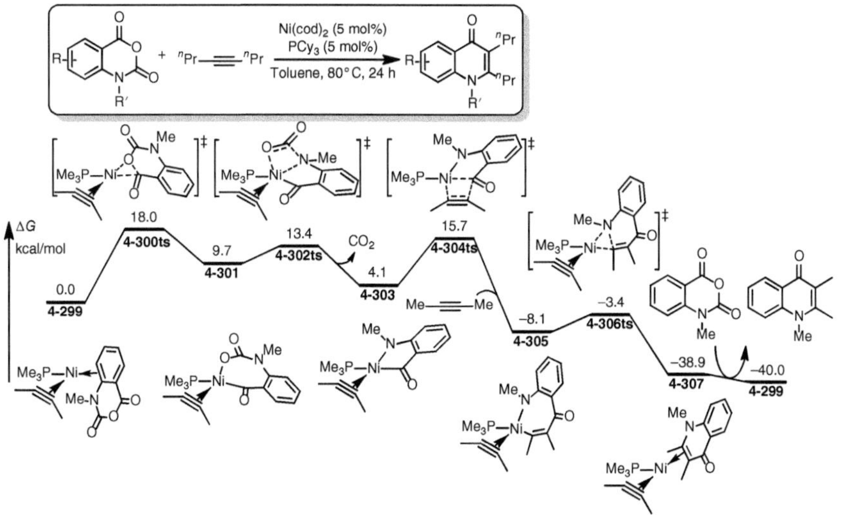

Figure 4.38 The free-energy profiles for the Ni-catalyzed cycloaddition ring substitution of oxazinediones. The energies were calculated at M06/6-311+G(2d,p) (SDD for Ni)/CPCM(toluene)//M06/6-31G(d) (LANL2TZ for Ni)/CPCM(toluene) level of theory. Source: Based on Jiang et al. [138].

As shown in Figure 4.38, when acetylene reacts with oxazinedione in the presence of Ni(0) catalyst, a pyridinone product is observed experimentally [158]. The reaction can be considered as the CO_2 moiety in oxazinedione is replaced by acetylene to afford pyridinone. The detailed mechanism of this reaction is studied by DFT calculations [159]. When oxazinedione coordinates with Ni(0) in intermediate **4-299**, the oxidative addition of C—O bond onto Ni takes place via a three-membered ring-type transition state **4-300ts** with an energy barrier of 18.0 kcal/mol to afford an acyl Ni(II) intermediate **4-301**. Then, a β-amino elimination takes place via transition state **4-302ts** to release a gaseous CO_2 and afford a nickela(II)cycle **4-303**. Subsequently, a migratory insertion of coordinated alkyne into C(acyl)—Ni bond forms vinyl Ni(II) intermediate **4-305** via transition state **4-304ts** with an energy barrier of 11.6 kcal/mol. Then, a C(vinyl)—N bond elimination takes place via transition state **4-306ts** with the coordination of another molecular acetylene. The ligand exchange with reacting oxazinedione can regenerate complex **4-299** with the release of pyridinone product.

In the Ni(0)-catalyzed reaction of phthalic anhydride and acetylene, after release of a gaseous CO, an isochromenone product is observed experimentally (Figure 4.39) [160]. A DFT study is carried out to reveal the mechanism of the substitution of CO by acetylene in phthalic anhydride [161]. The catalytic cycle starts from an oxidative addition of anhydride onto Ni(0) through transition state **4-309ts** with an energy barrier of only 5.1 kcal/mol to afford a nickela(II)cycle **4-310**. Then, a decarbonylation takes place via transition state **4-311ts** with an energy barrier of 13.7 kcal/mol to afford a phenyl Ni(II) intermediate **4-312**. Subsequently, alkyne insertion into C(aryl)—Ni bond occurs via transition state **4-313ts** to form a vinyl

Figure 4.39 The free-energy profiles for the Ni-catalyzed cycloaddition ring substitution of phthalic anhydrides. The energies were calculated at B3LYP/6-31G(d) (LANL2TZ(f) for Ni)/CPCM(acetonitrile)//B3LYP/6-31G(d) (LANL2TZ(f) for Ni) level of theory. Source: Based on Kajita et al. [160].

Ni(II) intermediate **4-314**. The reductive elimination yields isochromenone product and regenerates Ni(0) species. The rate-determining step of the whole catalytic cycle is considered to be the alkyne insertion process with an overall activation free energy of 26.1 kcal/mol.

4.7.3 Ni-Mediated Ring Extensions

In a Ni-mediated ring extension of a carbon ring (Scheme 4.15), the strain drives the oxidative addition of the carbon ring onto Ni(0) leading to the open of that ring.

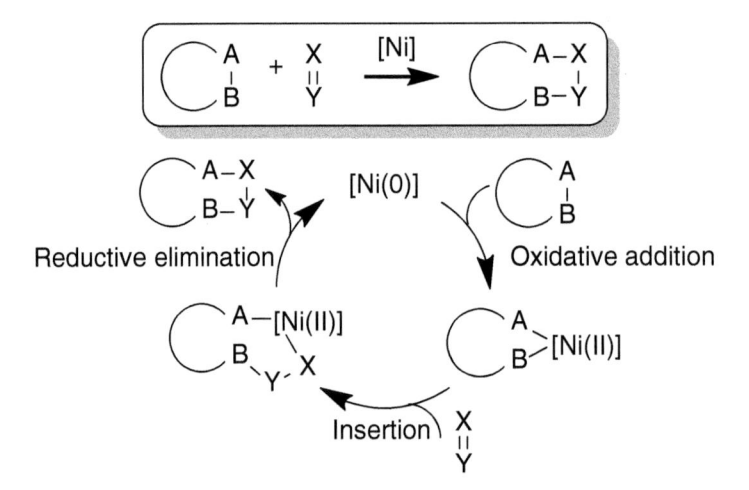

Scheme 4.15 The mechanism of Ni-mediated ring extensions.

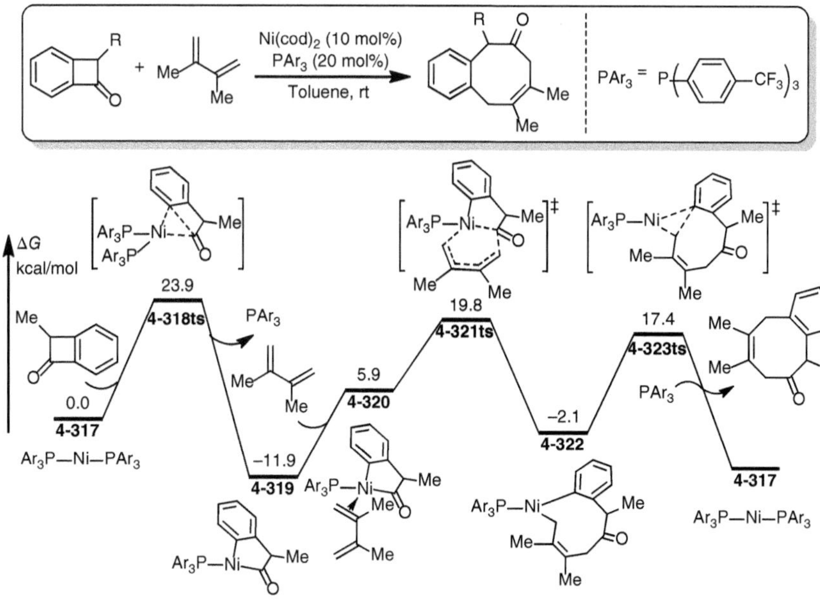

Figure 4.40 The free-energy profiles for the Ni-catalyzed ring extension of cyclobutanones. The energies were calculated at B3LYP/6-311G(d,p) (SDD(f) for Ni)/SMD(toluene)// B3LYP/6-31G(d) (LANL2TZ(f) for Ni) level of theory.

Then, an unsaturated compound insertion followed by reductive elimination regenerates a more stable larger ring to exclude ring strain.

As shown in Figure 4.40, in the presence of Ni-catalyst, cyclobutanone can react with diene to afford a ring-extended cyclooctenone product [162]. The details of the mechanism for this reaction are studied by DFT calculation. When cyclobutanone coordinates with Ni(0) **4-317**, an oxidative addition of C(aryl)—C(acyl) bond onto Ni(0) takes place via transition state **4-318ts** to afford a nickela(II)cycle **4-319**, which can be coordinated by diene to afford intermediate **4-320**. A 1,4-insertion of conjugated diene takes place via transition state **4-321ts** with an overall activation free energy of 31.7 kcal/mol to afford a metallacycle **4-322**. A rapid reductive elimination yields cyclooctenone product and regenerates Ni(PAr$_3$)$_2$ species **4-317** by the coordination of extra ligand PAr$_3$. In another DFT study, the calculated barrier for the diene insertion is considered to be only 25.7 kcal/mol, in which a different type of insertion transition state is observed theoretically.

The ring extension of vinylcyclopropane can be facilitated by Ni(0) catalyst. As shown in Figure 4.41, the mechanism of this reaction is revealed by DFT calculation [163]. Due to the strain of cyclopropane, the oxidative addition of C—C bond onto Ni(0) can take place via transition state **4-325ts** to afford a four-centered nickelacycle **4-326**, which can be isomerized to a six-centered nickelacycle **4-328** via transition state **4-327ts**. The reductive elimination affords cyclopentane coordinated Ni(0) species **4-330**. The ligand exchange yields the ring extended product and loads another molecular vinylcyclopropane substrate.

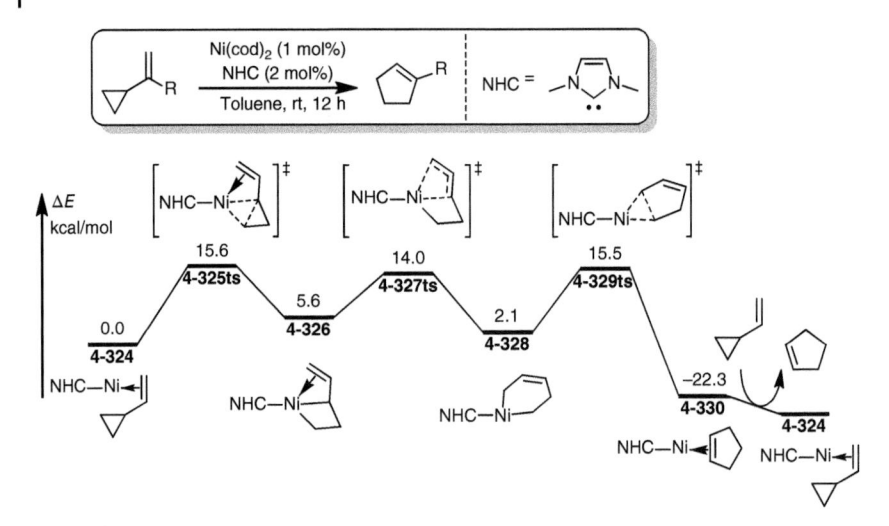

Figure 4.41 The free-energy profiles for the Ni-catalyzed ring extension of vinylcyclopropane. The energies were calculated at B3LYP/SZVP2+//B3LYP/LANL2DZ level of theory.

References

1. Ananikov, V.P. (2015). Nickel: the "Spirited Horse" of transition metal catalysis. *ACS Catalysis* 5 (3): 1964–1971.
2. Bullock, R.M. and Helm, M.L. (2015). Molecular electrocatalysts for oxidation of hydrogen using earth-abundant metals: shoving protons around with proton relays. *Accounts of Chemical Research* 48 (7): 2017–2026.
3. Hoshimoto, Y., Ohashi, M., and Ogoshi, S. (2015). Catalytic transformation of aldehydes with nickel complexes through η^2 coordination and oxidative cyclization. *Accounts of Chemical Research* 48 (6): 1746–1755.
4. Jana, R., Pathak, T.P., and Sigman, M.S. (2011). Advances in transition metal (Pd, Ni, Fe)-catalyzed cross-coupling reactions using alkyl-organometallics as reaction partners. *Chemical Reviews* 111 (3): 1417–1492.
5. Kurahashi, T. and Matsubara, S. (2015). Nickel-catalyzed reactions directed toward the formation of heterocycles. *Accounts of Chemical Research* 48 (6): 1703–1716.
6. Li, Z. and Liu, L. (2015). Recent advances in mechanistic studies on Ni catalyzed cross-coupling reactions. *Chinese Journal of Catalysis* 36 (1): 3–14.
7. Standley, E.A., Tasker, S.Z., Jensen, K.L. et al. (2015). Nickel catalysis: synergy between method development and total synthesis. *Accounts of Chemical Research* 48 (5): 1503–1514.
8. Su, B., Cao, Z.C., and Shi, Z.J. (2015). Exploration of earth-abundant transition metals (Fe, Co, and Ni) as catalysts in unreactive chemical bond activations. *Accounts of Chemical Research* 48 (3): 886–896.
9. Tamao, K., Sumitani, K., and Kumada, M. (1972). Selective carbon–carbon bond formation by cross-coupling of Grignard reagents with organic halides.

Catalysis by nickel–phosphine complexes. *Journal of the American Chemical Society* 94 (12): 4374–4376.

10 Tasker, S.Z., Standley, E.A., and Jamison, T.F. (2014). Recent advances in homogeneous nickel catalysis. *Nature* 509 (7500): 299–309.

11 Thakur, A. and Louie, J. (2015). Advances in nickel-catalyzed cycloaddition reactions to construct carbocycles and heterocycles. *Accounts of Chemical Research* 48 (8): 2354–2365.

12 Tobisu, M. and Chatani, N. (2015). Cross-couplings using aryl ethers via C—O bond activation enabled by nickel catalysts. *Accounts of Chemical Research* 48 (6): 1717–1726.

13 Tollefson, E.J., Hanna, L.E., and Jarvo, E.R. (2015). Stereospecific nickel-catalyzed cross-coupling reactions of benzylic ethers and esters. *Accounts of Chemical Research* 48 (8): 2344–2353.

14 Wang, C., Luo, L., and Yamamoto, H. (2016). Metal-catalyzed directed regio- and enantioselective ring-opening of epoxides. *Accounts of Chemical Research* 49 (2): 193–204.

15 Wilke, G. (1988). Contributions to organo-nickel chemistry. *Angewandte Chemie, International Edition* 27 (1): 185–206.

16 Au, C.-T., Liao, M.-S., and Ng, C.-F. (1998). A detailed theoretical treatment of the partial oxidation of methane to syngas on transition and coinage metal (M) catalysts (M = Ni, Pd, Pt, Cu). *Journal of Physical Chemistry A* 102 (22): 3959–3969.

17 Bellina, F. and Rossi, R. (2010). Transition metal-catalyzed direct arylation of substrates with activated sp^3-hybridized C—H bonds and some of their synthetic equivalents with aryl halides and pseudohalides. *Chemical Reviews* 110 (2): 1082–1146.

18 Butschke, B. and Schwarz, H. (2010). Mechanistic study on the gas-phase generation of "Rollover"-cyclometalated [M(bipy − H)]+(M = Ni, Pd, Pt)†. *Organometallics* 29 (22): 6002–6011.

19 Butschke, B. and Schwarz, H. (2011). Thermal C—H bond activation of benzene, toluene, and methane with cationic [M(X)(bipy)]+(M = Ni, Pd, Pt; X = CH$_3$, Cl; bipy = 2,2′-bipyridine): a mechanistic study. *Organometallics* 30 (6): 1588–1598.

20 Granville, S.L., Welch, G.C., and Stephan, D.W. (2012). Ni, Pd, Pt, and Ru complexes of phosphine-borate ligands. *Inorganic Chemistry* 51 (8): 4711–4721.

21 He, J., Wasa, M., Chan, K.S.L. et al. (2017). Palladium-catalyzed transformations of alkyl C—H bonds. *Chemical Reviews* 117 (13): 8754–8786.

22 Huang, L., Arndt, M., Goossen, K. et al. (2015). Late transition metal-catalyzed hydroamination and hydroamidation. *Chemical Reviews* 115 (7): 2596–2697.

23 Labinger, J.A. (2017). Platinum-catalyzed C–H functionalization. *Chemical Reviews* 117 (13): 8483–8496.

24 Petrone, D.A., Ye, J., and Lautens, M. (2016). Modern transition-metal-catalyzed carbon–halogen bond formation. *Chemical Reviews* 116 (14): 8003–8104.

25 Ruiz-Castillo, P. and Buchwald, S.L. (2016). Applications of palladium-catalyzed C–N cross-coupling reactions. *Chemical Reviews* 116 (19): 12564–12649.

26 Vasu, V., Kim, J.-S., Yu, H.-S. et al. (2018). *Toward Butadiene-ATRP with Group 10 (Ni, Pd, Pt) Metal Complexes*, vol. 1284, 205–225. ACS Publications.

27 Ahrens, T., Kohlmann, J., Ahrens, M. et al. (2015). Functionalization of fluorinated molecules by transition-metal-mediated C—F bond activation to access fluorinated building blocks. *Chemical Reviews* 115 (2): 931–972.

28 Cherney, A.H., Kadunce, N.T., and Reisman, S.E. (2015). Enantioselective and enantiospecific transition-metal-catalyzed cross-coupling reactions of organometallic reagents to construct C—C bonds. *Chemical Reviews* 115 (17): 9587–9652.

29 Correa, A. and Martin, R. (2014). Ni-catalyzed direct reductive amidation via C—O bond cleavage. *Journal of the American Chemical Society* 136 (20): 7253–7256.

30 Ding, D., Lan, Y., Lin, Z. et al. (2019). Synthesis of gem-difluoroalkenes by merging Ni-catalyzed C—F and C—C bond ctivation in cross-electrophile coupling. *Organic Letters* 21 (8): 2723–2730.

31 Gao, M., Sun, D., and Gong, H. (2019). Ni-catalyzed reductive C—O bond arylation of oxalates derived from α-hydroxy esters with aryl halides. *Organic Letters* 21 (6): 1645–1648.

32 Gu, Y. and Martin, R. (2017). Ni-catalyzed stannylation of aryl esters via C—O bond cleavage. *Angewandte Chemie International Edition* 56 (12): 3187–3190.

33 Honeycutt, A.P. and Hoover, J.M. (2017). Nickel-catalyzed oxidative decarboxylative (hetero)arylation of unactivated C—H bonds: Ni and Ag synergy. *ACS Catalysis* 7 (7): 4597–4601.

34 Jiang, C., Lu, H., Xu, W.-H. et al. (2019). Ni-catalyzed 1,2-acyl migration reactions triggered by C—C bond activation of ketones. *ACS Catalysis* 10 (3): 1947–1953.

35 Keen, A.L. and Johnson, S.A. (2006). Nickel(0)-catalyzed isomerization of an aryne complex: formation of a dinuclear Ni(I) complex via C—H rather than C—F bond activation. *Journal of the American Chemical Society* 128 (6): 1806–1807.

36 Korch, K.M. and Watson, D.A. (2019). Cross-coupling of heteroatomic electrophiles. *Chemical Reviews* 119 (13): 8192–8228.

37 Liu, Y.H., Xia, Y.N., and Shi, B.F. (2020). Ni-catalyzed chelation-assisted direct functionalization of inert C—H bonds. *Chinese Journal of Chemistry* 38 (6): 635–662.

38 Rosen, B.M., Quasdorf, K.W., Wilson, D.A. et al. (2011). Nickel-catalyzed cross-couplings involving carbon–oxygen bonds. *Chemical Reviews* 111 (3): 1346–1416.

39 Yin, Y., Yue, X., Zhong, Q. et al. (2018). Ni-catalyzed C—F bond functionalization of unactivated aryl fluorides and corresponding coupling with oxazoles. *Advanced Synthesis and Catalysis* 360 (8): 1639–1643.

40 Bernardi, F., Bottoni, A., and Rossi, I. (1998). A DFT investigation of ethylene dimerization catalyzed by Ni(0) complexes. *Journal of the American Chemical Society* 120 (31): 7770–7775.

41 Chan, B., Luo, Y., and Kimura, M. (2018). Mechanism for three-component Ni-catalyzed carbonyl–ene reaction for CO_2 transformation: what practical lessons do we learn from DFT modelling? *Australian Journal of Chemistry* 71 (4): 272.

42 Davies, D.L., Macgregor, S.A., and McMullin, C.L. (2017). Computational studies of carboxylate-assisted C–H activation and functionalization at group 8-10 transition metal centers. *Chemical Reviews* 117 (13): 8649–8709.

43 Hong, X., Liang, Y., and Houk, K.N. (2014). Mechanisms and origins of switchable chemoselectivity of Ni-catalyzed C(aryl)–O and C(acyl)–O activation of aryl esters with phosphine ligands. *Journal of the American Chemical Society* 136 (5): 2017–2025.

44 Hou, C., Li, Y., Zhao, C. et al. (2019). A DFT study of Co(I) and Ni(II) pincer complex-catalyzed hydrogenation of ketones: intriguing mechanism dichotomy by ligand field variation. *Catalysis Science & Technology* 9 (1): 125–135.

45 Huang, C., He, R., Shen, W. et al. (2015). Mechanisms for the synthesis of conjugated enynes from diphenylacetylene and trimethylsilylacetylene catalyzed by a nickel(0) complex: DFT study of ligand-controlled selectivity. *Journal of Molecular Modeling* 21 (5): 135.

46 Jackson, E.P., Malik, H.A., Sormunen, G.J. et al. (2015). Mechanistic basis for regioselection and regiodivergence in nickel-catalyzed reductive couplings. *Accounts of Chemical Research* 48 (6): 1736–1745.

47 Ji, C.L., Xie, P.P., and Hong, X. (2018). Computational study of mechanism and thermodynamics of Ni/IPr-catalyzed amidation of esters. *Molecules* 23 (10): 2618.

48 Jiang, F. and Ren, Q. (2014). Theoretical investigation of the mechanisms of the biphenyl formation in Ni-catalyzed reductive cross-coupling system. *Journal of Organometallic Chemistry* 757: 72–78.

49 Kawamata, Y., Vantourout, J.C., Hickey, D.P. et al. (2019). Electrochemically driven, Ni-catalyzed aryl amination: scope, mechanism, and applications. *Journal of the American Chemical Society* 141 (15): 6392–6402.

50 Li, Z., Jiang, Y.Y., and Fu, Y. (2012). Theoretical study on the mechanism of Ni-catalyzed alkyl–alkyl Suzuki cross-coupling. *Chemistry–A European Journal* 18 (14): 4345–4357.

51 Ren, Q., Jiang, F., and Gong, H. (2014). DFT study of the single electron transfer mechanisms in Ni-catalyzed reductive cross-coupling of aryl bromide and alkyl bromide. *Journal of Organometallic Chemistry* 770: 130–135.

52 Sperger, T., Sanhueza, I.A., Kalvet, I. et al. (2015). Computational studies of synthetically relevant homogeneous organometallic catalysis involving Ni, Pd, Ir, and Rh: an overview of commonly employed DFT methods and mechanistic insights. *Chemical Reviews* 115 (17): 9532–9586.

53 Xu, L., Chung, L.W., and Wu, Y.-D. (2015). Mechanism of Ni–NHC catalyzed hydrogenolysis of aryl ethers: roles of the excess base. *ACS Catalysis* 6 (1): 483–493.

54 Zhang, S.Q., Taylor, B.L.H., Ji, C.L. et al. (2017). Mechanism and origins of ligand-controlled stereoselectivity of Ni-catalyzed Suzuki–Miyaura coupling with benzylic esters: a computational study. *Journal of the American Chemical Society* 139 (37): 12994–13005.

55 Zhang, T., Zhang, X., and Chung, L.W. (2018). Computational insights into the reaction mechanisms of nickel-catalyzed hydrofunctionalizations and nickel-dependent enzymes. *Asian Journal of Organic Chemistry* 7 (3): 522–536.

56 Yamazaki, K., Obata, A., Sasagawa, A. et al. (2018). Computational mechanistic study on the nickel-catalyzed C–H/N–H oxidative annulation of aromatic amides with alkynes: the role of the nickel(0) ate complex. *Organometallics* 38 (2): 248–255.

57 Omer, H.M. and Liu, P. (2017). Computational study of Ni-catalyzed C–H functionalization: factors that control the competition of oxidative addition and radical pathways. *Journal of the American Chemical Society* 139 (29): 9909–9920.

58 Omer, H.M. and Liu, P. (2019). Computational study of the Ni-catalyzed C–H oxidative cycloaddition of aromatic amides with alkynes. *ACS Omega* 4 (3): 5209–5220.

59 Zhang, T., Liu, S., Zhu, L. et al. (2019). Theoretical study of FMO adjusted C–H cleavage and oxidative addition in nickel catalysed C–H arylation. *Communications Chemistry* 2 (1): 31.

60 Guihaumé, J., Halbert, S., Eisenstein, O. et al. (2011). Hydrofluoroarylation of alkynes with Ni catalysts. C–H activation via ligand-to-ligand hydrogen transfer, an alternative to oxidative addition. *Organometallics* 31 (4): 1300–1314.

61 He, Y., Cai, Y., and Zhu, S. (2017). Mild and regioselective benzylic C–H functionalization: Ni-catalyzed reductive arylation of remote and proximal olefins. *Journal of the American Chemical Society* 139 (3): 1061–1064.

62 Rej, S., Ano, Y., and Chatani, N. (2020). Bidentate directing groups: an efficient tool in C—H bond functionalization chemistry for the expedient construction of C—C bonds. *Chemical Reviews* 120 (3): 1788–1887.

63 Wang, Y.X., Qi, S.L., Luan, Y.X. et al. (2018). Enantioselective Ni–Al bimetallic catalyzed exo-selective C–H cyclization of imidazoles with alkenes. *Journal of the American Chemical Society* 140 (16): 5360–5364.

64 Zhan, B., Hu, F., Shi, B. et al. (2015). Recent progress on nickel-catalyzed direct functionalization of unactivated C—H bonds. *Chinese Science Bulletin* 60 (31): 2907–2917.

65 Kumar, R., Katari, M., Choudhary, A. et al. (2017). Computational insight into the hydroamination of an activated olefin, As catalyzed by a 1,2,4-triazole-derived nickel(II) N-heterocyclic carbene complex. *Inorganic Chemistry* 56 (24): 14859–14869.

66 Haines, B.E., Yu, J.Q., and Musaev, D.G. (2018). The mechanism of directed Ni(II)-catalyzed C–H iodination with molecular iodine. *Chemical Science* 9 (5): 1144–1154.

67 Jiang, Y.-Y., Li, Z., and Shi, J. (2012). Mechanistic origin of regioselectivity in nickel-catalyzed olefin hydroheteroarylation through C–H activation. *Organometallics* 31 (11): 4356–4366.
68 Singh, V., Nakao, Y., Sakaki, S. et al. (2017). Theoretical study of nickel-catalyzed selective alkenylation of pyridine: reaction mechanism and crucial roles of Lewis acid and ligands in determining the selectivity. *Journal of Organic Chemistry* 82 (1): 289–301.
69 Yu, H. and Fu, Y. (2012). Mechanistic origin of cross-coupling selectivity in Ni-catalysed Tishchenko reactions. *Chemistry – A European Journal* 18 (52): 16765–16773.
70 Feng, Z., Xiao, Y.L., and Zhang, X. (2018). Transition-metal (Cu, Pd, Ni)-catalyzed difluoroalkylation via cross-coupling with difluoroalkyl halides. *Accounts of Chemical Research* 51 (9): 2264–2278.
71 Guan, H., Zhang, Q., Walsh, P.J. et al. (2020). Nickel/photoredox-catalyzed asymmetric reductive cross-coupling of racemic α-chloro esters with aryl iodides. *Angewandte Chemie International Edition* 132 (13): 5210–5215.
72 Kurandina, D., Chuentragool, P., and Gevorgyan, V. (2019). Transition-metal-catalyzed alkyl Heck-type reactions. *Synthesis* 51 (05): 985–1005.
73 Li, Y., Fan, Y., and Jia, Q. (2019). Recent advance in Ni-catalyzed reductive cross-coupling to construct C(sp^2)—C(sp^2) and C(sp^2)—C(sp^3) bonds. *Chinese Journal of Organic Chemistry* 39 (2): 350.
74 Li, Y., Luo, Y., Peng, L. et al. (2020). Reaction scope and mechanistic insights of nickel-catalyzed migratory Suzuki–Miyaura cross-coupling. *Nature Communications* 11 (1): 417.
75 Mills, L.R., Graham, J.M., Patel, P. et al. (2019). Ni-catalyzed reductive cyanation of aryl halides and phenol derivatives via transnitrilation. *Journal of the American Chemical Society* 141 (49): 19257–19262.
76 Qiu, H., Shuai, B., Wang, Y.Z. et al. (2020). Enantioselective Ni-catalyzed electrochemical synthesis of biaryl atropisomers. *Journal of the American Chemical Society* 142 (22): 9872–9878.
77 Sperger, T., Sanhueza, I.A., and Schoenebeck, F. (2016). Computation and experiment: a powerful combination to understand and predict reactivities. *Accounts of Chemical Research* 49 (6): 1311–1319.
78 Terao, J. and Kambe, N. (2008). Cross-coupling reaction of alkyl halides with grignard reagents catalyzed by Ni, Pd, or Cu complexes with π-carbon ligand(s). *Accounts of Chemical Research* 41 (11): 1545–1554.
79 Wang, K. and Kong, W. (2019). Enantioselective reductive diarylation of alkenes by Ni-catalyzed domino Heck cyclization/cross coupling. *Synlett* 30 (09): 1008–1014.
80 Wu, Y.N., Fu, M.C., Shang, R. et al. (2020). Nickel-catalyzed carboxylation of aryl iodides with lithium formate through catalytic CO recycling. *Chemical Communications* 56 (29): 4067–4069.
81 Yu, D.-G., Li, B.-J., and Shi, Z.-J. (2010). Exploration of new C–O electrophiles in cross-coupling reactions. *Accounts of Chemical Research* 43 (12): 1486–1495.

82 Zhuo, J., Zhang, Y., Li, Z. et al. (2020). Nickel-catalyzed direct acylation of aryl and alkyl bromides with acylimidazoles. *ACS Catalysis* 10 (6): 3895–3903.

83 Bottoni, A., Miscione, G.P., Novoa, J.J. et al. (2003). DFT computational study of the mechanism of allyl halides carbonylation catalyzed by nickel tetracarbonyl. *Journal of the American Chemical Society* 125 (34): 10412–10419.

84 Yin, G., Kalvet, I., Englert, U. et al. (2015). Fundamental studies and development of nickel-catalyzed trifluoromethylthiolation of aryl chlorides: active catalytic species and key roles of ligand and traceless MeCN additive revealed. *Journal of the American Chemical Society* 137 (12): 4164–4172.

85 Zeng, Z., Zhang, T., Yue, X. et al. (2018). Mechanism investigation for anoin-assisted nickel catalyzed C—F bond functionalization reaction: a DFT study. *Scientia Sinica Chimica* 48 (7): 736–742.

86 Phapale, V.B., Guisan-Ceinos, M., Bunuel, E. et al. (2009). Nickel-catalyzed cross-coupling of alkyl zinc halides for the formation of $C(sp^2)$—$C(sp^3)$ bonds: scope and mechanism. *Chemistry – A European Journal* 15 (46): 12681–12688.

87 Cheung, M.S., Sheong, F.K., Marder, T.B. et al. (2015). Computational insight into nickel-catalyzed carbon–carbon versus carbon–boron coupling reactions of primary, secondary, and tertiary alkyl bromides. *Chemistry – A European Journal* 21 (20): 7480–7488.

88 Zhang, X., Liu, Y., Chen, G. et al. (2017). Theoretical insight into $C(sp^3)$—F bond activations and origins of chemo- and regioselectivities of "Tunable" nickel-mediated/-catalyzed couplings of 2-trifluoromethyl-1-alkenes with alkynes. *Organometallics* 36 (19): 3739–3749.

89 Rull, S.G., Funes-Ardoiz, I., Maya, C. et al. (2018). Elucidating the mechanism of aryl aminations mediated by NHC-supported nickel complexes: evidence for a nonradical Ni(0)/Ni(II) pathway. *ACS Catalysis* 8 (5): 3733–3742.

90 Alvarez-Bercedo, P. and Martin, R. (2010). Ni-catalyzed reduction of inert C—O bonds: a new strategy for using aryl ethers as easily removable directing groups. *Journal of the American Chemical Society* 132 (49): 17352–17353.

91 Cao, Z.C., Luo, Q.Y., and Shi, Z.J. (2016). Practical cross-coupling between O-based electrophiles and ryl bromides via Ni catalysis. *Organic Letters* 18 (23): 5978–5981.

92 Guan, B.T., Wang, Y., Li, B.J. et al. (2008). Biaryl construction via Ni-catalyzed C–O activation of phenolic carboxylates. *Journal of the American Chemical Society* 130 (44): 14468–14470.

93 Li, Z., Zhang, S.L., Fu, Y. et al. (2009). Mechanism of Ni-catalyzed selective C—O bond activation in cross-coupling of aryl esters. *Journal of the American Chemical Society* 131 (25): 8815–8823.

94 Lu, Q., Yu, H., and Fu, Y. (2014). Mechanistic study of chemoselectivity in Ni-catalyzed coupling reactions between azoles and aryl carboxylates. *Journal of the American Chemical Society* 136 (23): 8252–8260.

95 Martin, R., Martin-Montero, R., Krolikowski, T. et al. (2017). Stereospecific nickel-catalyzed borylation of secondary benzyl pivalates. *Synlett* 28 (19): 2604–2608.

96 Muto, K., Yamaguchi, J., and Itami, K. (2012). Nickel-catalyzed C–H/C–O coupling of azoles with phenol derivatives. *Journal of the American Chemical Society* 134 (1): 169–172.
97 Xiao, J., Chen, T., and Han, L.B. (2015). Nickel-catalyzed direct C–H/C–O cross couplings generating fluorobenzenes and heteroarenes. *Organic Letters* 17 (4): 812–815.
98 Yan, X., Yang, F., Cai, G. et al. (2018). Nickel(0)-catalyzed inert C—O bond functionalization: organo rare-earth metal complex as the coupling partner. *Organic Letters* 20 (3): 624–627.
99 Yang, J., Xiao, J., Chen, T. et al. (2016). Nickel-catalyzed phosphorylation of phenol derivatives via C–O/P–H cross-coupling. *Journal of Organic Chemistry* 81 (9): 3911–3916.
100 Zarate, C., Nakajima, M., and Martin, R. (2017). A mild and ligand-free Ni-catalyzed silylation via C–OMe cleavage. *Journal of the American Chemical Society* 139 (3): 1191–1197.
101 Sawatlon, B., Wititsuwannakul, T., Tantirungrotechai, Y. et al. (2014). Mechanism of Ni N-heterocyclic carbene catalyst for C—O bond hydrogenolysis of diphenyl ether: a density functional study. *Dalton Transactions* 43 (48): 18123–18133.
102 Schwarzer, M.C., Konno, R., Hojo, T. et al. (2017). Combined theoretical and experimental studies of nickel-catalyzed cross-coupling of methoxyarenes with arylboronic esters via C—O bond cleavage. *Journal of the American Chemical Society* 139 (30): 10347–10358.
103 Wititsuwannakul, T., Tantirungrotechai, Y., and Surawatanawong, P. (2016). Density functional study of nickel N-heterocyclic carbene catalyzed C—O bond hydrogenolysis of methyl phenyl ether: the concerted β–H transfer mechanism. *ACS Catalysis* 6 (3): 1477–1486.
104 Quasdorf, K.W., Antoft-Finch, A., Liu, P. et al. (2011). Suzuki–Miyaura cross-coupling of aryl carbamates and sulfamates: experimental and computational studies. *Journal of the American Chemical Society* 133 (16): 6352–6363.
105 Hu, J., Sun, H., Cai, W. et al. (2016). Nickel-catalyzed borylation of aryl- and benzyltrimethylammonium salts via C—N bond cleavage. *Journal of Organic Chemistry* 81 (1): 14–24.
106 Liu, L.L., Chen, P., Sun, Y. et al. (2016). Mechanism of nickel-catalyzed selective C—N bond activation in Suzuki–Miyaura cross-coupling of amides: a theoretical investigation. *Journal of Organic Chemistry* 81 (23): 11686–11696.
107 Liu, S., Qi, X., Bai, R. et al. (2019). Theoretical study of Ni-catalyzed C–N radical–radical cross-coupling. *Journal of Organic Chemistry* 84 (6): 3321–3327.
108 Nagae, H., Xia, J., Kirillov, E. et al. (2020). Asymmetric allylic alkylation of β-ketoesters via C—N bond cleavage of N-allyl-N-methylaniline derivatives catalyzed by a nickel–diphosphine system. *ACS Catalysis* 10 (10): 5828–5839.
109 Ni, S., Li, C.X., Mao, Y. et al. (2019). Ni-catalyzed deaminative cross-electrophile coupling of Katritzky salts with halides via C horizontal line N bond activation. *Science Advances* 5 (6): eaaw9516.

110 Ni, S., Zhang, W., Mei, H. et al. (2017). Ni-catalyzed reductive cross-coupling of amides with aryl Iodide electrophiles via C—N bond activation. *Organic Letters* 19 (10): 2536–2539.

111 Pulikottil, F.T., Pilli, R., Suku, R.V. et al. (2020). Nickel-catalyzed cross-coupling of alkyl carboxylic acid derivatives with pyridinium salts via C—N bond cleavage. *Organic Letters* 22 (8): 2902–2907.

112 Skhiri, A. and Chatani, N. (2019). Nickel-catalyzed reaction of benzamides with bicylic alkenes: cleavage of C—H and C—N bonds. *Organic Letters* 21 (6): 1774–1778.

113 Walker, J.A. Jr.,, Vickerman, K.L., Humke, J.N. et al. (2017). Ni-catalyzed alkene carboacylation via amide C—N bond activation. *Journal of the American Chemical Society* 139 (30): 10228–10231.

114 Wang, H., Zhang, S.Q., and Hong, X. (2019). Computational studies on Ni-catalyzed amide C—N bond activation. *Chemical Communications* 55 (76): 11330–11341.

115 Yu, C.G. and Matsuo, Y. (2020). Nickel-catalyzed deaminative acylation of activated aliphatic amines with aromatic amides via C—N bond activation. *Organic Letters* 22 (3): 950–955.

116 Yu, H., Gao, B., Hu, B. et al. (2017). Charge-transfer complex promoted C—N bond activation for Ni-catalyzed carbonylation. *Organic Letters* 19 (13): 3520–3523.

117 Yue, H., Zhu, C., Shen, L. et al. (2019). Nickel-catalyzed C—N bond activation: activated primary amines as alkylating reagents in reductive cross-coupling. *Chemical Science* 10 (16): 4430–4435.

118 Ji, C.L. and Hong, X. (2017). Factors controlling the reactivity and chemoselectivity of resonance destabilized amides in Ni-catalyzed decarbonylative and nondecarbonylative Suzuki–Miyaura coupling. *Journal of the American Chemical Society* 139 (43): 15522–15529.

119 Hie, L., Fine Nathel, N.F., Shah, T.K. et al. (2015). Conversion of amides to esters by the nickel-catalysed activation of amide C—N bonds. *Nature* 524 (7563): 79–83.

120 Fan, C., Lv, X.Y., Xiao, L.J. et al. (2019). Alkenyl exchange of allylamines via nickel(0)-catalyzed C—C bond cleavage. *Journal of the American Chemical Society* 141 (7): 2889–2893.

121 Huang, L., Ji, T., and Rueping, M. (2020). Remote nickel-catalyzed cross-coupling arylation via proton-coupled electron transfer-enabled C—C bond cleavage. *Journal of the American Chemical Society* 142 (7): 3532–3539.

122 Kwiatkowski, M.R. and Alexanian, E.J. (2019). Transition-metal (Pd, Ni, Mn)-catalyzed C—C bond constructions involving unactivated alkyl halides and fundamental synthetic building blocks. *Accounts of Chemical Research* 52 (4): 1134–1144.

123 Lv, Y., Pu, W., Niu, J. et al. (2018). nBu$_4$NI-catalyzed C—C bond formation to construct 2-carbonyl-1,4-diketones under mild conditions. *Tetrahedron Letters* 59 (15): 1497–1500.

124 Nakai, K., Kurahashi, T., and Matsubara, S. (2015). Nickel-catalyzed dual C—C σ bond activation to construct carbocyclic skeletons. *Tetrahedron Letters* 71 (26–27): 4512–4517.

125 Ogata, K. and Fukuzawa, S.-i. (2012). Development of nickel-catalyzed three-component reactions via C—H or C—C bond activation. *Journal of Synthetic Organic Chemistry, Japan* 70 (1): 2–10.

126 Pan, B., Wang, C., Wang, D. et al. (2013). Nickel-catalyzed [3 + 2] cycloaddition of diynes with methyleneaziridines via C—C bond cleavage. *Chemical Communications* 49 (44): 5073–5075.

127 Penney, J.M. and Miller, J.A. (2004). Alkynylation of benzonitriles via nickel catalyzed C—C bond activation. *Tetrahedron Letters* 45 (25): 4989–4992.

128 Suginome, M., Matsuda, T., Yoshimoto, T. et al. (2002). Nickel-catalyzed silaboration of small-ring vinylcycloalkanes: regio- and stereoselective (*E*)-allylsilane formation via C—C bond cleavage. *Organometallics* 21 (8): 1537–1539.

129 Zhao, P.P., Wang, Y.C., Sheng, Y. et al. (2017). Theoretical study of Ni$^+$ assisted C—C and C—H bond activations of propionaldehyde in the gas phase. *Computational & Theoretical Chemistry* 1114: 140–145.

130 Zhao, T.T., Xu, W.H., Zheng, Z.J. et al. (2018). Directed decarbonylation of unstrained aryl ketones via nickel-catalyzed C—C bond cleavage. *Journal of the American Chemical Society* 140 (2): 586–589.

131 Nakao, Y., Oda, S., and Hiyama, T. (2004). Nickel-catalyzed arylcyanation of alkynes. *Journal of the American Chemical Society* 126 (43): 13904–13905.

132 Ohnishi, Y.-y., Nakao, Y., Sato, H. et al. (2009). A theoretical study of nickel(0)-catalyzed phenylcyanation of alkynes. Reaction mechanism and regioselectivity. *Organometallics* 28 (8): 2583–2594.

133 Ni, S.-F., Yang, T.-L., and Dang, L. (2017). Transfer hydrocyanation by nickel(0)/Lewis acid cooperative catalysis, mechanism investigation, and computational prediction of shuttle catalysts. *Organometallics* 36 (15): 2746–2754.

134 Staudaher, N.D., Arif, A.M., and Louie, J. (2016). Synergy between experimental and computational chemistry reveals the mechanism of decomposition of nickel–ketene complexes. *Journal of the American Chemical Society* 138 (42): 14083–14091.

135 Hratchian, H.P., Chowdhury, S.K., Gutiérrez-García, V.M. et al. (2004). Combined experimental and computational investigation of the mechanism of nickel-catalyzed three-component addition processes. *Organometallics* 23 (20): 4636–4646.

136 McCarren, P.R., Liu, P., Cheong, P.H. et al. (2009). Mechanism and transition-state structures for nickel-catalyzed reductive alkyne–aldehyde coupling reactions. *Journal of the American Chemical Society* 131 (19): 6654–6655.

137 Mifleur, A., Mérel, D.S., Mortreux, A. et al. (2017). Deciphering the mechanism of the nickel-catalyzed hydroalkoxylation reaction: a combined experimental and computational study. *ACS Catalysis* 7 (10): 6915–6923.

138 Jiang, J., Fu, M., Li, C. et al. (2017). Theoretical investigation on nickel-catalyzed hydrocarboxylation of alkynes employing formic acid. *Organometallics* 36 (15): 2818–2825.

139 Xie, H., Kuang, J., Wang, L. et al. (2017). A DFT study on palladium and nickel-catalyzed regioselective and stereoselective hydrosilylation of 1,3-disubstituted allenes. *Organometallics* 36 (17): 3371–3381.

140 Gao, W., Lv, H., Zhang, T. et al. (2017). Nickel-catalyzed asymmetric hydrogenation of β-acylamino nitroolefins: an efficient approach to chiral amines. *Chemical Science* 8 (9): 6419.

141 Yuan, R. and Lin, Z. (2014). Computational insight into the mechanism of nickel-catalyzed reductive carboxylation of styrenes using CO_2. *Organometallics* 33 (24): 7147–7156.

142 Chen, J., Wang, Y., Ding, Z. et al. (2020). Synthesis of bridged tricyclo[5.2.1.0(1,5)]decanes via nickel-catalyzed asymmetric domino cyclization of enynones. *Nature Communications* 11 (1): 1882.

143 Desrosiers, J.N., Hie, L., Biswas, S. et al. (2016). Construction of quaternary stereocenters by nickel-catalyzed Heck cyclization reactions. *Angewandte Chemie International Edition* 55 (39): 11921–11924.

144 Li, Y., Wang, K., Ping, Y. et al. (2018). Nickel-catalyzed domino Heck cyclization/Suzuki coupling for the synthesis of 3,3-disubstituted oxindoles. *Organic Letters* 20 (4): 921–924.

145 Montgomery, J. (2004). Nickel-catalyzed reductive cyclizations and couplings. *Angewandte Chemie International Edition* 43 (30): 3890–3908.

146 Qin, X., Lee, M.W.Y., and Zhou, J.S. (2017). Nickel-catalyzed asymmetric reductive Heck cyclization of aryl halides to afford indolines. *Angewandte Chemie International Edition* 56 (41): 12723–12726.

147 Rajesh, M., Singam, M.K.R., Puri, S. et al. (2018). Nickel catalyzed syn-selective aryl nickelation and cyclization of aldehyde/enone-tethered terminal alkynes with arylboronic acids. *Journal of Organic Chemistry* 83 (24): 15361–15371.

148 Shen, D., Zhang, W.B., Li, Z. et al. (2020). Nickel/NHC-catalyzed enantioselective cyclization of pyridones and pyrimidones with tethered alkenes. *Advanced Synthesis and Catalysis* 362 (5): 1125–1130.

149 Thavaselvan, S. and Parthasarathy, K. (2020). Nickel-catalyzed cyclization strategy for the synthesis of pyrroloquinolines, indoloquinolines, and indoloisoquinolines. *Organic Letters* 22 (10): 3810–3814.

150 Xu, Z., Tang, Y., Shen, C. et al. (2020). Nickel-catalyzed regio- and diastereoselective hydroarylative and hydroalkenylative cyclization of 1,6-dienes. *Chemical Communications* 16 (19): 4984–4987.

151 Xue, W., Xu, H., Liang, Z. et al. (2014). Nickel-catalyzed reductive cyclization of alkyl dihalides. *Organic Letters* 56: 7741–7744.

152 Zhang, H. and Lu, Z. (2018). Nickel-catalyzed enantioselective sequential Nazarov cyclization/decarboxylation. *Organic Chemistry Frontiers* 5 (11): 1763–1767.

153 Zhang, Z., Qin, Z., Chang, W. et al. (2020). Nickel-catalyzed decarboxylative cyclization of isatoic anhydrides with carbodiimides: synthesis of 2,3-dihydroquinazolin-4(1*H*)-ones. *Advanced Synthesis and Catalysis* 362 (14): 2864–2869. https://doi.org/10.1002/adsc.202000047.

154 Straub, B.F. and Gollub, C. (2004). Mechanism of Reppe's nickel-catalyzed ethyne tetramerization to cyclooctatetraene: a DFT study. *Chemistry – A European Journal* 10 (12): 3081–3090.

155 Hong, X., Holte, D., Gotz, D.C. et al. (2014). Mechanism, reactivity, and selectivity of nickel-catalyzed [4 + 4 + 2] cycloadditions of dienes and alkynes. *Journal of Organic Chemistry* 79 (24): 12177–12184.

156 Hong, X., Liu, P., and Houk, K.N. (2013). Mechanism and origins of ligand-controlled selectivities in [Ni(NHC)]-catalyzed intramolecular (5 + 2) cycloadditions and homo-ene reactions: a theoretical study. *Journal of the American Chemical Society* 135 (4): 1456–1462.

157 Kumar, P., Thakur, A., Hong, X. et al. (2014). Ni(NHC)-catalyzed cycloaddition of diynes and tropone: apparent enone cycloaddition involving an 8π insertion. *Journal of the American Chemical Society* 136 (51): 17844–17851.

158 Yoshino, Y., Kurahashi, T., and Matsubara, S. (2009). Nickel-catalyzed decarboxylative carboamination of alkynes with isatoic anhydrides. *Journal of the American Chemical Society* 131 (22): 7494–7495.

159 Guan, W., Sakaki, S., Kurahashi, T. et al. (2013). Theoretical mechanistic study of novel Ni(0)-catalyzed [6 − 2 + 2] cycloaddition reactions of isatoic anhydrides with alkynes: origin of facile decarboxylation. *Organometallics* 32 (24): 7564–7574.

160 Kajita, Y., Kurahashi, T., and Matsubara, S. (2008). Nickel-catalyzed decarbonylative addition of anhydrides to alkynes. *Journal of the American Chemical Society* 130 (51): 17226–17227.

161 Xie, H., Sun, Q., Ren, G. et al. (2014). Mechanisms and reactivity differences for cycloaddition of anhydride to alkyne catalyzed by palladium and nickel catalysts: insight from density functional calculations. *Journal of Organic Chemistry* 79 (24): 11911–11921.

162 Yang, S., Xu, Y., and Li, J. (2016). Theoretical study of nickel-catalyzed proximal C–C cleavage in benzocyclobutenones with insertion of 1,3-diene: origin of selectivity and role of ligand. *Organic Letters* 18 (24): 6244–6247.

163 Wang, S.C., Troast, D.M., Conda-Sheridan, M. et al. (2009). Mechanism of the Ni(0)-catalyzed vinylcyclopropane–cyclopentene rearrangement. *Journal of Organic Chemistry* 74 (20): 7822–7833.

5

Theoretical Study of Pd-Catalysis

106.42
Pd⁴⁶
Palladium
[Kr]4d¹⁰

In the field of organometallic chemistry, Pd-complexes is one of the earliest catalysts used, and have been most widely used in organometallic catalysis so far [1–14]. In Pd chemistry, the common oxidative states of Pd are 0 and +2, which are easily converted to each other [15–21]. Moreover, +4 valent Pd can also exist as an active intermediate in Pd-catalysis, which adds some variables to Pd-chemistry [22–26]. Therefore, there are majorly two efficient and easy ways to construct Pd—C bond, including oxidative addition and metalation [27–37].

When electrophilic carbon subtracts are used, they can react with low-valence Pd-species to afford a Pd—C bond, in which carbon atom transfers from an electrophile to a nucleophile. This process is often considered to be oxidative addition, while the oxidative state of corresponding Pd atom is increased. In another case, when a nucleophilic carbon compound is employed to react with Pd-species, a Pd—C bond can be formed through metalation process. In this chemistry, main-group organometallic complexes are often used as nucleophile, which leads to a transmetalation process to afford Pd—C bond. In particular, in a weak polar C—H bond, the carbon moiety also can be considered as a formal nucleophile, because the electronegativity of carbon is slightly larger than that of hydrogen. Therefore, C—H bond can react with Pd-species through a deprotonation–metalation process, which also can provide Pd—C bond. In metalation processes, the oxidative state of Pd usually remains invariable [38].

When a Pd—C bond is generated, the further transformations of that bond lead to the formation of new organic compounds. One possibility is to construct new C—C covalent bonds by inserting an unsaturated molecule into Pd—C bonds. The usually used unsaturated molecules can be alkene, alkyne, carbon monoxide, carbene, nitrene, i.e. the newly formed Pd—C bond can react with another carbolic

Computational Methods in Organometallic Catalysis: From Elementary Reactions to Mechanisms,
First Edition. Yu Lan.
© 2021 WILEY-VCH GmbH. Published 2021 by WILEY-VCH GmbH.

or heteroatomic nucleophile to afford a C—C bond or a C—hetero bond, which is considered to be reductive elimination. In this process, the oxidative state of Pd is decreased. Moreover, protonation and β-hydride elimination are two other possible ways for the cleavage of Pd—C bond.

Many books and reviews have been devoted to Pd-chemistry, which focus on both experiments and theory [39–52]. Therefore, in this chapter, the author only wants to discuss several common reaction types for Pd-catalysis. The main concerns are the general reaction mechanism and catalytic cycle obtained by theoretical calculations in homogeneous Pd-chemistry.

5.1 Pd-Catalyzed Cross-coupling Reactions

The cross-coupling reaction, in which a nucleophilic carbon and an electrophilic carbon are used to construct a new C—C bond, is one of the most powerful tools in organic synthesis (Scheme 5.1). In this area, Pd-complexes are usually used as catalyst to promote the cross-coupling between organohalides (electrophile) and main-group organometallic compounds (nucleophile). According to the theoretical and experimental studies [53–56], a widely accepted reaction mechanism of Pd-catalyzed cross-coupling is shown in Scheme 5.1. The catalytic cycle starts from a Pd(0) species, which can react with organohalide to afford a Pd—C(R^1) bond in a Pd(II) intermediate through an oxidative addition. Then, a transmetalation of the resulting complex from oxidative addition by main-group organometallic compound takes place to afford a C(R^2)—Pd—C(R^1) intermediate. The reductive elimination gives the final coupling product R^1–R^2 and regenerates Pd(0) species, which can join the next catalytic cycle.

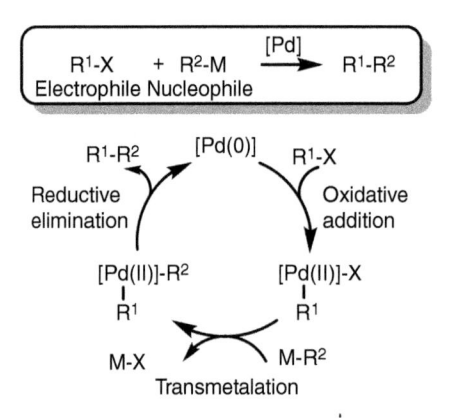

Scheme 5.1 General mechanism of Pd-catalyzed cross-coupling reactions.

Generally, a series of types for Pd-catalyzed cross-coupling reactions are commonly used, the difference of which is determined by the use of nucleophiles. Three of main developers of Pd-catalyzed cross-coupling reactions, R.F. Heck, E. Negishi, and A. Suzuki, were awarded with the Nobel prize in chemistry in 2010 for

5.1 Pd-Catalyzed Cross-coupling Reactions

their work in this field [57, 58]. In this chapter, mechanism discussion is classified by the reaction type, which involves Suzuki–Miyaura coupling, Negishi coupling, Stille–Kosugi–Migita coupling, Hiyama coupling, and Heck–Mizoroki reaction. Notably, a related C—C bond formation process in the Heck–Mizoroki reaction would be discussed in corresponding section, which has a mechanism slightly different from that shown in Scheme 5.1.

5.1.1 Suzuki–Miyaura Coupling

The Suzuki–Miyaura cross-coupling reaction consists of the coupling between an organoboron compound and an organic halide mediated by a Pd-catalyst in the presence of a base. The general catalytic cycle of Suzuki–Miyaura cross-coupling is exhibited in Scheme 5.2. The oxidative addition of organic halide onto Pd(0) species forms a R^1–Pd(II)–X species. In the presence of base, a transmetalation with organoboron can afford a R^1–Pd(II)–R^2 intermediate. Then, a reductive elimination yields R^1–R^2 coupling product and regenerates Pd(0) species.

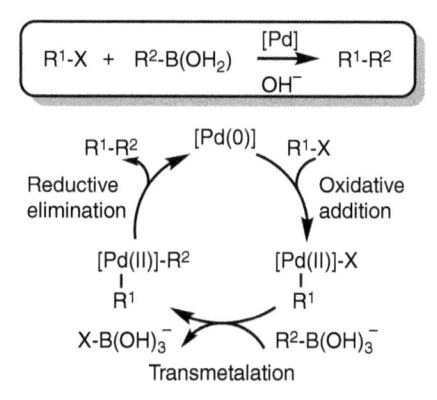

Scheme 5.2 The mechanism of Suzuki–Miyaura cross-coupling.

An early density functional theory (DFT) study, given in 2006, described the full free-energy profiles for the catalytic cycle of a model Suzuki–Miyaura cross-coupling reaction with vinyl bromide and vinyl boronic acid [59] (Figure 5.1). The reaction starts from a diphosphine-coordinated Pd(0) species **5-1**, which can react with vinyl bromide through an oxidative addition process via a three-membered ring-type transition state **5-3ts** with an energy barrier of 20.8 kcal/mol. The formed cis-vinyl Pd(II) intermediate **5-4** can easily isomerize to a trans-one **5-5**. In the presence of base, vinyl boronic acid can be activated to corresponding borate, which can coordinate onto Pd in intermediate **5-5** by release of bromide to afford intermediate **5-8**. The transmetalation can take place via transition state **5-9ts** with an energy barrier of 16.4 kcal/mol. After release of boric acid, a divinyl-Pd(II) intermediate **5-10** is formed with exergonic. The barrier of the following reductive elimination is only 14.4 kcal/mol via transition state **5-12ts**. Then, a C(vinyl)—C(vinyl) bond is formed with the regeneration of diphosphine Pd(0) species **5-1**. The computational

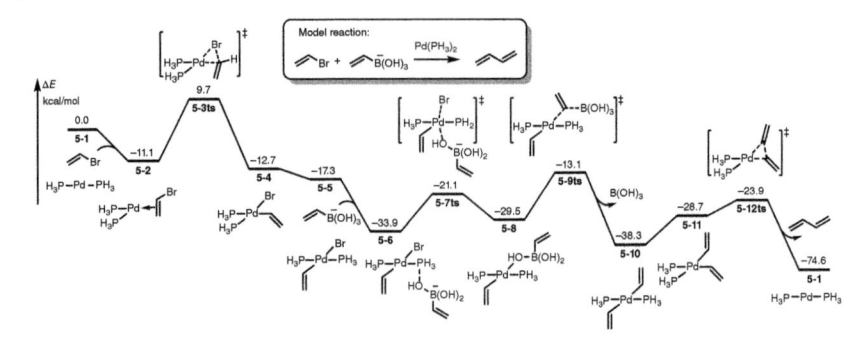

Figure 5.1 The energy profiles for a model reaction of Pd-catalyzed Suzuki–Miyaura cross-coupling. The energies were calculated at B3LYP/6-31G(d) (LANL2DZ for Pd and Br)/PCM(water)//B3LYP/6-31G(d) (LANL2DZ for Pd and Br) level of theory.

results show that the rate-determining step for the whole catalytic cycle is the transmetalation with base-activated vinyl borate.

In cases where the halogens are different, the observed selectivity is obviously related to the trend of C—I > C—Br > C—Cl > C—F. However, the situation is complicated in the case of cross-coupling reactions of polyhalogenated heterocycles bearing multiple identical halogens. As shown in Scheme 5.3a, Houk group predicted that the selectivity in this case is determined by both the strength of the carbon—halogen bond and the LUMO of heterocycle [60]. In a distortion–interaction analysis of this process (Scheme 5.3b), the total activation energy can be decomposed to distort isolated reactants to the transition-state geometry (the distortion energy) and the energy of interaction between these distorted reactants (the interaction energy). The distortion energy is related to the bond dissociation energy of reacting carbon—halogen bond, while the interaction energy is influenced by the secondary orbital interaction between occupied d orbital of Pd and π^* orbital in aryl-halide (Scheme 5.3c).

When a chloroaryl triflate is used in Suzuki–Miyaura cross-coupling, the regioselectivity is controlled by the ligand [61]. As shown in Scheme 5.4, tricyclohexylphosphine (PCy$_3$) gave coupling at the C—OTf bond, while *tert*-butylphosphine (P^{t}Bu$_3$) led to a remarkable reversal, with exclusive reaction at the C—Cl bond. The distortion–interaction analysis showed that in a monoligated Pd-mediated oxidative addition process, the regioselectivity is controlled by distortion energy; therefore, the weak C(aryl)—Cl bond reacts first due to its lower bond dissociation energy. In a biligated case, the regioselectivity is controlled by interaction energy. The lower LUMO of C(aryl)—O(OTf) bond reacts first.

Carboxylic anhydrides can play as an electrophile in a Pd-catalyzed Suzuki-type cross-coupling with arylboronic acids. DFT calculation was used to reveal the mechanism of this reaction [62]. As shown in Figure 5.2, the reaction starts from a biphosphine Pd(0) species **5-13**, which can react with acetic anhydride through an oxidative addition transition state **5-14ts** with an energy barrier of 18.0 kcal/mol to afford an acyl-Pd(II) intermediate **5-15**. After a ligand exchange with acetate, palladate **5-16**

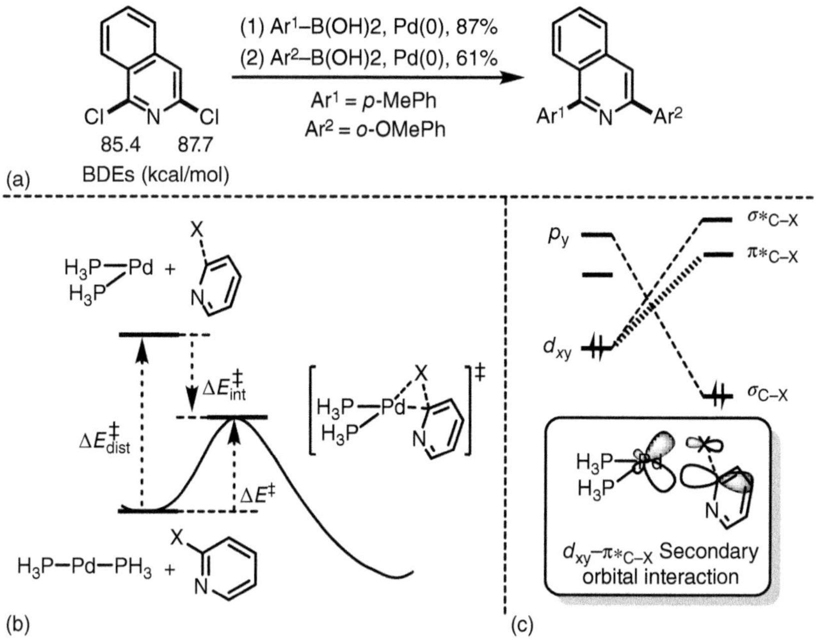

5.1 Pd-Catalyzed Cross-coupling Reactions | 185

(a)

(1) Ar¹–B(OH)2, Pd(0), 87%
(2) Ar²–B(OH)2, Pd(0), 61%

Ar¹ = p-MePh
Ar² = o-OMePh

Cl 85.4   Cl 87.7
BDEs (kcal/mol)

(b) Distortion–interaction diagram with $H_3P\text{—}Pd + $ and $H_3P\text{—}Pd\text{—}PH_3 +$ pathways, showing $\Delta E^{\ddagger}_{dist}$, $\Delta E^{\ddagger}_{int}$, $\Delta E^{\ddagger}$.

(c) Frontier MO diagram showing p_y, d_{xy}, σ^*_{C-X}, π^*_{C-X}, σ_{C-X} and $d_{xy}\text{–}\pi^*_{C-X}$ Secondary orbital interaction.

Scheme 5.3 (a) Pd-catalyzed tandem Suzuki–Miyaura cross-coupling with polyhalogenated heterocycles. (b) The distortion–interaction analysis. (c) The frontier molecular orbital analysis.

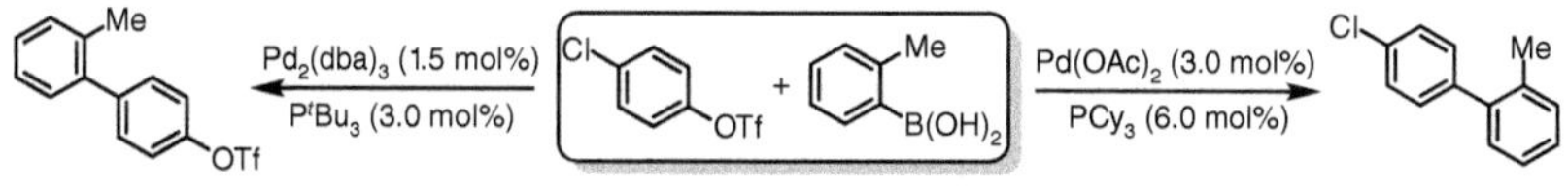

Pd₂(dba)₃ (1.5 mol%)
P^{t}Bu₃ (3.0 mol%)

Pd(OAc)₂ (3.0 mol%)
PCy₃ (6.0 mol%)

Reacting position	Pd species	$\Delta E^{\ddagger}$	$\Delta E^{\ddagger}_{dist}$ (Pd)	$\Delta E^{\ddagger}_{dist}$ (C–X)	$\Delta E^{\ddagger}_{int}$	$\Delta G^{\ddagger}$
C–C1	PdPMe₃	14.4	3.2	32.1	−20.9	13.7
C–OTf	PdPMe₃	19.9	4.3	52.6	−37.0	20.1
C–C1	Pd(PMe₃)₂	21.5	16.9	20.9	−16.3	35.1
C–OTf	Pd(PMe₃)₂	16.9	21.8	55.9	−60.8	31.8

Scheme 5.4 Regioselectivity of Pa-catalyzed Suzuki–Miyaura cross-coupling with chloroaryl triflate.

is formed. The coordination of phenylboronic acid forms intermediate **5-19**. Then, phenyl transfer from boronic acid onto Pd takes place through a transmetalation process. The calculated overall activation free energy of transmetalation is 26.8 kcal/mol via transition state **5-20ts**. After release of acetic boric anhydride, a phenyl acyl Pd(II) intermediate **5-21** is formed, the relative free energy of which is 5.5 kcal/mol higher than that of intermediate **5-16**. Finally, the reductive elimination occurs via transition state **5-22ts** with a barrier of 12.3 kcal/mol. The calculated overall activation free energy is 26.8 kcal/mol with the rate-determining step of transmetalation. DFT

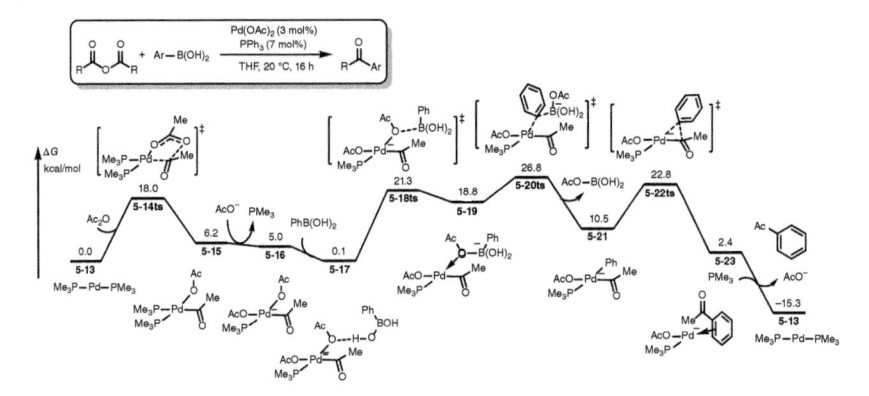

Figure 5.2 The free-energy profiles for Pd-catalyzed Suzuki–Miyaura cross-coupling with anhydride. The energies were calculated at BP86/6-31+G(d,p) (LANL2DZ for Pd)/CPCM(THF)//BP86/6-31G(d) (LANL2DZ for Pd) level of theory. THF; tetrahydrofuran.

calculation predicted that in the absence of base, the generation of less stable acetic boric anhydride leads to a higher relative free energy of aryl Pd(II) intermediate for this step.

When an N-heterocyclic carbene is used as ligand, aryl esters can be used as electrophile in Pd-catalyzed Suzuki–Miyaura type cross-coupling reactions [63]. As shown in Figure 5.3, DFT calculations showed that the ester's C(acyl)—O bond oxidative addition onto a monoligated Pd(0) species can occur via a three-membered ring-type transition state **5-25ts** with an energy barrier of 23.3 kcal/mol to afford an acyl Pd(II) intermediate **5-26**. With the activation of base, a transmetalation with phenylborate takes place via a four-membered ring-type transition state **5-28ts** to form a phenyl acyl Pd(II) intermediate **5-29**. The reductive elimination with a barrier of only 5.8 kcal/mol via transition state **5-30ts** yields benzophenone product and regenerates Pd(0) species **5-24** by substrate exchange.

In the presence of dialkylbiarylphosphine ligand, aryl sulfamates can be used as electrophile in Pd-catalyzed Suzuki–Miyaura cross-coupling reactions [64]. As shown in Figure 5.4. DFT calculation found that the biaryl moiety in phosphine ligand (XPhos) can coordinate onto Pd(0) in reacting species **5-32**. The calculated activation free energy for the oxidative addition step is 25.5 kcal/mol via a three-membered ring-type transition state **5-33ts**. Chelation-assisted oxidative additions via five-centered transition state **5-35ts**, **5-36ts**, **5-37ts**, or **5-38ts** are also considered; however, the calculated barriers for those cases are higher than those for the process via three-centered transition state **5-33ts**. In the absence of coordination with biaryl moiety, the barrier of oxidative is as high as 43.6 kcal/mol via transition state **5-34ts**.

5.1.2 Negishi Coupling

In a Pd-catalyzed Negishi cross-coupling reaction, organic zinc reagent is used as nucleophile to react with aryl halide, which is usually used to prepare alkyl or aryl arenes (Scheme 5.5).

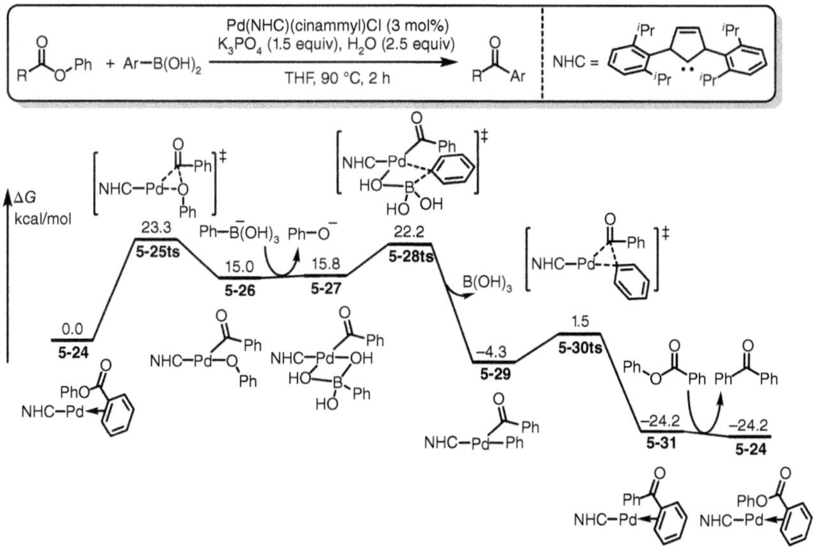

Figure 5.3 The free-energy profiles for Pd-catalyzed Suzuki–Miyaura cross-coupling with esters. The energies were calculated at M06/6-311+G(d,p) (SDD for Pd)/SMD(THF)//B3LYP/6-31G(d) (LANL2DZ for Pd) level of theory.

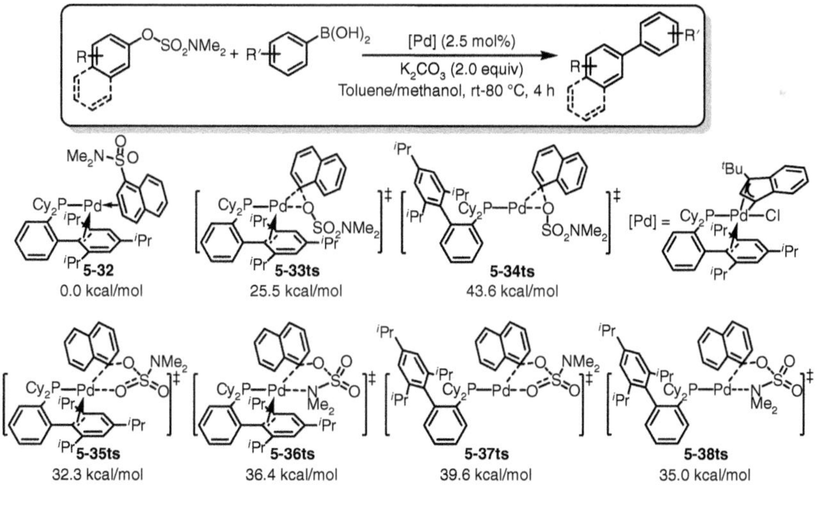

Figure 5.4 The calculated free energies for the key transition states of oxidative addition in Pd-catalyzed Suzuki–Miyaura cross-coupling with aryl sulfamates. The energies were calculated at M06/6-311G(d,p) (LANL2DZ(f) for Pd)/SMD(toluene)//M06-L/6-31G(d,p) (LANL2DZ(f) for Pd) level of theory.

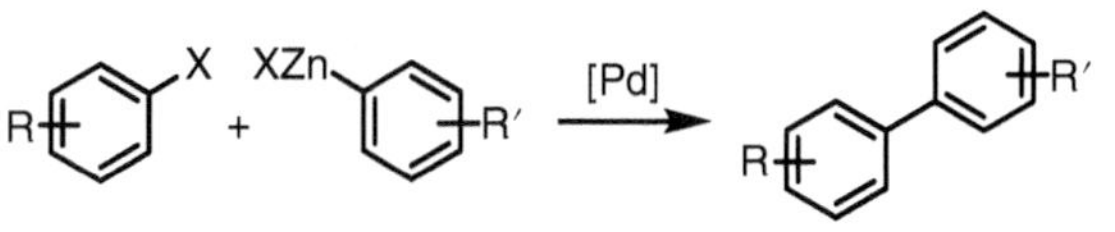

Scheme 5.5 Pd-catalyzed Negishi coupling. Source: Adapted from Ribagnac et al. [65].

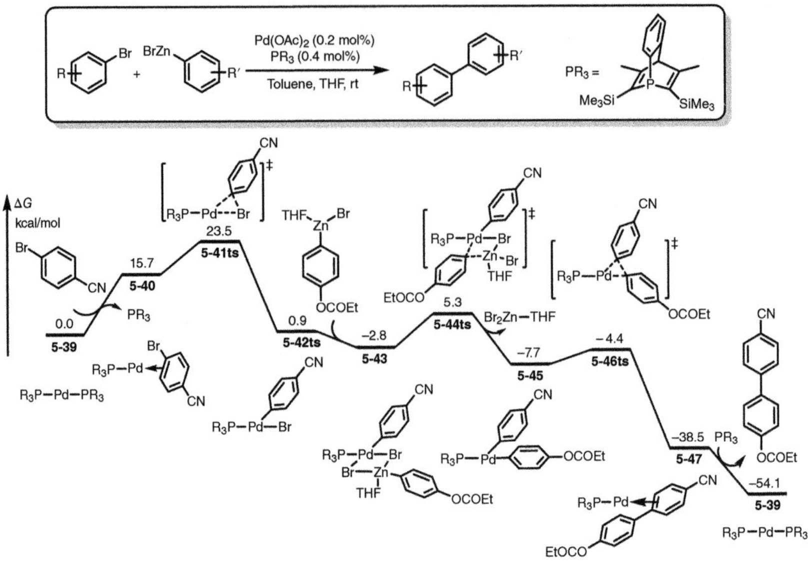

Figure 5.5 The free-energy profiles for a typical Pd-catalyzed Negishi cross-coupling. The energies were calculated at B3PW91/6-31G(d) (def2-QZVP for Pd and Zn)/PCM(THF)//B3PW91/6-31G(d) (def2-QZVP for Pd and Zn) level of theory.

The DFT-calculated free-energy profiles for a Negishi coupling using phosphine-coordinated Pd(0) species A are shown in Figure 5.5 [65]. The reaction starts from a ligand exchange with reacting aryl bromide to afford intermediate **5-40** by 15.7 kcal/mol endergonic. The oxidative addition takes place via transition state **5-41ts** with an overall activation free energy of 23.5 kcal/mol to form an aryl Pd(II) intermediate **5-42**, which is considered to be the rate-determining step for the whole catalytic cycle. In the presence of aryl zinc substrate, a transmetalation via a four-membered ring-type transition state **5-44ts** forms diaryl Pd(II) intermediate **5-45**. A rapid reductive elimination yields biaryl product and regenerates Pd(0) species **5-39**.

As shown in Figure 5.6, when an N-heterocyclic carbene is used as ligand, the barrier of oxidative addition is significantly reduced to only 10.0 kcal/mol via transition state **5-49ts** even when an alkyl bromide is used as electrophile [11]. The calculated free-energy profiles also revealed that the rate-determining step is changed to the following transmetalation via transition state **5-51ts**.

A detailed mechanistic study on the transmetalation step with alkyl zinc substrate is shown in Figure 5.7 [66]. The transmetalation can occur via transition state **5-56ts** with an energy barrier of 10.5 kcal/mol. The geometry information of that transition state is given in Figure 5.7. The lengths of calculated forming Pd—C(alkyl) bond and breaking Zn—C(alkyl) bond are 2.79 and 2.07 Å, respectively. Interestingly, the distance between Zn and Pd is only 3.01 Å, which revealed a significant interaction between those two metal atoms. Therefore, forward to this transition state, a methyl Pd(II) intermediate **5-57** is formed, in which a Zn—Pd bond is observed. The Zn—Pd

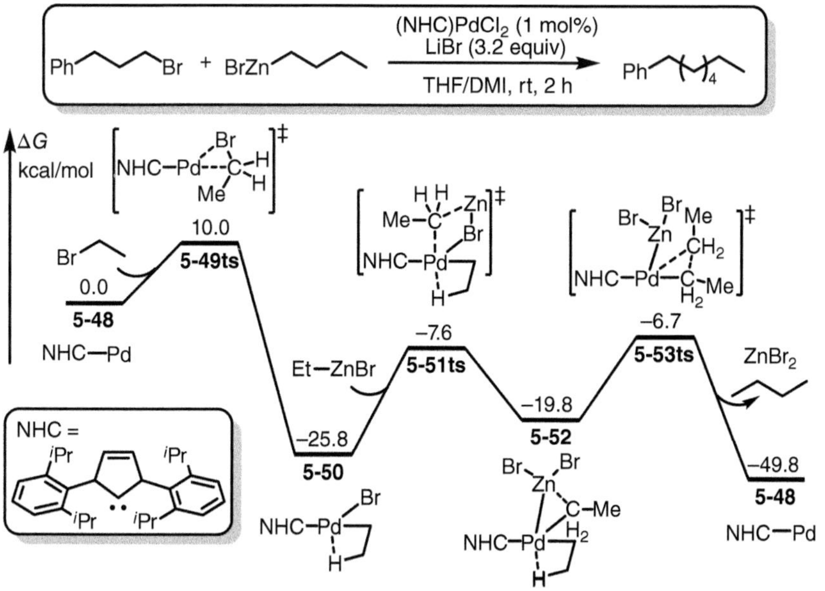

Figure 5.6 The free-energy profiles for a typical Pd-catalyzed Negishi cross-coupling with alkyl bromide. The energies were calculated at B3LYP/DZVP/PCM(THF)//B3LYP/DZVP level of theory.

bond interaction can be considered as the electron-rich Pd center that donates electron to the fairly positive Zn center [13].

When a chelation-involved electrophile (e.g. methyl *ortho*-iodobenzoate) is used in Pd-catalyzed Negishi coupling reaction, a second transmetalation is predicted by DFT calculation [67]. As shown in Figure 5.8, the transmetalation with phenyl zinc with aryl Pd(II) intermediate **5-59** generating from first transmetalation can take place via transition state **5-60ts** with an energy barrier of 18.3 kcal/mol, which is 1.5 kcal/mol lower than that of the competitive reductive elimination process via transition state **5-64ts**. Therefore, a homo-coupling biaryl product can be observed experimentally.

5.1.3 Stille Coupling

The catalytic cycle of a Pd-catalyzed Stille cross-coupling reaction involves oxidative addition, transmetalation, and reductive elimination, in which an organic tin reagent is used as nucleophile for transmetalation [68–70]. DFT study for the mechanism of Pd-catalyzed Stille cross-coupling with vinyl bromide and vinyl tin is shown in Figure 5.9 [68]. The oxidative addition of vinyl bromide onto Pd(0) species **5-65** takes place via transition state **5-66ts** with a free-energy barrier of 18.9 kcal/mol to afford a vinyl Pd(II) intermediate **5-67**. Then, a reversible ligand exchange with vinyl tin reagent forms intermediate **5-68** with 4.3 kcal/mol endergonic. A transmetalation takes place via a four-membered ring-type transition state **5-69ts** with an

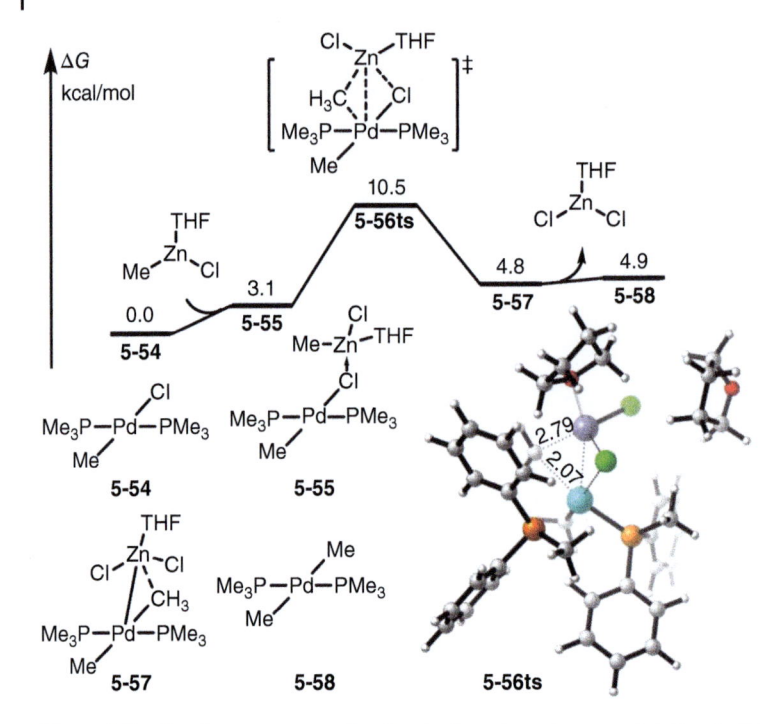

Figure 5.7 The free-energy profiles for the key step of transmetalation in Pd-catalyzed Negishi cross-coupling. The energies were calculated at M06/TZVP (DG-DZVP for Pd)/PCM(THF)//M06/6-31G(d) (LANL2DZ for Pd and Zn) level of theory. Source: Based on Fuentes et al. [66].

energy barrier of 20.4 kcal/mol, which is considered as the rate-determining step. Subsequently, reductive elimination releases a butadiene product and regenerates Pd(0) species **5-65**.

5.1.4 Hiyama Coupling

In organometallic chemistry, organosilicon compounds are available and usually stable in air. Therefore, if organosilicon compounds can be used in Pd-catalyzed cross-coupling reaction with organic halides, it would greatly increase the application field of this type of reaction. However, organosilicon compounds are not very reactive, probably because of the stable and strong C—Si bond. Fortunately, Hiyama and coworkers found that addition of fluoride anion can significantly accelerate this cross-coupling reaction and increase the product yield (Scheme 5.6).

Scheme 5.6 Pd-catalyzed Hiyama coupling.

Figure 5.8 The free-energy profiles for the second transmetalation in Negishi coupling reaction. The energies were calculated at B3LYP/6-31G(d) (SDD for Pd and Zn)/IEF-PCM(THF)//B3LYP/6-31G(d) (SDD for Pd and Zn) level of theory.

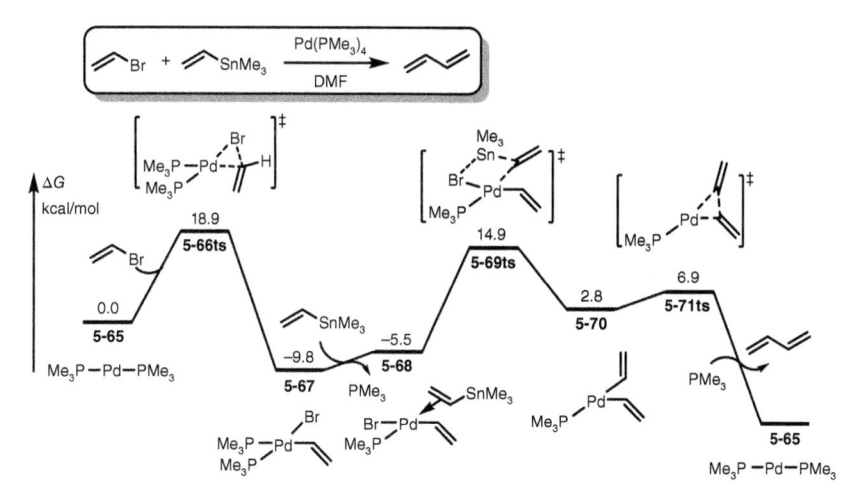

Figure 5.9 The free-energy profiles for a model reaction of Pd-catalyzed Stille coupling. The energies were calculated at B3LYP/6-31G(d) (SDD for Pd, Sn, and Br)/PCM(THF)//B3LYP/6-31G(d) (SDD for Pd, Sn, and Br) level of theory. Source: Based on Álvarez et al. [68].

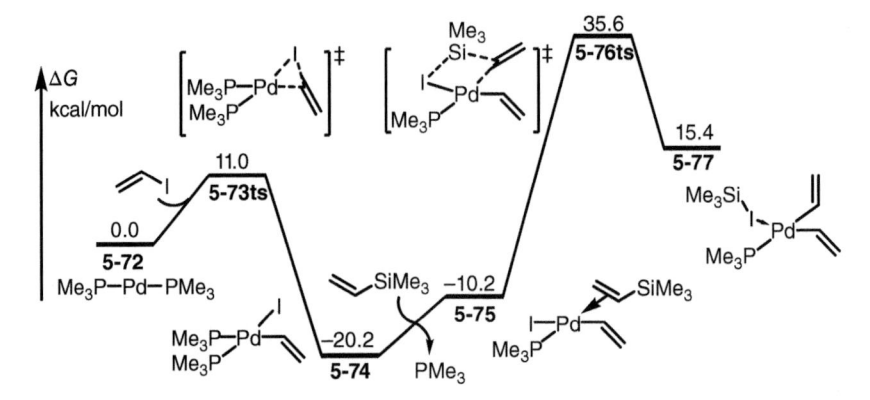

Figure 5.10 The energy profiles for a model reaction of Pd-catalyzed Hiyama coupling. The energies were calculated at B3LYP/6-311G(d) (LANL2DZ for Pd, Si, and I)/PCM(THF)//B3LYP/6-31G(d) (LANL2DZ for Pd, Si, and I) level of theory. Source: Based on Sugiyama et al. [71].

DFT calculations are taken to account the mechanism of Pd-catalyzed Hiyama coupling reaction [71]. As shown in Figure 5.10, a diphosphine-coordinated Pd(0) species **5-72** is set to relative zero of free-energy profiles. The oxidative of iodoethene takes place via transition state **5-73ts** to form a vinyl Pd(II) intermediate **5-74** with 20.2 kcal/mol exothermic. The coordination of vinyl silane forms a four-coordinated Pd(II) intermediate **5-75** with 10.0 kcal/mol endothermic with the release of phosphine ligand. Computational result showed that when the transmetalation carries out via a four-membered ring-type transition state **5-76ts**, the calculated barrier of this process is as high as 45.8 kcal/mol. Such a high barrier for the transmetalation can be attributed to the cleavage of a stable C—Si bond with the formation of a relatively unstable Si—I bond.

In Pd-catalyzed Hiyama coupling reaction, the aryl silane reagent can be generated in situ [73]. As shown in Figure 5.11, when an oxasilolane reacts with phenyl lithium, a phenyl silane can be generated in situ, which can play as nucleophile in Pd-catalyzed Hiyama-type cross-coupling reactions. DFT calculation is taken to reveal the mechanism of this reaction. When oxisilolane **5-82** reacts with phenyl lithium, transmetalation via a four-membered ring-type transition state **5-83ts** can afford a lithium silylphenyl methanolate intermediate **5-84**. The counterion exchange with Pd(II) species takes place through an associative pathway to afford intermediate **5-85**. With the intramolecular activation of oxygen atom, a transmetalation can take place via transition state TSb to afford phenyl Pd intermediate F. The calculated barrier of the transmetalation step is only 9.3 kcal/mol, which is much lower than that for the corresponding step in a typical Hiyama coupling reaction. Interestingly, oxasilolane A can be regenerated after transmetalation.

5.1.5 Heck–Mizoroki Reaction

The Heck–Mizoroki reaction, one of the most utilized cross-coupling reactions, is the palladium-catalyzed arylation of an olefin with an aryl halide under basic conditions

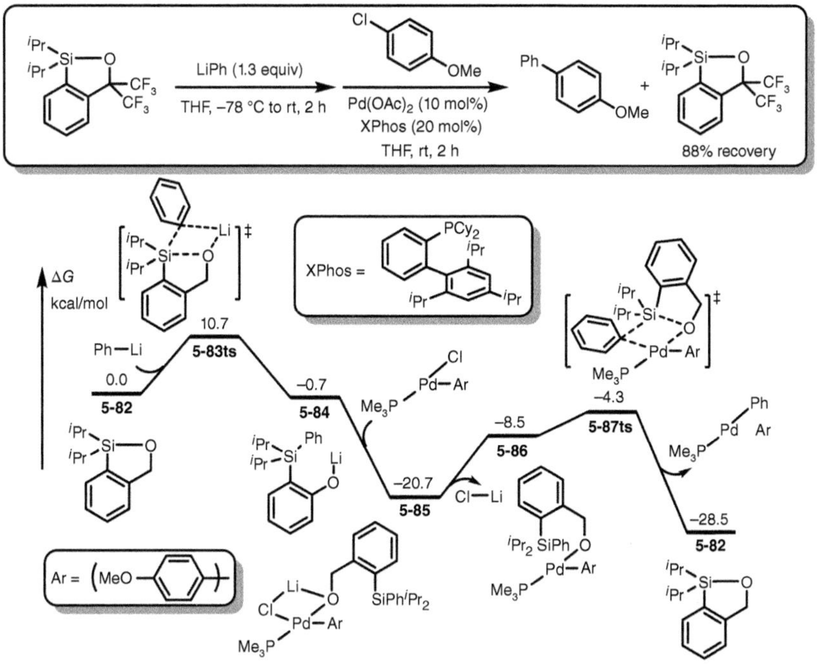

Figure 5.11 The free-energy profiles for a Pd-catalyzed Hiyama coupling with in situ-generated aryl silane. The energies were calculated at M06/6-311+G(d,p) (SDD for Pd)/CPCM(THF)//M06/6-31G(d) (SDD for Pd) level of theory. Source: Based on Albert et al. [72].

(Figure 5.11) [72]. Since its independent discovery in the early 1970s by Heck and Mizoroki, the Heck reaction has been widely used as a tool for organic synthesis because of its importance in the direct attachment of olefin groups to aromatic rings [57, 74–76]. In a Heck–Mizoroki reaction, olefins are usually used as nucleophile to react with organic halides. Therefore, the mechanism of Heck–Mizoroki reaction is slightly different from other Pd-catalyzed coupling reactions. A general catalytic cycle of Pd-catalyzed Heck–Mizoroki reaction is shown in Scheme 5.7, which also starts from a Pd(0) species. The oxidative addition of organic halide forms a C—Pd bond in a Pd(II) species. Then, alkene insertion into C—Pd bond leads to the formation of an alkyl Pd(II) intermediate. Subsequently, the β-hydride elimination takes place to yield olefin product and form a hydride Pd(II) species, which also can be considered to be a protonated palladate. In the presence of base, Pd(0) can regenerated by deprotonation.

DFT calculations are taken to reveal the mechanism of Heck–Mizoroki reaction [77]. As shown in Figure 5.12, the ligand exchange of Pd(0) species **5-88** with phenyl bromide takes place through a dissociative pathway. The oxidative addition of phenyl bromide onto Pd occurs via a three-membered ring-type transition state **5-89ts** with a total free-energy barrier of 21.0 kcal/mol to form a three-coordinated phenyl Pd(II) intermediate **5-90**. The coordination of ethylene forms intermediate **5-91** with 11.2 kcal/mol endergonic. Then, an alkene insertion takes place

Scheme 5.7 Mechanism of Heck–Mizoroki reaction.

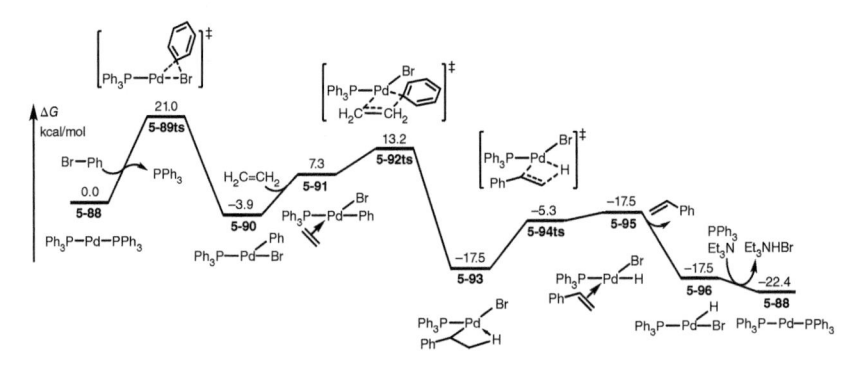

Figure 5.12 The free-energy profiles for a Pd-catalyzed Heck–Mizoroki reaction. The energies were calculated at B3LYP/6-31++G(d,p) (LANL2DZ+f for Pd, and LANL2DZ(d,p) for P and Br)/CPCM(DMSO)//PBE/6-31++G(d,p) (LANL2DZ+f for Pd, and LANL2DZ(d,p) for P and Br) level of theory. DMSO; dimethyl sulfoxide.

via a four-membered ring-type transition state **5-92ts** with an energy barrier of 5.9 kcal/mol. After alkene insertion, an agostic intermediate **5-93** is formed by exergonic. A β-hydride elimination occurs via transition state **5-94ts** with an energy barrier of 12.2 kcal/mol to afford an olefin-coordinated hydride Pd(II) intermediate **5-95**. After release of a molecule of olefin, a hydride Pd complex **5-96** can be generated. In the presence of base, the deprotonation with the coordination of phosphine ligand regenerates Pd(0) species **5-88**. DFT calculations found that the rate-determining step of the catalytic cycle for the Pd-catalyzed Heck–Mizoroki reaction is the oxidative addition step.

A theoretical study for the mechanism of Pd-catalyzed Heck–Mizoroki reaction with chelating bis-N-heterocyclic carbene ligand is also performed (Figure 5.13) [78]. The computed free-energy profiles for the catalytic cycle start from an alkene-coordinated Pd(0) species **5-97** with bis-carbene ligand. After substrate

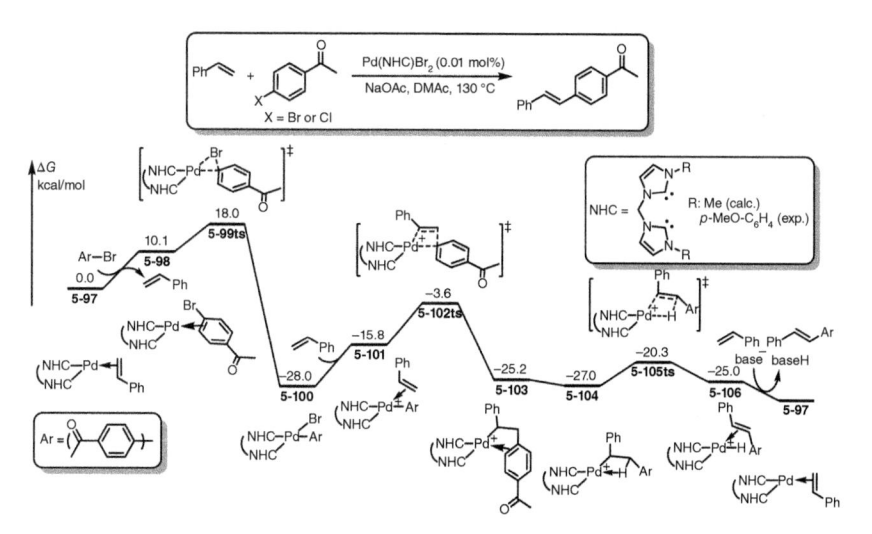

Figure 5.13 The free-energy profiles for a Pd-catalyzed Heck–Mizoroki reaction by using chelating bis-N-heterocyclic carbene ligand. The energies were calculated at M06/der2-TZVP (LANL2TZ(f) for Pd)/CPCM(DMSO)//B3LYP/6-31G(d) (LANL2DZ for Pd) level of theory. Source: Based on Allolio and Strassner [78].

loading in intermediate **5-98**, an oxidative addition of aryl bromide can occur via transition state **5-99ts** with an activation free energy of 18.0 kcal/mol to afford an aryl Pd(II) species **5-100** with 28.0 kcal/mol exergonic. After the ligand exchange with styrene substrate through a dissociative pathway, an alkene-coordinated cationic Pd(II) intermediate **5-101** is formed by 12.2 kcal/mol endergonic. The olefin insertion into Pd—C(aryl) bond takes place via a four-membered ring-type transition state **5-102ts** with an activation free energy of 24.4 kcal/mol to afford a benzylic Pd(II) intermediate **5-103**. The isomerization of benzylic Pd(II) **5-103** to agostic intermediate **5-104** leads to 1.8 kcal/mol exergonic. Then, a β-hydride elimination takes place via transition state **5-105ts** to afford a *trans*-olefin coordinated cationic Pd(II) complex **5-106**, which can be further deprotonated to regenerate Pd(0) species **5-97**. The computational results showed that the rate-determining step for the whole catalytic cycle is the olefin insertion process, when a chelating bis-N-heterocyclic carbene ligand is used. Moreover, the DFT study also found that the stereoselectivity is controlled by the β-hydride elimination step.

An intramolecular Heck–Mizoroki reaction can be used to construct ring-type coupling product [79]. As shown in Figure 5.14, in the presence of Pd-catalyst, an intramolecular transformation of *ortho*-bromodihydronaphthalenyl dihydropyran can yield a 9,10-dihydrophenanthrene product. The mechanism of this reaction is studied by DFT calculation. When diphosphine-coordinated Pd(0) complex **5-107** is set to relative zero of free-energy profiles, a substrate loading can form intermediate **5-108** through a dissociative pathway. The chelation-directed oxidative addition of vinyl bromide takes place via transition state **5-109ts** to form a vinyl Pd(II) intermediate **5-110**. Then, an olefin-coordinated Pd(II) intermediate **5-112** can be formed by intramolecular ligand exchange through associative replacement

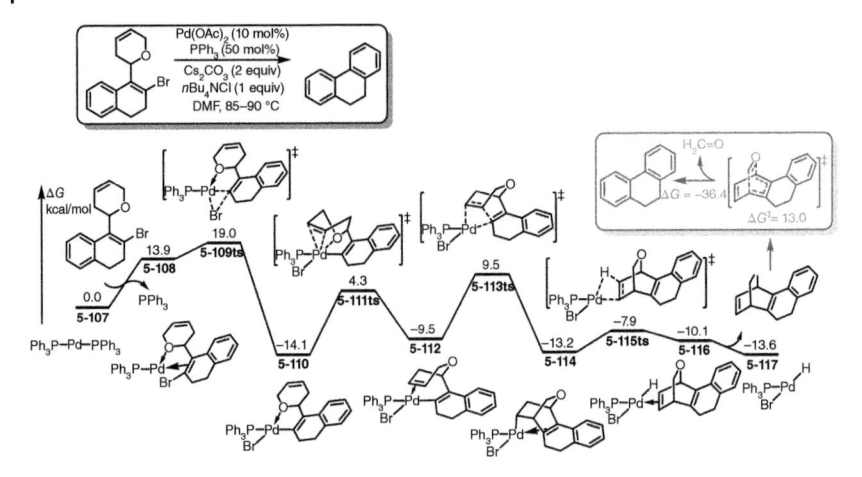

Figure 5.14 The free-energy profiles for a Pd-catalyzed intramolecular Heck–Mizoroki reaction. The energies were calculated at B3LYP/6-31G(d) (SDD for Pd and Br)/PCM(DMF)//B3LYP/6-31G(d) (SDD for Pd and Br) level of theory. DMF; N,N-dimethylformamide.

transition state **5-111ts**. Subsequently, an intramolecular olefin insertion occurs via transition state **5-113ts** with an energy barrier of 19.0 kcal/mol to afford alkyl Pd(II) intermediate **5-114**. A rapid β-hydride elimination takes place via transition state **5-115ts** with an energy barrier of only 5.3 kcal/mol to yield an oxabicyclo[2.2.2]octadiene molecule, which can undergo a retro-Diels–Alder reaction to afford 9,10-dihydrophenanthrene product.

5.2 Pd-Mediated C—Hetero Bond Formation

In Pd-catalyzed cross-coupling reactions, a new C—C bond can be formed between an electrophilic carbon and a nucleophilic carbon. In this case, an electrophilic carbon is often provided by organic halides, while the nucleophilic carbon comes from organic main-group metals. If the nucleophile is changed to a heteroatom, a C—heteroatom bond can be yielded from a cross-coupling reaction. In Pd-catalysis, C—B, C—S, C—I, and C—Si bonds can be formed following this idea. The general catalytic cycle of Pd-catalyzed C—heteroatom cross-coupling is shown in Scheme 5.8. The oxidative addition of organic halides onto Pd(0) species forms a Pd—C bond in a Pd(II) species. Then, a ligand exchange or metathesis with nucleophiles leads to the formation of Pd—heteroatom bond. The reductive elimination yields C—heteroatom cross-coupling product and regenerates Pd(0) species.

5.2.1 C—B Bond Formation

A Pd-catalyzed cross-coupling reaction of aryl halide with diboron was reported by Ishiyama and Miyaura [80]. This reaction receives considerable attention because this is very useful to synthesize a variety of organic boron compounds, which are

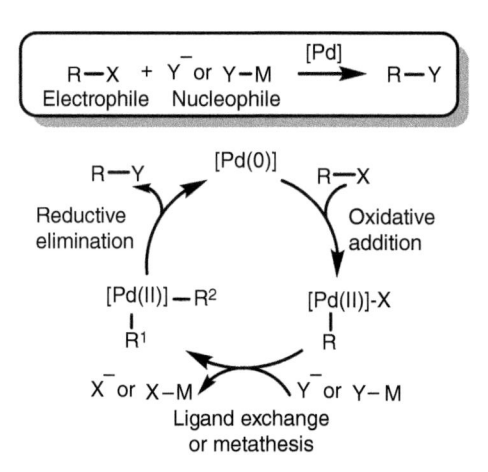

Scheme 5.8 Mechanism of Pd-mediated C—hetero bond formations.

important reagents for organic syntheses. A DFT calculation is focused on the mechanism of C—B bond formation process (Figure 5.15). In Sakaki's theoretical study, a hydroxide Pd(II) species **5-118** is set to relative zero in free-energy profiles [81], which can be generated by the oxidative addition of phenyl chloride followed by counterion exchange with base. The coordination of hydroxide in species **5-118** onto diboron forms intermediate **5-119** reversibly. Then, phosphine ligand dissociation takes place via transition state **5-120** to provide a vacancy in intermediate **5-121** for the following boronation. A transmetalation via a four-membered ring-type transition state **5-122ts** occurs with an overall activation free energy of 11.3 kcal/mol to form a boronated Pd(II) intermediate **5-123** irreversibly. A rapid aryl boryl elimination takes place via transition state **5-124ts** to yield aryl boron and generate Pd(0) species **5-125**.

5.2.2 C—S Bond Formation

The calculated zero-point energy profiles of a model reaction of phenyl chloride sulfuration are shown in Figure 5.16 [82]. The oxidative addition of phenyl chloride onto Pd(0) species **5-126** takes place via transition state **5-128ts** with an energy barrier of 13.7 kcal/mol to form a phenyl Pd(II) species **5-129**. In the presence of sulfide, a counterion exchange forms intermediate **5-130** with 13.8 kcal/mol exergonic. The reductive elimination of phenyl and sulfide would occur via transition state **5-131ts**, where the calculated barrier of this step is 21.6 kcal/mol.

When an anionic sulfonate is used as nucleophile, the barrier of reductive elimination for the formation of C—S bond can be significantly reduced. As shown in Figure 5.17 [83], benzylic sulfoxide can be used as sulfonate source, which can react with aryl bromide in the presence of Pd-catalyst. A DFT study is employed to reveal the mechanism of this reaction. As shown in Figure 5.17, the oxidative addition of phenyl bromide onto Pd(0) species **5-132** can occur via transition state **5-133ts** to afford a phenyl Pd(II) intermediate **5-134**. The ligand exchange with sulfonate anion forms oxygen-coordinated Pd(II) intermediate **5-135**, which can isomerize to

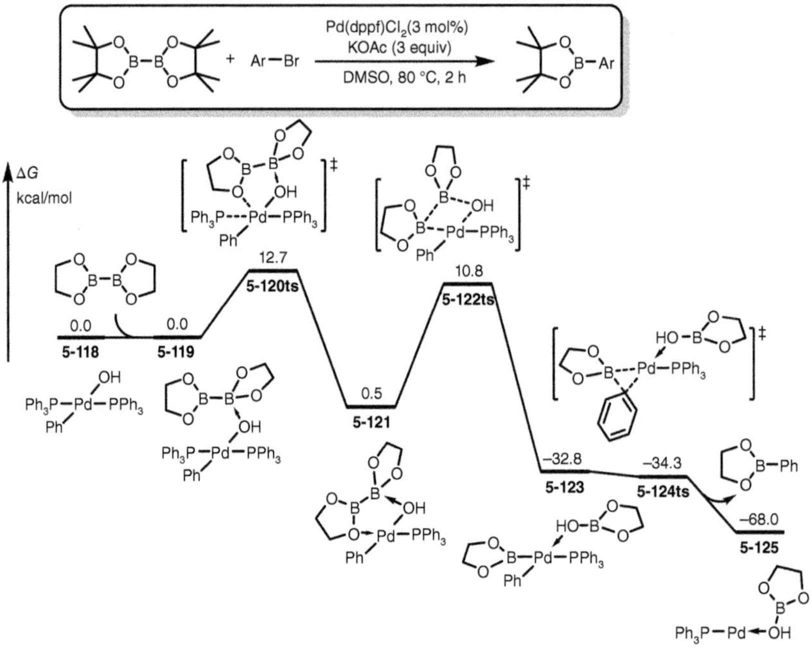

198 | *5 Theoretical Study of Pd-Catalysis*

Figure 5.15 The free-energy profiles for the key steps of Pd-catalyzed cross-coupling reaction of aryl halide with diboron. The energies were calculated at B3LYP/6-311G(d) (LANL2DZ for Pd and P)/PCM(DMSO)//B3LYP/6-31G(d) (LANL2DZ for Pd and P) level of theory.

sulfur-coordinated Pd(II) isomer **5-137** with a barrier of 20.7 kcal/mol. Then, a reductive elimination via a three-centered transition state takes place via transition state **5-138ts** with an energy barrier of only 8.4 kcal/mol to form a new C—S bond in sulfoxide product. The DFT calculations clearly revealed that the reductive elimination of aryl and sulfonate is much easier than that of aryl and sulfide in a regular Pd(0)–Pd(II) catalytic cycle.

To avoid the high barrier for the reductive elimination of aryl and sulfide from a Pd(II) species, a Pd(II)–Pd(IV) catalytic cycle is proposed in a Pd-catalyzed intramolecular σ-bond metathesis of S—C(aryl) and H—C(aryl) bonds for the synthesis of dibenzothiophene [84]. The calculated free-energy profiles for the main catalytic cycle are shown in Figure 5.18. The reaction starts from a Pd(II) species **5-140**. A C—H bond activation happens first through a concerted metalation–deprotonation (CMD) process via transition state **5-141ts** to form a sulfur-coordinated aryl Pd(II) intermediate **5-142**. Then, an intramolecular phenylsulfane oxidative addition onto Pd(II) takes place via a three-membered ring-type transition state **5-143ts** with an energy barrier of 24.2 kcal/mol to afford a sulfide Pd(IV) intermediate **5-144**. The following reductive elimination of another aryl group with sulfide occurs via transition state **5-145ts**. The calculated energy barrier of this step is only 12.4 kcal/mol, which can be attributed to the oxidability of Pd(IV) center. The generated phenyl Pd(II) intermediate **5-148** can be protonated by benzoic acid to regenerate Pd(II) species **5-140**.

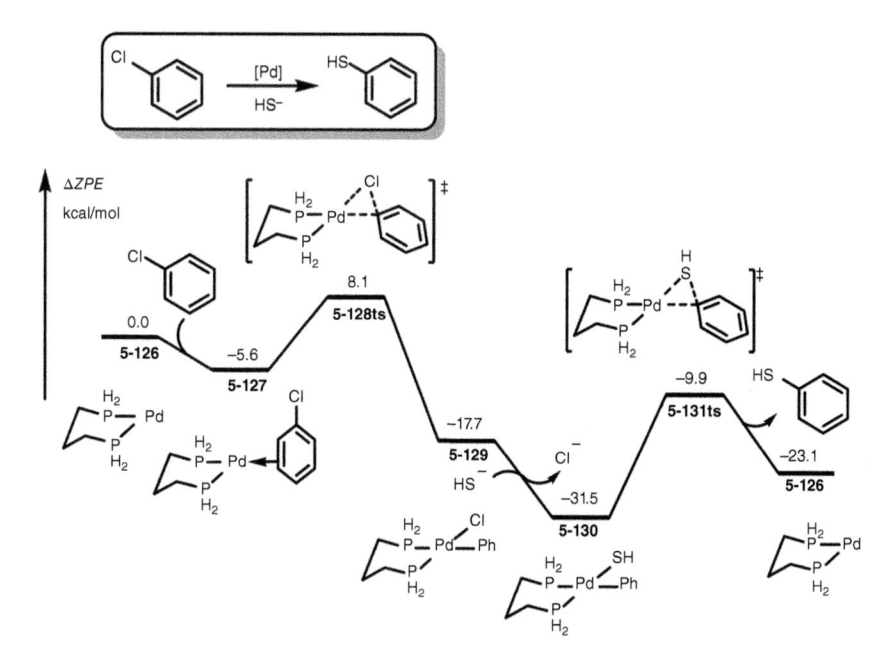

Figure 5.16 The zero-point energy profiles for a model reaction Pd-catalyzed sulfuration. The energies were calculated at B3LYP/6-31+G(d) (LACVP*+ for Pd)/PCM(THF) level of theory. Source: Based on Kozuch et al. [82].

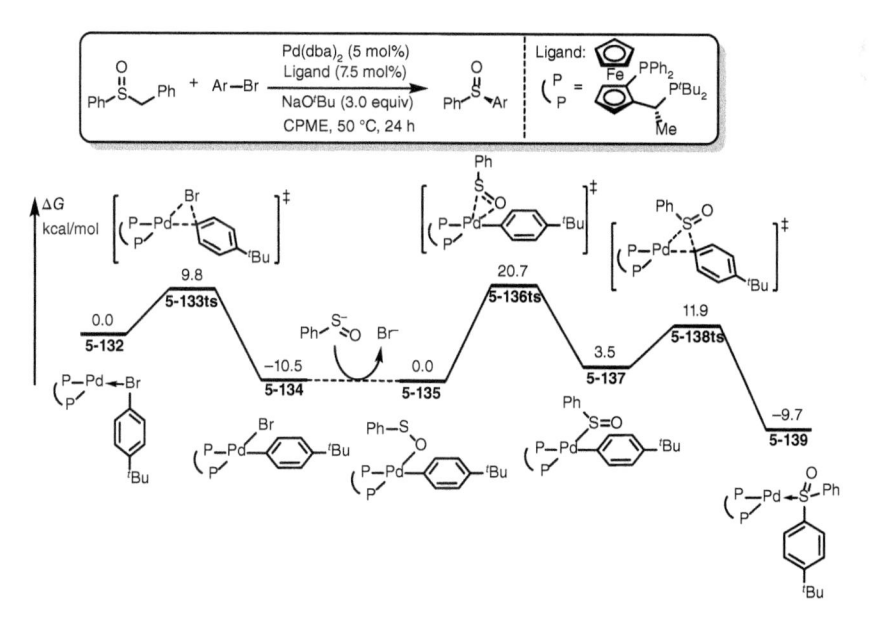

Figure 5.17 The free-energy profiles for a model reaction Pd-catalyzed sulfuration with benzylic sulfoxide. The energies were calculated at OLYP/def2-TZVP/COSMO(CPME) level of theory. CPME; cyclopentyl methyl ether.

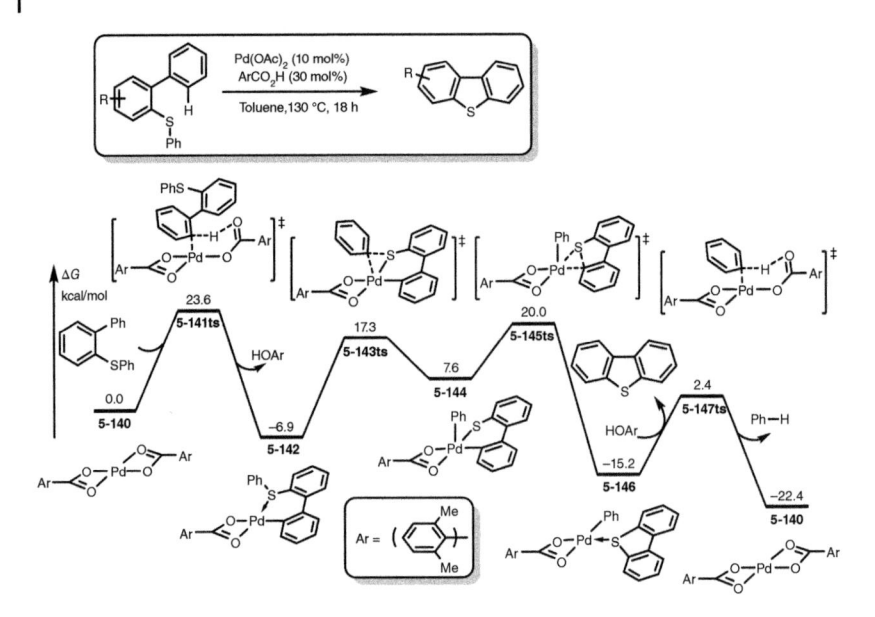

Figure 5.18 The free-energy profiles for intramolecular σ-bond metathesis of S—C(aryl) and H—C(aryl) bonds. The energies were calculated at M06-L/6-311+G(d,p) (SDD for Pd)/SMD(toluene)//B3LYP/6-31G(d) (SDD for Pd) level of theory.

5.2.3 C—I Bond Formation

Generally, the formation of C—I bond through reductive elimination from a Pd(II) species is both thermodynamically and kinetically unfavorable; therefore, driving force in other transformation is necessary for this process [85]. As shown in Figure 5.19, Lautens group reported a C(alkyl)—I bond formation reaction, involving an olefin insertion to form new C—C bond, which is considered to be the extra driving force for the whole transformation. DFT calculation is used to reveal the mechanism of this reaction. A diphosphine-coordinated Pd(0) species **5-148** is chosen as relative zero, which can be oxidized by phenyl iodide to form an aryl Pd(II) intermediate **5-151**. An intramolecular olefin insertion into Pd—C(aryl) bond occurs via transition state **5-152ts** with an energy barrier of 24.8 kcal/mol to form an alkyl Pd(II) species **5-153**. Then, a reductive elimination of alkyl and iodide could take place via a three-membered ring-type transition state **5-154ts**. The calculated energy barrier for this step is 24.9 kcal/mol. After ligand exchange, active catalyst **5-148** can be regenerated through a dissociative process. DFT calculations also predicted that the barriers for the corresponding reductive elimination for the formation of C—Br and C—Cl bonds are 27.4 and 27.9 kcal/mol, respectively. The higher barriers can be attributed to the stronger bond dissociation energies of Pd—Br and Pd—Cl bonds.

Xi group also provided another example for the conversion of aryl iodides to alkyl iodides. The DFT calculation results are shown in Figure 5.20 [85]. The oxidative addition of aryl iodide onto Pd(0) species **5-155** takes place via transition state **5-156ts** to afford an aryl Pd(II) intermediate **5-157**. The coordination of alkyne followed by its

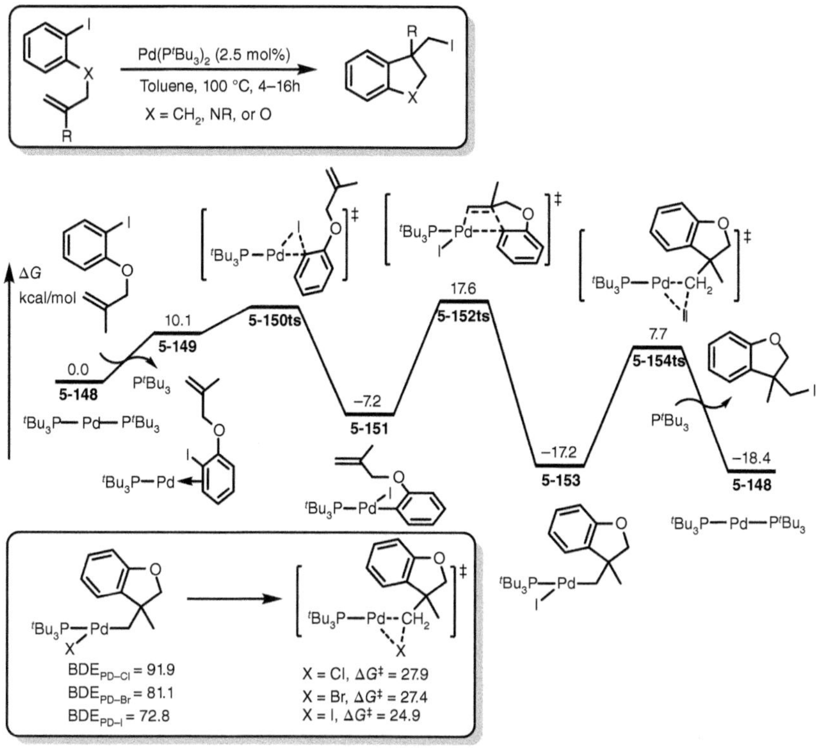

Figure 5.19 The free-energy profiles for a Pd-catalyzed C(alkyl)–I bond formation reaction. The energies were calculated at BP86/6-31+G(d) (SDD for Pd, Br, and I)/CPCM(toluene)//BP86/6-31+G(d) (SDD for Pd, Br, and I) level of theory.

insertion occurs via transition state **5-158ts** with a barrier of 17.8 kcal/mol to form a vinyl Pd(II) intermediate **5-159**. A σ bond metathesis between Pd—C(vinyl) and N—C(alkyl) bond takes place via a four-membered ring-type transition state **5-160ts** with an activation free energy of 38.7 kcal/mol to form an alkyl Pd(II) species **5-161** with the formation of indole moiety. The intramolecular metathesis is considered to be the rate-determining step for the whole catalytic cycle. The formation of indole leads to irreversible for the transformation. The coordination of phosphine ligand and *tert*-butoxide lithium forms a stable intermediate **5-162**. Then, a reductive elimination forms C(alkyl)–I via a three-membered ring-type transition state **5-163ts** with an energy barrier of 37.8 kcal/mol. The calculated high activation free energies for the catalytic cycle are coincident with a reaction temperature of 130 °C.

5.2.4 C—Si Bond Formation

The bis-silylation of unsaturated organic compounds, which can be formally represented as the insertion of unsaturated bond or atom into a Si—Si bond, is a particularly interesting transformation, which allows the simultaneous creation of two new Si—C bonds. The general catalytic cycle is shown in Scheme 5.9, the reaction starts

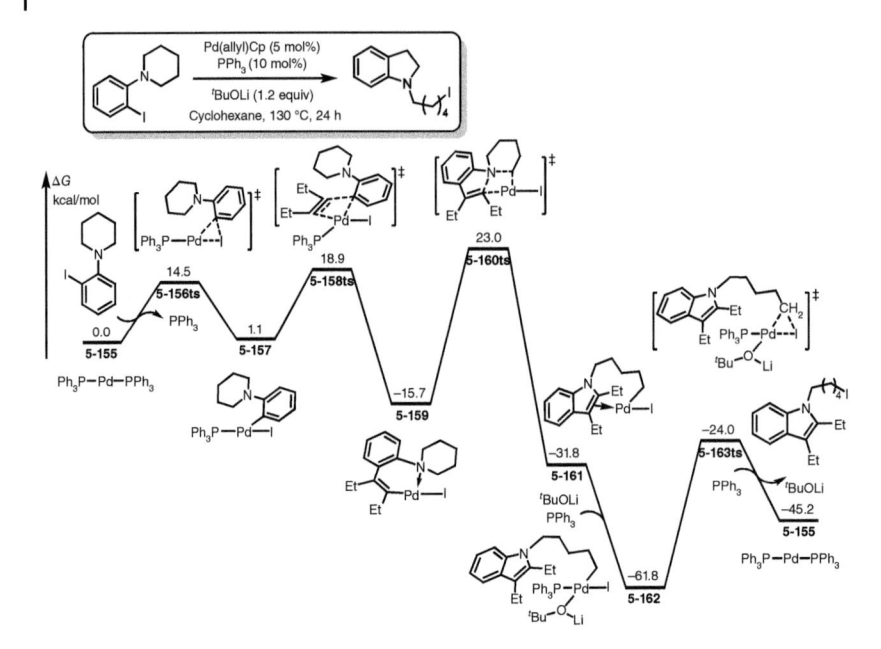

Figure 5.20 The free-energy profiles for a Pd-catalyzed conversion of aryl iodides to alkyl iodides. The energies were calculated at M06/6-311+G(d,p) (SDD for Pd and I)/PCM (cyclohexane)//B3LYP/6-31G(d) (LANL2DZ for Pd and I) level of theory.

from a Pd(0) species. Then, an oxidative addition of disilane affords a disilyl Pd(II) intermediate. After an unsaturated molecule insertion into Pd—Si bond, a C—Pd bond is formed. The reductive elimination would yield bis-silylation product and regenerate Pd(0) species.

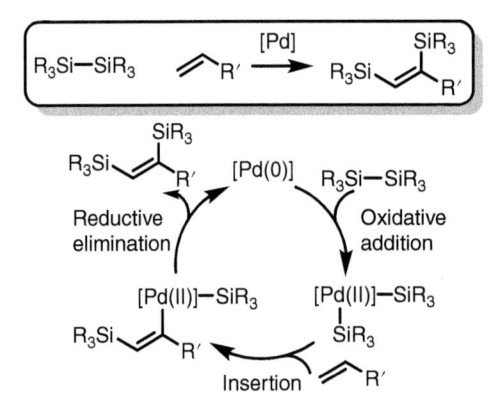

Scheme 5.9 Mechanism of Pd-catalyzed silyation of unsaturated bonds.

As shown in Scheme 5.9, when alkyne is used as unsaturated molecule, a bis-silylation can provide bis-silylethane derivatives in the presence of Pd-catalyst. DFT study is carried out to explore the mechanism of a model reaction [86]. As

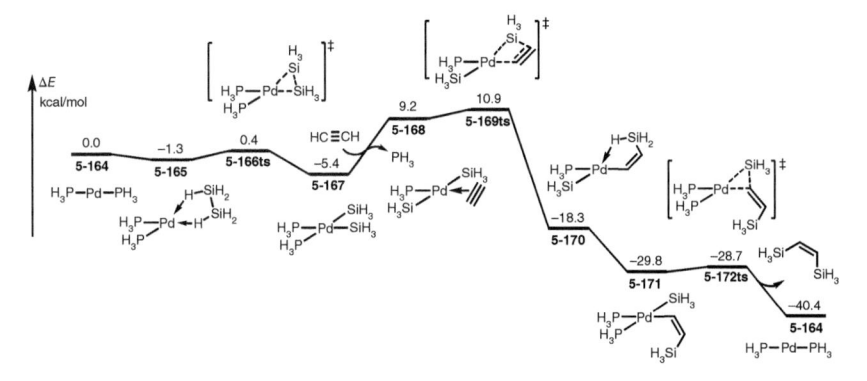

Figure 5.21 The energy profiles for a model reaction of Pd-catalyzed bis-silylation of alkynes. The energies were calculated at B3LYP/DZVP level of theory.

shown in Figure 5.21, diphosphine Pd(0) species **5-164** is set to relative zero, which can be coordinated by disilane to form intermediate **5-165** reversibly. The oxidative addition of disilane onto Pd takes place via a very early transition state **5-166ts** with an energy barrier of only 1.7 kcal/mol to afford disilyl Pd(II) intermediate **5-167**. The geometry information of transition state **5-166ts** is given in Figure 5.21. The lengths of forming Pd—Si bond and breaking Si—Si bond are 2.57 and 2.44 Å, respectively, which is close to them in coordination intermediate **5-165**. Therefore, an early transition state is observed. When disilyl Pd(II) intermediate **5-167** is formed, the coordination of acetylene in intermediate **5-168** leads to the dissociation of a phosphine ligand with 14.6 kcal/mol endergonic. Then, an acetylene insertion into Pd—Si bond takes place via transition state **5-169ts** with a barrier of only 1.7 kcal/mol to afford a vinyl Pd(II) intermediate **5-170**. The coordination of phosphine ligand generates intermediate **5-171**. Then, a rapid reductive elimination via a three-membered ring transition state **5-172ts** affords bis-silylethane product. The calculated rate-determining step is the acetylene insertion with an observed activation free energy of 21.5 kcal/mol.

Our group reported a theoretical study of Pd-catalyzed bis-silylation of carbene, which can provide a bis-silylmethane product correspondingly [87]. The reaction also starts from a diphosphine-coordinated Pd(0) species **5-173**. The calculated barrier of oxidative addition with disilane onto Pd(0) via transition state **5-174ts** is 12.4 kcal/mol to reversibly generate a disilyl Pd(II) species **5-175**. However, the calculated activation free energy for the following carbenation is as high as 31.3 kcal/mol via transition state **5-176ts**. Therefore, this pathway is excluded. Alternatively, a carbenation can take place first via transition state **5-178ts** with an activation free energy of 22.8 kcal/mol to afford a Pd–carbene complex **5-179**. Then, an oxidative addition of disilane happens via transition state **5-180ts** with a free-energy barrier of 14.7 kcal/mol to form a common disilyl Pd(II) species **5-177**. A carbene insertion into Pd—Si bond takes place via transition state **5-181ts** with an energy barrier of only 1.1 kcal/mol to afford an alkyl Pd(II) intermediate **5-182**. The reductive elimination of silyl and alkyl groups formed bis-silylmethane product via transition state **5-183ts** (Figure 5.22).

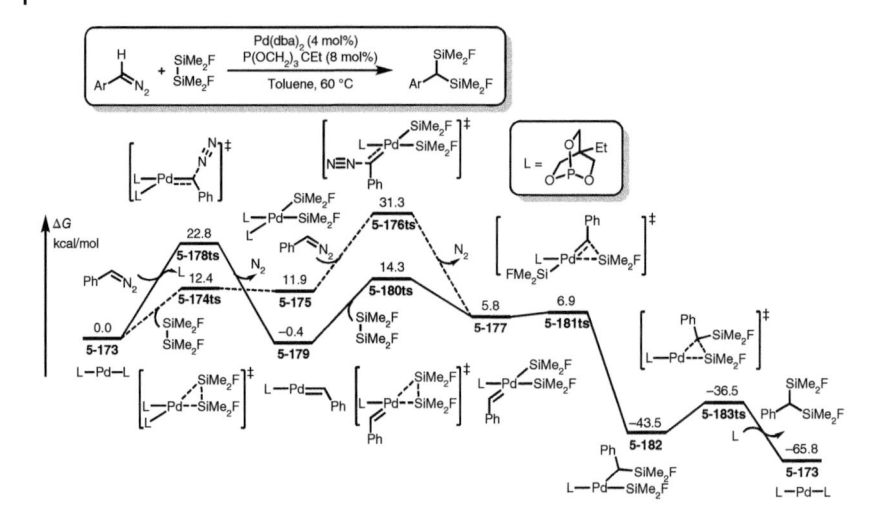

Figure 5.22 The free-energy profiles for Pd-catalyzed bis-silylation of carbenoids. The energies were calculated at M11-L/6-311+G(d,p) (LANL08f for Pd)/SMD(toluene)//B3LYP/6-31G(d) (LANL08f for Pd) level of theory.

Due to the low electronegativity of Si, silyl group can be considered to be a strong electron-donation group, which can be used for stable high-valence metal complex. Therefore, in the presence of Si, transition metals can be oxidized to form a formal high valence. As an example, in the theoretical study of Pd-catalyzed intermolecular σ-bond metathesis between cyclobutenones and silacyclobutanes, a Pd(0)–Pd(II)–Pd(IV) catalytic cycle is proposed and proved [88]. As shown in Figure 5.23, the oxidative addition of cyclobutenone onto Pd(0) species **5-184** can break the C(aryl)—C(acyl) bond to form a pallada(II)cycle **5-186** via transition state **5-185ts** with an energy barrier of 21.5 kcal/mol. Then, a further oxidative addition with silacyclobutane onto Pd(II) via transition state **5-187ts** leading to the cleavage of Si—C(alkyl) bond forms a Pd(IV) intermediate **5-188**. The calculated energy barrier of second oxidative addition is 23.3 kcal/mol. A silyl aryl reductive elimination occurs via transition state **5-190ts** to form a pallada(II)cycle **5-191**. Then, second reductive elimination of acyl and alkyl groups yields silacycle product.

5.3 Pd-Mediated C—H Activation Reactions

There is no doubt that transition-metal-catalyzed C—H activation and further functionalization is one of the most atom-economical reactions in organometallic catalysis, because all heavy atoms in the raw material are retained after the reaction. Various transition metal–involved catalysts revealed great reactivity in homogeneous C—H bond activation and functionalization [89–106]. Pd is undoubtedly the most commonly used transition metal in this type of reaction, which has attracted extensive attention in recent several decades [107–123]. In Pd-catalyzed C—H activation and functionalization, a C—H bond in arene reveals more reactivity than that

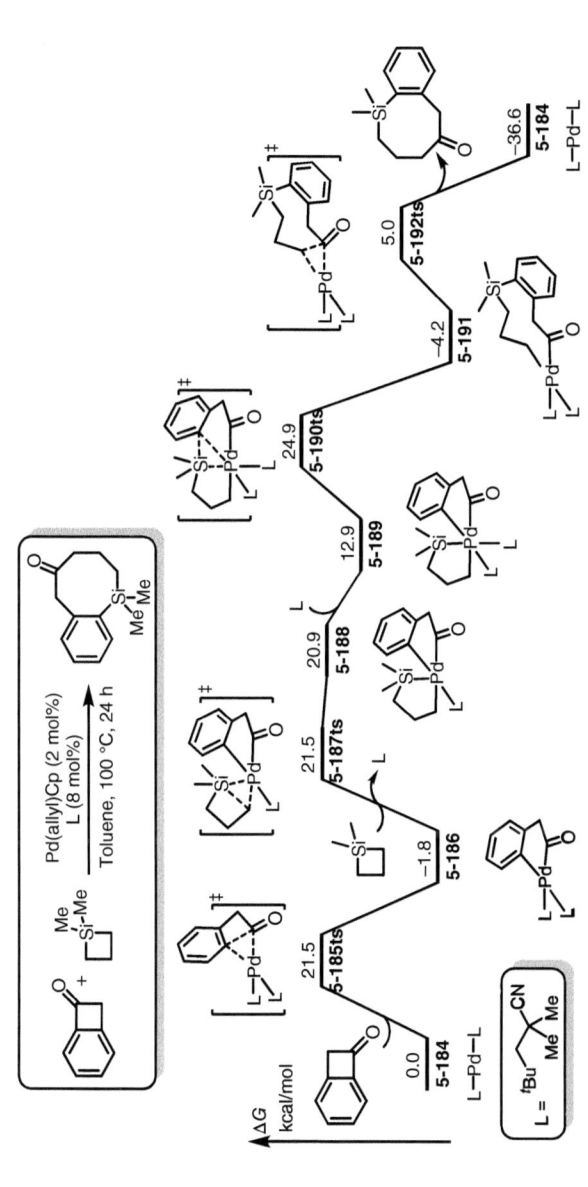

Figure 5.23 The free-energy profiles for Pd-catalyzed intermolecular σ-bond metathesis between cyclobutenones and silacyclobutanes. The energies were calculated at M06/6-311+G(d,p) (SDD for Pd)/SMD(toluene)//B3LYP/6-31G(d) (LANL2DZ for Pd) level of theory.

in alkanes. However, C(sp³)—H bond activation can also be obtained by choosing appropriate directing groups.

In a C—H bond, the electronegativity of carbon atom is slightly larger than that of hydrogen atom; therefore, it can be considered as a weak polar covalent bond, while carbon part represents a formal nucleophile. Following this idea, when an electrophile is used in Pd-catalyzed C—H activation and functionalization, a cross-coupling between nucleophile and electrophile can occur with formal redox neutral. Alternatively, if C—H bond reacts with another nucleophile, extra oxidant is necessary to take away undesired electrons, which is named oxidative coupling correspondingly.

A series of experimental and theoretical studies have shown that in most Pd-mediated C—H bond activation [124–135], the cleavage of C—H bond can proceed through a simultaneous metalation and intramolecular deprotonation, the so-called CMD mechanism. Moreover, some other mechanisms also can occur in Pd-catalyzed C—H activation. When a cationic Pd(II) reacts with an electron-rich arene, the coordination of arene onto Pd increases the acidity of C(aryl)—H, which can be deprotonated by exogenous base. This process can be considered as an electrophilic deprotonation. When a Pd—heteroatom σ-complex reacts with C—H bond, C—H bond cleavage can undergo a four-membered ring-type metathesis to form a C—Pd bond, which is named σ-complex-assisted metathesis mechanism. In a Pd(0)-mediated C—H bond cleavage, Pd atom inserts into C—H σ-bond to form a C—Pd(II)—H intermediate; therefore, oxidative addition is also one of the mechanisms for C—H activation.

In Pd-mediated concerted-metalation–deprotonation-type C—H bond activations, when a weak legate group (directing group) is involved in substrate, the reversible chelation of that group would partially cancel the entropy loss for an intermolecular combination with catalyst; therefore, the corresponding C—H bond cleavage is significantly accelerated. The commonly used directing groups can be divided into two categories, involving coordinative-directing groups and covalent-directing groups. Atoms with lone-pair electrons, such as nitrogen or oxygen, can coordinate with Pd before C—H activation to accelerate this process. Alternatively, some covalent bonds, involving Pd—C, Pd—N, Pd—O, can be formed reversibly and also can play as directing group to assist C—H activation. Assisted by directing group, the *ortho*-C—H bond in arenes usually demonstrates more reactivity. In some well-designed examples, *meta*-C—H bond in arenes can be also activated by specific directing groups. In the presence of directing group, even some C(sp³)—H bond can be activated by Pd-catalyst in mild reaction conditions. In this chapter, the author would try to discuss the mechanism of Pd-catalyzed C—H activation according to this classification.

5.3.1 Chelation-Free C(sp³)—H Activation

Methane C(sp³)—H bond is stable and unreactive. Therefore, the C(sp³)—H activation of methane is still a challenge for organometallic chemistry. A successful homogeneous approach of this process is shown in Scheme 5.10, in which a

5.3 Pd-Mediated C—H Activation Reactions | 207

relative harsh reaction condition is necessary. In the presence of bis-*N*-heterocyclic carbine-coordinated Pd(II) species and trifluoroacetic acid, methane can be oxidized by potassium peroxydisulfate to afford corresponding methyl trifluoroacetate.

Scheme 5.10 Pd-catalyzed C—H activation of methane.

The detailed mechanism study of methane C—H activation is reported by Strassner group (Figure 5.24) [136]. The reaction starts from a Pd(II) species **5-193**. The dissociation of trifluoroacetate forms a cationic Pd(II) intermediate, which can be coordinated by methane to afford intermediate **5-194** with 16.1 kcal/mol endergonic. The C—H cleavage of methane can occur via a CMD-type transition state **5-195ts**. The calculated overall activation free energy of this process is 39.5 kcal/mol, which is the rate-determining step for the whole catalytic cycle. After C—H bond cleavage, a methyl Pd(II) intermediate **5-197** is formed by release of a trifluoroacetic acid. The oxidation of intermediate **5-197** by peroxydisulfate affords a Pd(IV) species **5-198**. Then, a reductive elimination can yield methyl trifluoroacetate and regenerate Pd(II) species **5-193**.

In a similar example reported by the same group, the C—H activation of propane is studied by DFT calculations (Scheme 5.11) [132]. The results showed that the C—H bond cleavage undergoes a CMD mechanism. The calculated activation free energy

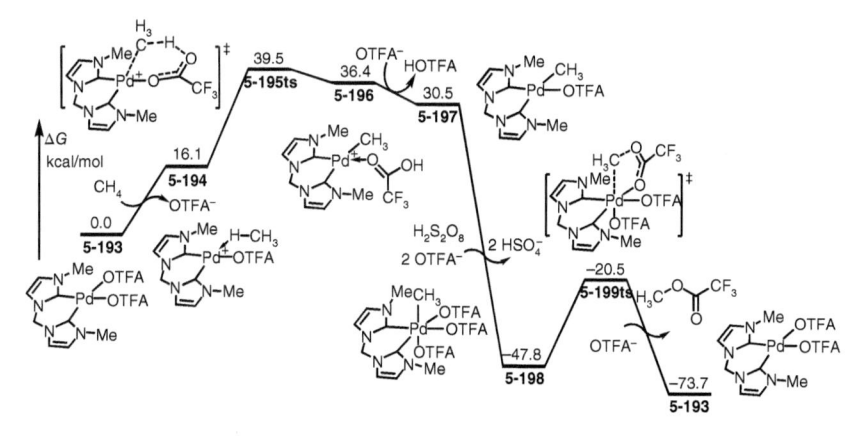

Figure 5.24 The free-energy profiles for Pd-catalyzed C—H activation of methane. The energies were calculated at B3LYP-D3/6-311++G(d,p) (LANL2DZ for Pd)/CPCM (trifluoroacetic acid)//B3LYP/6-31G(d) (LANL2DZ for Pd) level of theory. Source: Based on Munz and Strassner [136].

of primary C—H cleavage is 38.1 kcal/mol, which is 5.0 kcal/mol lower than that of secondary one.

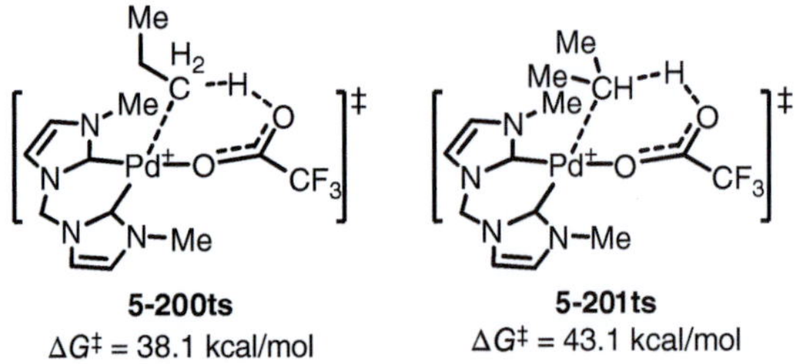

Scheme 5.11 The transition states of Pd-mediated C—H bond cleavage of propane. The energies were calculated at B3LYP-D3/6-311++G(d,p) (LANL2DZ for Pd)/CPCM (trifluoroacetic acid)//B3LYP/6-31G(d) (LANL2DZ for Pd) level of theory. Source: Based on Munz et al. [132].

5.3.2 Chelation-Free C(sp²)—H Activation

In the absence of chelated-directing group, the Pd-catalyzed C(sp²)—H bond activation is easier than the corresponding cases for C(sp³)—H bond. The reactivity of C(sp²)—H bonds for a series of arenes is studied by Fagnou group [33]. As shown in Scheme 5.12, an acetate Pd(II) species **5-202** was chosen as model substrate. All the arenes summarized in Scheme 5.13 are considered through a concerted-metalation–deprotonation process via transition state **5-203ts**. The geometry of that transition state is shown in Scheme 5.12. When arene coordinates with Pd, the C—H bond is activated by Pd, which leads to a deprotonation by the coordinated carboxylate. The lengths of forming Pd—C bond and breaking C—H bond are 2.24 and 1.42 Å, respectively, which exhibits a concerted process.

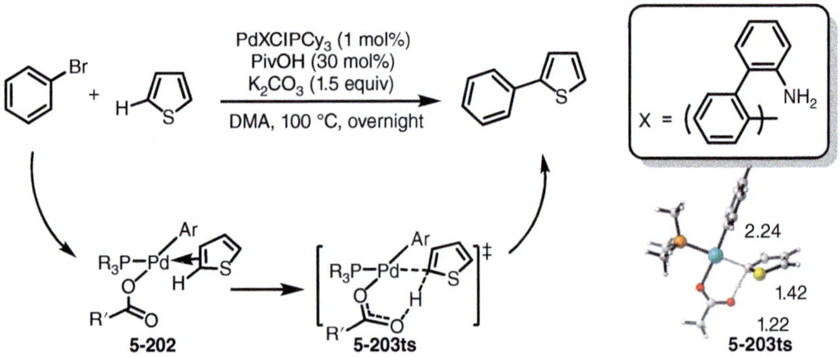

Scheme 5.12 Pd-catalyzed C—H activation/arylation.

The calculated activation free energies are labeled on corresponding C—H bonds. Generally, the reactivity of heteroarenes is higher than that of benzene. The electron-withdrawing group on arenes is favorable for C—H bond activation. The

5.3 Pd-Mediated C–H Activation Reactions | 209

Scheme 5.13 The calculated activation free energies for the Pd-mediated C–H activation.

distortion–interaction analysis revealed that the activation energies are correlated to the distortion energies in CMD-type transition states (Scheme 5.14).

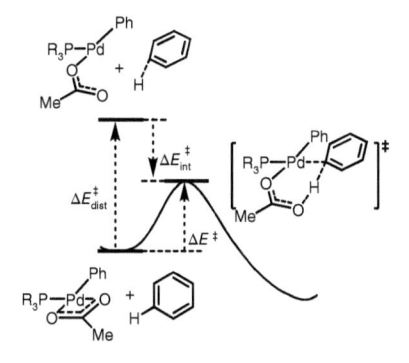

Scheme 5.14 Distortion–interaction analysis of the C–H bond cleavage step in Pd-catalyzed C–H functionalizations.

In the presence of AgOAc oxidant, the Pd-catalyzed oxidative coupling of benzene and chromone leads to the generation of arylation chromone product. DFT calculations for the detailed mechanism are shown in Figure 5.25 [137]. The coordination of benzene onto Pd(TFA)$_2$ leads to the formation of complex **5-205**. A CMD-type C—H activation takes place via transition state **5-206ts** with an energy barrier of 17.7 kcal/mol to form a phenyl Pd(II) intermediate **5-207**. There are two possible pathways for the subsequent phenylation of chromone. When this process occurs through chromone C—H activation followed by a reductive elimination through transition states **5-209ts** and **5-211ts**, the C3 selectivity is a favorable pathway

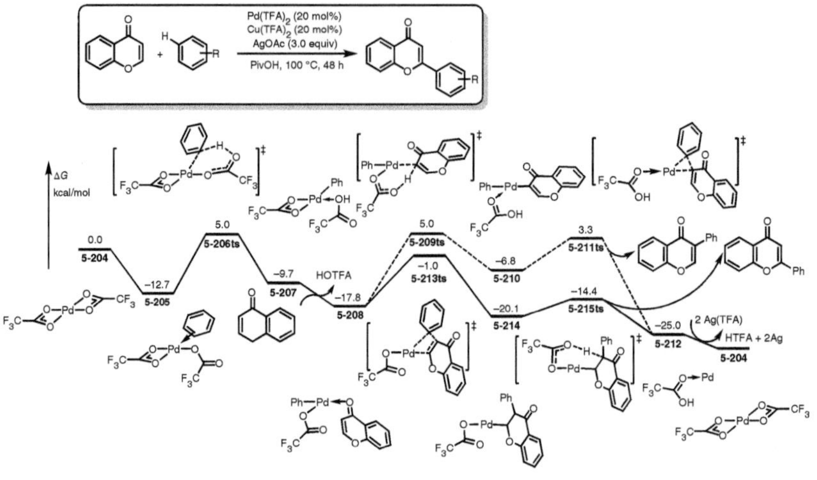

Figure 5.25 The free-energy profiles for Pd-catalyzed oxidative coupling of benzene and chromone. The energies were calculated at B3LYP-D3/6-311++G(d,p) (LANL2TZ(f) for Pd)/CPCM(acetic acid)//ωB97X-D/6-31G(d) (LANL2DZ for Pd) level of theory. Source: Based on Choi et al. [137].

with an energy barrier of 23.8 kcal/mol. Alternatively, olefin insertion followed by β-hydride elimination is also considered. The calculated barrier of olefin insertion is only 16.8 kcal/mol via transition state **5-213ts**. Then, the following rapid β-hydride deprotonation yields C2-phenylation product. The generated Pd(0) species can be oxidized by AgOAc to regenerate Pd(II) species. Therefore, the chromone insertion followed by β-hydride elimination is favorable.

In addition to arene, C(sp²)—H bond in aldehydes also can undergo a Pd-mediated chelation-free activation to form an acyl Pd intermediate [138]. As shown in Figure 5.26, in the presence of base, the Pd-catalyzed cross-coupling of aldehydes and aryl bromides can happen under a high temperature of 150 °C. The reaction starts from a carbonate Pd(II) species **5-216**. The coordination of benzaldehyde leads to 18.2 kcal/mol endergonic. The Pd-mediated C(acyl)—H bond activation takes place via transition state **5-218ts** with an energy barrier of only 4.7 kcal/mol to afford a hydrogen bond intermediate **5-219**. Then, the isomerization of **5-219** forms acyl Pd(II) intermediate **5-220ts**. The oxidative addition of phenyl bromide onto Pd(II) leads to the formation of a Pd(IV) intermediate **5-223**. Then, a rapid reductive elimination forms C(phenyl)—C(acyl) bond and regenerates Pd(II) active species **5-216** by the coordination of carbonate. The calculated overall activation free energy is 41.8 kcal/mol in the oxidative addition step.

5.3.3 Coordinative Chelation-Assisted *ortho*- C(aryl)—H Activation

As far as we know, the mechanism of chelation-assisted *ortho*- C(sp²)—H activation was first proposed by Davies and Macgregor in 2005 [139]. In their theoretical study, benzylamine is chosen as substrate to react with Pd(OAc)₂ (Figure 5.27). The amino group–coordinated Pd(II) species **5-226** is chosen as relative zero. Then,

Figure 5.26 The free-energy profiles for Pd-catalyzed cross-coupling of aldehyde and aryl bromide. The energies were calculated at M06-2X/6-31++G(d,p) (LACVP**++ for Pd) level of theory. Source: Based on Wakaki et al. [138].

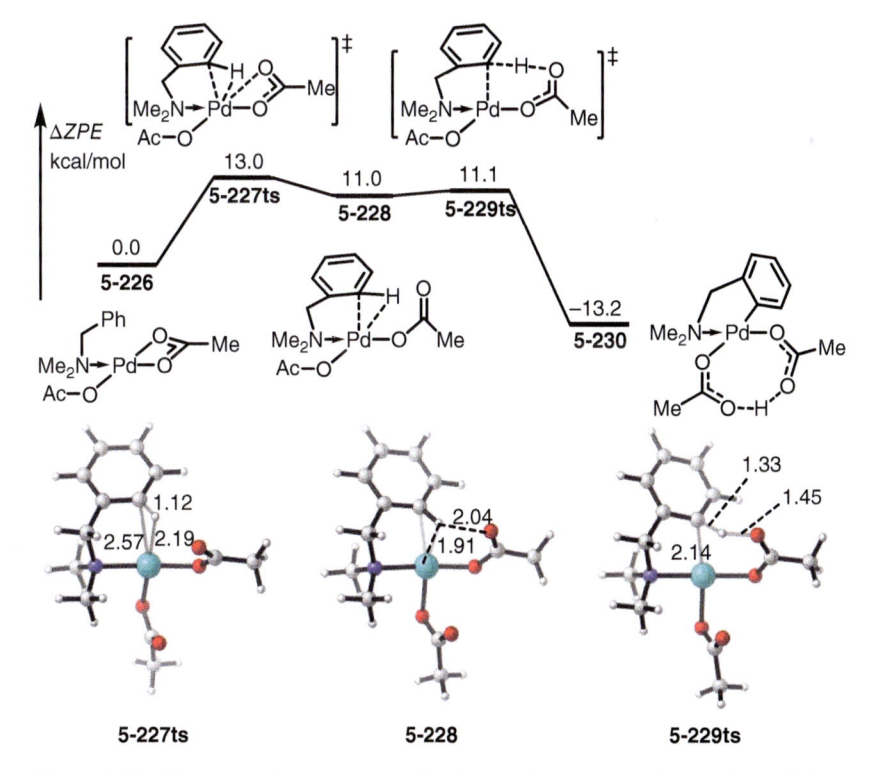

Figure 5.27 The zero-point energy profiles for the first-reported Pd-mediated CND-type C−H activation. The energies were calculated at BP86/6-31G(d,p) (RECP for Pd and P) level of theory.

a ligand exchange takes place via a substitution transition state **5-227ts** forming an agostic intermediate **5-228**, which is considered as the reacting precursor for C—H activation. In the geometry of intermediated **5-228**, the calculated distances of O(acetate)—H and Pd—H are 2.04 and 1.91 Å, respectively, which reveals strong interactions for the activation of this hydrogen. A concerted C(aryl)—H bond cleavage with the formation of C(aryl)—Pd bond formation takes place via a concerted six-membered ring-type transition state **5-229ts**. Thus, the amphiphilic palladium acetate provides electrophilic activation of a C—H bond and acts as an intramolecular base for the deprotonation. The acetate may also play a role in stabilizing the key agostic intermediate through hydrogen bonding. In the presence of amino-directing group, a cyclometalation intermediate **5-230** is afforded.

In another example, N-oxide of quinoline can play as a directing group, which can coordinate with Pd to assist a C8 selective C(sp^2)—H activation [128]. As shown in Scheme 5.15, when the reaction starts from a cationic Pd(II) species coordinated with quinoline N-oxide, an acetate-assisted C8—H activation can occur via a CMD-type transition state **5-231ts** with an energy barrier of 22.2 kcal/mol to afford cyclometalation intermediate. In a comparative case without coordination of N-oxide, the calculated energy barrier of C—H bond activation via corresponding CMD-type transition state **5-233** is 33.9 kcal/mol. Therefore, the coordinative chelation is favorable for CMD-type C—H bond activations.

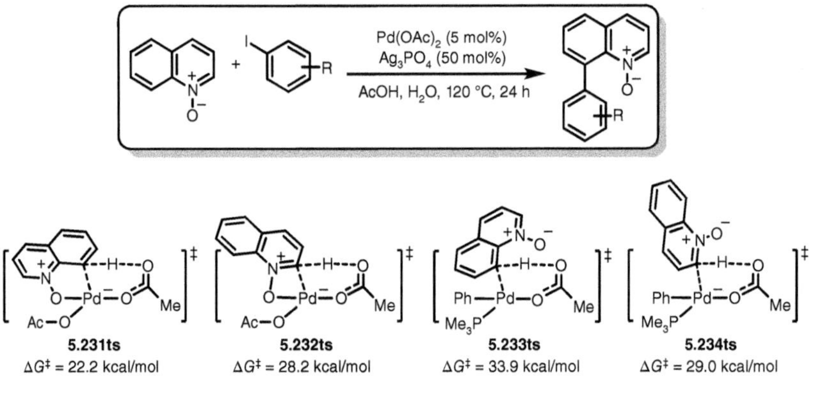

Scheme 5.15 The key transition states with (**5-231ts** and **5-232ts**) or without (**5-233ts** and **5-234ts**) the assistance of directing groups in Pd-catalyzed C8 selective arylation of quinoline N-oxide. The energies were calculated at B3LYP/6-311+G(d,p) (SDD for Pd and P)/SMD(DMF)//B3LYP/6-31G(d,p) (SDD for Pd and P) level of theory.

5.3.4 Covalent Chelation-Assisted *ortho*- C(aryl)—H Activation

When a covalent bond is formed between Pd and the directing group of reacting arene, the reactivity of Pd-mediated CMD type C—H bond activation may not be increased because of the strain of forming covalent bond with Pd [140]. As shown in Scheme 5.16, when a covalent bond is formed between aryl and Pd in intermediate **5-235**, the calculated barrier of CMD type C—H bond cleavage is 23.5 kcal/mol via a six-membered ring-type transition state **5-236ts**.

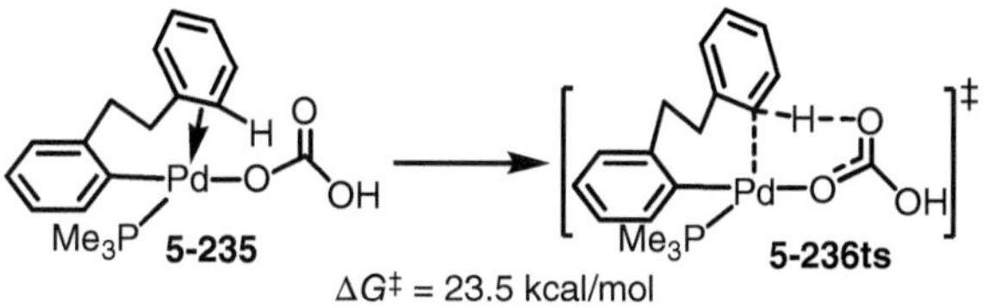

5.3 Pd-Mediated C–H Activation Reactions | 213

Scheme 5.16 Pd-mediated covalent chelation-assisted ortho- C(aryl)–H activation.

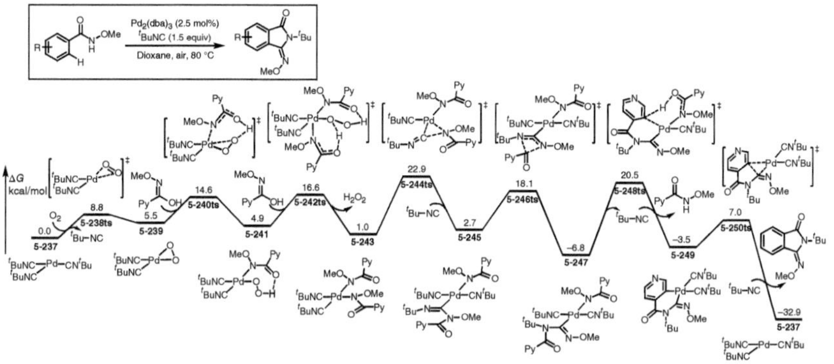

Figure 5.28 The free-energy profiles for Pd-catalyzed oxidative coupling of isonicotinamide and isonitrile. The energies were calculated at M06-L/6-311++G(d,p) (SDD for Pd)/SMD(dioxane)//B3LYP/6-31G(d,p) (SDD for Pd) level of theory. Source: Based on Dang et al. [124].

The mechanism for the formation of covalent bond between Pd and reacting arenes might be complicated [124]. As shown in Figure 5.28, in air condition, a Pd-catalyzed oxidative coupling of isonicotinamide and isonitrile leads to the formation of imino pyrrolone. The DFT calculations revealed the detailed mechanism of this reaction. As shown in Figure 5.28, in the presence of dioxygen, Pd(0) species **5-237** can be oxidized to a singlet cyclic dioxide Pd(II) intermediate **5-239** with 5.5 kcal/mol endergonic. After two-step reactions of deprotonation with isonicotinamide, a diamino Pd(II) intermediate **5-243** is given. An isocyanide insertion into Pd—N bond occurs via a three-membered ring-type transition state **5-244ts** to form a carbamimidoyl Pd(II) intermediate **5-245**, which can further be isomerized to its stable isomer **5-247** through acyl migration transition state **5-246ts**. Then, a CMD-type C—H bond cleavage takes place via a CMD-type transition state **5-248ts**. Although the covalent chelation was found in this step, the activation free energy of Pd-mediated C—H cleavage is as high as 27.3 kcal/mol. After C—H activation of pyridyl group, a rapid reductive elimination formed imino pyrrolone product with the regeneration of Pd(0) species **5-237**.

DFT calculation is used to reveal the mechanism of Pd/Cu co-catalyzed oxidative carbonylation of naphthyl amines to lactams (Figure 5.29) [141]. After oxidation by using Cu(II) species and gaseous dioxygen, an amino Pd(II) species **5-251** is formed. DFT calculation found that, in the presence of Pd—N covalent bond, the barrier of direct C8—H activation is as high as 26.6 kcal/mol via transition state **5-252ts**

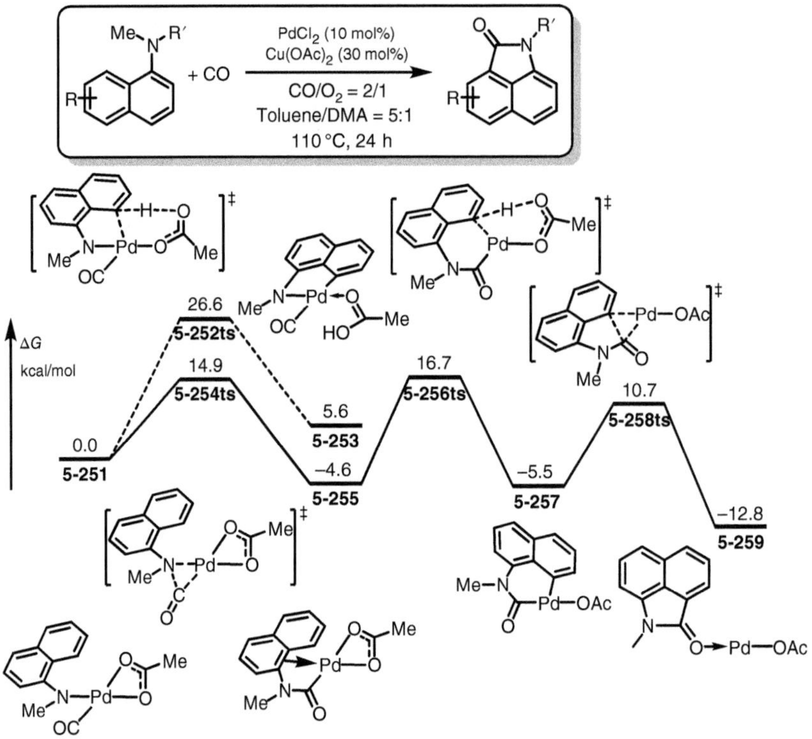

Figure 5.29 The free-energy profiles for the key step of Pd/Cu co-catalyzed oxidative carbonylation of naphthyl amines. The energies were calculated at M11-L/6-311++G(d) (SDD for Pd)/SMD(DMA)//B3LYP/6-31G(d,p) (SDD for Pd) level of theory. DMA; dimethylacetamide. Source: Based on Shi et al. [141].

to afford a five-membered palladacycle **5-253**. Alternatively, a carbonyl insertion into Pd—N bond can occur via transition state **5-254ts** with an energy barrier of only 14.9 kcal/mol to form a carbamoyl Pd(II) intermediate **5-255** reversibly. Then, a CMD-type C8—H bond cleavage takes place via transition state **5-256ts** with an energy barrier of 21.3 kcal/mol in the presence of Pd—C(acyl) covalent bond. Then, an acyl aryl reductive elimination leads to the formation of lactam product via transition state **5-258ts**.

5.3.5 Chelation-Assisted *meta*- C(aryl)—H Activation

In well-designed cases, a chelation-assisted *meta*-C(aryl)—H activation also can occur through a CMD-type pathway [142]. As shown in Scheme 5.17, when a nitrile-containing template is used, an oxidative coupling reaction between arenes and olefins is observed experimentally with a high regioselectivity of *meta*-position. DFT calculations revealed that the C—H activation takes place on a dimeric acetate Pd(II) complex. The calculated lowest barrier transition state is *meta*-**5-260ts** with a relative free energy of 24.8 kcal/mol. In the geometry of this transition state, one Pd linked with cyano group in substrate. The CMD-type C—H activation occurs on another Pd center. The corresponding activation barrier for *ortho-* or *para-* C(aryl)—H activations is 27.8 or 28.3 kcal/mol, which is higher than that of *meta-*C(aryl)—H activation. The case of mono-Pd-mediated C(aryl)—H activation was also considered theoretically. The calculated results showed that much higher energy barriers are observed through transition states **5-263ts, 5-264ts**, or **5-265ts**.

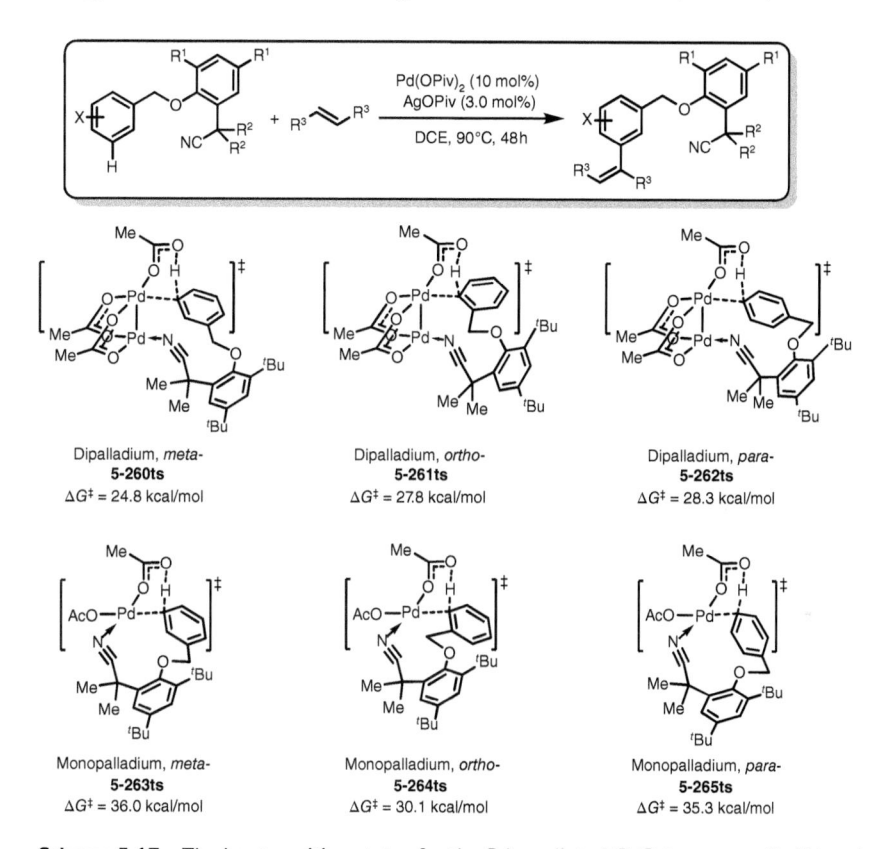

Dipalladium, *meta-*
5-260ts
ΔG‡ = 24.8 kcal/mol

Dipalladium, *ortho-*
5-261ts
ΔG‡ = 27.8 kcal/mol

Dipalladium, *para-*
5-262ts
ΔG‡ = 28.3 kcal/mol

Monopalladium, *meta-*
5-263ts
ΔG‡ = 36.0 kcal/mol

Monopalladium, *ortho-*
5-264ts
ΔG‡ = 30.1 kcal/mol

Monopalladium, *para-*
5-265ts
ΔG‡ = 35.3 kcal/mol

Scheme 5.17 The key transition states for the Pd-mediated CMD-type *meta-*C—H bond functionalizations. The energies were calculated at M06/6-311++G(d,p) (SDD for Pd)/SMD(DCE)//B3LYP/6-31G(d) (LANL2DZ+f for Pd) level of theory.

Interestingly, when amino acid is used as an additive in Pd(II)-mediated C—H activations, a *meta-*C(aryl)—H functionalized product is also given experimentally by using the same nitrile-containing template (Scheme 5.18). DFT calculations showed that, although the formation of neutral acetamido Pd(II) **5-266** is endergonic, it is the best way for substrate loading [143], while the C(aryl)—H activation mode is considered as **5-267ts**, in which acetamido group plays as base to assist deprotonation. In the presence of directing group, a mono-Pd-mediated CMD-type C(aryl)—H bond activation can occur via transition state **5-267ts** with an energy barrier of 23.6 kcal/mol, which is slight lower than that of **5-268ts** or **5-269ts** leading to the formation of *para-* or *ortho-*C(aryl)—H activation.

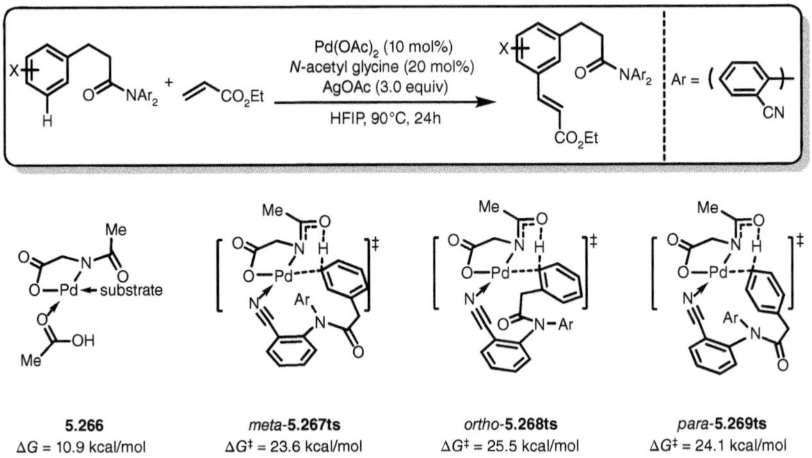

Scheme 5.18 The key transition states for the Pd-mediated CMD-type *meta*-C—H bond functionalizations assisted with *N*-acetyl glycine. The energies were calculated at M06/6-311++G(d,p) (SDD for Pd)/SMD(HFIP)//B3LYP/6-31G(d) (LANL2DZ+f for Pd) level of theory.

5.3.6 Coordinative Chelation-Assisted C(sp³)—H Activation

With the assistance of coordinative chelation group, a selective C(sp³)—H bond activation can occur through CMD-type mechanism [144]. As shown in Scheme 5.19, DFT calculation is used to reveal the mechanism of the C—H activation process in the Pd-catalyzed oxazolylmethyl C—H iodinations. The reaction starts from a trimer Pd₃(OAc)₆, which can be coordinated by oxazole to form a monomer Pd(II) species **5-270** with 6.7 kcal/mol endergonic. The C—H bond cleavage can undergo a CMD-type process via transition state **5-273ts** to form an alkyl Pd(II) intermediate by release of acetic acid. The calculated activation free energy for the C—H bond cleavage step is 26.2 kcal/mol. DFT calculation found that the diastereoselectivity is controlled by the position of leaving hydrogen atom in C—H activation transition state. As shown in Scheme 5.19, when the leaving hydrogen is close to the bulky group, the strain release would lead to a lower relative energy for that transition state.

5.3.7 Covalent Chelation-Assisted C(sp³)—H Activation

Consistent with the case of C(sp²)—H activation mentioned earlier, the covalent chelation cannot simply reduce the activation free energy of C(sp³)—H activation. However, the regioselectivity can be controlled by the covalent-directing group. As an early example, Fagnou reported an intramolecular cross-coupling of C(alkyl)—H bond with aryl bromide to synthesize benzofuran derivatives (Scheme 5.20) [145]. DFT calculations found that after an oxidative addition of aryl bromide followed by a counterion exchange, a CMD-type C(alkyl)—H cleavage can occur via transition state **5-275ts** with an energy barrier of 27.0 kcal/mol. The regioselectivity of different hydrogens was also considered. The primary hydrogen is more active than secondary

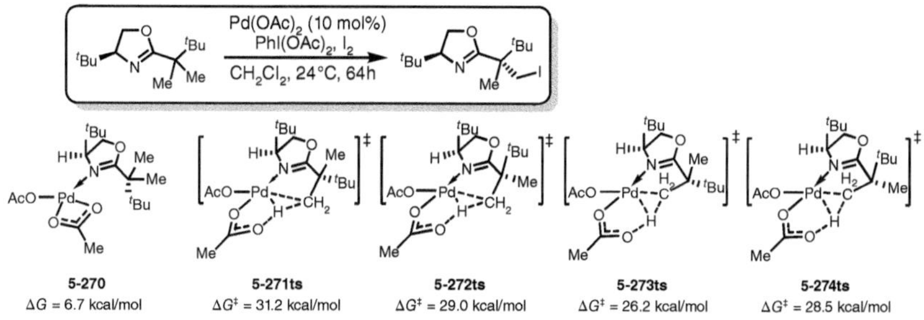

Scheme 5.19 The key transition states for the Pd-mediated CMD-type selective C(sp³)–H bond activation. The energies were calculated at M06/6-311++G(d,p) (SDD for Pd)/SMD(DCM)//B3LYP/6-31G(d) (SDD for Pd) level of theory.

hydrogen (**5-275ts** vs. **5-276ts**). The formation of six-centered palladacycle is more favorable than that of seven-centered one (**5-275ts** vs. **5-277ts**).

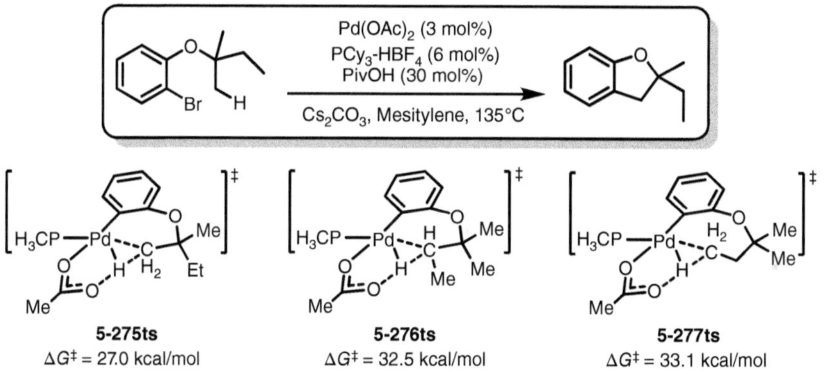

Scheme 5.20 The key transition states for the Pd-mediated CMD-type covalent chelation-assisted C(sp³)–H bond activation. The energies were calculated at B3LYP/DZVP (TZVP for Pd) level of theory. Source: Based on Lafrance et al. [145].

In another example, a cyclobutane can be prepared from a similar reaction, which also involves an intramolecular cross-coupling of C(alkyl)—H bond with aryl bromide. As shown in Figure 5.30, DFT calculations studied the mechanism for the formation of five-membered palladacycle [146]. When a P^{t}Bu$_3$ is chosen as ligand, the oxidative addition of aryl bromide can occur via transition state **5-279ts** to form an aryl Pd(II) intermediate **5-280**. The counterion exchange formed carbonate Pd(II) **5-281**. The formed C(aryl)—Pd bond can promote the following C(alkyl)—H bond activation via a CMD-type transition state **5-282ts**. The calculated activation free energy of this step is 33.6 kcal/mol, which causes a reaction temperature of 140 °C. The followed reductive elimination results in the generation of cyclobutene product.

Experimentally, the acetyl-protected aminoethyl quinoline additive can significantly reduce the reaction temperature of a Pd-mediated coordinative chelation-assisted C(sp³)—H activation. As shown in Figure 5.31, DFT calculations

Figure 5.30 The free-energy profiles of Pd-catalyzed intramolecular cross-coupling of C(alkyl)−H bond with aryl bromide. The energies were calculated at B3PW91/6-311++G(d,p) (SDD for Pd)/PCM(DMF)//B3PW91/6-31G(d,p) (SDD for Pd) level of theory.

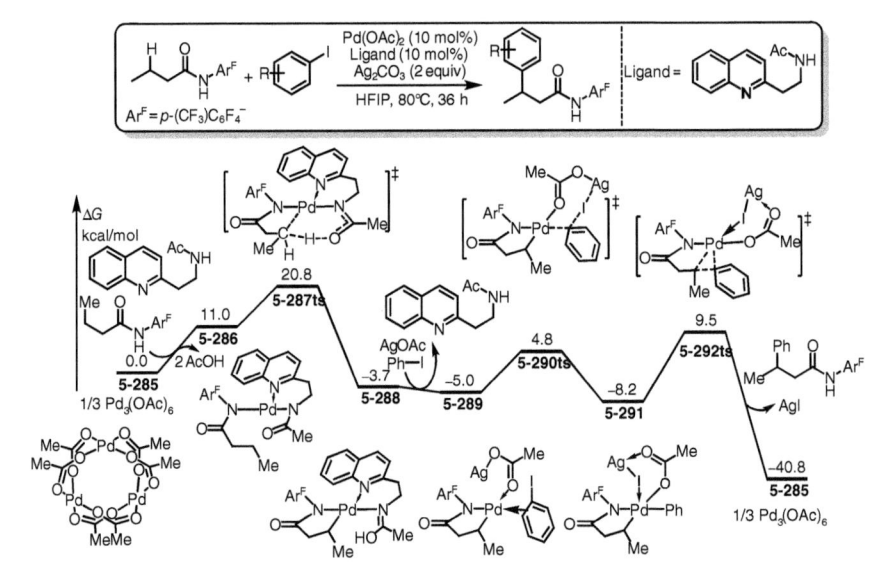

Figure 5.31 The free-energy profiles of Pd-catalyzed β-arylation of amides. The energies were calculated at M06/6-311++G(d,p) (LANL2DZ+f for Pd, I, and Ag)/SMD(HFIP)//B3PW91/6-31G(d) (LANL2DZ+f for Pd, I, and Ag) level of theory. Source: Based on Zhang et al. [147].

found that the amide additive can react with $Pd_3(OAc)_6$ first to form a chelated acetamido Pd(II) species **5-286** [148]. Although the formation of intermediate **5-286** is 11.0 kcal/mol endergonic, the alkalinity of acetamido group can promote a CMD-type C(alkyl)—H bond activation via transition state **5-287ts** to afford an alkyl Pd(II) species **5-288**. The calculated overall barrier of this process is only 20.8 kcal/mol, which is much lower than the above-mentioned cases. Then, the coordination of phenyl iodine leads to the dissociation of amide ligand and generates complex **5-289** with 1.3 kcal/mol exergonic. A silver-assisted oxidative addition takes place via transition state **5-290ts**, which results in a Pd(IV) intermediate **5-291** with 3.2 kcal/mol exergonic. In the presence of silver, the free-energy barrier of oxidative addition is only 9.8 kcal/mol. After this step, reductive elimination yields the β-arylation product and regenerates $Pd_3(OAc)_6$ catalyst.

5.3.8 C—H Bond Activation Through Electrophilic Deprotonation

When nucleophilic arenes are used in Pd-mediated C—H activations, the coordination of arenes onto Pd would activate the corresponding C(aryl)—H bond, which can bear an intermolecular deprotonation to afford C(aryl)—Pd bond for further transformations [149]. In the presence of chelation group, the corresponding Pd-mediated C—H activation can occur on C2 position of indoles. In the theoretical study for the mechanism of Pd-catalyzed indoles trifluoroethylation (Figure 5.32) [147], norbornene is necessary, which can play as a provisional covalent-directing group. In the catalytic cycle, an indolyl Pd(II) species **5-293** can be formed from a N—H bond cleavage. A reversible insertion of norbornene into Pd—N bond via transition state

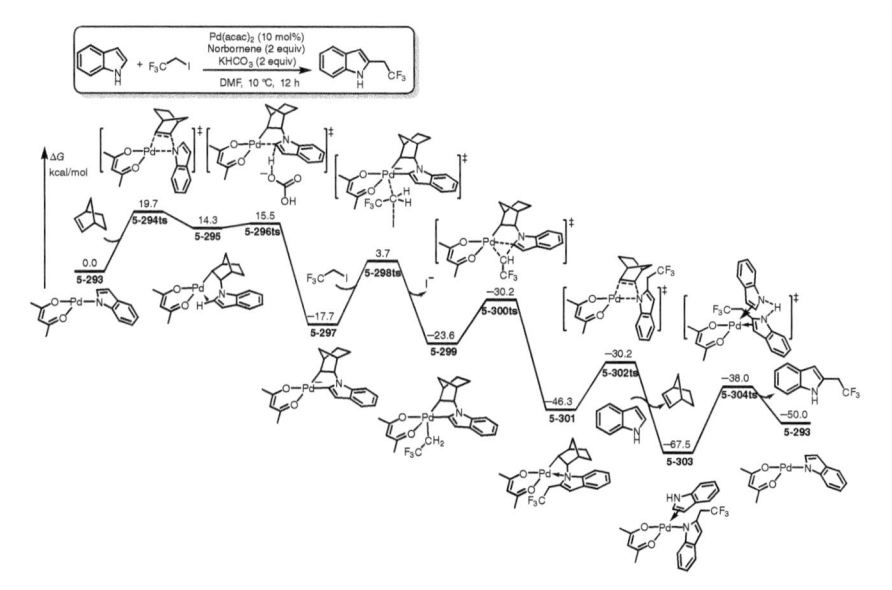

Figure 5.32 The free-energy profiles of norbornene-assisted Pd-catalyzed indoles trifluoroethylation. The energies were calculated at M11-L/6-311+G(d,p) (SDD for Pd)/SMD(DMF)//B3PW91/6-31G(d) (SDD for Pd) level of theory.

220 | 5 Theoretical Study of Pd-Catalysis

5-294ts affords an agostic complex **5-295**, in which the C2—H bond is activated by the electrophilic activation of Pd. Then, an outer-sphere deprotonation by bicarbonate leads to the formation of an anionic five-membered palladacycle **5-297** irreversibly. The calculated barrier of deprotonation via transition state **5-296ts** is only 1.2 kcal/mol. The formation of palladate increases the nucleophilicity; therefore, an S_N2-type oxidative addition with CF_3CH_2I can afford a neutral alkyl Pd(IV) species **5-299** via a linear transition state **5-298ts**. Subsequently, the C(aryl)—C(alkyl) reductive elimination formed C2-alkylated intermediate **5-301**. Then, a β-nitrogen elimination takes place via transition state **5-302ts** to regenerate norbornene and forms an alkylated indolyl Pd(II) species **5-303**. The proton transfer with another indole substrate via transition state **5-304ts** yields the final product and regenerates active species **5-293**. The calculated rate-determining step is the last proton transfer process with an energy barrier of 29.5 kcal/mol.

Increasing the oxidative state and positive charge of Pd can further enhance the electrophilicity, which is beneficial for electrophilic deprotonation process [150]. As shown in Figure 5.33, aryl Pd(II) species **5-305** can be oxidized by hypervalent iodine via transition state **5-306ts** to form an octahedral Pd(IV) complex **5-307**. The dissociation of chloride provides a vacancy, which can be occupied by aryl group to afford a cationic agostic intermediate **5-310**. The relative free energy of that complex is 16.8 kcal/mol higher than that of **5-308**. The activation of a cationic Pd(IV) leads

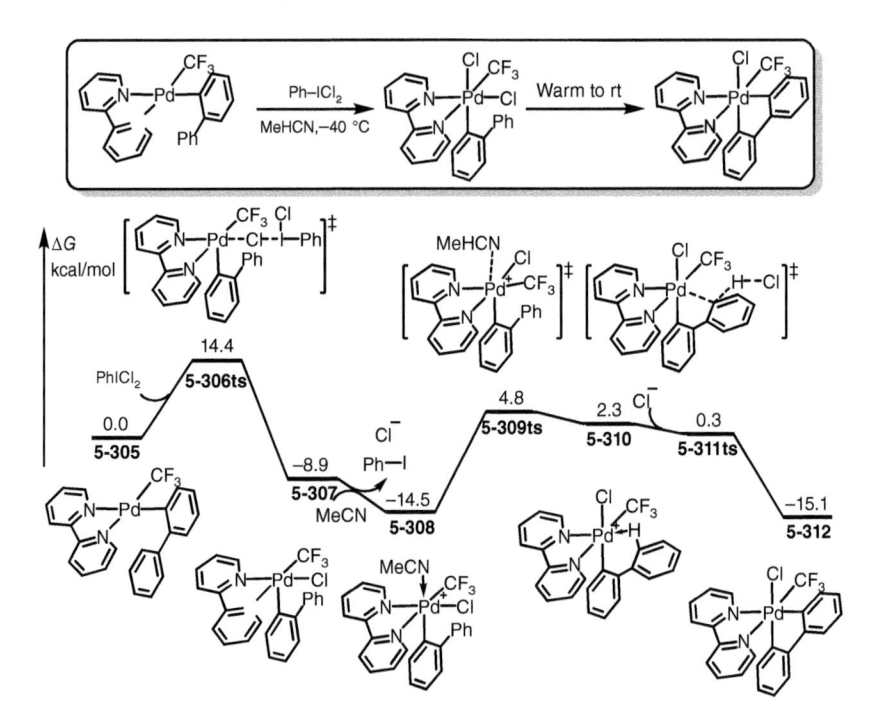

Figure 5.33 The free-energy profiles of the electrophilic deprotonation via a Pd(IV) intermediate. The energies were calculated at B3LYP-D3/DZVP/PCM(acetonitrile) level of theory.

to a rapid deprotonation in the absence of base via transition state **5-311ts** without barrier in solvent. Even chloride can play a proton accepter in this process to afford a neutral biaryl Pd(IV) species **5-312**.

5.3.9 C—H Bond Activation Through σ-Complex-Assisted Metathesis

When a C—H bond reacts with a Pd—heteroatom covalent bond, a σ-complex-assisted metathesis through four-membered ring-type transition state can be carried out to implement corresponding C—H activations. Usually, this type C—H activation often takes place with an alkoxyl Pd(II) species. In the mechanistic study of Pd-catalyzed azoles arylation reaction with aryl thioesters (Figure 5.34) [151], the oxidative addition of thioesters onto Pd(0) species can occur via transition state **5-314ts** with an energy barrier of 15.2 kcal/mol to form an aryl Pd(II) complex **5-315** with 7.2 kcal/mol exergonic. A counterion exchange with sodium *tert*-butoxide forms a *tert*-butoxy Pd(II) intermediate **5-316** with 15.5 kcal/mol endergonic. Then, a σ-bond metathesis between Pd—O and C(aryl)—H bond takes place via a four-membered ring-type transition state **5-317ts** to form an aryl Pd(II) species **5-318**. The calculated overall activation free energy of this step is 36.9 kcal/mol,

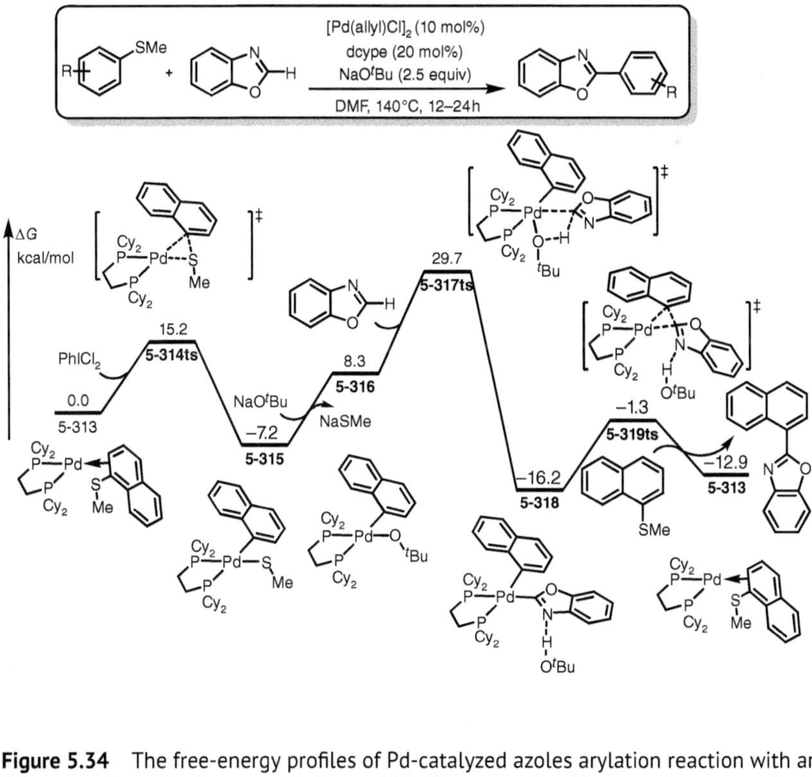

Figure 5.34 The free-energy profiles of Pd-catalyzed azoles arylation reaction with aryl thioesters through a key step of electrophilic deprotonation. The energies were calculated at M06/6-311++G(d,p) (LANL2DZ(f) for Pd)/SMD(DMF)//B3LYP/6-31G(d) (LANL2DZ(f) for Pd) level of theory. Source: Based on Yang et al. [151].

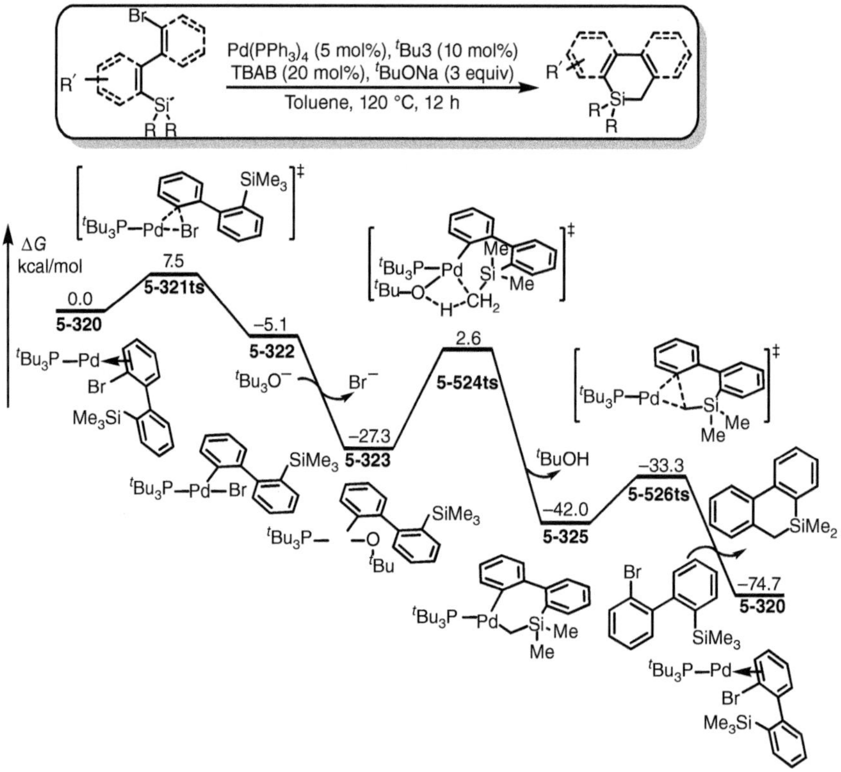

Figure 5.35 The free-energy profiles of Pd-catalyzed intramolecular cross-coupling of aryl bromide and C(alkyl)—H bond through a key step of electrophilic deprotonation. The energies were calculated at B3LYP/6-31G(d) (LANL2DZ(f) for Pd, Br, P, and Si)/CPCM (toluene)//B3LYP/6-31G(d) (LANL2DZ(f) for Pd, Br, P, and Si) level of theory. Source: Based on Yang et al. [151].

which is the rate-determining step for the whole catalytic cycle. The arylation product can be generated by reductive elimination with the regeneration of Pd(0) species **5-313**.

In the presence of chelation group, a σ-complex-assisted metathesis also can be carried out with C(sp³)—H bond. The calculated free-energy profiles for a Pd-catalyzed intramolecular cross-coupling of aryl bromide and C(alkyl)—H bond are shown in Figure 5.35 [152]. The reaction starts from an oxidative addition of aryl bromide onto Pd(0) species **5-320** to afford an aryl Pd(II) intermediate **5-322**. Then, a counterion exchange with tBuO⁻ forms a *tert*-butoxy Pd(II) intermediate **5-323** with 22.2 kcal/mol exergonic. Assisted by the chelation C(aryl)—Pd bond, a σ-bond metathesis takes place via a four-membered ring-type transition state **5-324ts** to form a seven-membered palladacycle **5-325**. The calculated energy barrier of this step is 29.9 kcal/mol, which is considered to be the rate-determining step for the whole catalytic cycle. Then, a rapid reductive elimination yields dibenzosiline product.

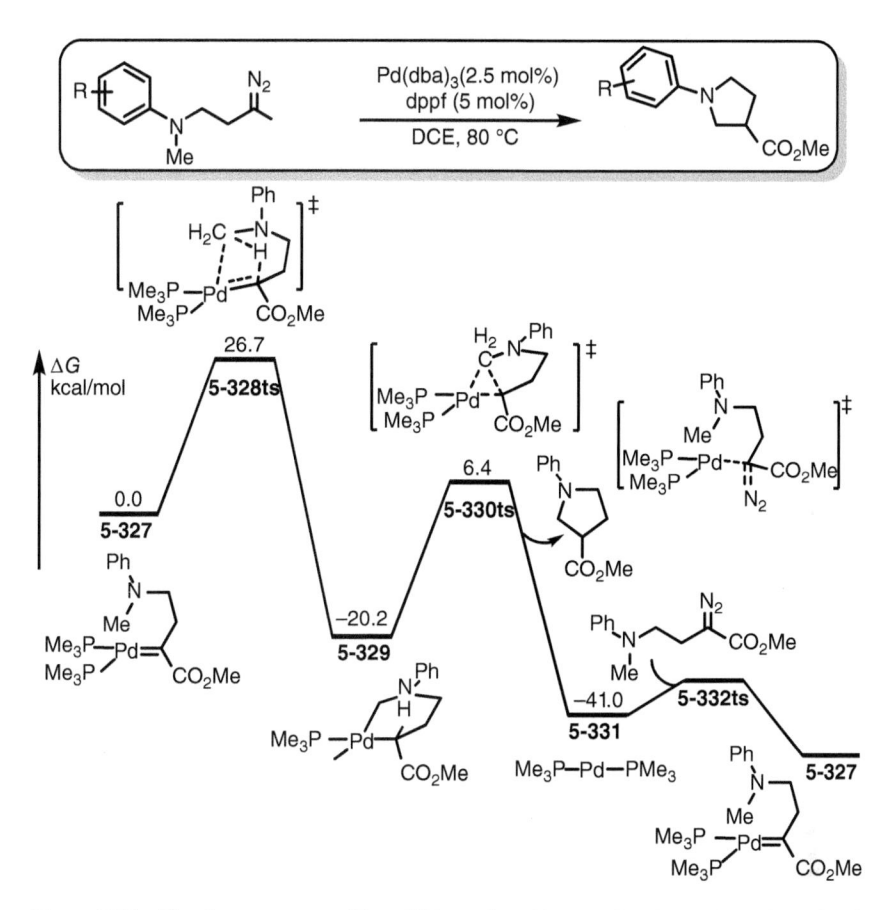

Figure 5.36 The free-energy profiles of Pd-catalyzed intramolecular carbene insertion into C(sp³)–H bond. The energies were calculated at M06-L/def2-TZVPP/PCM(chloroform)// B3LYP-D3/def2-SVP/PCM(chloroform) level of theory. Source: Based on Solé et al. [153].

In a unique example, an intramolecular σ-bond metathesis between C(alkyl)–H and Pd–carbene complex also can provide C–H metalation [153]. As shown in Figure 5.36, the mechanism of Pd-catalyzed intramolecular carbene insertion into C(sp³)–H bond is revealed by DFT calculations. When Pd–carbene **5-327** is formed, the intramolecular σ-bond metathesis between C(alkyl)–H and Pd–carbene complex takes place via transition state **5-328ts** with an energy barrier of 26.7 kcal/mol to form a six-membered palladacycle **5-329** with 20.2 kcal/mol exergonic. Then, an alkyl–alkyl reductive elimination forms cyclic product and regenerates Pd(0) species **5-331**, which can be carbonated by carbenoid.

5.3.10 C–H Bond Activation Through Oxidative Addition

The C–H bond activation through oxidative addition is rare in Pd-catalysis. One example is theoretical study for the mechanism of Pd-catalyzed deformylation [154]. As shown in Figure 5.37, DFT calculation found that when benzaldehyde

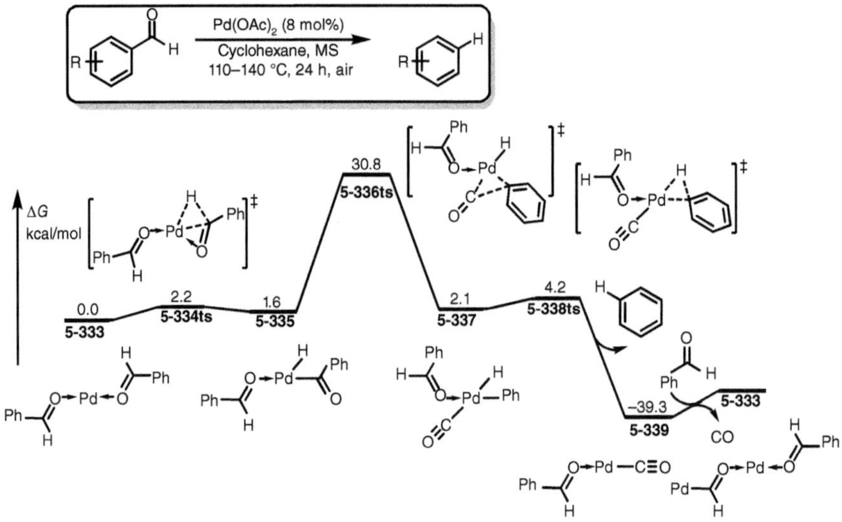

Figure 5.37 The free-energy profiles of Pd-catalyzed deformylation through an oxidative addition-type C(acyl)—H activation. The energies were calculated at M06/6-31+G(d) (SDD for Pd)/PCM(cyclohexane)//M06/6-31+G(d) (SDD for Pd) level of theory.

coordinates with Pd(0) species **5-333**, a C(acyl)—H bond oxidative addition onto Pd(0) can take place via transition state **5-334ts** with an energy barrier of only 2.2 kcal/mol to reversibly form a hydride acyl Pd(II) intermediate **5-335**. The oxidative addition also can be considered as an insertion of Pd into C(acyl)—H bond. Then, a decarbonylation takes place via an α-phenyl elimination transition state **5-336ts**. The calculated overall activation free energy of this step is 30.8 kcal/mol, which is considered to be the rate-determining step. Then, a rapid phenyl hydride reductive elimination yields deformylation product and regenerates Pd(0) species **5-333** with ligand exchange.

5.4 Pd-Mediated Activation of Unsaturated Molecules

As a late transition metal, Pd can be binding with a series of unsaturated molecules resulting in their activation [14, 155–164]. Generally, there are two ways to achieve this process. One way starts from a Pd(0) species, which is shown in Scheme 5.21a. The oxidative addition with electrophiles forms a Pd(II) species. Then, the coordination of unsaturated molecule results in a migratory insertion to afford a new C—Pd bond. The group exchange followed by reductive elimination yields bifunctionalized product and regenerates Pd(0) species. Alternatively, a non-neutral pathway also can provide functionalization of unsaturated molecules. As shown in Scheme 5.21b, the reaction starts from a Pd(II) species, which can react with a nucleophile to form a Pd—R (R = carbon or heteroatom) bond. Then, the coordination of unsaturated molecule also leads to a migratory insertion into the newly formed Pd—R bond. Finally, the electrophilic substitution regenerates Pd(II) species and yields functionalized product. Unsaturated molecules usually used as substrates in Pd-catalysis can be alkenes, alkynes, imines, carbonyls, carbenes, nitrenes, etc. The mechanistic details are selectively discussed in this section.

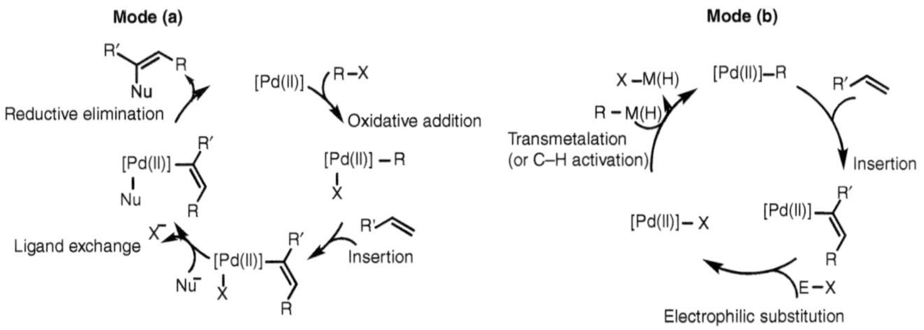

Scheme 5.21 General mechanism of Pd-catalyzed unsaturated bond activations: (a) redox process and (b) nonredox process.

5.4.1 Alkene Activation

The Pd-catalyzed olefin dihydrogenation can undergo a nonredox process [165]. As shown in Figure 5.38, the model reaction starts from a hydride Pd(II) species **5-340** with a pincer ligand. The ligand exchange with ethylene leads to the dissociation of one amine ligand moiety through an associative replacement pathway via transition state **5-341ts** to form an alkene-coordinated Pd(II) intermediate **5-342** by 7.7 kcal/mol endergonic. A migratory insertion of olefin into Pd—H bond occurs via a four-membered ring-type transition state **5-343ts** to form an agostic Pd(II) intermediate **5-344**. The coordination of dihydrogen molecule forms intermediate **5-345**. Then, a σ-bond metathesis between Pd—C(alkyl) and H—H bonds takes place via a concerted four-membered ring-type transition state **5-346ts** to form hydrolysis product ethylene and regenerate hydride Pd(II) species **5-340**.

The Pd-catalyzed olefin hydroarylation reaction also could bear a nonredox pathway [166]. As shown in Figure 5.39, the reaction starts from a cationic Pd(II) species **5-347**. The transmetalation with phenylboronic acid via transition state **5-348ts** affords a phenyl Pd(II) intermediate **5-349**. The coordination of enone forms intermediate **5-350**. Then, an olefin insertion into Pd—C(aryl) bond takes place via transition state **5-351ts** with an energy barrier of 21.3 kcal/mol to form an enolate Pd(II) intermediate **5-352**. The protonation with water molecule yields hydroarylation product and regenerates cationic Pd(II) species **5-347**. The rate-determining step is considered to be the olefin insertion step.

5.4.2 Alkyne Activation

The hydrogenation of alkyne also can be catalyzed by Pd in a homogeneous catalysis. The mechanistic details for this reaction are shown in Figure 5.40, which involved a nonredox catalytic cycle [167]. The coordination of alkyne onto a cationic

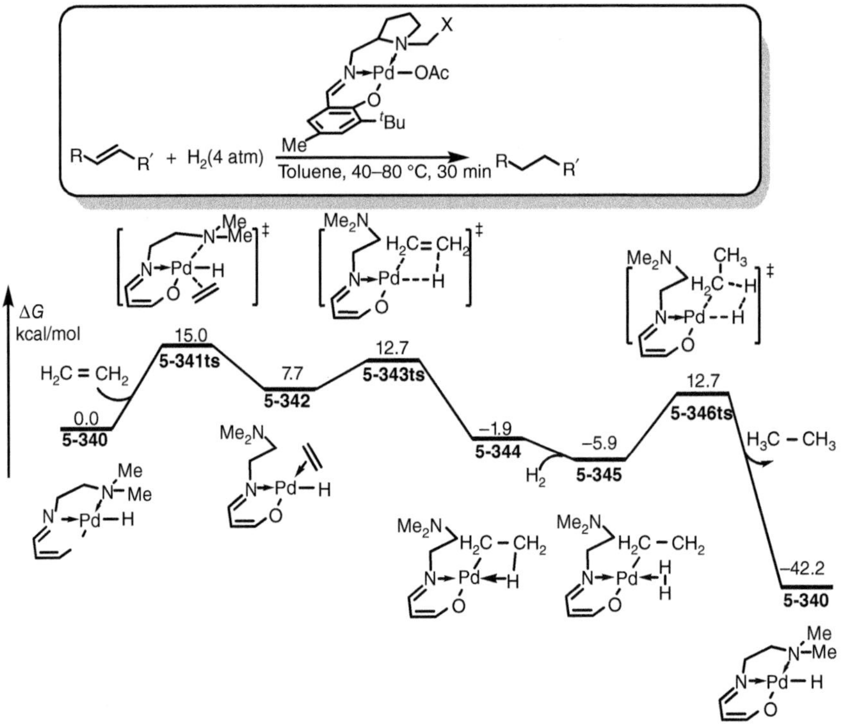

Figure 5.38 The free-energy profiles for a model reaction of Pd-catalyzed hydrogenation of olefins. The energies were calculated at B3LYP/6-31+G(d,p) (LANL2DZ for Pd)/PCM(ethanol) level of theory.

hydride Pd(II) species **5-355** forms intermediate **5-356** by 17.9 kcal/mol exothermic. The insertion of alkyne into Pd—H bond takes place via transition state **5-357ts** with an energy barrier of only 0.5 kcal/mol to form a vinyl Pd(II) intermediate **5-358**. Then, the coordination of dihydrogen molecule causes a σ-bond metathesis between Pd—C(vinyl) and H—H bonds via transition state **5-360ts** with an energy barrier of 8.6 kcal/mol. The formed alkene-coordinated hydride Pd can undergo an olefin insertion via transition state **5-362ts** to form an alkyl Pd(II) intermediate **5-363**. The second σ-bond metathesis between Pd—C(alkyl) and H—H bonds yields tetrahydrogenated product ethane and regenerates hydride Pd(II) species **5-355**.

5.4.3 Enyne Activation

In the presence of both alkenes and alkynes, a cascade insertion would cause the C—C bond formation from enynes [168]. As shown in Figure 5.41, a model reaction of Pd-catalyzed hydrovinylation of acetylene can undergo a nonredox catalytic cycle. When alkyne coordinates with hydride Pd(II) in **5-367**, a migratory insertion of alkyne into Pd—H bond takes place via transition state **5-368ts** with an overall free-energy barrier of 32.1 kcal/mol to form a vinyl Pd(II) intermediate **5-369**. Then, the coordination of alkene leads to an olefin insertion via transition state **5-370ts**

Figure 5.39 The free-energy profiles for a Pd-catalyzed hydroarylation of olefins. The energies were calculated at BP86/6-31G(d) (SDD for Pd)/CPCM(DCE) level of theory. DCE; dichloroethane.

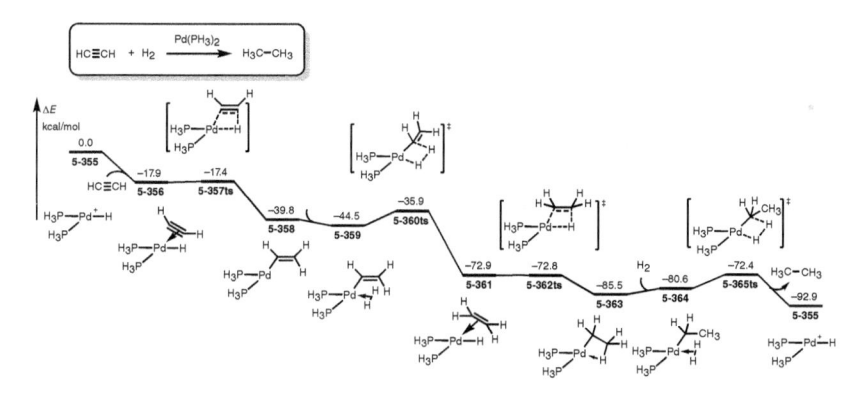

Figure 5.40 The energy profiles for a model reaction of Pd-catalyzed hydrogenation of acetylene. The energies were calculated at B3LYP/6-31G(d) (LANL2DZ(f) for Pd and P)/CPCM(DCE) level of theory.

with an energy barrier of 17.7 kcal/mol to form an alkyl Pd(II) species **5-371**. The β-hydride elimination yields hydrovinylated product and regenerates hydride Pd(II) species **5-366**.

As shown in Figure 5.42, the mechanism of Pd-catalyzed Pauson–Khand reaction with enynes is more complicated [169]. The oxidative cyclization with enyne onto Pd, which often appears in other transition metal–catalyzed Pauson–Khand reactions, can be excluded initially by the much higher barrier. DFT calculation results exhibited that the reaction starts from a *cis*-insertion of alkyne moiety into Pd—Cl bond via transition state **5-375ts** to form a vinyl Pd(II) intermediate **5-376**.

228 | *5 Theoretical Study of Pd-Catalysis*

Figure 5.41 The free-energy profiles for a model reaction of Pd-catalyzed hydrovinylation of acetylene. The energies were calculated at B3LYP/LACVP**/JPBS(DMF) level of theory.

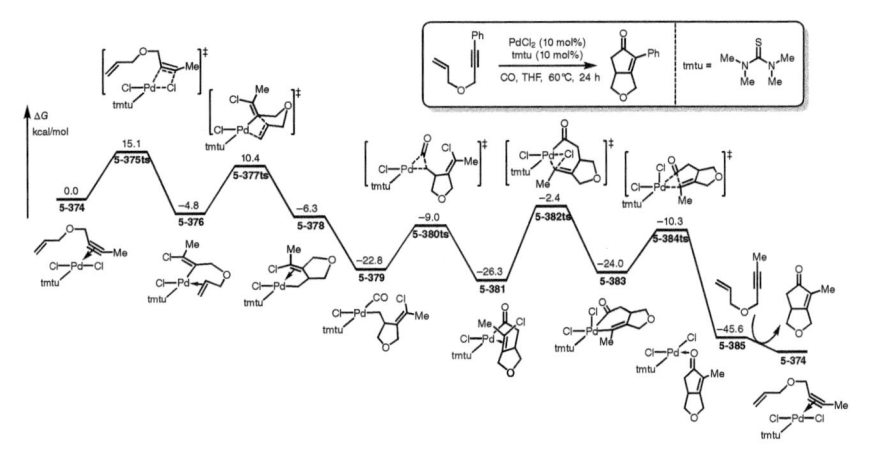

Figure 5.42 The free-energy profiles for Pd-catalyzed Pauson–Khand reaction. The energies were calculated at BP86/6-31+G(d) (SDD for Pd)/PCM(THF)//BP86/6-31+G(d) (SDD for Pd) level of theory. Source: Based on Lan et al. [169].

Then, the coordination of alkene leads to an intramolecular olefin insertion into Pd—C(vinyl) bond. The calculated energy barrier for this step is only 15.2 kcal/mol via transition state **5-377ts** to form an alkyl Pd(II) intermediate **5-378**. Then, the coordination of CO leads to a carbonyl insertion into Pd—C(alkyl) bond via transition state **5-380ts** to form an acyl Pd(II) intermediate **5-381**. An intramolecular vinyl chloride oxidative addition onto Pd(II) takes place via transition state **5-382ts** to form a six-membered pallada(IV)cycle **5-383**. The calculated energy barrier for this step is 23.9 kcal/mol, which is much lower than the commonly proposed enyne oxidative cyclization processes. Then, a rapid reductive elimination yields cyclopentenone product and regenerates a chloride Pd(II) species.

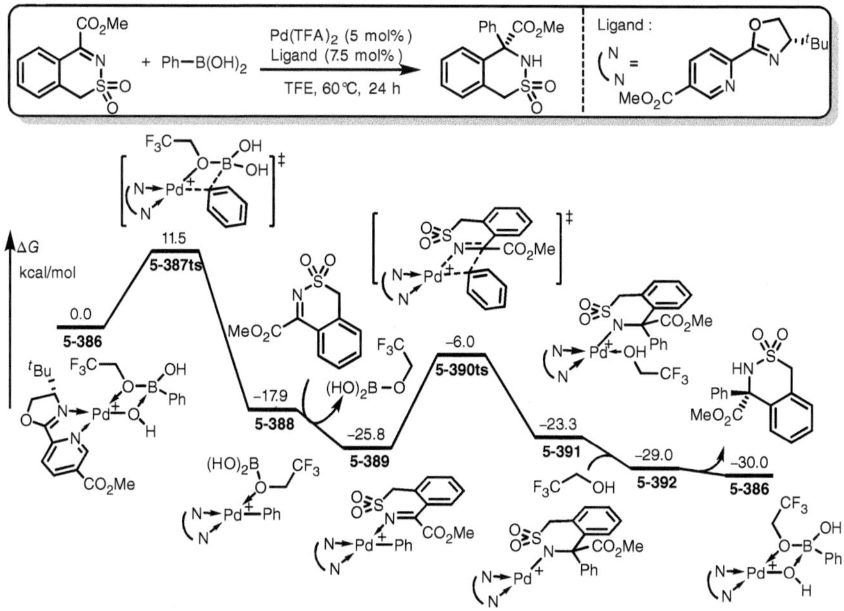

Figure 5.43 The free-energy profiles for Pd-catalyzed hydroarylation of imine. The energies were calculated at B3LYP/6-31G(d,p) (LANL2DZ for Pd)/CPCM(ethanol) level of theory. Source: Based on Quan et al. [170].

5.4.4 Imine Activation

In the presence of Pd catalyst, hydroarylation of imine can afford corresponding amine product through a nonredox pathway [170]. As shown in Figure 5.43, the reaction starts from a transmetalation with phenylboronic acid to form a cationic aryl Pd(II) species **5-388**, which can be coordinated by imine to form complex **5-389** by 7.9 kcal/mol exergonic. The C=N double bond insertion into Pd—C(aryl) bond occurs via transition state **5-390ts** with an energy barrier of 19.8 kcal/mol to form an amino Pd(II) intermediate **5-391**, which can be protonated by alcohol solvent to regenerate active species **5-386** and release arylated product.

5.4.5 CO Activation

Carbon monoxide is a useful carbonyl source, which is often employed to construct esters and amides in Pd-catalysis. The mechanism of Pd-catalyzed alkoxycarbonylation and aminocarbonylation of alkynes is studied by DFT calculation.

As shown in Figure 5.44, the alkoxycarbonylation starts from a hydride Pd(II) species **5-393**, where Pd-hydride is coordinated by alkyne [171]. Subsequently, alkyne inserts into Pd—H bond via transition state **5-394ts** with an energy barrier of 26.0 kcal/mol to form a vinyl Pd(II) intermediate **5-396** by the coordination of CO. Then, CO insertion occurs via transition state **5-397ts** via an energy barrier of 25.2 kcal/mol to form an acyl Pd(II) species **5-398**. The coordination of methanol

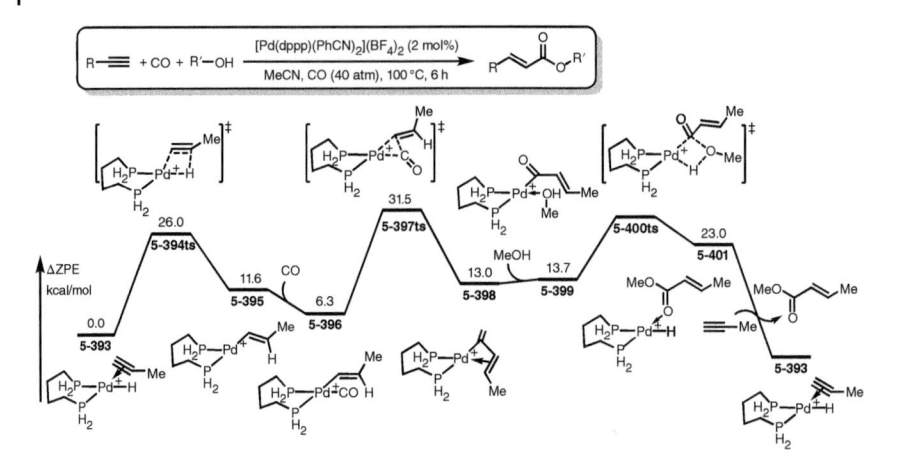

Figure 5.44 The zero-point energy profiles for Pd-catalyzed alkoxycarbonylation. The energies were calculated at B3LYP/6-31G(d) (LANL2DZ for Pd)/PCM(acetonitrile) level of theory. Source: Based on Suleiman et al. [171].

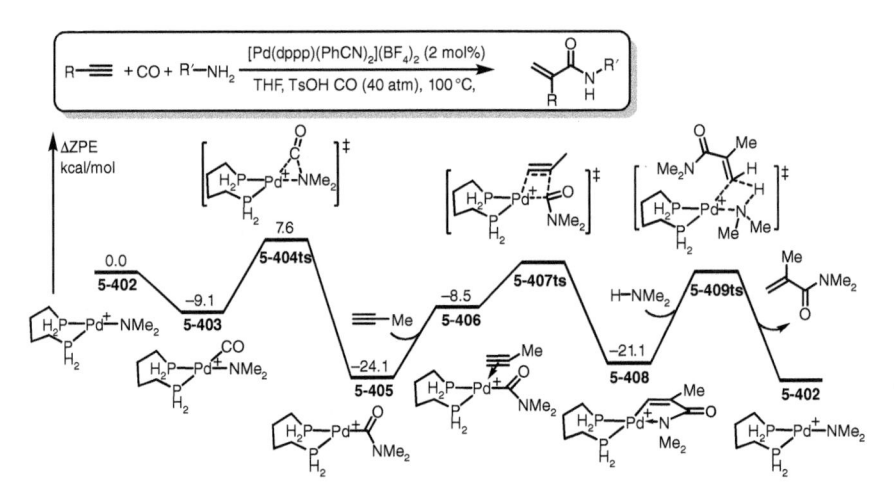

Figure 5.45 The zero-point energy profiles for Pd-catalyzed aminocarbonylation. The energies were calculated at B3LYP/6-31G(d) (LANL2DZ for Pd)/PCM(acetonitrile) level of theory. Source: Based on El Ali et al. [172].

forms complex **5-399**. Then, a σ-bond metathesis between Pd—C(acyl) and O—H bonds yields ester product and regenerates Pd-hydride species **5-393**.

As shown in Figure 5.45, aminocarbonylation starts from an amino Pd(II) species **5-402**, which can be coordinated by CO to form complex **5-403** with 9.1 kcal/mol exergonic [172]. Then, a carbonyl insertion takes place via transition state **5-404ts** with an energy barrier of 16.7 kcal/mol to form a carbamoyl Pd(II) intermediate **5-405**. The coordination of alkyne leads to a migratory insertion of alkyne into Pd—C(carbamoyl) bond via transition state **5-407ts**. Then, the coordination of amine followed by proton transfer achieves aminolysis to form amide product and regenerate amino Pd(II) species **5-402** via transition state **5-409ts**.

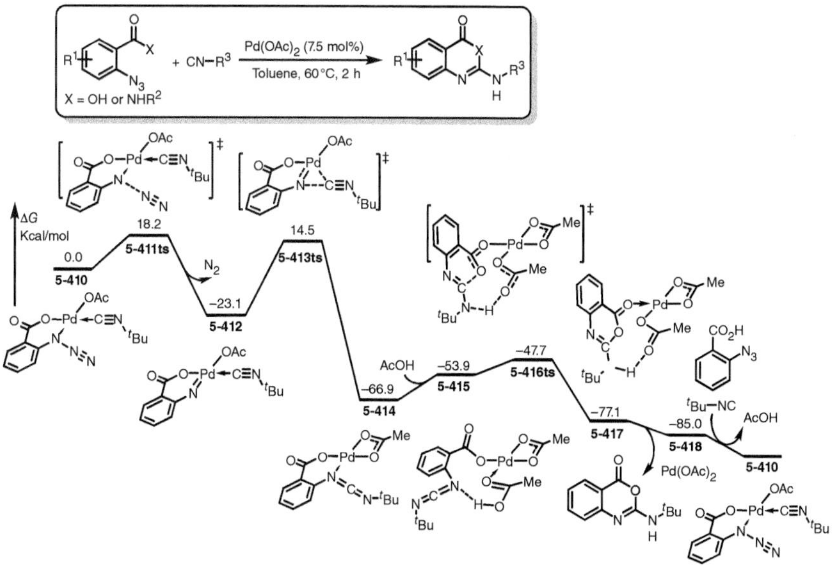

Figure 5.46 The free-energy profiles for Pd-catalyzed tandem azide-isocyanide cross-coupling and cyclization. The energies were calculated at B3LYP/6-31G(d,p) (LANL2DZ for Pd) level of theory.

5.4.6 Isocyanide Activation

Isocyanide can be considered to be the isoelectronic species of carbon monoxide, in which an oxygen atom is replaced by nitrogen. Therefore, isocyanide also can be activated by Pd through an insertion process. DFT calculations are used to reveal the mechanism of Pd-catalyzed tandem azide-isocyanide cross-coupling and cyclization [173]. As shown in Figure 5.46, the calculated free-energy profiles start from an azide-coordinated Pd(II) complex **5-410**. The energy barrier of denitrogenation is 18.2 kcal/mol via transition state **5-411ts**. Then, a Pd–nitrene complex **5-412** is formed. The coordinated isocyanide inserts into Pd—N(nitrene) bond via a three-membered ring-type transition state **5-413ts** with an energy barrier of 37.6 kcal/mol, which is considered to be the rate-determining step. The formation of carbodiimide-coordinated Pd(II) complex **5-414** is a strong exergonic process. Then, an intramolecular nucleophilic addition with carbonate onto carbodiimide takes place via transition state **5-416ts** to generate oxazinone in intermediate **5-417**. Finally, proton transfer with acetic acid yields oxazinone product and regenerates Pd(II) active species **5-410**.

5.4.7 Carbene Activation

Carbene is an important organic intermediate that contains two unshared valence electrons and a neutral carbon atom with a valence of 2 [174–178]. The center carbon atom in carbene has six valence electrons, which is two electrons less than that allowed by the octet rule; therefore, carbenes are classified as either singlet or triplet

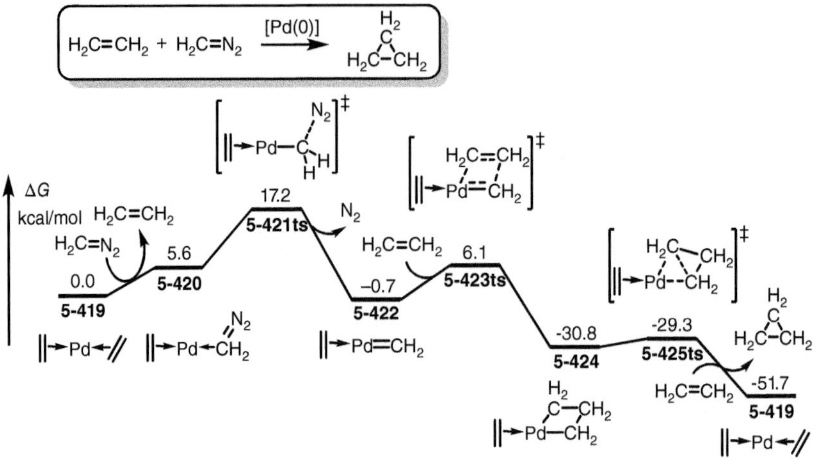

Figure 5.47 The free-energy profiles for a model reaction of Pd-catalyzed carbonation of olefins. The energies were calculated at B3LYP/LACV3P**++//B3LYP/LACV3P* level of theory. Source: Based on Straub [195].

depending upon their electronic behavior. Pd–carbene complexes usually play as active intermediates in further functionalizations [179–194].

As an example, Pd–carbene can react with alkene to construct cyclopropane derivatives. The mechanism of this reaction is studied by DFT calculation, which is shown in Figure 5.47 [195]. In a ligand-free reaction condition, diethylene-coordinated Pd(0) species **5-419** is chosen as relative zero in free-energy profiles. A ligand exchange with diazo forms complex **5-420** with 5.6 kcal/mol endergonic. The denitrogenation forms a Pd–carbene complex **5-422** with an energy barrier of 11.6 kcal/mol. The intermolecular π-bond metathesis via a four-membered ring-type transition state **5-423ts** forms a four-membered pallada(II)cycle **5-424**. Then, a rapid reductive elimination yields cyclopropane product and regenerates complex **5-419** by the coordination of another ethylene.

Interestingly, the mechanism of Pd-catalyzed alkylation of carbenoid can undergo various processes [196]. As shown in Figure 5.48, a redox-neutral process reaction between diazo compounds was proposed by DFT calculation. The coordination of diazo compound onto Pd(OAc)$_2$ species **5-426** leads to a nucleophilic substitution via transition state **5-427ts** to afford an acetoxyalkyl Pd(II) intermediate **5-428** with release of dinitrogen molecule. The calculated activation free energy is only 0.9 kcal/mol, which is much lower than typical metal-assisted decomposition of diazo compound. Then, an intermolecular aliphatic deprotonation by *tert*-butoxide via transition state **5-429ts** generates a vinyl acetate–coordinated palladate(0) species **5-430**. The sequential oxidative addition takes place via a five-membered ring-type transition state **5-431ts** to afford a vinyl palladate **5-432**. Then, alkene reactant inserts into Pd—C(vinyl) bond via transition state **5-434ts** to form an enolate Pd(II) species **5-435**. A water-assisted protonation yields alkylation product and regenerates Pd(OAc)$_2$ species **5-426**, which is considered to be the rate-determining step for the whole catalytic cycle.

5.4 Pd-Mediated Activation of Unsaturated Molecules | 233

Figure 5.48 The free-energy profiles for a model reaction of a redox-neutral Pd-catalyzed alkylation of carbenoid. The energies were calculated at B3LYP/6-31G(d,p) (LANL2DZ(f) for Pd)/CPCM(acetonitrile)//B3LYP/6-31G(d,p) (LANL2DZ(f) for Pd) level of theory.

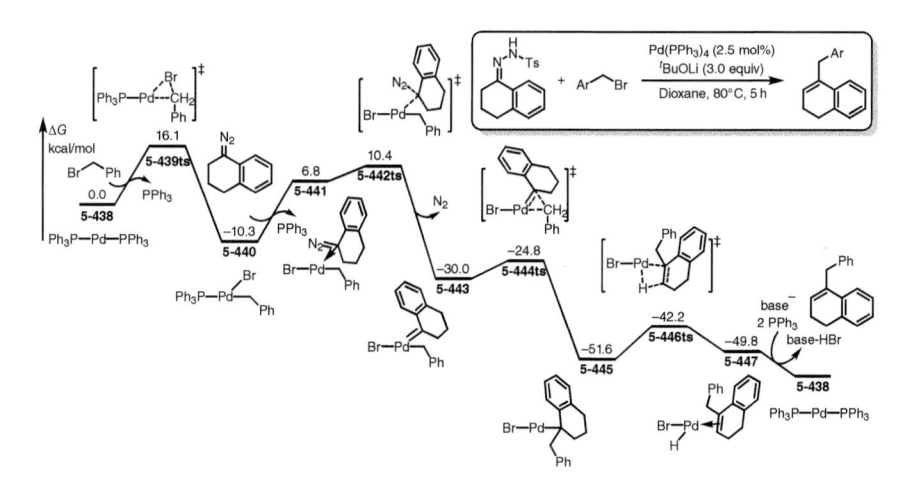

Figure 5.49 The free-energy profiles for a model reaction of a redox-involved Pd-catalyzed alkylation of carbenoid. The energies were calculated at B3LYP/6-31G(d,p) (LANL2DZ(f) for Pd)/CPCM(acetonitrile)//B3LYP/6-31G(d,p) (LANL2DZ(f) for Pd) level of theory. Source: Based on Sun et al. [197].

An alternative redox process for Pd-catalyzed alkylation of carbenoid is shown in Figure 5.49 [197]. The oxidative addition onto Pd(0) species with benzyl bromide takes place via transition state **5-439ts** to form a benzylic Pd(II) intermediate **5-440** with 10.3 kcal/mol exergonic. The coordination of diazo forms complex **5-441**. The sequential denitrogenation forms a Pd–carbene complex **5-443**. A migratory insertion of carbene into Pd—C(alkyl) bond generates an alkyl Pd(II) species **5-445**. A rapid β-hydride elimination yields olefin product and forms a hydride Pd species, which can further transfer to Pd(0) species **5-438** in the presence of extra base.

5.5 Allylic Pd Complex

Allylic Pd complex is a useful intermediate, which has wide potential applications for the construction of new C—C and C—heteroatom bonds in allylic compound [198–212]. Formally, allyl group in organometallic complexes is a nucleophile; however, in an allylic Pd complex, the η^3-coordination of allyl group onto a cationic Pd(II) results in an electrophilicity of allyl group, which can react with an even weak nucleophile in a formal reductive reaction to construct new covalent bond with the generation of Pd(0) species. As shown in Scheme 5.22, allylic Pd(II) can be generated from an oxidative addition (model (a)) or substitution (model (b)) with an electrophilic allyl compound. The electrophilic attack onto allene-coordinated Pd(0) species also can afford allylic Pd (model (c)). In another nonredox process, the allylic C—H activation (model (d)) or transmetalation (model (e)) with Pd(II) species is an alternative pathway for the construction of allylic Pd.

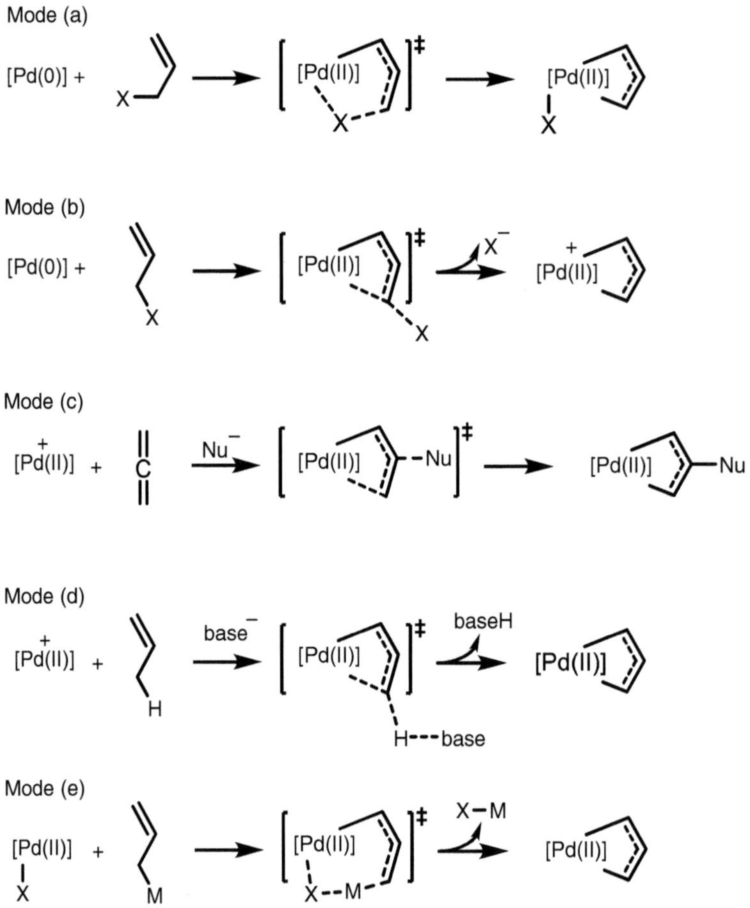

Scheme 5.22 The generation of allylic palladium. (a) Oxidative addition, (b) nucleophilic substitution, (c, d) electrophilic deprotonation, and (e) transmetalation.

A typical catalytic cycle with a key intermediate of allylic Pd is shown in Scheme 5.23. The nucleophilicity of Pd(0) leads to a nucleophilic substitution of allyl compound by release of leaving group X to form a cationic allylic Pd(II) species. Then, a nucleophilic attack with extra nucleophile onto allyl group leads to the formation of new allyl compound and regenerates Pd(0) species.

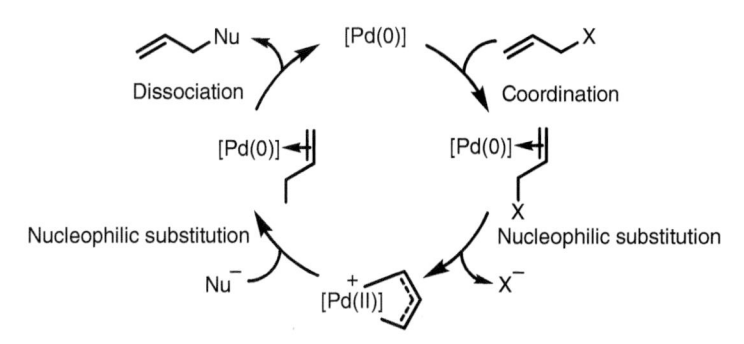

Scheme 5.23 A typical mechanism of Pd-mediated allylic substitution.

5.5.1 Formation from Allylic Oxidative Addition

An inner-sphere oxidative addition of allylic halides onto Pd(0) species is a direct way to form allylic Pd(II) complex [213]. As shown in Figure 5.50, when a benzylic chloride is used as substrate in a dearomatized Stille-type cross-coupling reaction, a η^3-allylic η^3-benzylic Pd(II) species **5-448** was considered as the key point on the calculated free-energy profiles. The nucleophilic substitution of phosphine ligand occurs via transition state **5-449ts** to afford a η^3-allylic η^1-benzylic Pd(II) intermediate **5-450** reversibly, which can be further isomerized to an unstable η^1-allylic η^3-benzylic Pd(II) intermediate **5-452** with 13.6 kcal/mol endergonic. Then, a reductive elimination takes place via a nine-membered ring-type transition state **5-453ts** with overall activation free energy of 20.7 kcal/mol. This transition state can be considered as a nucleophilic allyl group attack onto benzylic Pd(II). The

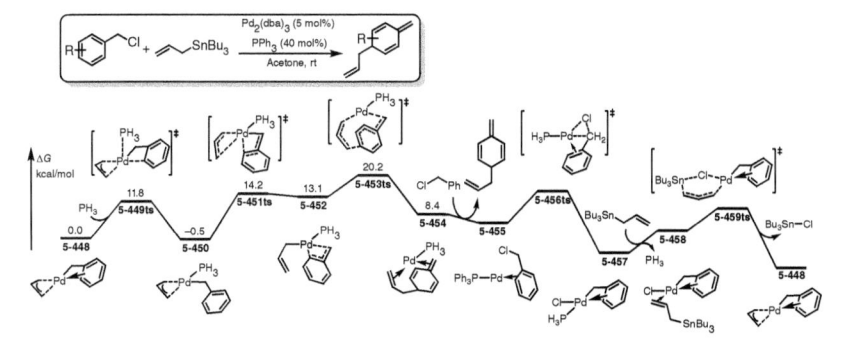

Figure 5.50 The free-energy profiles for Pd-catalyzed Stille coupling with benzylic chloride through a key intermediate of allylic Pd(II). The energies were calculated at B3LYP/6-31G(d) (LANL2DZ for Pd) level of theory.

release of allylation product forms intermediate **5-455**; however, the computational information is absent after this step. The coordination of benzylic chloride leads to an oxidative addition via transition state **5-456ts** to result in a η^3-benzylic Pd(II) intermediate **5-457**, which can react with allylic Sn through transmetalation via transition state **5-459ts** to regenerate the active species **5-448**.

5.5.2 Formation from Allylic Nucleophilic Substitution

A low-valence transition metal often reveals electron-rich character, which can represent a nucleophile to react with allyl compound by nucleophilic substitution. After release of an anionic-leaving group, a cationic allylic Pd(II) complex can be formed for further transformations.

DFT calculations were used to reveal the mechanism of amination of allylic alcohols by using Pd(II) catalyst, which involves an allylic Pd(II) complex as key intermediate [214]. The calculated free-energy profiles are shown in Figure 5.51. The reaction starts from a cationic allylic Pd(II) species **5-460**. The intermolecular nucleophilic attack from the back of allyl by amine takes place via transition state **5-461ts** with an energy barrier of 8.9 kcal/mol. The formation of allylic amine coordinated Pd(0) complex **5-462** is 9.5 kcal/mol exergonic. The olefin exchange with propenol forms a π-coordinated Pd(0) complex **5-464** with 6.1 kcal/mol endergonic by release of propenaminium. Assisted with the hydrogen bond forming between propenol and propenaminium, a nucleophilic substitution of Pd(0) can regenerate allylic Pd(II) species **5-460** by release of water. The calculated allylation of Pd(0) is the rate-determining step for the whole catalytic cycle with an energy barrier of 19.7 kcal/mol. A similar catalytic cycle is also present in Pd-catalyzed amination of allylic ethers.

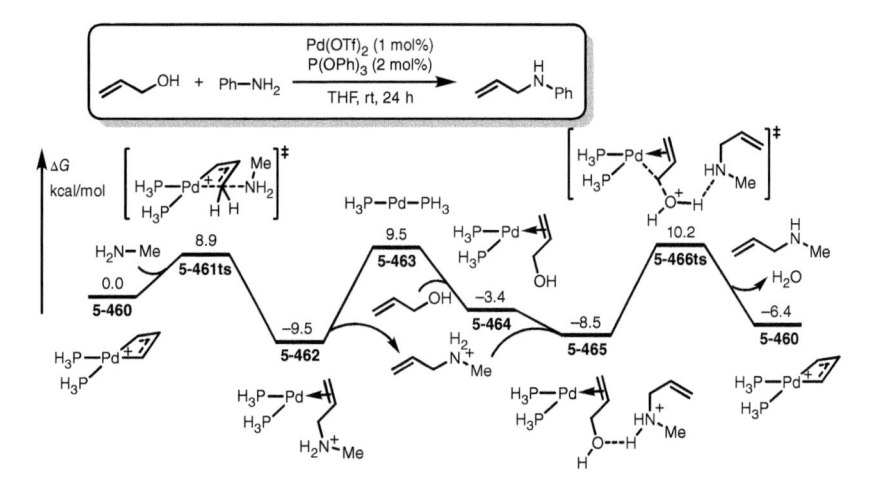

Figure 5.51 The free-energy profiles for Pd-catalyzed amination of allylic alcohols. The energies were calculated at B3PW91/6-31+G(d) (LANL2DZ for Pd)/PCM(THF)// B3PW91/6-31+G(d) (LANL2DZ for Pd) level of theory.

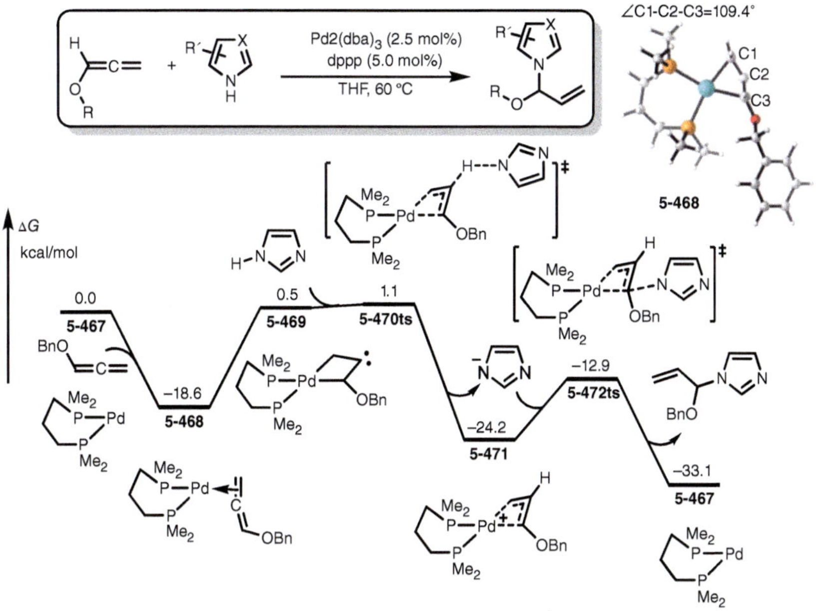

Figure 5.52 The free-energy profiles for Pd-catalyzed hydroamination of allenes. The energies were calculated at M06/6-311+G(d,p) (SDD for Pd)/CPCM(acetonitrile) level of theory.

5.5.3 Formation from the Nucleophilic Attack onto Allene

According to the coordinative activation of Pd(0), the middle carbon of allene plays nucleophilicity, which can react with electrophile to form an allylic Pd(II) complex from an oxidative process. This transformation was considered as the key step of Pd-catalyzed hydroamination of allenes [215]. As shown in Figure 5.52, the reaction starts from an allene-coordinated Pd(0) complex **5-468**, which can isomerize to a four-membered pallada(II)cycle **5-469** by 19.1 kcal/mol endergonic. In the geometry of intermediate **5-469**, the bond angle of C1—C2—C3 from allene moiety is 109.4°, which clearly reveals carbene character. The calculated frontier molecular orbitals of intermediate **5-469** confirmed this structure. Because of the nucleophilicity of carbene moiety, an intermolecular proton transfer from imidazole to middle carbon takes place via transition state **5-470ts** with an overall activation free energy of 19.7 kcal/mol to form a cationic allylic Pd(II) complex **5-471**. Then, a nucleophilic attack of imidazolide takes place via transition state **5-472ts** to afford hydroamination product and regenerate Pd(0) species **5-467**.

5.5.4 Formation from Allylic C—H Activation

The allylic C—H activation is a nonredox way for the generation of allylic Pd(II) complex. A combination of theoretical and experimental studies found that the aluminum salt can significantly accelerate Pd-catalyzed isomerization of allylbenzene, in which an allylic Pd(II) complex is key species formed by C—H activation [216].

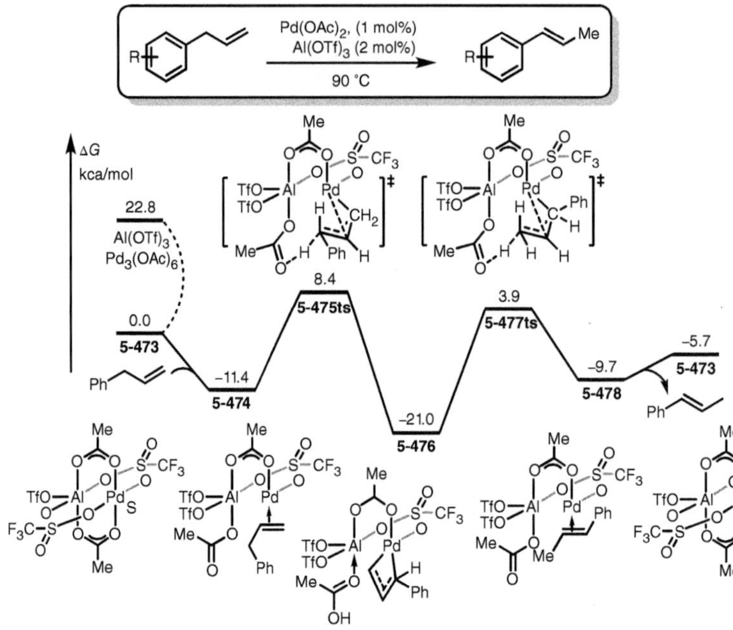

Figure 5.53 The free-energy profiles for Pd-catalyzed isomerization of allylbenzene in the presence of aluminum salt. The energies were calculated at B3LYP-D3/def2-TZVPPD (SDD for Pd)/SMD(acetonitrile)//B3LYP/def2-SVP (SDD for Pd) level of theory.

As shown in Figure 5.53, DFT calculations found that trimer $Pd_3(OAc)_6$ can be stabilized by the $Al(OTf)_3$ additive to form a $Pd(OAc)_2$–$Al(OTf)_3$ complex **5-473** with 22.8 kcal/mol exergonic. Then, the coordination of allylbenzene forms **5-474** with a further 11.4 kcal/mol exergonic. An acetate-assisted allylic C—H activation takes place via transition state **5-475ts** with an energy barrier of 19.8 kcal/mol to form an allylic Pd(II) complex **5-476**. This complex can be protonated by acetic acid via transition state **5-477ts** to yield vinylbenzene product and regenerate $Pd(OAc)_2$–$Al(OTf)_3$ complex **5-473**.

5.5.5 Formation from Allene Insertion

Allylic Pd can be formed from the insertion of allene into a Pd—C bond, which is a nonredox process [217]. As shown in Figure 5.54, the mechanism of Pd-catalyzed carbocyclization of bisallene involves this step. The reaction starts from a bisallene-coordinated Pd(II) complex **5-479**, which can undergo an allylic C—H activation via transition state **5-480ts** with an energy barrier of 14.9 kcal/mol to afford a η^1-allylic Pd(II) complex **5-481**. Then, an intramolecular allene insertion into Pd—C(allyl) bond takes place via transition state **5-482ts** to form a η^1-allylic Pd(II) complex **5-483**, which can isomerize to η^3-coordinated complex **5-485** via transition state **5-484ts**. The barrier of η^1- to η^3-isomerization is considered to be 19.5 kcal/mol. Subsequently, an acetate-assisted β-hydride elimination takes place

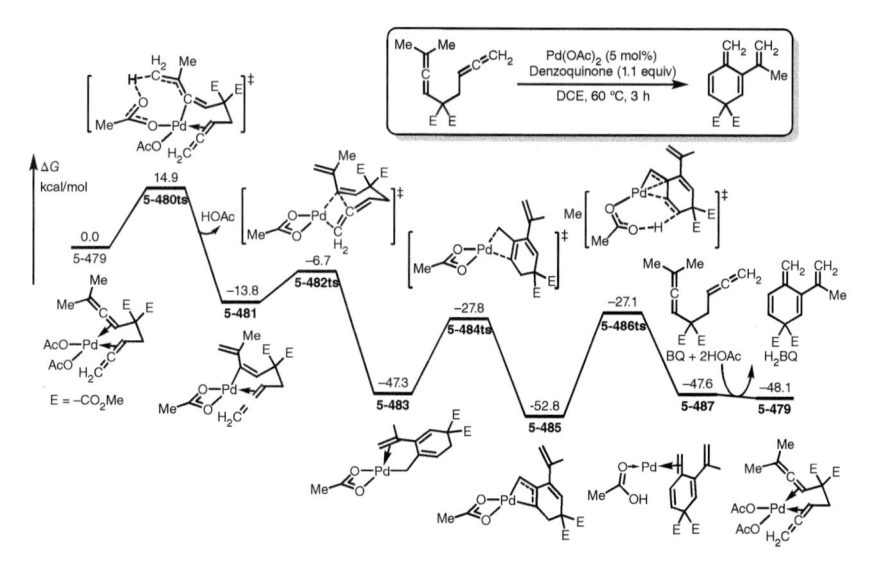

Figure 5.54 The free-energy profiles for Pd-catalyzed carbocyclization of bisallene. The energies were calculated at M06/6-311++G(d,p) (SDD for Pd)/SMD(DCE)//B3LYP/6-31G(d,p) (LANL2DZ for Pd) level of theory. Source: Based on Xie et al. [217].

via transition state **5-486ts** to yield carbocyclization product. The generated Pd(0) species **5-487** can be oxidized by quinone to regenerate active species **5-479**.

References

1 Zanardi, A., Mata, J.Á., and Peris, E. (2009). Well-defined Ir/Pd complexes with a triazolyl-diylidene bridge as catalysts for multiple tandem reactions. *Journal of the American Chemical Society* 131: 14531–14537.

2 Brunel, P., Monot, J., Kefalidis, C.E. et al. (2017). Valorization of CO_2: preparation of 2-oxazolidinones by metal–ligand cooperative catalysis with SCS indenediide Pd complexes. *ACS Catalysis* 7 (4): 2652–2660.

3 Liversedge, I.A., Higgins, S.J., Giles, M. et al. (2006). Suzuki route to regioregular polyalkylthiophenes using Ir-catalysed borylation to make the monomer, and Pd complexes of bulky phosphanes as coupling catalysts for polymerisation. *Tetrahedron Letters* 47 (29): 5143–5146.

4 Scarel, A., Rosa Axet, M., Amoroso, F. et al. (2008). Subtle balance of steric and electronic effects for the synthesis of atactic polyketones catalyzed by Pd complexes with meta-substituted aryl-BIAN ligands. *Organometallics* 27: 1486–1494.

5 Ding, B., Zhang, Z., Xu, Y. et al. (2013). P-stereogenic PCP pincer? Pd complexes: synthesis and application in asymmetric addition of diarylphosphines to nitroalkenes. *Organic Letters* 15: 5476–5479.

6 Biffis, A., Centomo, P., Del Zotto, A. et al. (2018). Pd metal catalysts for cross-couplings and related reactions in the 21st century: a critical review. *Chemical Reviews* 118 (4): 2249–2295.

7 Schnyder, A., Indolese, A.F., Studer, M. et al. (2002). A new generation of air stable, highly active Pd complexes for C–C and C–N coupling reactions with aryl chlorides. *Angewandte Chemie International Edition* 114: 3820–3823.

8 Iyer, S., Kulkarni, G.M., and Ramesh, C. (2004). Mizoroki–Heck reaction, catalysis by nitrogen ligand Pd complexes and activation of aryl bromides. *Tetrahedron* 60 (9): 2163–2172.

9 Haneda, S., Gan, Z., Eda, K. et al. (2007). Ligand effects of 2-(2-pyridyl)benzazole-Pd complexes on the X-ray crystallographic structures,1H NMR spectra, and catalytic activities in Mizoroki–Heck reactions. *Organometallics* 26: 6551–6555.

10 Hamashima, Y., Takano, H., Hotta, D. et al. (2003). Immobilization and reuse of Pd complexes in ionic liquid: efficient catalytic asymmetric fluorination and Michael reactions with β-ketoesters. *Organic Letters* 5: 3225–3228.

11 Chass, G.A., O'Brien, C.J., Hadei, N. et al. (2009). Density functional theory investigation of the alkyl–alkyl Negishi cross-coupling reaction catalyzed by N-heterocyclic carbene (NHC)-Pd complexes. *Chemistry* 15 (17): 4281–4288.

12 Tauchman, J., Cisarova, I., and Stepnicka, P. (2011). Chiral phosphinoferrocene carboxamides with amino acid substituents as ligands for Pd-catalysed asymmetric allylic substitutions. Synthesis and structural characterisation of catalytically relevant Pd complexes. *Dalton Transactions* 40 (44): 11748–11757.

13 Ding, B., Zhang, Z., Liu, Y. et al. (2013). Chemoselective transfer hydrogenation of α,β-unsaturated ketones catalyzed by pincer-Pd complexes using alcohol as a hydrogen source. *Organic Letters* 15: 3690–3693.

14 Selander, N. and Szabó, K.J. (2011). Catalysis by palladium pincer complexes. *Chemical Reviews* 111 (3): 2048–2076.

15 Bheeter, C.B., Chen, L., Soulé, J.-F. et al. (2016). Regioselectivity in palladium-catalysed direct arylation of 5-membered ring heteroaromatics. *Catalysis Science & Technology* 6 (7): 2005–2049.

16 Chen, C., Luo, Y., Fu, L. et al. (2018). Palladium-catalyzed intermolecular ditrifluoromethoxylation of unactivated alkenes: CF_3O-palladation initiated by Pd(IV). *Journal of the American Chemical Society* 140 (4): 1207–1210.

17 Gadge, S.T. and Bhanage, B.M. (2014). Recent developments in palladium catalysed carbonylation reactions. *RSC Advances* 4 (20): 10367.

18 McDonald, R.I., Liu, G., and Stahl, S.S. (2011). Palladium(II)-catalyzed alkene functionalization via nucleopalladation: stereochemical pathways and enantioselective catalytic applications. *Chemical Reviews* 111 (4): 2981–3019.

19 Sather, A.C. and Buchwald, S.L. (2016). The evolution of Pd(0)/Pd(II)-catalyzed aromatic fluorination. *Accounts of Chemical Research* 49 (10): 2146–2157.

20 Topczewski, J.J., Cabrera, P.J., Saper, N.I. et al. (2016). Palladium-catalysed transannular C–H functionalization of alicyclic amines. *Nature* 531 (7593): 220–224.

21 Ye, J. and Lautens, M. (2015). Palladium-catalysed norbornene-mediated C–H functionalization of arenes. *Nature Chemistry* 7 (11): 863–870.

22 Abada, E., Zavalij, P.Y., and Vedernikov, A.N. (2017). Reductive C(sp(2))-N elimination from isolated Pd(IV) amido aryl complexes prepared using H_2O_2 as oxidant. *Journal of the American Chemical Society* 139 (2): 643–646.

23 Malacria, M. and Maestri, G. (2013). Palladium/norbornene catalytic system: chelation as a tool to control regioselectivity of Pd(IV) reductive elimination. *Journal of Organic Chemistry* 78 (4): 1323–1328.

24 Narbonne, V., Retailleau, P., Maestri, G. et al. (2014). Diastereoselective synthesis of dibenzoazepines through chelation on palladium(IV) intermediates. *Organic Letters* 16 (2): 628–631.

25 Plata, R.E., Hill, D.E., Haines, B.E. et al. (2017). A role for Pd(IV) in catalytic enantioselective C–H functionalization with monoprotected amino acid ligands under mild conditions. *Journal of the American Chemical Society* 139 (27): 9238–9245.

26 Xu, L.M., Li, B.J., Yang, Z. et al. (2010). Organopalladium(IV) chemistry. *Chemical Society Reviews* 39 (2): 712–733.

27 Ahlquist, M. and Norrby, P.-O. (2007). Oxidative addition of aryl chlorides to monoligated palladium(0): a DFT-SCRF study. *Organometallics* 26 (3): 550–553.

28 Anand, M. and Sunoj, R.B. (2011). Palladium(II)-catalyzed direct alkoxylation of arenes: evidence for solvent-assisted concerted metalation deprotonation. *Organic Letters* 13 (18): 4802–4805.

29 Banerjee, M. and Roy, S. (2003). Palladium(0) catalyzed regioselective carbonyl propargylation across tetragonal tin(II) oxide via redox transmetallation. Electronic supplementary information (ESI) available: general method and experimental procedure and spectroscopic data for compounds 3–10. *Chemical Communications* 4: 534–535.

30 Gazvoda, M., Virant, M., Pinter, B. et al. (2018). Mechanism of copper-free Sonogashira reaction operates through palladium–palladium transmetallation. *Nature Communications* 9 (1): 4814.

31 Goossen, L.J., Koley, D., Hermann, H.L. et al. (2005). Mechanistic pathways for oxidative addition of aryl halides to palladium(0) complexes: a DFT study. *Organometallics* 24 (10): 2398–2410.

32 Gorelsky, S.I., Lapointe, D., and Fagnou, K. (2008). Analysis of the concerted metalation-deprotonation mechanism in palladium-catalyzed direct arylation across a broad range of aromatic substrates. *Journal of the American Chemical Society* 130 (33): 10848–10849.

33 Gorelsky, S.I., Lapointe, D., and Fagnou, K. (2011). Analysis of the palladium-catalyzed (aromatic)C–H bond metalation–deprotonation mechanism spanning the entire spectrum of arenes. *Journal of Organic Chemistry* 77 (1): 658–668.

34 Lei, A. and Zhang, X. (2002). A novel palladium-catalyzed homocoupling reaction initiated by transmetallation of palladium enolates. *Tetrahedron Letters* 43 (14): 2525–2528.

35 Roy, A.H. and Hartwig, J.F. (2003). Oxidative addition of aryl tosylates to palladium(0) and coupling of unactivated aryl tosylates at room temperature. *Journal of the American Chemical Society* 125 (29): 8704–8705.

36 Senn, H.M. and Ziegler, T. (2004). Oxidative addition of aryl halides to palladium(0) complexes: a density-functional study including solvation. *Organometallics* 23 (12): 2980–2988.

37 Zeni, G. and Larock, R.C. (2006). Synthesis of heterocycles via palladium-catalyzed oxidative addition. *Chemical Reviews* 106 (11): 4644–4680.

38 Garcia-Melchor, M., Braga, A.A., Lledos, A. et al. (2013). Computational perspective on Pd-catalyzed C–C cross-coupling reaction mechanisms. *Accounts of Chemical Research* 46 (11): 2626–2634.

39 Shi, S., Nolan, S.P., and Szostak, M. (2018). Well-defined palladium(II)-NHC precatalysts for cross-coupling reactions of amides and esters by selective N–C/O–C cleavage. *Accounts of Chemical Research* 51 (10): 2589–2599.

40 Minami, Y. and Hiyama, T. (2016). Synthetic transformations through alkynoxy-palladium interactions and C–H activation. *Accounts of Chemical Research* 49 (1): 67–77.

41 He, G., Wang, B., Nack, W.A. et al. (2016). Syntheses and transformations of alpha-amino acids via palladium-catalyzed auxiliary-directed sp(3) C–H functionalization. *Accounts of Chemical Research* 49 (4): 635–645.

42 Baudoin, O. (2017). Ring construction by palladium(0)-catalyzed C(sp(3))–H activation. *Accounts of Chemical Research* 50 (4): 1114–1123.

43 Fihri, A., Meunier, P., and Hierso, J.-C. (2007). Performances of symmetrical achiral ferrocenylphosphine ligands in palladium-catalyzed cross-coupling reactions: a review of syntheses, catalytic applications and structural properties. *Coordination Chemistry Reviews* 251 (15–16): 2017–2055.

44 Khan, F., Dlugosch, M., Liu, X. et al. (2018). The palladium-catalyzed Ullmann cross-coupling reaction: a modern variant on a time-honored process. *Accounts of Chemical Research* 51 (8): 1784–1795.

45 Yin, G., Mu, X., and Liu, G. (2016). Palladium(II)-catalyzed oxidative difunctionalization of alkenes: bond forming at a high-valent palladium center. *Accounts of Chemical Research* 49 (11): 2413–2423.

46 Cannon, J.S. and Overman, L.E. (2016). Palladium(II)-catalyzed enantioselective reactions using COP catalysts. *Accounts of Chemical Research* 49 (10): 2220–2231.

47 Wang, D., Weinstein, A.B., White, P.B. et al. (2018). Ligand-promoted palladium-catalyzed aerobic oxidation reactions. *Chemical Reviews* 118 (5): 2636–2679.

48 Yang, Y.F., Hong, X., Yu, J.Q. et al. (2017). Experimental-computational synergy for selective Pd(II)-catalyzed C–H activation of aryl and alkyl groups. *Accounts of Chemical Research* 50 (11): 2853–2860.

49 Yang, B., Qiu, Y., and Backvall, J.E. (2018). Control of selectivity in palladium(II)-catalyzed oxidative transformations of allenes. *Accounts of Chemical Research* 51 (6): 1520–1531.

References

50 Sperger, T., Sanhueza, I.A., Kalvet, I. et al. (2015). Computational studies of synthetically relevant homogeneous organometallic catalysis involving Ni, Pd, Ir, and Rh: an overview of commonly employed DFT methods and mechanistic insights. *Chemical Reviews* 115 (17): 9532–9586.

51 Yu, J.-Q. and Shi, Z. (2010). *C–H Activation*. Berlin, Heidelberg: Springer.

52 Li, J.J. and Gribble, G.W. (2007). *Palladium in Heterocyclic Chemistry: A Guide for the Synthetic Chemist*. Elsevier.

53 Lan, Y., Liu, P., Newman, S.G. et al. (2012). Theoretical study of Pd(0)-catalyzed carbohalogenation of alkenes: mechanism and origins of reactivities and selectivities in alkyl halide reductive elimination from Pd(II) species. *Chemical Science* 3 (6): 1987.

54 Liu, L., Zhang, A.A., Wang, Y. et al. (2015). Asymmetric synthesis of P-stereogenic phosphinic amides via Pd(0)-catalyzed enantioselective intramolecular C–H arylation. *Organic Letters* 17 (9): 2046–2049.

55 Rousseaux, S., Gorelsky, S.I., Chung, B.K. et al. (2010). Investigation of the mechanism of C(sp^3)–H bond cleavage in Pd(0)-catalyzed intramolecular alkane arylation adjacent to amides and sulfonamides. *Journal of the American Chemical Society* 132 (31): 10692–10705.

56 Xue, L. and Lin, Z. (2010). Theoretical aspects of palladium-catalysed carbon–carbon cross-coupling reactions. *Chemical Society Reviews* 39 (5): 1692–1705.

57 Beletskaya, I.P. and Cheprakov, A.V. (2000). The Heck reaction as a sharpening stone of palladium catalysis. *Chemical Reviews* 100 (8): 3009–3066.

58 Negishi, E. and Anastasia, L. (2003). Palladium-catalyzed alkynylation. *Chemical Reviews* 103 (5): 1979–2017.

59 Braga, A.A.C., Ujaque, G., and Maseras, F. (2006). A DFT study of the full catalytic cycle of the Suzuki–Miyaura cross-coupling on a model system. *Organometallics* 25: 3647–3658.

60 Legault, C.Y., Garcia, Y., Merlic, C.A. et al. (2007). Origin of regioselectivity in palladium-catalyzed cross-coupling reactions of polyhalogenated heterocycles. *Journal of the American Chemical Society* 129 (42): 12664–12665.

61 Schoenebeck, F. and Houk, K.N. (2010). Ligand-controlled regioselectivity in palladium-catalyzed cross coupling reactions. *Journal of the American Chemical Society* 132 (8): 2496–2497.

62 Goossen, L.J., Koley, D., Hermann, H.L. et al. (2005). The palladium-catalyzed cross-coupling reaction of carboxylic anhydrides with arylboronic acids: a DFT study. *Journal of the American Chemical Society* 127 (31): 11102–11114.

63 Ben Halima, T., Zhang, W., Yalaoui, I. et al. (2017). Palladium-catalyzed Suzuki–Miyaura coupling of aryl esters. *Journal of the American Chemical Society* 139 (3): 1311–1318.

64 Melvin, P.R., Nova, A., Balcells, D. et al. (2017). DFT investigation of Suzuki–Miyaura reactions with aryl sulfamates using a dialkylbiarylphosphine-ligated palladium catalyst. *Organometallics* 36 (18): 3664–3675.

65 Ribagnac, P., Blug, M., Villa-Uribe, J. et al. (2011). Room-temperature palladium-catalyzed Negishi-type coupling: a combined experimental and theoretical study. *Chemistry* 17 (51): 14389–14393.

66 Fuentes, B., Garcia-Melchor, M., Lledos, A. et al. (2010). Palladium round trip in the Negishi coupling of *trans*-[PdMeCl(PMePh$_2$)$_2$] with ZnMeCl: an experimental and DFT study of the transmetalation step. *Chemistry* 16 (29): 8596–8599.

67 Liu, Q., Lan, Y., Liu, J. et al. (2009). Revealing a second transmetalation step in the Negishi coupling and its competition with reductive elimination: improvement in the interpretation of the mechanism of biaryl syntheses. *Journal of the American Chemical Society* 131 (29): 10201–10210.

68 Álvarez, R.A., Faza, O.N., López, C.S. et al. (2006). Computational characterization of a complete palladium-catalyzed cross-coupling process: the associative transmetalation in the Stille reaction. *Organic Letters* 8 (1): 35–38.

69 Faza, R.Á.O.N., de Lera, A.R., and Cárdenas, D.J. (2007). A density functional theory study of the Stille cross-coupling via associative transmetalation. The role of ligands and coordinating solvents. *Advanced Synthesis and Catalysis* 349 (6): 887–906.

70 Álvarez, R., Pérez, M., Faza, O.N. et al. (2008). Associative transmetalation in the Stille cross-coupling reaction to form dienes: theoretical insights into the open pathway. *Organometallics* 27: 3378–3389.

71 Sugiyama, A., Ohnishi, Y.-y., Nakaoka, M. et al. (2008). Why does fluoride anion accelerate transmetalation between vinylsilane and palladium(II)-vinyl complex? Theoretical study. *Journal of the American Chemical Society* 130: 12975–12985.

72 Albert, K., Gisdakis, P., and Rösch, N. (1998). On C–C coupling by carbene-stabilized palladium catalysts: a density functional study of the Heck reaction. *Organometallics* 17: 1608–1616.

73 Martinez-Solorio, D., Melillo, B., Sanchez, L. et al. (2016). Design, synthesis, and validation of an effective, reusable silicon-based transfer agent for room-temperature Pd-catalyzed cross-coupling reactions of aryl and heteroaryl chlorides with readily available aryl lithium reagents. *Journal of the American Chemical Society* 138 (6): 1836–1839.

74 Biffis, A., Zecca, M., and Basato, M. (2001). Palladium metal catalysts in Heck C–C coupling reactions. *Journal of Molecular Catalysis A: Chemical* 173 (1–2): 249–274.

75 Xu, L., Chen, W., and Xiao, J. (2000). Heck reaction in ionic liquids and the in situ identification of *N*-heterocyclic carbene complexes of palladium. *Organometallics* 19 (6): 1123–1127.

76 Trzeciak, A.M. and Ziółkowski, J.J. (2007). Monomolecular, nanosized and heterogenized palladium catalysts for the Heck reaction. *Coordination Chemistry Reviews* 251 (9–10): 1281–1293.

77 Surawatanawong, P., Fan, Y., and Hall, M.B. (2008). Density functional study of the complete pathway for the Heck reaction with palladium diphosphines. *Journal of Organometallic Chemistry* 693 (8–9): 1552–1563.

78 Allolio, C. and Strassner, T. (2014). Palladium complexes with chelating bis-NHC ligands in the Mizoroki–Heck reaction-mechanism and electronic effects, a DFT study. *Journal of Organic Chemistry* 79 (24): 12096–12105.

79 Surawatanawong, P. and Hall, M.B. (2008). Theoretical study of alternative pathways for the Heck reaction through dipalladium and "ligand-free" palladium intermediates. *Organometallics* 27: 6222–6232.

80 Takagi, J., Takahashi, K., Ishiyama, T. et al. (2002). Palladium-catalyzed cross-coupling reaction of bis(pinacolato)diboron with 1-alkenyl halides or triflates: convenient synthesis of unsymmetrical 1,3-dienes via the borylation-coupling sequence. *Journal of the American Chemical Society* 124 (27): 8001–8006.

81 Sumimoto, M., Iwane, N., Takahama, T. et al. (2004). Theoretical study of trans-metalation process in palladium-catalyzed borylation of iodobenzene with diboron. *Journal of the American Chemical Society* 126: 10457–11047.

82 Kozuch, S., Amatore, C., Jutand, A. et al. (2005). What makes for a good catalytic cycle? A theoretical study of the role of an anionic palladium(0) complex in the cross-coupling of an aryl halide with an anionic nucleophile. *Organometallics* 24 (10): 2319–2330.

83 Jia, T., Zhang, M., McCollom, S.P. et al. (2017). Palladium-catalyzed enantioselective arylation of aryl sulfenate anions: a combined experimental and computational study. *Journal of the American Chemical Society* 139 (24): 8337–8345.

84 Xu, D., Qi, X., Duan, M. et al. (2017). Thiolate–palladium(IV) or sulfonium–palladate(0)? A theoretical study on the mechanism of palladium-catalyzed C–S bond formation reactions. *Organic Chemistry Frontiers* 4 (6): 943–950.

85 Hao, W., Wei, J., Chi, Y. et al. (2016). A DFT study on the conversion of aryl iodides to alkyl iodides: reductive elimination of R–I from alkylpalladium iodide complexes with accessible β-hydrogens. *Chemistry - A European Journal* 22 (10): 3422–3429.

86 Bottoni, A., Higueruelo, A.P., and Miscione, G.P. (2002). A DFT computational study of the bis-silylation reaction of acetylene catalyzed by palladium complexes. *Journal of the American Chemical Society* 124: 5506–5513.

87 Xu, Z.-Y., Zhang, S.-Q., Liu, J.-R. et al. (2018). Mechanism and origins of chemo- and regioselectivities of Pd-catalyzed intermolecular σ-bond exchange between benzocyclobutenones and silacyclobutanes: a computational study. *Organometallics* 37 (4): 592–602.

88 Yue, X., Shan, C., Qi, X. et al. (2018). Insights into disilylation and distannation: sequence influence and ligand/steric effects on Pd-catalyzed difunctionalization of carbenes. *Dalton Transactions* 47 (6): 1819–1826.

89 Iwai, T. and Sawamura, M. (2015). Transition-metal-catalyzed site-selective C–H functionalization of quinolines beyond C2 selectivity. *ACS Catalysis* 5 (9): 5031–5040.

90 Ackermann, L., Vicente, R., and Kapdi, A.R. (2009). Transition-metal-catalyzed direct arylation of (hetero)arenes by C–H bond cleavage. *Angewandte Chemie International Edition* 48 (52): 9792–9826.

91 Hummel, J.R., Boerth, J.A., and Ellman, J.A. (2017). Transition-metal-catalyzed C–H bond addition to carbonyls, imines, and related polarized π bonds. *Chemical Reviews* 117 (13): 9163–9227.

92 Lee, D. and Otte, R.D. (2004). Transition-metal-catalyzed aldehydic C–H activation by azodicarboxylates. *Journal of Organic Chemistry* 69 (10): 3569–3571.

93 Huang, Z., Lim, H.N., Mo, F. et al. (2015). Transition metal-catalyzed ketone-directed or mediated C–H functionalization. *Chemical Society Reviews* 44 (21): 7764–7786.

94 Qiu, G. and Wu, J. (2015). Transition metal-catalyzed direct remote C–H functionalization of alkyl groups via C(sp^3)–H bond activation. *Organic Chemistry Frontiers* 2 (2): 169–178.

95 Bellina, F. and Rossi, R. (2010). Transition metal-catalyzed direct arylation of substrates with activated sp^3-hybridized C–H bonds and some of their synthetic equivalents with aryl halides and pseudohalides. *Chemical Reviews* 110 (2): 1082–1146.

96 Sandtorv, A.H. (2015). Transition metal-catalyzed C–H activation of indoles. *Advanced Synthesis and Catalysis* 357 (11): 2403–2435.

97 Chen, Z., Wang, B., Zhang, J. et al. (2015). Transition metal-catalyzed C–H bond functionalizations by the use of diverse directing groups. *Organic Chemistry Frontiers* 2 (9): 1107–1295.

98 Giri, R., Shi, B.F., Engle, K.M. et al. (2009). Transition metal-catalyzed C–H activation reactions: diastereoselectivity and enantioselectivity. *Chemical Society Reviews* 38 (11): 3242–3272.

99 Gadge, S.T., Gautam, P., and Bhanage, B.M. (2016). Transition metal-catalyzed carbonylative C–H bond functionalization of arenes and C(sp(3))–H bond of alkanes. *Chemical Record* 16 (2): 835–856.

100 Ramirez, T.A., Zhao, B., and Shi, Y. (2012). Recent advances in transition metal-catalyzed sp^3 C–H amination adjacent to double bonds and carbonyl groups. *Chemical Society Reviews* 41 (2): 931–942.

101 Lopez, L.A. and Lopez, E. (2015). Recent advances in transition metal-catalyzed C–H bond functionalization of ferrocene derivatives. *Dalton Transactions* 44 (22): 10128–10135.

102 Cho, S.H., Kim, J.Y., Kwak, J. et al. (2011). Recent advances in the transition metal-catalyzed twofold oxidative C–H bond activation strategy for C–C and C–N bond formation. *Chemical Society Reviews* 40 (10): 5068–5083.

103 Guo, T., Huang, F., Yu, L. et al. (2015). Indole synthesis through transition metal-catalyzed C–H activation. *Tetrahedron Letters* 56 (2): 296–302.

104 Jazzar, R., Hitce, J., Renaudat, A. et al. (2010). Functionalization of organic molecules by transition-metal-catalyzed C(sp^3)–H activation. *Chemistry* 16 (9): 2654–2672.

References

105 Ackermann, L. (2011). Carboxylate-assisted transition-metal-catalyzed C–H bond functionalizations: mechanism and scope. *Chemical Reviews* 111 (3): 1315–1345.

106 Gandeepan, P. and Cheng, C.H. (2016). Advancements in the synthesis and applications of cationic *N*-heterocycles through transition metal-catalyzed C–H activation. *Chemistry - An Asian Journal* 11 (4): 448–460.

107 Majhi, B., Kundu, D., Ahammed, S. et al. (2014). *tert*-Butyl nitrite mediated regiospecific nitration of (*E*)-azoarenes through palladium-catalyzed directed C–H activation. *Chemistry* 20 (32): 9862–9866.

108 Karthikeyan, J. and Cheng, C.H. (2011). Synthesis of phenanthridinones from *N*-methoxybenzamides and arenes by multiple palladium-catalyzed C–H activation steps at room temperature. *Angewandte Chemie International Edition* 50 (42): 9880–9883.

109 Thirunavukkarasu, V.S., Parthasarathy, K., and Cheng, C.H. (2008). Synthesis of fluorenones from aromatic aldoxime ethers and aryl halides by palladium-catalyzed dual C–H activation and Heck cyclization. *Angewandte Chemie International Edition* 47 (49): 9462–9465.

110 Preciado, S., Mendive-Tapia, L., Albericio, F. et al. (2013). Synthesis of C-2 arylated tryptophan amino acids and related compounds through palladium-catalyzed C–H activation. *Journal of Organic Chemistry* 78 (16): 8129–8135.

111 Chaumontet, M., Piccardi, R., Audic, N. et al. (2008). Synthesis of benzocyclobutenes by palladium-catalyzed C–H activation of methyl groups: method and mechanistic study. *Journal of the American Chemical Society* 130 (45): 15157–15166.

112 Calleja, J., Pla, D., Gorman, T.W. et al. (2015). A steric tethering approach enables palladium-catalysed C–H activation of primary amino alcohols. *Nature Chemistry* 7 (12): 1009–1016.

113 Houlden, C.E., Hutchby, M., Bailey, C.D. et al. (2009). Room-temperature palladium-catalyzed C–H activation: *ortho*-carbonylation of aniline derivatives. *Angewandte Chemie International Edition* 48 (10): 1830–1833.

114 Piou, T., Bunescu, A., Wang, Q. et al. (2013). Palladium-catalyzed through-space C(sp(3))–H and C(sp(2))–H bond activation by 1,4-palladium migration: efficient synthesis of [3,4]-fused oxindoles. *Angewandte Chemie International Edition* 52 (47): 12385–12389.

115 Duan, P., Yang, Y., Ben, R. et al. (2014). Palladium-catalyzed benzo[d]isoxazole synthesis by C–H activation/[4+1] annulation. *Chemical Science* 5 (4): 1574–1578.

116 Lee, T.H., Jayakumar, J., Cheng, C.H. et al. (2013). One pot synthesis of bioactive benzopyranones through palladium-catalyzed C–H activation and CO insertion into 2-arylphenols. *Chemical Communications* 49 (100): 11797–11799.

117 Maleckis, A., Kampf, J.W., and Sanford, M.S. (2013). A detailed study of acetate-assisted C–H activation at palladium(IV) centers. *Journal of the American Chemical Society* 135 (17): 6618–6625.

118 Baber, R.A., Bedford, R.B., Betham, M. et al. (2006). Chiral palladium bis(phosphite)PCP-pincer complexes via ligand C–H activation. *Chemical Communications* 37: 3880–3882.

119 Piou, T., Neuville, L., and Zhu, J. (2012). Activation of a C(sp^3)–H bond by a transient sigma-alkylpalladium(II) complex: synthesis of spirooxindoles through a palladium-catalyzed domino carbopalladation/C(sp^3)–C(sp^3) bond-forming process. *Angewandte Chemie International Edition* 51 (46): 11561–11565.

120 Watanabe, T., Oishi, S., Fujii, N. et al. (2008). Palladium-catalyzed sp(3) CH activation of simple alkyl groups: direct preparation of indoline derivatives from N-alkyl-2-bromoanilines. *Organic Letters* 10 (9): 1759–1762.

121 Qian, B., Guo, S., Shao, J. et al. (2010). Palladium-catalyzed benzylic addition of 2-methyl azaarenes to N-sulfonyl aldimines via C–H bond activation. *Journal of the American Chemical Society* 132 (11): 3650–3651.

122 Nishikata, T., Abela, A.R., Huang, S. et al. (2010). Cationic palladium(II) catalysis: C–H activation/Suzuki–Miyaura couplings at room temperature. *Journal of the American Chemical Society* 132 (14): 4978–4979.

123 Gandeepan, P., Parthasarathy, K., and Cheng, C.H. (2010). Synthesis of phenanthrone derivatives from sec-alkyl aryl ketones and aryl halides via a palladium-catalyzed dual C–H bond activation and enolate cyclization. *Journal of the American Chemical Society* 132 (25): 8569–8571.

124 Dang, Y., Deng, X., Guo, J. et al. (2016). Unveiling secrets of overcoming the "heteroatom problem" in palladium-catalyzed aerobic C–H functionalization of heterocycles: a DFT mechanistic study. *Journal of the American Chemical Society* 138 (8): 2712–2723.

125 Wang, J.-R., Yang, C.-T., Liu, L. et al. (2007). Pd-catalyzed aerobic oxidative coupling of anilides with olefins through regioselective C–H bond activation. *Tetrahedron Letters* 48 (31): 5449–5453.

126 Katayev, D., Larionov, E., Nakanishi, M. et al. (2014). Palladium-N-heterocyclic carbene (NHC)-catalyzed asymmetric synthesis of indolines through regiodivergent C(sp^3)–H activation: scope and DFT study. *Chemistry - A European Journal* 20 (46): 15021–15030.

127 Ke, Z. and Cundari, T.R. (2010). Palladium-catalyzed C–H activation/C–N bond formation reactions: DFT study of reaction mechanisms and reactive intermediates. *Organometallics* 29 (4): 821–834.

128 Stephens, D.E., Lakey-Beitia, J., Atesin, A.C. et al. (2014). Palladium-catalyzed C8-selective C–H arylation of quinoline N-oxides: insights into the electronic, steric, and solvation effects on the site selectivity by mechanistic and DFT computational studies. *ACS Catalysis* 5 (1): 167–175.

129 Wang, G.-W. and Yuan, T.-T. (2010). Palladium-catalyzed alkoxylation of N-methoxybenzamides via direct sp^2 C–H bond activation. *Journal of Organic Chemistry* 75 (2): 476–479.

130 Paul, P., Sengupta, P., and Bhattacharya, S. (2013). Palladium mediated C–H bond activation of thiosemicarbazones: catalytic application of organopalladium complexes in C–C and C–N coupling reactions. *Journal of Organometallic Chemistry* 724: 281–288.

131 Justicia, J., Oltra, J.E., and Cuerva, J.M. (2004). Palladium mediated C–H activation in the field of terpenoids: synthesis of rostratone. *Tetrahedron Letters* 45 (22): 4293–4296.

132 Munz, D., Meyer, D., and Strassner, T. (2013). Methane CH activation by palladium complexes with chelating bis(NHC) ligands: a DFT study. *Organometallics* 32 (12): 3469–3480.

133 Strassner, T., Muehlhofer, M., Zeller, A. et al. (2004). The counterion influence on the CH-activation of methane by palladium(II) biscarbene complexes – structures, reactivity and DFT calculations. *Journal of Organometallic Chemistry* 689 (8): 1418–1424.

134 Wang, D.-H., Hao, X.-S., Wu, D.-F. et al. (2006). Palladium-catalyzed oxidation of Boc-protected N-methylamines with IOAc as the oxidant: ABoc-directed sp^3 C–H bond activation. *Organic Letters* 8 (15): 3387–3390.

135 Aguilar, D., Navarro, R., Soler, T. et al. (2010). Regioselective functionalization of iminophosphoranes through Pd-mediated C–H bond activation: C–C and C–X bond formation. *Tetrahedron Letters* 39 (43): 10422.

136 Munz, D. and Strassner, T. (2014). On the mechanism of the palladium bis(NHC) complex catalyzed CH functionalization of propane: experiment and DFT calculations. *Chemistry* 20 (45): 14872–14879.

137 Choi, H., Min, M., Peng, Q. et al. (2016). Unraveling innate substrate control in site-selective palladium-catalyzed C–H heterocycle functionalization. *Chemical Science* 7 (6): 3900–3909.

138 Wakaki, T., Togo, T., Yoshidome, D. et al. (2018). Palladium-catalyzed synthesis of diaryl ketones from aldehydes and (hetero)aryl halides via C–H bond activation. *ACS Catalysis* 8: 3123–3128.

139 Davies, D.L., Donald, S.M.A., and Macgregor, S.A. (2005). Computational study of the mechanism of cyclometalation by palladium acetate. *Journal of the American Chemical Society* 127: 13754–13755.

140 Garcia-Cuadrado, D., Braga, A.A., Maseras, F. et al. (2006). Proton abstraction mechanism for the palladium-catalyzed intramolecular arylation. *Journal of the American Chemical Society* 128 (4): 1066–1067.

141 Shi, R., Lu, L., Xie, H. et al. (2016). C8–H bond activation vs. C2–H bond activation: from naphthyl amines to lactams. *Chemical Communications* 52 (90): 13307–13310.

142 Yang, Y.-F., Cheng, G.-J., Liu, P. et al. (2013). Palladium-catalyzed meta-selective C–H bond activation with a nitrile-containing template: computational study on mechanism and origins of selectivity. *Journal of the American Chemical Society* 136 (1): 344–355.

143 Cheng, G.-J., Yang, Y.-F., Liu, P. et al. (2014). Role of N-acyl amino acid ligands in Pd(II)-catalyzed remote C–H activation of tethered arenes. *Journal of the American Chemical Society* 136 (3): 894–897.

144 Giri, R., Lan, Y., Liu, P. et al. (2012). Understanding reactivity and stereoselectivity in palladium-catalyzed diastereoselective sp^3 C–H bond activation: intermediate characterization and computational studies. *Journal of the American Chemical Society* 134 (34): 14118–14126.

145 Lafrance, M., Gorelsky, S.I., and Fagnou, K. (2007). High-yielding palladium-catalyzed intramolecular alkane arylation: reaction development and mechanistic studies. *Journal of the American Chemical Society* 129 (47): 14570–14571.

146 Kefalidis, C.E., Baudoin, O., and Clot, E. (2010). DFT study of the mechanism of benzocyclobutene formation by palladium-catalysed C(sp^3)–H activation: role of the nature of the base and the phosphine. *Tetrahedron Letters* 39 (43): 10528–10535.

147 Zhang, H., Wang, H.-Y., Luo, Y. et al. (2018). Regioselective palladium-catalyzed C–H bond trifluoroethylation of indoles: exploration and mechanistic insight. *ACS Catalysis* 8 (3): 2173–2180.

148 Yang, Y.F., Chen, G., Hong, X. et al. (2017). The origins of dramatic differences in five-membered vs. six-membered chelation of Pd(II) on efficiency of C(sp(3))–H bond activation. *Journal of the American Chemical Society* 139 (25): 8514–8521.

149 Zhang, S., Chen, Z., Qin, S. et al. (2016). Non-redox metal ion promoted oxidative coupling of indoles with olefins by the palladium(II) acetate catalyst through dioxygen activation: experimental results with DFT calculations. *Organic & Biomolecular Chemistry* 14 (17): 4146–4157.

150 Xing, Y.-M., Zhang, L., and Fang, D.-C. (2015). DFT studies on the mechanism of palladium(IV)-mediated C–H activation reactions: oxidant effect and regioselectivity. *Organometallics* 34 (4): 770–777.

151 Yang, Y.-M., Dang, Z.-M., and Yu, H.-Z. (2016). Density functional theory investigation on Pd-catalyzed cross-coupling of azoles with aryl thioethers. *Organic & Biomolecular Chemistry* 14 (19): 4499–4506.

152 Xie, H., Zhang, H., and Lin, Z. (2013). DFT studies on the mechanisms of palladium-catalyzed intramolecular arylation of a silyl C(sp^3)–H bond. *New Journal of Chemistry* 37 (9): 2856.

153 Solé, D., Amenta, A., Mariani, F. et al. (2017). Transition metal-catalysed intramolecular carbenoid C–H insertion for pyrrolidine formation by decomposition of α-diazoesters. *Advanced Synthesis and Catalysis* 359 (20): 3654–3664.

154 Modak, A., Rana, S., Phukan, A.K. et al. (2017). Palladium-catalyzed deformylation reactions with detailed experimental and in silico mechanistic studies. *European Journal of Organic Chemistry* 2017 (28): 4168–4174.

155 Mo, J., Xu, L., Ruan, J. et al. (2006). Regioselective Heck arylation of unsaturated alcohols by palladium catalysis in ionic liquid. *Chemical Communications* 34: 3591–3593.

156 Ye, J. and Ma, S. (2014). Palladium-catalyzed cyclization reactions of allenes in the presence of unsaturated carbon–carbon bonds. *Accounts of Chemical Research* 47 (4): 989–1000.

157 Minami, Y., Anami, T., and Hiyama, T. (2014). Palladium-catalyzed annulation of 2-substituted silylethynyloxybiaryls through δ-C–H activation. *Chemistry Letters* 43 (11): 1791–1793.

158 Yasui, Y. and Takemoto, Y. (2008). Intra- and intermolecular amidation of C–C unsaturated bonds through palladium-catalyzed reactions of carbamoyl derivatives. *Chemical Record* 8 (6): 386–394.

159 Ozaki, T., Kotani, M., Kusano, H. et al. (2011). Highly regioselective hydroselenation and double-bond isomerization of terminal alkynes with benzeneselenol catalyzed by bis(triphenylphosphine)palladium(II) dichloride. *Journal of Organometallic Chemistry* 696 (1): 450–455.

160 Fu, C.F., Lee, C.C., Liu, Y.H. et al. (2010). Biscarbene palladium(II) complexes reactivity of saturated versus unsaturated N-heterocyclic carbenes. *Inorganic Chemistry* 49 (6): 3011–3018.

161 Liu, G. and Lu, X. (2003). Palladium(II)-catalyzed coupling of allenoic acids and α,β-unsaturated carbonyl compounds through tandem intramolecular oxypalladation and conjugate addition reactions. *Tetrahedron Letters* 44 (1): 127–130.

162 Liu, J., Liu, Q., Franke, R. et al. (2015). Ligand-controlled palladium-catalyzed alkoxycarbonylation of allenes: regioselective synthesis of alpha,beta- and beta,gamma-unsaturated esters. *Journal of the American Chemical Society* 137 (26): 8556–8563.

163 Wu, W. and Jiang, H. (2012). Palladium-catalyzed oxidation of unsaturated hydrocarbons using molecular oxygen. *Accounts of Chemical Research* 45 (10): 1736–1748.

164 Yang, Y., Chen, L., Zhang, Z. et al. (2011). Palladium-catalyzed oxidative C–H bond and C horizontal line C double bond cleavage: C-3 acylation of indolizines with alpha,beta-unsaturated carboxylic acids. *Organic Letters* 13 (6): 1342–1345.

165 Comasvives, A., Gonzalezarellano, C., Boronat, M. et al. (2008). Mechanistic analogies and differences between gold- and palladium-supported Schiff base complexes as hydrogenation catalysts: a combined kinetic and DFT study. *Journal of Catalysis* 254 (2): 226–237.

166 Holder, J.C., Zou, L., Marziale, A.N. et al. (2013). Mechanism and enantioselectivity in palladium-catalyzed conjugate addition of arylboronic acids to beta-substituted cyclic enones: insights from computation and experiment. *Journal of the American Chemical Society* 135 (40): 14996–15007.

167 López-Serrano, J., Lledós, A., and Duckett, S.B. (2008). A DFT study on the mechanism of palladium-catalyzed alkyne hydrogenation: neutral versus cationic pathways. *Organometallics* 27 (1): 43–52.

168 Henriksen, S.T., Tanner, D., Skrydstrup, T. et al. (2010). DFT investigation of the palladium-catalyzed ene-yne coupling. *Chemistry* 16 (31): 9494–9501.

169 Lan, Y., Deng, L., Liu, J. et al. (2009). On the mechanism of the palladium catalyzed intramolecular Pauson–Khand-type reaction. *The Journal of Organic Chemistry* 74: 5049–5058.

170 Quan, M., Yang, G., Xie, F. et al. (2015). Pd(II)-catalyzed asymmetric addition of arylboronic acids to cyclic N-sulfonyl ketimine esters and a DFT study of its mechanism. *Organic Chemistry Frontiers* 2 (4): 398–402.

171 Suleiman, R., Ibdah, A., and El Ali, B. (2011). A DFT study of the mechanism of palladium-catalyzed alkoxycarbonylation and aminocarbonylation of alkynes: hydride versus amine pathways. *Journal of Organometallic Chemistry* 696 (11–12): 2355–2363.

172 El Ali, B., Tijani, J., and El-Ghanam, A.M. (2002). Total regioselective control of the carbonylative coupling of 1-heptyne with aniline and N-methyl aniline catalyzed by palladium(II) and phosphine ligand. *Journal of Molecular Catalysis A: Chemical* 187: 17–33.

173 Ansari, A.J., Pathare, R.S., Maurya, A.K. et al. (2018). Synthesis of diverse nitrogen heterocycles via palladium-catalyzed tandem azide-isocyanide cross-coupling/cyclization: mechanistic insight using experimental and theoretical studies. *Advanced Synthesis and Catalysis* 360 (2): 290–297.

174 Che, C., Ho, C., and Huang, J. (2007). Metal–carbon multiple bonded complexes carbene, vinylidene and allenylidene complexes of ruthenium and osmium supported by macrocyclic ligands. *Coordination Chemistry Reviews* 251 (17–20): 2145–2166.

175 Hahn, F.E. (2018). Introduction: carbene chemistry. *Chemical Reviews* 118 (19): 9455–9456.

176 Ryan, S.J., Candish, L., and Lupton, D.W. (2013). Acyl anion free N-heterocyclic carbene organocatalysis. *Chemical Society Reviews* 42 (12): 4906–4917.

177 Herndon, J.W. (2000). Applications of carbene complexes toward organic synthesis. *Coordination Chemistry Reviews* 206–207: 237–262.

178 Curran, D.P., Solovyev, A., Makhlouf Brahmi, M. et al. (2011). Synthesis and reactions of N-heterocyclic carbene boranes. *Angewandte Chemie International Edition* 50 (44): 10294–10317.

179 Schneider, S.K., Roembke, P., Julius, G.R. et al. (2006). Pyridin-, quinolin- and acridinylidene palladium carbene complexes as highly efficient C–C coupling catalysts. *Advanced Synthesis and Catalysis* 348 (14): 1862–1873.

180 Herrmann, W.A., Öfele, K., Preysing, D.v. et al. (2003). Phospha-palladacycles and N-heterocyclic carbene palladium complexes: efficient catalysts for CC-coupling reactions. *Journal of Organometallic Chemistry* 687 (2): 229–248.

181 Khlebnikov, V., Meduri, A., Mueller-Bunz, H. et al. (2012). Palladium carbene complexes for selective alkene di- and oligomerization. *Organometallics* 31 (3): 976–986.

182 Baker, M.V., Skelton, B.W., White, A.H. et al. (2001). Palladium carbene complexes derived from imidazolium-linked *ortho*-cyclophanes. *Journal of the Chemical Society, Dalton Transactions* 2: 111–120.

183 Altenhoff, G., Würtz, S., and Glorius, F. (2006). The first palladium-catalyzed Sonogashira coupling of unactivated secondary alkyl bromides. *Tetrahedron Letters* 47 (17): 2925–2928.

184 Clement, N.D., Routaboul, L., Grotevendt, A. et al. (2008). Development of palladium-carbene catalysts for telomerization and dimerization of 1,3-dienes: from basic research to industrial applications. *Chemistry* 14 (25): 7408–7420.

185 Comanescu, C.C. and Iluc, V.M. (2015). C–H activation reactions of a nucleophilic palladium carbene. *Organometallics* 34 (19): 4684–4692.

186 Albeniz, A.C., Espinet, P., Manrique, R. et al. (2005). Aryl palladium carbene complexes and carbene-aryl coupling reactions. *Chemistry* 11 (5): 1565–1573.

187 Selvakumar, K., Zapf, A., and Beller, M. (2002). New palladium carbene catalysts for the Heck reaction of aryl chlorides in ionic liquids. *Organic Letters* 4 (18): 3031–3033.

188 Marion, N. and Nolan, S.P. (2008). Well-defined N-heterocyclic carbenes-palladium(II) precatalysts for cross-coupling reactions. *Accounts of Chemical Research* 41 (11): 1440–1449.

189 Kim, J.H., Kim, J.W., Shokouhimehr, M. et al. (2005). Polymer-supported N-heterocyclic carbene-palladium complex for heterogeneous Suzuki cross-coupling reaction. *Journal of Organometallic Chemistry* 70 (17): 6714–6720.

190 Karimi, B. and Enders, D. (2006). New N-heterocyclic carbene palladium complex/ionic liquid matrix immobilized on silica: application as recoverable catalyst for the Heck reaction. *Organic Letters* 8 (6): 1237–1240.

191 Devine, S.K. and Van Vranken, D.L. (2007). Palladium-catalyzed carbene insertion into vinyl halides and trapping with amines. *Organic Letters* 9 (10): 2047–2049.

192 Batey, R.A., Shen, M., and Lough, A.J. (2002). Carbamoyl-substituted N-heterocyclic carbene complexes of palladium(II): application to Sonogashira cross-coupling reactions. *Organic Letters* 4 (9): 1411–1414.

193 Schönfelder, D., Fischer, K., Schmidt, M. et al. (2005). Poly(2-oxazoline)s functionalized with palladium carbene complexes: soluble, amphiphilic polymer supports for C–C coupling reactions in water. *Macromolecules* 38 (2): 254–262.

194 Albéniz, A.C., Espinet, P., Manrique, R. et al. (2002). Observation of the direct products of migratory insertion in aryl palladium carbene complexes and their subsequent hydrolysis. *Angewandte Chemie* 114 (13): 2469–2472.

195 Straub, B.F. (2002). Pd(0) mechanism of palladium-catalyzed cyclopropanation of alkenes by CH_2N_2: a DFT study. *Journal of the American Chemical Society* 124: 14195–14201.

196 Li, B., Bi, S., Liu, Y. et al. (2014). Role of acetate and water in the water-assisted $Pd(OAc)_2$-catalyzed cross-coupling of alkenes with N-tosyl hydrazones: a DFT study. *Organometallics* 33 (13): 3453–3463.

197 Sun, Z., Du, C., Liu, P. et al. (2018). Highly selective β-hydride elimination in the Pd-catalyzed cross-coupling of N-tosylhydrazones with benzyl bromides. *ChemistrySelect* 3 (3): 900–903.

198 Jensen, T. and Fristrup, P. (2009). Toward efficient palladium-catalyzed allylic C–H alkylation. *Chemistry* 15 (38): 9632–9636.

199 Hazari, A., Gouverneur, V., and Brown, J.M. (2009). Palladium-catalyzed substitution of allylic fluorides. *Angewandte Chemie International Edition* 48 (7): 1296–1299.

200 Hollingworth, C., Hazari, A., Hopkinson, M.N. et al. (2011). Palladium-catalyzed allylic fluorination. *Angewandte Chemie International Edition* 50 (11): 2613–2617.

201 Huo, X., Yang, G., Liu, D. et al. (2014). Palladium-catalyzed allylic alkylation of simple ketones with allylic alcohols and its mechanistic study. *Angewandte Chemie International Edition* 53 (26): 6776–6780.

202 You, S.L. and Dai, L.X. (2006). Enantioselective palladium-catalyzed decarboxylative allylic alkylations. *Angewandte Chemie International Edition* 45 (32): 5246–5248.

203 Wallner, O.A. and Szabo, K.J. (2004). Palladium pincer complex-catalyzed allylic stannylation with hexaalkylditin reagents. *Organic Letters* 6 (11): 1829–1831.

204 Usui, I., Schmidt, S., Keller, M. et al. (2008). Allylation of *N*-heterocycles with allylic alcohols employing self-assembling palladium phosphane catalysts. *Organic Letters* 10 (6): 1207–1210.

205 Uozumi, Y. and Shibatomi, K. (2001). Catalytic asymmetric allylic alkylation in water with a recyclable amphiphilic resin-supported P,N-chelating palladium complex. *Journal of the American Chemical Society* 123 (12): 2919–2920.

206 Trost, B.M., Xu, J., and Schmidt, T. (2009). Palladium-catalyzed decarboxylative asymmetric allylic alkylation of enol carbonates. *Journal of the American Chemical Society* 131 (51): 18343–18357.

207 Trost, B.M. and Thaisrivongs, D.A. (2008). Strategy for employing unstabilized nucleophiles in palladium-catalyzed asymmetric allylic alkylations. *Journal of the American Chemical Society* 130 (43): 14092–14093.

208 Trost, B.M. and Brennan, M.K. (2006). Palladium asymmetric allylic alkylation of prochiral nucleophiles: horsfiline. *Organic Letters* 8 (10): 2027–2030.

209 Trost, B.M. and Xu, J. (2005). Palladium-catalyzed asymmetric allylic alpha-alkylation of acyclic ketones. *Journal of the American Chemical Society* 127 (49): 17180–17181.

210 Faller, J.W. and Wilt, J.C. (2005). Palladium/BINAP(S)-catalyzed asymmetric allylic amination. *Organic Letters* 7 (4): 633–636.

211 Evans, D.A., Campos, K.R., Tedrow, J.S. et al. (2000). Application of chiral mixed phosphorus/sulfur ligands to palladium-catalyzed allylic substitutions. *Journal of the American Chemical Society* 122 (33): 7905–7920.

212 Kazmaier, U. and Zumpe, F.L. (2000). Palladium-catalyzed allylic alkylations without isomerization—dream or reality? *Angewandte Chemie International Edition* 39 (4): 802–804.

213 Ariafard, A. and Lin, Z. (2006). DFT studies on the mechanism of allylative dearomatization catalyzed by palladium. *Journal of the American Chemical Society* 128: 13010–13016.

214 Piechaczyk, O., Thoumazet, C., Jean, Y. et al. (2006). DFT study on the palladium-catalyzed allylation of primary amines by allylic alcohol. *Journal of the American Chemical Society* 128: 14306–14317.

215 Bernar, I., Fiser, B., Blanco-Ania, D. et al. (2017). Pd-catalyzed hydroamination of alkoxyallenes with azole heterocycles: examples and mechanistic proposal. *Organic Letters* 19 (16): 4211–4214.

216 Senan, A.M., Qin, S., Zhang, S. et al. (2016). Nonredox metal-ion-accelerated olefin isomerization by palladium(II) catalysts: density functional theory (DFT) calculations supporting the experimental data. *ACS Catalysis* 6 (7): 4144–4148.

217 Xie, H., Zhang, H., and Lin, Z. (2013). DFT studies on the palladium-catalyzed dearomatization reaction between chloromethylnaphthalene and the cyclic amine morpholine. *Organometallics* 32 (8): 2336–2343.

6

Theoretical Study of Pt-Catalysis

195.078
Pt[78]
Platinum
[Xe]4f^{14}5d^{9}6s^1

Group 10 transition metals, majorly including Ni, Pd, and Pt, exhibit significant efficacy for catalyzing organometallic coupling reactions and easy access to the synthetic chemistry in both industrial and academic fields [1–11]. In particular, platinum has played a central role in the organometallic chemistry by soluble species over the past several decades, in both fundamental understanding and approaches to the construction of new C—C and C—heteroatom bonds [12–16]. As an element of the same group, Pt-complexes often display the same reactivity with Pd, while the mechanism of Pt-catalysis also could be analogous to Pd-chemistry [17]. The elementary reactions, such as oxidative addition, insertion, transmetallation, and reductive elimination, which are popular in Pd-chemistry, are also present in Pt-catalyzed organometallic transformations [18–24]. However, there are important elemental properties that affect the reactivity and catalytic potential of Pt-complex that are not present in Pd-chemistry [25–27]. Possibly, the most significant differences between those two group 10 metals are the lanthanide contraction and relativistic effects present in the former [28–30], both of which contribute to more diffuse 5d orbitals that enhance Pt's "soft" character [31]. This, coupled with a slower rate of ligand substitution for Pt(II), allows for the development of catalytic cycles that rely on alternative pathways for M—C bond cleavage, such as protonolysis, cyclopropanation, and heteroatom insertion [32–34].

Alternatively, as a late transition metal, complexes and salts derived from Pt have shown an exceptional ability to promote a variety of organometallic transformations of unsaturated precursors, which also revealed a similar reactivity of the compounds of Au [35–41]. These processes result from the peculiar Lewis acid property of Pt, whose alkynophilic character promotes the nucleophilic attack of unsaturated reactants by the π-acid activation [42, 43]. Moreover, the Lewis acid property of Pt also allows the formation of stable carbenoid complexes, which would be used in the

Computational Methods in Organometallic Catalysis: From Elementary Reactions to Mechanisms,
First Edition. Yu Lan.
© 2021 WILEY-VCH GmbH. Published 2021 by WILEY-VCH GmbH.

6 Theoretical Study of Pt-Catalysis

further transformations for the construction of new C—C bonds [44–52]. Unlike the Au-catalysis, the chemical property of Pt is also similar with the other group 10 transition metals, i.e. Ni and Pd [53–62]. Therefore, the alkyl, vinyl, and aryl Pt species play a significant stability, which could be used for further functionalization by insertion, isomerization, or reductive elimination [63–67].

The chemical properties and reactivity of Pt-catalysis could analogous consider a combination of Pd- and Au-chemistry, which lead to the complexity for the mechanism of Pt-catalyzed transformations [68–78]. Indeed, the mechanism in Pt-catalysis could be different from corresponding Pd- or Au-catalysis [79–82]. Therefore, a series of mechanistic studies around Pt-catalysis have been performed to reveal the glamor of Pt chemistry. This chapter is organized to reveal the recent advances of theoretical study in Pt-catalyzed organic transformations, which would be sorted by the activated species, majorly including C—H activation, alkyne activation, alkene activation, and allene activation.

6.1 Mechanism of Pt-Catalyzed C—H Activation

C—H activation could be defined as a facile C—H bond cleavage process with an "M–X" species that proceeds by coordination of a hydrocarbon leading to an M—C intermediate [84–87]. In well-accepted Pd-chemistry, the mechanism of most Pd^{II}-mediated C—H activation could be described as a concerted metalation–deprotonation (CMD), in which the C—H bond cleavage and C—Pd bond formation take place synchronously [88–98]. Also, as a group 10 element, Pt-catalysts were also found to be active in C—H activation and functionalization reactions. However, there are rare examples that affirmed a CMD process in Pt-mediated C—H activation process [83]. Alternatively (Scheme 6.1), an oxidative addition of C—H bond is often proposed in Pt-catalysis instead of CMD (model a). Besides, electrophilic deprotonation also has been proposed in Pt-mediated C—H activations (model b). Moreover, carbenoid Pt complexes are easily prepared by

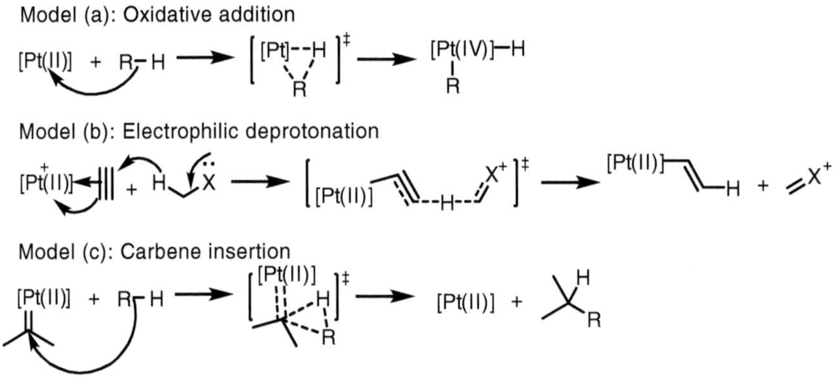

Scheme 6.1 The possible mechanisms of Pt-mediated C—H activations. Source: Based on Webb et al. [83].

the activation of acetylene derivatives, which contain a Pt=C double bond. The migratory insertion of carbenoid into C—H bond also could be considered to be a C—H bond activation in Pt-chemistry (model c) [99–101].

6.1.1 Oxidative Addition of C—H Bond

The controllable and selective C—H activation and conversion of methane remains a grand challenge for both academia and industry [102–106]. In 1998, Periana et al. reported a promising system using Pt(bpym)Cl$_2$ as catalyst to convert methane to methyl sulfate in concentrated sulfuric acid, which acts as both the solvent and the oxidant [107].

A density functional theory (DFT) calculation is performed to reveal the mechanism of C—H activation process (Figure 6.1) [108]. The calculated free-energy profiles showed that the coordination of methane to form a square-plane Pt-methane σ-complex **6-2** is 16.1 kcal/mol endothermic. The oxidative addition could take place via a three-membered ring-type transition state **6-3ts** with a free-energy barrier of 8.0 kcal/mol to form σ-complex **6-4**. When a quadrangular pyramid Pt(IV) intermediate **6-4** is formed, an outer-sphere reductive deprotonation could occur via transition state **6-5ts** with a free-energy barrier of 9.0 kcal/mol to form methyl Pt(II) complex **6-6**. Then, the oxidation by concentrated sulfuric acid results in a Pt(IV) complex **6-8**, which can undergo a reductive elimination via transition state **6-9ts** to yield methyl sulfate product and regenerate the active catalyst **6-1**.

6.1.2 Electrophilic Dehydrogenation

A high oxidation state Pt(IV)I$_4$ species could be used as a π-acid in some of C—H activation reactions, in which acetylene could be activated by the coordinated Pt^{IV}I$_4$ to reduce the electron density of unsaturated bond to activate an

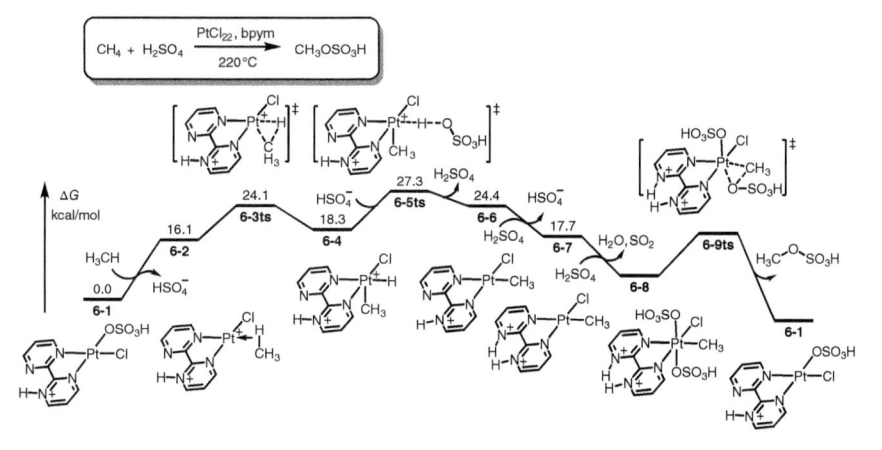

Figure 6.1 Free-energy profiles for the Pt-catalyzed oxidation of methane in concentrated sulfuric acid. The energies were calculated at B3LYP/LACV3P**++/PBF(sulfuric acid)// B3LYP/LACVP**/PBF(sulfuric acid) level of theory.

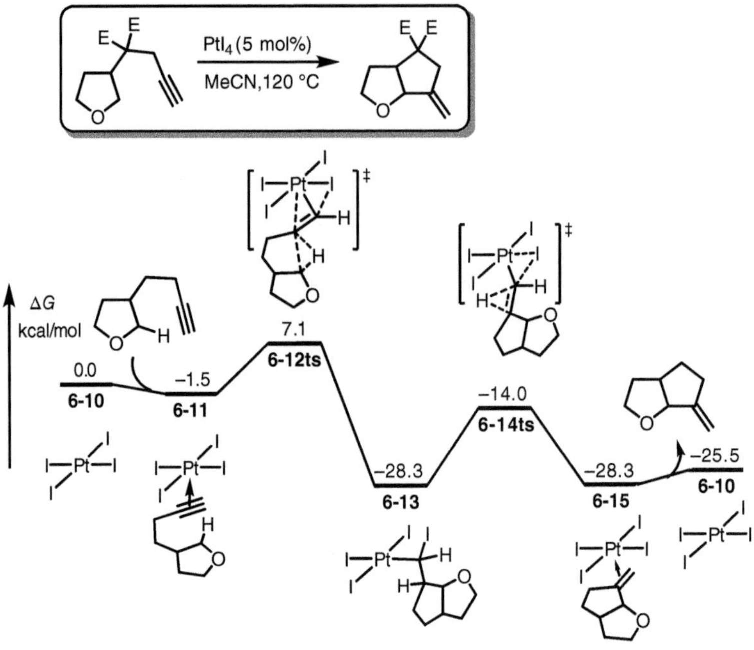

Figure 6.2 Free-energy profiles for the PtI$_4$-catalyzed intramolecular C—H annulation of tetrahydrofuran. The energies were calculated at M06/6-311++G(d,p) (LANL2DZ+f for Pt)//IEF-PCM(acetonitrile)//B3LYP/6-31+G(d,p) (LANL2DZ+f for Pt) level of theory. Source: Based on Jin et al. [116].

electron-rich hydrogen in hydrocarbon [109–114]. In 2009, Sames group reported a Pt(IV)I$_4$-catalyzed C(sp^3)—H activation of tetrahydrofuran, in which the following intramolecular annulation of acetylene formed a pure five-membered product in good yield in acetonitrile at 120 °C, while Au-catalysts offered complex mixtures of many products and gave only small amounts of the desired product [115].

DFT calculations were performed by Zhao group to reveal the mechanism of PtI$_4$-catalyzed tetrahydrofuran C—H annulation reaction [116]. As shown in Figure 6.2, PtI$_4$ **6-10** is set to relative zero in free-energy profiles, which could coordinate with acetylene to form a π-acid complex **6-11** with 1.5 kcal/mol exoergic. With the activation of high oxidation state Pt, the electrophilicity of acetylene moiety is significantly increased. Then, an intramolecular [1,5]-hydride transfer could occur via a six-membered ring-type transition state **6-12ts** irreversibly. The activation barrier of this step was determined to be only 8.6 kcal/mol, indicating a rather rapid process. Intrinsic reaction coordinate (IRC) calculation revealed that the expectant oxonium ylide was not observed in gas phase. Alternatively, the C—C bond formation occurs followed by hydride transfer to generate a C—C coupling intermediate **6-13**. An intramolecular [1,2]-hydrogen shift could take place via transition state **6-14ts** with a barrier of 14.3 kcal/mol to form an olefin-coordinated PtI$_4$ intermediate **6-15**, which is considered to be the rate-determining step in the whole catalytic cycle. The ligand exchange with reactant regenerates active catalyst **6-10** and releases annulation product.

6.1.3 Carbene Insertion into C—H Bonds

It is well established that terminal alkynes on treatment with transition metals, such as Pt, Au, and Ru, lead to the formation of vinylidene–metal complexes, which reveal a carbenoid character in the following transformations [117, 118]. In particular, when a vinylidene Pt(II) intermediate is generated, a series of following processes with C—H bond in hydrocarbon could occur to cleavage C—H bond [119, 120]. Formally, the above-mentioned transformations could be considered as the migratory insertion of carbenoid into C—H bond. However, the intrinsic mechanism of this process is far more complicated than superficial equation, which could include concerted carbene migratory insertion into C—H bond (model a), Friedel–Crafts-type electrophilic substitution (model b), hydride transfer followed by electrophilic substitution (model c), and Pt-carbine-involved pericyclic [1,5]-hydrogen shift (model d) (Scheme 6.2).

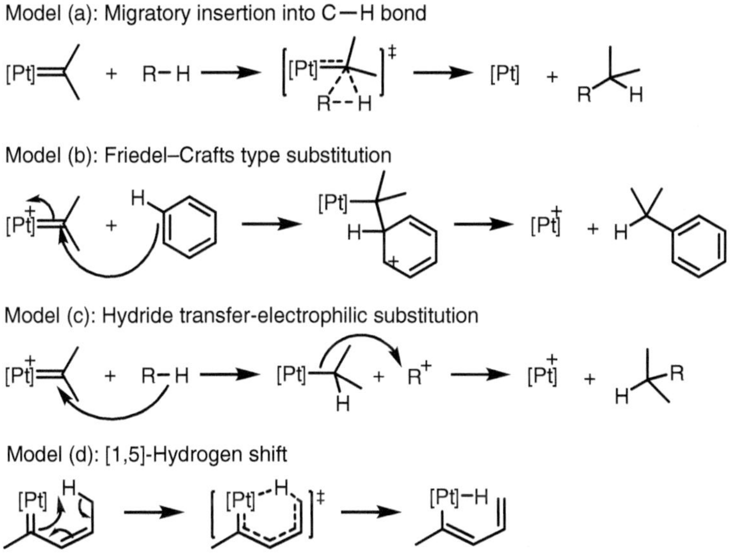

Model (a): Migratory insertion into C—H bond

Model (b): Friedel–Crafts type substitution

Model (c): Hydride transfer-electrophilic substitution

Model (d): [1,5]-Hydrogen shift

Scheme 6.2 Possible mechanisms of Pt-mediated C—H bond cleavage.

In 2006, Yamamoto group reported a PtBr$_2$-catalyzed transformation of ethynyl–benzylalcohol into indene, where a C(sp^3)—H activation takes place by formal carbene insertion into the C—H bond of the vinylidene Pt complex [122]. A DFT calculation revealed the detailed reaction mechanism of this reaction (Figure 6.3). Enyne-coordinated Pt species **6-16** is set to relative zero in free-energy profiles. The 1,2-hydrogen shift could take place via transition state **6-17ts** with a free-energy barrier of 33.4 kcal/mol, which is considered to be the rate-determining step. The relative free energy of the generated Pt–vinylidene complex **6-18** is only 2.2 kcal/mol higher than that of active catalyst **6-16**. A concerted transition state of carbene migratory insertion into C—H bond was located at **6-19ts** with a free-energy barrier of only 3.1 kcal/mol. The geometry information of transition

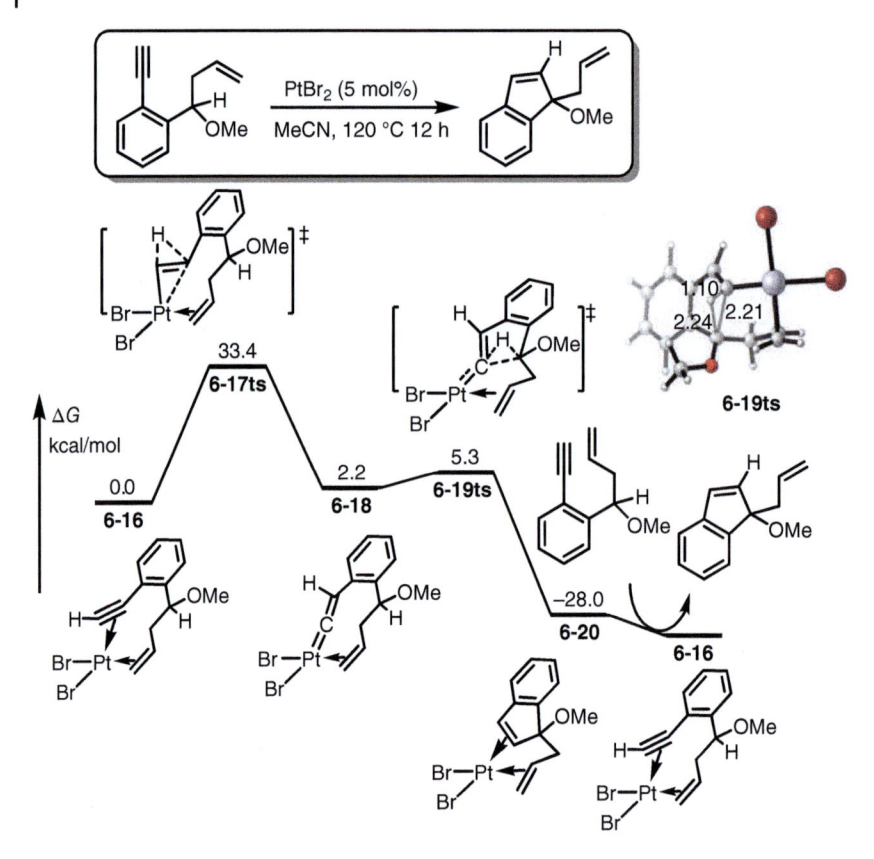

Figure 6.3 Free-energy profiles for the PtBr$_2$-catalyzed isomerization of ethynyl–benzylalcohol into indene involving a C(sp^3)–H activation. The energies were calculated at B3LYP/SDD level of theory. Source: Based on Wang et al. [121].

state **6-19ts** is shown in Figure 6.3. The lengths of forming C—C bond, forming C—H bond, and breaking C—H bond are 2.21, 2.24, and 1.10 Å, respectively, which clearly revealed a concerted process.

When vinylidene Pt species is formed, the electron deficiency of vinylidene moiety could urge it to eliminate a hydride from another C—H bond to form a vinyl platinate species. In 2009, Chatani group reported a PtCl$_2$-catalyzed cyclization of *ortho*-alkyl ethynylbenzenes under relatively mild conditions leading to benzylic C—H bond functionalization products [101]. The DFT study of this reaction was performed by Yu group [121]. As shown in Figure 6.4, the ligand exchange loads the ethynylbenzene reactant with an endergonic of 6.1 kcal/mol in intermediate **6-22**. The 1,2-hydrogen shift could occur via transition state **6-23ts** with a free-energy barrier of 16.3 kcal/mol, which was considered to be the rate limit. When a vinylidene Pt intermediate **6-24** is formed, an intramolecular 1,5-hydrogen shift of benzylic hydrogen could occur via transition state **6-25ts** with a free-energy barrier of only 4.1 kcal/mol. A zwitterionic intermediate **6-26** was possible, where the positive

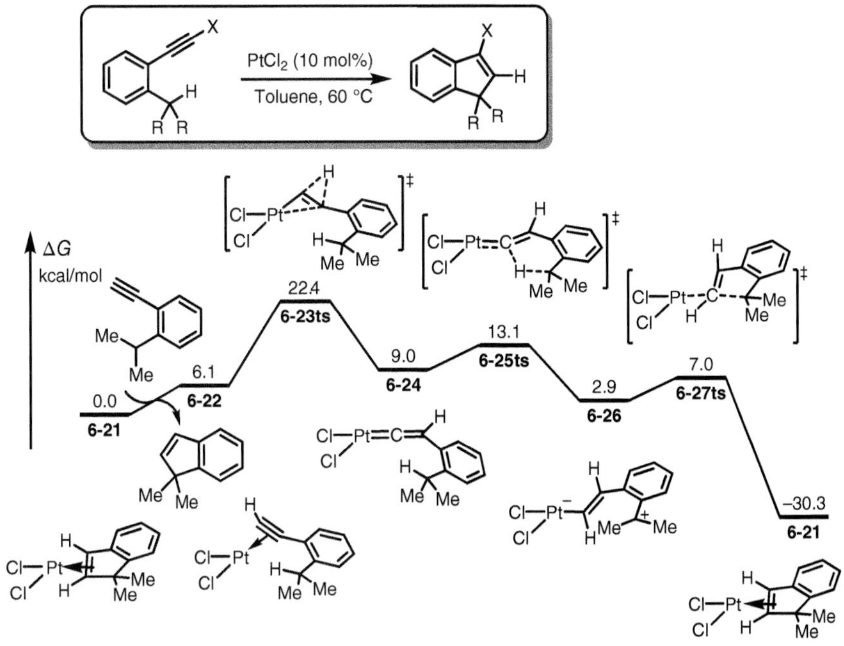

Figure 6.4 Free-energy profiles for the PtCl$_2$-catalyzed cyclization of *ortho*-alkyl ethynylbenzenes. The energies were calculated at M06/6-31G(d) (LANL2DZ for Pt)/CPCM(toluene)//B3LYP/6-31G(d) (LANL2DZ for Pt) level of theory.

charge is located on benzylic position, and the negative charge concentrates upon platinate moiety. The subsequent intramolecular electrophilic substitution with benzyl carbocation onto vinyl platinate could take place via transition state **6-27ts** to form an indene-coordinated PtCl$_2$ **6-21**.

Interestingly, when a butenylidene Pt intermediate is present in Pt-catalysis, it could be considered as an equivalent of pentadiene, which could undergo a pericyclic 1,5-hydrogen shift to afford a hydride Pt species [124]. The example for this process is presented in a Pt-mediated isomerization of 1,6-enyne with propargylic ester group. The process furnished a complex mixture from which the triene product was characterized and isolated in only 19% yield [125]. As shown in Figure 6.5, a DFT calculation by Soriano et al. is taken to reveal the mechanism for the Pt-mediated isomerization of 1,6-enyne [120]. With the coordination onto PtCl$_2$, the electron density of acetylene would be reduced by the Lewis acidity of Pt in complex **6-28**. Therefore, an intramolecular nucleophilic attack by the oxygen atom of ester group could occur via transition state **6-29ts** with a barrier of only 4.0 kcal/mol to form a five-membered ring-type oxonium ylide **6-30**. Subsequently, the initial C—O bond cleavage affords a butenylidene Pt intermediate **6-32** through transition state **6-31ts** with a barrier of only 8.3 kcal/mol. The key step of this reaction is pericyclic 1,5-hydrogen shift, which could take place via a concerted six-membered ring-type transition state **6-33ts** with a free-energy barrier of 15.6 kcal/mol. After a rapid reductive elimination, triene product yielded with the regeneration of catalyst **6-28**.

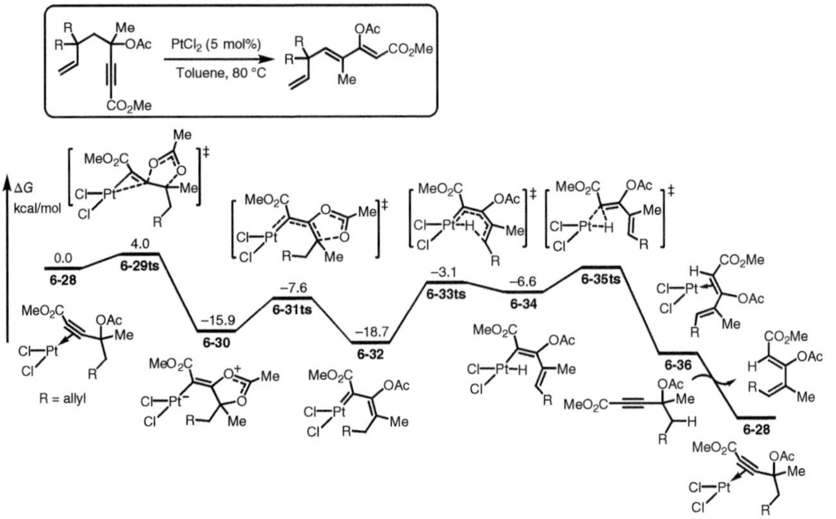

Figure 6.5 Free-energy profiles for the PtCl$_2$-catalyzed isomerization of 1,6-enyne. The energies were calculated at B3LYP/6-31G(d) (LANL2DZ for Pt)/PCM(toluene)//B3LYP/6-31G(d) (LANL2DZ for Pt) level of theory. Source: Based on Pujanauski et al. [123].

6.2 Mechanism of Pt-Catalyzed Alkyne Activation

As a soft and polarizable late transition metal, Pt is found to be applied to active unsaturated bonds, which reveals a carbophilic character. In essence, the coordination of alkyne onto Pt deprives the unsaturated bond of electron density to the extent that it becomes susceptible to an outer-sphere attack by an adequate nucleophile [126]. Although this basic mechanism of carbophilic activation with Pt-catalysts is certainly oversimplified and exceptions do exist, it is worth noting that it illustrates that no metal-based redox steps are involved in whole catalytic cycle. Actually, the catalytic behavior of Pt-catalysts in carbophilic activation is analogous to the Au(I)-catalysis, which also plays a π-acidic feature in alkyne activations [127, 128].

Alternatively, as a group 10 transition metal, the oxidative cycloadditions of Pt species with 1,6-enynes were usually proposed as a competing pathway as a contrast to corresponding carbophilic activation pathway, which is well known in Pt and Ni chemistry. However, it is often energetically disadvantageous in Pt-catalysis. The unfrequent examples of Pt-mediated oxidative cycloadditions would occur with activated enynes [129].

6.2.1 Nucleophilic Additions

With the activation of the coordinated π-acidic catalyst Pt(II) species, the electron density in acetylene is significantly deprived to allow an outer-sphere nucleophilic attack by appropriate nucleophiles. Generally, an intramolecular nucleophilic attack is recommended in Pt-catalysis, in which both heteroatoms and heterocycles could

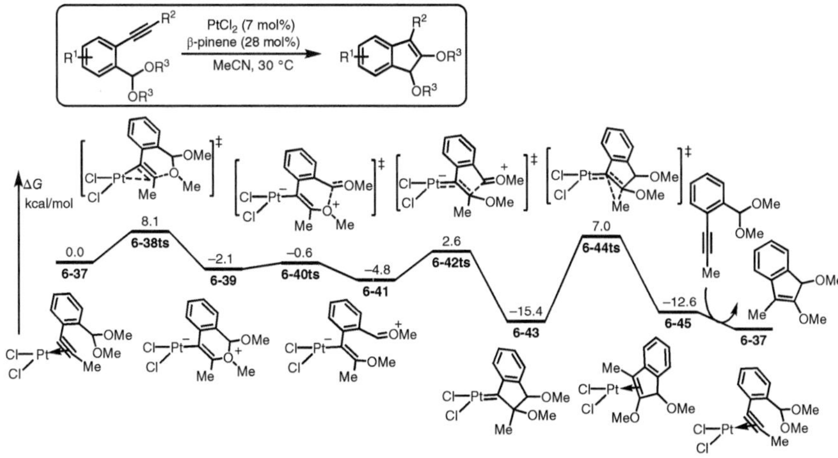

Figure 6.6 Free-energy profiles for the PtCl$_2$-catalyzed cyclization of o-alkynylbenzacetal. The energies were calculated at M06/6-311++G(2d,p) (SDD for Pt)/PCM(acetonitrile)//M06/ 6-31+G(d) (LANL2DZ for Pt) level of theory.

act as adequate nucleophile. Generally, the oxidative state of Pt remains unchanged, which means metal-based redox steps are excluded in this process [130].

Heteroatoms, such as oxygen, which contains at least one lone pair of electrons, play an important role in nucleophilic attack of π-acidic Pt(II)-activated acetylenes. In 2002, Yamamoto group reported an experimental study for the cyclizations of o-alkynylbenzacetal derivatives catalyzed by Pt halides, which yielded indene derivatives with a good isolated yield under a temperature of 30 °C in acetonitrile solvent [131]. A DFT study is taken out to further reveal the mechanism for the cyclization of o-alkynylbenzacetal derivatives [132]. The calculated free-energy profiles are shown in Figure 6.6. The coordination of o-alkynylbenzacetal with PtCl$_2$ forms intermediate **6-37**, which is considered as relative zero in free-energy profiles. The 6-*endo*-dig cyclization could occur via transition state **6-38ts** with a free-energy barrier of 8.1 kcal/mol to for isochromenylium intermediate **6-39**. Subsequently, the C—O bond cleavage of oxonium takes place via transition state **6-40ts** with a free-energy barrier of only 1.5 kcal/mol to form a zwitterionic intermediate **6-41**. The C—C bond formation via transition state **6-42ts** with a free-energy barrier of 7.4 kcal/mol forms a Pt–carbene complex **6-43**. The computational results showed that the 1,2-transfer of methyl group via transition state **6-44ts** is energetically favorable compared with the corresponding 1,2-transfer of methoxyl group. Therefore, indene derivative is observed as the major product, which would release from generated complex **6-45** by ligand exchange.

In another case, the propargylic ester group could act as nucleophile in the nucleophilic attack of π-acidic Pt(II)-activated acetylene [123]. As shown in Figure 6.7, an isomerization of oxiranyl propargylic esters catalyzed by Pt salt could afford pyran derivatives, which could undergo a further pentannulation to yield cyclopentenone product with a moderate-to-high yield under a temperature of 100 °C after 8 hours. DFT studies taken by Lera group revealed the mechanism of acetylene activation

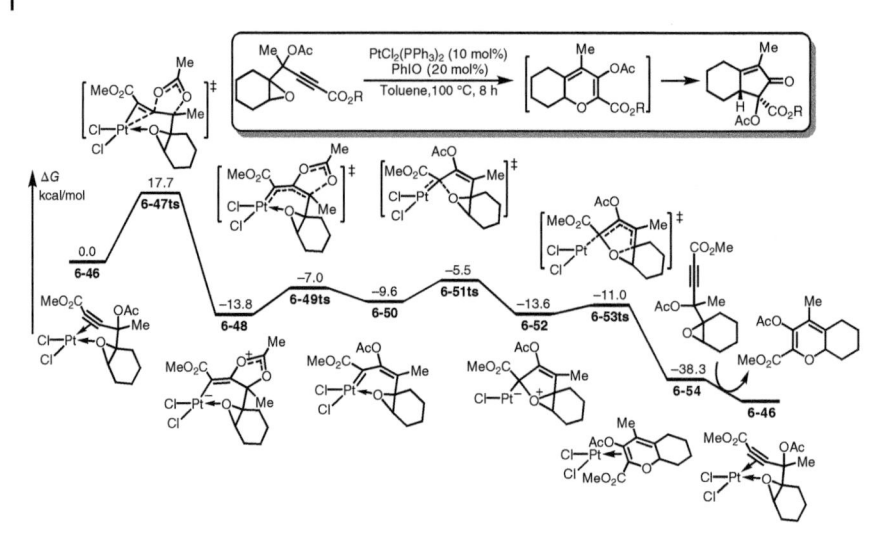

Figure 6.7 Free-energy profiles for the PtCl$_2$-catalyzed isomerization of oxiranyl propargylic esters. The energies were calculated at M06/6-31++G(d,p) (LANL2DZ for Pt)/PCM(toluene)//M06/6-31++G(d,p) (LANL2DZ for Pt) level of theory.

process [133]. As shown in Figure 6.7, the acetylene-coordinated PtCl$_2$ species **6-46** is set to relative zero. An intramolecular nucleophilic attack by propargylic ester group occurs via transition state **6-47ts** with a free-energy barrier of 17.7 kcal/mol to form a dioxolium intermediate **6-48** irreversibly. The following original C—O bond cleavage leading to the rearrangement of unsaturated bonds takes place via transition state **6-49ts** to afford an allylidene Pt intermediate **6-50**. Subsequently, an intramolecular nucleophilic attack of oxiranyl onto Pt–carbene occurs via transition state **6-51ts** with a barrier of 4.1 kcal/mol to form an oxonium intermediate **6-52**. Then, the C—O bond cleavage in oxiranyl moiety could take place via transition state **6-53ts** to afford a pyran-coordinated Pt species **6-54** irreversibly. The computational study showed that the rate-determining step for the whole catalytic cycle is the first nucleophilic attack of acetylene with propargylic ester group.

6.2.2 Cyclopropanation

Pt-catalyzed cycloisomerization of enynes has emerged as a powerful tool for the synthesis of carbocycles. Among this area, when 1,5-enynes or 1,6-enynes are employed as substrate, the cycloisomerization process usually goes through a bicycle-[3.1.0]hexene or [4.1.0]heptene intermediate, respectively [134, 135].

As shown in Figure 6.8, a wide experimental study by Marco–Contelles group aimed at establishing the scope and generality of the PtCl$_2$-catalyzed cycloisomerization of 1,6-enynes to afford bicycle[4.1.0]heptene product with moderate isolated yields [136]. The combination of theoretical and experimental studies by the same group revealed the mechanism of this cycloisomerization reaction [137]. As shown in Figure 6.8, with the activation of Pt, the π-electrons of alkene moiety in complex **6-55** could act as nucleophile to attack acetylene moiety. This process could occur

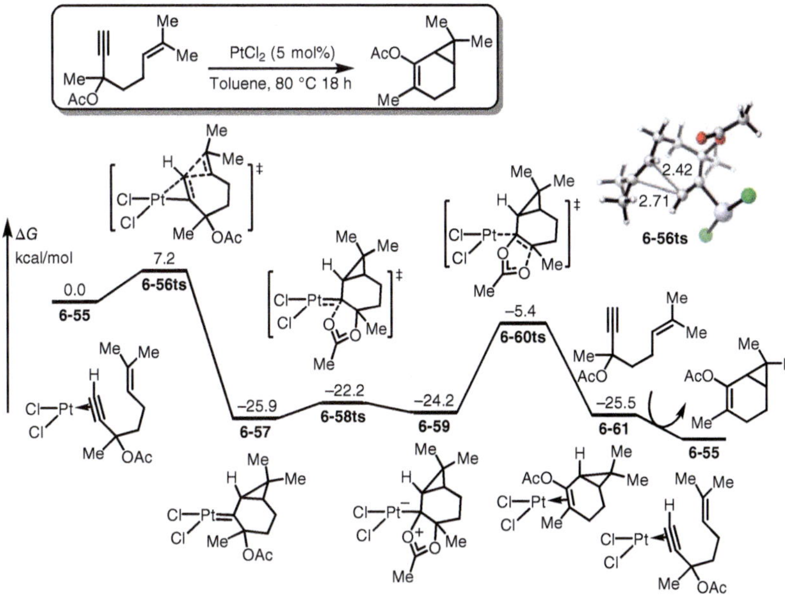

Figure 6.8 Free-energy profiles for the PtCl$_2$-catalyzed cyclopropanation of 1,6-enyne. The energies were calculated at B3LYP/6-311+G(2d,p) (LANL2DZ for Pt)//B3LYP/6-31G(d) (LANL2DZ for Pt) level of theory.

via a concerted three-membered ring-type transition state **6-56ts** with a free-energy barrier of only 7.2 kcal/mol. In the geometry of transition state **6-56ts**, length of two forming C—C bond is 2.42 and 2.71 Å, respectively, which indicate that the formation of two C—C bond is simultaneous. A bicycle[4.1.0]heptanylidene Pt intermediate **6-57** is formed with 25.9 kcal/mol exergonic. Then, an intramolecular migration of ester group could go through a dioxolium intermediate **6-59** to afford bicycle[4.1.0]heptane-coordinated Pt species **6-61**.

In another case, Li, Liu, Xu, and coworkers reported a PtI$_2$-catalyzed cycloisomerization of 1,5-enynes to afford terphenyl through a bicycle[3.1.0]hexene intermediate [138] (Figure 6.9). An additional theoretical calculation revealed the mechanism of this process. As shown in Figure 6.9, the intramolecular nucleophilic attack by ethylene moiety could take place via transition state **6-64ts** with a free-energy barrier of 18.9 kcal/mol. The geometry of this transition state also revealed a concerted process. When bicycle[3.1.0]hexanylidene Pt intermediate **6-65** is formed, the inner-sphere acetate acid elimination could occur via transition state **6-66ts** to afford bicycle[3.1.0]hexenylidene Pt intermediate **6-67**. Then, a methylene migration takes place via transition state **6-68ts** with an energy barrier of 22.4 kcal/mol to afford a zwitterionic intermediate **6-69**. This step is considered the rate limit around the whole catalytic cycle. After this step, an intramolecular C—C bond formation via transition state **6-70ts** forms a zwitterionic bicycle[3.1.0]-intermediate **6-71**. Further C—C bond cleavage followed by 1,2-hydrogen shift results in the generation of benzene product.

268 | 6 Theoretical Study of Pt-Catalysis

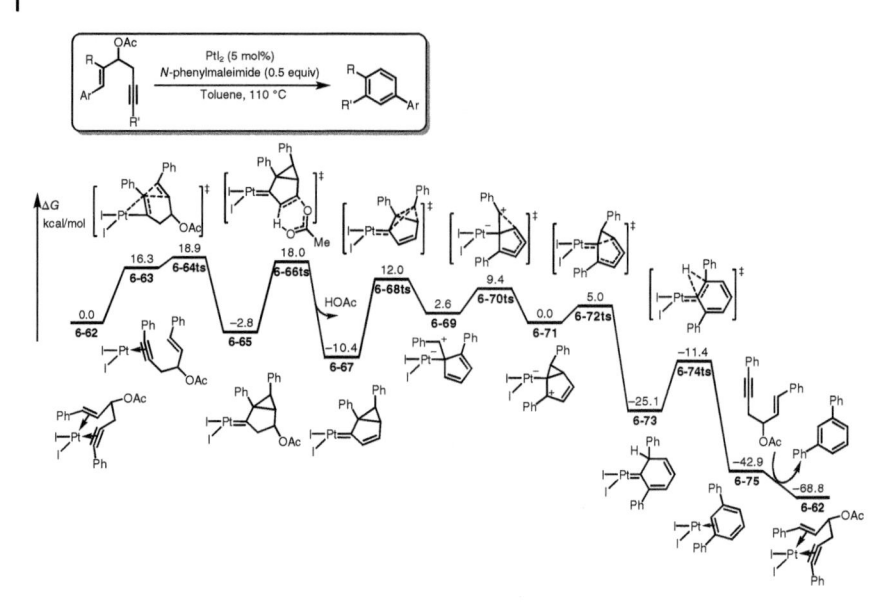

Figure 6.9 Free-energy profiles for the PtI$_2$-catalyzed cycloisomerization of 1,5-enyne. The energies were calculated at B3LYP-D3/6-311+G(d,p) (SDD for Pt and I)/SMD(toluene) //B3LYP/6-31G(d) (SDD for Pt and I) level of theory.

6.2.3 Oxidative Cycloaddition

Metal-involved intramolecular Alder-ene type reaction is one of the most effective ways for the construction of carbocycles, which could be catalyzed by a series of transition metals, such as RhI, TiII, PdII, and RuII species [140–146]. In 2001, Echavarren group reported a PtCl$_2$-catalyzed Alder-ene type cycloaddition of simple 1,6-enynes [147]. High-yield diene products could be generated under a temperature of 70 °C after 17–20 hours [139] (Scheme 6.3a. The mechanism of this transformation is proposed by the same group, which is shown in Scheme 6.3c. The Alder-ene type isomerization of 1,6-enyne probably takes place by the coordination of Pt(II) species to both unsaturated bonds. The formation of key Pt(IV) metalacycle intermediate would suffer from oxidative cycloaddition. The following β-hydrogen elimination from an alkyl group takes place to form a hydride vinyl Pt(IV) species. Finally, a reductive elimination leads to cycloisomerized products and regenerates catalytically active Pt(II) species. In another case, when a more active 1,6-allenyne is used in Pt-mediated Alder-ene type reaction (Scheme 6.3c), the corresponding cycloisomerization could occur in more moderate conditions.

As shown in Figure 6.10, allenyne-coordinated PtCl$_2$ species **6-76** is set to relative zero in calculated free-energy profiles [149, 150]. An oxidative cycloaddition could occur via transition state **6-77ts** with a barrier of only 16.4 kcal/mol, which is much lower than the corresponding step with simple enynes. The generation of cyclometalation Pt(IV) intermediate **6-78** is 28.5 kcal/mol exergonic. The following β-hydride elimination occurs via transition state **6-79ts** with a barrier of only 16.0 kcal/mol. The following reductive elimination takes place from hydride Pt(IV) intermediate

6.2 Mechanism of Pt-Catalyzed Alkyne Activation | 269

Scheme 6.3 Pt-mediated annulation of enynes through a key step of oxidative cycloadditions (a and b) and the common mechanism. Source: Based on Muñoz et al. [139].

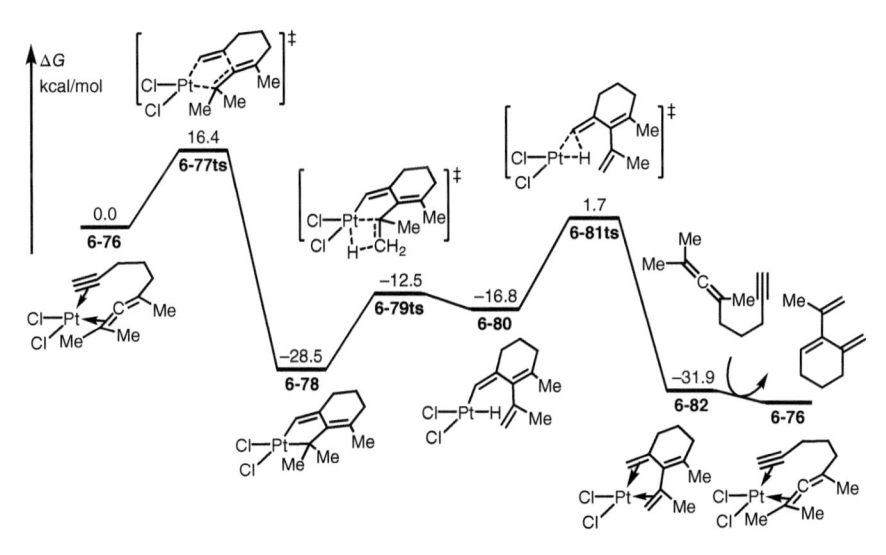

Figure 6.10 Free-energy profiles for the PtCl$_2$-catalyzed Alder-ene type cycloisomerization of 1,6-enyne. The energies were calculated at B3LYP/6-31G(d,p) (LANL2DZ for Pt) level of theory. Source: Based on Kefalidis and Tsipis [148].

6-80 via transition state **6-81ts** to generate diene product–coordinated Pt(II) species **6-82**. The overall activation free energy of the reductive elimination step is considered to be 30.2 kcal/mol from intermediate **6-78** to transition state **6-81ts**. Therefore, the reductive elimination step would be rate limiting in the whole catalytic cycle.

6.3 Mechanism of Pt-Catalyzed Alkene Activation

As a group 10 transition metal, Pt-species also exhibits an analogous catalytic activity in comparison with corresponding Pd-species. This propriety is centrally presented in Pt-catalyzed alkene functionalization reactions. The general mechanism of Pt-catalyzed alkene functionalization reactions could be inducted as a catalytic cycle shown in Scheme 6.4. The reaction starts from a Pt(II) intermediate. Then, an alkene insertion into Pt–hetero bond forms an alkyl Pt intermediate. An oxidative addition of hetero bond affords a Pt(IV) species, which could undergo reductive elimination to release the alkane product and regenerate Pt(II) species [151–153].

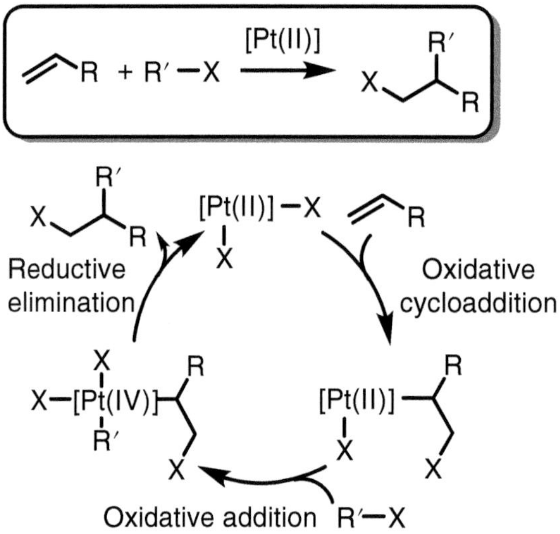

Scheme 6.4 The general mechanism of Pt-mediated alkene functionalizations.

6.3.1 Hydroamination of Alkenes

Hydroamination reaction is an atom economical process that results in the N—H addition of an amine or ammonia to an unsaturated bond to produce a higher amine [155–161]. In particular, a number of works have been donated onto the Pt-catalyzed hydroamination of ethylene and its derivatives [162–165]. A DFT calculation is performed by Poli group to reveal the proposed Pt(II)–Pt(IV) catalytic cycle of hydroamination reaction [166]. As shown in Figure 6.11, ethylene-coordinated platinate(II) complex **6-83** is set to relative of free-energy profiles. The intermolecular nucleophilic attack by aniline takes place via transition state **6-84ts** with a free-energy

6.3 Mechanism of Pt-Catalyzed Alkene Activation | 271

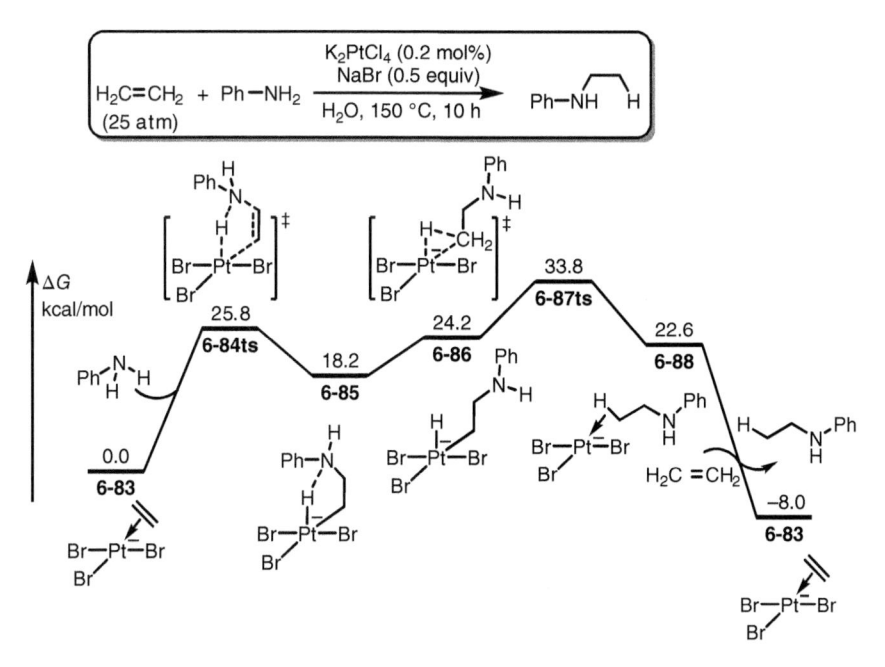

Figure 6.11 Free-energy profiles for the K_2PtCl_4-NaBr-catalyzed hydroamination of acetylene. The energies were calculated at B3LYP/6-31+G(d) (LANL2TZ(f) for Pt)/CPCM (aniline)//B3LYP/6-31+G(d) (LANL2TZ(f) for Pt) level of theory. Source: Based on Dias et al. [154].

barrier of 25.8 kcal/mol to afford an internal salt **6-85**, where the oxidative state of Pt is increased to +4 by a concomitant hydrogen transfer from ammonia to Pt. When the N—H hydrogen bond is broken in intermediate **6-86**, a subsequent reductive elimination forms C—H bond via transition state **6-87ts**. The overall activation free energy is as high as 33.8 kcal/mol, which is considered to be the rate-determining step in the whole catalytic cycle. Active catalyst **6-83** could be regenerated by the ligand exchange with ethylene substrate.

As a debate, alternative nonredox catalytic cycles for amino hydride Pt(II)-catalyzed hydroamination of ethylene are proposed by Tsipis and Kefalidis (Scheme 6.5) [148]. As shown in Figure 6.12, hydride amino Pt(II) species **6-89** is set to relative zero in free-energy profiles, which could be coordinated with ethylene to form intermediate **6-90** reversibly. An intermolecular nucleophilic attack by ammonia forms an internal salt **6-92** with 6.3 kcal/mol endergonic; however, the transition state **6-91ts** for this step was not located. The following proton transfer could occur via transition state **6-93ts** to form an alkyl Pt(II) intermediate **6-94** with an activation free energy of 33.9 kcal/mol. Then, the protonation of alkyl group by coordinated ammonia could occur via transition state **6-95ts** with an activation free energy of 33.7 kcal/mol to regenerate the active catalyst **6-89**. The calculated overall activation free energy of turnover-determining step is 35.0 kcal/mol, which indicates a turnover frequency of $2.1*10^{-11}$ h^{-1}. These results parallel the Poli results for anionic Pt complex, which are in support of this mechanism.

272 | 6 Theoretical Study of Pt-Catalysis

Scheme 6.5 PtBr$_2$-catalyzed hydroamination of olefins. Source: Based Kefalidis and Tsipis [148].

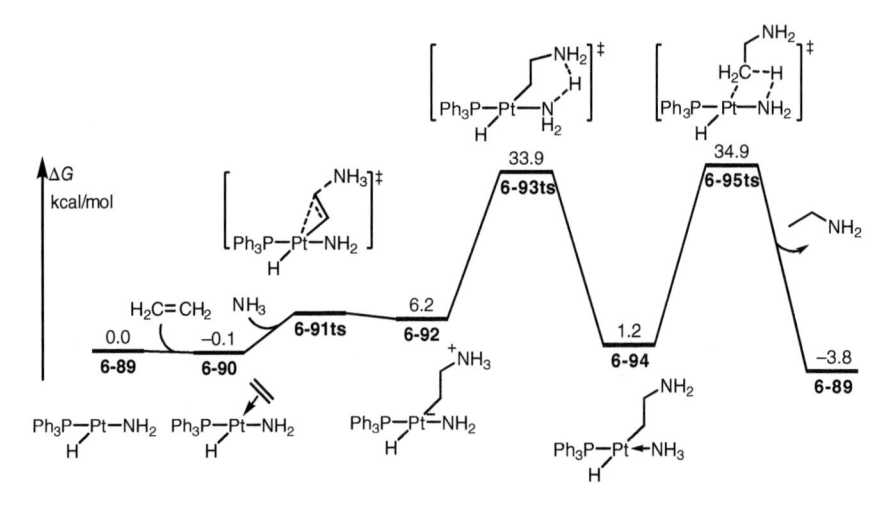

Figure 6.12 Free-energy profiles for an amino-Pt(II)-catalyzed hydroamination of acetylene. The energies were calculated at M06-L/6-311+G(d,p) (LANL2TZ(f) for Pt)/CPCM (DCM)//M06-L/6-31G(d,p) (LANL2TZ(f) for Pt) level of theory. Source: Based on Liu et al. [167].

6.3.2 Hydroformylation of Alkenes

Transition-metal-catalyzed hydroformylation of olefins represents a versatile way for the production of commercially important aldehydes, which are difficult to obtain by conventional synthetic routes [168–173]. In previous experimental studies, a variety of transition-metal-catalysts, including Co, Rh, and Pd species, have been used for this transformation. PtCl$_2$ was also found to be active in this process in the presence of SnCl$_2$ and phosphine ligand [174–182]. DFT calculations were performed by Rocha and coworker [154]. As shown in Figure 6.13, hydride Pt(II) active catalyst **6-96** is set to relative zero. The coordination of propene forms intermediate **6-97** with 6.0 kcal/mol endergonic due to entropy loss. The olefin insertion could occur via transition state **6-98ts** with a free-energy barrier of 14.2 kcal/mol

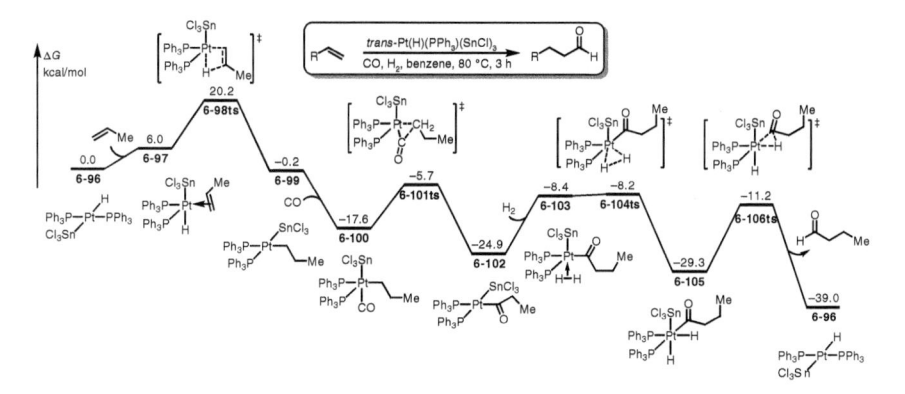

Figure 6.13 Free-energy profiles for an amino-Pt(II)-catalyzed hydroformylation of olefins. The energies were calculated at BP86/6-31G(d) (LANL2DZ for Pt and Sn)/PCM(benzene)// BP86/6-31G(d) (LANL2DZ for Pt and Sn) level of theory.

to form an alkyl Pt(II) species **6-99**. The coordination of carbon monoxide affords intermediate **6-100** endergonically. The following carbonyl insertion could occur via a three-membered ring-type transition state **6-101ts** with a free-energy barrier of 11.9 kcal/mol to form an acetyl Pt(II) complex **6-102** irreversibly. The coordination of dihydrogen followed by oxidative addition affords a hydride Pt(IV) species **6-105** via transition state **6-104ts** with a free-energy barrier of 16.7 kcal/mol. Subsequently, a reductive elimination takes place via transition state **6-106ts** to afford butyral product.

6.3.3 Isomerization of Cyclopropenes

Cyclopropenes are highly strained but readily accessible carbocyclic molecules and therefore exhibit remarkable activities that extend far beyond typical reactions of simple olefins, alkynes, and allenes. The metal-catalyzed cyclopropene rearrangement to allenes has attracted much attention, because this kind of reaction can be achieved under mild reaction conditions. In 2011, Lee and coworkers developed the PtCl$_2$-catalyzed rearrangement of silylated cyclopropenes to allenes in dichloromethane at 50 °C for 2 hours in high yields to 95% [183].

DFT calculations were performed by Bi group in 2012 on the mechanism of Pt-catalyzed isomerization of cyclopropene [167]. As shown in Figure 6.14, the calculated pathway originated from silylated cyclopropene coordination onto the PtCl$_2$, which would result in the generation of π-complex **6-107**. This π complex combines with the second reactant molecule to afford the adduct **6-108**. Subsequent Pt would insert into the C—C bond of the three-membered ring of second reactant molecule to form **6-110** via the transition state **6-109ts** with a free-energy barrier of only 0.8 kcal/mol, from which the reactant molecule initially coordinated with the PtCl$_2$ unit is expelled. Then, a reductive elimination of chloride is in order to afford a vinyl chloride moiety in the generated intermediate **6-113**, where the reacting chloride still coordinates onto Pt. Then a 1,2-silyl shift occurs via transition state **6-114ts** with an energy barrier of 11.1 kcal/mol. Simultaneously, the C—Cl bond is

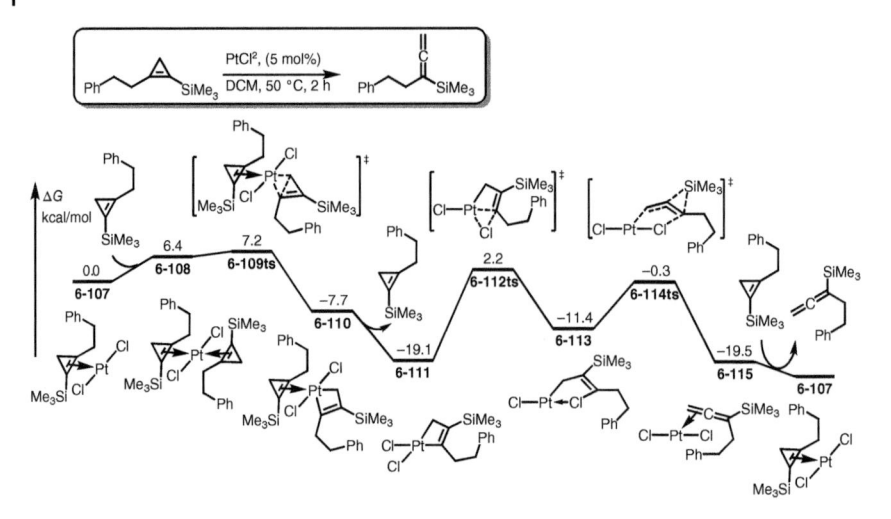

Figure 6.14 Free-energy profiles for an amino-Pt(II)-catalyzed isomerization of cyclopropenes. The energies were calculated at BP86/6-31G(d) (LANL2DZ for Pt and Sn)/PCM(DCM)//BP86/6-31G(d) (LANL2DZ for Pt) level of theory.

also broken to achieve the formation of allene species, which coordinates onto Pt in complex **6-115**. Finally, ligand exchange with cyclopropene reactant releases allene product and regenerates the active Pt(II) catalyst **6-107**.

References

1 Amatore, C. and Jutand, A. (2000). Anionic Pd(0) and Pd(II) intermediates in palladium-catalyzed Heck and cross-coupling reactions. *Accounts of Chemical Research* 33 (5): 314–321.
2 Cho, H.Y. and Morken, J.P. (2014). Catalytic bismetallative multicomponent coupling reactions: scope, applications, and mechanisms. *Chemical Society Reviews* 43 (13): 4368–4380.
3 Colacot, T.J. (2011). The 2010 Nobel prize in chemistry: palladium-catalysed cross-coupling. *Platinum Metals Review* 55 (2): 84–90.
4 Han, F.S. (2013). Transition-metal-catalyzed Suzuki–Miyaura cross-coupling reactions: a remarkable advance from palladium to nickel catalysts. *Chemical Society Reviews* 42 (12): 5270–5298.
5 Jana, R., Pathak, T.P., and Sigman, M.S. (2011). Advances in transition metal (Pd, Ni, Fe)-catalyzed cross-coupling reactions using alkyl-organometallics as reaction partners. *Chemical Reviews* 111 (3): 1417–1492.
6 Lyons, T.W. and Sanford, M.S. (2010). Palladium-catalyzed ligand-directed C–H functionalization reactions. *Chemical Reviews* 110 (2): 1147–1169.
7 Munoz, M.P. (2014). Silver and platinum-catalysed addition of O—H and N—H bonds to allenes. *Chemical Society Reviews* 43 (9): 3164–3183.

8 Negishi, E. (1982). Palladium- or nickel-catalyzed cross coupling. A new selective method for carbon–carbon bond formation. *Accounts of Chemical Research* 15 (11): 340–348.

9 Phapale, V.B. and Cárdenas, D.J. (2009). Nickel-catalysed Negishi cross-coupling reactions: scope and mechanisms. *Chemical Society Reviews* 38 (6): 1598.

10 Vedernikov, A.N. (2011). Direct functionalization of M–C (M = Pt^{II}, Pd^{II}) bonds using environmentally benign oxidants, O$_2$ and H$_2$O$_2$. *Accounts of Chemical Research* 45 (6): 803–813.

11 Xu, L.M., Li, B.J., Yang, Z. et al. (2010). Organopalladium(IV) chemistry. *Chemical Society Reviews* 39 (2): 712–733.

12 Felix, R.J., Munro-Leighton, C., and Gagne, M.R. (2014). Electrophilic Pt(II) complexes: precision instruments for the initiation of transformations mediated by the cation-olefin reaction. *Accounts of Chemical Research* 47 (8): 2319–2331.

13 Fürstner, A. (2014). From understanding to prediction: gold- and platinum-based π-acid catalysis for target oriented synthesis. *Accounts of Chemical Research* 47 (3): 925–938.

14 Labinger, J.A. (2017). Platinum-catalyzed C–H functionalization. *Chemical Reviews* 117 (13): 8483–8496.

15 Lersch, M. and Tilset, M. (2005). Mechanistic aspects of C–H activation by Pt complexes. *Chemical Reviews* 105 (6): 2471–2526.

16 Vigalok, A. (2015). Electrophilic halogenation–reductive elimination chemistry of organopalladium and -platinum complexes. *Accounts of Chemical Research* 48 (2): 238–247.

17 Khan, M.S., Haque, A., Al-Suti, M.K. et al. (2015). Recent advances in the application of group-10 transition metal based catalysts in C–H activation and functionalization. *Journal of Organometallic Chemistry* 793: 114–133.

18 Balcells, D., Clot, E., and Eisenstein, O. (2010). C—H bond activation in transition metal species from a computational perspective. *Chemical Reviews* 110 (2): 749–823.

19 Crabtree, R.H. (1985). The organometallic chemistry of alkanes. *Chemical Reviews* 85 (4): 245–269.

20 Jia, C., Kitamura, T., and Fujiwara, Y. (2001). Catalytic functionalization of arenes and alkanes via C—H bond activation. *Accounts of Chemical Research* 34 (8): 633–639.

21 Koga, N. and Morokuma, K. (1991). Ab initio molecular orbital studies of catalytic elementary reactions and catalytic cycles of transition-metal complexes. *Chemical Reviews* 91 (5): 823–842.

22 Niu, S. and Hall, M.B. (2000). Theoretical studies on reactions of transition–metal complexes. *Chemical Reviews* 100 (2): 353–406.

23 Sehnal, P., Taylor, R.J., and Fairlamb, I.J. (2010). Emergence of palladium(IV) chemistry in synthesis and catalysis. *Chemical Reviews* 110 (2): 824–889.

24 Shilov, A.E. and Shul'pin, G.B. (1997). Activation of C—H bonds by metal complexes. *Chemical Reviews* 97 (8): 2879–2932.

25 Garcia-Melchor, M., Braga, A.A., Lledos, A. et al. (2013). Computational perspective on Pd-catalyzed C–C cross-coupling reaction mechanisms. *Accounts of Chemical Research* 46 (11): 2626–2634.

26 Guo, W., Kuniyil, R., Gomez, J.E. et al. (2018). A domino process toward functionally dense quaternary carbons through Pd-catalyzed decarboxylative C(sp^3)–C(sp^3) bond formation. *Journal of the American Chemical Society* 140 (11): 3981–3987.

27 Toledo, A., Funes-Ardoiz, I., Maseras, F. et al. (2018). Palladium-catalyzed aerobic homocoupling of alkynes: full mechanistic characterization of a more complex oxidase-type behavior. *ACS Catalysis* 8 (8): 7495–7506.

28 Furstner, A. and Davies, P.W. (2007). Catalytic carbophilic activation: catalysis by platinum and gold π acids. *Angewandte Chemie International Edition* 46 (19): 3410–3449.

29 Heinemann, C., Hertwig, R.H., Wesendrup, R. et al. (1995). Relativistic effects on bonding in cationic transition-metal-carbene complexes: a density-functional study. *Journal of the American Chemical Society* 117 (1): 495–500.

30 Schwarz, H. (2003). Relativistische effekte in der Gasphasenchemie von ionen aus experimenteller Sicht. *Angewandte Chemie International Edition* 115 (37): 4580–4593.

31 Jansen, M. (2005). Effects of relativistic motion of electrons on the chemistry of gold and platinum. *Solid State Sciences* 7 (12): 1464–1474.

32 Bhanu Prasad, B.A., Yoshimoto, F.K., and Sarpong, R. (2005). Pt-catalyzed pentannulations from in situ generated metallo-carbenoids utilizing propargylic esters. *Journal of the American Chemical Society* 127 (36): 12468–12469.

33 Fernandez-Alvarez, V.M., Ho, S.K.Y., Britovsek, G.J.P. et al. (2018). A DFT-based mechanistic proposal for the light-driven insertion of dioxygen into Pt(II)—C bonds. *Chemical Science* 9 (22): 5039–5046.

34 Helm, L. and Merbach, A.E. (2005). Inorganic and bioinorganic solvent exchange mechanisms. *Chemical Reviews* 105 (6): 1923–1959.

35 Cariou, K., Ronan, B., Mignani, S. et al. (2007). From PtCl$_2$– and acid-catalyzed to uncatalyzed cycloisomerization of 2-propargyl anilines: access to functionalized indoles. *Angewandte Chemie International Edition* 119 (11): 1913–1916.

36 de Orbe, M.E., Amenos, L., Kirillova, M.S. et al. (2017). Cyclobutene vs 1,3-diene formation in the gold-catalyzed reaction of alkynes with alkenes: the complete mechanistic picture. *Journal of the American Chemical Society* 139 (30): 10302–10311.

37 Hashmi, A.S.K., Kurpejović, E., Frey, W. et al. (2007). Gold catalysis contra platinum catalysis in hydroarylation contra phenol synthesis. *Tetrahedron* 63 (26): 5879–5885.

38 Li, Z., Brouwer, C., and He, C. (2008). Gold-catalyzed organic transformations. *Chemical Reviews* 108 (8): 3239–3265.

39 Liu, Y., Zhang, D., Bi, S. et al. (2013). Theoretical investigation on Pt(II)- and Au(I)-mediated cycloisomerizations of propargylic 3-indoleacetate: [3 + 2]-

versus [2 + 2]-cycloaddition products. *Organic & Biomolecular Chemistry* 11 (2): 336–343.
40 Mamane, V., Gress, T., Krause, H. et al. (2004). Platinum- and gold-catalyzed cycloisomerization reactions of hydroxylated enynes. *Journal of the American Chemical Society* 126 (28): 8654–8655.
41 Taduri, B.P., Sohel, S.M., Cheng, H.M. et al. (2007). Pt- and Au-catalyzed oxidative cyclization of 2-ethenyl-1-(prop-2′-yn-1′-ol)benzenes to naphthyl aldehydes and ketones: catalytic oxidation of metal–alkylidene intermediates using H_2O and H_2O_2. *Chemical Communications* 24: 2530–2532.
42 Furstner, A. (2009). Gold and platinum catalysis – a convenient tool for generating molecular complexity. *Chemical Society Reviews* 38 (11): 3208–3221.
43 Yamamoto, Y. (2007). From σ- to π-electrophilic Lewis acids. Application to selective organic transformations. *Journal of Organic Chemistry* 72 (21): 7817–7831.
44 Beletskaya, I. and Moberg, C. (1999). Element–element addition to alkynes catalyzed by the group 10 metals. *Chemical Reviews* 99 (12): 3435–3462.
45 Beletskaya, I. and Moberg, C. (2006). Element–element additions to unsaturated carbon–carbon bonds catalyzed by transition metal complexes. *Chemical Reviews* 106 (6): 2320–2354.
46 Cardin, D.J., Cetinkaya, B., and Lappert, M.F. (1972). Transition metal–carbene complexes. *Chemical Reviews* 72 (5): 545–574.
47 Doyle, M.P. (1986). Catalytic methods for metal carbene transformations. *Chemical Reviews* 86 (5): 919–939.
48 Luh, T.Y., Leung Mk, M.K., and Wong, K.T. (2000). Transition metal-catalyzed activation of aliphatic C—X bonds in carbon–carbon bond formation. *Chemical Reviews* 100 (8): 3187–3204.
49 Ren, T. (2008). Peripheral covalent modification of inorganic and organometallic compounds through C—C bond formation reactions. *Chemical Reviews* 108 (10): 4185–4207.
50 Saito, K., Sogou, H., Suga, T. et al. (2011). Platinum(II)-catalyzed generation and [3+2] cycloaddition reaction of α, β-unsaturated carbene complex intermediates for the preparation of polycyclic compounds. *Journal of the American Chemical Society* 133 (4): 689–691.
51 Wang, B., Qiu, D., Zhang, Y. et al. (2016). Recent advances in $C(sp^3)$—H bond functionalization via metal–carbene insertions. *Beilstein Journal of Organic Chemistry* 12: 796–804.
52 Zhang, J., Shan, C., Zhang, T. et al. (2019). Computational advances aiding mechanistic understanding of silver-catalyzed carbene/nitrene/silylene transfer reactions. *Coordination Chemistry Reviews* 382: 69–84.
53 Coombs, J.R., Haeffner, F., Kliman, L.T. et al. (2013). Scope and mechanism of the Pt-catalyzed enantioselective diboration of monosubstituted alkenes. *Journal of the American Chemical Society* 135 (30): 11222–11231.
54 de Aguirre, A., Funes-Ardoiz, I., and Maseras, F. (2019). Four oxidation states in a single photoredox nickel-based catalytic cycle: a computational study. *Angewandte Chemie International Edition* 58 (12): 3898–3902.

55 Ishiyama, T. and Miyaura, N. (2004). Metal-catalyzed reactions of diborons for synthesis of organoboron compounds. *Chemical Record* 3 (5): 271–280.

56 Jover, J., Miloserdov, F.M., Benet-Buchholz, J. et al. (2014). On the feasibility of nickel-catalyzed trifluoromethylation of aryl halides. *Organometallics* 33 (22): 6531–6543.

57 Oestreich, M., Hartmann, E., and Mewald, M. (2013). Activation of the Si—B interelement bond: mechanism, catalysis, and synthesis. *Chemical Reviews* 113 (1): 402–441.

58 Oost, R., Neuhaus, J.D., Misale, A. et al. (2018). Catalyst-dependent selectivity in sulfonium ylide cycloisomerization reactions. *Chemical Science* 9 (35): 7091–7095.

59 Rull, S.G., Funes-Ardoiz, I., Maya, C. et al. (2018). Elucidating the mechanism of aryl aminations mediated by NHC-supported nickel complexes: evidence for a nonradical Ni(0)/Ni(II) pathway. *ACS Catalysis* 8 (5): 3733–3742.

60 Sperger, T., Sanhueza, I.A., Kalvet, I. et al. (2015). Computational studies of synthetically relevant homogeneous organometallic catalysis involving Ni, Pd, Ir, and Rh: an overview of commonly employed DFT methods and mechanistic insights. *Chemical Reviews* 115 (17): 9532–9586.

61 Suginome, M. (2010). Catalytic carboborations. *Chemical Record* 10 (5): 348–358.

62 Yoshida, H. (2016). Borylation of alkynes under base/coinage metal catalysis: some recent developments. *ACS Catalysis* 6 (3): 1799–1811.

63 Dong, Z., Ren, Z., Thompson, S.J. et al. (2017). Transition-metal-catalyzed C–H alkylation using alkenes. *Chemical Reviews* 117 (13): 9333–9403.

64 Levin, E., Ivry, E., Diesendruck, C.E. et al. (2015). Water in *N*-heterocyclic carbene-assisted catalysis. *Chemical Reviews* 115 (11): 4607–4692.

65 Newkome, G.R., Puckett, W.E., Gupta, V.K. et al. (1986). Cyclometalation of the platinum metals with nitrogen and alkyl, alkenyl, and benzyl carbon donors. *Chemical Reviews* 86 (2): 451–489.

66 Shiotsuki, M., White, P.S., Brookhart, M. et al. (2007). Mechanistic studies of platinum(II)-catalyzed ethylene dimerization: determination of barriers to migratory insertion in diimine Pt(II) hydrido ethylene and ethyl ethylene intermediates. *Journal of the American Chemical Society* 129 (13): 4058–4067.

67 Williams, B.S. and Goldberg, K.I. (2001). Studies of reductive elimination reactions to form carbon–oxygen bonds from Pt(IV) complexes. *Journal of the American Chemical Society* 123 (11): 2576–2587.

68 Abu Sohel, S.M. and Liu, R.S. (2009). Carbocyclisation of alkynes with external nucleophiles catalysed by gold, platinum and other electrophilic metals. *Chemical Society Reviews* 38 (8): 2269–2281.

69 Barluenga, J., Dieguez, A., Fernandez, A. et al. (2006). Gold- or platinum-catalyzed tandem cycloisomerization/prins-type cyclization reactions. *Angewandte Chemie International Edition* 45 (13): 2091–2093.

70 Dorel, R. and Echavarren, A.M. (2015). Gold(I)-catalyzed activation of alkynes for the construction of molecular complexity. *Chemical Reviews* 115 (17): 9028–9072.

71 Ghosh, A., Basak, A., Chakrabarty, K. et al. (2018). Au-catalyzed hexannulation and Pt-catalyzed pentannulation of propargylic ester bearing a 2-alkynyl-phenyl substituent: a comparative DFT study. *ACS Omega* 3 (1): 1159–1169.

72 Hashmi, A.S. (2007). Gold-catalyzed organic reactions. *Chemical Reviews* 107 (7): 3180–3211.

73 Munoz, M.P. (2012). Transition metal-catalysed intermolecular reaction of allenes with oxygen nucleophiles: a perspective. *Organic & Biomolecular Chemistry* 10 (18): 3584–3594.

74 Nevado, C. and Echavarren, A.M. (2005). Intramolecular hydroarylation of alkynes catalyzed by platinum or gold: mechanism and endo selectivity. *Chemistry – A European Journal* 11 (10): 3155–3164.

75 Raynal, M., Ballester, P., Vidal-Ferran, A. et al. (2014). Supramolecular catalysis. Part 1: non-covalent interactions as a tool for building and modifying homogeneous catalysts. *Chemical Society Reviews* 43 (5): 1660–1733.

76 Villa, A., Dimitratos, N., Chan-Thaw, C.E. et al. (2016). Characterisation of gold catalysts. *Chemical Society Reviews* 45 (18): 4953–4994.

77 Xia, Y. and Huang, G. (2010). Mechanisms of the Au- and Pt-catalyzed intramolecular acetylenic Schmidt reactions: a DFT study. *Journal of Organic Chemistry* 75 (22): 7842–7854.

78 Zhdanko, A. and Maier, M.E. (2014). Gold(I)-, palladium(II)-, platinum(II)-, and mercury(II)-catalysed spirocyclization of 1,3-enynediols: reaction scope. *European Journal of Organic Chemistry* 2014 (16): 3411–3422.

79 Evans, E.J., Li, H., Yu, W.Y. et al. (2017). Mechanistic insights on ethanol dehydrogenation on Pd–Au model catalysts: a combined experimental and DFT study. *Physical Chemistry Chemical Physics* 19 (45): 30578–30589.

80 Kumar, C.H.V., Jagadeesh, R.V., Shivananda, K.N. et al. (2010). Catalysis and mechanistic studies of Ru(III), Os(VIII), Pd(II), and Pt(IV) metal ions on oxidative conversion of folic acid. *Industrial & Engineering Chemistry Research* 49 (4): 1550–1560.

81 McKeown, B.A., Gonzalez, H.E., Friedfeld, M.R. et al. (2011). Mechanistic studies of ethylene hydrophenylation catalyzed by bipyridyl Pt(II) complexes. *Journal of the American Chemical Society* 133 (47): 19131–19152.

82 Yu, W., Porosoff, M.D., and Chen, J.G. (2012). Review of Pt-based bimetallic catalysis: from model surfaces to supported catalysts. *Chemical Reviews* 112 (11): 5780–5817.

83 Webb, J.R., Munro-Leighton, C., Pierpont, A.W. et al. (2011). Pt(II) and Pt(IV) amido, aryloxide, and hydrocarbyl complexes: synthesis, characterization, and reaction with dihydrogen and substrates that possess C–H bonds. *Inorganic Chemistry* 50 (9): 4195–4211.

84 Cundari, T.R., Grimes, T.V., and Gunnoe, T.B. (2007). Activation of carbon–hydrogen bonds via 1,2-addition across M–X (X = OH or NH_2) bonds of d6 transition metals as a potential key step in hydrocarbon functionalization: a computational study. *Journal of the American Chemical Society* 129 (43): 13172–13182.

85 Davies, D.L., Macgregor, S.A., and McMullin, C.L. (2017). Computational studies of carboxylate-assisted C–H activation and functionalization at group 8–10 transition metal centers. *Chemical Reviews* 117 (13): 8649–8709.

86 Haibach, M.C. and Seidel, D. (2014). C–H bond functionalization through intramolecular hydride transfer. *Angewandte Chemie International Edition* 53 (20): 5010–5036.

87 Webb, J.R., Burgess, S.A., Cundari, T.R. et al. (2013). Activation of carbon–hydrogen bonds and dihydrogen by 1,2-CH-addition across metal–heteroatom bonds. *Dalton Transactions* 42 (48): 16646.

88 Chaumontet, M., Piccardi, R., Audic, N. et al. (2008). Synthesis of benzocyclobutenes by palladium-catalyzed C–H activation of methyl groups: method and mechanistic study. *Journal of the American Chemical Society* 130 (45): 15157–15166.

89 Chempath, S. and Bell, A.T. (2006). Density functional theory analysis of the reaction pathway for methane oxidation to acetic acid catalyzed by Pd^{2+} in sulfuric acid. *Journal of the American Chemical Society* 128 (14): 4650–4657.

90 Giri, R., Shi, B.-F., Engle, K.M. et al. (2009). Transition metal-catalyzed C–H activation reactions: diastereoselectivity and enantioselectivity. *Chemical Society Reviews* 38 (11): 3242.

91 Gorelsky, S.I. (2013). Origins of regioselectivity of the palladium-catalyzed (aromatic)CH bond metalation–deprotonation. *Coordination Chemistry Reviews* 257 (1): 153–164.

92 Gorelsky, S.I., Lapointe, D., and Fagnou, K. (2008). Analysis of the concerted metalation–deprotonation mechanism in palladium-catalyzed direct arylation across a broad range of aromatic substrates. *Journal of the American Chemical Society* 130 (33): 10848–10849.

93 Han, H., Zhang, T., Yang, S.D. et al. (2019). Palladium-catalyzed enantioselective C–H aminocarbonylation: synthesis of chiral isoquinolinones. *Organic Letters* 21 (6): 1749–1754.

94 He, J., Wasa, M., Chan, K.S.L. et al. (2017). Palladium-catalyzed transformations of alkyl C–H bonds. *Chemical Reviews* 117 (13): 8754–8786.

95 Lafrance, M., Gorelsky, S.I., and Fagnou, K. (2007). High-yielding palladium-catalyzed intramolecular alkane arylation: reaction development and mechanistic studies. *Journal of the American Chemical Society* 129 (47): 14570–14571.

96 Musaev, D.G., Figg, T.M., and Kaledin, A.L. (2014). Versatile reactivity of Pd-catalysts: mechanistic features of the mono-*N*-protected amino acid ligand and cesium–halide base in Pd-catalyzed C–H bond functionalization. *Chemical Society Reviews* 43 (14): 5009–5031.

97 Xu, D., Qi, X., Duan, M. et al. (2017). Thiolate–palladium(IV) or sulfonium–palladate(0)? A theoretical study on the mechanism of palladium-catalyzed C–S bond formation reactions. *Organic Chemistry Frontiers* 4 (6): 943–950.

98 Zhang, X., Chung, L.W., and Wu, Y.D. (2016). New mechanistic insights on the selectivity of transition-metal-catalyzed organic reactions: the role of computational chemistry. *Accounts of Chemical Research* 49 (6): 1302–1310.

99 Jiang, Y.-Y., Man, X., and Bi, S. (2016). Advances in theoretical study on transition-metal-catalyzed C–H activation. *Science China Chemistry* 59 (11): 1448–1466.

100 Rivada-Wheelaghan, O., Rosello-Merino, M., Ortuno, M.A. et al. (2014). Reactivity of coordinatively unsaturated bis(*N*-heterocyclic carbene) Pt(II) complexes toward H_2. Crystal structure of a 14-electron Pt(II) hydride complex. *Inorganic Chemistry* 53 (8): 4257–4268.

101 Tobisu, M., Nakai, H., and Chatani, N. (2009). Platinum and ruthenium chloride-catalyzed cycloisomerization of 1-alkyl-2-ethynylbenzenes: interception of π-activated alkynes with a benzylic C–H bond. *Journal of Organic Chemistry* 74 (15): 5471–5475.

102 Chen, K., Zhang, G., Chen, H. et al. (2012). Spin-orbit coupling and outer-core correlation effects in Ir- and Pt-catalyzed C–H activation. *Journal of Chemical Theory and Computation* 8 (5): 1641–1645.

103 Gunsalus, N.J., Koppaka, A., Park, S.H. et al. (2017). Homogeneous functionalization of methane. *Chemical Reviews* 117 (13): 8521–8573.

104 Johansson, L., Ryan, O.B., and Tilset, M. (1999). Hydrocarbon activation at a cationic platinum(II) diimine aqua complex under mild conditions in a hydroxylic solvent. *Journal of the American Chemical Society* 121 (9): 1974–1975.

105 Jones, C.J., Taube, D., Ziatdinov, V.R. et al. (2004). Selective oxidation of methane to methanol catalyzed, with C–H activation, by homogeneous, cationic gold. *Angewandte Chemie International Edition* 43 (35): 4626–4629.

106 Xiao, L. and Wang, L. (2007). Methane activation on Pt and Pt_4: a density functional theory study. *Journal of Physical Chemistry B* 111 (7): 1657–1663.

107 Periana, R.A., Taube, D.J., Gamble, S. et al. (1998). Platinum catalysts for the high-yield oxidation of methane to a methanol derivative. *Science* 280 (5363): 560.

108 Ahlquist, M., Periana, R.A., and Goddard, W.A. (2009). C–H activation in strongly acidic media. The co-catalytic effect of the reaction medium. *Chemical Communications* (17): 2373–2375.

109 Chang, H.K., Liao, Y.C., and Liu, R.S. (2007). Diversity in platinum-catalyzed hydrative cyclization of trialkyne substrates to form tetracyclic ketones. *Journal of Organic Chemistry* 72 (21): 8139–8141.

110 Cooper, L., Alonso, J.M., Eagling, L. et al. (2018). Synthesis of a novel type of 2,3'-BIMs via platinum-catalysed reaction of indolylallenes with indoles. *Chemistry – A European Journal* 24 (23): 6105–6114.

111 Ghosh, S., Khamarui, S., Gayen, K.S. et al. (2013). ArCH(OMe)$_2$ – a Pt^{IV}-catalyst originator for diverse annulation catalysis. *Scientific Reports* 3 (1): 2987.

112 Harrak, Y., Blaszykowski, C., Bernard, M. et al. (2004). $PtCl_2$-catalyzed cycloisomerizations of 5-En-1-yn-3-ol systems. *Journal of the American Chemical Society* 126 (28): 8656–8657.

113 Pastine, S.J., Youn, S.W., and Sames, D. (2003). Pt^{IV}-catalyzed cyclization of arene–alkyne substrates via intramolecular electrophilic hydroarylation. *Organic Letters* 5 (7): 1055–1058.

114 Ponte, F., Russo, N., and Sicilia, E. (2018). Insights from computations on the mechanism of reduction by ascorbic acid of Pt(IV) prodrugs with asplatin and its chlorido and bromido analogues as model systems. *Chemistry – A European Journal* 24 (38): 9572–9580.

115 Vadola, P.A. and Sames, D. (2009). C–H bond functionalization via hydride transfer: direct coupling of unactivated alkynes and sp^3 C–H bonds catalyzed by platinum tetraiodide. *Journal of the American Chemical Society* 131 (45): 16525–16528.

116 Jin, L., Wu, Y., and Zhao, X. (2012). PtI^4-catlyzed C–H bond functionalization in alkynyl ether: density functional theory survey. *Organometallics* 31 (8): 3065–3073.

117 Shan, C., Zhu, L., Qu, L.B. et al. (2018). Mechanistic view of Ru-catalyzed C–H bond activation and functionalization: computational advances. *Chemical Society Reviews* 47 (20): 7552–7576.

118 Xia, Y., Qiu, D., and Wang, J. (2017). Transition-metal-catalyzed cross-couplings through carbene migratory insertion. *Chemical Reviews* 117 (23): 13810–13889.

119 Ortega-Moreno, L., Peloso, R., Lopez-Serrano, J. et al. (2017). A cationic unsaturated platinum(II) complex that promotes the tautomerization of acetylene to vinylidene. *Angewandte Chemie International Edition* 56 (10): 2816–2819.

120 Soriano, E. and Marco-Contelles, J. (2007). DFT-based mechanism for the unexpected formation of dienes in the PtCl$_2$ isomerization of propargylic acetates: examples of inhibition of the rautenstrauch process. *Journal of Organic Chemistry* 72 (4): 1443–1448.

121 Wang, Y., Liao, W., Huang, G. et al. (2014). Mechanisms of the PtCl$_2$-catalyzed intramolecular cyclization of o-isopropyl-substituted aryl alkynes for the synthesis of indenes and comparison of three sp^3 C–H bond activation modes. *Journal of Organic Chemistry* 79 (12): 5684–5696.

122 Bajracharya, G.B., Pahadi, N.K., Gridnev, I.D. et al. (2006). PtBr$_2$-catalyzed transformation of allyl(o-ethynylaryl)carbinol derivatives into functionalized indenes. Formal sp^3 C–H bond activation. *Journal of Organic Chemistry* 71 (16): 6204–6210.

123 Pujanauski, B.G., Bhanu Prasad, B.A., and Sarpong, R. (2006). Pt-catalyzed tandem epoxide fragmentation/pentannulation of propargylic esters. *Journal of the American Chemical Society* 128 (21): 6786–6787.

124 Oh, C.H., Lee, J.H., Lee, S.M. et al. (2009). Divergent insertion reactions of Pt–carbenes generated from [3+2] cyclization of platinum-bound pyrylliums. *Chemistry – A European Journal* 15 (1): 71–74.

125 Marco-Contelles, J., Arroyo, N., Anjum, S. et al. (2006). PtCl$_2$- and PtCl$_4$-catalyzed cycloisomerization of polyunsaturated precursors. *European Journal of Organic Chemistry* 2006 (20): 4618–4633.

126 Yang, S., Li, Z., Jian, X. et al. (2009). Platinum(II)-catalyzed intramolecular cyclization of o-substituted aryl alkynes through sp^3 C–H activation. *Angewandte Chemie International Edition* 48 (22): 3999–4001.

127 Cordonnier, M.-C., Blanc, A., and Pale, P. (2008). Gold(I)-catalyzed rearrangement of alkynyloxiranes: a mild access to divinyl ketones. *Organic Letters* 10 (8): 1569–1572.

128 Jin, S., Jiang, C., Peng, X. et al. (2016). Gold(I)-catalyzed angle strain controlled strategy to furopyran derivatives from propargyl vinyl ethers: insight into the regioselectivity of cycloisomerization. *Organic Letters* 18 (4): 680–683.

129 Nakamura, I., Sato, Y., and Terada, M. (2009). Platinum-catalyzed dehydroalkoxylation–cyclization cascade via N–O bond cleavage. *Journal of the American Chemical Society* 131 (12): 4198–4199.

130 Shu, D., Winston-McPherson, G.N., Song, W. et al. (2013). Platinum-catalyzed tandem indole annulation/arylation for the synthesis of diindolylmethanes and indolo[3,2-b]carbazoles. *Organic Letters* 15 (16): 4162–4165.

131 Nakamura, I., Bajracharya, G.B., Wu, H. et al. (2004). Catalytic cyclization of o-alkynylbenzaldehyde acetals and thioacetals. Unprecedented activation of the platinum catalyst by olefins. Scope and mechanism of the reaction. *Journal of the American Chemical Society* 126 (47): 15423–15430.

132 Mi, Y., Zhou, T., Wang, K.-P. et al. (2015). Mechanistic understanding of the divergent cyclizations of o-alkynylbenzaldehyde acetals and thioacetals catalyzed by metal halides. *Chemistry – A European Journal* 21 (48): 17256–17268.

133 Gonzáles-Pérez, A.B., Vaz, B., Faza, O.N. et al. (2012). Mechanistic and sterochemical insights on the Pt-catalyzed rearrangement of oxiranylpropargylic esters to cyclopentenones. *Journal of Organic Chemistry* 77 (19): 8733–8743.

134 Luzung, M.R., Markham, J.P., and Toste, F.D. (2004). Catalaytic isomerization of 1,5-enynes to bicyclo[3.1.0]hexenes. *Journal of the American Chemical Society* 126 (35): 10858–10859.

135 Ye, L., Chen, Q., Zhang, J. et al. (2009). PtCl$_2$-catalyzed cycloisomerization of 1,6-enynes for the synthesis of substituted bicyclo[3.1.0]hexanes. *Journal of Organic Chemistry* 74 (24): 9550–9553.

136 Mainetti, E., Mouriès, V., Fensterbank, L. et al. (2002). The effect of a hydroxy protecting group on the PtCl$_2$-catalyzed cyclization of dienynes – a novel, efficient, and selective synthesis of carbocycles. *Angewandte Chemie International Edition* 41 (12): 2132–2135.

137 Soriano, E., Ballesteros, P., and Marco-Contelles, J. (2004). A theoretical investigation on the mechanism of the PtCl$_2$-mediated cycloisomerization of heteroatom-tethered 1,6-enynes. *Journal of Organic Chemistry* 69 (23): 8018–8023.

138 Huang, K., Ke, X., Wang, H. et al. (2015). PtI$_2$-catalyzed cyclization of 3-acyloxy-1,5-enynes with the elimination of HOAc and a benzyl shift: synthesis of unsymmetrical m-terphenyls. *Organic & Biomolecular Chemistry* 13 (15): 4486–4493.

139 Muñoz, M.P., Adrio, J., Carretero, J.C. et al. (2005). Ligand effects in gold- and platinum-catalyzed cyclization of enynes: chiral gold complexes for enantioselective alkoxycyclization. *Organometallics* 24 (6): 1293–1300.

140 Echavarren, A.M. and Nevado, C. (2004). Non-stabilized transition metal carbenes as intermediates in intramolecular reactions of alkynes with alkenes. *Chemical Society Reviews* 33 (7): 431–436.

141 Liu, S., Zhang, T., Zhu, L. et al. (2018). Retro-metal-ene versus retro-Aldol: mechanistic insight into Rh-catalysed formal [3+2] cycloaddition. *Chemical Communications* 54 (96): 13551–13554.

142 Long, R., Huang, J., Shao, W. et al. (2014). Asymmetric total synthesis of (−)-lingzhiol via a Rh-catalysed [3+2] cycloaddition. *Nature Communications* 5: 5707.

143 Qi, X., Liu, S., Zhang, T. et al. (2016). Effective chirality transfer in [3+2] reaction between allenyl–rhodium and enal: mechanistic study based on DFT calculations. *Journal of Organic Chemistry* 81 (18): 8306–8311.

144 Shan, C., Zhong, K., Qi, X. et al. (2018). Long distance unconjugated agostic-assisted 1,5-H shift in a Ru-mediated Alder-ene type reaction: mechanism and stereoselectivity. *Organic Chemistry Frontiers* 5 (21): 3178–3185.

145 Zhang, J., Shan, C., Lv, K. et al. (2019). Mechanistic insight into palladium-catalyzed carbocyclization–functionalization of bisallene: a computational study. *ChemCatChem* 11 (4): 1228–1237.

146 Zhu, L., Qi, X., and Lan, Y. (2016). Rhodium-catalyzed hetero-(5 + 2) cycloaddition of vinylaziridines and alkynes: a theoretical view of the mechanism and chirality transfer. *Organometallics* 35 (5): 771–777.

147 Méndez, M., Muñoz, M.P., Nevado, C. et al. (2001). Cyclizations of enynes catalyzed by $PtCl_2$ or other transition metal chlorides: divergent reaction pathways. *Journal of the American Chemical Society* 123 (43): 10511–10520.

148 Kefalidis, C.E. and Tsipis, C.A. (2012). DFT study of the mechanism of hydroamination of ethylene with ammonia catalyzed by diplatinum(II) complexes: inner- or outer-sphere? *Journal of Computational Chemistry* 33 (20): 1689–1700.

149 Bernoud, E., Lepori, C., Mellah, M. et al. (2015). Recent advances in metal free- and late transition metal-catalysed hydroamination of unactivated alkenes. *Catalysis Science & Technology* 5 (4): 2017–2037.

150 Soriano, E. and Marco-Contelles, J. (2005). A DFT-based theoretical investigation of the mechanism of the $PtCl_2$-mediated cycloisomerization of allenynes. *Chemistry – A European Journal* 11 (2): 521–533.

151 Brunet, J.-J., Chu, N.-C., and Rodriguez-Zubiri, M. (2007). Platinum-catalyzed intermolecular hydroamination of alkenes: halide-anion-promoted catalysis. *European Journal of Inorganic Chemistry* 2007 (30): 4711–4722.

152 Lavery, C.B., Ferguson, M.J., and Stradiotto, M. (2010). Platinum-catalyzed alkene cyclohydroamination: evaluating the utility of bidentate P,N/P,P ligation and phosphine-free catalyst systems. *Organometallics* 29 (22): 6125–6128.

153 Tsipis, C.A. and Kefalidis, C.E. (2007). Hydrosilylation, hydrocyanation, and hydroamination of ethene catalyzed by bis(hydrido-bridged)diplatinum complexes: added insight and predictions from theory. *Journal of Organometallic Chemistry* 692 (23): 5245–5255.

154 Dias, R.P. and Rocha, W.R. (2011). DFT study of the homogeneous hydroformylation of propene promoted by a heterobimetallic Pt–Sn catalyst. *Organometallics* 30 (16): 4257–4268.

155 Chen, J. and Lu, Z. (2018). Asymmetric hydrofunctionalization of minimally functionalized alkenes via earth abundant transition metal catalysis. *Organic Chemistry Frontiers* 5 (2): 260–272.

156 Chen, Q.A., Chen, Z., and Dong, V.M. (2015). Rhodium-catalyzed enantioselective hydroamination of alkynes with indolines. *Journal of the American Chemical Society* 137 (26): 8392–8395.

157 Coman, S.M. and Parvulescu, V.I. (2015). Nonprecious metals catalyzing hydroamination and C–N coupling reactions. *Organic Process Research & Development* 19 (10): 1327–1355.

158 Guo, S., Yang, J.C., and Buchwald, S.L. (2018). A practical electrophilic nitrogen source for the synthesis of chiral primary amines by copper-catalyzed hydroamination. *Journal of the American Chemical Society* 140 (46): 15976–15984.

159 Huang, L., Arndt, M., Goossen, K. et al. (2015). Late transition metal-catalyzed hydroamination and hydroamidation. *Chemical Reviews* 115 (7): 2596–2697.

160 Strom, A.E., Balcells, D., and Hartwig, J.F. (2016). Synthetic and computational studies on the rhodium-catalyzed hydroamination of aminoalkenes. *ACS Catalysis* 6: 5651–5665.

161 Thomas, A.A., Speck, K., Kevlishvili, I. et al. (2018). Mechanistically guided design of ligands that significantly improve the efficiency of CuH-catalyzed hydroamination reactions. *Journal of the American Chemical Society* 140 (42): 13976–13984.

162 Bender, C.F., Hudson, W.B., and Widenhoefer, R.A. (2008). Sterically hindered mono(phosphines) as supporting ligands for the platinum-catalyzed hydroamination of amino alkenes. *Organometallics* 27 (10): 2356–2358.

163 Béthegnies, A., Daran, J.-C., and Poli, R. (2013). Platinum-catalyzed hydroamination of ethylene: study of the catalyst decomposition mechanism. *Organometallics* 32 (2): 673–681.

164 Dub, P.A., Daran, J.-C., Levina, V.A. et al. (2011). Modeling the platinum-catalyzed intermolecular hydroamination of ethylene: the nucleophilic addition of $HNEt_2$ to coordinated ethylene in trans-$PtBr_2(C_2H_4)(HNEt_2)$. *Journal of Organometallic Chemistry* 696 (6): 1174–1183.

165 Rodriguez-Zubiri, M., Anguille, S., Brunet, J.-J. et al. (2013). Pt-catalysed intermolecular hydroamination of non-activated olefins using a novel family of catalysts: arbuzov-type phosphorus metal complexes. *Journal of Molecular Catalysis A: Chemical* 379: 103–111.

166 Dub, P.A., Béthegnies, A., and Poli, R. (2011). DFT and experimental studies on the PtX_2/X^--catalyzed olefin hydroamination: effect of halogen, amine

basicity, and olefin on activity, regioselectivity, and catalyst deactivation. *Organometallics* 31 (1): 294–305.

167 Liu, Y., Zhang, D., and Bi, S. (2012). Theoretical insight into PtCl$_2$-catalyzed isomerization of cyclopropenes to allenes. *Organometallics* 31 (13): 4769–4778.

168 Agbossou, F., Carpentier, J.-F., and Mortreux, A. (1995). Asymmetric hydroformylation. *Chemical Reviews* 95 (7): 2485–2506.

169 Breit, B. (2003). Synthetic aspects of stereoselective hydroformylation. *Accounts of Chemical Research* 36 (4): 264–275.

170 Franke, R., Selent, D., and Borner, A. (2012). Applied hydroformylation. *Chemical Reviews* 112 (11): 5675–5732.

171 Haumann, M. and Riisager, A. (2008). Hydroformylation in room temperature ionic liquids (RTILs): catalyst and process developments. *Chemical Reviews* 108 (4): 1474–1497.

172 Luo, X., Bai, R., Liu, S. et al. (2016). Mechanism of rhodium-catalyzed formyl activation: a computational study. *Journal of Organic Chemistry* 81 (6): 2320–2326.

173 Paulik, F.E. (1972). Recent developments in hydroformylation catalysis. *Catalysis Reviews* 6 (1): 49–84.

174 Dias, A.d.O., Augusti, R., dos Santos, E.N. et al. (1997). Convenient one-pot synthesis of 4,8-dimethyl-bicyclo[3.3.1]non-7-en-2-ol via platinum/tin catalyzed hydroformylation/cyclization of limonene. *Tetrahedron Letters* 38 (1): 41–44.

175 Gusevskaya, E.V., dos Santos, E.N., Augusti, R. et al. (2000). Platinum/tin catalyzed hydroformylation of naturally occurring monoterpenes. *Journal of Molecular Catalysis A: Chemical* 152 (1): 15–24.

176 Kistamurthy, D., Otto, S., Moss, J.R. et al. (2010). Synthesis and characterization of [MCl2(^{n}Pr-N(Ph$_2$P)$_2$)] (M = Pt, Pd) complexes and hydroformylation of 1-octene using [PtCl$_2$(R-N(Ph$_2$P)$_2$)] (R = benzyl, 2-picolyl, nPr) complexes. *Transition Metal Chemistry* 35 (5): 633–637.

177 Kollár, L., Sándor, P., and Szalontai, G. (1991). Temperature dependence of the enantioselective hydroformylation with PtCl$_2$[(S)-BINAP] + SnCl$_2$ catalyst and the dynamic NMR study of the catalytic precursor. *Journal of Molecular Catalysis* 67 (2): 191–198.

178 Papp, T., Kollár, L., and Kégl, T. (2013). Mechanism of the platinum/tin-catalyzed asymmetric hydroformylation of styrene: a detailed computational investigation of the chiral discrimination. *Organometallics* 32 (13): 3640–3650.

179 Pongrácz, P. and Kollár, L. (2016). Enantioselective hydroformylation of 2- and 4-substituted styrenes with PtCl$_2$[(R)-BINAP] + SnCl$_2$ 'in situ' catalyst. *Journal of Organometallic Chemistry* 824: 118–123.

180 Pongrácz, P., Kollár, L., and Mika, L.T. (2016). A step towards hydroformylation under sustainable conditions: platinum-catalysed enantioselective hydroformylation of styrene in γ-valerolactone. *Green Chemistry* 18 (3): 842–847.

181 Pongrácz, P., Kostas, I.D., and Kollár, L. (2013). Platinum complexes of P,N- and P,N,P-ligands and their application in the hydroformylation of styrene. *Journal of Organometallic Chemistry* 723: 149–153.

182 Pongrácz, P., Petőcz, G., Shaw, M. et al. (2010). Platinum complexes of 2-diphenylphosphinobenzaldehyde-derived P-alkene ligands and their application in the hydroformylation of styrene. *Journal of Organometallic Chemistry* 695 (22): 2381–2384.

183 Li, J., Sun, C., Demerzhan, S. et al. (2011). Metal-catalyzed rearrangement of cyclopropenes to allenes. *Journal of the American Chemical Society* 133 (33): 12964–12967.

7

Theoretical Study of Co-Catalysis

58.9332
Co[27]
Cobalt
[Ar]3d^{7}4s^2

Undoubtedly, cobalt (Co) is an earth-abundant first-row transition metal, which has been successfully used as a homogeneous catalyst in organometallic chemistry [1–5]. The oxidative states of Co vary greatly in Co-catalysis. The common oxidation states of Co are 0, +1, +2, +3, +5, etc., [6–10]. It is precisely because of the variety of oxidative states that the redox processes of Co involved in the catalytic cycle also have various channels, which can be redox with double-electron transfer or single-electron transfer (SET) [11–14]. This characteristic leads to the mechanistic complexity of Co-catalysis, which involves not only the common elementary reactions, but also the variety of different spin configurations and the intersection of potential energy surfaces with different spin states.

Generally, the low-valence Co can be stabilized by π-acidic ligand, which leads to the activation of unsaturated bonds. As an example, Pauson–Khand reaction, the coupling of enyne with carbon monoxide to afford cyclopentenones, can be catalyzed by low-valence Co_2CO_8 species [15–17]. When σ-donor ligands are used in Co-catalysis, high-valence Co-species can be further stabilized. Therefore, C—C bond can be formed in Co-catalyzed coupling reactions. In this chapter, the mechanism of some Co-catalyzed typical reactions is discussed, involving C—H bond functionalizations, coupling reactions, hydrogenations, hydroformylations, and some unsaturated molecule activations.

7.1 Co-Mediated C—H Bond Activation

Co-catalyzed C—H bond activation/functionalization has become an increasingly important strategy for the construction of complex organic products, and

Computational Methods in Organometallic Catalysis: From Elementary Reactions to Mechanisms,
First Edition. Yu Lan.
© 2021 WILEY-VCH GmbH. Published 2021 by WILEY-VCH GmbH.

remarkable progress has been achieved for a wide range of potential applications [1, 18–21]. As the earth-abundant element, the 3d transition metal cobalt has significantly lower electronegativity than that of the 4d transition metals (such as Rh); thus, the C—Co bond in organometallic cobalt intermediates would usually reveal more nucleophilicity and exhibit unique reactivity [22–26]. In the field of cobalt-catalyzed C—H bond activation and functionalization, the groups of Brookhart [27–30], Nakamura [31–33], Yoshikai [32, 34–43], Ackermann [44–53], and others [54–59] have pioneered low-valent-Co-catalyzed selective C—H activation/coupling reactions with a wide range of substrates under mild reaction conditions, and the oxidative C—H activation would be involved in these low-valent cobalt catalytic systems. Since 2013, Matsunaga, Kanai, and coworkers reported the first example of the Cp*Co(III) catalysis [60, 61], which exhibits particular reactivities and higher activities than the analogous Cp*Rh(III) catalysis, and it has revealed widely potential application of C—H functionalizations. Over the past years, the Co(III) (i.e. Cp*Co^{III}) catalyst systems have witnessed major progress in the area of C—H activation, which can be tracked in the reports by the groups of Kanai [62–66], Matsunaga [62–69], Glorius [70–74], Ackermann [44–53], Ellman [25, 75–80], Chang [24, 54, 81], Li [26, 82–86], Yu [87–89], and others [54–56, 58, 59]. Noteworthy, the oxidant additive cooperated with the cheaper cobalt catalysts, such as Co^{II}(OAc)$_2$, Co^{II}C$_2$O$_4$, Co^{II}(acac)$_2$, Co^{III}(acac)$_3$, and so on [90–94], and has attracted more and more attention since Daugulis and Grigorjeva reported the amazing strategies in 2014 [58]. Actually, the precatalyst Co^{II} can be oxidized to Co^{III}, and the high-valent Co^{III} complexes would initiate the catalytic cycle by the concerted metalation–deprotonation (CMD) or SET/proton concerted electron transfer (PCET) pathway in the latter two cases.

As depicted in Scheme 7.1, a series of activation models for Co-catalyzed C—H bond cleavage have been proposed based on extensive theoretical and experimental studies of cobalt chemistry [11, 95–100]. The direct oxidative addition of a C—H bond onto cobalt can take place via a three-membered ring transition state for the low-valent cobalt species (model a), which also could be considered to be cobalt insertion into C—H bond [101–104]. For the high-valent cobalt species, CMD mechanism is commonly involved when there is an inner-sphere carboxylate (model b) [14, 105–107]. Alternatively, when an exogenous base is used, an electrophilic deprotonation also can access the C—H bond cleavage (model c) [31, 108, 109].

Different from 4d transition metals (such as Rh(III)), the electronegativity of Co(III) is much lower and thus can easily transfer one electron to oxidant for the formation of Co(IV)-R radical complex. Therefore, the intermolecular SET [14, 110, 111] and PCET [112–114] can also result in cobalt-mediated C—H activation (models d and e). As shown in Scheme 7.1d, the intermolecular SET can occur between oxidant and the hydrocarbon substrate to provide a radical intermediate. Then, C—H bond cleavage could take place through deprotonation. Alternatively, the C—H bond activation can also occur via the intramolecular PCET transition state through spin crossover process.

The C—H cleavage usually provides a new Co—C bond, which can undergo further transformations for the construction of new organic molecules. In

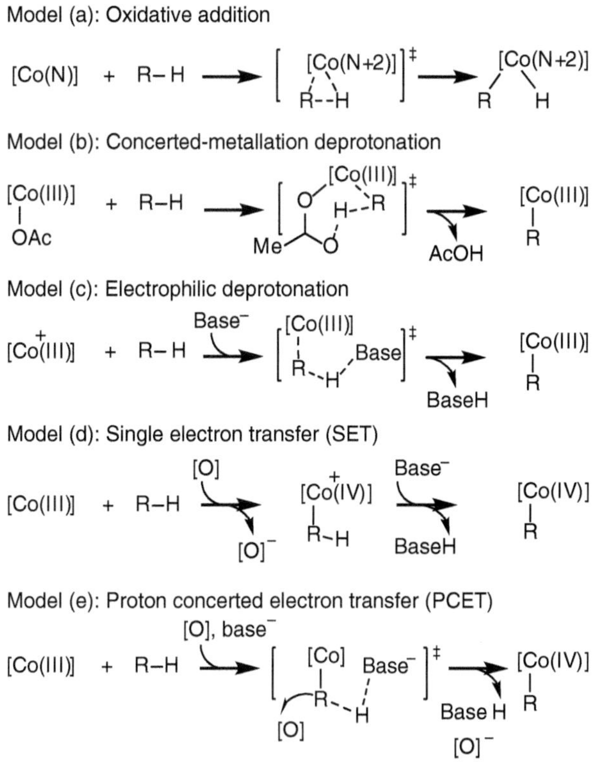

Model (a): Oxidative addition

$$[Co(N)] + R-H \longrightarrow \left[\begin{array}{c} [Co(N+2)] \\ | \\ R--H \end{array} \right]^{\ddagger} \longrightarrow [Co(N+2)]\begin{array}{c} \diagup \diagdown \\ R \quad H \end{array}$$

Model (b): Concerted-metallation deprotonation

$$\underset{\underset{OAc}{|}}{[Co(III)]} + R-H \longrightarrow \left[\begin{array}{c} [Co(III)] \\ O \diagup \diagdown H--R \\ Me \diagdown \diagup O \end{array} \right]^{\ddagger} \xrightarrow[AcOH]{} \underset{\underset{R}{|}}{[Co(III)]}$$

Model (c): Electrophilic deprotonation

$$[\overset{+}{Co(III)}] + R-H \xrightarrow{Base^-} \left[\begin{array}{c} [Co(III)] \\ | \\ R_{\cdot\cdot}_{H'}Base \end{array} \right]^{\ddagger} \xrightarrow[BaseH]{} \underset{\underset{R}{|}}{[Co(III)]}$$

Model (d): Single electron transfer (SET)

$$[Co(III)] + R-H \xrightarrow[{[O]^-}]{[O]} \underset{\underset{R-H}{|}}{[\overset{+}{Co(IV)}]} \xrightarrow[BaseH]{Base^-} \underset{\underset{R}{|}}{[Co(IV)]}$$

Model (e): Proton concerted electron transfer (PCET)

$$[Co(III)] + R-H \xrightarrow{[O], base^-} \left[\begin{array}{c} [Co] \quad Base^- \\ | \quad\quad\, / \\ R_{\cdot\cdot}H \\ [O] \end{array} \right]^{\ddagger} \xrightarrow[\underset{[O]^-}{Base\,H}]{} \underset{\underset{R}{|}}{[Co(IV)]}$$

Scheme 7.1 Possible models of cobalt-mediated C—H activation. Source: Chen et al. [31]; Wei et al. [14]; Liang and Jiao [105]; Santhoshkumar and Cheng [106]; Yamazaki et al. [107]; Roslan et al. [108]; Shin et al. [109].

Co-chemistry, those transformations were often considered to be the insertion of unsaturated bonds into newly formed Co—C bonds. Therefore, in this section, the mechanism of Co-mediated C—H activation/functionalization was organized by the insertion molecules.

7.1.1 Hydroarylation of Alkenes

In Co-catalyzed arene functionalizations, the olefin insertion is one of the most important ways for the alkylation (Scheme 7.2a) [115]. The mechanism for this type of reactions could be classified by the valence of Co catalyst. Generally, two possible pathways for the alkene functionalizations through Co-mediated C—H activation have been proposed involving redox and nonredox cases. In a redox process (Scheme 7.2b), when low-valent Co(0) species is used as catalyst, there are three steps involved in the catalytic cycle of one catalytic cycle, including the oxidative addition with C—H bond, olefin insertion into newly formed Co—H bond, reductive elimination for the regeneration of cobalt catalyst, and the formation of new C(aryl)—C(alkyl) bond. It should be mentioned that the valence of cobalt during the catalytic cycle is varied from Co(0) to Co(II), and finally returned to Co(0).

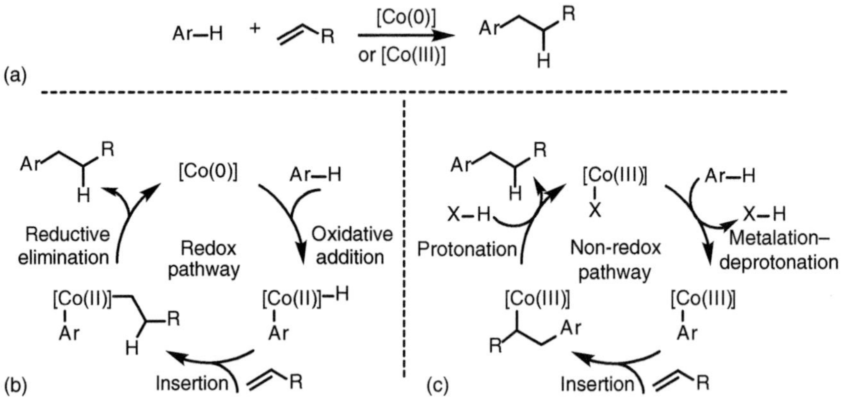

Scheme 7.2 (a) Co-catalyzed C—H alkylations through olefin insertion and the catalytic cycles through redox pathway (b) or nonredox pathway (c). Source: (a) Based on Yang et al. [115].

Alternatively, when high-valent Co(III) species is used as catalyst (Scheme 7.2c), the CMD type C—H bond cleavage would form an aryl–Co(III) complex. Then, the olefin insertion into Co(III)—C(aryl) bond forms new C(aryl)—C(alkyl) bond in alkyl cobalt(III) intermediate. The protonation yields the alkyl arene product and regenerates the Co(III) catalyst. The whole catalytic cycle is a nonredox process.

As an example for low-valent case, Yoshikai group developed a cobalt-catalyzed hydroarylation of styrene with *ortho*-phenylpyridine [116]. To gain insight into the mechanism of this reaction, a density functional theory (DFT) study was made by Fu and coworkers [115] (Figure 7.1). In their theoretical calculations, the Co(0)-mediated C—H bond cleavage could take place via a three-membered ring-type oxidative addition transition state **7-2ts** with an energy barrier of 17.7 kcal/mol to form an aryl hydride–Co(II) complex **7-3** reversibly. Then, the insertion of styrene into Co—H bond occurs via a four-membered ring transition state **7-4ts** with an energy barrier of 19.2 to generate the related linear alkyl aryl Co(II) intermediate **7-5**. Reductive elimination of aryl and alkyl groups occurs via a three-membered ring-type transition state **7-6ts** to yield alkylated arene and regenerate the active catalyst **7-1**. The computational results found that the rate-determining step for the catalytic cycle is the reductive elimination step with an overall observed activation free energy of 25.3 kcal/mol.

Alternatively, high valent Co(III)-catalyzed alkylation of arene could take place through a nonredox process by a CMD type C—H activation. As shown in Figure 7.2, Whiteoak and coworkers reported a Cp*Co(III)-catalyzed coupling reaction of benzamides with α,β-unsaturated carbonyl compounds in preparing aliphatic ketones [117]. The DFT computations revealed the mechanism of this reaction. In their theoretical study, neutral active catalyst Cp*Co(III) diacetate **7-7** is set to the relative zero in free-energy profiles. The ligand exchange with benzamide forms a cationic Co(III) intermediate **7-8** reversibly with the release of an acetate. A CMD type C—H bond cleavage takes place via a six-membered ring transition state **7-9ts** with an energy barrier of 21.3 kcal/mol to generate aryl Co(III) **7-10**. The olefin insertion into

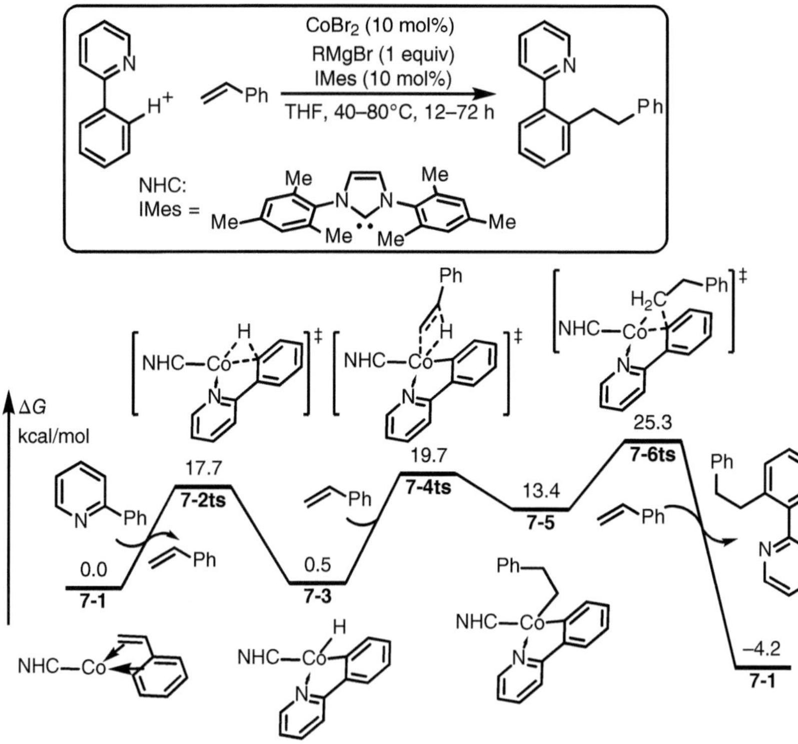

Figure 7.1 Free-energy profiles for the low-valent Co(0)-catalyzed arene C–H bond alkylation reaction. The energies were calculated at B3P86/6-311G(d, p) (SDD for Co)/CPCM(THF)//B3P86/6-31G(d) (LANL2DZ(f) for Co) level of theory. Source: Based on Yang et al. [115].

C(aryl)—Co(III) bond occurs via a four-membered ring transition state **7-11ts** with an energy barrier of only 6.4 kcal/mol to afford alkyl–Co(III) intermediate **7-12**. The protonation of enolate moiety via transition state **7-13ts** forms complex **7-14** with an energy barrier of 22.5 kcal/mol, which is considered to be the rate-determining step for the whole nonredox catalytic cycle.

7.1.2 Hydroarylation of Allenes

A series of successful ortho-alkenylations with allenes have been reported in cobalt catalysis. The high-valent Co-catalyzed ortho-alkenylation of arenes is mainly by the insertion of allenes into the Co—C bond rather than Co—H bond. Commonly, ortho-alkenylation of arene can be achieved by high-valent Cp*Co(III) catalyst. In this field, allene could be considered as electrophile, in which the C—C π bond acts as an inner-sphere oxidant to maintain the redox neutrality.

In 2017, Ackermann and coworkers developed an example of Co(III)-catalyzed hydroarylation of allenes with decorated arenes/indole derivatives to synthesize highly chemo- and regio-selective alkenylated heteroarene products [118]. A DFT study by the same group to reveal the mechanism of this reaction is illustrated in

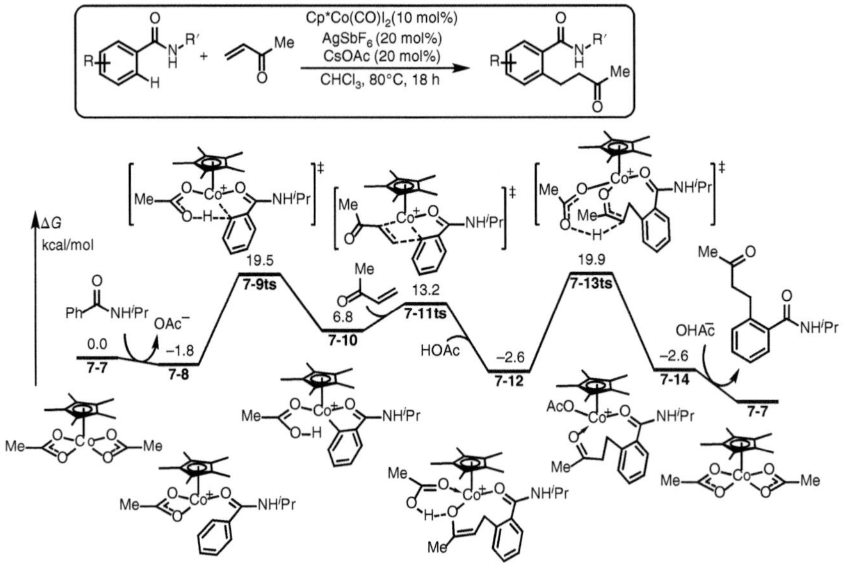

Figure 7.2 Free-energy profiles for the high-valent Co(III)-catalyzed arene C—H bond alkylation reaction. The energies were calculated at M06/def2-TZVP/COSMO(DCM)// BP86-D3BJ/def2-TZVP level of theory.

Figure 7.3. Aryl Co(III) species **7-15** can be formed through electrophilic deprotonation, which was considered as relative zero in free-energy profiles. The allene insertion into Co—C bond forms a vinyl Co(III) intermediate **7-17** via transition state **7-16ts** with an energy barrier of 6.0 kcal/mol. Subsequently, isomerization of vinyl-Co(III) to allylic one **7-21** can undergo sequential β-hydride elimination and alkene insertion via transition state **7-18ts** and **7-20ts** with an observed energy barrier of 16.2 kcal/mol. The generated allylic Co(III) complex **7-21** can be protonated by pyrimidinium via transition state **7-22ts**. The ligand exchange of dicationic intermediate **7-23** with indole reactant releases the hydroarylation product. The indolyl Co(III) species **7-15** can be regenerated by a further electrophilic deprotonation.

7.1.3 Hydroarylation of Alkynes

The high-valent Co(III)-catalyzed hydroarylation of alkyenes can also be achieved through the insertion of alkyne into the Co—C bond after a CMD-type C—H activation. For instance, an efficient, atom-economical, and regioselective Co(III)-catalyzed alkenylation of indoles with terminal alkyne was reported under mild conditions to prepare linear olefins in 2017 [86]. As illustrated in Figure 7.4, Co(III)-mediated C—H bond cleavage in the substrate-coordinated Co(III) cation **7-27** takes place through CMD process via transition state **7-28ts** with an energy barrier of 13.6 kcal/mol. Then, alkyne inserts into the Co—C bond via transition state **7-30ts** affording a vinyl Co(III) intermediate **7-31**. The overall activation free energy starting from intermediate **7-27** is 29.2 kcal/mol,

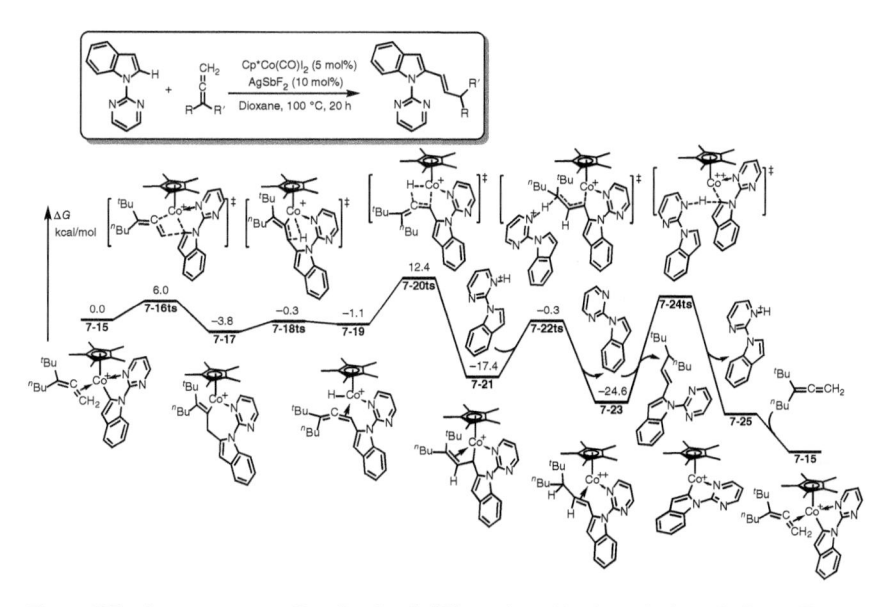

Figure 7.3 Free-energy profiles for the Co(III)-catalyzed hydroarylation of allene. The energies were calculated at BP86-D3BJ/def2-TZVP/COSMO(DCE)//BP86-D3BJ/def2-SVP level of theory.

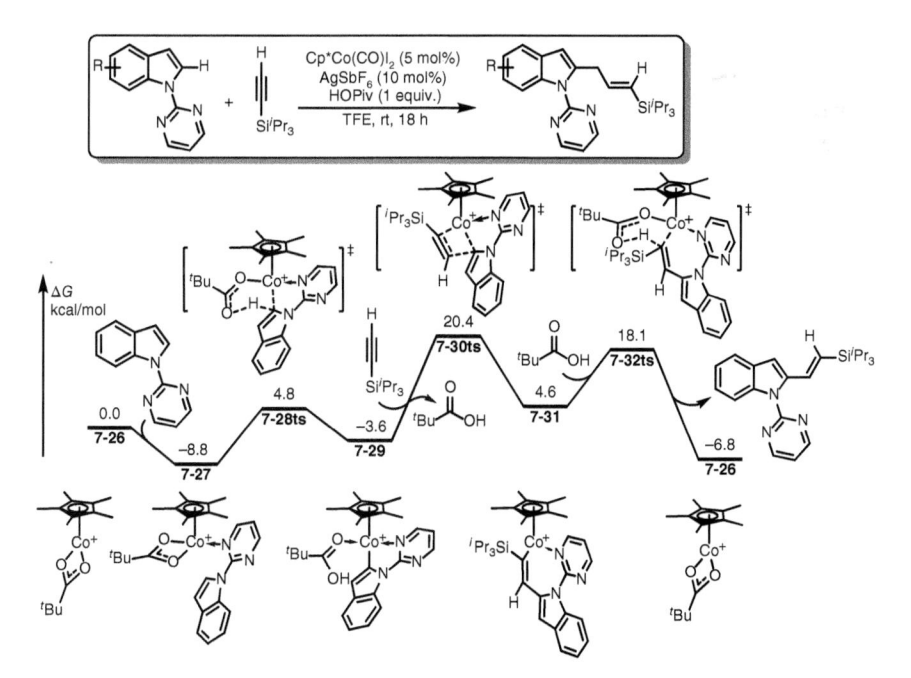

Figure 7.4 Free-energy profiles for the Co(III)-catalyzed hydroarylation of alkyne. The energies were calculated at M11-L/6-311+G(d,p) (SDD for Co)/SMD(TFE)//B3LYP/6-31+G(d) (SDD for Co) level of theory. Source: Based on Khand et al. [120].

which is considered to be the rate-determining step for the whole catalytic cycle. The protonation of substrate by extra pivalate acid takes place in regenerating Co(III) catalyst **7-26** and producing the hydroarylated product via transition states **7-32ts**.

7.1.4 Hydroarylation of Nitrenoid

From a mechanistic point of view, hydroarylation of nitrenoid can be considered to be ortho-amination through the insertion of a nitrene or formal one into a C(aryl)—H bond. For example, Whiteoak group disclosed a Cp*Co(III)-catalyzed C(aryl)—H activation/amination insertion, which provides desired acetamidobenzamide derivatives in both experiment and theory [119]. As presented in Figure 7.5, the calculated energy profiles reveal the catalytic cycle of this reaction. The ligand exchange with benzamide substrate leads to a C(sp^2)—H activation via a CMD-type transition state **7-35ts** with an energy barrier of 21.5 kcal/mol for generating aryl-Co(III) intermediate **7-36**. The coordination of dioxazolone leads to a sequential nucleophilic substitution with aryl group on cobalt via transition state **7-38ts**, which can be considered to be a formal nitrene insertion into C(aryl)—Co bond. The release of carbon dioxide results in a six-membered cobalt cycle **7-39** irreversibly. The calculated overall activation free energy in this step is as high as 30.6 kcal/mol from reference of **7-34**. Then, the protonation by acetic acid via transition state **7-40ts** results in the acetamidobenzamide product and regenerates the active catalyst **7-33**.

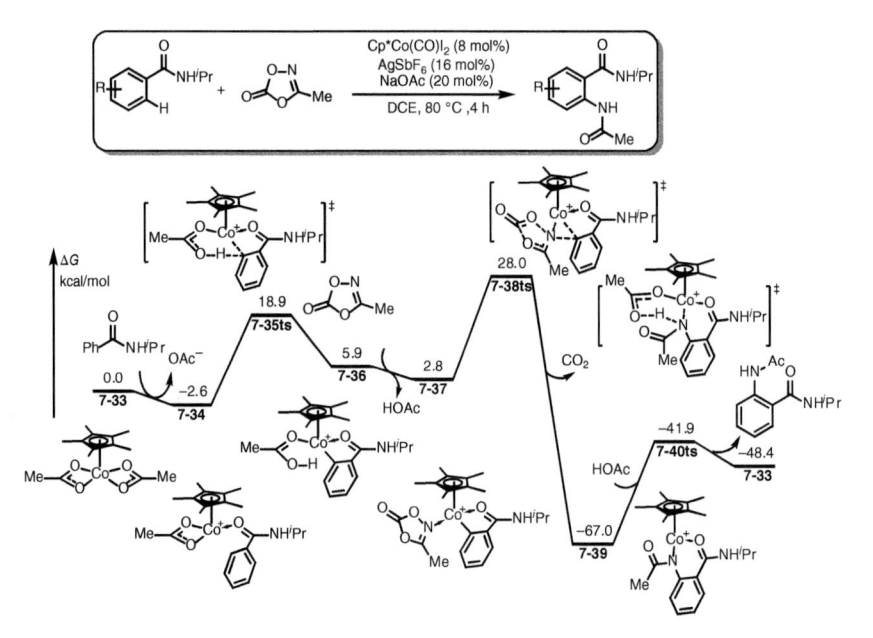

Figure 7.5 Free-energy profiles for the Co(III)-catalyzed hydroarylation of nitrenoid. The energies were calculated at M06/def2-TZVP/COSMO(DCE)//BP86-D3BJ/def2-SVP level of theory.

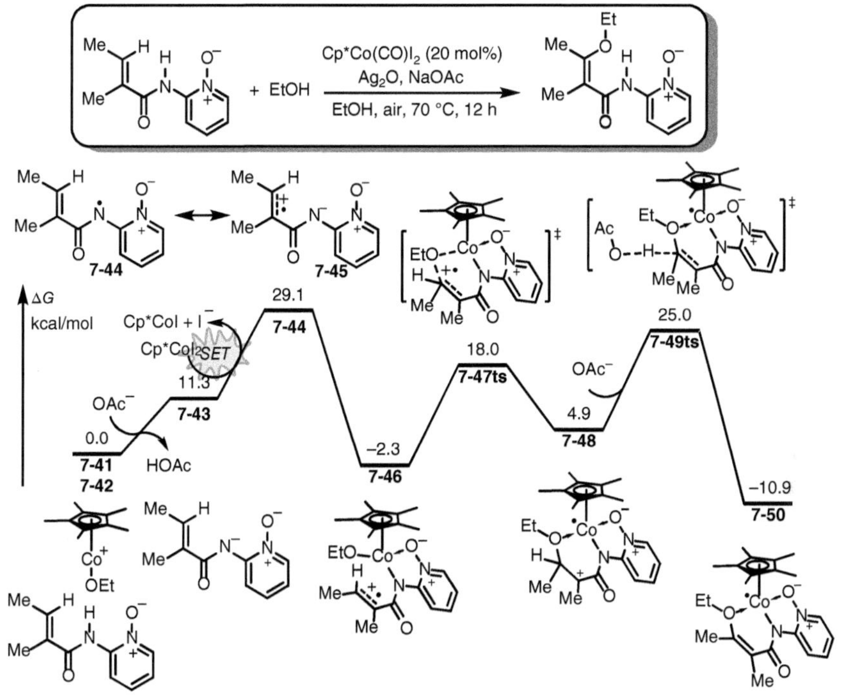

Figure 7.6 Free-energy profiles for the Co(III)-catalyzed oxidative alkoxylation of alkenes. The energies were calculated at M06-L/6-311++G(2df, 2pd) (SDD for Co)/SMD(ethanol)//M06-L/6-31G(d) (LANL2DZ for Co)/SMD(ethanol) level of theory.

7.1.5 Oxidative C—H Alkoxylation

The oxidative coupling of alkene and alcohols can undergo a SET-assisted C—H activation in the presence of Co(III) catalyst, where the oxidative state of cobalt is reduced in C—H bond cleavage step [11]. As shown in Figure 7.6, the mechanism of Co(III)-catalyzed oxidative alkoxylation of alkenes starts from the deprotonation of aryl amide. The generated anion intermediate **7-43** can undergo a SET with Co(III) to afford an amino radical **7-44**, which also can be considered as its resonance structure **7-45**. The coordination of **7-44** onto cationic Co(III) species **7-41** results in complex **7-46**. Then, an intramolecular nucleophilic attack of ethoxy onto cationic radical C=C bond occurs via transition state **7-47ts** to form a low-valence Co(II) intermediate **7-48**. An intermolecular deprotonation takes place via transition state **7-49ts** to form a neutral Co(II) intermediate **7-50**. In further process, Co(III) catalyst can be regenerated by the oxidation with dioxygen and ligand exchange.

7.2 Co-Mediated Cycloadditions

The low-valence Co-catalyst can be coordinated with π-acidic compounds, which leads to their activation. Usually, the following oxidative addition with alkenes

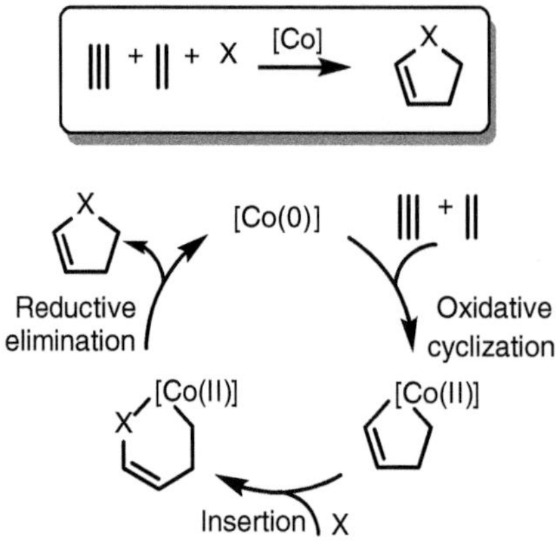

Scheme 7.3 General mechanism of Co-catalyzed cycloadditions. Source: Based on Khand et al. [120].

and/or alkynes leads to the formation of corresponding cobaltic cycle. Then, further transformation could yield cyclic products through insertion and reductive elimination steps. The general reaction pathway of Co-mediated cycloadditions can be summarized in Scheme 7.3 [120]. The most famous Co-mediated cycloaddition is Pauson–Khand reaction, which is used to construct cyclopentenones from enynes and carbon monoxide [121–123]. Moreover, some other cyclizations of alkynes are also important in Co-catalysis.

7.2.1 Co-Mediated Pauson–Khand Reaction

Pauson–Khand reaction was first reported in 1973 [120], which represents a one-step synthesis of the cyclopentenone ring through [2+2+1] assembly of one molecule each of alkene, alkyne, and carbon monoxide (Figure 7.7). In the early reports, $Co_2(CO)_8$ complex was necessary, as both catalyst and carbon monoxide source [15–17].

The first DFT study on the mechanism of Co-mediated Pauson–Khand reaction was reported by Nakamura and Yamanaka, in which a simple model of the reaction with ethylene and acetylene was used [123]. The calculated free-energy profiles for the bimetallic pathway are shown in Figure 7.7. The $Co_2(CO)_8$ complex **7-51** is set to relative zero, which can react with acetylene to afford an alkyne-bridged bicobalt complex **7-52** with 6.9 kcal/mol endergonic. A ligand exchange with ethylene through associate pathway forms intermediate **7-53**. Then, an oxidative cyclization by the coordinated alkene and alkyne leads to the formation of five-membered cobaltcycle **7-55** via transition state **7-54ts**. The calculated overall activation energy of this step is 36.2 kcal/mol, which can be majorly attributed to the high energy of substrates loading. Then, a carbonyl insertion into C(alkyl)—Co bond occurs via transition state **7-56ts** to form a six-membered ring intermediate **7-57**. Finally, the coordination of carbon monoxide leads to the reductive elimination to yield cyclopentenone product via transition state **7-58ts**. The rate-determining step of this reaction is the oxidative cyclization with alkene and alkyne. The strong bonding of

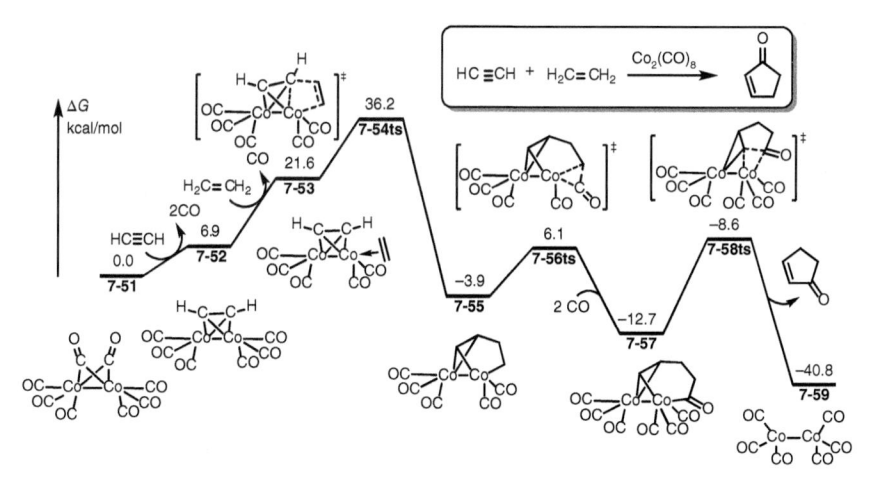

Figure 7.7 Free-energy profiles for a model Pauson–Khand reaction. The energies were calculated at B3LYP/6-311+G(d) (SDD for Co)//B3LYP/6-31G(d) (LANL2DZ for Co) level of theory.

carbon monoxide onto low-valence Co-restricting ligand exchange results in such a high activation energy of this step.

7.2.2 Co-Catalyzed [4+2] Cyclizations

A Co_2CO_8-catalyzed intramolecular [4+2] cyclization of alkyne and cyclobutanone can be achieved to synthesize phenols. The mechanism of this reaction is revealed by DFT calculation (Figure 7.8) [124]. Similar to Pauson–Khand reaction, this transformation also starts from an alkyne-bridged dicobalt complex **7-60**. The dissociation of carbon monoxide forms intermediate **7-61** with 13.3 kcal/mol endergonic. Then, an intramolecular oxidative addition of cyclobutanone onto one of Co takes place via a three-membered ring-type transition state **7-62ts** to form a Co(II) intermediate **7-63**. Subsequently, alkyne insertion into Co—C(aryl) bond generates a seven-membered ring-type Co(II) complex **7-65** via transition state **7-64ts**. The calculated overall activation free energy of this step is 36.6 kcal/mol. A rapid reductive elimination via transition state **7-66ts** forms a cyclohexadienone-coordinated Co(0) complex **7-67**. The phenol product can release from that complex by ligand exchange.

7.2.3 Co-Catalyzed [2+2+2] Cyclizations

In the presence of low-valence Co-catalyst, the cyclization of three-component unsaturated bonds would form six-membered product. Generally, two molecules of acetylene are necessary, while the other one can be either acetylene, ethylene, or C—heteroatom unsaturated bonds [125–127].

As shown in Figure 7.9, CpCo(I) species is often used as catalyst in [2+2+2] cyclization reactions [127]. In the mechanism study of acetylene trimerization reactions, the reaction starts from an oxidative cyclization with two molecules of

7 Theoretical Study of Co-Catalysis

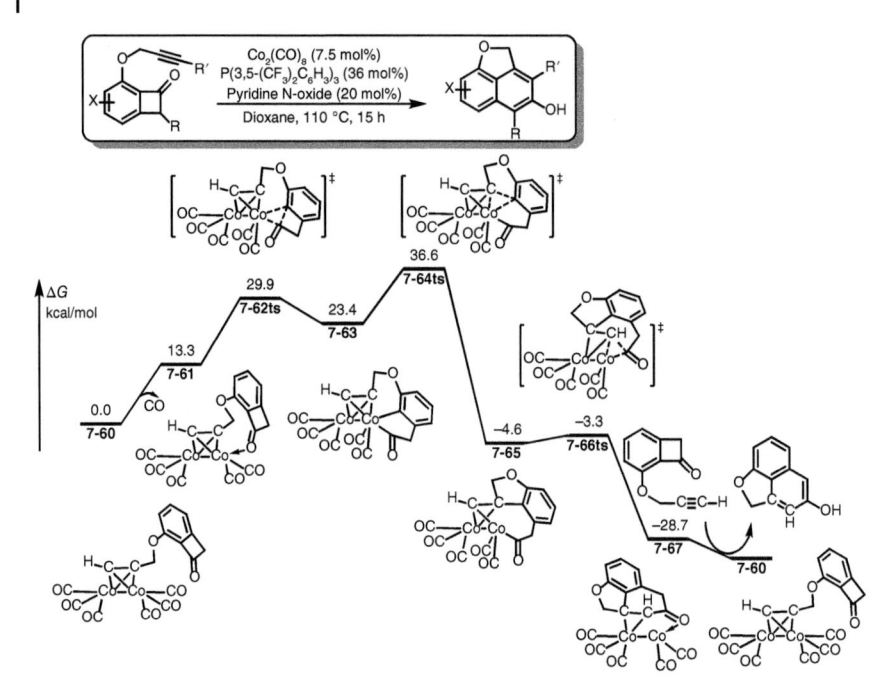

Figure 7.8 Free-energy profiles for a Co$_2$(CO)$_8$-catalyzed intramolecular [4+2] cycloaddition. The energies were calculated at M06/6-311+G(d,p)/SMD(dioxane)//B3LYP/6-31G(d) (LANL2DZ for Co) level of theory. Source: Based on Zhu et al. [124].

ethylene via transition state **7-69ts** with an energy barrier of 11.9 kcal/mol to form a cobaltacyclopentadiene compound **7-70**, which can be isolated experimentally. The coordination of another molecule acetylene forms a singlet Co(III) complex **7-71**. Then, acetylene inserts into Co—C bond via transition state **7-72ts** to form a seven-membered cyclic Co(III) intermediate **7-73**. The reductive elimination yields cyclohexdiene product and regenerates active species **7-68** by ligand exchange.

7.2.4 Co-Catalyzed [2+2] Cyclizations

In the absence of extra ligands and substrates, cobaltacyclopentadiene intermediate can be further transformed. As an example (Figure 7.10), when alkynes are added to CpCo(I) species, a cyclobutadiene-coordinated Co(I) complex is observed experimentally in the refluxed toluene [128]. Mechanism of this reaction is studied by DFT calculations. The reaction starts from a bialkyne-coordinated Co(I) species **7-76**. The oxidative cyclization can take place via transition state **7-77ts** with an energy barrier of only 12.0 kcal/mol to form a singlet cobaltacyclopentadiene intermediate **7-78**. The followed reductive elimination via a singlet transition state **7-79ts** yields cyclobutadiene-coordinated Co(I) complex **7-80** with an energy barrier of only 4.5 kcal/mol. In further theoretical studies, when phenyl is removed from alkyne reactant, the corresponding triplet state of cobaltacyclopentadiene is very stable, which restricted the following reductive elimination.

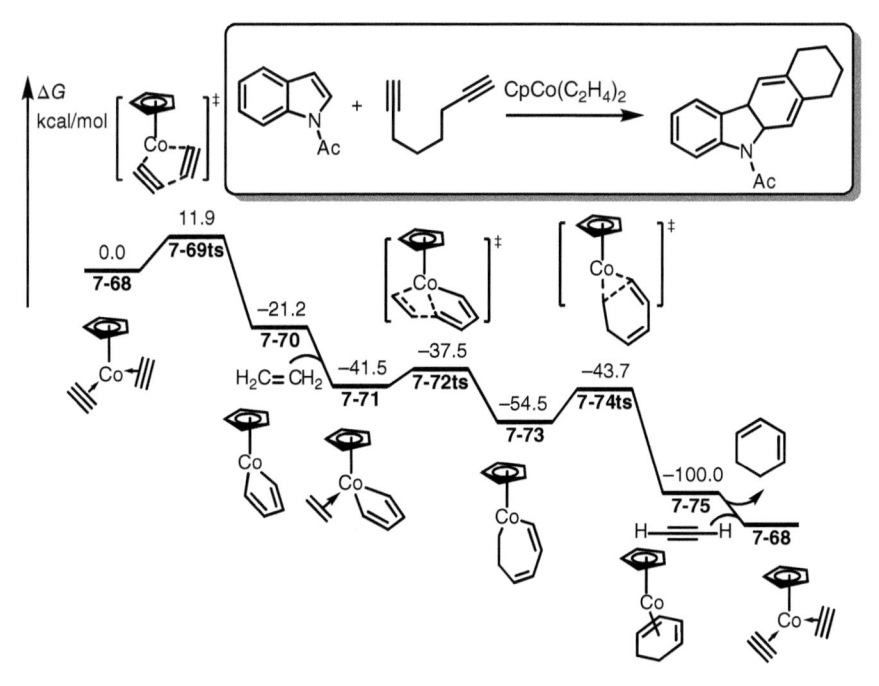

Figure 7.9 Free-energy profiles for a model reaction of CpCo(I)-catalyzed [2+2+2] cycloaddition. The energies were calculated at B3LYP5/6-31G(d,p) (LANL2DZ for Co) level of theory.

7.3 Co-Catalyzed Hydrogenation

In Co-catalysis, a Co–H species is familiar and can provide nucleophilic hydride to achieve hydrogenation of unsaturated bonds. Uniquely, the mechanism for Co-catalyzed hydrogenation of carbon dioxide is well studied by Yang [129–132] and Ke groups [133, 134] individually. Moreover, the mechanism studies of Co-catalyzed hydrogenation of alkenes, alkynes, and carbonyls were also reported.

The general catalytic cycle of Co-catalyzed hydrogenation is shown in Scheme 7.4. The hydride Co species can be considered as nucleophile, which can react with unsaturated bonds to form a new X—Co bond. Then, a concerted or stepwise metathesis with dihydrogen regenerates hydride Co species and yields dihydrogenation products.

7.3.1 Hydrogenation of Carbon Dioxide

The reutilization of carbon dioxide as an abundant, inexpensive, and nontoxic carbon source for the synthesis of valuable chemicals has attracted increasing attention in recent decades. To overcome the low activity of carbon dioxide, transition metal-catalyzed hydrogenation of carbon dioxide to formic acid and its derivatives has been studied extensively.

302 | *7 Theoretical Study of Co-Catalysis*

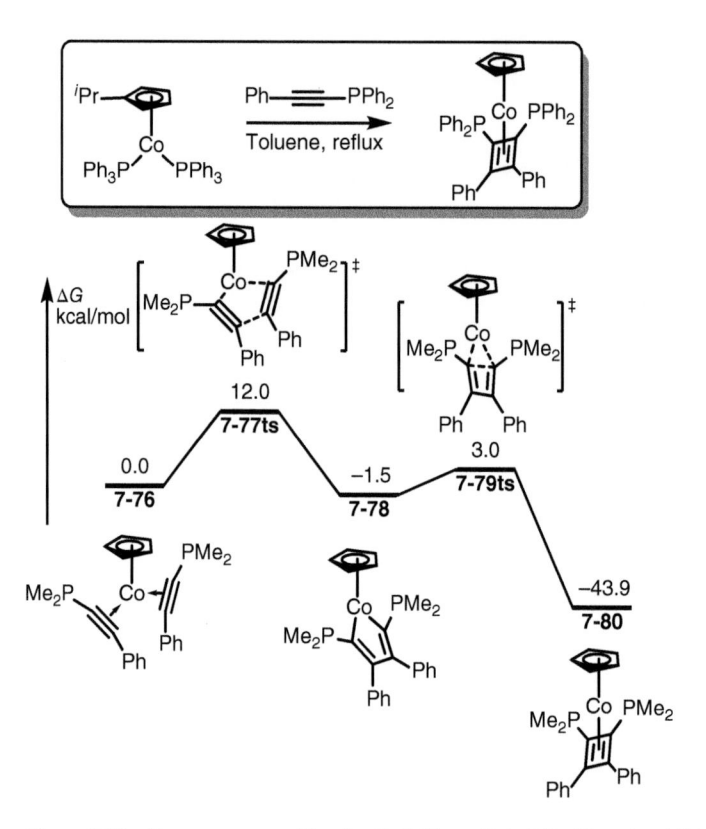

Figure 7.10 Free-energy profiles for CpCo(I)-mediated [2+2] cycloaddition of acetylenes. The energies were calculated at M06/6-311+G(2d,2p) (LANL2DZ for Co)/PCM(toluene)//M06/6-31G(d) (LANL2DZ for Co) level of theory.

Scheme 7.4 The general mechanism of Co-catalyzed hydrogenation reactions.

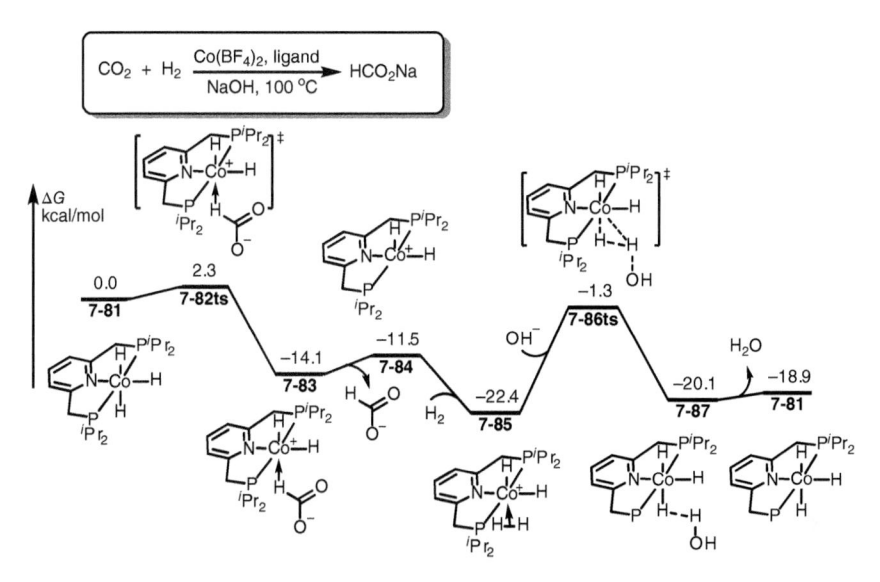

Figure 7.11 Free-energy profiles for Co-catalyzed hydrogenation of carbon dioxide in the presence of PNP-pincer ligand. The energies were calculated at ωB97X/6-31++G(d,p)/IEF-PCM(water) level of theory.

As shown in Figure 7.11, the detailed mechanisms of Co-catalyzed formate formation from H_2 and CO_2 in the presence of base are studied by DFT [132]. The reaction starts from a hydride Co(III) species **7-81**, in which hydride represents nucleophile. An intermolecular nucleophilic attack of hydride onto carbon dioxide takes place via transition state **7-82ts** with an energy barrier of only 2.3 kcal/mol leading to the release of formate and the formation of cationic Co(III) species **7-84**. The coordination of dihydrogen forms intermediate **7-85** with 10.9 kcal/mol exergonic. Then, a base-assisted H—H bond cleavage takes place via transition state **7-86ts** to form a hydrogen bond complex **7-87**. After release of water molecule, hydride Co(III) species **7-81** is regenerated. The computational results showed that the rate-determining step is the H—H bond cleavage with an energy barrier of 21.1 kcal/mol.

When biphosphine ligand is used in Co-catalyzed dihydrogenation of carbon dioxide, an inner-sphere pathway is proposed by Ke and further proved in DFT calculations [134]. As shown in Figure 7.12, the reaction starts from a cationic Co(I) species **7-88**. The oxidative addition of dihydrogen onto Co(I) takes place rapidly via transition state **7-89ts** to form a bihydride Co(III) cation **7-90**. The deprotonation of **7-90** by formate via transition state **7-91ts** forms a neutral hydride Co(I) intermediate **7-92**. Then, the coordination of carbon dioxide takes place via transition state **7-93ts** to form a C(CO$_2$)—Co bond in intermediate **7-94**. The calculated overall barrier of this step is 20.9 kcal/mol, which is considered to be the rate-determining step for the whole catalytic cycle. Then, a migratory insertion of carbon dioxide into Co—H bond takes place via transition state **7-95ts** to yield formate and regenerate cationic Co(I) species **7-88**.

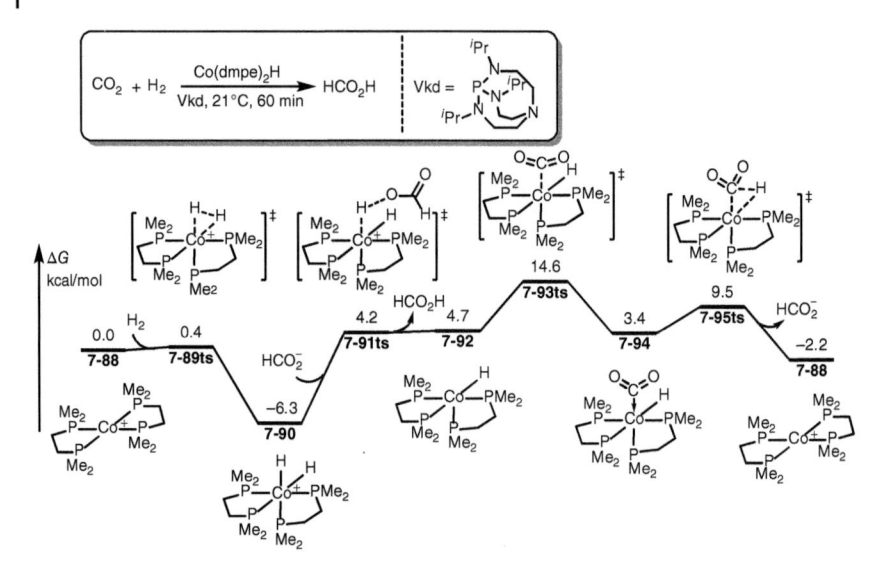

Figure 7.12 Free-energy profiles for Co-catalyzed hydrogenation of carbon dioxide in the presence of biphosphine ligand. The energies were calculated at ωB97X-D/6-31++G(d,p)/SMD(THF) level of theory.

7.3.2 Hydrogenation of Alkenes

In the presence of hydride Co species, the addition of gaseous dihydrogen onto alkenes would result in alkane product. The nonredox mechanism for Co-pincer complex-catalyzed dihydrogenation of alkenes was studied by Yang group, in which a nonredox pathway is proposed [135]. As shown in Figure 7.13, the reaction starts from a cationic hydride Co(II) species **7-96**. Alkene insertion into Co—H bond takes place via transition state **7-97ts** with an energy barrier of 19.8 kcal/mol to form an alkyl Co(II) intermediate **7-98**. In next step, the metathesis between Co—C(alkyl) bond and H—H bond in dihydrogen occurs via transition state **7-99ts**. The energy barrier of the second step is 26.7 kcal/mol, which was considered to be the rate-determining step. After this step, hydride Co(II) species **7-96** is regenerated with the release of alkane product.

When diphosphine ligand is used in Co-catalyzed dihydrogenation reaction, a redox pathway was proposed by Lei and Ma [136]. As shown in Figure 7.14, the reaction starts from a planar dihydride Co(II) species **7-100**. The coming alkene inserts into Co—H bond via transition state **7-101ts** to form an agostic Co(II) complex **7-102**. The rotation of hydride via transition state **7-103ts** generates an alkyl Co(II) intermediate **7-104**. Then, a reductive elimination of alkyl and hydride via transition state **7-105ts** yields alkane product and forms a diphosphine-coordinated Co(0) complex **7-106**. The oxidative addition of dihydrogen onto Co(0) complex **7-106** is a barrierless process, which can regenerate dihydride Co(II) species **7-100** by 48.0 kcal/mol exergonic.

Alternatively, an anionic Co(−I)–Co(I) catalytic cycle for alkene dihydrogenation is proposed by Wu et al. [137]. As shown in Figure 7.15, the reaction starts

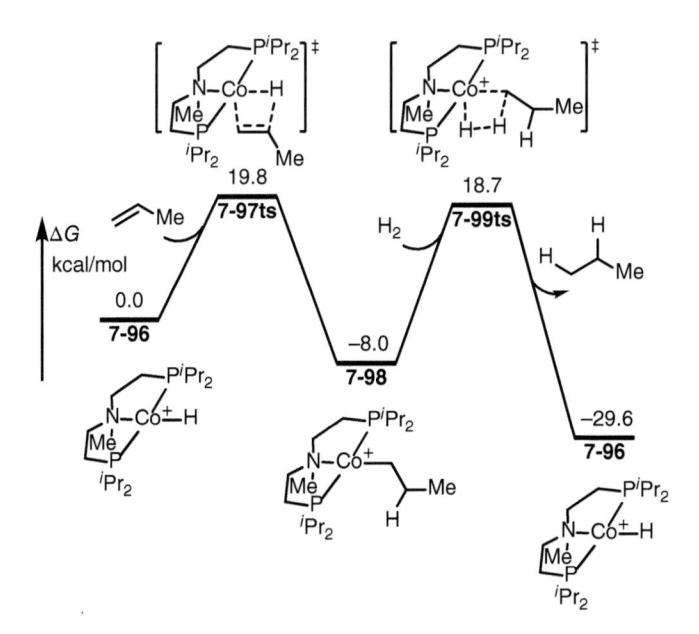

Figure 7.13 Free-energy profiles for Co-catalyzed hydrogenation of alkene through a nonredox pathway in the presence of PNP-pincer ligand. The energies were calculated at ωB97X/6-31++G(d,p)/IEF-PCM(water) level of theory.

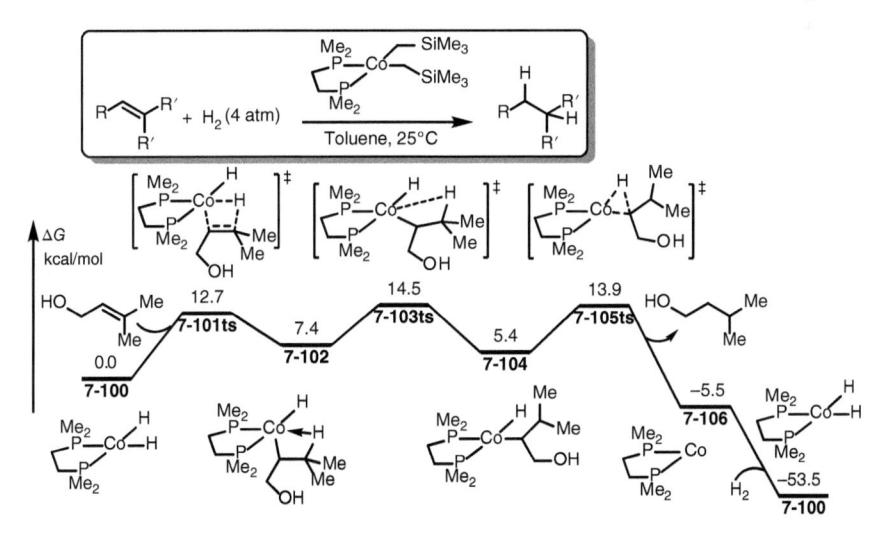

Figure 7.14 Free-energy profiles for Co-catalyzed hydrogenation of alkene through a redox pathway. The energies were calculated at B3LYP/6-31G(d) (LANL2DZ for Co)/PCM(toluene)//B3LYP/6-31G(d) (LANL2DZ for Co) level of theory.

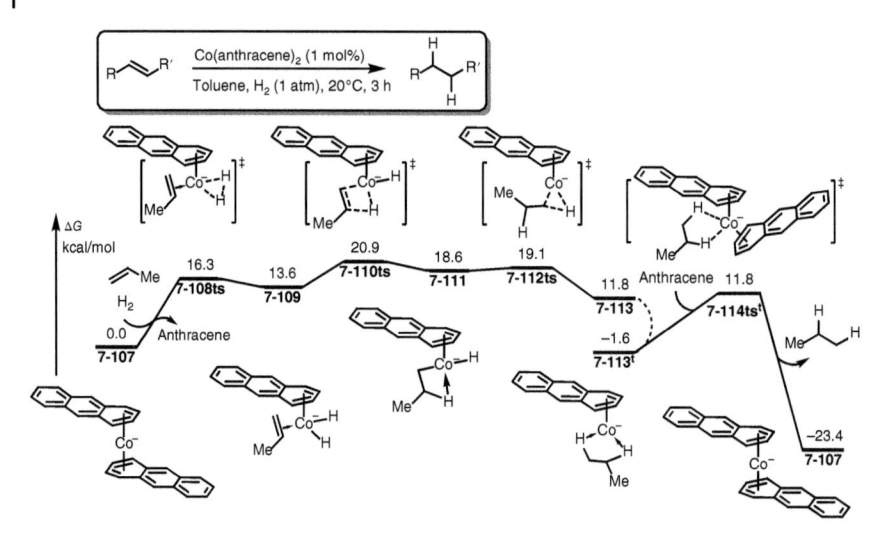

Figure 7.15 Free-energy profiles for Co(−I)-catalyzed hydrogenation of alkene through a redox pathway. The energies were calculated at B3LYP-D3/6-31G(d) (def2-TZVP for Co) level of theory.

from an anionic η⁴-anthracene-coordinated Co(−I) species **7-107**. The oxidative addition of dihydrogen takes place via transition state **7-108ts** to afford an anionic dihydride Co(I) intermediate **7-109**. The insertion of coordinated alkene occurs via transition state **7-110ts** to form an agostic Co(I) complex **7-111**. The calculated overall activation free energy of this step is 20.9 kcal/mol, which is represented as the rate-determining step. Then, a rapid reductive elimination via transition state **7-112ts** results in a singlet state Co(−I) complex **7-113**, which can easily transfer to its triplet isomer **7-113ᵗ** through spin cross-over. The coordination of anthracene ligand leads to the dissociation of alkane product via transition state **7-114tsᵗ**. After release of alkane product, η⁴-anthracene-coordinated Co(−I) species **7-107** is regenerated.

7.3.3 Hydrogenation of Alkynes

When pincer P–N–P type ligand is used, a Co-catalyzed hydrogenation of alkynes is reported experimentally, in which a Lewis-acid–base pair of ammonia and borane is used as hydrogen source [138]. As shown in Figure 7.16, the mechanism of this reaction is studied by DFT calculation. When hydride Co(I) complex **7-115ᵗ** is used as active catalyst, the coordination of alkyne leads to an insertion into Co—H bond via transition state **7-116ts** with an activation free energy of only 4.5 kcal/mol. When vinyl Co(I) complex **7-117ᵗ** is formed, protonation by methanol solvent takes place via transition state **7-118tsᵗ** with an energy barrier of 16.5 kcal/mol. In this transition state, methanol molecule is activated by the hydrogen bond with amino moiety in ligand. After release of (Z)-alkene, an amino Co(I) complex **7-119ᵗ** is formed, which can be hydrided by ammonia-coordinated borane without barrier to regenerate hydride

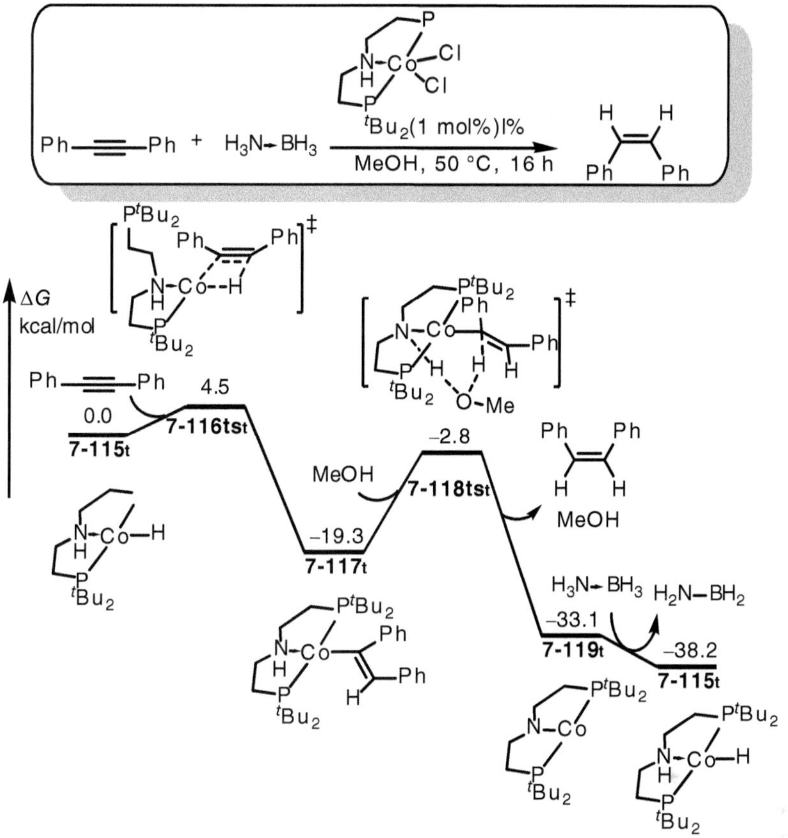

Figure 7.16 Free-energy profiles for Co-catalyzed hydrogenation of alkyne. The energies were calculated at M06/6-311+G(d,p) (SDD for Co)/SMD(methanol)//B3LYP/6-31G(d) (SDD for Co) level of theory.

Co(I) species **7-115ᵗ**. DFT calculation found that the whole catalytic cycle runs in a triplet spin state.

7.4 Co-Catalyzed Hydroformylation

Hydroformylation can be formally considered as that where olefin reacts with dihydrogen and carbon monoxide to form a new aldehyde with one more carbon in comparison with the raw olefin. In Co-catalysis (Scheme 7.5), H_2 and CO can be used as acyl source to achieve the addition onto olefins, which can be named as direct hydroformylation. Alternatively, other aldehydes can be used as acyl source in this type of reactions, which can be considered as a formal acyl transfer from one olefin to another. Therefore, it can be named as transfer hydroformylation.

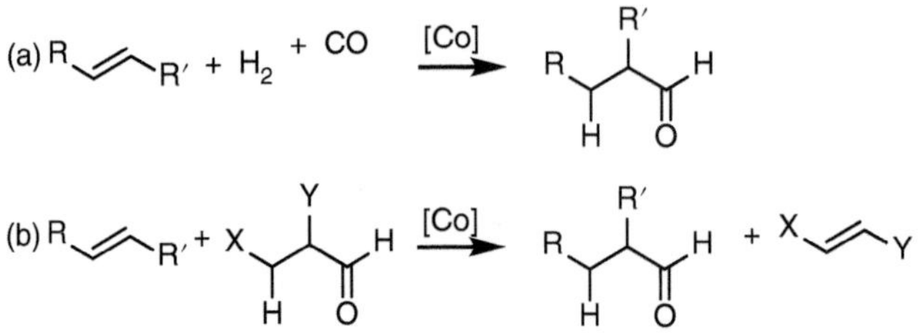

308 | *7 Theoretical Study of Co-Catalysis*

Scheme 7.5 Co-catalyzed direct (a) and transfer (b) hydroformylations.

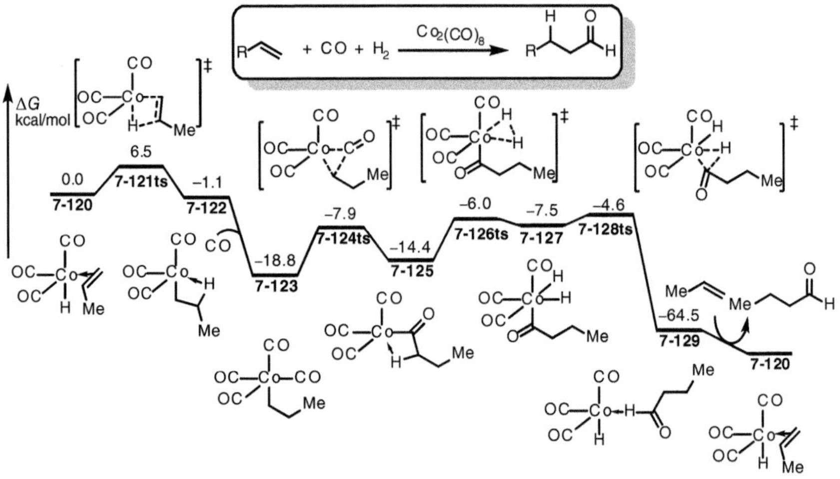

Figure 7.17 Free-energy profiles for a model reaction of Co-catalyzed hydroformylation of alkenes. The energies were calculated at B3LYP/6-311+G(d) level of theory.

7.4.1 Direct Hydroformylation by H_2 and CO

In the presence of hydride Co catalyst, a three-component reaction with olefins, H_2, and CO can provide aldehyde product, in which the acyl group comes from CO reactant. The mechanism of this type reaction is revealed by DFT calculation [139]. As shown in Figure 7.17, the olefin-coordinated hydride Co(I) complex **7-120** is set to relative zero in free-energy profiles. A migratory insertion of olefin into Co—H bond occurs via transition state **7-121ts** with an energy barrier of 6.5 kcal/mol to form an agostic Co(I) intermediate **7-122**. The coordination of extra CO molecule forms an alkyl Co(I) intermediate **7-123** with 18.8 kcal/mol endergonic. Then, a carbonyl insertion takes place via transition state **7-124ts** with an energy barrier of 10.9 kcal/mol to form an acyl Co(I) intermediate **7-125**. The oxidative addition by dihydrogen occurs via transition state **7-126ts** to afford a dihydride Co(III) intermediate **7-127**. Then, a rapid reductive elimination of acyl and hydride via transition state **7-128ts** yields aldehyde product and regenerates hydride Co(I) active species **7-120** through ligand exchange.

7.4.2 Transfer Hydroformylation

When an aldehyde is used as reactant, the dehydroformylation can formally afford a H_2 and a CO, which can be employed as acyl source in hydroformylation with another olefin to achieve a transfer hydroformylation. The full reaction can be considered as an aldehyde reacts with olefin to obtain another olefin and aldehyde. Although Rh- and Ir-catalyzed transfer hydroformylation has been reported experimentally, there are rare examples for the Co-catalyzed ones. Nevertheless, theoretical study is ahead of this reaction, which revealed the mechanism of this reaction with the calculation of whole catalytic cycle [140]. In computational study, Xantphos is chosen as a bidentate ligand. As shown in Figure 7.18, the reaction starts from a benzoate Co(I) species **7-130**. The oxidative addition of C(acyl)—H bond in propanal

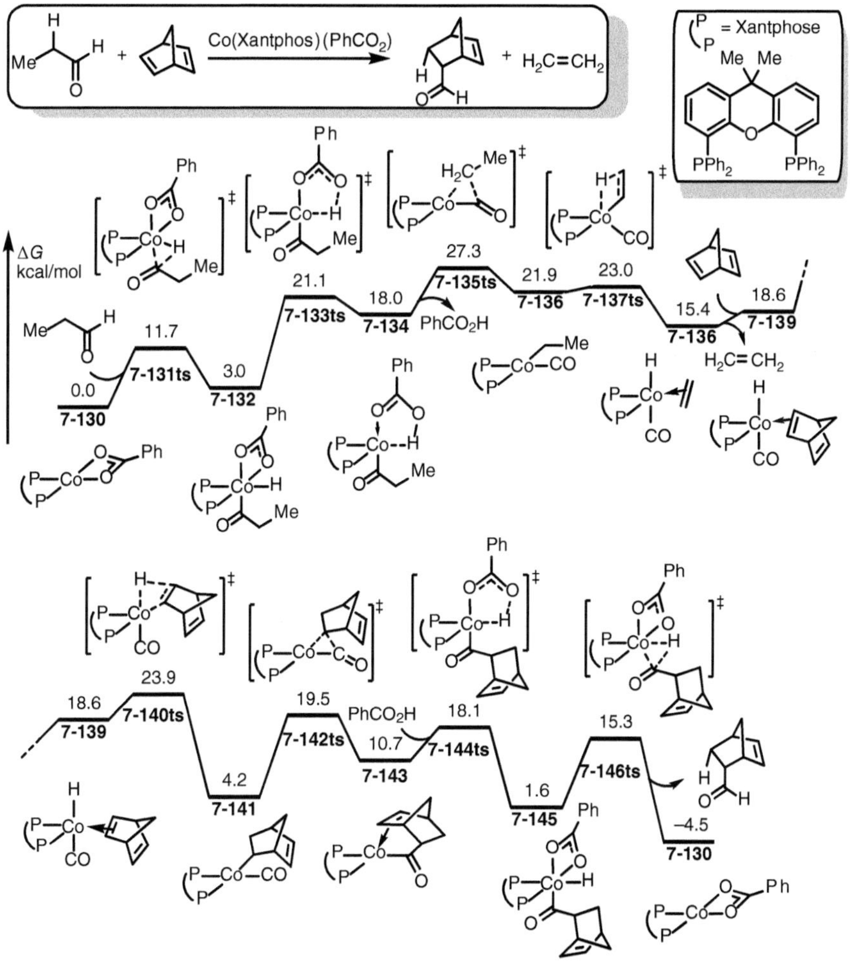

Figure 7.18 Free-energy profiles for a model reaction of Co-catalyzed transfer hydroformylation of alkenes. The energies were calculated at B3LYP/6-311+G(d) level of theory.

onto Co via transition state **7-131ts** forms an acyl hydride Co(III) intermediate **7-132**. Then, a reductive elimination gives an acyl Co(I) intermediate **7-134** by releasing benzoic acid. An α-alkyl elimination takes place via transition state **7-135ts** to form a carbonyl-coordinated alkyl Co(I) complex **7-136**. The calculated overall barrier of this step is 27.3 kcal/mol, which is considered to be the rate-determining step for the whole catalytic cycle. Subsequently, a β-hydride elimination generates ethylene product and forms a carbonyl-coordinated hydride Co(I) species **7-138**. After alkene exchange, the double bond in norbornadiene inserts into Co—H bond via transition state **7-140ts** to form another alkyl Co(I) intermediate **7-141**. The sequential carbonyl insertion into Co—C(alkyl) bond via transition state **7-142ts** forms an acyl Co(I) intermediate **7-143**. Finally, a two-step protonation with benzoic acid yields aldehyde product and regenerates Co(I) species **7-130**.

7.5 Co-Mediated Carbene Activation

Similar to other transition metals, Co can be used in carbene precursor functionalizations. In fact, recent theoretical studies have revealed some unique reaction modes and mechanisms of Co-catalyzed carbene functionalizations.

7.5.1 Arylation of Carbene

In organometallic catalysis (Scheme 7.6), diazo compounds are often used as carbene precursor, which can react with transition metal to form a metal–carbene complex with release of gaseous dinitrogen. The formed metal–carbene can undergo further transformations to achieve the functionalization of carbene. However, in Co(III)-catalyzed arylation of diazo compounds, a concerted substitutive denitrogenation pathway was found to be more favorable in comparison with the commonly accepted stepwise process [73, 141, 142].

The calculated free-energy profiles for the Cp*Co(III)-catalyzed annulation of phenyl pyridine and diazo compound [141] is given in Figure 7.19 involving two catalytic cycles, both of which start from a cationic acetate Cp*Co(III) species **7-147**. The first catalytic cycle initials form a C—H activation with phenyl pyridine via

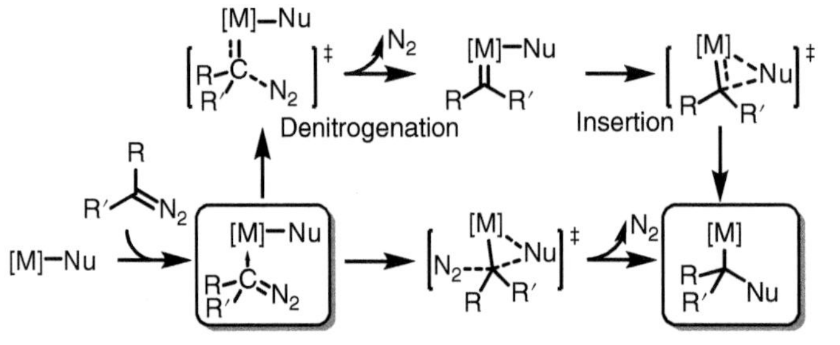

Scheme 7.6 Mechanism of transition metal-mediated diazo transformation.

7.5 Co-Mediated Carbene Activation

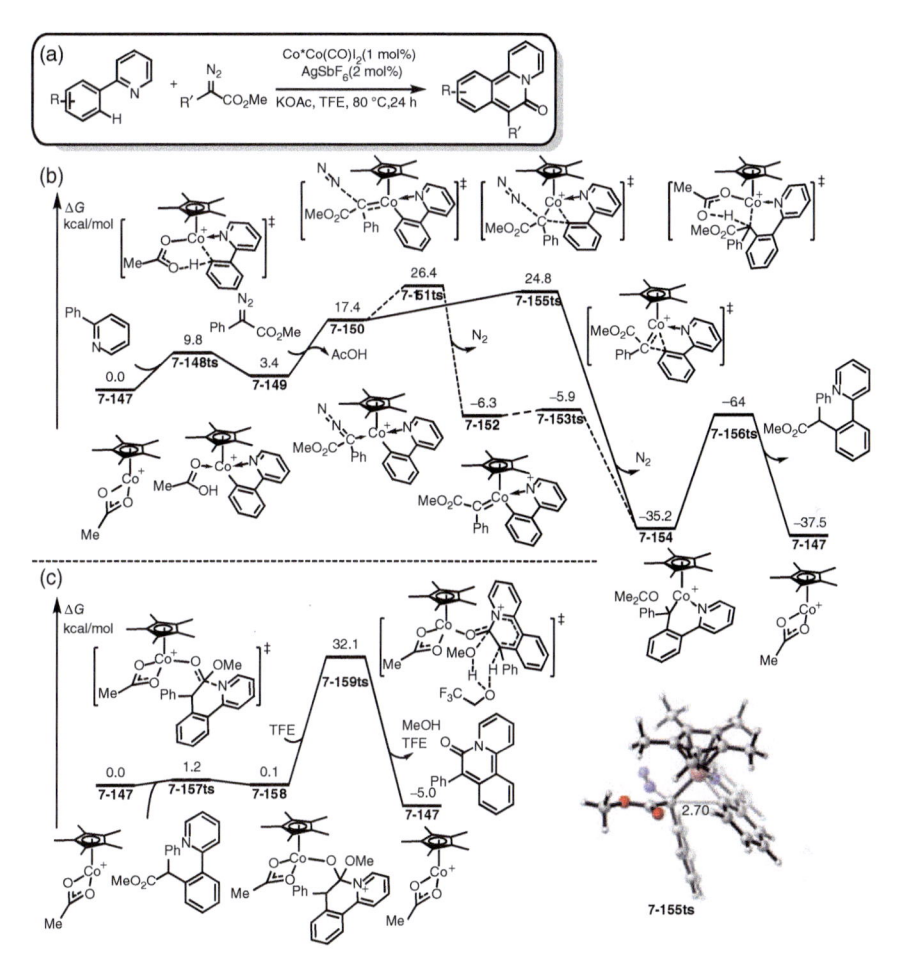

Figure 7.19 Free-energy profiles for the Cp*Co(III)-catalyzed annulation of phenyl pyridine and diazo compound. The energies were calculated at M06-L/def2-TZVPP/SMD(TFE)// M06-L/6-31G(d,p) level of theory.

a CMD-type transition state **7-148ts** to form an aryl Co(III) intermediate **7-149**. The coordination of diazo compound forms intermediate **7-150**. If the reaction occurs through a stepwise process, a denitrogenation can happen via transition state **7-151ts** to give a Co(III)–carbene complex **7-152**. Then, a rapid carbene insertion into Co—C(aryl) bond takes place via transition state **7-153ts** to form an alkyl Co(III) intermediate **7-154**. Alternatively, when diazo compound has been coordinated onto Co in complex **7-150**, a nucleophilic substitution by aryl group on Co can occur via transition state **7-155ts**, which leads to the leaving of dinitrogen. The geometry information of transition state **7-155ts** is given in Figure 7.19. The length of forming C(aryl)—C(diazo) bond is 2.70 Å, which shows a weak interaction between those two atoms. The calculated relative free energy of transition state **7-155ts** is 24.8 kcal/mol, which is 1.6 kcal/mol lower than that of **7-151ts**. Although such an energy penalty cannot exclude the stepwise pathway, it at least indicates that the concerted mechanism is also possible. When intermediate **7-154** is formed, the protonation of that species by acetic acid can afford an alkylation intermediate product and regenerate acetate Cp*Co(III) species **7-147**.

In the second catalytic cycle, with the activation of cationic Co **7-147**, an intramolecular nucleophilic attack with pyridyl onto ester group via transition state **7-157ts** reversibly forms intermediate **7-158**. Subsequently, a solvent-assisted 1,2-elimination of methanol takes place via transition state **7-159ts** with an energy barrier of 32.0 kcal/mol and yields final annulation product with the regeneration of active catalyst **7-147**.

7.5.2 Carboxylation of Carbene

When diazo compounds react with equivalent $Co_2(CO)_8$, a good isolated yield (μ^2-η^1)-coordinated Co–carbene complex **7-160** can be isolated experimentally [143]. In the further carbonylation of this complex, a ketene product can be yielded. The mechanism of carboxylation is shown in Figure 7.20. In a (μ^2-η^1)-coordinated Co–carbene complex **7-160**, the bond lengths of two Co—C(carbene) bonds are 1.99 Å, which can be considered as two Co—C single bonds. Therefore, a carbonyl insertion into that bond can occur via transition state **7-161ts** to form intermediate **7-162** with an energy barrier of only 8.9 kcal/mol. The geometry of complex **7-162**, given in Figure 7.20, is an interesting example between the (μ^1-η^2) and (μ^2-η^2) coordination modes as the CH moiety is bound stronger to Co_1; however, the strength of the interaction to Co_2 is not negligible, suggested by the 2.207 Å C—Co bond length. Then, the coordination of a CO molecule forms complex **7-164** with

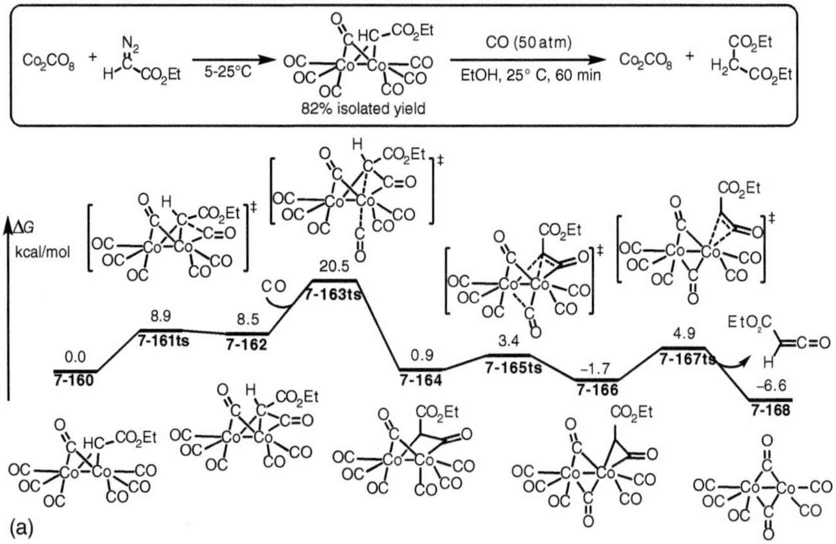

Figure 7.20 Free-energy profiles for the key step of Co-mediated carboxylation of diazo compounds. The energies were calculated at BP86/6-31G(d,p) (SDD for Co) level of theory.

(b) 4TS   5

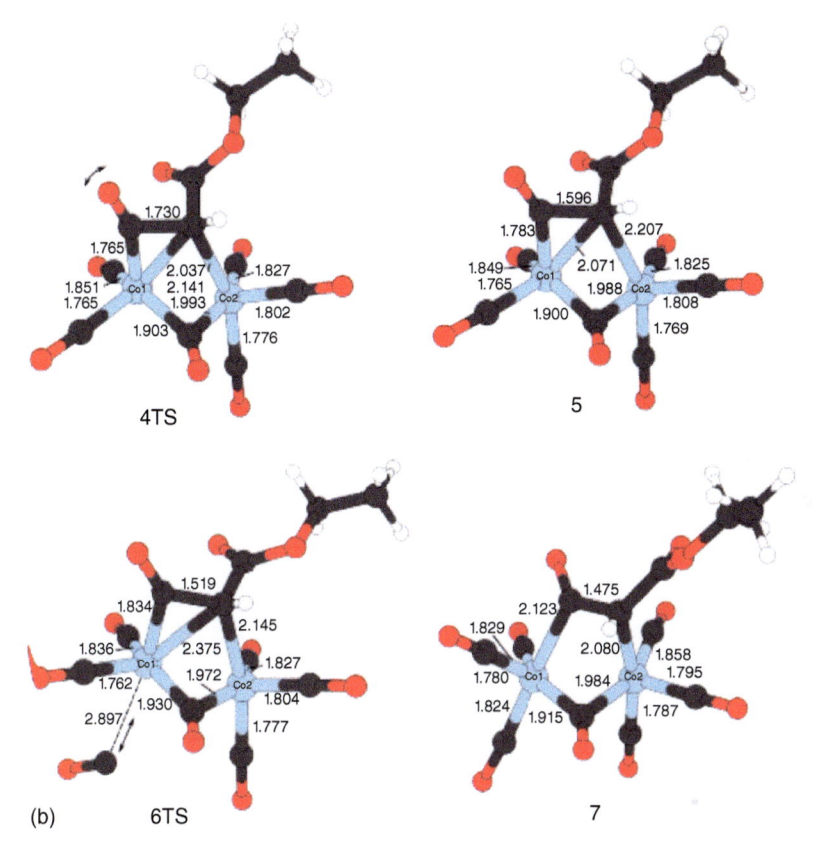

6TS   7

Figure 7.20 (Continued)

an energy barrier of 12.0 kcal/mol via transition state **7-163ts**, which is considered as the rate-determining step. Unlike **7-162**, complex **7-163ts** can be characterized as a true (μ^2-η^2) ketene complex. After further ligand exchange, ketene product can be released from metal with the formation of $Co_2(CO)_7$ complex **7-168** with an overall exergonic of 6.6 kcal/mol.

7.6 Co-Mediated Nitrene Activation

Organic azides can react with Co(II) species by releasing dinitrogen to afford a Co–nitrene complex. Interestingly, the redox non-innocence of the nitrene moiety further adds to the complexity of the mechanism of this chemistry [144]. As shown in Scheme 7.7, when azides react with Co(II), the formed Co–nitrene complex can be considered as a classical metal–nitrene complex with an unpaired electron on Co(II) (left in Scheme 7.7a). Alternatively, an internal SET can afford a Co(III)—N single bond in a resonance structure (right in Scheme 7.7a), where

314 | *7 Theoretical Study of Co-Catalysis*

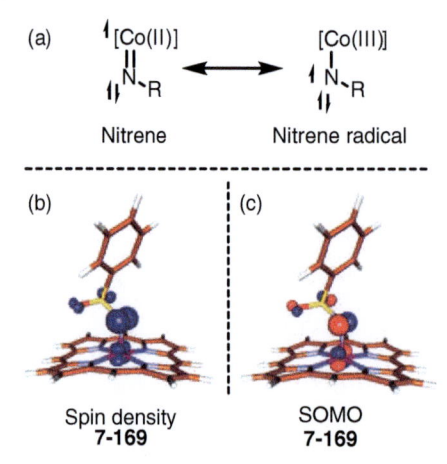

Scheme 7.7 The resonance structures of Co–nitrene complex (a); Calculated spin density map (b) and SOMO (c) for Co–nitrene complex **7-169**.

the unpaired electron locates on nitrogen atom. As an example, DFT calculations for the molecular orbitals of a Co–nitrene complex **7-169** found that both the spin density and SOMO for this complex are located on nitrogen atom. Therefore, the carbene presents more radical character. Following this idea, Co–nitrene complex can react with olefins, isonitriles, or C—H bonds through radical processes to achieve corresponding aminations.

7.6.1 Aziridination of Olefins

When a porphyrin-coordinated Co(II) is used as catalyst, the aziridination of olefins by azides can undergo a Co(III)–nitrene radical intermediate. DFT calculation is used to reveal this mechanism [145]. As shown in Figure 7.21, the coordination of azide onto porphyrin-coordinated Co(II) species **7-170** forms a Co(II) intermediate **7-171** with 7.4 kcal/mol endergonic. The N—N bond cleavage occurs via transition state **7-172ts** with an activation free energy of 22.7 kcal/mol to afford a Co(III)–nitrene radical intermediate **7-173** by 21.8 kcal/mol exergonic. An intermolecular radical addition with nitrogen radical onto styrene substrate takes place via transition state **7-174ts** and forms a benzylic radical **7-175**. Subsequently, another C—N bond can be formed by radical addition via transition state **7-176ts** to yield aziridine product and regenerate Co(II) species **7-170**.

7.6.2 Amination of Isonitriles

The Co—N bond in Co(III)–nitrene radical plays as a single bond; therefore, isonitrile can insert into this bond for further transformations [146]. As shown in Figure 7.22, when an oxalate Co(II) is used as catalyst, a three-component coupling with amine, isonitriles, and azides can be used to synthesize guanidine

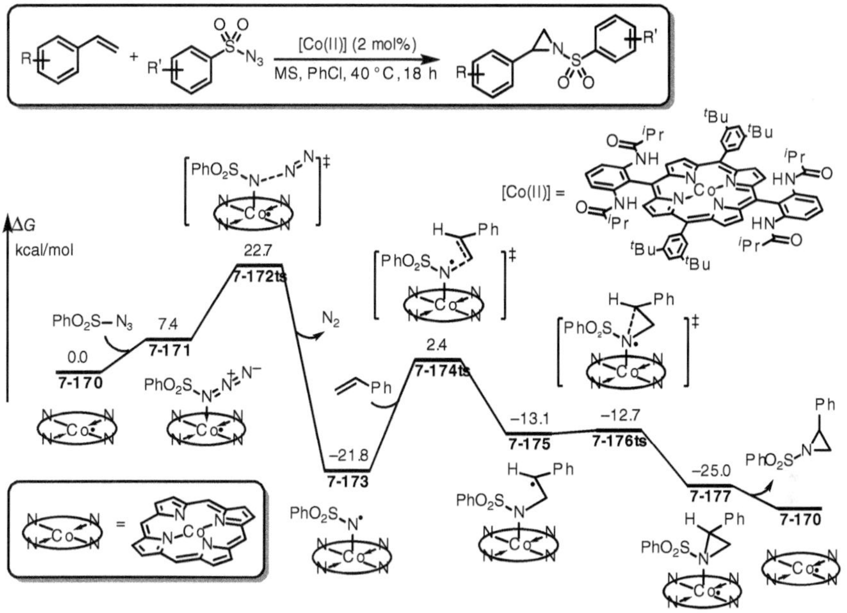

Figure 7.21 Free-energy profiles for the Co-catalyzed aziridination of olefins via a Co–nitrene intermediate. The energies were calculated at BP86/TZVP level of theory.

derivatives. The DFT calculated free-energy profiles for the catalytic cycle is given in the same figure. The coordination of azide forms complex **7-179** with 19.3 kcal/mol endergonic. The N—N bond cleavage takes place via transition state **7-180ts** with a further energy barrier of 19.9 kcal/mol to form a Co(III)–nitrene radical **7-181** with 3.1 kcal/mol overall exergonic. Then, a coordinated-isonitrile insertion into Co—N(nitrene) bond occurs via transition state **7-182ts** to afford a methanediimine-coordinated Co(II) complex **7-183**. Finally, an intermolecular nucleophilic attack of aniline via transition state **7-184ts** yields guanidine product and regenerates Co(II) species **7-178**.

7.6.3 Amination of C—H Bonds

In the absence of any other active species, the Co(III)–nitrene radical can even react with C—H bond to achieve amination [97]. As shown in Figure 7.23, the reaction starts from the coordination of azide onto Co(II) species **7-186** to form complex **7-187**. Then, the N—N bond cleavage takes place via transition state **7-188ts** with an enthalpy barrier of 14.1 kcal/mol to form a Co(III)–nitrene radical **7-189**. An intermolecular radical substitution with the benzylic-H of ethylbenzene occurs via transition state **7-190ts** with an enthalpy barrier of only 6.3 kcal/mol. The release of benzylic radical forms an amino Co(III) intermediate **7-191**. Then, another radical substitution by benzylic radical takes place without enthalpy barrier forming new C—N bond and regenerating Co(II) species **7-186**.

316 | *7 Theoretical Study of Co-Catalysis*

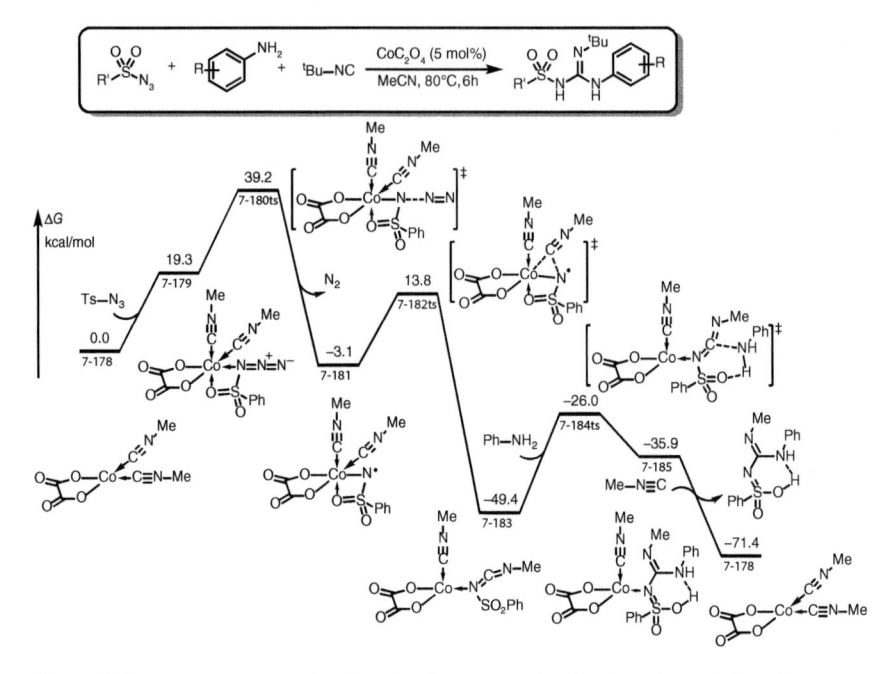

Figure 7.22 Free-energy profiles for the Co-catalyzed amination of isonitriles via a Co–nitrene intermediate. The energies were calculated at B3LYP/6-311++G(d,p) (LANL2TZ(f) for Co)/SMD(acetonitrile)//B3LYP/6-31G(d) (LANL2DZ for Co) level of theory.

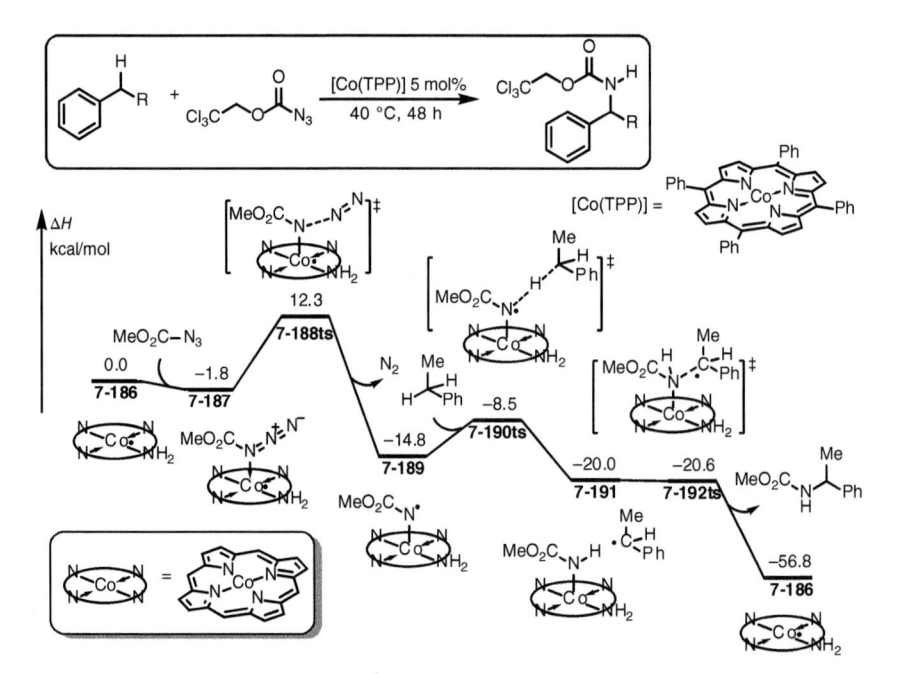

Figure 7.23 Relative enthalpy profiles for the Co-mediated nitrene insertion into C–H bond. The energies were calculated at BP86/def-TZVP//BP86/SV(P) level of theory.

References

1. Gao, K. and Yoshikai, N. (2014). Low-valent cobalt catalysis: new opportunities for C–H functionalization. *Accounts of Chemical Research* 47 (4): 1208–1219.
2. Su, B., Cao, Z.C., and Shi, Z.J. (2015). Exploration of earth-abundant transition metals (Fe, Co, and Ni) as catalysts in unreactive chemical bond activations. *Accounts of Chemical Research* 48 (3): 886–896.
3. Yu, D.G., Wang, X., Zhu, R.Y. et al. (2012). Direct arylation/alkylation/magnesiation of benzyl alcohols in the presence of Grignard reagents via Ni-, Fe-, or Co-catalyzed sp^3 C—O bond activation. *Journal of the American Chemical Society* 134 (36): 14638–14641.
4. Obligacion, J.V. and Chirik, P.J. (2018). Earth-abundant transition metal catalysts for alkene hydrosilylation and hydroboration: opportunities and assessments. *Nature Reviews Chemistry* 2 (5): 15–34.
5. Du, P. and Eisenberg, R. (2012). Catalysts made of earth-abundant elements (Co, Ni, Fe) for water splitting: recent progress and future challenges. *Energy and Environmental Science* 5 (3): 6012–6021.
6. Beale, A.M., Sankar, G., Catlow, C.R. et al. (2005). Towards an understanding of the oxidation state of cobalt and manganese ions in framework substituted microporous aluminophosphate redox catalysts: an electron paramagnetic resonance and X-ray absorption spectroscopy investigation. *Physical Chemistry Chemical Physics* 7 (8): 1856–1860.
7. Brik, Y., Kacimi, M., Ziyad, M. et al. (2001). Titania-supported cobalt and cobalt–phosphorus catalysts: characterization and performances in ethane oxidative dehydrogenation. *Journal of Catalysis* 202 (1): 118–128.
8. Dillard, J.G., Crowther, D.L., and Murray, J.W. (1982). The oxidation states of cobalt and selected metals in Pacific ferromanganese nodules. *Geochimica et Cosmochimica Acta* 46 (5): 755–759.
9. Kadish, K.M., Shen, J., Fremond, L. et al. (2008). Clarification of the oxidation state of cobalt corroles in heterogeneous and homogeneous catalytic reduction of dioxygen. *Inorganic Chemistry* 47 (15): 6726–6737.
10. Masset, A.C., Michel, C., Maignan, A. et al. (2000). Misfit-layered cobaltite with an anisotropic giant magnetoresistance: $Ca_3Co_4O_9$. *Physical Review B* 62 (1): 166–175.
11. Guo, X.K., Zhang, L.B., Wei, D. et al. (2015). Mechanistic insights into cobalt(II/III)-catalyzed C–H oxidation: a combined theoretical and experimental study. *Chemical Science* 6 (12): 7059–7071.
12. Kenion, R.L. and Ananth, N. (2016). Direct simulation of electron transfer in the cobalt hexammine(II/III) self-exchange reaction. *Physical Chemistry Chemical Physics* 18 (37): 26117–26124.
13. Mei, R., Fang, X., He, L. et al. (2020). Cobaltaelectro-catalyzed oxidative allene annulation by electro-removable hydrazides. *Chemical Communications* 56 (9): 1393–1396.

14 Wei, D., Zhu, X., Niu, J.-L. et al. (2016). High-valent-cobalt-catalyzed C–H functionalization based on concerted metalation–deprotonation and single-electron-transfer mechanisms. *ChemCatChem* 8 (7): 1242–1263.

15 Belanger, D.B. and Livinghouse, T. (1998). Hexacarbonyldicobalt–alkyne complexes as convenient $Co_2(CO)_8$ surrogates in the catalytic Pauson–Khand reaction. *Tetrahedron Letters* 39 (42): 7641–7644.

16 Belanger, D.B., O'Mahony, D.J.R., and Livinghouse, T. (1998). Thermal promotion of the cobalt catalyzed intramolecular Pauson–Khand reaction – an alternative experimental protocol for cyclopentenone synthesis. *Tetrahedron Letters* 39 (42): 7637–7640.

17 Geis, O. and Schmalz, H.-G. (1998). New developments in the Pauson–Khand reaction. *Angewandte Chemie International Edition* 37 (7): 911–914.

18 Ackermann, L. (2020). Metalla-electrocatalyzed C–H activation by earth-abundant 3d metals and beyond. *Accounts of Chemical Research* 53 (1): 84–104.

19 Cahiez, G. and Moyeux, A. (2010). Cobalt-catalyzed cross-coupling reactions. *Chemical Reviews* 110 (3): 1435–1462.

20 Dethe, D.H., C, B.N., and Bhat, A.A. (2020). Cp*Co(III)-catalyzed ketone-directed ortho-C–H activation for the synthesis of indene derivatives. *Journal of Organic Chemistry* 85 (11): 7565–7575.

21 Moselage, M., Li, J., and Ackermann, L. (2015). Cobalt-catalyzed C–H activation. *ACS Catalysis* 6 (2): 498–525.

22 Hyster, T.K. (2014). High-valent Co(III)- and Ni(II)-catalyzed C–H activation. *Catalysis Letters* 145 (1): 458–467.

23 Khan, B., Dwivedi, V., and Sundararaju, B. (2020). Cp*Co(III)-catalyzed o-amidation of benzaldehydes with dioxazolones using transient directing group strategy. *Advanced Synthesis and Catalysis* 362 (5): 1195–1200.

24 Park, J. and Chang, S. (2015). Comparative catalytic activity of group 9 [Cp*M(III)] complexes: cobalt-catalyzed C–H amidation of arenes with dioxazolones as amidating reagents. *Angewandte Chemie International Edition* 54 (47): 14103–14107.

25 Scamp, R.J., deRamon, E., Paulson, E.K. et al. (2020). Cobalt(III)-catalyzed C–H amidation of dehydroalanine for the site-selective structural diversification of thiostrepton. *Angewandte Chemie International Edition* 59 (2): 890–895.

26 Wang, F., Jin, L., Kong, L. et al. (2017). Cobalt(III)- and rhodium(III)-catalyzed C–H amidation and synthesis of 4-quinolones: C–H activation assisted by weakly coordinating and functionalizable enaminone. *Organic Letters* 19 (7): 1812–1815.

27 Lenges, C.P. and Brookhart, M. (1997). Co(I)-catalyzed inter- and intramolecular hydroacylation of olefins with aromatic aldehydes. *Journal of the American Chemical Society* 119 (13): 3165–3166.

28 Lenges, C.P., Brookhart, M., and Grant, B.E. (1997). H/D exchange reactions between C_6D_6 and $C_5Me_5Co(CH_2=CHR)_2$ (R = H, $SiMe_3$): evidence for oxidative addition of bonds to the $[C_5Me_5(L)Co]$ moiety. *Journal of Organometallic Chemistry* 528 (1–2): 199–203.

29 Lenges, C.P., White, P.S., and Brookhart, M. (1998). Mechanistic and synthetic studies of the addition of alkyl aldehydes to vinylsilanes catalyzed by Co(I) complexes. *Journal of the American Chemical Society* 120 (28): 6965–6979.

30 Bolig, A.D. and Brookhart, M. (2007). Activation of sp^3 C—H bonds with cobalt(I): catalytic synthesis of enamines. *Journal of the American Chemical Society* 129 (47): 14544–14545.

31 Chen, Q., Ilies, L., and Nakamura, E. (2011). Cobalt-catalyzed ortho-alkylation of secondary benzamide with alkyl chloride through directed C—H bond activation. *Journal of the American Chemical Society* 133 (3): 428–429.

32 Chen, Q., Ilies, L., Yoshikai, N. et al. (2011). Cobalt-catalyzed coupling of alkyl Grignard reagent with benzamide and 2-phenylpyridine derivatives through directed C—H bond activation under air. *Organic Letters* 13 (12): 3232–3234.

33 Ilies, L., Chen, Q., Zeng, X. et al. (2011). Cobalt-catalyzed chemoselective insertion of alkene into the ortho C—H bond of benzamide. *Journal of the American Chemical Society* 133 (14): 5221–5223.

34 Ding, Z. and Yoshikai, N. (2013). Cobalt-catalyzed intramolecular olefin hydroarylation leading to dihydropyrroloindoles and tetrahydropyridoindoles. *Angewandte Chemie International Edition* 52 (33): 8574–8578.

35 Gao, K., Lee, P.S., Long, C. et al. (2012). Cobalt-catalyzed ortho-arylation of aromatic imines with aryl chlorides. *Organic Letters* 14 (16): 4234–4237.

36 Gao, K., Paira, R., and Yoshikai, N. (2014). Cobalt-catalyzed ortho-C–H alkylation of 2-arylpyridines via ring-opening of aziridines. *Advanced Synthesis and Catalysis* 356 (7): 1486–1490.

37 Gao, K. and Yoshikai, N. (2013). Cobalt-catalyzed ortho alkylation of aromatic imines with primary and secondary alkyl halides. *Journal of the American Chemical Society* 135 (25): 9279–9282.

38 Lee, P.S., Fujita, T., and Yoshikai, N. (2011). Cobalt-catalyzed, room-temperature addition of aromatic imines to alkynes via directed C—H bond activation. *Journal of the American Chemical Society* 133 (43): 17283–17295.

39 Lee, P.S. and Yoshikai, N. (2015). Cobalt-catalyzed enantioselective directed C–H alkylation of indole with styrenes. *Organic Letters* 17 (1): 22–25.

40 Tan, B.-H., Dong, J., and Yoshikai, N. (2012). Cobalt-catalyzed addition of arylzinc reagents to alkynes to form *ortho*-alkenylarylzinc species through 1,4-cobalt migration. *Angewandte Chemie International Edition* 124 (38): 9748–9752.

41 Yamakawa, T. and Yoshikai, N. (2013). Annulation of α,β-unsaturated imines and alkynes via cobalt-catalyzed olefinic C–H activation. *Organic Letters* 15 (1): 196–199.

42 Yang, J., Seto, Y.W., and Yoshikai, N. (2015). Cobalt-catalyzed intermolecular hydroacylation of olefins through chelation-assisted imidoyl C–H activation. *ACS Catalysis* 5 (5): 3054–3057.

43 Yang, J. and Yoshikai, N. (2014). Cobalt-catalyzed enantioselective intramolecular hydroacylation of ketones and olefins. *Journal of the American Chemical Society* 136 (48): 16748–16751.

44 Sauermann, N., Meyer, T.H., Tian, C. et al. (2017). Electrochemical cobalt-catalyzed C–H oxygenation at room temperature. *Journal of the American Chemical Society* 139 (51): 18452–18455.

45 Mei, R. and Ackermann, L. (2016). Cobalt-catalyzed C–H functionalizations by imidate assistance with aryl and alkyl chlorides. *Advanced Synthesis and Catalysis* 358 (15): 2443–2448.

46 Ackermann, L. (2014). Cobalt-catalyzed C–H arylations, benzylations, and alkylations with organic electrophiles and beyond. *Journal of Organic Chemistry* 79 (19): 8948–8954.

47 Punji, B., Song, W., Shevchenko, G.A. et al. (2013). Cobalt-catalyzed C—H bond functionalizations with aryl and alkyl chlorides. *Chemistry* 19 (32): 10605–10610.

48 Mei, R., Wang, H., Warratz, S. et al. (2016). Cobalt-catalyzed oxidase C–H/N–H alkyne annulation: mechanistic insights and access to anticancer agents. *Chemistry* 22 (20): 6759–6763.

49 Li, J. and Ackermann, L. (2015). Cobalt-catalyzed C–H arylations with weakly-coordinating amides and tetrazoles: expedient route to angiotensin-II-receptor blockers. *Chemistry* 21 (15): 5718–5722.

50 Song, W. and Ackermann, L. (2012). Cobalt-catalyzed direct arylation and benzylation by C–H/C–O cleavage with sulfamates, carbamates, and phosphates. *Angewandte Chemie International Edition* 124 (33): 8376–8379.

51 Li, J. and Ackermann, L. (2015). Cobalt-catalyzed C–H cyanation of arenes and heteroarenes. *Angewandte Chemie International Edition* 127 (12): 3706–3709.

52 Mei, R., Sauermann, N., Oliveira, J.C.A. et al. (2018). Electroremovable traceless hydrazides for cobalt-catalyzed electro-oxidative C–H/N–H activation with internal alkynes. *Journal of the American Chemical Society* 140 (25): 7913–7921.

53 Li, J. and Ackermann, L. (2015). Cobalt(III)-catalyzed aryl and alkenyl C–H aminocarbonylation with isocyanates and acyl azides. *Angewandte Chemie International Edition* 54 (29): 8551–8554.

54 Pawar, A.B. and Chang, S. (2015). Cobalt-catalyzed C–H cyanation of (hetero)arenes and 6-arylpurines with N-cyanosuccinimide as a new cyanating agent. *Organic Letters* 17 (3): 660–663.

55 Zhang, L.B., Hao, X.Q., Zhang, S.K. et al. (2015). Cobalt-catalyzed $C(sp^2)$–H alkoxylation of aromatic and olefinic carboxamides. *Angewandte Chemie International Edition* 54 (1): 272–275.

56 Obligacion, J.V., Semproni, S.P., and Chirik, P.J. (2014). Cobalt-catalyzed C–H borylation. *Journal of the American Chemical Society* 136 (11): 4133–4136.

57 Gandeepan, P., Rajamalli, P., and Cheng, C.-H. (2016). Diastereoselective [3+2] annulation of aromatic/vinylic amides with bicyclic alkenes through cobalt-catalyzed C–H activation and intramolecular nucleophilic addition. *Angewandte Chemie International Edition* 128 (13): 4380–4383.

58 Grigorjeva, L. and Daugulis, O. (2014). Cobalt-catalyzed, aminoquinoline-directed $C(sp^2)$—H bond alkenylation by alkynes. *Angewandte Chemie International Edition* 126 (38): 10373–10376.

References

59 Kalsi, D. and Sundararaju, B. (2015). Cobalt catalyzed C—H and N—H bond annulation of sulfonamide with terminal and internal alkynes. *Organic Letters* 17 (24): 6118–6121.

60 Yoshino, T., Ikemoto, H., Matsunaga, S. et al. (2013). A cationic high-valent Cp*Co(III) complex for the catalytic generation of nucleophilic organometallic species: directed C—H bond activation. *Angewandte Chemie International Edition* 52 (8): 2207–2211.

61 Yoshino, T., Ikemoto, H., Matsunaga, S. et al. (2013). Cp*Co(III)-catalyzed C_2-selective addition of indoles to imines. *Chemistry* 19 (28): 9142–9146.

62 Andou, T., Saga, Y., Komai, H. et al. (2013). Cobalt-catalyzed C_4-selective direct alkylation of pyridines. *Angewandte Chemie International Edition* 125 (11): 3295–3298.

63 Suzuki, Y., Sun, B., Sakata, K. et al. (2015). Dehydrative direct C–H allylation with allylic alcohols under [Cp*Co^{III}] catalysis. *Angewandte Chemie International Edition* 127 (34): 10082–10085.

64 Suzuki, Y., Sun, B., Yoshino, T. et al. (2015). Cp*Co(III)-catalyzed oxidative C–H alkenylation of benzamides with ethyl acrylate. *Tetrahedron* 71 (26–27): 4552–4556.

65 Yamamoto, S., Saga, Y., Andou, T. et al. (2014). Cobalt-catalyzed C-4 selective alkylation of quinolines. *Advanced Synthesis and Catalysis* 356 (2–3): 401–405.

66 Yoshino, T., Ikemoto, H., Matsunaga, S. et al. (2013). A cationic high-valent Cp*Co^{III} complex for the catalytic generation of nucleophilic organometallic species: directed C—H bond activation. *Angewandte Chemie International Edition* 125 (8): 2263–2267.

67 Tanaka, R., Ikemoto, H., Kanai, M. et al. (2016). Site- and regioselective monoalkenylation of pyrroles with alkynes via Cp*Co(III) catalysis. *Organic Letters* 18 (21): 5732–5735.

68 Yoshino, T. and Matsunaga, S. (2018). Cobalt-catalyzed C(sp^3)–H functionalization reactions. *Asian Journal of Organic Chemistry* 7 (7): 1193–1205.

69 Yoshino, T., Satake, S., and Matsunaga, S. (2020). Diverse approaches for enantioselective C–H functionalization reactions using group 9 Cp^x M^{III} catalysts. *Chemistry* 26 (33): 7346–7357.

70 Kim, J.H., Gressies, S., and Glorius, F. (2016). Cooperative Lewis acid/Cp*Co(III) catalyzed C—H bond activation for the synthesis of isoquinolin-3-ones. *Angewandte Chemie International Edition* 55 (18): 5577–5581.

71 Gensch, T., Vasquez-Cespedes, S., Yu, D.G. et al. (2015). Cobalt(III)-catalyzed directed C–H allylation. *Organic Letters* 17 (15): 3714–3717.

72 Yu, D.G., Gensch, T., de Azambuja, F. et al. (2014). Co(III)-catalyzed C–H activation/formal SN-type reactions: selective and efficient cyanation, halogenation, and allylation. *Journal of the American Chemical Society* 136 (51): 17722–17725.

73 Zhao, D., Kim, J.H., Stegemann, L. et al. (2015). Cobalt(III)-catalyzed directed C–H coupling with diazo compounds: straightforward access towards extended π-systems. *Angewandte Chemie International Edition* 54 (15): 4508–4511.

74 Gensch, T., Klauck, F.J., and Glorius, F. (2016). Cobalt-catalyzed C–H thiolation through dehydrogenative cross-coupling. *Angewandte Chemie International Edition* 55 (37): 11287–11291.

75 Shen, Z., Li, C., Mercado, B.Q. et al. (2019). Cobalt(III)-catalyzed diastereoselective three-component C—H bond addition to butadiene and activated ketones. *Synthesis* 52 (08): 1239–1246.

76 Boerth, J.A., Maity, S., Williams, S.K. et al. (2018). Selective and synergistic cobalt(III)-catalyzed three-component C—H bond addition to dienes and aldehydes. *Nature Catalysis* 1: 673–679.

77 Hummel, J.R. and Ellman, J.A. (2015). Cobalt(III)-catalyzed synthesis of indazoles and furans by C—H bond functionalization/addition/cyclization cascades. *Journal of the American Chemical Society* 137 (1): 490–498.

78 Boerth, J.A. and Ellman, J.A. (2017). A convergent synthesis of functionalized alkenyl halides through cobalt(III)-catalyzed three-component C—H bond addition. *Angewandte Chemie International Edition* 129 (33): 10108–10112.

79 Hummel, J.R. and Ellman, J.A. (2015). Cobalt(III)-catalyzed C—H bond amidation with isocyanates. *Organic Letters* 17 (10): 2400–2403.

80 Boerth, J.A., Hummel, J.R., and Ellman, J.A. (2016). Highly stereoselective cobalt(III)-catalyzed three-component C—H bond addition cascade. *Angewandte Chemie International Edition* 55 (41): 12650–12654.

81 Patel, P. and Chang, S. (2015). Cobalt(III)-catalyzed C–H amidation of arenes using acetoxycarbamates as convenient amino sources under mild conditions. *ACS Catalysis* 5 (2): 853–858.

82 Jia, Q., Kong, L., and Li, X. (2019). Cobalt(III)-catalyzed C–H amidation of weakly coordinating sulfoxonium ylides and α-benzoylketene dithioacetals. *Organic Chemistry Frontiers* 6 (6): 741–745.

83 Kong, L., Yu, S., Tang, G. et al. (2016). Cobalt(III)-catalyzed C–C coupling of arenes with 7-oxabenzonorbornadiene and 2-vinyloxirane via C–H activation. *Organic Letters* 18 (15): 3802–3805.

84 Li, L., Wang, H., Yu, S. et al. (2016). Cooperative Co(III)/Cu(II)-catalyzed C–N/N–N coupling of imidates with anthranils: access to 1H-indazoles via C–H activation. *Organic Letters* 18 (15): 3662–3665.

85 Wang, F., Wang, Q., Bao, M. et al. (2016). Cobalt-catalyzed redox-neutral synthesis of isoquinolines: C–H activation assisted by an oxidizing N—S bond. *Chinese Journal of Catalysis* 37 (8): 1423–1430.

86 Zhou, X., Luo, Y., Kong, L. et al. (2017). Cp*Co^{III}-catalyzed branch-selective hydroarylation of alkynes via C–H activation: efficient access to α-gem-vinylindoles. *ACS Catalysis* 7 (10): 7296–7304.

87 Wang, S., Hou, J.T., Feng, M.L. et al. (2016). Cobalt(III)-catalyzed alkenylation of arenes and 6-arylpurines with terminal alkynes: efficient access to functional dyes. *Chemical Communications* 52 (13): 2709–2712.

88 Ye, Y.-H., Zhang, J., Wang, G. et al. (2011). Cobalt-catalyzed benzylic C–H amination via dehydrogenative-coupling reaction. *Tetrahedron* 67 (25): 4649–4654.

89 Wang, S., Chen, S.Y., and Yu, X.Q. (2017). C–H functionalization by high-valent Cp*Co(III) catalysis. *Chemical Communications* 53 (22): 3165–3180.

90 Bai, L., Wyrwalski, F., Lamonier, J.-F. et al. (2013). Effects of β-cyclodextrin introduction to zirconia supported-cobalt oxide catalysts: from molecule-ion associations to complete oxidation of formaldehyde. *Applied Catalysis B: Environmental* 138-139: 381–390.

91 Prakash, S., Muralirajan, K., and Cheng, C.-H. (2016). Cobalt-catalyzed oxidative annulation of nitrogen-containing arenes with alkynes: an atom-economical route to heterocyclic quaternary ammonium salts. *Angewandte Chemie International Edition* 128 (5): 1876–1880.

92 Mukaiyama, T., Isayama, S., Inoki, S. et al. (1989). Oxidation–reduction hydration of olefins with molecular oxygen and 2-propanol catalyzed by bis(acetylacetonato)cobalt(II). *Chemistry Letters* 18 (3): 449–452.

93 Harrison, P.G., Ball, I.K., Daniell, W. et al. (2003). Cobalt catalysts for the oxidation of diesel soot particulate. *Chemical Engineering Journal* 95 (1–3): 47–55.

94 Wang, F.-S., Yang, T.-Y., Hsu, C.-C. et al. (2016). The mechanism and thermodynamic studies of CMRP: different control mechanisms demonstrated by Co^{II}(TMP), Co^{II}(salen*), and Co^{II}(acac)$_2$ mediated polymerization, and the correlation of reduction potential, equilibrium constant, and control mechanism. *Macromolecular Chemistry and Physics* 217 (3): 422–432.

95 Sakata, K., Eda, M., Kitaoka, Y. et al. (2017). Cp*Co(III)-catalyzed C–H alkenylation/annulation reactions of indoles with alkynes: a DFT study. *Journal of Organic Chemistry* 82 (14): 7379–7387.

96 Wang, Q., Huang, F., Jiang, L. et al. (2018). Comprehensive mechanistic insight into cooperative Lewis acid/Cp*Co(III)-catalyzed C–H/N–H activation for the synthesis of isoquinolin-3-ones. *Inorganic Chemistry* 57 (5): 2804–2814.

97 Lyaskovskyy, V., Suarez, A.I., Lu, H. et al. (2011). Mechanism of cobalt(II) porphyrin-catalyzed C–H amination with organic azides: radical nature and H-atom abstraction ability of the key cobalt(III)–nitrene intermediates. *Journal of the American Chemical Society* 133 (31): 12264–12273.

98 Wang, Y., Du, C., Wang, Y. et al. (2018). High-valent cobalt-catalyzed C–H activation/annulation of 2-benzamidopyridine 1-oxide with terminal alkyne: a combined theoretical and experimental study. *Advanced Synthesis and Catalysis* 360 (14): 2668–2677.

99 Ghorai, J. and Anbarasan, P. (2019). Developments in Cp*Co^{III}-catalyzed C—H bond functionalizations. *Asian Journal of Organic Chemistry* 8 (4): 430–455.

100 Martinez, A.M., Rodriguez, N., Gomez-Arrayas, R. et al. (2017). Cobalt-catalyzed ortho-C–H functionalization/alkyne annulation of benzylamine derivatives: access to dihydroisoquinolines. *Chemistry* 23 (48): 11669–11676.

101 Gao, K., Lee, P.S., Fujita, T. et al. (2010). Cobalt-catalyzed hydroarylation of alkynes through chelation-assisted C—H bond activation. *Journal of the American Chemical Society* 132 (35): 12249–12251.

102 Yoshikai, N. (2014). Development of cobalt-catalyzed C—H bond functionalization reactions. *Bulletin of the Chemical Society of Japan* 87 (8): 843–857.

103 Klein, H.-F., Helwig, M., Koch, U. et al. (1993). Coordination and reactions of diazenes in trimethylphosphinecobalt(I). Complexes – syntheses and structures of complexes containing μ²-(N,N′)-Benzo[c]cinnoline and η²-azobenzene ligands. *Zeitschrift für Naturforschung Part B* 48 (6): 778–784.

104 Kakiuchi, F. and Kochi, T. (2008). Transition-metal-catalyzed carbon–carbon bond formation via carbon–hydrogen bond cleavage. *Synthesis* 2008 (19): 3013–3039.

105 Liang, Y. and Jiao, N. (2016). Cationic cobalt(III) catalyzed indole synthesis: the regioselective intermolecular cyclization of N-nitrosoanilines and alkynes. *Angewandte Chemie International Edition* 128 (12): 4103–4107.

106 Santhoshkumar, R. and Cheng, C.H. (2018). Hydroarylations by cobalt-catalyzed C–H activation. *Beilstein Journal of Organic Chemistry* 14: 2266–2288.

107 Yamazaki, K., Kommagalla, Y., Ano, Y. et al. (2019). A computational study of cobalt-catalyzed C–H iodination reactions using a bidentate directing group with molecular iodine. *Organic Chemistry Frontiers* 6 (4): 537–543.

108 Roslan, I.I., Sun, J., Chuah, G.-K. et al. (2015). Cobalt(II)-catalyzed electrophilic alkynylation of 1,3-dicarbonyl compounds to form polysubstituted Furans via π–π activation. *Advanced Synthesis and Catalysis* 357 (4): 719–726.

109 Shin, B., Sutherlin, K.D., Ohta, T. et al. (2016). Reactivity of a cobalt(III)-hydroperoxo complex in electrophilic reactions. *Inorganic Chemistry* 55 (23): 12391–12399.

110 Whiteoak, C.J., Planas, O., Company, A. et al. (2016). A first example of cobalt-catalyzed remote C–H functionalization of 8-aminoquinolines operating through a single electron transfer mechanism. *Advanced Synthesis and Catalysis* 358 (10): 1679–1688.

111 Adams, D.M. and Hendrickson, D.N. (1996). Pulsed laser photolysis and thermodynamics studies of intramolecular electron transfer in valence tautomeric cobalto–quinone complexes. *Journal of the American Chemical Society* 118 (46): 11515–11528.

112 Cowley, R.E., Bontchev, R.P., Sorrell, J. et al. (2007). Formation of a cobalt(III) imido from a cobalt(II) amido complex. Evidence for proton-coupled electron transfer. *Journal of the American Chemical Society* 129 (9): 2424–2425.

113 Costentin, C., Passard, G., Robert, M. et al. (2013). Concertedness in proton-coupled electron transfer cleavages of carbon–metal bonds illustrated by the reduction of an alkyl cobalt porphyrin. *Chemical Science* 4 (2): 819–823.

114 Symes, M.D., Surendranath, Y., Lutterman, D.A. et al. (2011). Bidirectional and unidirectional PCET in a molecular model of a cobalt-based oxygen-evolving catalyst. *Journal of the American Chemical Society* 133 (14): 5174–5177.

115 Yang, Z., Yu, H., and Fu, Y. (2013). Mechanistic study on ligand-controlled cobalt-catalyzed regioselectivity-switchable hydroarylation of styrenes. *Chemistry* 19 (36): 12093–12103.

116 Gao, K. and Yoshikai, N. (2011). Regioselectivity-switchable hydroarylation of styrenes. *Journal of the American Chemical Society* 133 (3): 400–402.

117 Chirila, P.G., Adams, J., Dirjal, A. et al. (2018). Cp*Co(III)-catalyzed coupling of benzamides with α,β-unsaturated carbonyl compounds: preparation of aliphatic ketones and azepinones. *Chemistry* 24 (14): 3584–3589.

118 Nakanowatari, S., Mei, R., Feldt, M. et al. (2017). Cobalt(III)-catalyzed hydroarylation of allenes via C–H activation. *ACS Catalysis* 7 (4): 2511–2515.

119 Chirila, P.G., Skibinski, L., Miller, K. et al. (2018). Towards a sequential one-pot preparation of 1,2,3-benzotriazin-4(3*H*)-ones employing a key Cp*Co(III)-catalyzed C–H amidation step. *Advanced Synthesis and Catalysis* 360 (12): 2324–2332.

120 Khand, I.U., Knox, G.R., Pauson, P.L. et al. (1973). Organocobalt complexes. Part II. Reaction of acetylenehexacarbonyldicobalt complexes, (R^{1}C$_2$R^2)Co$_2$(CO)$_6$, with norbornene and its derivatives. *Journal of the Chemical Society, Perkin Transactions 1* 2 (0): 977–981.

121 Lesage, D., Milet, A., Memboeuf, A. et al. (2014). The Pauson–Khand mechanism revisited: origin of CO in the final product. *Angewandte Chemie International Edition* 53 (7): 1939–1942.

122 Liu, S., Shen, H., Yu, Z. et al. (2014). What controls stereoselectivity and reactivity in the synthesis of a trans-decalin with a quaternary chiral center via the intramolecular Pauson–Khand reaction: a theoretical study. *Organometallics* 33 (22): 6282–6285.

123 Yamanaka, M. and Nakamura, E. (2001). Density functional studies on the Pauson–Khand reaction. *Journal of the American Chemical Society* 123 (8): 1703–1708.

124 Zhu, Z., Li, X., Chen, S. et al. (2018). Cobalt-catalyzed intramolecular alkyne/benzocyclobutenone coupling: C—C bond cleavage via a tetrahedral dicobalt intermediate. *ACS Catalysis* 8 (2): 845–849.

125 Agenet, N., Gandon, V., Vollhardt, K.P. et al. (2007). Cobalt-catalyzed cyclotrimerization of alkynes: the answer to the puzzle of parallel reaction pathways. *Journal of the American Chemical Society* 129 (28): 8860–8871.

126 Dahy, A.A. and Koga, N. (2014). A computational study on the formation of pyridin-2(1*H*)-one and pyridine-2(1*H*)-thione from the reaction of cobaltacyclopentadiene with isocyanate and isothiocyanate. *Journal of Organometallic Chemistry* 770: 101–115.

127 Gandon, V., Agenet, N., Vollhardt, K.P. et al. (2006). Cobalt-mediated cyclic and linear 2:1 cooligomerization of alkynes with alkenes: a DFT study. *Journal of the American Chemical Society* 128 (26): 8509–8520.

128 Weng, C.-M. and Hong, F.-E. (2011). DFT studies on the reaction of CpCo(PPh$_3$)$_2$ with diphenylphosphinoalkynes: formation of cobaltacycles and cyclobutadiene-substituted CpCoCb diphosphines. *Organometallics* 30 (14): 3740–3748.

129 Chen, X., Ge, H., and Yang, X. (2017). Newly designed manganese and cobalt complexes with pendant amines for the hydrogenation of CO$_2$ to methanol: a DFT study. *Catalysis Science and Technology* 7 (2): 348–355.

130 Ge, H., Chen, X., and Yang, X. (2016). A mechanistic study and computational prediction of iron, cobalt and manganese cyclopentadienone complexes for hydrogenation of carbon dioxide. *Chemical Communications* 52 (84): 12422–12425.

131 Ge, H., Chen, X., and Yang, X. (2017). Hydrogenation of carbon dioxide to methanol catalyzed by iron, cobalt, and manganese cyclopentadienone complexes: mechanistic insights and computational design. *Chemistry* 23 (37): 8850–8856.

132 Ge, H., Jing, Y., and Yang, X. (2016). Computational design of cobalt catalysts for hydrogenation of carbon dioxide and dehydrogenation of formic acid. *Inorganic Chemistry* 55 (23): 12179–12184.

133 Hou, C., Jiang, J., Zhang, S. et al. (2014). Hydrogenation of carbon dioxide using half-sandwich cobalt, rhodium, and iridium complexes: DFT study on the mechanism and metal effect. *ACS Catalysis* 4 (9): 2990–2997.

134 Zhang, Z., Li, Y., Hou, C. et al. (2018). DFT study of CO_2 hydrogenation catalyzed by a cobalt-based system: an unexpected formate anion-assisted deprotonation mechanism. *Catalysis Science and Technology* 8 (2): 656–666.

135 Jing, Y., Chen, X., and Yang, X. (2015). Computational mechanistic study of the hydrogenation and dehydrogenation reactions catalyzed by cobalt pincer complexes. *Organometallics* 34 (24): 5716–5722.

136 Ma, X. and Lei, M. (2017). Mechanistic insights into the directed hydrogenation of hydroxylated alkene catalyzed by bis(phosphine)cobalt dialkyl complexes. *Journal of Organic Chemistry* 82 (5): 2703–2712.

137 Wu, S.B., Zhang, T., Chung, L.W. et al. (2019). A missing piece of the mechanism in metal-catalyzed hydrogenation: Co(−I)/Co(0)/Co(+I) catalytic cycle for co(−I)-catalyzed hydrogenation. *Organic Letters* 21 (2): 360–364.

138 Qi, X., Liu, X., Qu, L.-B. et al. (2018). Mechanistic insight into cobalt-catalyzed stereodivergent semihydrogenation of alkynes: the story of selectivity control. *Journal of Catalysis* 362: 25–34.

139 Huo, C.-F., Li, Y.-W., Beller, M. et al. (2003). HCo(CO)$_3$-catalyzed propene hydroformylation. Insight into detailed mechanism. *Organometallics* 22 (23): 4665–4677.

140 Dong, C., Ji, M., Yang, X. et al. (2017). Mechanisms of the transfer hydroformylation catalyzed by rhodium, cobalt, and iridium complexes: insights from density functional theory study. *Journal of Organometallic Chemistry* 833: 71–79.

141 Qu, S. and Cramer, C.J. (2017). Mechanistic study of Cp*Co(III)/Rh(III)-catalyzed directed C–H functionalization with diazo compounds. *Journal of Organic Chemistry* 82 (2): 1195–1204.

142 Planas, O., Roldan-Gomez, S., Martin-Diaconescu, V. et al. (2018). Mechanistic insights into the S_N2-type reactivity of aryl-Co(III) masked-carbenes for C–C bond forming transformations. *Chemical Science* 9 (26): 5736–5746.

143 Ungvári, N., Fördős, E., Kégl, T. et al. (2010). Mechanism of the cobalt-catalyzed carbonylation of ethyl diazoacetate. *Inorganica Chimica Acta* 363 (9): 2016–2028.

144 Suarez, A.I., Jiang, H., Zhang, X.P. et al. (2011). The radical mechanism of cobalt(II) porphyrin-catalyzed olefin aziridination and the importance of cooperative H-bonding. *Dalton Transactions* 40 (21): 5697–5705.

145 Ruppel, J.V., Jones, J.E., Huff, C.A. et al. (2008). A highly effective cobalt catalyst for olefin aziridination with azides: hydrogen bonding guided catalyst design. *Organic Letters* 10 (10): 1995–1998.

146 Gu, Z.-Y., Liu, Y., Wang, F. et al. (2017). Cobalt(II)-catalyzed synthesis of sulfonyl guanidines via nitrene radical coupling with isonitriles: a combined experimental and computational study. *ACS Catalysis* 7 (6): 3893–3899.

8
Theoretical Study of Rh-Catalysis

102.906
Rh⁴⁵
Rhodium
[Kr]4d⁸5s¹

In organometallic catalysis, the most popular metal would be palladium [1–9], while rhodium might be the best alternative of palladium [10–16], which has also been applied as catalyst for the construction of new C—C/C—hetero bonds and the organic transformation of functional groups. In Rh-catalysis, the familiar oxidative state of Rh are +1 and +3. In some recent researches, the oxidative state Rh was also observed as active intermediates by both experiment and theoretical study [17–22].

When Rh reacts with organic compounds, a Rh—C bond can be easily formed, which exhibits relatively stability [23–27]. Therefore, it plays the role of an efficient nucleophile, which can be further transferred for the construction of new C—C/C—hetero bonds. Rh can be coordinated with some unsaturated molecules, such as alkene, alkyne, carbon monoxide, nitrene, and carbene, which leads to the activation of those molecules. Subsequently, those unsaturated molecules can insert into Rh—C bond to construct new covalent bonds [28–30]. In the presence of both nucleophile and electrophiles, a Rh-catalyzed redox-neutral cross-coupling reaction can take place. After a catalytic cycle, the oxidative state of Rh remains constant. In Rh-catalyzed oxidative coupling between two nucleophiles, the low-valence Rh is obtained after a catalytic cycle, which can be oxidized by extra metallic or organic oxidant [31–38]. Interestingly, some endogenous oxidants, such as quinoline-N-oxide, oxime ethers, or oxime esters, can be used to keep redox-neutral, which are also employed as reactant [39–44]. Moreover, some inert covalent bonds can be activated by Rh, which leads to an insertion of unsaturated molecule into this inert covalent bond [45–47].

In Rh-chemistry, ligand is usually necessary to stabilize Rh in catalytic cycle. Analogous to Pd, neutral phosphine and carbene ligands are commonly suitable for Rh

Computational Methods in Organometallic Catalysis: From Elementary Reactions to Mechanisms,
First Edition. Yu Lan.
© 2021 WILEY-VCH GmbH. Published 2021 by WILEY-VCH GmbH.

in a series of organometallic reactions [48–54]. Moreover, bidentate olefin ligands, such as norbornadiene derivatives, also play ligative activation [55–62]. However, the Rh atom in catalytic cycle should be cationic; thus, an anionic ligand can skillfully improve the catalytic activity of Rh. For instance, cyclepentadiene derivatives act an important ligand role in Rh-catalysis, which can coordinate with Rh to form a η^6-complex to further stabilize Rh [63–70]. Interestingly, dimeric Rh(II) can be stabilized by tetracarboxylate ligand, which reveals unique catalytic property in carbene or nitrene activations [71–76].

According to the above understanding, the mechanism of Rh-catalysis would be very concise. However, due to the stability of Rh—C bond, a series of covalent bonds can be formed in one catalytic cycle, which leads to the complexity of Rh-catalysis. During the past several decades, a series of theoretical studies focusing on Rh-catalysis have been made to reveal their mechanisms. In this chapter, the first three sections discuss some Rh-mediated inert bond activation and functionalization, including C—H, C—C, and C—hetero bond activations. Then, the functionalizations and transformations of some unsaturated molecules are summarized, including the reactions with alkenes, alkynes, carbonyls, imines, carbenes, and nitrenes. The last section would discuss the mechanism of Rh-catalyzed cyclizations.

8.1 Rh-Mediated C—H Activation Reactions

A broad range of experimental transformations can be achieved through Rh-catalyzed C—H bond functionalization. Notably, cyclopentadienyl (Cp) ligands play an important role in Rh-catalyzed C—H activation reactions [11]. Use of a chelating-directing group could increase the reactivity of substrates, thereby broadening the functional group tolerance and increasing the efficiency [77, 78]. Along with the explosive progress witnessed from an experimental aspect, the computational study of Rh-catalyzed C—H functionalization has also achieved significant accomplishments [79, 80]. Although the mechanism details may vary from case to case, the catalytic cycle generally consists of three main steps: C—H bond cleavage, C—Rh bond transformation, and regeneration of the active catalyst (Scheme 8.1) [81].

8.1.1 Rh-Catalyzed Arylation of C—H Bond

Rh-catalyzed C—H bond arylations can be categorized as redox-neutral cross-coupling reactions with electrophilic aryl reagents or oxidative coupling reactions with nucleophilic aryl reagents.

In 2005, Sames and coworkers reported a [Rh(coe)$_2$Cl]$_2$-catalyzed (coe = cyclooctene) C2-selective arylation of indoles in the presence of an electron-poor phosphine ligand [P(p-CF$_3$-C$_6$H$_4$)$_3$] and cesium pivalate using aryl iodides as electrophiles [82]. The reaction temperature was optimized to 120 °C, which indicated a high activation free energy.

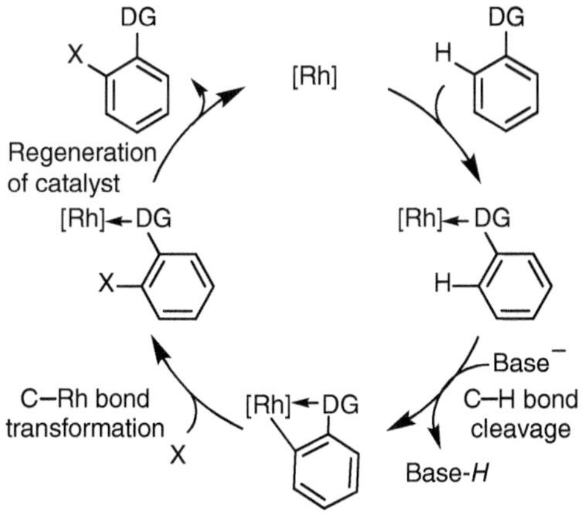

Scheme 8.1 General mechanism of Rh-catalyzed C−H activation and functionalizations. Source: Modified from Qi et al. [81].

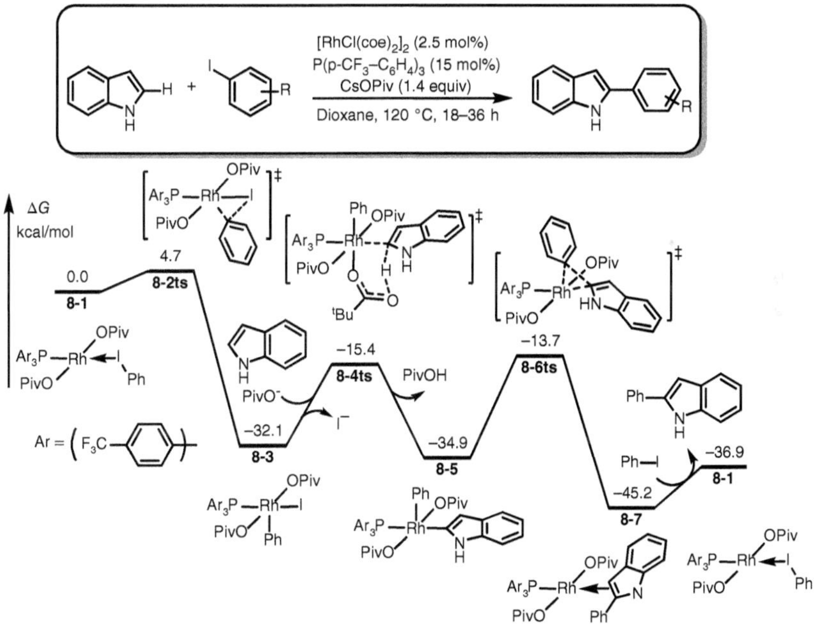

Figure 8.1 Free-energy profiles for Rh(I)-catalyzed C2-selective C−H bond activation and arylation of indoles. The energies were calculated at B3LYP-D2/6-311+G(2d,2p) (LNAL2DZ for Rh and I)/SMD(dioxane)//B3LYP/6-31G(d,p) (LANL2DZ for Rh and I) level of theory.

Santoro and Himo investigated the mechanism of the above arylation reaction using a density functional theory (DFT) study [83]. As shown in Figure 8.1, the catalytic cycle begins with four-coordinate anionic Rh(I) complex **8-1**, in which phenyl iodide is coordinated with Rh(I). The oxidation addition of phenyl iodide to Rh(I) occurs via transition state **8-2ts** to generate five-coordinate phenyl–Rh(III) intermediate **8-3**. In the following step, concerted metalation-deprotonation

(CMD)-type C—H bond cleavage at the C2 position then occurs via transition state **8-4ts**, generating aryl–Rh(III) intermediate **8-5** with an energy barrier of 16.5 kcal/mol. Reductive elimination of diaryl–Rh(III) complex **8-5** then gives C2-arylated indole intermediate **8-7** via transition state **8-6ts**. After the release of arylation product and coordination of phenyl iodide, Rh(I) complex **8-1** is regenerated. The calculated results indicated that reductive elimination is the rate-determining step in this catalytic cycle.

In another example, Glorius and coworkers reported the first Rh(III)-catalyzed oxidative C—H/C—H cross-coupling of aromatic compounds in 2012 [84]. In this reaction, benzamides smoothly reacted with solvent amounts of haloarenes to give the desired products in good yields. Bromides, chlorides, and iodides were all compatible with this Rh-catalyzed system.

Zhao, et al. performed DFT calculations to investigate the mechanism of this Rh(III)-catalyzed oxidative C—H/C—H cross-coupling reaction [85]. The calculated free-energy profile for the rational reaction pathway is shown in Figure 8.2. Amide-directed C—H bond activation occurs via CMD-type transition state **8-9ts**, in which acetate acts as the base to deprotonate the ortho-aromatic proton, with concomitant formation of aryl–Rh(III) intermediate **8-10**. The activation barrier for this C—H bond cleavage via transition state **8-9ts** is 20.0 kcal/mol. After ligand exchange of PivO$^-$ with PivOH, subsequent C—H activation of aryl bromide occurs via transition state **8-11ts**, in which PivO$^-$ also acts as the base for deprotonation. This electrophilic deprotonative metalation was calculated as the rate-determining step of the catalytic cycle, with an overall activation free energy of 29.3 kcal/mol. The C(aryl)—C(aryl) reductive elimination of diaryl–Rh(III) intermediate **8-12** gives arylation-product-coordinated Rh(I) complex **8-14**. The activation barrier for the reductive elimination step is 28.0 kcal/mol, indicating that reductive elimination is also a slow process.

8.1.2 Rh-Catalyzed Alkylation of C—H Bond

Rh-catalyzed C—H bond alkylation is an efficient method for constructing C(sp^2)—C(sp^3) bonds [46, 86–88]. Olefins [89–92], alkynes [93–95], and carbenoids [96, 97] are usually used as alkyl sources in such reactions. Mechanistically, the key step in C(aryl)—C(alkyl) bond formation is often the insertion of an unsaturated molecule into a newly formed C(aryl)—Rh bond.

In 2012, Chang, Jung, and coworkers reported a Rh-catalyzed C—H alkylation of bipyridines using olefins (Figure 8.3) [98]. In this transformation, Rh(acac)$_3$ was identified as the most suitable catalyst precursor, and significantly increased reaction efficiency and selectivity were achieved when IMes·HCl was employed. Various simple terminal olefins were used as alkylating reagents to give the bisalkylated products. Kinetic isotope effect (KIE) experiments ($k_{H/D}$ = 1.01) suggested that C—H bond cleavage was excluded from the rate-determining step in the catalytic cycle.

The same group also performed DFT calculations to study the mechanism of this C—H alkylation reaction [98]. The calculated free-energy profiles for the first alkylation cycle are shown in Figure 8.3. The catalytic cycle starts with

8.1 Rh-Mediated C–H Activation Reactions | 333

Figure 8.2 Free-energy profiles for Rh(III)-catalyzed oxidative C–H/C–H cross-coupling reaction. The energies were calculated at PBE0-D3BJ/6-311+G(d,p) (SDD for Rh and Br)/SMD(bromobenzene)//PBE0-D3BJ/6-31G(d) (SDD for Rh and Br) level of theory.

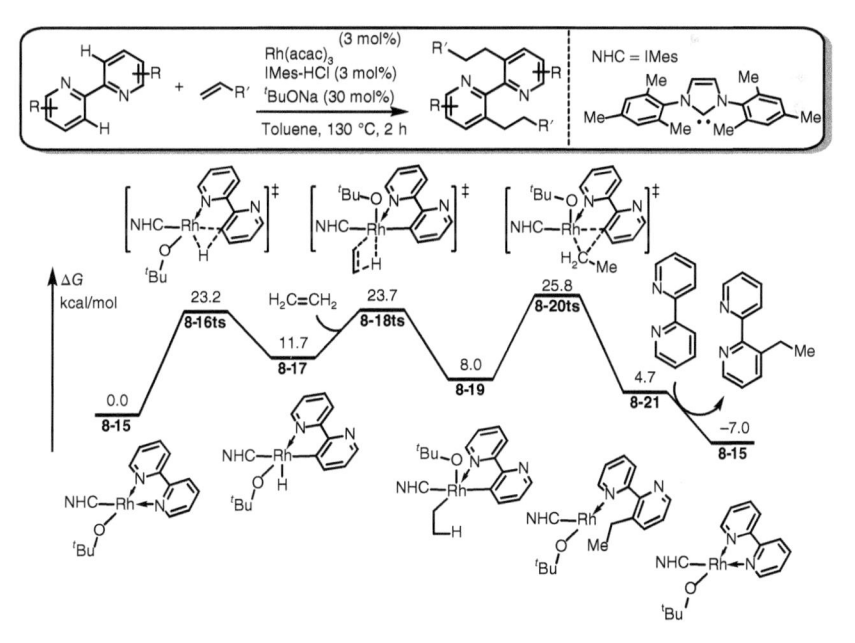

Figure 8.3 Free-energy profiles for Rh(III)-catalyzed bipyridine C–H alkylation by using olefins. The energies were calculated at B3LYP-D3/6-31G(d) (SRSC-ECP for Rh)//B3LYP/6-31G(d) (SRSC-ECP for Rh) level of theory. Source: Based on Kwak et al. [98].

chelated Rh(I)–NHC (N-heterocyclic carbene) complex **8-15**. C—H bond activation occurs through an oxidative addition-type transition state **8-16ts** to generate Rh(III)-hydride species **8-17**. Subsequent coordination of the olefin substrate leads to insertion into the Rh—H bond occurring relatively easily via transition state **8-18ts**, with an activation barrier of 23.7 kcal/mol, to afford alkyl–Rh(III) intermediate **8-19**. The C(aryl)—C(alkyl) reductive elimination of complex **8-19** occurs via transition state **8-20ts** to generate alkylated bipyridine-coordinated Rh(I) intermediate **8-21**, which would undergo recomplexation to deliver Rh(I) species **8-15**.

In 2014, Dong and coworkers developed a ketone α-alkylation reaction using a metal–organic cooperative catalysis strategy [99, 100] comprising a secondary amine and Rh(I) complex [101]. Notably, the reaction only afforded the 5-position monoalkylation product with complete regioselectivity, with overalkylation completely avoided. In this reaction, azaindoline reacts with the ketone to afford an enamine intermediate, which can be further functionalized regioselectively controlled by the chelated directing effect.

In 2015, Wang and coworkers performed DFT calculations to further investigate the mechanism and origins of regioselectivity in the above reaction [79]. As shown in Figure 8.4, the reaction between azaindoline and ketones affords corresponding enamine, which can react with azaindoline-coordinated Rh(I) complex **8-22** through ligand exchange to give complex **8-23**. Chelation by the azaindoline cocatalyst allows oxidation addition of the C—H bond onto the Rh(I) center via transition state **8-24ts**, with a free energy barrier of 21.5 kcal/mol, to give vinyl Rh(III)–hydride complex **8-25**. Subsequent olefin coordination leads to

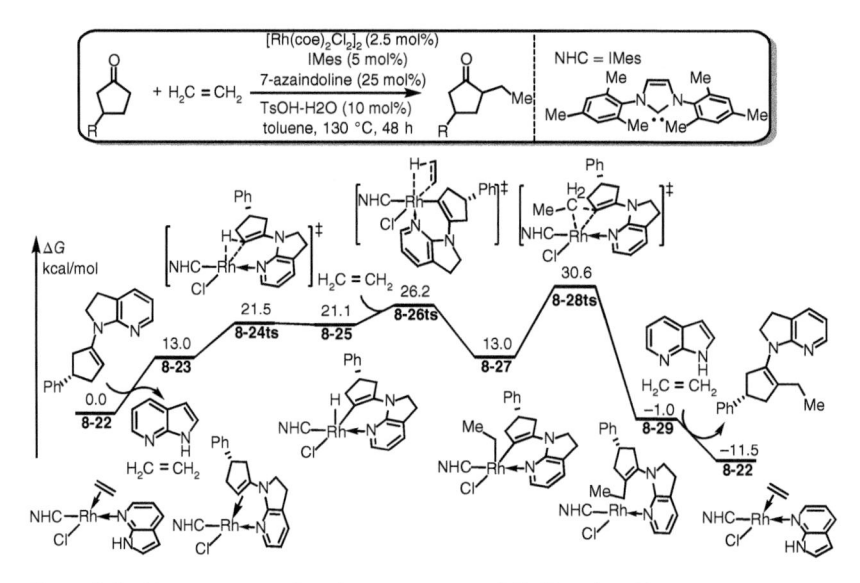

Figure 8.4 Free-energy profiles for the key step of Rh/7-azaindoline co-catalyzed ketone α-alkylation. The energies were calculated at M06/6-31G(d,p) (SDD for Rh)/SMD(toluene) //B3LYP/6-31G(d,p) (SDD for Rh) level of theory.

sequential insertion into the Rh(III)–hydride bond via four-membered ring-type transition state **8-26ts** to form an alkyl Rh(III) intermediate **8-27**. The subsequent reductive elimination occurs via transition state **8-28ts** with an overall activation energy barrier of 30.6 kcal/mol to afford alkylated enamine-coordinated Rh(I) complex **8-29**. The calculated results showed that the reductive elimination was the rate-determining step in the catalytic cycle. The release of alkylation enamine through ligand exchange with 7-azaindoline and ethylene regenerates active Rh(I) catalyst **8-22** to complete the catalytic cycle. Finally, hydrolysis of alkylated enamine can yield the α-alkylated ketone product and regenerate the 7-azaindoline cocatalyst.

Alkynes can also be used as the alkyl source in Rh-catalyzed C—H bond alkylations through more complicated transformations [93–95]. In 2014, Li and Chang independently reported Cp*Rh(III)-catalyzed C—H bond alkylations of quinoline N-oxide using alkynes (Figure 8.5) [93–95]. In these transformations, the N-oxide moiety served as a directing group, leading to regioselective C8—H bond activation. Notably, the N-oxide moiety also acted as an endogenous oxidant to keep the reaction redox-neutral. Furthermore, O-atom transfer played an important role in completing the formal alkylation reaction.

The free-energy profile of the preferred pathway for this reaction, obtained from DFT calculations by the Lan group, is shown in Figure 8.5 [102]. Substrate-coordinated cationic Cp*Rh(III) acetate complex **8-30** was set as relative zero for the reaction pathway. The N-oxide-directed CMD-type C—H bond cleavage occurs via six-membered ring-type transition state **8-31ts**, with an activation free energy of 23.3 kcal/mol affording aryl–Rh(III) intermediate **8-32**. Subsequent insertion of the coordinated acetylene into the C(aryl)—Rh bond occurs via transition state **8-33ts**, with an energy barrier of 16.5 kcal/mol, to generate rhodacycle **8-34**. Then C—O bond reductive elimination, which is considered the rate-determining step, leads to the formation of oxazinoquinolinium-coordinated Rh(I) complex **8-36**. The overall activation energy for this process is 27.0 kcal/mol. Oxidation addition of the N—O bond to Rh(I) occurs via transition state **8-37ts** to give enolate–Rh(III) complex **8-38** in an exothermic process. Finally, protonation of intermediate **8-38** gives an enol complex **8-40**, which undergoes isomerization to give alkylation product and regenerate active catalyst **8-30**.

8.1.3 Rh-Catalyzed Alkenylation of C—H Bond

Olefins can act as nucleophiles to undergo oxidative coupling with arene C—H bonds using a Rh(III) catalyst in the presence of an external oxidant, which can be mechanistically considered an oxidative Heck-type coupling. In 2011, Liu and coworkers reported a Rh(III)-catalyzed oxidative Heck-type coupling reaction of protected phenols with olefins (Figure 8.6) [103]. In this reaction, electron-deficient olefins, such as acrylates, can be used as the alkenyl source, with [Cp*RhCl$_2$]$_2$ selected as the catalyst and Cu(OAc)$_2$ as the external oxidant. The KIE ($k_H/k_D = 3.1$) was observed experimentally, indicating that C—H bond cleavage was most likely involved in the rate-determining step of the catalytic cycle.

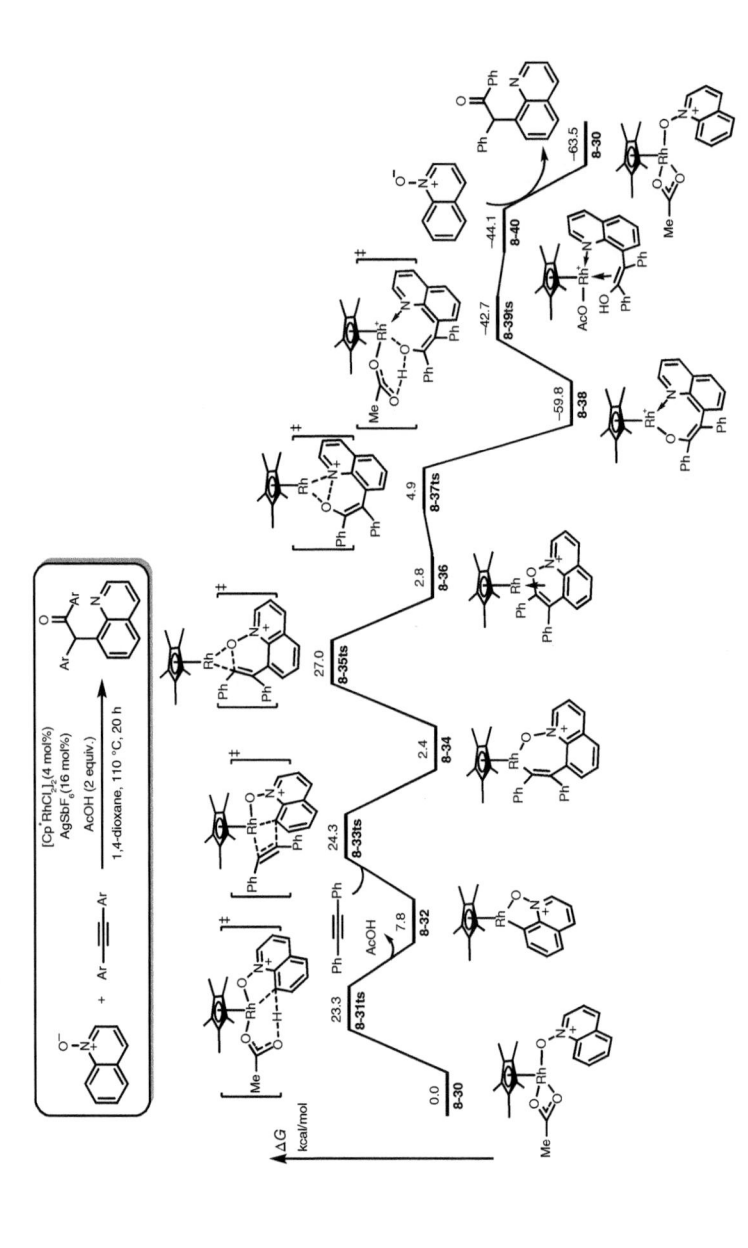

Figure 8.5 Free-energy profiles for Cp*Rh(III)-catalyzed C–H bond alkylations of quinoline N-oxide using alkynes. The energies were calculated at M11-L/6-311+G(d) (SDD for Rh)/SMD(dioxane)//B3LYP/6-311G(d) (LANL2DZ for Rh) level of theory. Source: Sharma et al. [93]; Zhang et al. [94]; Dateer and Chang [95]; Li et al. [102].

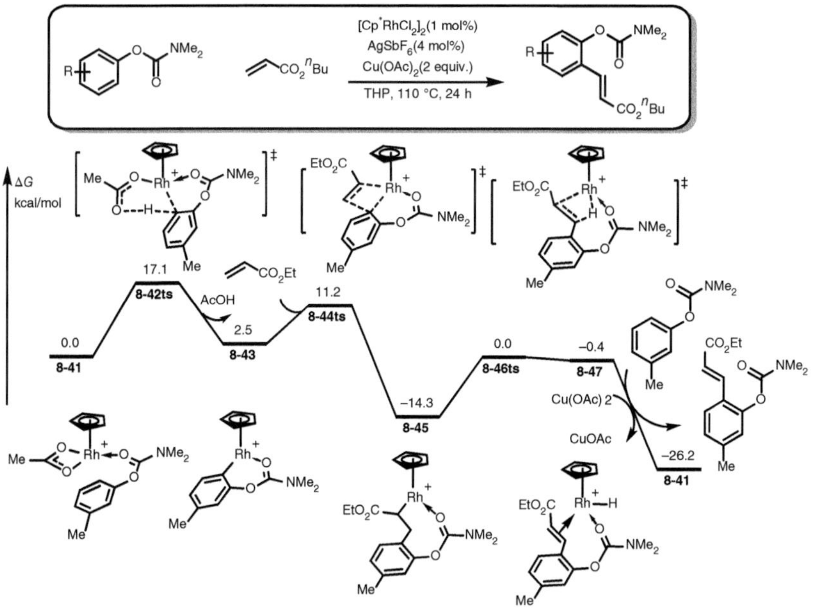

Figure 8.6 Free-energy profiles for Rh(III)-catalyzed oxidative Heck-type coupling reaction. The energies were calculated at M06/6-311+G(d,p) (LANL2DZ(f) for Rh and Cu)/SMD(THF)//B3LYP/6-31G(d) (LANL2DZ for Rh and Cu) level of theory.

Fu and coworkers conducted DFT calculations to determine the detailed mechanism of this reaction [104]. As shown in Figure 8.6, cationic Cp*Rh(III) complex **8-41** was selected as relative zero in the calculated free-energy profiles. C—H bond activation occurs through a CMD-type mechanism via transition state **8-42ts**, with a free-energy barrier of 17.1 kcal/mol, to form aryl–Rh(III) intermediate **8-43**. The calculated results indicated that the CMD-type C—H bond cleavage was the rate-determining step in this catalytic cycle, which was consistent with experimental observations. Subsequent migratory insertion of the C=C double bond into the Rh—C(aryl) bond occurs via transition state **8-44ts** to form alkyl–Rh(III) intermediate **8-45**, which undergoes β-H elimination through four-membered ring transition state **8-46ts** to give Rh(III)–hydride intermediate **8-47**. After releasing alkenylated product, active catalytic complex **8-41** is regenerated by oxidation with Cu(OAc)$_2$.

Alkynes can be used as coupling partners in Rh-catalyzed C—H bond alkenylation reactions [105–107]. This transformation can be considered a nonredox formal alkyne insertion into an aryl C—H bond. Wang and coworkers reported a Rh(III)-catalyzed alkenylation reaction of 8-methylquinolines with alkynes through C(sp^3)—H bond activation (Figure 8.7) [107]. The KIE was found to be $k_H/k_D = 4.0$, indicating that C—H bond cleavage was involved in the rate-determining step.

In 2015, Morokuma and coworkers conducted a detailed theoretical study to investigate the mechanism of the above reaction [108]. As shown in Figure 8.7, the coordination of 8-methylquinoline onto Rh(III) in complex **8-48** leads to activation of the

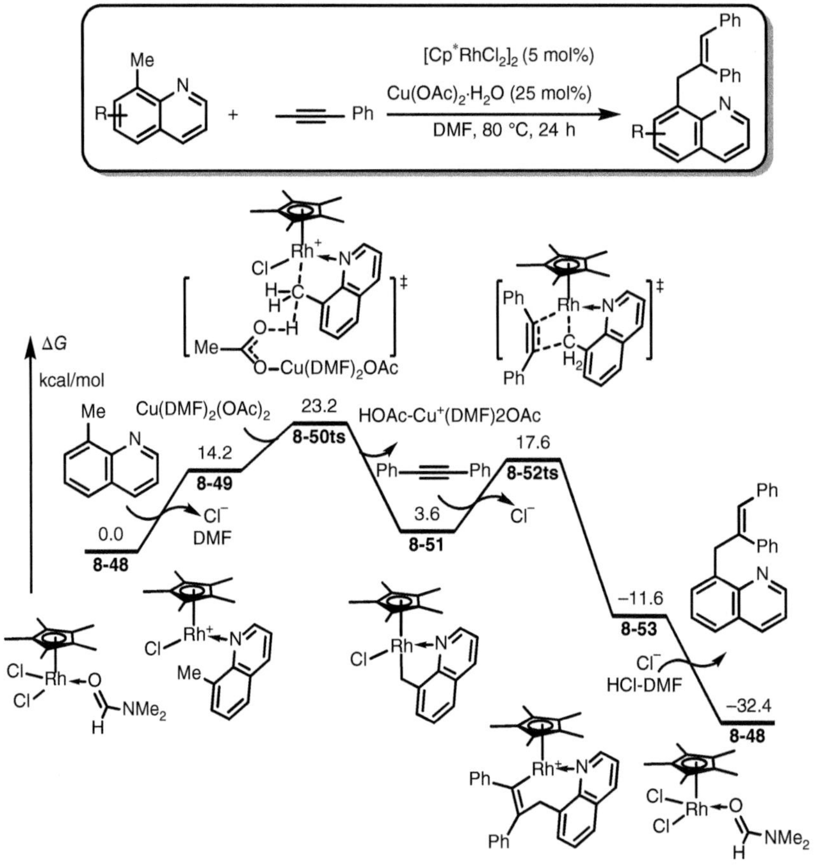

338 | *8 Theoretical Study of Rh-Catalysis*

Figure 8.7 Free-energy profiles for Rh(III)-catalyzed alkenylation of 8-methylquinolines through C(sp³)—H bond activation. The energies were calculated at B3LYP-D3/6-311G(d,p) (SDD for Rh)/IEF-PCM(DMF)//B3LYP-D3/6-311G(d,p) (SDD for Rh) level of theory. Source: Based on Liu et al. [107].

benzylic C—H bond, which can form a hydrogen bond with the incoming copper acetate complex. C—H activation proceeds through intermolecular deprotonation by copper acetate complex as base via transition state **8-50ts**, with an overall activation free energy of 23.2 kcal/mol to give an alkyl–Rh(III) intermediate **8-51**. Subsequent acetylene insertion occurs via transition state **8-52ts** with an activation energy barrier of 14.0 kcal/mol giving alkenyl–Rh(III) complex **8-53** after chloride dissociation. Finally, protonation and ligand exchange yield alkenylation product and regenerate active species **8-48**. The oxidation state of the Rh(III) center is retained throughout the catalytic cycle.

8.1.4 Rh-Catalyzed Amination of C—H Bond

Rh–nitrene complexes play important roles in C—N bond formation reactions following Rh-catalyzed C—H bond activation [109]. A series of nitrene precursors

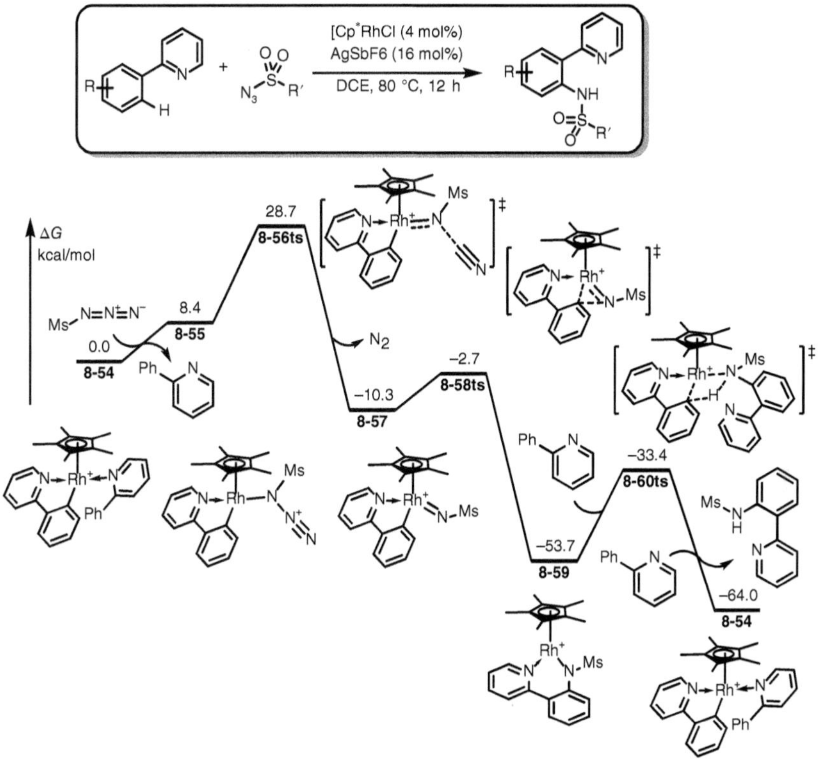

Figure 8.8 Free-energy profiles for Rh(III)-catalyzed intermolecular C—H bond activation and amination of arenes with azides. The energies were calculated at B3LYP-D3/6-31G(d) (SRSC-ECP for Rh)/IEF-PCM(DCE)//B3LYP/6-31G(d) (SRSC-ECP for Rh) level of theory. DCE; dichloroethane. Source: Kim et al. [117]; Ryu et al. [118]; Shin et al. [119]; Park et al. [120].

can be used to construct Rh–nitrene complexes, including azides [19], dioxazolones [110–112], anthranils [113–115], and N-phenoxyacetamides [116]. Azides are frequently used nitrene precursors in Rh-catalyzed arene aminations. Chang and coworkers made a significant contribution to Rh(III)-catalyzed intermolecular arene C—H bond amination, using organic azides as the amino source (Figure 8.8) [117–119]. These reactions do not require an external oxidant and release nitrogen as the only byproduct. A series of directing groups are effective in this reaction, furnishing the desired ortho-aminated products in good yields with excellent selectivity.

The authors also performed DFT studies to investigate the detailed mechanism of nitrene insertion, which leads to C—N bond formation (Figure 8.8) [120]. Five-membered rhodacycle complex **8-54**, which was separated and proven to be the active intermediate in the catalytic cycle, is set as the relative zero point in the calculated free-energy profiles. Ligand exchange of azide with 2-phenylpyridine in **8-54** gives sterically matched Rh–azide species **8-55**. Subsequent denitrogenation occurs via transition state **8-56ts**, which was calculated to be the rate-determining step in the catalytic cycle with an overall activation free energy of 28.7 kcal/mol to

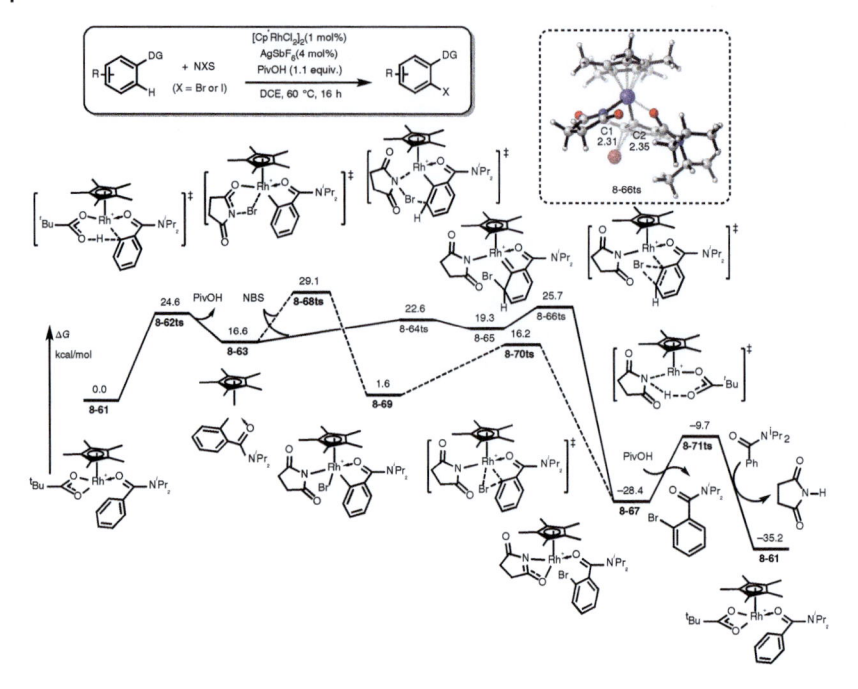

Figure 8.9 Free-energy profiles for Rh(III)-catalyzed ortho-C−H bond activation and bromination of arenes using NBS. The energies were calculated at M11-L/6-311+G(d) (LANL08-f for Rh)/SMD(acetic acid)//B3LYP/6-31+G(d) (SDD for Rh) level of theory. Source: Based on Lied et al. [122].

give Rh(V)–nitrene complex **8-57**. Nitrene reductive insertion into the C(aryl)—Rh bond occurs via three-centered transition state **8-58ts** with a low-energy barrier of 7.6 kcal/mol. Generated six-membered amino-Rh(III) species **8-59**, considered a Brønsted base, can undergo σ-bond metathesis with 2-phenylpyridine via transition state **8-60ts**, with an energy barrier of 20.3 kcal/mol, to afford amination product and regenerate Rh(III) active species **8-54** through coordination with a 2-phenylpyridine substrate.

8.1.5 Rh-Catalyzed Halogenation of C−H Bond

For Rh-catalyzed C−H bond activation and C—halogen bond formation reactions, among the best strategies involves using NXS (N-halosuccinimide) (X = Br or I) as an electrophilic halogen source to cross-couple with C−H bond [121, 122]. Previous theoretical studies have strongly indicated that these transformations usually favor a nonredox mechanism involving an X transfer pathway, rather than the alternative oxidative addition of NXS following reductive elimination through a Rh(V) species, when an aryl–Rh(III) species is formed. In 2012, Glorius and coworkers reported the first Rh(III)-catalyzed ortho-bromination/iodination of arenes with high yields (Figure 8.9) [122]. This strategy was compatible with various highly useful directing groups. In this transformation, N-bromosuccinimide (NBS) and N-iodosuccinimide (NIS) were used as efficient brominating and iodinating reagents, respectively.

DFT calculations were performed to study the mechanism and evaluate the feasibility of Rh(V) species formation in this Rh(III)-catalyzed bromination reaction [123]. As shown in Figure 8.9, amide-directed C—H bond cleavage occurs via transition state **8-62ts**, with an energy barrier of 24.6 kcal/mol, to form aryl–Rh(III) intermediate **8-63**. Subsequent intermolecular Br transfer from NBS to the aryl moiety affords bromonium complex **8-65** via transition state **8-64ts**, with a small barrier of 6.0 kcal/mol. The Br shift from C2 to C1 occurs with C(aryl)—Rh bond cleavage via transition state **8-66ts**, with an overall activation free energy of 25.7 kcal/mol, to give bromination product-coordinated Rh(III) complex **8-67**. In transition state **8-66ts**, the lengths of the breaking C2—Br and forming C2—Br bonds are 2.31 and 2.35 Å, respectively. The release of bromination product and protonation regenerate active catalyst **8-61** to complete the catalytic cycle. In this pathway, the Rh oxidation state remains at +3, indicating a nonredox process. Alternatively, the oxidative addition–reductive elimination pathway was ruled out owing to an unfavorable activation energy of 29.1 kcal/mol via a key transition state of **8-68ts**. However, DFT calculations found that the Rh(III)/Rh(V) catalytic cycle, which involves a Rh(V) intermediate generated by oxidative addition of NBS, becomes favorable when a strong electron-withdrawing group is installed on the arene.

8.2 Rh-Catalyzed C—C Bond Activations and Transformations

Due to the bond strength, the selective cleavage of a C—C bond in organic compound still remains a challenge in homogeneous catalysis. In the presence of Rh-catalyst, some relatively active C—C bond can be broken in a mild reaction condition through oxidative additions or eliminations (Scheme 8.2) [50, 124–128].

8.2.1 Strain-driven Oxidative Addition

The ring strain of benzocyclobutanone leads to a rather weak C—C bond, which can react with low-valence Rh species to break the four-membered ring. As shown

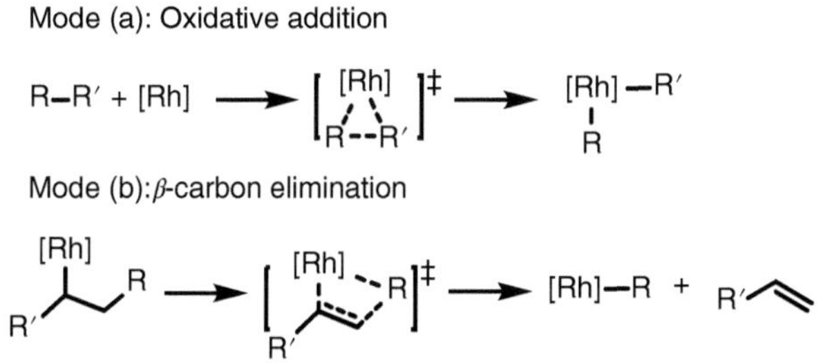

Scheme 8.2 The general pathways for the Rh mediated C—C bond cleavage. Source: Jiao et al. [50]; Shaw et al. [124]; Zhao et al. [125]; Xu and Dong [126]; Xu and Dong [127]; Fu et al. [128].

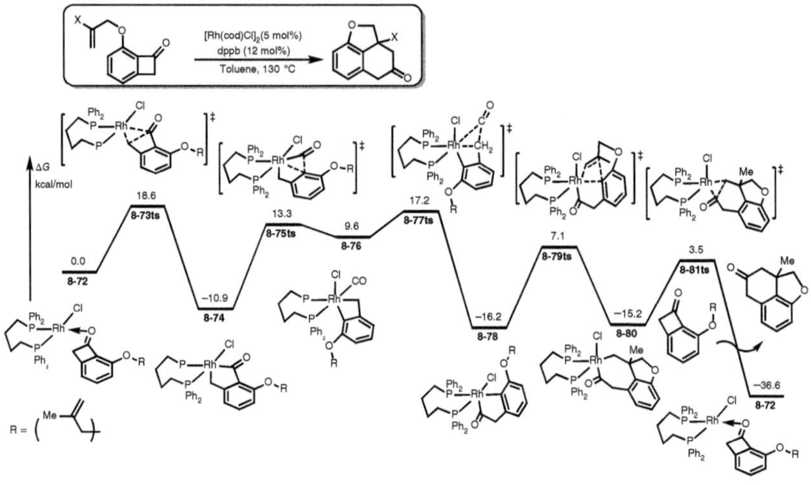

Figure 8.10 Free-energy profiles for Rh-catalyzed intramolecular ring expansion reaction of benzocyclobutanone. The energies were calculated at M06/6-311+G(d,p) (SDD for Rh)/SMD(toluene)//B3LYP/6-31G(d) (LANL2DZ for Rh) level of theory.

in Figure 8.10, Liu studied the mechanism of an intramolecular ring expansion reaction of benzocyclobutanone [129]. The computational result showed that the oxidative addition of benzocyclobutanone onto Rh(I) species **8-72** occurs with C(acyl)—C(alkyl) bond preferentially via transition state **8-73ts** with an energy barrier of 18.6 kcal/mol. Then, a carbonyl shift takes place through decarbonylation followed by insertion to afford a rhoda(III)cycle **8-78**. Then, an intramolecular alkene insertion takes place via transition state **8-79ts** with an energy barrier of 23.3 kcal/mol, which is considered to be the rate-determining step for the whole catalytic cycle. When the seven-membered rhoda(III)cycle **8-80** is formed, the sequential reductive elimination yields ring-expansion product and regenerates Rh(I) species **8-72**.

8.2.2 The Carbon–Cyano Bond Activation

The cyano group usually exhibits halogen character in organic chemistry. The C(aryl)—C(cyano) bond can be activated by the oxidative addition onto low-valence Rh species [130]. The mechanism of nitrile borylation was revealed by DFT calculations [131]. As shown in Figure 8.11, boryl Rh(I) species **8-82** is considered as active catalyst, which is coordinated with benzonitrile. A cyano insertion into Rh—B bond takes place via transition state **8-83ts** to form a boryliminoacyl intermediate **8-84**, in which the C(aryl)—C(cyano) bond is further activated by the formed N—B bond. A phenyl α-elimination takes place via transition state **8-85ts** with an energy barrier of 12.3 kcal/mol. Then, a borylisocyanide-coordinated phenyl Rh(I) complex **8-86** is formed. The dissociation of borylisocyanide leads to the formation of phenyl Rh(I) intermediate **8-87** with 21.3 kcal/mol endergonic. Then, the oxidative addition of diboron onto **8-87** via transition state **8-92ts** forms a Rh(III) intermediate **8-93**.

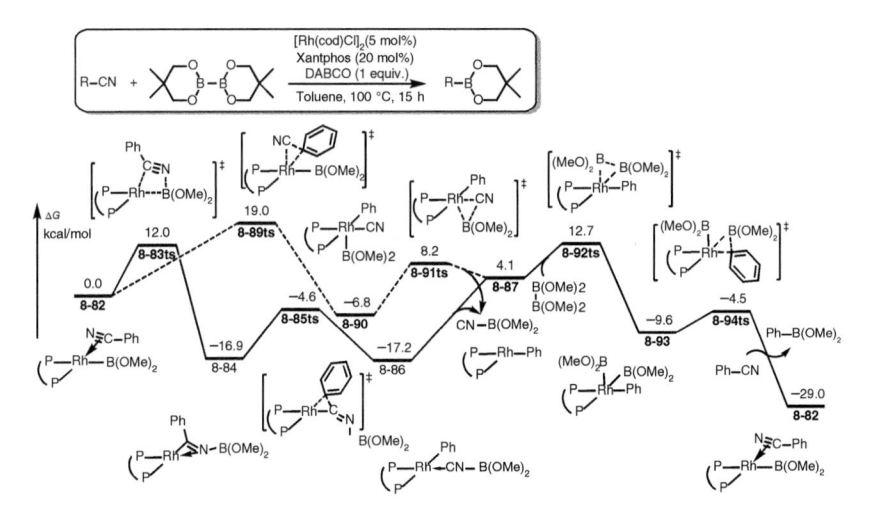

Figure 8.11 Free-energy profiles for Rh-catalyzed borylation of nitriles. The energies were calculated at B3LYP-D/6-311+G(d,p) (SDD for Rh)/PCM(toluene)//B3LYP/6-31G(d) (SDD for Rh) level of theory.

After a rapid reductive elimination, boryl Rh(I) species **8-82** can be regenerated by the release of borylated product. The overall barrier of oxidative addition is 29.9 kcal/mol, which would be the rate limit in catalytic cycle. A direct oxidative addition of benzonitrile onto Rh(I) species **8-82** is also considered, which could take place via transition state **8-89ts**. However, the relative free energy of **8-89ts** is 7.0 kcal/mol higher than that of **8-83ts**. Therefore, the direct oxidative addition can be excluded.

8.2.3 β-Carbon Elimination

In the presence of ring-strain driving force, a β-aryl elimination can cause the cleavage of C—C bond [132]. DFT calculations are used to reveal the mechanism of Rh-catalyzed ring-opening reactions of cyclobutanol [133]. As shown in Figure 8.12, the reaction starts from a cyclobutanolate Rh(I) species **8-95**. Assisted by the ring strain, a β-aryl elimination takes place via transition state **8-96ts** with an energy barrier of only 7.5 kcal/mol to form a carbonyl-coordinated aryl Rh(I) intermediate **8-97**. Then, another cyclobutanol coordinate with Rh forms intermediate **8-98** with 14.6 kcal/mol endergonic. A proton transfer takes place via transition state **8-99ts** to regenerate cyclobutanolate Rh(I) species **8-95** by release of ring-opening product.

8.3 Rh-Mediated C—Hetero Bond Activations

In the presence of Rh-catalyst, some unique C—N or C—O bond can be activated for further transformations. When low-valence Rh is used as catalyst, both oxidative addition–type and metathesis-type activations are possible [134–139].

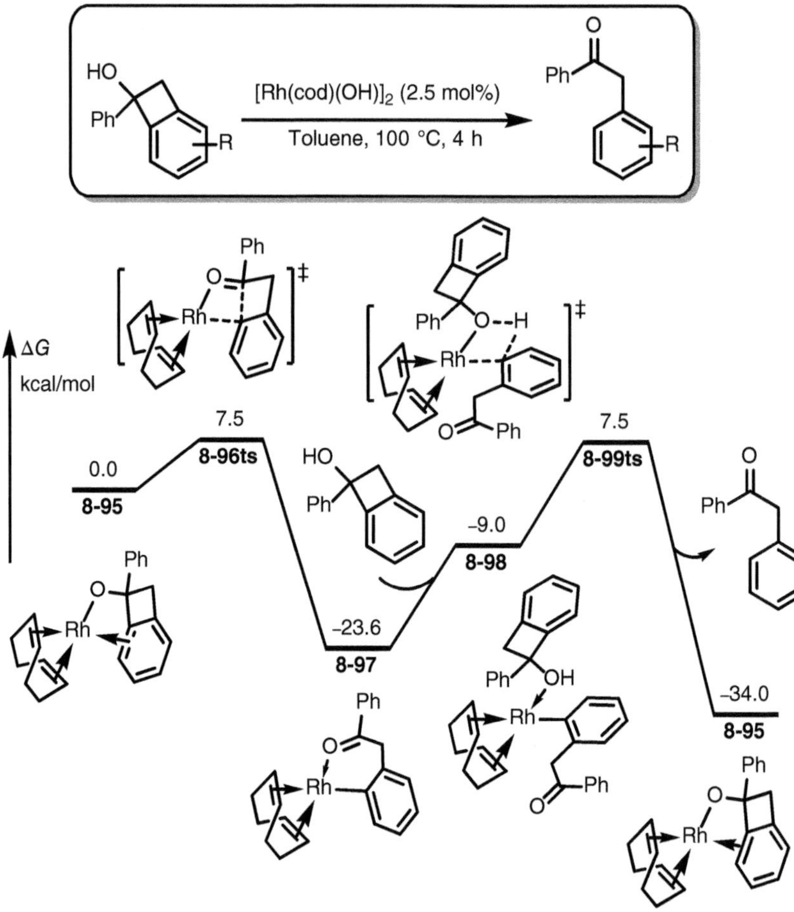

Figure 8.12 Free-energy profiles for Rh-catalyzed ring-opening reactions of cyclobutanol. The energies were calculated at B3LYP/6-31G(d) (LANL2DZ(f) for Rh)/CPCM(toluene)// B3LYP/6-31G(d) (LANL2DZ(f) for Rh) level of theory.

8.3.1 C—N Bond Activation

When a strain C—N bond is present in aziridine compounds, a Rh(I)-catalyzed carbonyl insertion reaction can expand the ring of aziridine to provide corresponding β-lactams (Figure 8.13) [140]. DFT calculation is used to reveal the mechanism of this ring-expanding reaction [141]. The coordination of aziridine onto Rh(I) specie forms **8-100**. The oxidative addition of C—N bond onto Rh takes place via a four-membered ring-type transition state **8-101ts** with an energy barrier of 23.9 kcal/mol. When a rhoda(III)cycle **8-102** is formed, a carbonyl insertion into Rh—N bond occurs via transition state **8-103ts** to afford a five-membered rhoda(III)cycle **8-104**. The coordination of carbon monoxide advances the reductive elimination to yield β-lactam via transition state **8-106ts** with an energy barrier of 23.5 kcal/mol.

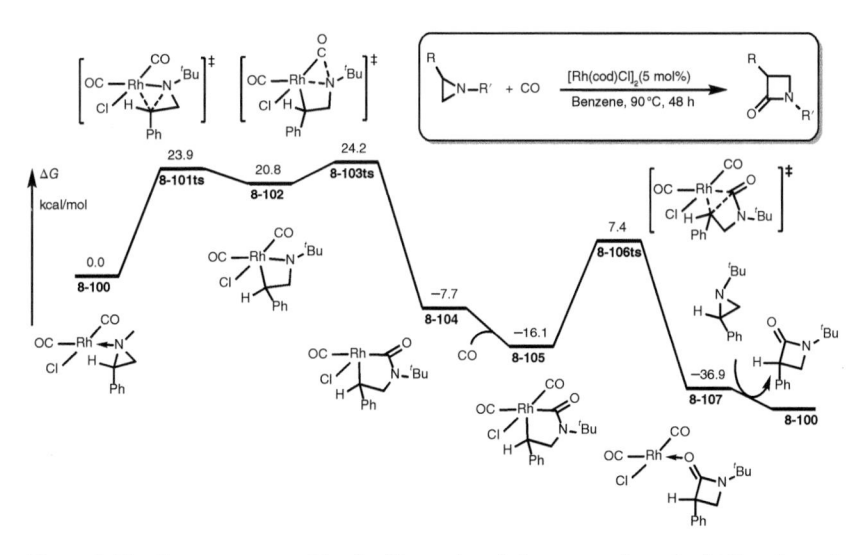

Figure 8.13 Free-energy profiles for Rh-catalyzed ring-expansion of aziridine through carbonyl insertion. The energies were calculated at B3LYP/6-31G(d) (LANL2DZ(f) for Rh)/PCM(benzene)//B3LYP/6-31G(d) (LANL2DZ(f) for Rh) level of theory. Source: Based on Lu and Alper [140].

8.3.2 C—O Bond Activation

In the theoretical study of Rh(I)-catalyzed C—O bond cleavage reaction, a stepwise metathesis mechanism was proposed [142]. As shown in Figure 8.14, when a N-hetero carbene IMes[Me] is used as ligand, a boryl Rh(I) species **8-108** is considered as the active species in the whole catalytic cycle. The nucleophilic addition of pyridyl group in aryl pyridyl ether substrate onto boryl group forms complex **8-109** with 11.1 kcal/mol exergonic. The B—N coordination leads to the activation of C(pyridyl)—O bond in complex **8-109**. Then, an aryl group migration from ether to Rh via transition state **8-110ts** affords an aryl Rh(I) intermediate **8-111** with 16.5 kcal/mol exergonic. The calculated energy barrier of this step is only 12.7 kcal/mol. Therefore, transition state **8-110ts** can be considered as a S_N2_{Ar}-type substitution. When intermediate **8-111** is formed, it can undergo an oxidative addition with diboryl via transition state **8-112ts** to afford a bisboryl Rh(III) intermediate **8-113**. The calculated activation free energy for this transformation is 26.0 kcal/mol, resulting in the rate-determining step of oxidative addition. Finally, a rapid reductive elimination yields boryl arene product and regenerates boryl Rh(I) active species **8-108**.

8.4 Rh-Catalyzed Alkene Functionalizations

Rh-catalyzed alkene functionalizations often proceed through a redox-involved pathway starting from a low-valence Rh(I) species [143–149]. As shown in Scheme 8.3, the oxidative addition of another reactant results in a high-valence

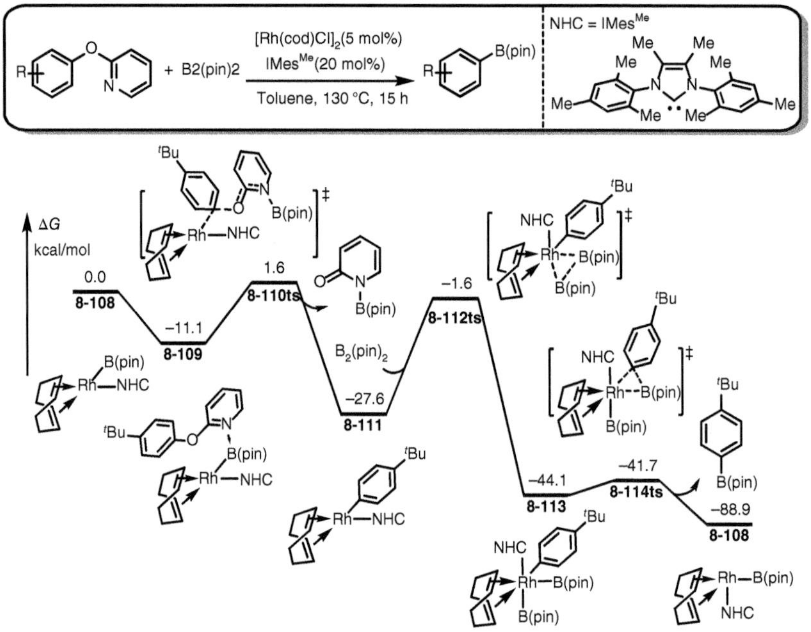

Figure 8.14 Free-energy profiles for Rh-catalyzed pyridinyl-assisted C−O bond activation and borylation. The energies were calculated at M06/6-311++G(d,p) (SDD for Rh)/SMD(toluene)//B3LYP/6-31G(d,p) (LANL2DZ for Rh) level of theory.

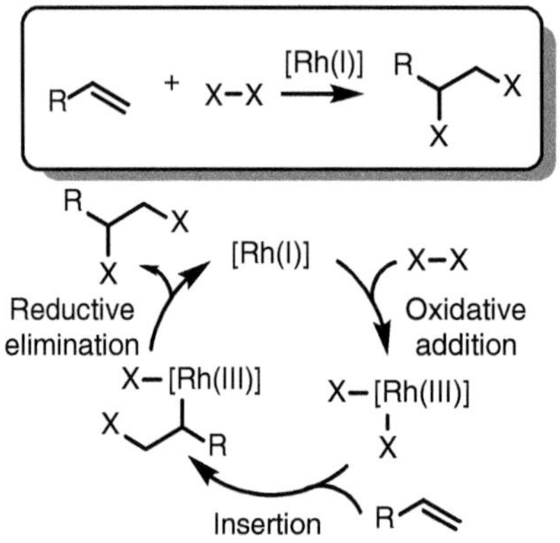

Scheme 8.3 General mechanism of Rh-catalyzed alkene functionalizations.

Rh(III) intermediate. Then, alkene insertion takes place to afford an alkyl Rh(III) intermediate, which can undergo a rapid reductive elimination to regenerate the Rh(I) active species.

8.4.1 Hydrogenation of Alkene

When hydride in a Rh—H complex is chosen as nucleophile, dihydrogenation of alkenes can give corresponding alkane product. As shown in Figure 8.15,

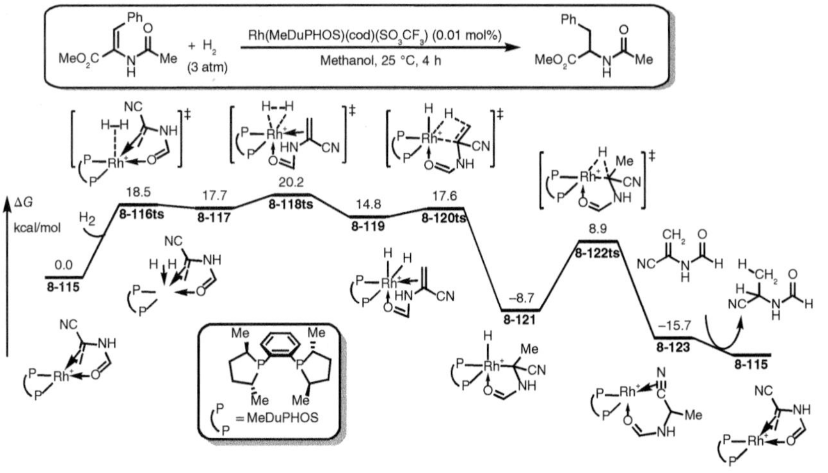

Figure 8.15 Free-energy profiles for Rh-catalyzed hydrogenation of alkene. The energies were calculated at ONIOM(B2LYP/6-31G(d,p) (LANL2DZ for Rh):HF/LANL2MB:UFF) level of theory.

the mechanism of cationic Rh(I)-catalyzed dihydrogenation of alkenes was revealed by DFT calculation [150]. In the presence of acetyl chelation group, the alkene-coordinated cationic Rh(I) species **8-115** can be coordinated by dihydrogen molecule to form a complex **8-117** with 17.7 kcal/mol endergonic. Then, the oxidative addition of dihydrogen onto Rh takes place via transition state **8-118ts** to form a dihydride Rh(III) intermediate **8-119**. The calculated overall activation free energy of this process is 20.2 kcal/mol. The sequential alkene insertion occurs via transition state **8-120ts** with an energy barrier of only 2.8 kcal/mol to give an alkyl Rh(III) species **8-121**. Finally, alkane product can be yielded by reductive elimination with the regeneration of Rh(I) catalyst via transition state **8-122ts**.

In another example, sulfoxide can act as chelation group in Rh-catalyzed alkene dihydrogenation [151]. As shown in Figure 8.16, chelating by oxygen atom, an alkene-coordinated Rh(I) species **8-124** is chosen as relative zero in free-energy profiles. The oxidative addition of dihydrogen molecule can occur via transition state **8-127ts** to form a dihydride Rh(III) species **8-128**. Then, alkene insertion via transition state **8-129ts** affords an alkyl Rh(III) intermediate **8-130**. After a rapid reductive elimination, alkane product is yielded by the coordination of another molecule substrate with the regeneration of Rh(I) species **8-124**. The calculated overall activation free energy is 18.4 kcal/mol, which is higher than that for the Rh-catalyzed racemization of allylic sulfoxide substrate. Therefore, when a racemic allylic sulfoxide was used as substrate, chiral product could be observed experimentally.

8.4.2 Diboration of Alkene

In the presence of Rh-catalyst, alkenes can react with bisboryl to afford diborated alkanes. The mechanism of alkene diboration is revealed by DFT calculation

348 | *8 Theoretical Study of Rh-Catalysis*

Figure 8.16 Free-energy profiles for Rh-catalyzed hydrogenation of alkene. The energies were calculated at M06/6-311+G(d,p) (SDD for Rh)/SMD(methanol)//M06/6-311+G(d,p) (SDD for Rh) level of theory.

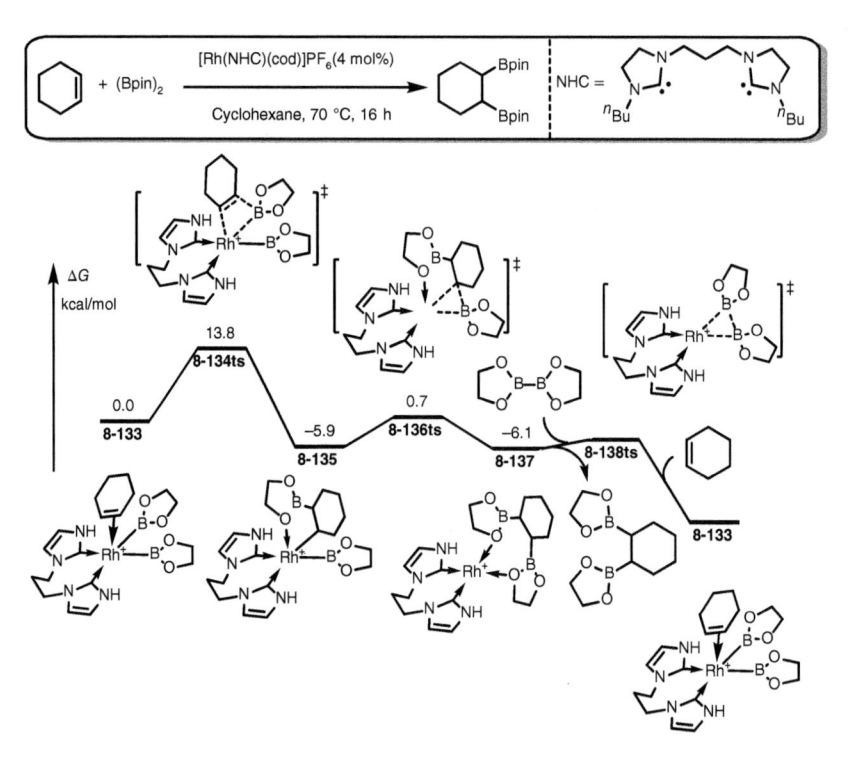

Figure 8.17 Free-energy profiles for Rh-catalyzed diboration of alkene. The energies were calculated at VWN level of theory. Source: Based on Pubill-Ulldemolins et al. [152].

(Figure 8.17) [152]. The oxidative addition with bisboryl can afford a cationic biboryl Rh(III) species, which can be coordinated by alkene to form intermediate **8-133**. Subsequently, alkene inserts into Rh—B bond via transition state **8-134ts**. Then, a reductive elimination takes place via transition state **8-136ts** to generate another C—B bond. Diboryl alkane can release from Rh(I) species **8-137**, while it can be oxidized by bisboryl reactant to regenerate Rh(III) species **8-133**.

8.5 Rh-Catalyzed Alkyne Functionalizations

In the presence of Rh-catalyst, the addition of unsaturated bond in alkynes with nucleophiles and electrophiles can afford corresponding bifunctionalized alkene product [153–157]. The mechanism of this type reaction can be roughly divided into alkyne activation and nucleophile activation by the initial step [158–161]. The common nucleophiles in Rh-catalyzed alkyne functionalizations are acyl, amino, alkoxyl, thiol, acetoxyl, and aryl groups [162–165].

8.5.1 Hydroacylation of Alkynes

Cyclopentenone derivatives can be synthesized by an intramolecular Rh-catalyzed hydrocylation of alkynes. The mechanism of this reaction was revealed by Wu group [166]. As shown in Figure 8.18, the reaction starts from the chelation of alkyne group in Rh(I) species **8-139**. The oxidative addition of C(acyl)—H bond onto Rh via transition state **8-140ts** with an energy barrier of 20.8 kcal/mol affords an acyl hydride Rh(III) intermediate **8-141**. Then, an intramolecular alkyne insertion into Rh—H forms a six-membered rhoda(III)cycle **8-143**. The energy barrier of direct reductive elimination via transition state **8-144ts** is as high as 18.8 kcal/mol. Alternatively, an α-alkyl elimination via transition state **8-146ts** would provide a carbonyl-coordinated five-membered rhoda(III)cycle **8-147**. Then, a carbonyl insertion into Rh—C(vinyl) bond can form intermediate **8-149** via transition state **8-148ts**. Finally, the reductive elimination from intermediate **8-149** can yield the common cyclopentenone product and regenerate Rh(I) species **8-145**. The calculated relative free energy of transition state **8-150ts** is 5.3 kcal/mol, lower than that of **8-144ts**.

8.5.2 Hydroamination of Alkynes

The DFT study of a Rh-catalyzed alkyne hydroamination found that the alkyne substrate is activated first to transfer to allylic Rh species, then it can be functionalized by amine [167]. As shown in Figure 8.19, the reaction starts from an oxidative addition of benzoic acid onto Rh(I) species **8-146** to afford a hydride Rh(III) intermediate **8-148** via transition state **8-147ts**. Then, alkyne substrate inserts into Rh—H bond via transition state **8-149ts** with an energy barrier of 16.9 kcal/mol to form a vinyl Rh(III) species **8-150**. Starting from an allylic Rh(III) species **8-154** in previous catalytic cycle, the calculated overall activation free energy for this

350 | 8 Theoretical Study of Rh-Catalysis

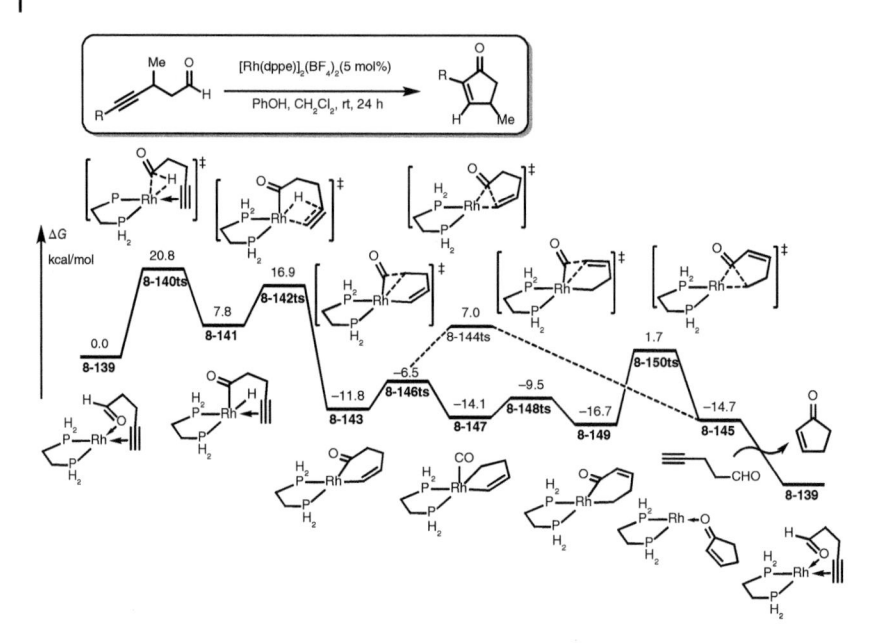

Figure 8.18 Free-energy profiles for Rh-catalyzed intramolecular hydrocylation of alkynes. The energies were calculated at MP2/6-31+G(d) (LANL2DZ for Rh and P)/CPCM(DCM)//MP2/6-31G(d) (LANL2DZ for Rh and P) level of theory.

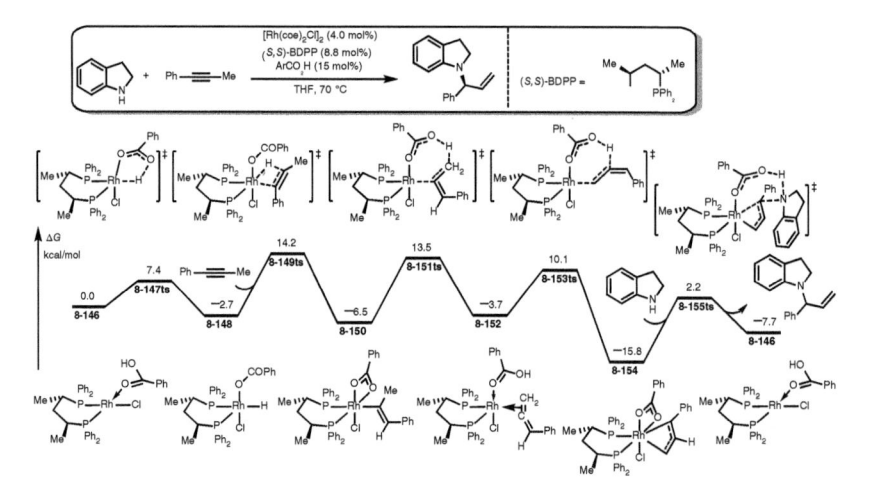

Figure 8.19 Free-energy profiles for Rh-catalyzed hydroamination of alkynes. The energies were calculated at B3LYP-D3/6-311G(d,p) (SDD for Rh)/SMD(THF)//B3LYP-D3/6-31G(d,p) (SDD for Rh) level of theory.

step is as high as 22.3 kcal/mol. Therefore, the alkyne insertion can be considered as the rate-determining step for the whole catalytic cycle. A benzoate-assisted reductive deprotonation via transition state **8-151ts** can form an allene-coordinated Rh(I) intermediate **8-152**. The calculated barrier of this step is 20.0 kcal/mol. The followed protonation of allene can generate an allylic Rh(III) intermediate **8-154**

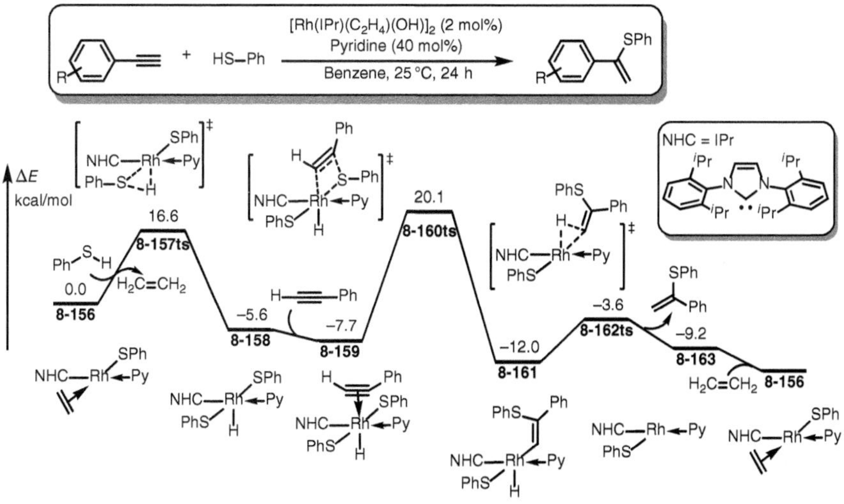

Figure 8.20 Energy profiles for Rh-catalyzed hydrothiolation of alkynes. The energies were calculated at B3LYP/6-31G(d,p) (SDD for Rh) level of theory.

via transition state **8-153ts**. An intermolecular nucleophilic attack by amine takes place via transition state **8-155ts** with an energy barrier of 18.0 kcal/mol to yield hydroamination product and regenerate active catalyst **8-146**.

8.5.3 Hydrothiolation of Alkynes

The mechanism of Rh-catalyzed direct hydrothiolation of alkynes was revealed by Castarlenas and Oro [168]. As shown in Figure 8.20, when the reaction starts from a thiolate Rh(I) species **8-156**, the S—H bond activation can be achieved by an oxidative addition of S—H bond onto Rh via transition state **8-157ts** with an energy barrier of 16.6 kcal/mol to afford a hydride Rh(III) intermediate **8-158**. Then, the coordination of alkyne leads to the insertion into Rh—S bond via transition state **8-160ts**. The calculated energy barrier of this step is 27.8 kcal/mol. When the thiovinyl hydride Rh(III) species **8-161** is formed, a reductive elimination yields the thioalkene product and regenerates Rh(I) species **8-156**.

8.5.4 Hydroacetoxylation of Alkynes

When a carboxylate is used as nucleophile, in Rh-catalyzed hydrofunctionalization of alkynes, an allylic ester can be yielded experimentally. The mechanism of this reaction was revealed by Breit group [169]. As shown in Figure 8.21, the reaction starts from an alkyne-coordinated Rh(I) species **8-164**. With the activation of Rh, coordinated alkyne can be protonated by acetic acid via transition state **8-166ts** with an energy barrier of only 1.6 kcal/mol to afford a vinyl Rh(III) species **8-167**. Then, a β-hydride elimination via transition state **8-168ts** can form an allene-coordinated hydride-Rh(III) **8-169**. The following olefin insertion via transition state **8-170ts** can

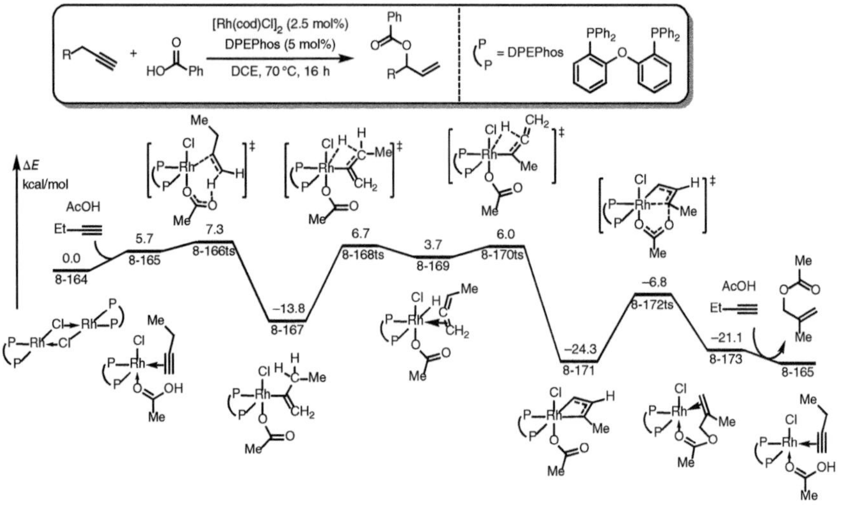

352 | *8 Theoretical Study of Rh-Catalysis*

Figure 8.21 Energy profiles for Rh-catalyzed hydroacetoxylation of alkynes. The energies were calculated at M06/def2-SVP/IEF-PCM(DCE)//BP86/def2-SVP level of theory.

afford an allylic Rh(III) **8-171**. After a nucleophilic attack with acetate, the allylic ester product can be yielded with the regeneration of Rh(I) species **8-165**.

8.6 Rh-Catalyzed Addition Reactions of Carbonyl Compounds

The coordination of carbonyl group onto Rh leads to the activation of C—O π bond, which can undergo a nucleophilic attack to afford an alkoxyl Rh species with nucleophiles [170–176]. However, the detailed mechanism of this type reactions might contain complexity. The DFT studies for the mechanism of Rh-catalyzed addition reactions of carbonyl compounds are summarized in this section, majorly involving the transformations of ketones or carbon dioxide.

8.6.1 Hydrogenation of Ketones

The Rh-catalyzed direct hydrogenation of ketones by gaseous dihydrogen can be used to synthesize corresponding secondary alcohols. The mechanism of this reaction was revealed by DFT calculations [177]. As shown in Figure 8.22, the reaction starts from a cationic Rh(I) species **8-174**. The oxidative addition of dihydrogen molecule onto Rh via transition state **8-175ts** leads to the formation of dihydride Rh(III) intermediate **8-176** with an energy barrier of 10.4 kcal/mol. Then, an intermolecular carbonyl insertion into Rh—H bond takes place via a four-membered transition state **8-177ts** with an energy barrier of 16.4 kcal/mol to reversibly form an alkoxyl Rh(III) intermediate **8-178**. The water-assisted reductive elimination via seven-membered ring-type transition state **8-180ts** generates propanol product with

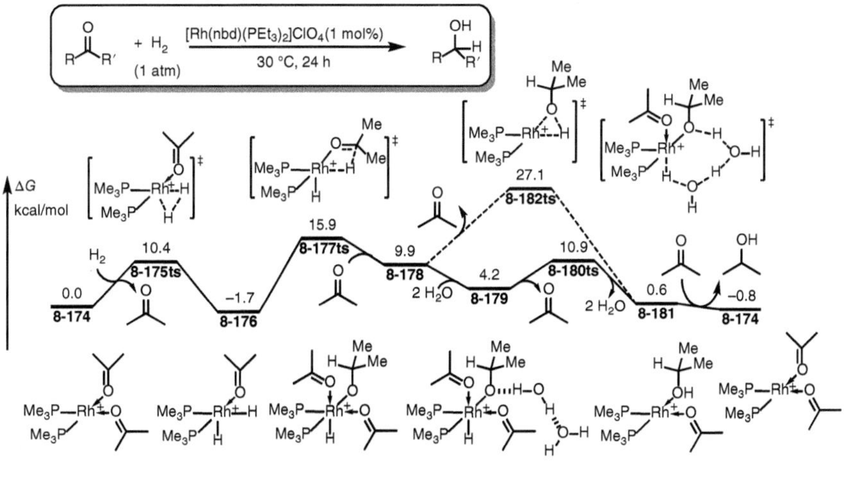

Figure 8.22 Free-energy profiles for Rh-catalyzed hydrogenation of ketones. The energies were calculated at B3LYP-D3BJ/def2-DZVP/PCM(acetone)//B3LYP-D3BJ/def2-SVP level of theory.

an energy barrier of only 6.7 kcal/mol. As a contrast, the energy barrier of reductive elimination is detected to be 17.2 kcal/mol in the absence of water-assist.

8.6.2 Hydrogenation of Carbon Dioxide

In the presence of pincer PCP ligand-coordinated Rh-complex, carbon dioxide can be reduced by gaseous dihydrogen to form equivalent formate Rh(III) hydride complex. The mechanism of this reaction was revealed by Huang and Fujita [178]. As shown in Figure 8.23, the oxidative addition of dihydrogen onto pincer ligand-coordinated Rh(I) specie **8-183** takes place without energy barrier to form a dihydride Rh(III) species **8-184**. Then, carbonyl of carbon dioxide inserts into Rh—H bond via a four-membered ring-type transition state **8-185ts** to form a formate Rh(III) intermediate **8-186**. The calculated overall activation free energy of this step is 17.8 kcal/mol. The possible reductive elimination via a five-membered ring-type transition state **8-187ts** could yield formic acid product **8-188**; however, the calculated reaction energy is 9.0 kcal/mol endergonic, which is thermodynamically unfavorable.

8.6.3 Hydroacylation of Ketones

The Rh-catalyzed hydroacylation of ketones leads to the formation of corresponding ester product. Mechanism of an intramolecular hydroacylation for the construction of lactones was revealed by DFT calculations [172]. As shown in Figure 8.24, the chelation of ketone moiety leads to the coordination of formyl group in Rh(I) species **8-189**. The oxidative addition of C—H bond in formyl group takes place via transition state **8-190ts** to afford an acyl hydride Rh(III) intermediate **8-191**. Then, the coordinated carbonyl group inserts into Rh—H bond via transition state **8-192ts** to

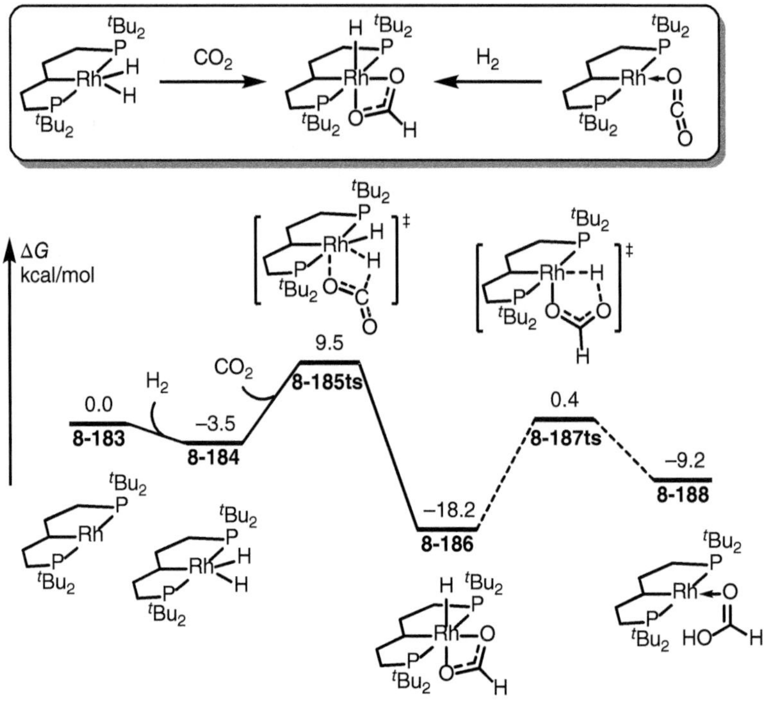

Figure 8.23 Free-energy profiles for Rh-mediated hydrogenation of carbon dioxide. The energies were calculated at KMLYP/6-31+G(d) (LANL2DZ for Rh) level of theory.

afford an rhoda(III)cycle **8-193**. The calculated activation free energy for this step is 22.7 kcal/mol, which is considered as rate-determining step. After a reductive elimination via transition state **8-194ts**, lactone product can be yielded by the regeneration of Rh(I) species **8-189**.

8.7 Rh-Catalyzed Carbene Transformations

In Rh-catalysis, Fischer-type Rh–carbene complex can be easily achieved by the denetrogenation of diazo compounds [179–182]. In Rh–carbene complex, carbon atom in carbenoid plays electrophilicity, which can react with a series of nucleophiles, including C—H bonds, alkenes, alcohols, electron-rich arenes, and carbon monoxide [183–188]. As shown in Scheme 8.4, the bifunctionalization of carbenoid can be processed through either a concerted process (Path [a]) or a stepwise process (Path [b]) [189–192].

8.7.1 Carbene Insertion into C—H Bonds

When a Fischer-type Rh–carbene complex is formed, carbon atom in carbenoid plays a remarkable electrophilicity. The σ-orbital of a C—H bond in alkanes

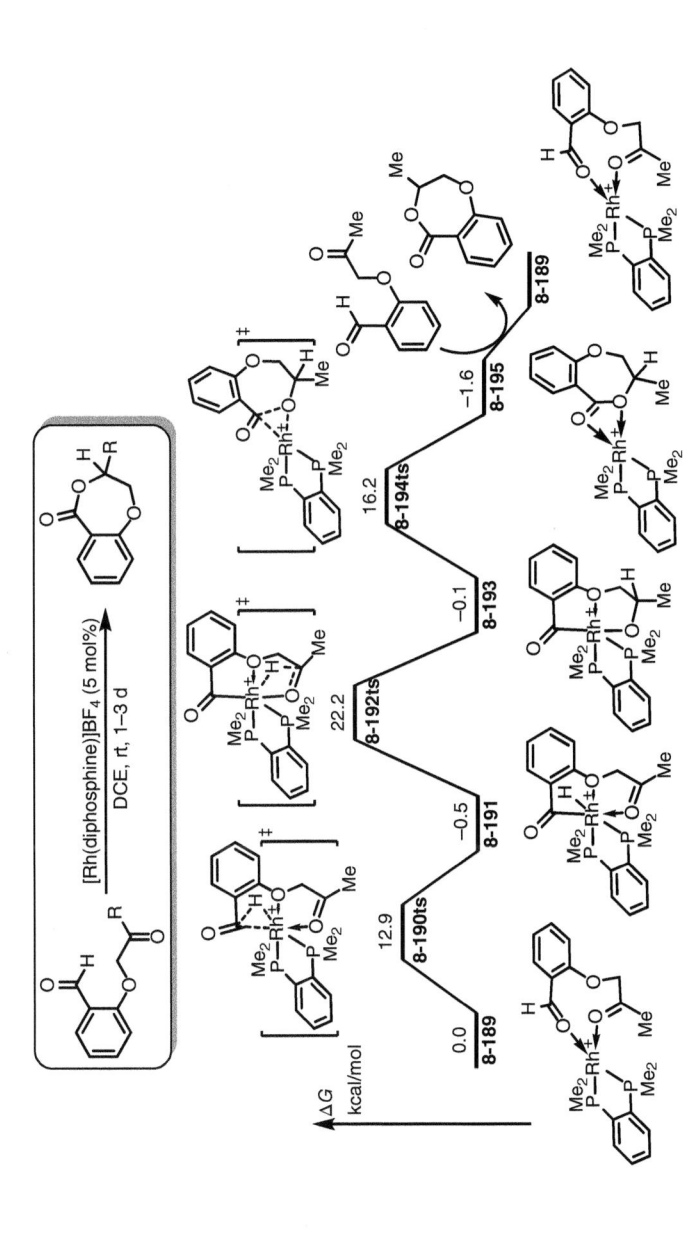

Figure 8.24 Free-energy profiles for Rh-catalyzed hydroacylation of ketones. The energies were calculated at B3LYP/LACV3P** level of theory.

356 | *8 Theoretical Study of Rh-Catalysis*

Scheme 8.4 General mechanism of Rh-mediated carbene transformations. (a) Concerted pathway. (b) Stepwise pathway.

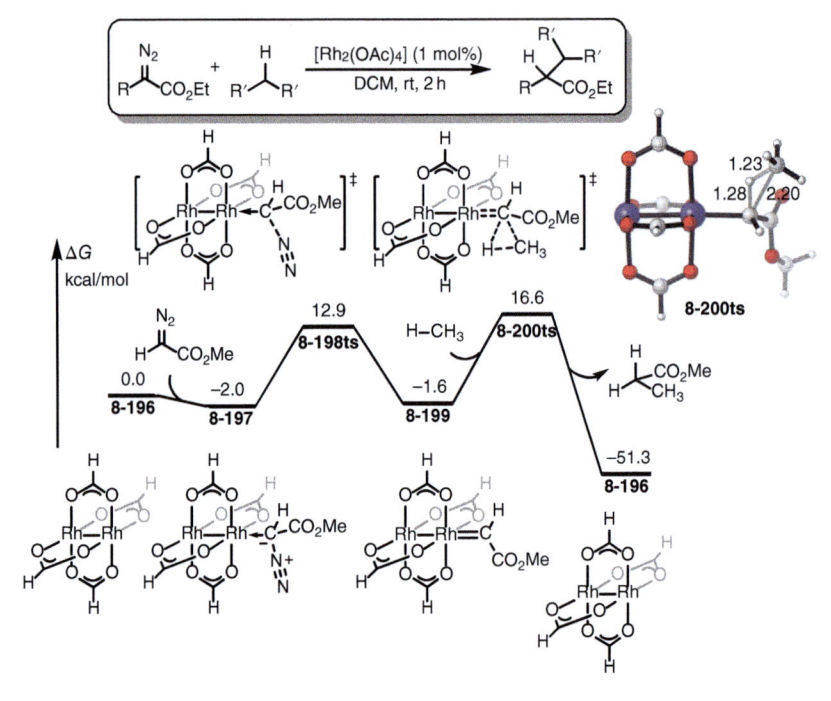

Figure 8.25 Free-energy profiles for dirhodium-catalyzed carbene insertion into C(alkyl)−H bond. The energies were calculated at B3LYP/6-31G(d) (LANL2DZ for Rh)/PCM(DCM)// B3LYP/6-31G(d) (LANL2DZ for Rh) level of theory.

can be employed as a nucleophile to react with Rh–carbene complex for the hydroalkylation of carbenoid [193–196]. In this type of reaction, dirhodium tetracarboxylate complexes are commonly used catalysts. The detailed reaction mechanism is exhibited by DFT calculations [197]. As shown in Figure 8.25, the reaction starts from a dirhodium tetracarboxylate complex **8-196**, which can be coordinated by diazo compound to afford complex **8-197** with 2.0 kcal/mol exergonic. The C—N bond cleavage via a linear transition state **8-198ts** gives a Rh–carbene complex **8-199** with an energy barrier of 14.9 kcal/mol. Then, a Rh-assisted carbene insertion into C(alkyl)—H bond takes place via a three-membered ring-type transition state **8-200ts** to yield hydroalkylated product with the regeneration of dirhodium complex **8-196**. The geometry information of transition state **8-200ts** shows that the lengths of forming C(carbene)—C(alkyl), C(carbene)—H bonds, and breaking C(alkyl)—H bond are 2.20, 1.28, and 1.23 Å, respectively, which reveal a concerted process for the C(carbene)—C(alkyl)/C(carbene)—H bonds formation and C(alkyl)—H bond cleavage.

8.7.2 Arylation of Carbenes

When Rh–carbene complex reacts with electron-rich arenes, the carbene insertion into C(aryl)—H bond can be achieved through a stepwise process. In this reaction, nucleophilic attack of arene occurs first. Then, proton transfer provides the hydroarylated product. DFT calculations are used to study the Rh-catalyzed hydroarylation of carbenoids by using phenylpyrrolidine as nucleophile [198]. As shown in Figure 8.26, the coordination of diazo onto dirhodium species **8-201** forms complex **8-202** with 3.9 kcal/mol endergonic. Then, a denitrogenation occurs via transition state **8-203ts** to afford a Rh–carbene complex **8-204**. An intermolecular nucleophilic attack with the phenyl π-bond in aminobenzene takes place via transition state **8-205ts** with an energy barrier of only 10.5 kcal/mol to generate a zwitterionic iminium rhodate intermediate **8-206**.

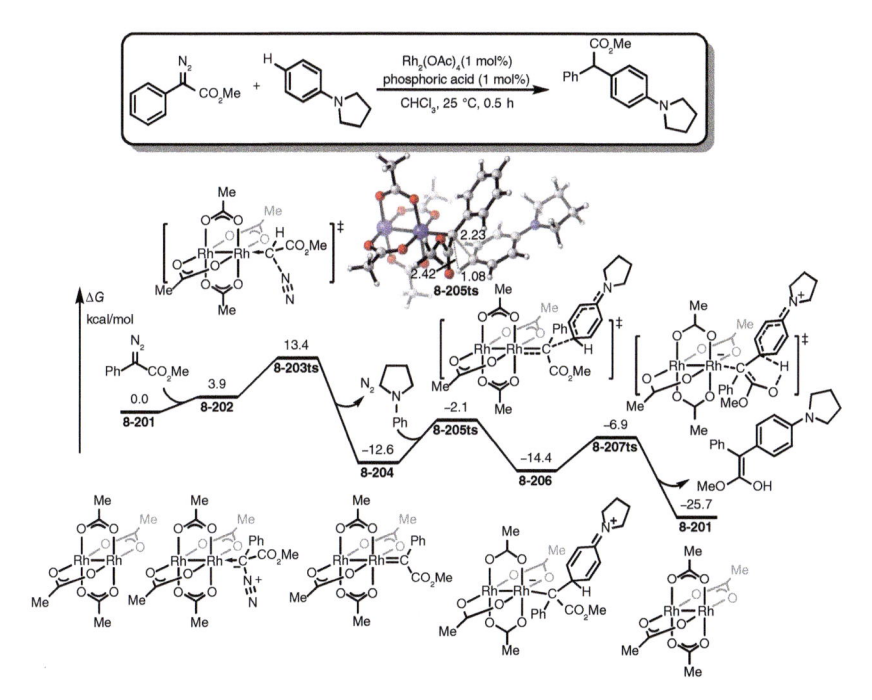

Figure 8.26 Free-energy profiles for dirhodium-catalyzed carbene insertion into C(aryl)—H bond. The energies were calculated at M06/6-311+G(d,p) (LANL2DZ for Rh)/SMD (chloroform)//B3LYP/6-31G(d) (LANL2DZ for Rh) level of theory.

358 | *8 Theoretical Study of Rh-Catalysis*

The geometry information of transition state BTS4 shows that the length of forming C(carbene)—C(aryl) bond is 2.23 Å. Moreover, the length of C(aryl)—H bond and the distance of C(carbene)—H are 1.08 and 2.42 Å, respectively, which clearly reveal a stepwise process for the formation of only C(carbene)—C(aryl) bond. After the nucleophilic attack, an intramolecular proton transfer occurs via a five-membered ring-type transition state **8-207ts** to regenerate dirhodium species **8-201**. The generated enol can easily transfer to corresponding ester product.

8.7.3 Cyclopropanation of Carbenes

When a Fischer-type Rh–carbene complex reacts with alkenes, a (2 + 1) cycloaddition would provide corresponding cyclopropanes through a concerted process. DFT calculations were used to reveal the mechanism of this reaction [199]. As shown in Figure 8.27, the coordination of diazo compound onto dirhodium species **8-208** affords complex **8-209** with 12.8 kcal/mol exothermic. The denitrogenation via transition state **8-210ts** with an energy barrier of 15.3 kcal/mol generates Rh–carbene complex **8-211**. An intermolecular nucleophilic attack by styrene takes place via transition state **8-212ts** resulting in the formation of cyclopropane product. The geometry information of transition state **8-212ts** shows that the lengths for the two forming C(carbene)—C(alkene) bonds are 3.30 and 3.70 Å, respectively, which reveal a very early coming transition state with the synergetic formation of two C(carbene)—C(alkene) bonds.

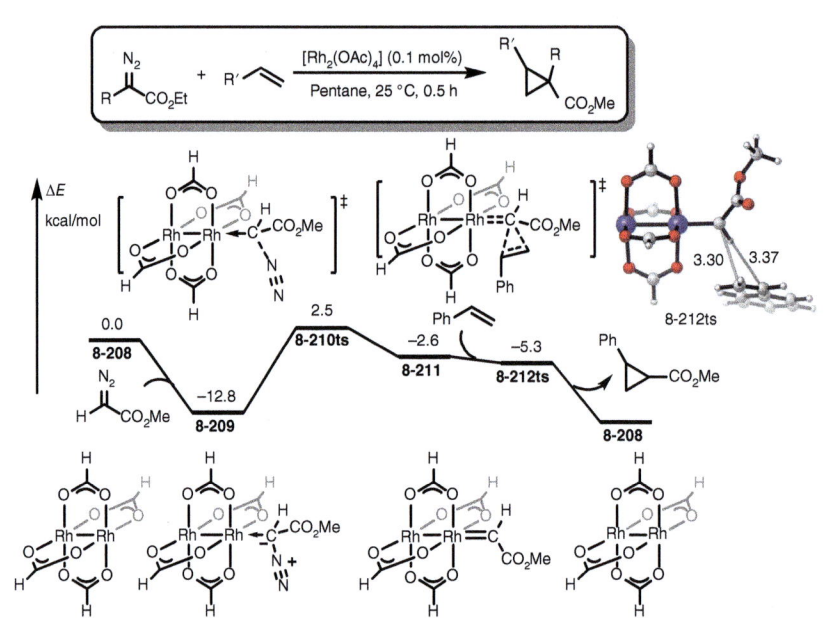

Figure 8.27 Energy profiles for dirhodium-catalyzed cyclopropanation of carbenes. The energies were calculated at B3LYP/6-31G(d) (LANL2DZ for Rh) level of theory.

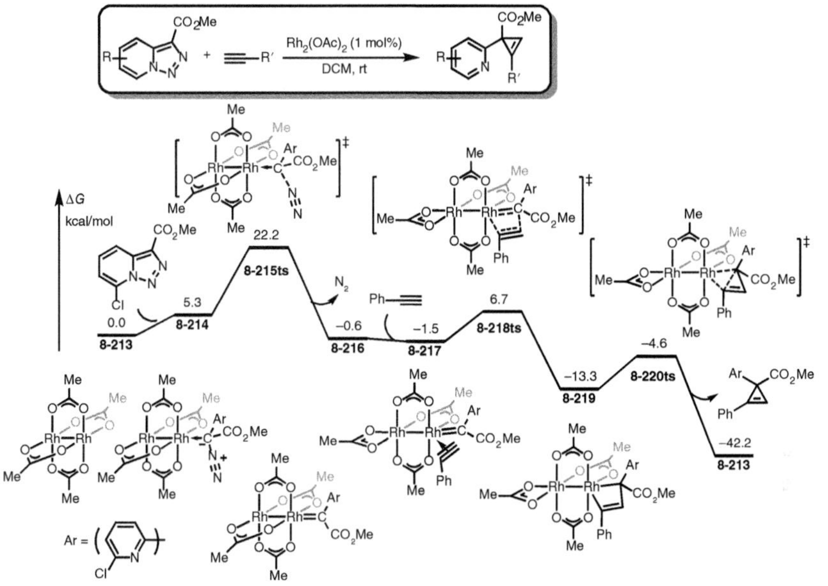

Figure 8.28 Free-energy profiles for dirhodium-catalyzed cyclopropenation of carbenes. The energies were calculated at ωB97XD/6-311++G(d,p) (LANL2TZ(f) for Rh). SMD(DCM) //ωB97XD/6-31G(d) (LANL2DZ for Rh) level of theory. Source: Based on Wang et al. [200].

8.7.4 Cyclopropenation of Carbenes

When alkyne is used as nucleophile to react with Fischer-type Rh–carbene complex, corresponding cyclopropene product can be observed experimentally. The mechanism of this reaction was revealed by Bao group (Figure 8.28) [200]. When pyridotriazole is used as substrate, the tautomerization can form an aryl diazo isomer with an energy barrier of 17.5 kcal/mol. The coordination of this aryl diazo isomer onto dirhodium species **8-213** leads to the denitrogenation via transition state **8-215ts** with an overall activation free energy of 22.2 kcal/mol to form a Rh–carbene complex **8-216**. The coordination of acetylene onto one of the rhodium forms complex **8-217** reversibly. Then, an alkyne insertion takes place via a four-membered ring-type transition state **8-218ts** to afford a four-membered rhodacycle **8-219**. The calculated overall activation free energy for this step is only 7.3 kcal/mol. After a rapid reductive elimination via transition state **8-220ts**, cyclopropane product is yielded with the regeneration of dirhodium catalyst **8-213**.

8.8 Rh-Catalyzed Nitrene Transformations

Usually, metal–nitrene complexes can be considered as isoelectronic species of corresponding metal–carbene complexes. However, high-spin state of metal–nitrene complexes can provide unique reactivities in this chemistry [201]. Moreover, the lower electronegativity of nitrogen atom in Rh–nitrene complexes leads to nucleophilicity of nitrene, which also can make some difference when compared with corresponding Rh–carbene complexes [28, 86].

8.8.1 Nitrene Insertion into C—H Bonds

The mechanism of nitrene insertion into C—H bonds is determined by the usage of Rh catalyst (Figure 8.29). When a tetradentate liganted Rh(III) catalyst is used, a stepwise insertion pathway was proposed by DFT calculations [201]. When Rh(III) species reacts with azide compound, a high-spin triplet Rh(III)–nitrene complex **8-221** is formed, in which nitrogen atom exhibits radical character. A radical-type substitution with C(alkyl)—H bond takes place via transition state **8-222ts** with an energy barrier of 10.8 kcal/mol forming amino Rh(III) complex **8-223** and an alkyl

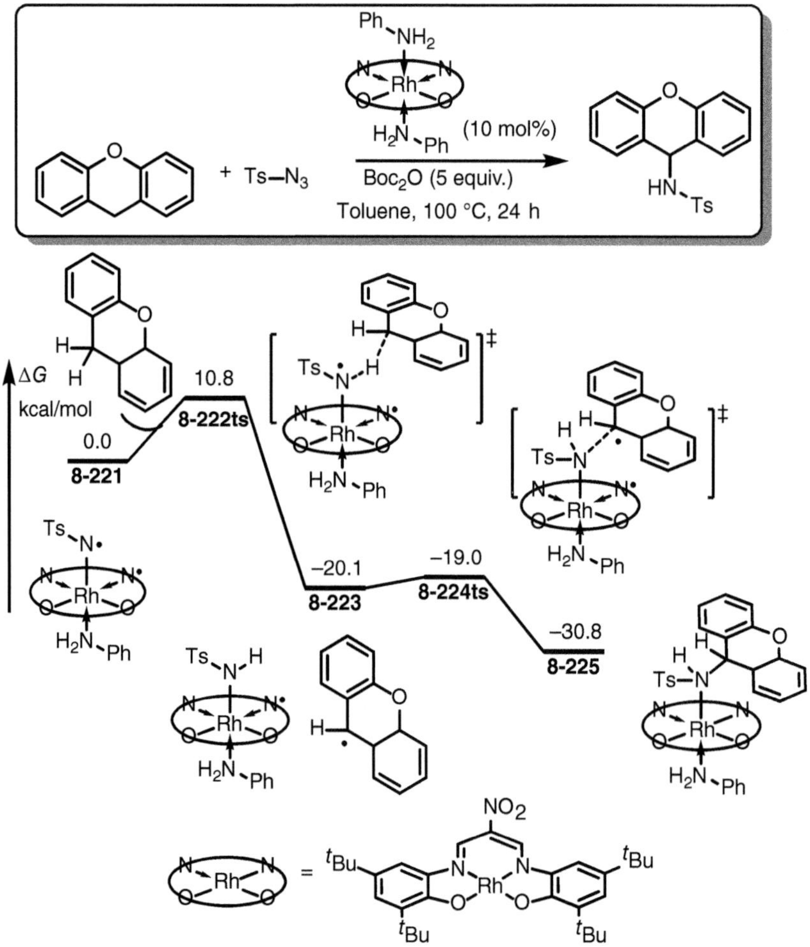

Figure 8.29 Free-energy profiles for rhodium-mediated nitrene insertion into C(alkyl)–H bond through a radical substitution. The energies were calculated at B3LYP/D95** (SDD for Rh) level of theory.

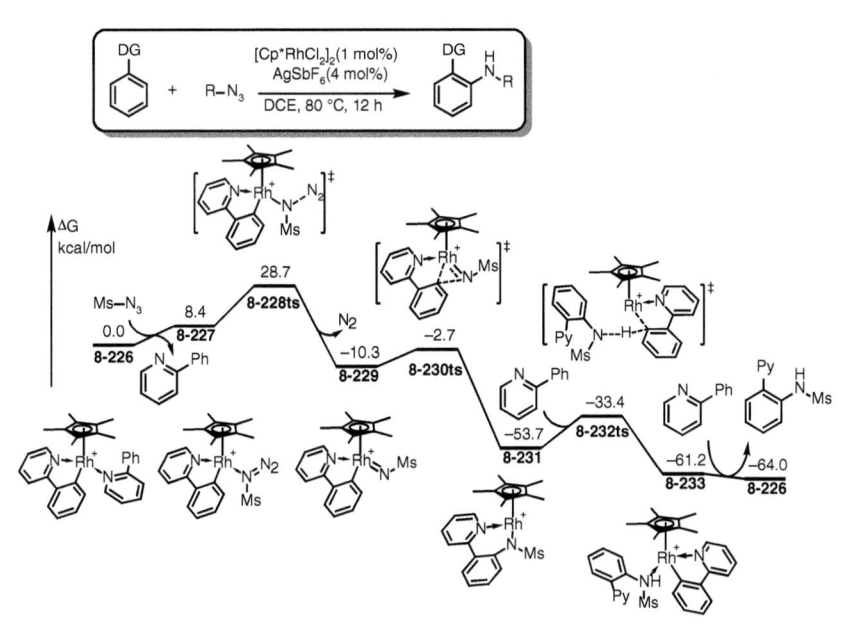

Figure 8.30 Free-energy profiles for rhodium-catalyzed nitrene insertion into C(aryl)—H bond through C—H activation. The energies were calculated at B3LYP/D95** (SDD for Rh) level of theory. Source: Based on Park et al. [120].

radical. Then, a rapid radical coupling occurs via transition state **8-224** generated hydroalkylation product **8-225**. In the whole process, the oxidative state of Rh remains at +3, while the ligand can exhibit radical character.

When a CpRh(III) species is used as catalyst in nitrene insertion into C—H bond, the reaction can undergo a low-spin process through C—H activation, nitrene insertion, and protonation (Figure 8.30) [120]. The catalytic cycle starts to form an aryl Rh(III) species **8-226**. The coordination of azide compound in complex **8-227** leads to the N—N bond cleavage via transition state **8-228ts**. The calculated overall activation free energy of this step is 28.7 kcal/mol, which is considered as rate limit in the whole catalytic cycle. When aryl Rh–nitrene complex **8-229** is formed, a nitrene insertion into Rh—C(aryl) bond can occur via a three-membered ring-type transition state **8-230ts** with an energy barrier of only 7.6 kcal/mol to form a six-membered rhodacycle **8-231**. The σ-bond metathesis-type C—H activation assisted by pyridyl directing group takes place via a four-membered ring-type transition state **8-232ts** to release hydroalrylation product and regenerate aryl Rh(III) active species **8-226**.

8.8.2 Aziridination of Nitrenes

In the presence of dirhodium catalyst, nitrenoid can react with alkene to offer aziridine product. The mechanism of this reaction was studied by DFT calculations [202]. As shown in Figure 8.31, when iminoiodane is used as nitrene source, the coordination onto dirhodium species leads to the formation of complex **8-235^s** with 9.6 kcal/mol exergonic. The decomposition of iminoiodane can undergo a

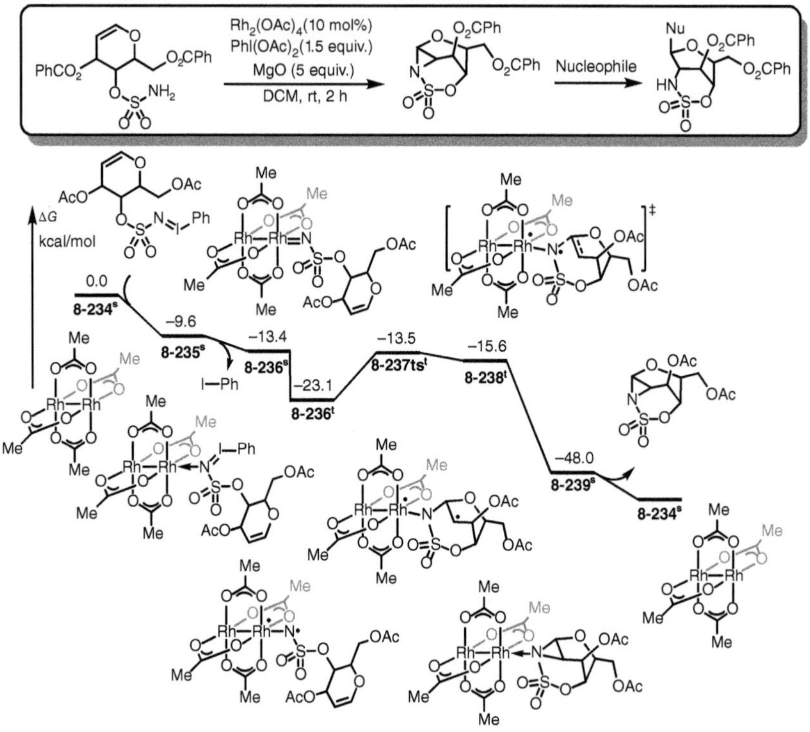

Figure 8.31 Free-energy profiles for rhodium-catalyzed intramolecular aziridination of nitrenes. The energies were calculated at BPW91/6-31G(d) (LANL2DZ for Rh and LANL2DZ(dp) for I)/PCM(DCM) level of theory.

barrierless process to afford a singlet Rh–nitrene complex **8-236^s**; however, the triplet state **8-236^t** of this complex is 9.7 kcal/mol more stable. Then, an intramolecular radical addition onto C=C double bond occurs via transition state **8-237ts^t** with an energy barrier of 9.6 kcal/mol in a triplet potential energy surface. Then, a barrierless radical coupling generates aziridine product through spin crossover.

8.9 Rh-Catalyzed Cycloadditions

Small-to-medium-sized cyclic organic compounds (3–9 carbon/heteroatoms) have attracted increasing interest in natural product synthesis and pharmaceutical chemistry, while developing new strategies for their construction is an important and challenging task [203–207]. Generally, there are two ways for the synthesis of carboncycles from chain-type organic compounds, involving ring-closure reactions and cycloadditions [208–213]. New bond between the two terminals of a chain-type organic compound can be formed to construct organic ring molecule; however, the atomic number of the newly formed ring cannot exceed that of the original chain. Alternatively, in a cycloaddition reaction, the combination of two or more fragments forms carboncycle, whose atomic number can be increased by using more reactants.

The common organic sources for cycloaddition reactions are unsaturated molecules, which can be categorized by the atomic number of carbon and heteroatoms used in cycloadditions [214–216]. The one-atom sources can be unsaturated atoms, such as carbon monoxide, isocyanide, carbene, and nitrene. The two-atom sources are usually unsaturated bonds, such as alkene, alkyne, and carbonyl. 1,3-dipoles can be used as three-atom sources in cycloadditions, which are often used to construct heterocycles. The strain in cyclopropane derivatives provides more reactivity, which can be used as three-atom sources. Allylic ions are also frequently used three-atom sources in the synthesis of carboncycles. The conjugative unsaturated molecules offer four or more atom sources for cycloadditions.

It is remarkable that the cycloaddition reactions often involved two or more unsaturated molecules. In a noncatalytic process, it should bear harsh reaction conditions with unacceptable reaction yield. Effectively, transition metals are often used as catalyst, which can significantly accelerate the cycloaddition with two or more unsaturated molecules [217, 218]. In the past several decades, a series of transition metals, such as Ni, Pd, Pt, Co, Rh, Ir, Cu, and Au, had been used as catalyst, which provide many powerful cycloaddition reactions for the construction of carbon(heteroatom)cycles [219–222]. Among them, Rh-catalysis revealed the best reactivity and efficiency, because almost all for the unsaturated molecules can be well activated by Rh-species. Indeed, three to nine-membered carbon(hetero)rings can be constructed by using Rh-catalysis [223–228]. In Rh-catalyzed cycloadditions, more than two unsaturated molecules would be activated, while their reacting sequences and activation modes are variable. Therefore, the mechanism of Rh-catalyzed cycloadditions would be very complicated.

In a mechanistic point of view, the catalytic cycle of Rh-catalyzed cycloadditions can be separated into three major processes, including oxidative addition of Rh(I) species to afford a rhoda(III)cycle, ring extension by unsaturated molecule insertion, and reductive elimination of the generation of cycloadducts (Scheme 8.5) [229–234]. Generally, the oxidative cyclization by enynes or diynes with Rh(I) species is often used in the initiations of cycloaddition. The oxidative addition by using vinylcyclopropane is an alternative choice for this step. When a rhoda(III)cycle is formed, carbonyl, alkenes, or alkynes can insert into Rh—C bond to enlarge rhodacycle. After a reductive elimination, cycloadduct product can be released from Rh(I) species. However, the detailed mechanism for a specific reaction would be very complicated.

During the past two decades, DFT calculations provided a powerful tool for the understanding reaction mechanism. A series of DFT explorations are used to reveal the mechanism of Rh-catalyzed cycloaddition reactions. In this section, the previous reported mechanistic studies for Rh-catalyzed cycloadditions are summarized, which would lead to new ideas for designing new type of cycloadditions in future. Majorly, this section is organized by the atomic number of carbon/heteroatoms in each reactant.

364 | 8 Theoretical Study of Rh-Catalysis

Scheme 8.5 General mechanism of Rh-catalyzed cycloadditions. Source: Padwa et al. [229]; Iron et al. [230]; Dalton et al. [231]; Haraburda et al. [232]; Zheng et al. [233]; Zhang and Davies [234].

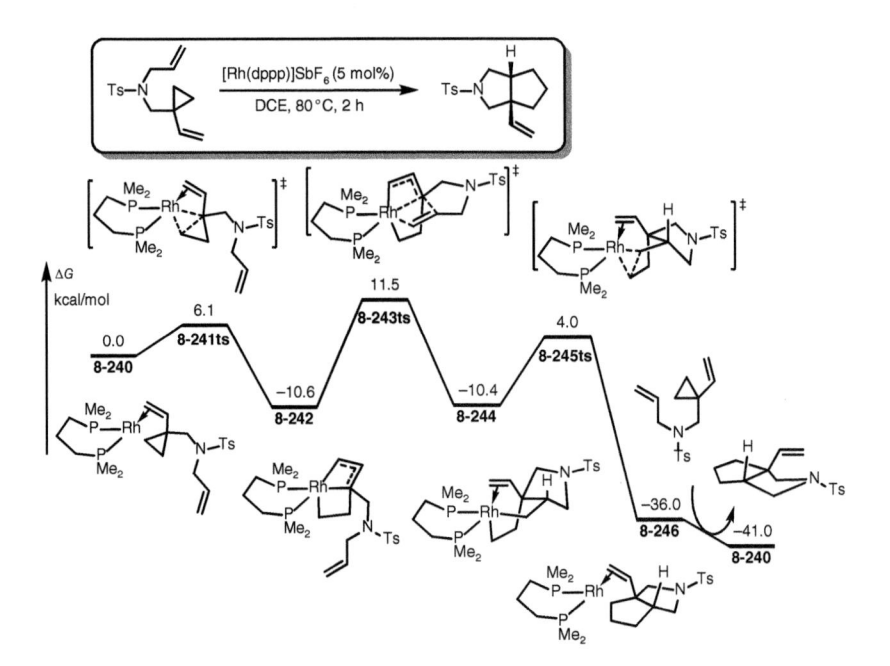

Figure 8.32 Free-energy profiles for rhodium-catalyzed intramolecular (3+2) cycloadditions of cyclopropanes and alkenes. The energies were calculated at B3LYP/6-31G(d) (LANL2DZ for Rh)/CPCM(DCE)//B3LYP/6-31G(d) (LANL2DZ for Rh) level of theory. Source: Based on Lin et al. [235].

8.9.1 (3+2) Cycloadditions

The (3+2) cycloadditions of cyclopropanes and alkenes can be used to synthesize cyclopentane, in which cyclopropanes act as three-carbon source. However, the reactivity of normal cyclopropanes is still defective in Rh-catalysis. Therefore, vinyl cyclopropanes can be used as substrate, in which vinyl group is used to activate the strained three-membered ring. The mechanism of Rh-catalyzed intramolecular (3+2) cycloadditions of vinyl cyclopropanes and alkenes is shown in Figure 8.32 [235]. The reaction starts from a vinyl cyclopropane-coordinated Rh(I) species **8-240**. Assisted by the vinyl group, the oxidative addition of three-membered ring onto Rh occurs via transition state **8-241ts** with an energy barrier of only 6.1 kcal/mol to afford an allylic Rh(III) intermediate **8-242**. Then, an intramolecular alkene insertion into Rh—C(allyl) bond takes place via transition state **8-243ts** to form a six-membered rhodacycle **8-244**. The calculated activation free energy for this step is 22.1 kcal/mol, which is considered as rate-determining step. After a reductive elimination via transition state **8-245ts**, cyclopentane product is generated. The mechanism of (3+2) cycloadditions of cyclopropanes and alkynes for the synthesis of cyclopentenes is similar to above.

In a more complicated transformations, cyclopentenes can be generated by a formal (3+2) cycloaddition of enynol in the presence of Rh(I)-catalyst [236].

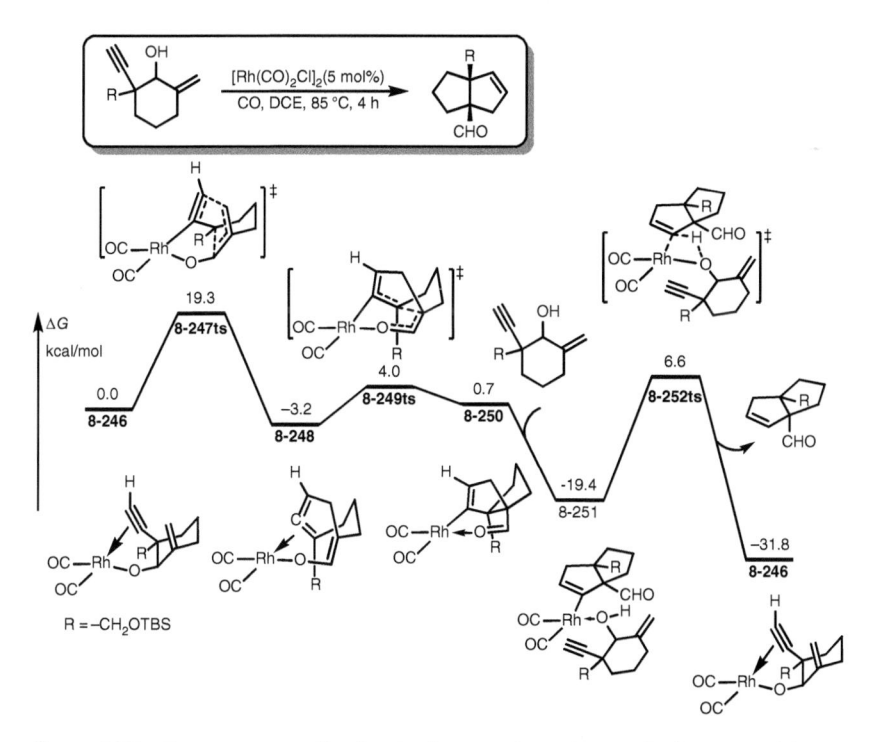

Figure 8.33 Free-energy profiles for rhodium-catalyzed formal (3+2) cycloadditions of enynols. The energies were calculated at M11-L/6-311+G(d) (LANL08 for Rh)/SMD(DCE)// M11-L/6-311+G(d) (LANL08 for Rh) level of theory.

366 | *8 Theoretical Study of Rh-Catalysis*

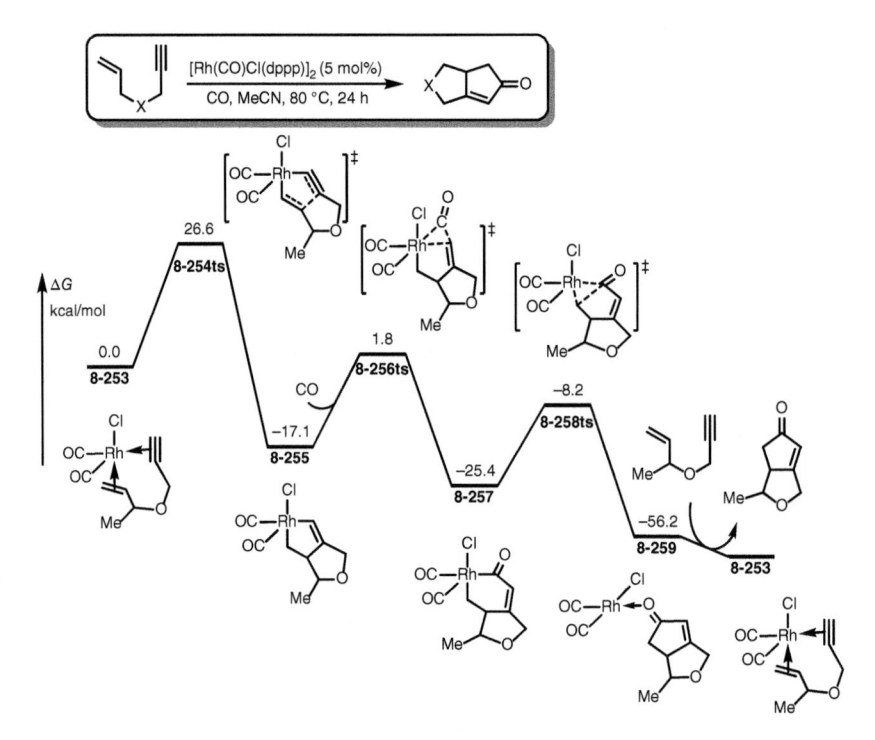

Figure 8.34 Free-energy profiles for rhodium-catalyzed Pauson–Khand reaction. The energies were calculated at B3LYP/cc-pVTZ(-f) level of theory.

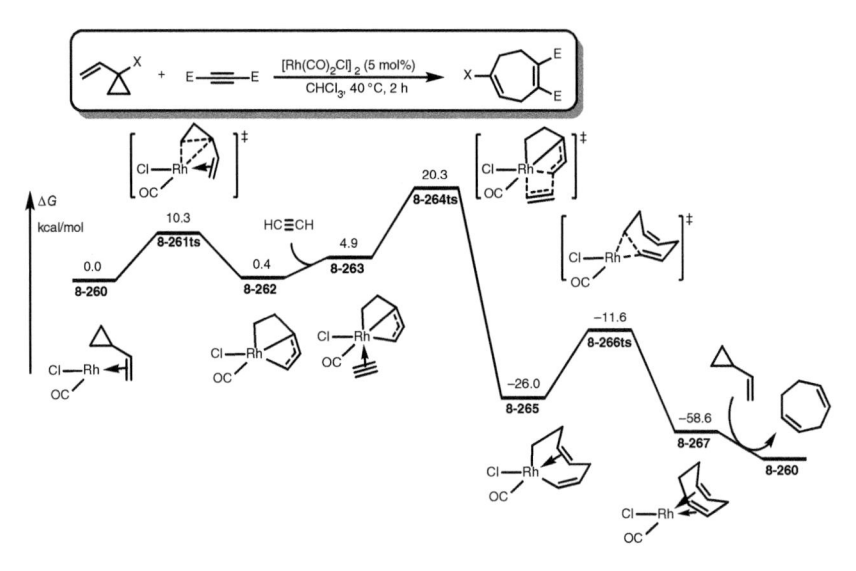

Figure 8.35 Free-energy profiles for rhodium-catalyzed intermolecular (5+2) cycloaddition of vinyl cyclopropane. The energies were calculated at B3LYP/6-31G(d) (LANL2DZ for Rh) level of theory.

8.9 Rh-Catalyzed Cycloadditions

DFT calculations found that when complex **8-246** is formed from alcoholysis, a Rh-assisted intramolecular cope-type 3,3-sigmatropic migration can occur via transition state **8-247ts** with an energy barrier of 19.3 kcal/mol to afford an allene intermediate **8-248**. Then, a Conia-ene type addition takes place via transition state **8-249ts** that is formed cyclopentene framework in intermediate **8-250**. The alcoholysis releases product and regenerates active species **8-251** (Figure 8.33).

8.9.2 Pauson–Khand-type (2+2+1) Cycloadditions

In a Rh-catalyzed Pauson–Khand-type reaction, the combination of alkene, alkyne, and carbon monoxide results in cyclopentenone product leading to a three-component (2+2+1) cycloaddition [237]. As shown in Figure 8.34, DFT calculations found that when enyne coordinates onto Rh(I) in species **8-253**, the oxidative cyclization takes place via a five-membered ring-type transition state **8-254ts** with an energy barrier of 26.6 kcal/mol to form a rhoda(III)cycle **8-255**. Then, the coordinated carbonyl inserts into Rh—C(alkyl) bond via a three-membered ring-type transition state **8-256ts** to form a six-membered rhodacycle **8-257**. After reductive elimination, cyclopentenone product is yielded with the regeneration of Rh(I) species **8-253**.

8.9.3 (5+2) Cycloadditions

During the past two decades, a series of Rh-catalyzed (5+2) cycloadditions have been developed by Wender, Houk, and coworkers [238–241]. In these reactions,

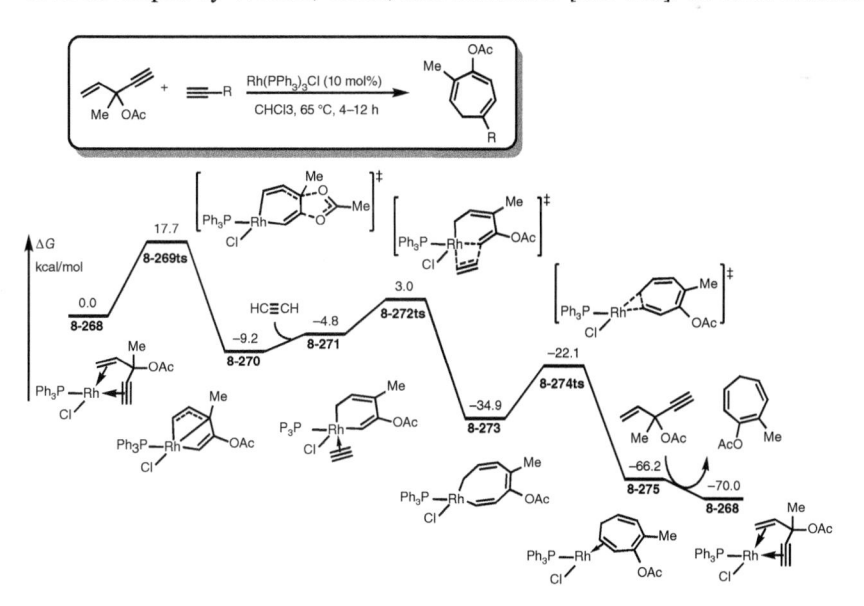

Figure 8.36 Free-energy profiles for rhodium-catalyzed intermolecular (5+2) cycloaddition of 3-acyloxy-1,4-enynes. The energies were calculated at M06/6-311+G(d,p) (SDD for Rh)/SMD(chloroform)//B3LYP/6-31G(d) (SDD for Rh)/SMD(chloroform) level of theory. Source: Xu et al. [242]; Wang et al. [243]; Liang et al. [244].

vinylcyclopropanes were most frequently used C5 sources, which can be activated by low-valence Rh(I) complex. The pioneering theoretical study for the simplest model reaction found that the coordination of vinylcyclopropane onto Rh(I) in complex **8-260** leads to a rapid oxidative addition via transition state **8-261ts** to afford an allylic Rh(III) complex **8-262**. Then, an intermolecular alkyne insertion into Rh—C(allyl) bond occurs via transition state **8-264ts** to form an eight-membered rhoda(III)cycle **8-265**. The following reductive elimination takes place via transition state **8-266ts** to form cycloheptadiene product and regenerate Rh(I) species. The calculated rate-determining step is the alkyne insertion with an overall activation free energy of 20.3 kcal/mol (Figure 8.35).

Moreover, 3-acyloxy-1,4-enynes could be used as an effective alternative in Rh-catalyzed cycloadditions (Figure 8.36) [242]. DFT calculations found that when 3-acyloxy-1,4-enyne coordinates with Rh(I) in complex **8-268**, a 1,2-acyloxy migration leads to the formation of an allylic Rh(III) complex **8-270** with an energy barrier of 17.7 kcal/mol. Then, the coordination of alkyne leads to an intermolecular alkyne insertion into Rh—C(alkenyl) bond via transition state **8-272ts** to form an eight-membered rhoda(III)cycle **8-273**. After a reductive

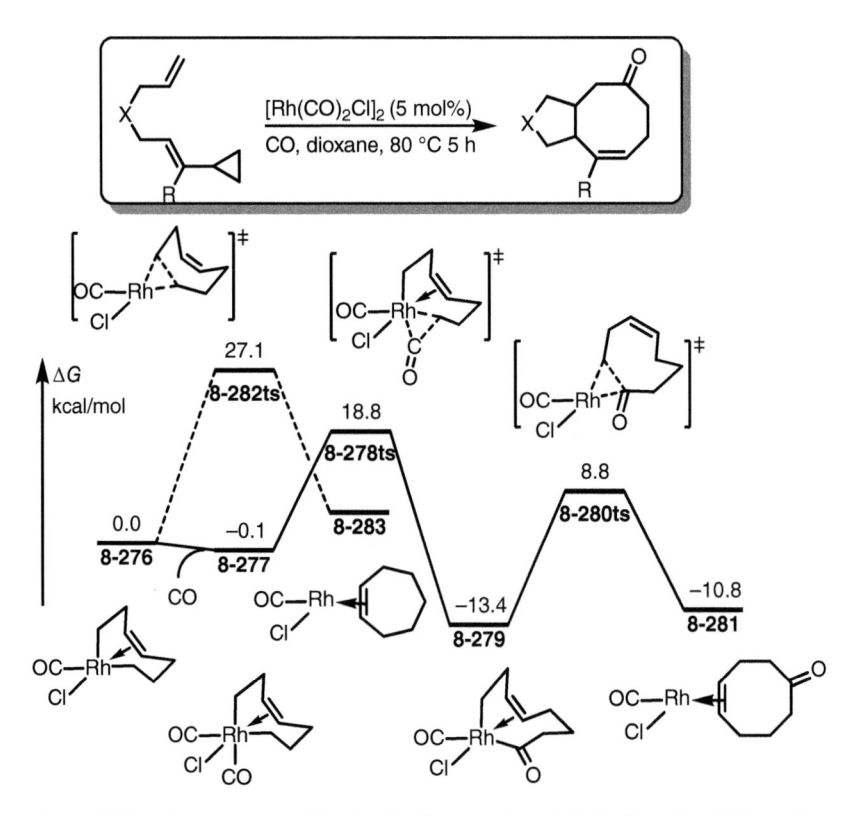

Figure 8.37 Free-energy profiles for rhodium-catalyzed (5+2+1) cycloaddition of ene-vinylcyclopropane. The energies were calculated at M06/6-311+G(d,p) (SDD for Rh)/SMD(chloroform)//B3LYP/6-31G(d) (SDD for Rh)/SMD(chloroform) level of theory.

elimination, cycloheptatriene product can be yielded with the regeneration of Rh(I) species **8-268**.

8.9.4 (5+2+1) Cycloadditions

Yu group provided a good example for computationally designed organometallic catalysis (Figure 8.37) [243, 244]. DFT calculations found that when a seven-membered rhoda(III)cycle **8-276** is formed by the reaction of Rh(I) with ene-vinylcyclopropane, the coordination of carbon monoxide can form complex **8-277** reversibly. Then, a carbonyl insertion takes place via transition state **8-278ts** to afford nine-membered rhoda(III)cycle **8-279**. After reductive elimination through transition state **8-280**, (5+2+1) cycloadduct can be generated. Compared with corresponding (5+2) cycloaddition, the relative free energy of (5+2+1) transition state **8-278ts** is 8.3 kcal/mol lower than that of transition state **8-282ts** leading to (5+2) cycloaddition. Therefore, (5+2+1) cycloaddition would be a favorable pathway in the presence of carbon monoxide. The followed experiments proved this prediction by the same group.

References

1. Widenhoefer, R.A. (2002). Synthetic and mechanistic studies of the cycloisomerization and cyclization/hydrosilylation of functionalized dienes catalyzed by cationic palladium(II) complexes. *Accounts of Chemical Research* 35 (10): 905–913.
2. Zeni, G., Braga, A.L., and Stefani, H.A. (2003). Palladium-catalyzed coupling of sp^2-hybridized tellurides. *Accounts of Chemical Research* 36 (10): 731–738.
3. Sigman, M.S. and Jensen, D.R. (2006). Ligand-modulated palladium-catalyzed aerobic alcohol oxidations. *Accounts of Chemical Research* 39 (3): 221–229.
4. Marion, N. and Nolan, S.P. (2008). Well-defined N-heterocyclic carbenes–palladium(II) precatalysts for cross-coupling reactions. *Accounts of Chemical Research* 41 (11): 1440–1449.
5. Guo, L.N., Duan, X.H., and Liang, Y.M. (2011). Palladium-catalyzed cyclization of propargylic compounds. *Accounts of Chemical Research* 44 (2): 111–122.
6. Neufeldt, S.R. and Sanford, M.S. (2012). Controlling site selectivity in palladium-catalyzed C—H bond functionalization. *Accounts of Chemical Research* 45 (6): 936–946.
7. Ruiz-Castillo, P. and Buchwald, S.L. (2016). Applications of palladium-catalyzed C–N cross-coupling reactions. *Chemical Reviews* 116 (19): 12564–12649.
8. He, J., Wasa, M., Chan, K.S.L. et al. (2017). Palladium-catalyzed transformations of alkyl C–H bonds. *Chemical Reviews* 117 (13): 8754–8786.
9. Wang, D., Weinstein, A.B., White, P.B. et al. (2018). Ligand-promoted palladium-catalyzed aerobic oxidation reactions. *Chemical Reviews* 118 (5): 2636–2679.

10 Fagnou, K. and Lautens, M. (2003). Rhodium-catalyzed carbon–carbon bond forming reactions of organometallic compounds. *Chemical Reviews* 103 (1): 169–196.

11 Hirner, J.J., Shi, Y., and Blum, S.A. (2011). Organogold reactivity with palladium, nickel, and rhodium: transmetalation, cross-coupling, and dual catalysis. *Accounts of Chemical Research* 44 (8): 603–613.

12 Gillingham, D. and Fei, N. (2013). Catalytic X–H insertion reactions based on carbenoids. *Chemical Society Reviews* 42 (12): 4918–4931.

13 Wang, Y. and Yu, Z.X. (2015). Rhodium-catalyzed [5+2+1] cycloaddition of ene-vinylcyclopropanes and CO: reaction design, development, application in natural product synthesis, and inspiration for developing new reactions for synthesis of eight-membered carbocycles. *Accounts of Chemical Research* 48 (8): 2288–2296.

14 Koschker, P. and Breit, B. (2016). Branching out: rhodium-catalyzed allylation with alkynes and allenes. *Accounts of Chemical Research* 49 (8): 1524–1536.

15 Blaszczyk, S.A., Glazier, D.A., and Tang, W. (2020). Rhodium-catalyzed (5+2) and (5+1) cycloadditions using 1,4-enynes as five-carbon building blocks. *Accounts of Chemical Research* 53 (1): 231–243.

16 Zhu, W. and Gunnoe, T.B. (2020). Advances in rhodium-catalyzed oxidative arene alkenylation. *Accounts of Chemical Research* 53 (4): 920–936.

17 Ball, Z.T. (2013). Designing enzyme-like catalysts: a rhodium(II) metallopeptide case study. *Accounts of Chemical Research* 46 (2): 560–570.

18 Zhang, G., Yang, L., Wang, Y. et al. (2013). An efficient Rh/O$_2$ catalytic system for oxidative C–H activation/annulation: evidence for Rh(I) to Rh(III) oxidation by molecular oxygen. *Journal of the American Chemical Society* 135 (24): 8850–8853.

19 Shin, K., Kim, H., and Chang, S. (2015). Transition-metal-catalyzed C–N bond forming reactions using organic azides as the nitrogen source: a journey for the mild and versatile C–H amination. *Accounts of Chemical Research* 48 (4): 1040–1052.

20 Sperger, T., Sanhueza, I.A., Kalvet, I. et al. (2015). Computational studies of synthetically relevant homogeneous organometallic catalysis involving Ni, Pd, Ir, and Rh: an overview of commonly employed DFT methods and mechanistic insights. *Chemical Reviews* 115 (17): 9532–9586.

21 Yu, S., Liu, S., Lan, Y. et al. (2015). Rhodium-catalyzed C–H activation of phenacyl ammonium salts assisted by an oxidizing C–N bond: a combination of experimental and theoretical studies. *Journal of the American Chemical Society* 137 (4): 1623–1631.

22 Liu, S., Qi, X., Qu, L.-B. et al. (2018). C–H bond cleavage occurring on a Rh(V) intermediate: a theoretical study of Rh-catalyzed arene azidation. *Catalysis Science and Technology* 8 (6): 1645–1651.

23 Chen, P.H., Billett, B.A., Tsukamoto, T. et al. (2017). "Cut and sew" transformations via transition-metal-catalyzed carbon–carbon bond activation. *ACS Catalysis* 7 (2): 1340–1360.

24 Desnoyer, A.N. and Love, J.A. (2017). Recent advances in well-defined, late transition metal complexes that make and/or break C–N, C–O and C–S bonds. *Chemical Society Reviews* 46 (1): 197–238.

25 Deng, L., Fu, Y., Lee, S.Y. et al. (2019). Kinetic resolution via Rh-catalyzed C–C activation of cyclobutanones at room temperature. *Journal of the American Chemical Society* 141 (41): 16260–16265.

26 Luo, Y., Shan, C., Liu, S. et al. (2019). Oxidative addition promoted C–C bond cleavage in Rh-mediated cyclopropenone activation: a DFT study. *ACS Catalysis* 9 (12): 10876–10886.

27 Long, Y., Su, Z., Zheng, Y. et al. (2020). Rhodium-catalyzed transarylation of benzamides: C–C bond vs C–N bond activation. *ACS Catalysis* 10 (5): 3398–3403.

28 Ritleng, V., Sirlin, C., and Pfeffer, M. (2002). Ru-, Rh-, and Pd-catalyzed C–C bond formation involving C–H activation and addition on unsaturated substrates: reactions and mechanistic aspects. *Chemical Reviews* 102 (5): 1731–1770.

29 Patureau, F.W. and Glorius, F. (2010). Rh catalyzed olefination and vinylation of unactivated acetanilides. *Journal of the American Chemical Society* 132 (29): 9982–9983.

30 Kim, D.S., Park, W.J., and Jun, C.H. (2017). Metal-organic cooperative catalysis in C–H and C–C bond activation. *Chemical Reviews* 117 (13): 8977–9015.

31 Chan, J., Baucom, K.D., and Murry, J.A. (2007). Rh(II)-catalyzed intermolecular oxidative sulfamidation of aldehydes: a mild efficient synthesis of *N*-sulfonylcarboxamides. *Journal of the American Chemical Society* 129 (46): 14106–14107.

32 Stuart, D.R., Bertrand-Laperle, M., Burgess, K.M. et al. (2008). Indole synthesis via rhodium catalyzed oxidative coupling of acetanilides and internal alkynes. *Journal of the American Chemical Society* 130 (49): 16474–16475.

33 Rakshit, S., Grohmann, C., Besset, T. et al. (2011). Rh(III)-catalyzed directed C–H olefination using an oxidizing directing group: mild, efficient, and versatile. *Journal of the American Chemical Society* 133 (8): 2350–2353.

34 Brasse, M., Campora, J., Ellman, J.A. et al. (2013). Mechanistic study of the oxidative coupling of styrene with 2-phenylpyridine derivatives catalyzed by cationic rhodium(III) via C–H activation. *Journal of the American Chemical Society* 135 (17): 6427–6430.

35 Xing, L., Fan, Z., Hou, C. et al. (2014). Synthesis of pyrazolo[1,2-a]cinnolinesviaa rhodium-catalyzed oxidative coupling approach. *Advanced Synthesis and Catalysis* 356 (5): 972–976.

36 Okada, T., Sakai, A., Hinoue, T. et al. (2018). Rhodium(III)-catalyzed oxidative coupling of *N*-phenylindole-3-carboxylic acids with alkenes and alkynes via C4-H and C2-H/C2'-H bond cleavage. *The Journal of Organic Chemistry* 83 (10): 5639–5649.

37 Meng, R., Bi, S., Jiang, Y.Y. et al. (2019). C–H activation versus ring opening and inner- versus outer-sphere concerted metalation–deprotonation in Rh(III)-catalyzed oxidative coupling of oxime ether and cyclopropanol: a density functional theory study. *The Journal of Organic Chemistry* 84 (17): 11150–11160.

38 Rej, S. and Chatani, N. (2019). Rhodium-catalyzed C(sp^2)- or C(sp^3)–H bond functionalization assisted by removable directing groups. *Angewandte Chemie International Edition* 58 (25): 8304–8329.

39 Guimond, N., Gorelsky, S.I., and Fagnou, K. (2011). Rhodium(III)-catalyzed heterocycle synthesis using an internal oxidant: improved reactivity and mechanistic studies. *Journal of the American Chemical Society* 133 (16): 6449–6457.

40 Zhao, D., Shi, Z., and Glorius, F. (2013). Indole synthesis by rhodium(III)-catalyzed hydrazine-directed C–H activation: redox-neutral and traceless by N–N bond cleavage. *Angewandte Chemie International Edition* 52 (47): 12426–12429.

41 Wu, S., Zeng, R., Fu, C. et al. (2015). Rhodium-catalyzed C–H functionalization-based approach to eight-membered lactams. *Chemical Science* 6 (4): 2275–2285.

42 Wang, Q., Li, Y., Qi, Z. et al. (2016). Rhodium(III)-catalyzed annulation between N-sulfinyl ketoimines and activated olefins: C–H activation assisted by an oxidizing N–S bond. *ACS Catalysis* 6 (3): 1971–1980.

43 Yang, Y.F., Houk, K.N., and Wu, Y.D. (2016). Computational exploration of Rh(III)/Rh(V) and Rh(III)/Rh(I) catalysis in rhodium(III)-catalyzed C–H activation reactions of N-phenoxyacetamides with alkynes. *Journal of the American Chemical Society* 138 (21): 6861–6868.

44 Wu, J.Q., Zhang, S.S., Gao, H. et al. (2017). Experimental and theoretical studies on rhodium-catalyzed coupling of benzamides with 2,2-difluorovinyl tosylate: diverse synthesis of fluorinated heterocycles. *Journal of the American Chemical Society* 139 (9): 3537–3545.

45 Lewis, J.C., Bergman, R.G., and Ellman, J.A. (2008). Direct functionalization of nitrogen heterocycles via Rh-catalyzed C–H bond activation. *Accounts of Chemical Research* 41 (8): 1013–1025.

46 Colby, D.A., Tsai, A.S., Bergman, R.G. et al. (2012). Rhodium catalyzed chelation-assisted C–H bond functionalization reactions. *Accounts of Chemical Research* 45 (6): 814–825.

47 To, C.T. and Chan, K.S. (2017). Selective aliphatic carbon–carbon bond activation by rhodium porphyrin complexes. *Accounts of Chemical Research* 50 (7): 1702–1711.

48 Jun, C.-H. and Hong, J.-B. (1999). Catalytic transformation of aldimine to ketimine by Wilkinson's complex through transimination. *Organic Letters* 1 (6): 887–889.

49 Brak, K. and Ellman, J.A. (2009). Asymmetric synthesis of alpha-branched allylic amines by the Rh(I)-catalyzed addition of alkenyltrifluoroborates to N-tert-butanesulfinyl aldimines. *Journal of the American Chemical Society* 131 (11): 3850–3851.

50 Jiao, L., Lin, M., and Yu, Z.X. (2010). Rh(I)-catalyzed intramolecular [3+2] cycloaddition reactions of 1-ene-, 1-yne- and 1-allene-vinylcyclopropanes. *Chemical Communications* 46 (7): 1059–1061.

51 Sun, Z.M., Zhang, J., Manan, R.S. et al. (2010). Rh(I)-catalyzed olefin hydroarylation with electron-deficient perfluoroarenes. *Journal of the American Chemical Society* 132 (20): 6935–6937.

52 Oonishi, Y., Hosotani, A., and Sato, Y. (2011). Rh(I)-catalyzed formal [6+2] cycloaddition of 4-allenals with alkynes or alkenes in a tether. *Journal of the American Chemical Society* 133 (27): 10386–10389.

53 Dong, K., Wang, Z., and Ding, K. (2012). Rh(I)-catalyzed enantioselective hydrogenation of α-substituted ethenylphosphonic acids. *Journal of the American Chemical Society* 134 (30): 12474–12477.

54 Kou, K.G., Le, D.N., and Dong, V.M. (2014). Rh(I)-catalyzed intermolecular hydroacylation: enantioselective cross-coupling of aldehydes and ketoamides. *Journal of the American Chemical Society* 136 (26): 9471–9476.

55 Tobisu, M., Kita, Y., and Chatani, N. (2006). Rh(I)-catalyzed silylation of aryl and alkenyl cyanides involving the cleavage of C—C and Si—Si bonds. *Journal of the American Chemical Society* 128 (25): 8152–8153.

56 Harada, Y., Nakanishi, J., Fujihara, H. et al. (2007). Rh(I)-catalyzed carbonylative cyclization reactions of alkynes with 2-bromophenylboronic acids leading to indenones. *Journal of the American Chemical Society* 129 (17): 5766–5771.

57 Xia, Y., Liu, Z., Liu, Z. et al. (2014). Formal carbene insertion into C—C bond: Rh(I)-catalyzed reaction of benzocyclobutenols with diazoesters. *Journal of the American Chemical Society* 136 (8): 3013–3015.

58 Syu, J.F., Wang, Y.T., Liu, K.C. et al. (2016). Rh(I)-catalyzed 1,4-conjugate addition of alkenylboronic acids to a cyclopentenone useful for the synthesis of prostaglandins. *Journal of Organic Chemistry* 81 (22): 10832–10844.

59 Zhao, C., Liu, L.C., Wang, J. et al. (2016). Rh(I)-catalyzed insertion of allenes into C—C bonds of benzocyclobutenols. *Organic Letters* 18 (2): 328–331.

60 Liu, Y., Zhou, C., Xiong, M. et al. (2018). Asymmetric Rh(I)-catalyzed functionalization of the 3-C(sp^3)—H bond of benzofuranones with α-diazoesters. *Organic Letters* 20 (18): 5889–5893.

61 Haito, A. and Chatani, N. (2019). Rh(I)-catalyzed [3+2] annulation reactions of cyclopropenones with amides. *Chemical Communications* 55 (40): 5740–5742.

62 Nie, Y., Chen, J., and Zhang, W. (2019). Rh(I)-catalyzed ring-opening of cyclobutanols via C—C bond activation: synthesis of cis-olefin with a remote aldehyde. *Tetrahedron* 75 (39): 130563.

63 Ueura, K., Satoh, T., and Miura, M. (2007). An efficient waste-free oxidative coupling via regioselective C—H bond cleavage: Rh/Cu-catalyzed reaction of benzoic acids with alkynes and acrylates under air. *Organic Letters* 9 (7): 1407–1409.

64 Uto, T., Shimizu, M., Ueura, K. et al. (2008). Rhodium-catalyzed oxidative coupling of triarylmethanols with internal alkynes via successive C—H and C—C bond cleavages. *Journal of Organic Chemistry* 73 (1): 298–300.

65 Shibata, Y. and Tanaka, K. (2011). Catalytic [2+2+1] cross-cyclotrimerization of silylacetylenes and two alkynyl esters to produce substituted silylfulvenes. *Angewandte Chemie International Edition* 50 (46): 10917–10921.

66 Wodrich, M.D., Ye, B., Gonthier, J.F. et al. (2014). Ligand-controlled regiodivergent pathways of rhodium(III)-catalyzed dihydroisoquinolone synthesis: experimental and computational studies of different cyclopentadienyl ligands. *Chemistry – A European Journal* 20 (47): 15409–15418.

67 Ye, B. and Cramer, N. (2015). Chiral cyclopentadienyls: enabling ligands for asymmetric Rh(III)-catalyzed C–H functionalizations. *Accounts of Chemical Research* 48 (5): 1308–1318.

68 Takahama, Y., Shibata, Y., and Tanaka, K. (2016). Heteroarene-directed oxidative sp^2 C—H bond allylation with aliphatic alkenes catalyzed by an (electron-deficient η^5-cyclopentadienyl)rhodium(III) complex. *Organic Letters* 18 (12): 2934–2937.

69 Hong, S.Y., Jeong, J., and Chang, S. (2017). [4+2] or [4+1] annulation: changing the reaction pathway of a rhodium-catalyzed process by tuning the Cp ligand. *Angewandte Chemie International Edition* 56 (9): 2408–2412.

70 Piou, T. and Rovis, T. (2018). Electronic and steric tuning of a prototypical piano stool complex: Rh(III) catalysis for C–H functionalization. *Accounts of Chemical Research* 51 (1): 170–180.

71 Xu, F., Zhang, S., Wu, X. et al. (2006). 2,3-migration in Rh(II)-catalyzed reactions of β-trifluoroacetamido α-diazocarbonyl compounds. *Organic Letters* 8 (15): 3207–3210.

72 Liao, M., Peng, L., and Wang, J. (2008). Rh(II)-catalyzed Sommelet–Hauser rearrangement. *Organic Letters* 10 (5): 693–696.

73 Marcoux, D., Azzi, S., and Charette, A.B. (2009). TfNH$_2$ as achiral hydrogen-bond donor additive to enhance the selectivity of a transition metal catalyzed reaction. Highly enantio- and diastereoselective rhodium-catalyzed cyclopropanation of alkenes using α-cyano diazoacetamide. *Journal of the American Chemical Society* 131 (20): 6970–6972.

74 Chuprakov, S., Kwok, S.W., and Fokin, V.V. (2013). Transannulation of 1-sulfonyl-1,2,3-triazoles with heterocumulenes. *Journal of the American Chemical Society* 135 (12): 4652–4655.

75 Panish, R., Selvaraj, R., and Fox, J.M. (2015). Rh(II)-catalyzed reactions of diazoesters with organozinc reagents. *Organic Letters* 17 (16): 3978–3981.

76 Wu, J., Li, X., Qi, X. et al. (2019). Site-selective and stereoselective O-alkylation of glycosides by Rh(II)-catalyzed carbenoid insertion. *Journal of the American Chemical Society* 141 (50): 19902–19910.

77 Lewis, L.N. and Smith, J.F. (1986). Catalytic C—C bond formation via ortho-metalated complexes. *Journal of the American Chemical Society* 108 (10): 2728–2735.

78 Zhang, F. and Spring, D.R. (2014). Arene C–H functionalisation using a removable/modifiable or a traceless directing group strategy. *Chemical Society Reviews* 43 (20): 6906–6919.

79 Dang, Y., Qu, S., Tao, Y. et al. (2015). Mechanistic insight into ketone α-alkylation with unactivated olefins via C–H activation promoted by metal-organic cooperative catalysis (MOCC): enriching the MOCC chemistry. *Journal of the American Chemical Society* 137 (19): 6279–6291.

80 Neufeldt, S.R., Jiménez-Osés, G., Huckins, J.R. et al. (2015). Pyridine N-oxide vs pyridine substrates for Rh(III)-catalyzed oxidative C–H bond functionalization. *Journal of the American Chemical Society* 137 (31): 9843–9854.
81 Qi, X., Li, Y., Bai, R. et al. (2017). Mechanism of rhodium-catalyzed C–H functionalization: advances in theoretical investigation. *Accounts of Chemical Research* 50 (11): 2799–2808.
82 Wang, X., Lane, B.S., and Sames, D. (2005). Direct C-arylation of free (NH)-indoles and pyrroles catalyzed by Ar–Rh(III) complexes assembled in situ. *Journal of the American Chemical Society* 127 (14): 4996–4997.
83 Santoro, S. and Himo, F. (2018). Mechanism and selectivity of rhodium-catalyzed C—H bond arylation of indoles. *International Journal of Quantum Chemistry* 118 (9): e25526.
84 Wencel-Delord, J., Nimphius, C., Patureau, F.W. et al. (2012). [Rh^{III}Cp*]-catalyzed dehydrogenative aryl–aryl bond formation. *Angewandte Chemie International Edition* 51 (9): 2247–2251.
85 Zhao, D., Li, X., Han, K. et al. (2015). Theoretical investigations on Rh(III)-catalyzed cross-dehydrogenative aryl–aryl coupling via C—H bond activation. *Journal of Physical Chemistry A* 119 (12): 2989–2997.
86 Colby, D.A., Bergman, R.G., and Ellman, J.A. (2010). Rhodium-catalyzed C—C bond formation via heteroatom-directed C—H bond activation. *Chemical Reviews* 110 (2): 624–655.
87 Song, G., Wang, F., and Li, X. (2012). C—C, C—O and C—N bond formation via rhodium(III)-catalyzed oxidative C–H activation. *Chemical Society Reviews* 41 (9): 3651–3678.
88 Song, G. and Li, X. (2015). Substrate activation strategies in rhodium(III)-catalyzed selective functionalization of arenes. *Accounts of Chemical Research* 48 (4): 1007–1020.
89 Lewis, J.C., Bergman, R.G., and Ellman, J.A. (2007). Rh(I)-catalyzed alkylation of quinolines and pyridines via C—H bond activation. *Journal of the American Chemical Society* 129 (17): 5332–5333.
90 Huang, L., Wang, Q., Qi, J. et al. (2013). Rh(III)-catalyzed ortho-oxidative alkylation of unactivated arenes with allylic alcohols. *Chemical Science* 4 (6): 2665.
91 Shibata, K. and Chatani, N. (2014). Rhodium-catalyzed alkylation of C—H bonds in aromatic amides with α, β-unsaturated esters. *Organic Letters* 16 (19): 5148–5151.
92 Deng, H., Li, H., and Wang, L. (2015). A unique alkylation of azobenzenes with allyl acetates by Rh^{III}-catalyzed C–H functionalization. *Organic Letters* 17 (10): 2450–2453.
93 Sharma, U., Park, Y., and Chang, S. (2014). Rh(III)-catalyzed traceless coupling of quinoline N-oxides with internal diarylalkynes. *The Journal of Organic Chemistry* 79 (20): 9899–9906.
94 Zhang, X., Qi, Z., and Li, X. (2014). Rhodium(III)-catalyzed C–C and C–O coupling of quinoline N-oxides with alkynes: combination of C–H activation with O-atom transfer. *Angewandte Chemie International Edition* 53 (40): 10794–10798.

95 Dateer, R.B. and Chang, S. (2015). Selective cyclization of arylnitrones to indolines under external oxidant-free conditions: dual role of Rh(III) catalyst in the C–H activation and oxygen atom transfer. *Journal of the American Chemical Society* 137 (15): 4908–4911.

96 Chen, X., Zheng, G., Li, Y. et al. (2017). Rhodium-catalyzed site-selective coupling of indoles with diazo esters: C4-alkylation versus C2-annulation. *Organic Letters* 19 (22): 6184–6187.

97 Wu, Y., Chen, Z., Yang, Y. et al. (2018). Rh(III)-catalyzed redox-neutral unsymmetrical C–H alkylation and amidation reactions of *N*-phenoxyacetamides. *Journal of the American Chemical Society* 140 (1): 42–45.

98 Kwak, J., Ohk, Y., Jung, Y. et al. (2012). Rollover cyclometalation pathway in rhodium catalysis: dramatic NHC effects in the C—H bond functionalization. *Journal of the American Chemical Society* 134 (42): 17778–17788.

99 Park, Y.J., Park, J.W., and Jun, C.H. (2008). Metal-organic cooperative catalysis in C—H and C—C bond activation and its concurrent recovery. *Accounts of Chemical Research* 41 (2): 222–234.

100 Ko, H.M. and Dong, G. (2014). Cooperative activation of cyclobutanones and olefins leads to bridged ring systems by a catalytic [4+2] coupling. *Nature Chemistry* 6 (8): 739–744.

101 Mo, F. and Dong, G. (2014). C—H bond activation. Regioselective ketone α-alkylation with simple olefins via dual activation. *Science* 345 (6192): 68–72.

102 Li, Y., Liu, S., Qi, Z. et al. (2015). The mechanism of N—O bond cleavage in rhodium-catalyzed C—H bond functionalization of quinoline *N*-oxides with alkynes: a computational study. *Chemistry – A European Journal* 21 (28): 10131–10137.

103 Gong, T.J., Xiao, B., Liu, Z.J. et al. (2011). Rhodium-catalyzed selective C–H activation/olefination of phenol carbamates. *Organic Letters* 13 (12): 3235–3237.

104 Zhang, Q., Yu, H.Z., Li, Y.T. et al. (2013). Computational study on mechanism of Rh(III)-catalyzed oxidative Heck coupling of phenol carbamates with alkenes. *Dalton Transactions* 42 (12): 4175–4184.

105 Liu, G., Shen, Y., Zhou, Z. et al. (2013). Rhodium(III)-catalyzed redox-neutral coupling of *N*-phenoxyacetamides and alkynes with tunable selectivity. *Angewandte Chemie International Edition* 52 (23): 6033–6037.

106 Fukui, Y., Liu, P., Liu, Q. et al. (2014). Tunable arylative cyclization of 1,6-enynes triggered by rhodium(III)-catalyzed C–H activation. *Journal of the American Chemical Society* 136 (44): 15607–15614.

107 Liu, B., Zhou, T., Li, B. et al. (2014). Rhodium(III)-catalyzed alkenylation reactions of 8-methylquinolines with alkynes by C(sp^3)–H activation. *Angewandte Chemie International Edition* 53 (16): 4191–4195.

108 Jiang, J., Ramozzi, R., and Morokuma, K. (2015). Rh(III)-catalyzed C(sp^3)—H bond activation by an external base metalation/deprotonation mechanism: a theoretical study. *Chemistry – A European Journal* 21 (31): 11158–11164.

109 Wang, F., Yu, S., and Li, X. (2016). Transition metal-catalysed couplings between arenes and strained or reactive rings: combination of C–H activation and ring scission. *Chemical Society Reviews* 45 (23): 6462–6477.

110 Park, J. and Chang, S. (2015). Comparative catalytic activity of group 9 [Cp*M^{III}] complexes: cobalt-catalyzed C–H amidation of arenes with dioxazolones as amidating reagents. *Angewandte Chemie International Edition* 54 (47): 14103–14107.

111 Park, Y., Park, K.T., Kim, J.G. et al. (2015). Mechanistic studies on the Rh(III)-mediated amido transfer process leading to robust C–H amination with a new type of amidating reagent. *Journal of the American Chemical Society* 137 (13): 4534–4542.

112 Wang, H., Tang, G., and Li, X. (2015). Rhodium(III)-catalyzed amidation of unactivated C(sp^3)—H bonds. *Angewandte Chemie International Edition* 54 (44): 13049–13052.

113 Lerchen, A., Knecht, T., Daniliuc, C.G. et al. (2016). Unnatural amino acid synthesis enabled by the regioselective cobalt(III)-catalyzed intermolecular carboamination of alkenes. *Angewandte Chemie International Edition* 55 (48): 15166–15170.

114 Yu, S., Li, Y., Zhou, X. et al. (2016). Access to structurally diverse quinoline-fused heterocycles via rhodium(III)-catalyzed C–C/C–N coupling of bifunctional substrates. *Organic Letters* 18 (12): 2812–2815.

115 Yu, S., Tang, G., Li, Y. et al. (2016). Anthranil: an aminating reagent leading to bifunctionality for both C(sp^3)–H and C(sp^2)–H under rhodium(III) catalysis. *Angewandte Chemie International Edition* 55 (30): 8696–8700.

116 Piou, T. and Rovis, T. (2015). Rhodium-catalysed syn-carboamination of alkenes via a transient directing group. *Nature* 527 (7576): 86–90.

117 Kim, J.Y., Park, S.H., Ryu, J. et al. (2012). Rhodium-catalyzed intermolecular amidation of arenes with sulfonyl azides via chelation-assisted C—H bond activation. *Journal of the American Chemical Society* 134 (22): 9110–9113.

118 Ryu, J., Shin, K., Park, S.H. et al. (2012). Rhodium-catalyzed direct C–H amination of benzamides with aryl azides: a synthetic route to diarylamines. *Angewandte Chemie International Edition* 51 (39): 9904–9908.

119 Shin, K., Baek, Y., and Chang, S. (2013). Direct C–H amination of arenes with alkyl azides under rhodium catalysis. *Angewandte Chemie International Edition* 52 (31): 8031–8036.

120 Park, S.H., Kwak, J., Shin, K. et al. (2014). Mechanistic studies of the rhodium-catalyzed direct C–H amination reaction using azides as the nitrogen source. *Journal of the American Chemical Society* 136 (6): 2492–2502.

121 Schroder, N., Wencel-Delord, J., and Glorius, F. (2012). High-yielding, versatile, and practical [Rh(III)Cp*]-catalyzed ortho bromination and iodination of arenes. *Journal of the American Chemical Society* 134 (20): 8298–8301.

122 Lied, F., Lerchen, A., Knecht, T. et al. (2016). Versatile Cp*Rh(III)-catalyzed selective ortho-chlorination of arenes and heteroarenes. *ACS Catalysis* 6 (11): 7839–7843.

123 Zhang, T., Qi, X., Liu, S. et al. (2017). Computational investigation of the role played by rhodium(V) in the rhodium(III)-catalyzed ortho-bromination of arenes. *Chemistry – A European Journal* 23 (11): 2690–2699.

124 Shaw, M.H., Croft, R.A., Whittingham, W.G. et al. (2015). Modular access to substituted azocanes via a rhodium-catalyzed cycloaddition-fragmentation strategy. *Journal of the American Chemical Society* 137 (25): 8054–8057.

125 Zhao, P., Incarvito, C.D., and Hartwig, J.F. (2006). Direct observation of β-aryl eliminations from Rh(I) alkoxides. *Journal of the American Chemical Society* 128 (10): 3124–3125.

126 Xu, T. and Dong, G. (2012). Rhodium-catalyzed regioselective carboacylation of olefins: a C—C bond activation approach for accessing fused-ring systems. *Angewandte Chemie International Edition* 51 (30): 7567–7571.

127 Xu, T. and Dong, G. (2014). Coupling of sterically hindered trisubstituted olefins and benzocyclobutenones by C–C activation: total synthesis and structural revision of cycloinumakiol. *Angewandte Chemie International Edition* 53 (40): 10733–10736.

128 Fu, X.F., Xiang, Y., and Yu, Z.X. (2015). Rh(I)-catalyzed benzo/[7+1] cycloaddition of cyclopropyl-benzocyclobutenes and CO by merging thermal and metal-catalyzed C—C bond cleavages. *Chemistry – A European Journal* 21 (11): 4242–4246.

129 Lu, G., Fang, C., Xu, T. et al. (2015). Computational study of Rh-catalyzed carboacylation of olefins: ligand-promoted rhodacycle isomerization enables regioselective C—C bond functionalization of benzocyclobutenones. *Journal of the American Chemical Society* 137 (25): 8274–8283.

130 Tobisu, M., Kinuta, H., Kita, Y. et al. (2012). Rhodium(I)-catalyzed borylation of nitriles through the cleavage of carbon–cyano bonds. *Journal of the American Chemical Society* 134 (1): 115–118.

131 Kinuta, H., Takahashi, H., Tobisu, M. et al. (2014). Theoretical studies of rhodium-catalyzed borylation of nitriles through cleavage of carbon–cyano bonds. *Bulletin of the Chemical Society of Japan* 87 (6): 655–669.

132 Ishida, N., Sawano, S., Masuda, Y. et al. (2012). Rhodium-catalyzed ring opening of benzocyclobutenols with site-selectivity complementary to thermal ring opening. *Journal of the American Chemical Society* 134 (42): 17502–17504.

133 Li, Y. and Lin, Z. (2013). Theoretical studies of ring-opening reactions of phenylcyclobutabenzenol and its reactions with alkynes catalyzed by rhodium complexes. *Journal of Organic Chemistry* 78 (22): 11357–11365.

134 Calet, S., Urso, F., and Alper, H. (1989). Enantiospecific and stereospecific rhodium(I)-catalyzed carbonylation and ring expansion of aziridines. Asymmetric synthesis of β-lactams and the kinetic resolution of aziridines. *Journal of the American Chemical Society* 111 (3): 931–934.

135 Ikeda, S., Chatani, N., and Murai, S. (1992). Rhodium-catalyzed reaction of N,N-acetals with a hydrosilane and carbon monoxide. *Organometallics* 11 (11): 3494–3495.

136 Gandelman, M. and Milstein, D. (2000). Homogeneously catalyzed, chelate assisted hydrogenolysis of an amine C—N bond. *Chemical Communications* (17): 1603–1604.

137 Sharma, S., Han, S., Kim, M. et al. (2014). Rh-catalyzed oxidative C—C bond formation and C—N bond cleavage: direct access to C2-olefinated free

(NH)-indoles and pyrroles. *Organic and Biomolecular Chemistry* 12 (11): 1703–1706.

138 Kinuta, H., Tobisu, M., and Chatani, N. (2015). Rhodium-catalyzed borylation of aryl 2-pyridyl ethers through cleavage of the carbon–oxygen bond: borylative removal of the directing group. *Journal of the American Chemical Society* 137 (4): 1593–1600.

139 Ouyang, K., Hao, W., Zhang, W.X. et al. (2015). Transition-metal-catalyzed cleavage of C–N single bonds. *Chemical Reviews* 115 (21): 12045–12090.

140 Lu, S.-M. and Alper, H. (2004). Carbonylative ring expansion of aziridines to β-lactams with rhodium-complexed dendrimers on a resin. *Journal of Organic Chemistry* 69 (10): 3558–3561.

141 Ardura, D., Lopez, R., and Sordo, T.L. (2006). A theoretical study of rhodium(I) catalyzed carbonylative ring expansion of aziridines to β-lactams: crucial activation of the breaking C—N bond by hyperconjugation. *Journal of Organic Chemistry* 71 (19): 7315–7321.

142 Liu, G.-D., Du, C.-Y., Zheng, X.-W. et al. (2017). Theoretical insight into the mechanisms and regioselectivity for the borylation reactions of aryl 2-pyridyl ethers catalyzed by rhodium. *Journal of Organometallic Chemistry* 830: 175–180.

143 Landis, C.R. and Steven, F. (2000). A simple model for the origin of enantioselection and the anti "Lock-and-Key" motif in asymmetric hydrogenation of enamides as catalyzed by chiral diphosphine complexes of Rh(I). *Angewandte Chemie International Edition* 39 (16): 2863–2866.

144 Coulter, M.M., Dornan, P.K., and Dong, V.M. (2009). Rh-catalyzed intramolecular olefin hydroacylation: enantioselective synthesis of seven- and eight-membered heterocycles. *Journal of the American Chemical Society* 131 (20): 6932–6933.

145 Willis, M.C. (2010). Transition metal catalyzed alkene and alkyne hydroacylation. *Chemical Reviews* 110 (2): 725–748.

146 Lutz, J.P., Rathbun, C.M., Stevenson, S.M. et al. (2011). Rate-limiting step of the Rh-catalyzed carboacylation of alkenes: C—C bond activation or migratory insertion? *Journal of the American Chemical Society* 134 (1): 715–722.

147 Crozet, D., Kefalidis, C.E., Urrutigoïty, M. et al. (2014). Hydroaminomethylation of styrene catalyzed by rhodium complexes containing chiral diphosphine ligands and mechanistic studies: why is there a lack of asymmetric induction? *ACS Catalysis* 4 (2): 435–447.

148 Dong, Z., Ren, Z., Thompson, S.J. et al. (2017). Transition-metal-catalyzed C–H alkylation using alkenes. *Chemical Reviews* 117 (13): 9333–9403.

149 Vickerman, K.L. and Stanley, L.M. (2017). Catalytic, enantioselective synthesis of polycyclic nitrogen, oxygen, and sulfur heterocycles via Rh-catalyzed alkene hydroacylation. *Organic Letters* 19 (19): 5054–5057.

150 Gridnev, I.D., Liu, Y., and Imamoto, T. (2013). Mechanism of asymmetric hydrogenation of β-dehydroamino acids catalyzed by rhodium complexes: large-scale experimental and computational study. *ACS Catalysis* 4 (1): 203–219.

151 Dornan, P.K., Kou, K.G., Houk, K.N. et al. (2014). Dynamic kinetic resolution of allylic sulfoxides by Rh-catalyzed hydrogenation: a combined theoretical and experimental mechanistic study. *Journal of the American Chemical Society* 136 (1): 291–298.

152 Pubill-Ulldemolins, C., Poyatos, M., Bo, C. et al. (2013). Rhodium–NHC complexes mediate diboration versus dehydrogenative borylation of cyclic olefins: a theoretical explanation. *Dalton Transactions* 42 (3): 746–752.

153 Willis, M.C., Randell-Sly, H.E., Woodward, R.L. et al. (2006). Rhodium-catalyzed intermolecular chelation controlled alkene and alkyne hydroacylation: synthetic scope of β-S-substituted aldehyde substrates. *Journal of Organic Chemistry* 71 (14): 5291–5297.

154 Gonzalez-Rodriguez, C., Pawley, R.J., Chaplin, A.B. et al. (2011). Rhodium-catalyzed branched-selective alkyne hydroacylation: a ligand-controlled regioselectivity switch. *Angewandte Chemie International Edition* 50 (22): 5134–5138.

155 Zhang, X., Li, Y., Shi, H. et al. (2014). Rhodium(III)-catalyzed intramolecular amidoarylation and hydroarylation of alkyne via C–H activation: switchable synthesis of 3,4-fused tricyclic indoles and chromans. *Chemical Communications* 50 (55): 7306–7309.

156 Hooper, J.F., Seo, S., Truscott, F.R. et al. (2016). α-Amino aldehydes as readily available chiral aldehydes for Rh-catalyzed alkyne hydroacylation. *Journal of the American Chemical Society* 138 (5): 1630–1634.

157 Coxon, T.J., Fernandez, M., Barwick-Silk, J. et al. (2017). Exploiting carbonyl groups to control intermolecular rhodium-catalyzed alkene and alkyne hydroacylation. *Journal of the American Chemical Society* 139 (29): 10142–10149.

158 Di Giuseppe, A., Castarlenas, R., Perez-Torrente, J.J. et al. (2012). Ligand-controlled regioselectivity in the hydrothiolation of alkynes by rhodium N-heterocyclic carbene catalysts. *Journal of the American Chemical Society* 134 (19): 8171–8183.

159 Yu, Z., Qi, X., Li, Y. et al. (2016). Mechanism, chemoselectivity and enantioselectivity for the rhodium-catalyzed desymmetric synthesis of hydrobenzofurans: a theoretical study. *Organic Chemistry Frontiers* 3 (2): 209–216.

160 Deng, J., Wen, X., and Li, J. (2017). Mechanistic investigation of Rh(I)-catalyzed alkyne–isatin decarbonylative coupling. *Organic Chemistry Frontiers* 4 (7): 1304–1312.

161 Barwick-Silk, J., Hardy, S., Willis, M.C. et al. (2018). Rh(DPEPhos)-catalyzed alkyne hydroacylation using β-carbonyl-substituted aldehydes: mechanistic insight leads to low catalyst loadings that enables selective catalysis on gram-scale. *Journal of the American Chemical Society* 140 (23): 7347–7357.

162 Alonso, F., Beletskaya, I.P., and Yus, M. (2004). Transition-metal-catalyzed addition of heteroatom–hydrogen bonds to alkynes. *Chemical Reviews* 104 (6): 3079–3160.

163 Song, G., Chen, D., Pan, C.-L. et al. (2010). Rh-catalyzed oxidative coupling between primary and secondary benzamides and alkynes: synthesis of polycyclic amides. *Journal of Organic Chemistry* 75 (21): 7487–7490.

164 Wang, H., Grohmann, C., Nimphius, C. et al. (2012). Mild Rh(III)-catalyzed C–H activation and annulation with alkyne MIDA boronates: short, efficient synthesis of heterocyclic boronic acid derivatives. *Journal of the American Chemical Society* 134 (48): 19592–19595.

165 Palacios, L., Artigas, M.J., Polo, V. et al. (2013). Hydroxo–rhodium–N-heterocyclic carbene complexes as efficient catalyst precursors for alkyne hydrothiolation. *ACS Catalysis* 3 (12): 2910–2919.

166 Chung, L.W., Wiest, O., and Wu, Y.D. (2008). A theoretical study on the trans-addition intramolecular hydroacylation of 4-alkynals catalyzed by cationic rhodium complexes. *Journal of Organic Chemistry* 73 (7): 2649–2655.

167 Athira, C., Changotra, A., and Sunoj, R.B. (2018). Rhodium catalyzed asymmetric hydroamination of internal alkynes with indoline: mechanism, origin of enantioselectivity, and role of additives. *Journal of Organic Chemistry* 83 (5): 2627–2639.

168 Palacios, L., Di Giuseppe, A., Artigas, M.J. et al. (2016). Mechanistic insight into the pyridine enhanced α-selectivity in alkyne hydrothiolation catalysed by quinolinolate–rhodium(I)–N-heterocyclic carbene complexes. *Catalysis Science and Technology* 6 (24): 8548–8561.

169 Gellrich, U., Meißner, A., Steffani, A. et al. (2014). Mechanistic investigations of the rhodium catalyzed propargylic CH activation. *Journal of the American Chemical Society* 136 (3): 1097–1104.

170 Hutschka, F., Dedieu, A., Eichberger, M. et al. (1997). Mechanistic aspects of the rhodium-catalyzed hydrogenation of CO_2 to formic acid – a theoretical and kinetic study. *Journal of the American Chemical Society* 119 (19): 4432–4443.

171 Shen, Z., Khan, H.A., and Dong, V.M. (2008). Rh-catalyzed carbonyl hydroacylation: an enantioselective approach to lactones. *Journal of the American Chemical Society* 130 (10): 2916–2917.

172 Shen, Z., Dornan, P.K., Khan, H.A. et al. (2009). Mechanistic insights into the rhodium-catalyzed intramolecular ketone hydroacylation. *Journal of the American Chemical Society* 131 (3): 1077–1091.

173 Khan, H.A., Kou, K.G.M., and Dong, V.M. (2011). Nitrogen-directed ketone hydroacylation: enantioselective synthesis of benzoxazecinones. *Chemical Science* 2 (3): 407–410.

174 Yu, S. and Li, X. (2014). Mild synthesis of chalcones via rhodium(III)-catalyzed C–C coupling of arenes and cyclopropenones. *Organic Letters* 16 (4): 1220–1223.

175 Kou, K.G., Longobardi, L.E., and Dong, V.M. (2015). Rhodium(I)-catalyzed intermolecular hydroacylation of α-keto amides and isatins with non-chelating aldehydes. *Advanced Synthesis and Catalysis* 357 (10): 2233–2237.

176 Hummel, J.R., Boerth, J.A., and Ellman, J.A. (2017). Transition-metal-catalyzed C—H bond addition to carbonyls, imines, and related polarized π bonds. *Chemical Reviews* 117 (13): 9163–9227.

177 Polo, V., Schrock, R.R., and Oro, L.A. (2016). A DFT study of the role of water in the rhodium-catalyzed hydrogenation of acetone. *Chemical Communications* 52 (96): 13881–13884.

178 Huang, K.-W., Han, J.H., Musgrave, C.B. et al. (2007). Carbon dioxide reduction by pincer rhodium η²-dihydrogen complexes: hydrogen-binding modes and mechanistic studies by density functional theory calculations. *Organometallics* 26 (3): 508–513.

179 Vigalok, A. and Milstein, D. (2000). Direct synthesis of thermally stable PCP-type rhodium carbenes. *Organometallics* 19 (11): 2061–2064.

180 Cohen, R., Rybtchinski, B., Gandelman, M. et al. (2003). Metallacarbenes from diazoalkanes: an experimental and computational study of the reaction mechanism. *Journal of the American Chemical Society* 125 (21): 6532–6546.

181 Werle, C., Goddard, R., Philipps, P. et al. (2016). Structures of reactive donor/acceptor and donor/donor rhodium carbenes in the solid state and their implications for catalysis. *Journal of the American Chemical Society* 138 (11): 3797–3805.

182 Xia, Y., Qiu, D., and Wang, J. (2017). Transition-metal-catalyzed cross-couplings through carbene migratory insertion. *Chemical Reviews* 117 (23): 13810–13889.

183 Tsoi, Y.T., Zhou, Z., and Yu, W.Y. (2011). Rhodium-catalyzed cross-coupling reaction of arylboronates and diazoesters and tandem alkylation reaction for the synthesis of quaternary α,α-heterodiaryl carboxylic esters. *Organic Letters* 13 (19): 5370–5373.

184 Yu, X., Yu, S., Xiao, J. et al. (2013). Rhodium(III)-catalyzed azacycle-directed intermolecular insertion of arene C–H bonds into α-diazocarbonyl compounds. *Journal of Organic Chemistry* 78 (11): 5444–5452.

185 Yada, A., Fujita, S., and Murakami, M. (2014). Enantioselective insertion of a carbenoid carbon into a C—C bond to expand cyclobutanols to cyclopentanols. *Journal of the American Chemical Society* 136 (20): 7217–7220.

186 Ghorai, J. and Anbarasan, P. (2015). Rhodium catalyzed arylation of diazo compounds with aryl boronic acids. *Journal of Organic Chemistry* 80 (7): 3455–3461.

187 Ng, F.N., Lau, Y.F., Zhou, Z. et al. (2015). [Rh(III)(Cp*)]-catalyzed cascade arylation and chlorination of α-diazocarbonyl compounds with arylboronic acids and *N*-chlorosuccinimide for facile synthesis of α-aryl–α-chloro carbonyl compounds. *Organic Letters* 17 (7): 1676–1679.

188 Feng, S., Mo, F., Xia, Y. et al. (2016). Rhodium(I)-catalyzed C—- bond activation of siloxyvinylcyclopropanes with diazoesters. *Angewandte Chemie International Edition* 55 (49): 15401–15405.

189 Chan, W.W., Lo, S.F., Zhou, Z. et al. (2012). Rh-catalyzed intermolecular carbenoid functionalization of aromatic C—H bonds by α-diazomalonates. *Journal of the American Chemical Society* 134 (33): 13565–13568.

190 Xia, Y., Feng, S., Liu, Z. et al. (2015). Rhodium(I)-catalyzed sequential C(sp)–C(sp³) and C(sp³)–C(sp³) bond formation through migratory carbene insertion. *Angewandte Chemie International Edition* 54 (27): 7891–7894.

191 Xia, Y., Liu, Z., Feng, S. et al. (2015). Rh(I)-catalyzed cross-coupling of α-diazoesters with arylsiloxanes. *Organic Letters* 17 (4): 956–959.

192 Lu, Y.S. and Yu, W.Y. (2016). Cp*Rh(III)-catalyzed cross-coupling of alkyltrifluoroborate with alpha-diazomalonates for C(sp^3)–C(sp^3) bond formation. *Organic Letters* 18 (6): 1350–1353.

193 Pirrung, M.C., Liu, H., and Morehead, A.T. Jr., (2002). Rhodium chemzymes: Michaelis–Menten kinetics in dirhodium(II) carboxylate-catalyzed carbenoid reactions. *Journal of the American Chemical Society* 124 (6): 1014–1023.

194 Yoshikai, N. and Nakamura, E. (2003). Theoretical studies on diastereo- and enantioselective rhodium-catalyzed cyclization of diazo compoundvia intramolecular C–H bond insertion. *Advanced Synthesis and Catalysis* 345 (910): 1159–1171.

195 Xie, Q., Song, X.-S., Qu, D. et al. (2015). DFT study on the rhodium(II)-catalyzed C–H functionalization of indoles: enol versus oxocarbenium ylide. *Organometallics* 34 (13): 3112–3119.

196 Zhou, T., Guo, W., and Xia, Y. (2015). RhV-nitrenoid as a key intermediate in RhIII-catalyzed heterocyclization by C–H activation: a computational perspective on the cycloaddition of benzamide and diazo compounds. *Chemistry – A European Journal* 21 (25): 9209–9218.

197 Nakamura, E., Yoshikai, N., and Yamanaka, M. (2002). Mechanism of C—H bond activation/C—C bond formation reaction between diazo compound and alkane catalyzed by dirhodium tetracarboxylate. *Journal of the American Chemical Society* 124 (24): 7181–7192.

198 Liu, H., Duan, J.-X., Qu, D. et al. (2016). Mechanistic insights into asymmetric C–H insertion cooperatively catalyzed by a dirhodium(II) complex and chiral phosphoric acid. *Organometallics* 35 (11): 2003–2009.

199 Nowlan, D.T. 3rd,, Gregg, T.M., Davies, H.M. et al. (2003). Isotope effects and the nature of selectivity in rhodium-catalyzed cyclopropanations. *Journal of the American Chemical Society* 125 (51): 15902–15911.

200 Wang, C., Zhou, Y., and Bao, X. (2017). Mechanistic insights into the Rh-catalyzed transannulation of pyridotriazole with phenylacetylene and benzonitrile: a DFT study. *Journal of Organic Chemistry* 82 (7): 3751–3759.

201 Fujita, D., Sugimoto, H., Shiota, Y. et al. (2017). Catalytic C–H amination driven by intramolecular ligand-to-nitrene one-electron transfer through a rhodium(III) Centre. *Chemical Communications* 53 (35): 4849–4852.

202 Lorpitthaya, R., Xie, Z.-Z., Kuo, J.-L. et al. (2008). Stereocontrolled Intramolecular aziridination of glycals: ready access to aminoglycosides and mechanistic insights from DFT studies. *Chemistry – A European Journal* 14 (5): 1561–1570.

203 Molander, G.A. (1998). Diverse methods for medium ring synthesis. *Accounts of Chemical Research* 31 (10): 603–609.

204 Yet, L. (2000). Metal-mediated synthesis of medium-sized rings. *Chemical Reviews* 100 (8): 2963–3008.

205 Deiters, A. and Martin, S.F. (2004). Synthesis of oxygen- and nitrogen-containing heterocycles by ring-closing metathesis. *Chemical Reviews* 104 (5): 2199–2238.

206 Martí-Centelles, V., Pandey, M.D., Burguete, M.I. et al. (2015). Macrocyclization reactions: the importance of conformational, configurational,

and template-induced preorganization. *Chemical Reviews* 115 (16): 8736–8834.
207 Remy, R. and Bochet, C.G. (2016). Arene–alkene cycloaddition. *Chemical Reviews* 116 (17): 9816–9849.
208 Illuminati, G. and Mandolini, L. (2002). Ring closure reactions of bifunctional chain molecules. *Accounts of Chemical Research* 14 (4): 95–102.
209 Coldham, I. and Hufton, R. (2005). Intramolecular dipolar cycloaddition reactions of azomethine ylides. *Chemical Reviews* 105 (7): 2765–2810.
210 Kurteva, V.B. and Afonso, C.A. (2009). Synthesis of cyclopentitols by ring-closing approaches. *Chemical Reviews* 109 (12): 6809–6857.
211 Monfette, S. and Fogg, D.E. (2009). Equilibrium ring-closing metathesis. *Chemical Reviews* 109 (8): 3783–3816.
212 Gilmore, K. and Alabugin, I.V. (2011). Cyclizations of alkynes: revisiting Baldwin's rules for ring closure. *Chemical Reviews* 111 (11): 6513–6556.
213 Johansson, J.R., Beke-Somfai, T., Said Stalsmeden, A. et al. (2016). Ruthenium-catalyzed azide alkyne cycloaddition reaction: scope, mechanism, and applications. *Chemical Reviews* 116 (23): 14726–14768.
214 Gothelf, K.V. and Jorgensen, K.A. (1998). Asymmetric 1,3-dipolar cycloaddition reactions. *Chemical Reviews* 98 (2): 863–910.
215 Ylijoki, K.E. and Stryker, J.M. (2013). [5 + 2] Cycloaddition reactions in organic and natural product synthesis. *Chemical Reviews* 113 (3): 2244–2266.
216 Chen, J.R., Hu, X.Q., Lu, L.Q. et al. (2015). Formal [4+1] annulation reactions in the synthesis of carbocyclic and heterocyclic systems. *Chemical Reviews* 115 (11): 5301–5365.
217 Wender, P.A., Gamber, G.G., Hubbard, R.D. et al. (2002). Three-component cycloadditions: the first transition metal-catalyzed [5+2+1] cycloaddition reactions. *Journal of the American Chemical Society* 124 (12): 2876–2877.
218 Heller, B. and Hapke, M. (2007). The fascinating construction of pyridine ring systems by transition metal-catalysed [2 + 2 + 2] cycloaddition reactions. *Chemical Society Reviews* 36 (7): 1085–1094.
219 Schore, N.E. (1988). Transition metal-mediated cycloaddition reactions of alkynes in organic synthesis. *Chemical Reviews* 88 (7): 1081–1119.
220 Wender, P.A., Takahashi, H., and Witulski, B. (1995). Transition metal catalyzed [5 + 2] cycloadditions of vinylcyclopropanes and alkynes: a homolog of the Diels–Alder reaction for the synthesis of seven-membered rings. *Journal of the American Chemical Society* 117 (16): 4720–4721.
221 Lautens, M., Klute, W., and Tam, W. (1996). Transition metal-mediated cycloaddition reactions. *Chemical Reviews* 96 (1): 49–92.
222 Chopade, P.R. and Louie, J. (2006). [2+2+2] Cycloaddition reactions catalyzed by transition metal complexes. *Advanced Synthesis and Catalysis* 348 (16–17): 2307–2327.
223 Padwa, A., Carter, S.P., Nimmesgern, H. et al. (1988). Rhodium(II) acetate induced intramolecular dipolar cycloadditions of o-carboalkoxy-α-diazoacetophenone derivatives. *Journal of the American Chemical Society* 110 (9): 2894–2900.

224 Murakami, M., Itami, K., and Ito, Y. (1997). Rhodium-catalyzed asymmetric [4 + 1] cycloaddition. *Journal of the American Chemical Society* 119 (12): 2950–2951.

225 Wender, P.A., Fuji, M., Husfeld, C.O. et al. (1999). Rhodium-catalyzed [5 + 2] cycloadditions of allenes and vinylcyclopropanes: asymmetric total synthesis of (+)-dictamnol. *Organic Letters* 1 (1): 137–140.

226 Tanaka, K., Takeishi, K., and Noguchi, K. (2006). Enantioselective synthesis of axially chiral anilides through rhodium-catalyzed [2+2+2] cycloaddition of 1,6-diynes with trimethylsilylynamides. *Journal of the American Chemical Society* 128 (14): 4586–4587.

227 Reddy, R.P. and Davies, H.M.L. (2007). Asymmetric synthesis of tropanes by rhodium-catalyzed [4 + 3] cycloaddition. *Journal of the American Chemical Society* 129 (34): 10312–10313.

228 Hyster, T.K. and Rovis, T. (2010). Rhodium-catalyzed oxidative cycloaddition of benzamides and alkynes via C–H/N–H activation. *Journal of the American Chemical Society* 132 (30): 10565–10569.

229 Padwa, A., Fryxell, G.E., and Zhi, L. (1990). Tandem cyclization–cycloaddition reaction of rhodium carbenoids. Scope and mechanistic details of the process. *Journal of the American Chemical Society* 112 (8): 3100–3109.

230 Iron, M.A., Martin, J.M., and van der Boom, M.E. (2003). Cycloaddition reactions of metalloaromatic complexes of iridium and rhodium: a mechanistic DFT investigation. *Journal of the American Chemical Society* 125 (38): 11702–11709.

231 Dalton, D.M., Oberg, K.M., Yu, R.T. et al. (2009). Enantioselective rhodium-catalyzed [2 + 2 + 2] cycloadditions of terminal alkynes and alkenyl isocyanates: mechanistic insights lead to a unified model that rationalizes product selectivity. *Journal of the American Chemical Society* 131 (43): 15717–15728.

232 Haraburda, E., Torres, O., Parella, T. et al. (2014). Stereoselective rhodium-catalysed [2+2+2] cycloaddition of linear allene-ene/yne-allene substrates: reactivity and theoretical mechanistic studies. *Chemistry* 20 (17): 5034–5045.

233 Zheng, W.-F., Sun, G.-J., Chen, L. et al. (2018). Enantioselective synthesis of trans-vicinal diamines via rhodium-catalyzed [2+2] cycloaddition of allenamides. *Advanced Synthesis and Catalysis* 360 (9): 1790–1794.

234 Zhang, B. and Davies, H.M.L. (2020). Rhodium-catalyzed enantioselective [4+2] cycloadditions of vinylcarbenes with dienes. *Angewandte Chemie International Edition* 59 (12): 4937–4941.

235 Lin, M., Kang, G.-Y., Guo, Y.-A. et al. (2011). Asymmetric Rh(I)-catalyzed intramolecular [3 + 2] cycloaddition of 1-Yne-vinylcyclopropanes for bicyclo[3.3.0] compounds with a chiral quaternary carbon stereocenter and density functional theory study of the origins of enantioselectivity. *Journal of the American Chemical Society* 134 (1): 398–405.

236 Qi, X., Liu, S., Zhang, T. et al. (2016). Effective chirality transfer in [3+2] reaction between allenyl–rhodium and enal: mechanistic study based on DFT calculations. *Journal of Organic Chemistry* 81 (18): 8306–8311.

237 Baik, M.H., Mazumder, S., Ricci, P. et al. (2011). Computationally designed and experimentally confirmed diastereoselective rhodium-catalyzed Pauson–Khand reaction at room temperature. *Journal of the American Chemical Society* 133 (20): 7621–7623.

238 Wender, P.A., Rieck, H., and Fuji, M. (1998). The transition metal-catalyzed intermolecular [5+2] cycloaddition: the homologous Diels–Alder reaction. *Journal of the American Chemical Society* 120 (42): 10976–10977.

239 Yu, Z.X., Wender, P.A., and Houk, K.N. (2004). On the mechanism of [Rh(CO)$_2$Cl]$_2$-catalyzed intermolecular (5 + 2) reactions between vinylcyclopropanes and alkynes. *Journal of the American Chemical Society* 126 (30): 9154–9155.

240 Liu, P., Sirois, L.E., Cheong, P.H. et al. (2010). Electronic and steric control of regioselectivities in Rh(I)-catalyzed (5 + 2) cycloadditions: experiment and theory. *Journal of the American Chemical Society* 132 (29): 10127–10135.

241 Xu, X., Liu, P., Lesser, A. et al. (2012). Ligand effects on rates and regioselectivities of Rh(I)-catalyzed (5 + 2) cycloadditions: a computational study of cyclooctadiene and dinaphthocyclooctatetraene as ligands. *Journal of the American Chemical Society* 134 (26): 11012–11025.

242 Xu, X., Liu, P., Shu, X.Z. et al. (2013). Rh-catalyzed (5+2) cycloadditions of 3-acyloxy-1,4-enynes and alkynes: computational study of mechanism, reactivity, and regioselectivity. *Journal of the American Chemical Society* 135 (25): 9271–9274.

243 Wang, Y., Wang, J., Su, J. et al. (2007). A computationally designed Rh(I)-catalyzed two-component [5+2+1] cycloaddition of ene-vinylcyclopropanes and CO for the synthesis of cyclooctenones. *Journal of the American Chemical Society* 129 (33): 10060–10061.

244 Liang, Y., Jiang, X., Fu, X.F. et al. (2012). Total synthesis of (+)-asteriscanolide: further exploration of the rhodium(I)-catalyzed [(5+2)+1] reaction of ene-vinylcyclopropanes and CO. *Chemistry – An Asian Journal* 7 (3): 593–604.

9

Theoretical Study of Ir-Catalysis

192.217
Ir[77]
Iridium
[Xe]4f[14]5d[7]6s[2]

As a group 9 transition metal, Ir is relatively seldom applied in homogeneous catalysis when compared with the same-group transition metals Rh and Co. In this area, Ir-complexes are often used in hydrogenation and hydrofunctionalization reactions because of the stability of Ir—H bonds [1–3]. Moreover, boryl Ir species also play an important role in Ir-mediated homogeneous catalysis, which can provide a series of borylation products [4–6]. The common oxidative states of Ir are +1 and +3 [7–9]; however, a formal +5 valence state frequently appears in hydride or boryl Ir species [10, 12]. The mechanism of Ir-catalysis is relatively easy, and theoretical studies are often focused on Ir-catalyzed hydrogenations, hydrofunctionalizations, borylations, and some cross-coupling reactions [12–16].

9.1 Ir-Catalyzed Hydrogenations

Ir-mediated dihydrogenation was adopted as an important method to produce chiral high-value-added bulk and fine chemicals [17–20]. From the mechanistic point of view, the hydrogenation of a nonpolar C—C double or triple bond can be considered as that where unsaturated bond inserts into H—H single bond to achieve hydrogenation [21, 22]. In this case, one hydrogen atom added to the unsaturated bond would come from the migratory insertion of Ir—H complexes. The other one would originate by reductive elimination of Ir—H. For the hydrogenation of a polar C—hetero bond, one hydrogen atom could be considered as a "hydride" as nucleophile; the other acts as a "proton," which would be electrophile [23, 24]. In some cases, the other hydrogen could also come from protonation by protonated ligand or solvent [3, 25].

Computational Methods in Organometallic Catalysis: From Elementary Reactions to Mechanisms,
First Edition. Yu Lan.
© 2021 WILEY-VCH GmbH. Published 2021 by WILEY-VCH GmbH.

As shown in Scheme 9.1, gaseous dihydrogen is an ideal hydrogen source, which can realize the atomic economic hydrogenation [26, 27]. On the other hand, some reductive organics, such as an alcohol with α-hydrogen, also can be employed as dihydrogen source, which is named transfer hydrogenation [28, 29]. The substrates with unsaturated bond of Ir-mediated dihydrogenation could be olefins [30–33], carbonyl compounds [34–37], or imines [38–41], which have been studied both experimentally and theoretically. The hydrogen source in this process could originate from hydrogen molecule or some alcohols with α-H. In this section, the discussion of reaction mechanism obtained by density functional theory (DFT) methods would be organized by the type of substrates.

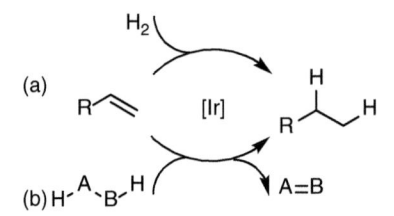

Scheme 9.1 Ir-catalyzed hydrogenations. (a) Direct hydrogenation. (b) Transfer hydrogenation.

9.1.1 Hydrogenation of Alkenes

As illustrated in Scheme 9.2, three possible reaction pathways have been proposed for Ir-promoted direct hydrogenation of olefins based on experimental observations and DFT calculations [43, 44]. Both the Ir(I)–Ir(III) and Ir(III)–Ir(V) catalytic cycles are possible in individual Ir catalysis, and the specific cycle depends on the ligands and detailed reaction conditions [44]. Catalytic cycle A shows the Ir(III)–Ir(V) pathway, which starts from Ir(III) species **9-1** with coordination of the olefin and dihydrogen. Subsequent oxidative addition of dihydrogen to Ir leads to simultaneous insertion of the olefin into an H—Ir covalent bond owing to the coordination number limit of Ir, which forms alkyl Ir(V) intermediate **9-2**. Reductive elimination gives the alkane product and forms Ir(III) species **9-3**, which undergoes ligand exchange with the olefin and dihydrogen to regenerate active catalyst **9-1**. Alternatively, the other Ir(III)–Ir(V) catalytic cycle could also occur (catalytic cycle B).

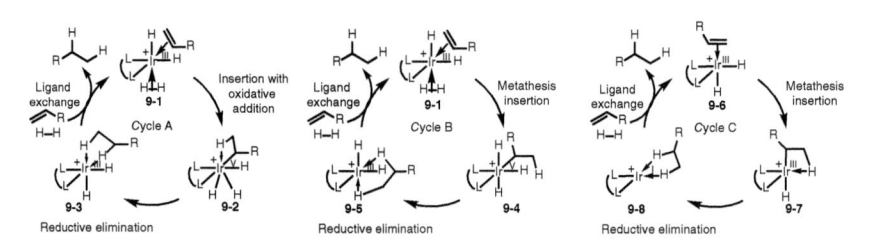

Scheme 9.2 Mechanism of Ir-catalyzed hydrogenation of alkenes. Source: Modified from Mazet et al. [42].

In this cycle, when common Ir(III) species **9-1** forms, σ-bond metathesis of the coordinated dihydrogen and olefin occurs to form alkyl Ir(V) complex **9-4**, which can also undergo reductive elimination to give the alkane product. An Ir(I)–Ir(III) catalytic cycle C is also possible. Oxidative addition of dihydrogen to the Ir(I) species leads to formation of Ir(III) dihydride complex **9-6**. Migratory insertion of the olefin into the H—Ir(III) bond then leads to formation of alkyl Ir(III) intermediate **9-7**. Subsequently, hydride–alkyl reductive elimination occurs to give the alkane product with regeneration of Ir(I) species **9-8** via coordination of the olefin and dihydrogen.

Among the three possible catalytic cycles, cycle A is the most commonly accepted pathway for unfunctionalized alkenes and Ir-catalytic systems with a phosphinooxazoline ligand. For the Ir(PHOX) complex shown in the chemical reaction equation in Figure 9.1, Mazet et al. [42] reported Ir-catalyzed asymmetric hydrogenation of (E)-1,2-diphenyl-1-propene and calculated the energy differences between the various stereoisomers by DFT methods. Dihydrogen-coordinated Ir(III) dihydride catalyst **9-9** forms by oxidation addition of a H_2 molecule and subsequent coordination of a second H_2 molecule to the truncated Ir(I)-(PHOX) catalyst, which is endergonic by 6.6 kcal/mol. Olefin insertion into the H—Ir(III) bond below the basal plane then simultaneously occurs with oxidative addition of H_2 via transition state **9-11ts**. In the geometry of this transition state, the double bond of the olefin is aligned parallel to the axial ligands. The coordination number of Ir decreases with insertion of the olefin into the Ir—H bond, which leads to simultaneous oxidative addition of the second coordinated H_2 molecule to afford alkyl Ir(V) trihydride intermediate **9-12**. This step has an energy barrier of 12.0 kcal/mol. **9-12** is seven-coordinated and possesses an agostic bond to the hydrogenation carbon center. The second hydrogenation step then occurs by reductive elimination via

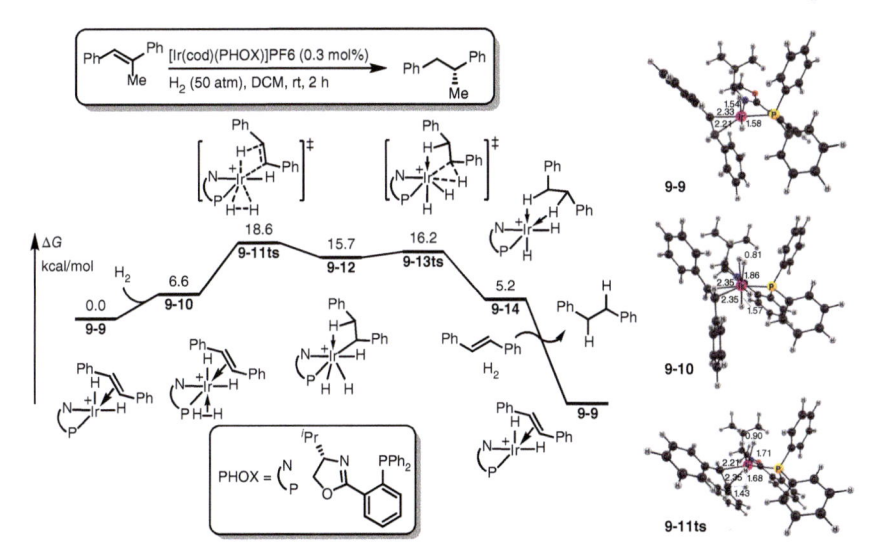

Figure 9.1 The free-energy profiles for the Ir-catalyzed hydrogenation of alkenes through an Ir(III)–Ir(V) catalytic cycle. The energies were calculated at B3LYP/6-31G(d) (LANL2DZ for Ir) level of theory.

9 Theoretical Study of Ir-Catalysis

transition state **9-13ts** to form alkane-coordinated Ir(III) complex **9-14**, which has an energy barrier of only 0.5 kcal/mol. The following ligand exchange with reactants and dihydrogen molecule regenerates active catalyst **9-9**. The computational study revealed that the rate-determining step is the concerted olefin insertion with dihydrogen oxidative addition step, which has an overall activation free energy of 18.6 kcal/mol.

Catalytic cycle B is a possible alternative to cycle A, and the path is determined by the steric and electronic properties of the ligands. In this case, a typical example may elucidate such effects. For N-heterocyclic carbene (NHC)-oxazole ligands in Ir catalysis, Hall and coworkers [44] performed a DFT study of the mechanism of Ir-catalyzed hydrogenation of phenylbutene using the full-atomic oxazole-carbene ligand. They found that a metathesis–insertion type Ir(III)–Ir(V) catalytic cycle B is preferred. The catalytic cycle starts from ethylene and dihydrogen-coordinated Ir(III) complex **9-15** (Figure 9.2), where the Ir center is hexa-coordinated. Consequently, oxidation addition cannot occur unless σ-bond metathesis between the coordinated ethylene and dihydrogen occurs via transition state **9-16ts** with

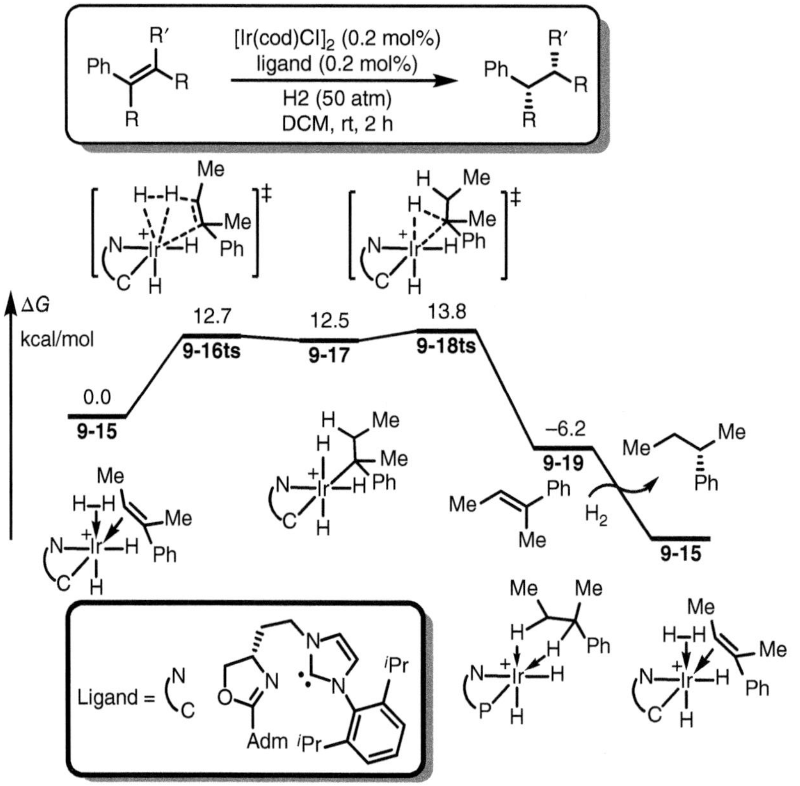

Figure 9.2 The free-energy profiles for the Ir-catalyzed hydrogenation of alkenes using a chiral NHC-oxazole ligand. The energies were calculated at PBE/D95v (LANL2DZ for Ir) level of theory.

a free-energy barrier of 12.7 kcal/mol, where the coordination number remains unchanged. Alkyl Ir(V) complex **9-17** then forms. Butylbenzene-coordinated Ir(III) complex **9-19** then forms by reductive elimination of the alkyl and hydride, with a small energy barrier of 1.3 kcal/mol via transition state **9-18ts**. This step is an exergonic process with a free energy of 6.2 kcal/mol. Dihydrogen and phenylbutene then coordinate to catalyst along with the release of butylbenzene by ligand exchange to regenerate active catalyst **9-15**, which releases a large amount of free energy. The calculated overall activation free for the whole catalytic cycle B is 13.8 kcal/mol at the reductive elimination step, which is favorable compared with concerted olefin insertion with dihydrogen oxidative addition in the truncated model of cycle A.

9.1.2 Hydrogenation of Carbonyl Compounds

Hydrogenation of carbonyl compounds has been a topic of considerable research interest in organic synthesis, where the stereospecific centers of chiral alcohols can be constructed by asymmetric reduction of the C=O bond [45–47]. Mechanistically, the carbonyl group is a typical polar covalent bond, in which the carbon atom is electrophilic, while the oxygen atom can be considered to be a nucleophile. Based on the hydrogen source, carbonyl compound hydrogenation can be divided into two possible processes: direct hydrogenation and transfer hydrogenation. In direct hydrogenation, the hydride source for formation of the C—H bond is gaseous dihydrogen, which pre-reacts with the Ir species to form a hydride–Ir complex [23, 24, 48]. The proton source for formation of the O—H bond can formally be a polar O—H or N—H bond, which also comes from gaseous dihydrogen. Alternatively, hydrogenation can occur with a reductive organic molecule, such as an alcohol with an α-hydrogen atom, which is called transfer hydrogenation [3, 49, 50]. In this type of reaction, hydride transfer occurs to form a new C—H bond, while the O—H bond can be formed by proton transfer with the dihydrogen source [51].

Zhang and coworkers [23, 24, 52] reported a series of enantioselective direct hydrogenation reactions of ketones under mild conditions with moderate-pressure hydrogen gas. A common mechanism of this reaction type is shown in Scheme 9.3. In this reaction, alkoxy or amino Ir(III) complex **9-20** is used as a catalyst, which includes an Ir—X (X = N or O) polar bond. The coordinated hydrogen molecule can be polarly activated by this Ir—X moiety to formally result in hydride coordinating to Ir and a proton coordinating to the X atom in complex **9-21**. The formed X—H moiety is then used to activate additional carbonyl compounds, which can further polarize the C=O double bond in intermediate **9-22**. By this activation, hydride transfer occurs to reduce the carbonyl group to an alkoxy group via key transition state **9-23ts**. Finally, release of alcohol leads to regeneration of the X—Ir bond in species **9-20** to accomplish the catalytic cycle.

The free-energy profiles for the Ir-catalyzed hydrogenation of a carbonyl compound with dihydrogen are shown in Figure 9.3, where the lowest energy neutral Ir(III) complex **9-25** is set to the relative zero free energy. Coordination of one hydrogen molecule to complex **9-25** generates octahedral intermediate **9-26**, which is endergonic by 14.0 kcal/mol. The hydrogen molecule can be activated by the

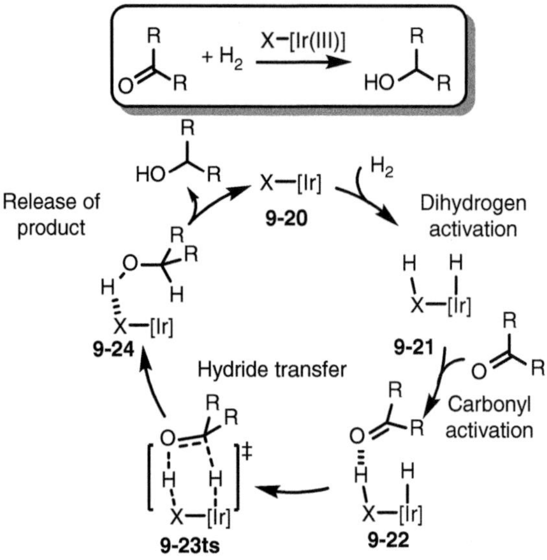

Scheme 9.3 General mechanism of Ir-catalyzed hydrogenation of a carbonyl compound with dihydrogen.

polar Ir—N covalent bond via four-membered-ring metathesis transition state **9-27ts** to form Ir(III) hydride complex **9-28** with the coordinated amine moiety. The reacting acetophenone can be activated by hydrogen bond interaction with the coordinated amine moiety, which undergoes concerted proton–hydride transfer via transition state **9-29ts** to afford phenethylol and regenerate active catalyst **9-25**. The computational results revealed that the rate-determining step for this catalytic cycle is cleavage of the H—H bond in the hydrogen molecule, so 20 atm pressure of hydrogen gas is necessary.

N—H bond cleavage/formation is the unique characteristic of tractional Noyori-type hydrogenation, which enables chemoselectivity by donating/accepting a proton [55, 56]. Dub's group [57, 59] proposed a revised mechanism for Noyori-type catalysis, in which they suggest that the N—H bond is not accessed but only plays a role in stabilizing the transition state of the reaction-determining step via formation of a strong N—H···O hydrogen bond. As an example, the energy profiles of Ir-catalyzed hydrogenation of esters by the outer-sphere pathway are given in Figure 9.4. The catalytic cycle starts by outer-sphere hydride transfer from Ir(III) dihydride complex **9-30** to the reacting trifluoroacetate, which needs to overcome an energy barrier of 16.3 kcal/mol and gives more active anionic complex **9-32**. Dissociation of methoxyethanolate accompanied by further H_2 coordination gives H_2 ion-pair complex **9-33**. Proton transfer from the coordinated H_2 to the anionic methoxyethanolate through **9-34ts** then gives the dihydrogenated ester and completes the catalytic cycle by overcoming an energy barrier of 2.0 kcal/mol.

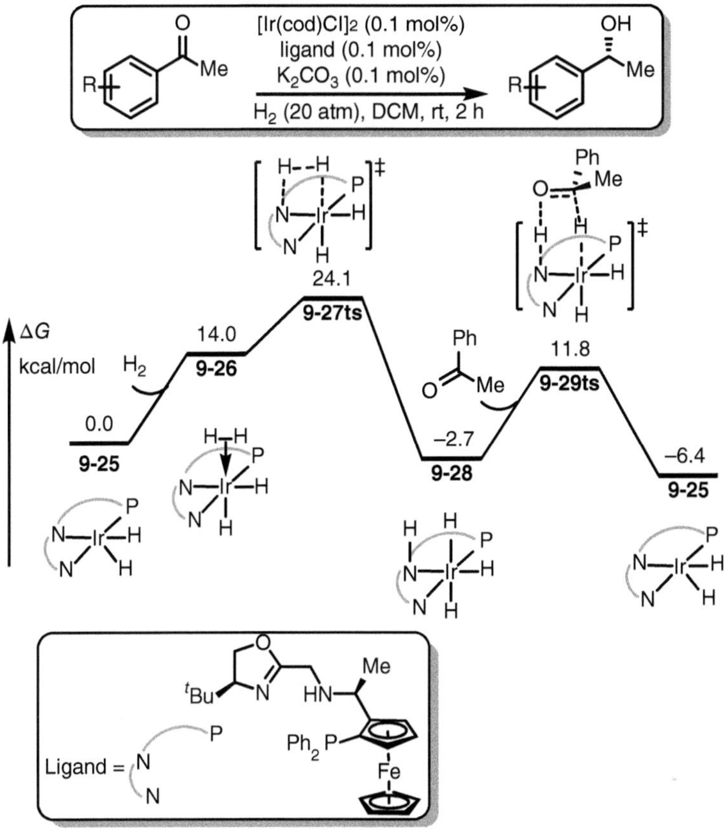

Figure 9.3 The free-energy profiles for the Ir-catalyzed hydrogenation of ketones using a chiral pincer ligand. The energies were calculated at M05-2X/6-311+G(d,p)/IEF-PCM(DCM) //B3LYP/6-31G(d) (LANL08(f) for Ir) level of theory. DCM; dichloromethane. Source: Dobereiner et al. [53]; Hopmann and Bayer [54].

9.1.3 Hydrogenation of Imines

Ir-catalyzed hydrogenation of imines is one of the most powerful approaches to prepare amine derivatives [60–62]. Mechanistically, the C=N double bond in imines is a polar bond, similar to the C=O double bond in carbonyl compounds [63, 64]. Consequently, a C—H bond can be formed between the carbon atom in an imine and a hydride source, while a N—H bond can be formed between the nitrogen atom in an imine and a proton source. However, the alkalinity of imines is significantly stronger than that of carbonyl compounds. Therefore, deprotonation by the imine to give an iminium intermediate is a favorable process, which would occur as a matter of priority. Accordingly, hydrogenation of imines can also be categorized as direct hydrogenation by gaseous dihydrogen or transfer hydrogenation by an alcohol as the hydrogen source.

According to the alkalinity of imines, proton transfer leading to formation of an iminium ion is an energetically favorable process, which is crucial in Ir-catalyzed

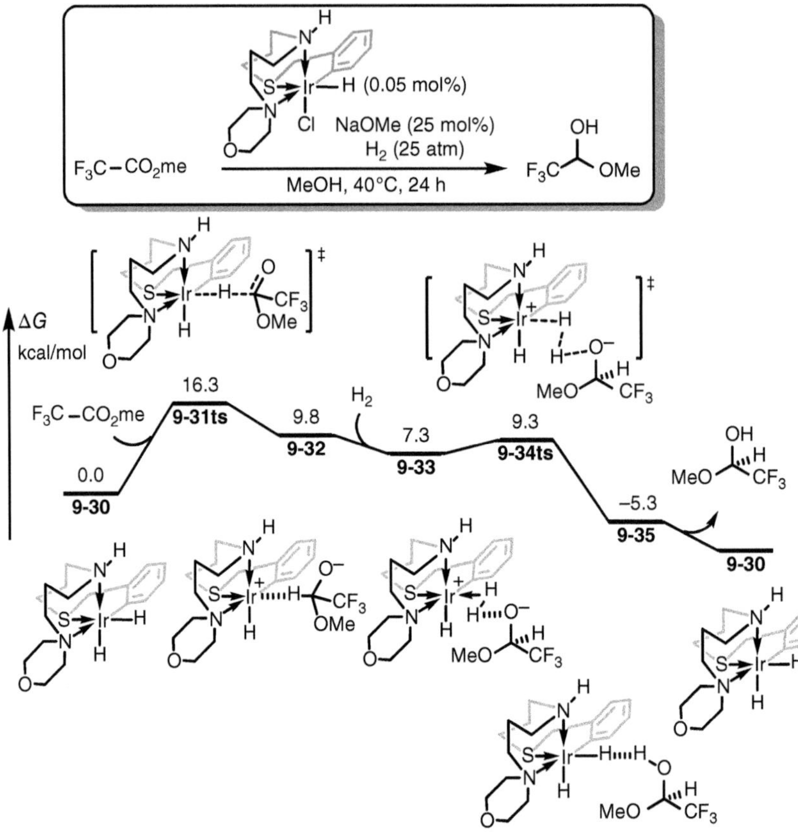

Figure 9.4 The free-energy profiles for the Ir-catalyzed hydrogenation of esters through an outer-sphere mechanism. The energies were calculated at ωB97XD/6-311++G(d,p) (ECP60MDF_VTZ for Ir)/SMD(methanol) level of theory.

direct hydrogenation of imines. Various mechanisms have been proposed for different Ir-based catalysts on the basis of experimental observations [65–67]. The well-accepted catalytic cycle for the outer-sphere mechanism is shown in Scheme 9.4 [53, 54]. When dihydrogen coordinates to the cationic Ir(III) species in **9-36**, the H—H covalent bond is polarized by the Ir center. In the presence of an imine, heterolysis of the H—H bond gives a cationic iminium ion and forms neutral Ir(III) hydride complex **9-37**. Formation of an iminium cation increases the electrophilicity of the C=N double bond. Therefore, intermolecular hydride transfer via the key transition state **9-38ts** generates an amine product. The active catalyst **9-36** can be regenerated by coordination of a hydrogen molecule.

The outer-sphere catalytic cycle for Ir-(PHOX)-mediated imine hydrogenation has been studied by DFT calculations [54]. The free-energy profile of the catalytic cycle is shown in Figure 9.5. Replacement of dihydrogen with equatorially coordinated diphenylethanimine in complex **9-40** is an endothermic process with an energy cost of 10.1 kcal/mol. Proton transfer then occurs from the coordinated H_2 to the imine

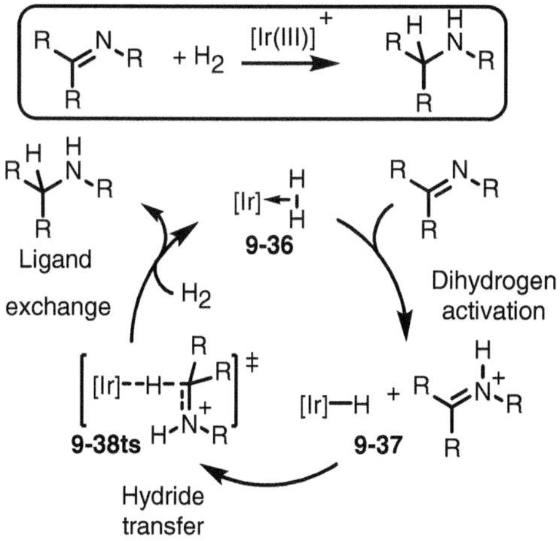

9.1 Ir-Catalyzed Hydrogenations

Scheme 9.4 General mechanism of Ir-catalyzed hydrogenation of imines.

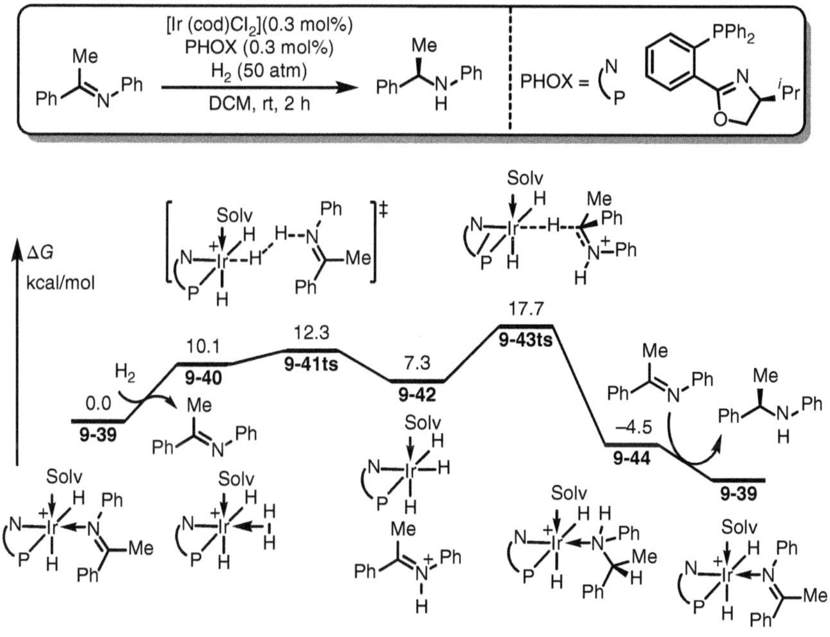

Figure 9.5 The free-energy profiles for the Ir-catalyzed hydrogenation of imines. The energies were calculated at B3LYP-D3/6-311G(d,p) (LANL2DZ(f) for Ir)/IEF-PCM(DCM) //B3LYP/6-311G(d,p) (LANL2DZ(f) for Ir) level of theory.

carbon atom of diphenylethanimine via **9-41ts**, which needs to overcome an energy barrier of 12.3 kcal/mol. Transfer of the hydride to the cationic iminium then leads to formation of amine product via **9-43ts** with an energy barrier of 10.4 kcal/mol, which is an endothermic process. With the release of the amine product, the empty coordination site can bind a new imine reactant. Hydride transfer to the iminium carbon atom is the rate-limiting and enantioselectivity-determining step for this reaction.

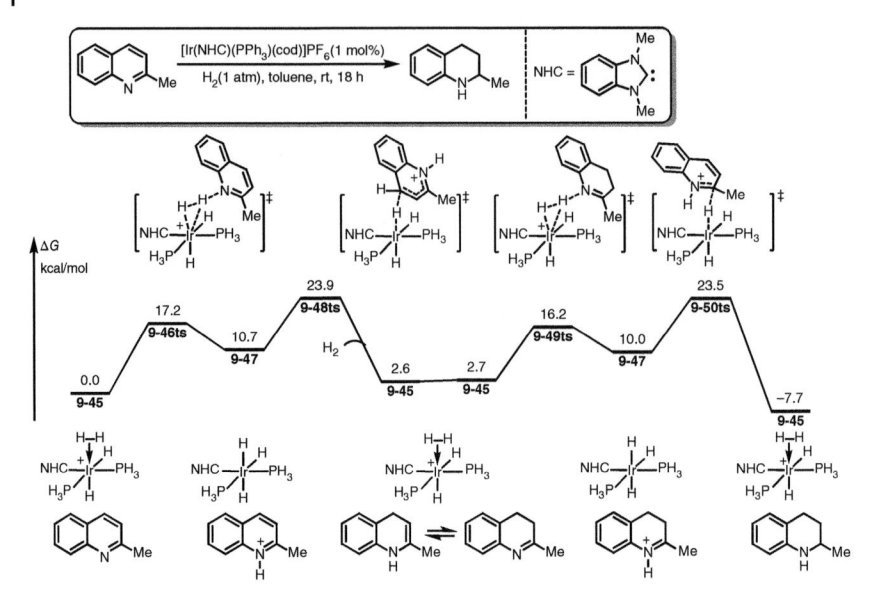

Figure 9.6 The free-energy profiles for the Ir-catalyzed hydrogenation of quinolines. The energies were calculated at B3PW91/6-311+G(d,p) (SDDALL for Ir and P)/SMD(toluene) //B3PW91/6-31G(d,p) (SDDALL for Ir and P) level of theory.

9.1.4 Hydrogenation of Quinolines

Eisenstein and coworkers [53] reported phosphine-coordinated Ir-catalyzed tetrahydrogenation of quinoline (Figure 9.6). Tetrahydroquinolines can be formed with up to 95% yield under rather mild reaction conditions at room temperature with only 1 atm pressure of gaseous dihydrogen. Inspired by the experimentally proposed outer-sphere mechanism for Ru-catalyzed hydrogenation of quinolines, Eisenstein and coworkers [68] also proposed an outer-sphere catalytic cycle supported by DFT calculations. The calculated free-energy profile is shown in Figure 9.6, in which dihydrogen-coordinated cationic Ir(III) complex **9-45** is set to the relative zero energy in the calculated free-energy profiles. Based on activation of Ir, intermolecular deprotonation of dihydrogen by the quinoline occurs via transition state **9-46ts** with an energy barrier of 17.2 kcal/mol, leading to heterolysis of a H—H bond to form cationic quinolinium intermediate and neutral Ir(III) trihydride complex **9-47**. Subsequently, intermolecular hydride transfer occurs from Ir(III) trihydride complex **9-47** to quinolinium via line-type transition state **9-48ts** with an overall activation free energy of 23.9 kcal/mol, which is considered to be the rate-determining step of the whole catalytic cycle. The enamine type 1,4-dihydrogenated intermediate is provided for the second hydrogenation process, which can transfer to its imine-type isomer reversibly. Then, dihydrogen-coordinated cationic Ir(III) complex **9-45** can participate in the second hydrogenation of the imine type 3,4-dihydroquinoline coming from isomerization of enamine type 1,4-dihydroquinoline via a H—H bond cleavage transition state **9-49ts**. The following hydride transfer process needs to overcome an overall activation free energy of 23.5 kcal/mol via transition state **9-50ts**. The whole catalytic cycle is 7.7 kcal/mol for the generation of tetrahydrogenation of quinoline.

9.2 Ir-Catalyzed Hydrofunctionalizations

In Ir-catalyzed dihydrogenations, polarization of dihydrogen coming from gaseous molecule or other hydrogen sources, which is usually considered as a nucleophilic hydride and an electrophilic proton, can be added onto an unsaturated bond to afford corresponding saturated bond. Moreover, other functional groups, such as amino, silyl, and aryl groups, can be used instead of hydride as nucleophile individually in the addition reactions, which can be named as hydroamination [69, 72], hydrosilylation [73–76], and hydroarylation [77–80], correspondingly. In Ir-catalysis, the mechanism of those reactions could also be related to the dihydrogenation; therefore, they are discussed in this section.

9.2.1 Ir-Catalyzed Hydroaminations

Hydroamination, N—H bond addition to an unsaturated C=C double bond, is an atomic efficient transformation for the synthesis of amine derivatives [15, 69, 81]. In this process, amino group often plays as nucleophile to add to unsaturated bond. Moreover, the hydrogen, which is added on unsaturated bond, acts as electrophile in this process.

As shown in Figure 9.7, Tobisch, Stradiotto, and coworkers [69] reported a combination of experimental and theoretical study of an Ir-catalyzed intramolecular hydroamination. In DFT calculations, animo olefin-coordinated IrI species **9-51** is set to relative zero in free-energy profiles. Assisted by the coordination onto Ir, an intramolecular nucleophilic attack of amino group onto olefin takes place via transition state **9-52ts** with an energy barrier of 16.9 kcal/mol to afford an ammonium iridate species **9-53**. An intramolecular proton transfer from ammonium to iridate occurs via transition state **9-54ts** affording a hydride Ir(III) complex **9-55**. In fact, the formation of H—Ir covalent bond also can be considered the oxidation of Ir(I) by ammonium. The followed reductive elimination of hydride and alkyl takes place via transition state **9-56ts** to afford an amine-coordinated Ir(I) complex **9-57**. The relative free energy of this transition state is calculated to be 24.6 kcal/mol, which indicates the rate-determining step here. The ligand exchange with amino alkene reactant yields annulation amine product with the regeneration of active catalyst **9-51**.

9.2.2 Ir-Catalyzed Hydroarylations

The C—H bond in arenes could be activated by low oxidation state Ir(I) complex to afford alkyl hydride Ir(III) species, which could be applied to hydroarylation of C=C

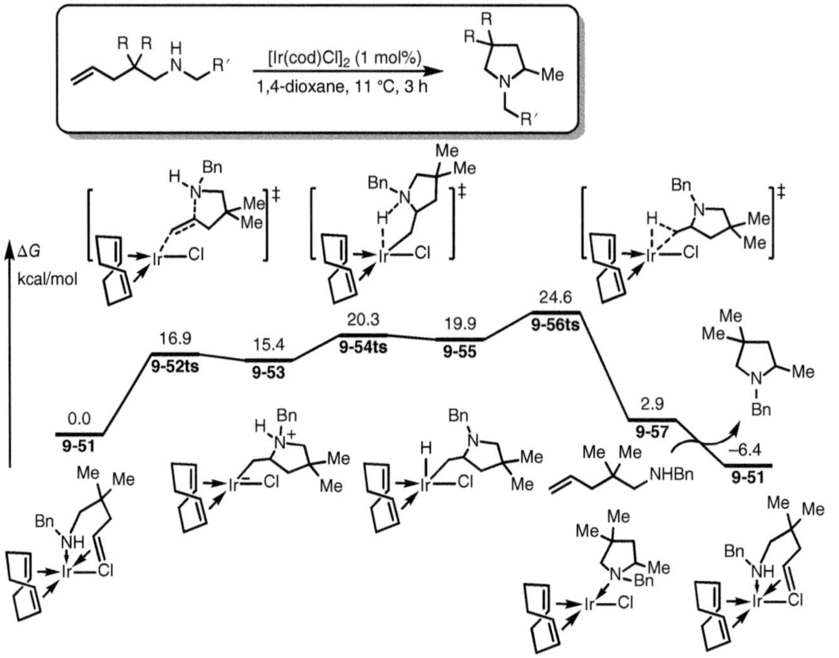

Figure 9.7 The free-energy profiles for the Ir-catalyzed intramolecular hydroamination of alkenes. The energies were calculated at TPSS(COSMO)/TZVP (SDD for Ir)/CPCM (1,4-dioxane) level of theory. Source: Modified from Hesp et al. [69].

double bond [82]. Huang and Liu [78] reported a DFT study for the hydroarylation of styrene. As shown in Scheme 9.5, a modified Chalk–Harrod pathway is considered as major process, which involves an oxidative addition of C—H bond onto Ir, olefin insertion into Ir—C(aryl) bond, and C(alkyl)—H reductive elimination. However, the normal Chalk–Harrod catalytic cycle was excluded, which undergoes olefin insertion into Ir—H bond and C(aryl)—C(alkyl) reductive elimination.

The DFT level calculated free-energy profiles for the catalytic cycle of IrIII-catalyzed hydroarylation of styrene and is shown in Figure 9.8 An acyl-directed oxidative addition of C(aryl)—H bond onto Ir(I) species **9-58** takes place via transition state **9-59ts** with a free-energy barrier of 18.8 kcal/mol to afford an aryl hydride Ir(III) complex **9-60**. In a modified Chalk–Harrod pathway, olefin insertion into C—Ir bond via transition state **9-61ts** forms an alkyl Ir(III) complex **9-62**. The subsequent reductive elimination of alkyl and hydride group occurs via transition state **9-63ts** to afford alkylation product. The rate-determining step of this catalytic cycle is considered to be the olefin insertion into aryl—Ir bond with an activation free energy of 25.9 kcal/mol. Moreover, the original Chalk–Harrod pathway is also considered by DFT calculations. Although a reversible olefin insertion into Ir—H bond can occur rapidly, the barrier of following reductive elimination of aryl and alkyl group is as high as 34.9 kcal/mol. Therefore, the Chalk–Harrod pathway can be excluded.

Scheme 9.5 General mechanism of Ir-catalyzed hydroarylation of unsaturated bonds.

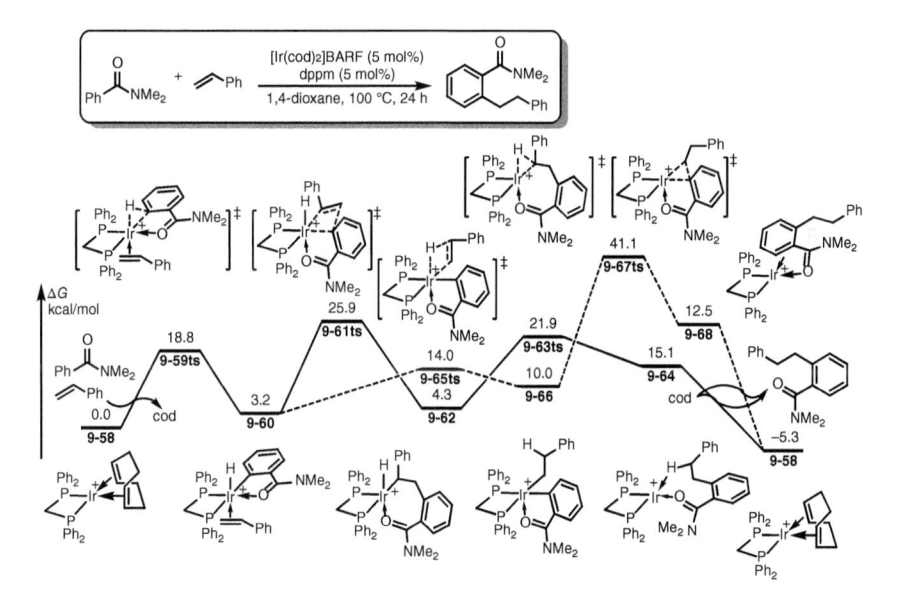

Figure 9.8 The free-energy profiles for the Ir-catalyzed hydroarylation of alkenes. The energies were calculated at TPSS(COSMO)/TZVP (SDD for Ir)/CPCM(1,4-dioxane) level of theory.

9.2.3 Ir-Catalyzed Hydrosilylations

Iridium catalyst is one of the powerful tools to facilitate hydrosilylation of unsaturated bond, presumably due to its exceptional capability in silane Si—H bond activation [83]. As known, the electronegativity of silicon is slightly lower than that of

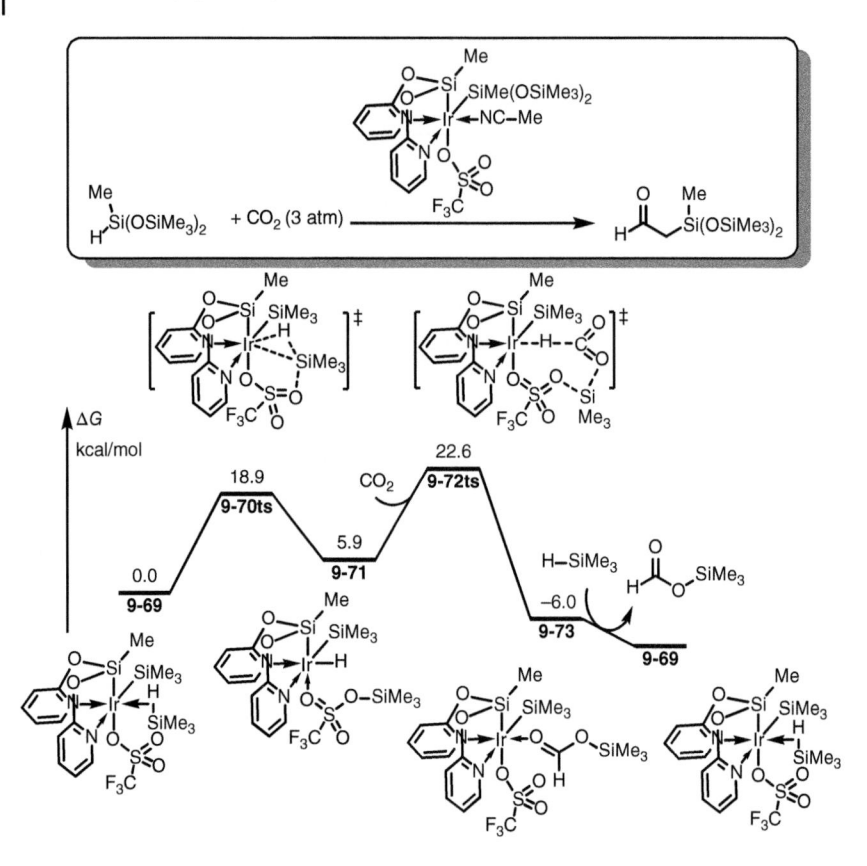

Figure 9.9 The free-energy profiles for the Ir-catalyzed hydrosilylation of carbon dioxide. The energies were calculated at M06/6-31G(d) (SDD(f) for Ir) level of theory.

hydrogen. Therefore, silyl often plays as electrophile in this process; also, hydrogen is always used as nucleophile. In Ir-catalyzed hydrosilylation, both the Ir—H and Ir—SiR$_3$ species can play as key intermediate in catalytic cycle.

In 2012, Alvarez and Oro [83] reported an Ir(III)-catalyzed hydrosilylation of carbon dioxide to afford silyl formate derivatives. As shown in Figure 9.9, DFT studies are used to reveal the mechanism of this reaction. The coordination of silane forms Ir(III) complex **9-69** was considered as active species. A triflate-assisted metathesis of coordinated H—Si bond forms a silyltriflate-coordinated hydride Ir(III) complex **9-71** via transition state **9-70ts** with a free-energy barrier of 18.9 kcal/mol. By the activation of Ir, a concerted hydrosilylation can occur via an eight-membered ring-type transition state **9-72ts** to afford silyl formate product. In this step, hydride provides nucleophilicity to attack oxygen atom in CO$_2$, while silyl group acts as electrophile.

Sun group [76] reported an Ir(I)-catalyzed hydrosilylation of acetylenes, which can highly yield prepared Z-configuration ethylenes in room temperature. Zhang and Wu have done a DFT study on the mechanism of this reaction (Figure 9.10). In the calculated catalytic cycle, an acetylene-coordinated hydride Ir(I) species **9-74**

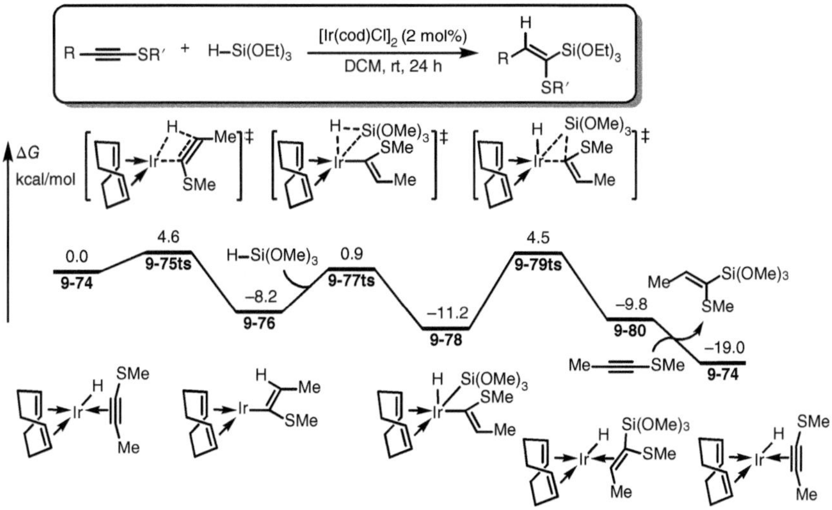

Figure 9.10 The free-energy profiles for the Ir-catalyzed hydrosilylation of alkynes. The energies were calculated at M06/6-311++G(3df,3pd) (def2-TZVP for Ir)/SMD(DCM)//B3LYP/6-31G(d) (LANL2DZ+f for Ir) level of theory.

is considered to be the active catalyst. The insertion of coordinated acetylene into Ir—H bond takes place via transition state **9-75ts** with a barrier of only 4.6 kcal/mol to form a vinyl Ir(I) complex **9-76**. An intermolecular oxidative addition of Si—H bond in silane occurs via transition state **9-77ts**. Then, the reductive elimination forms new C—Si bond in silyl acetylene product via transition state **9-79ts**. DFT calculations revealed that the rate-determining step of the whole catalytic cycle is the reductive elimination of silyl and alkyl groups with a free-energy barrier of only 15.7 kcal/mol. Therefore, this reaction can occur in mild conditions.

Alternatively, a silyl Ir(III) complex is considered as the key species in an Ir(III)-catalyzed hydrosilylation without redox processes. A DFT study is taken to prove the proposed Ir(III) catalytic cycle. As shown in Figure 9.11, an acetylene-coordinated silyl Ir(III) complex **9-81** is set to relative zero in free-energy profiles. A migratory insertion of acetylene into Ir—Si bond occurs via transition state **9-82ts** with an energy barrier of 23.2 kcal/mol, which is considered as the rate limit of the whole catalytic cycle. The coordination of silane forms vinyl Ir(III) complex **9-84**. Then, a σ-bond metathesis yields an olefin-coordinated silyl Ir(III) complex **9-86**. Active catalyst **9-81** can be regenerated by ligand exchange with alkyne reactant. In the whole catalytic cycle, the oxidative state of Ir keeps at +3.

9.3 Ir-Catalyzed Borylations

As a group 9 transition metal, low-valence Ir-species can be used in the C—H activation to afford corresponding alkyl or aryl Ir species through an oxidative addition

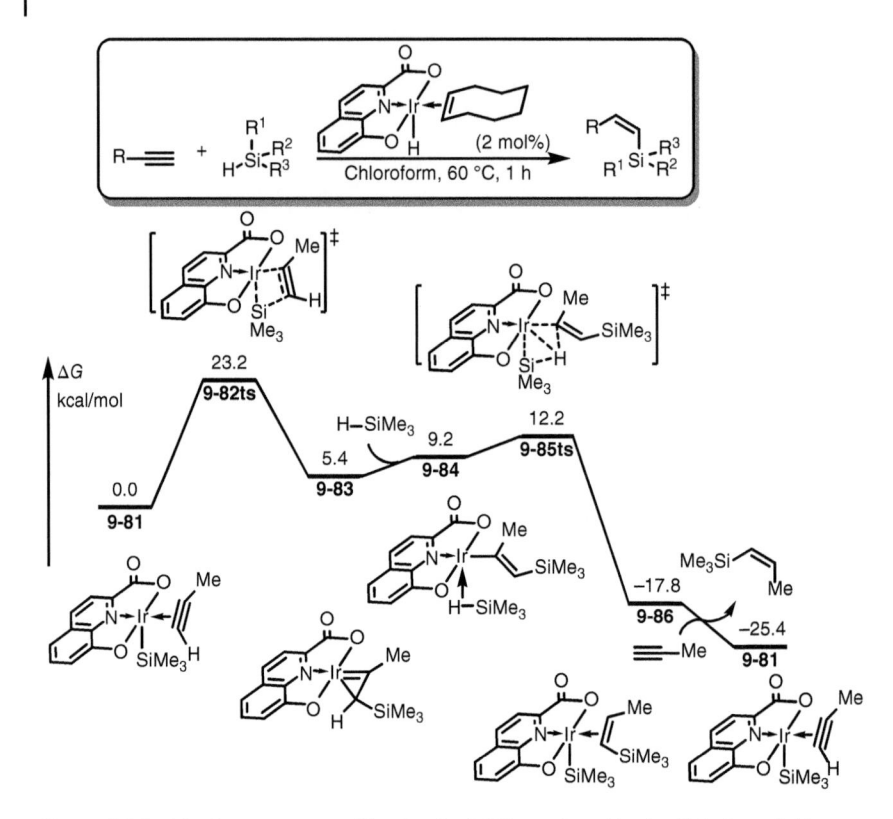

Figure 9.11 The free-energy profiles for the Ir(III)-catalyzed hydrosilylation of alkynes. The energies were calculated at M06/6-311++G(3df,3pd) (def2-TZVP for Ir)/SMD(DCM)//B3LYP/6-31G(d) (LANL2DZ+f for Ir) level of theory.

process. When diboron compounds are used, the generated alkyl or aryl Ir species can be borylated to provide alkyl or aryl borane product. However, the mechanism of Ir-catalyzed borylation reactions still remains under debate.

9.3.1 Borylation of Alkanes

When an active methyl group is involved in methylsilane compound, an Ir-catalyzed borylation can take place under 80 °C [85, 86]. As shown in Figure 9.12, DFT calculations at M06 level of theory found that Ir(III) species **9-88** is 9.3 kcal/mol more unstable than Ir(V) species **9-87**. A boryl-assisted C—H bond oxidative addition onto Ir(III) species **9-88** can occur via transition state **9-89ts** with an overall activation free energy of 28.3 kcal/mol to afford an alkyl Ir(V) intermediate **9-90**. Then, a C(alkyl)—B(boryl) reductive elimination takes place via transition state **9-91ts** to yield borylated product and form a hydride Ir(III) intermediate **9-92**. A rapid oxidative addition with diboron B$_2$pin$_2$ via transition state **9-93ts** forms a Ir(V) intermediate **9-94**. A B(boryl)—H(hydride) elimination regenerates active species **9-88**.

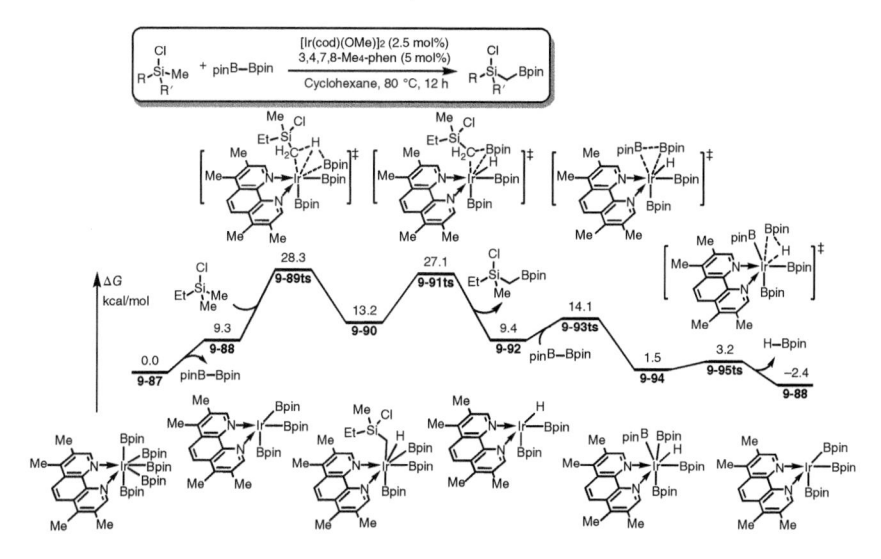

Figure 9.12 The free-energy profiles for the Ir-catalyzed borylation of alkanes. The energies were calculated at M06/6-311+G(2d,2p) (LANL2TZ(f) for Ir)/PCM(cyclohexane)// B3LYP/6-31G(d,p) (LANL2DZ for Ir) level of theory.

9.3.2 Borylation of Arenes

In Ir-catalyzed borylation of arenes, the C(aryl)—H bond activation is the key step, which can undergo an oxidative addition process [87]. In this step, both Ir(I)–Ir(III) and Ir(III)–Ir(V) transformations are possible, which would be determined by the steric effect of ligands.

When small-hindrance bipyridine was used as a bidentate ligand in Ir-catalysis, 3-borylation of pyridine can be performed through an Ir(III)–Ir(V) catalytic cycle [88]. The calculated free-energy profiles for the full catalytic cycle is shown in Figure 9.13. Starting from an Ir(III) species **9-96**, an oxygen-directed oxidative addition of ortho C—H bond can occur via transition state **9-97ts** with an energy barrier of 16.4 kcal/mol to form an Ir(V) intermediate **9-98**. In this key intermediate, the oxidative state of Ir is +5, and the coordination number of Ir is 7. Then, a C(aryl)—B reductive elimination via transition state **9-99ts** can yield *ortho*-borylated arene product and form a hydride Ir(III) intermediate **9-100**, which can react with diboron reactant to regenerate Ir(III) species **9-96** through sequential oxidative addition and reductive elimination via transition states **9-101ts** and **9-103ts**, respectively.

When a large diphosphine ligand Xyl-MeO-BIPHEP was used in Ir-catalyzed *para*-borylation of silyl-benzene, DFT calculation found that the steric hindrance of ligand leads to a large activation free energy of 61.3 kcal/mol for the formation of Ir(V) intermediate **9-104**. Therefore, an alternative Ir(I)–Ir(III) catalytic cycle was proposed by DFT calculation [10]. As shown in Figure 9.14, the catalytic cycle starts from a boryl Ir(I) species **9-102**, which can react with silyl-benzene via an oxidative addition transition state **9-105ts** to form a boryl aryl hydride Ir(III) intermediate **9-106** with an energy barrier of 14.7 kcal/mol. Then, a reductive

Figure 9.13 The free-energy profiles for the Ir-catalyzed borylation of arenes through an Ir(III)–Ir(V) catalytic cycle. The energies were calculated at PBE-D3/6-311+G(d,p) (cc-pVTZ-PP for Ir)/SMD(octane)//PBE/6-31G(d) (SDD for Ir)/SMD(octane) level of theory.

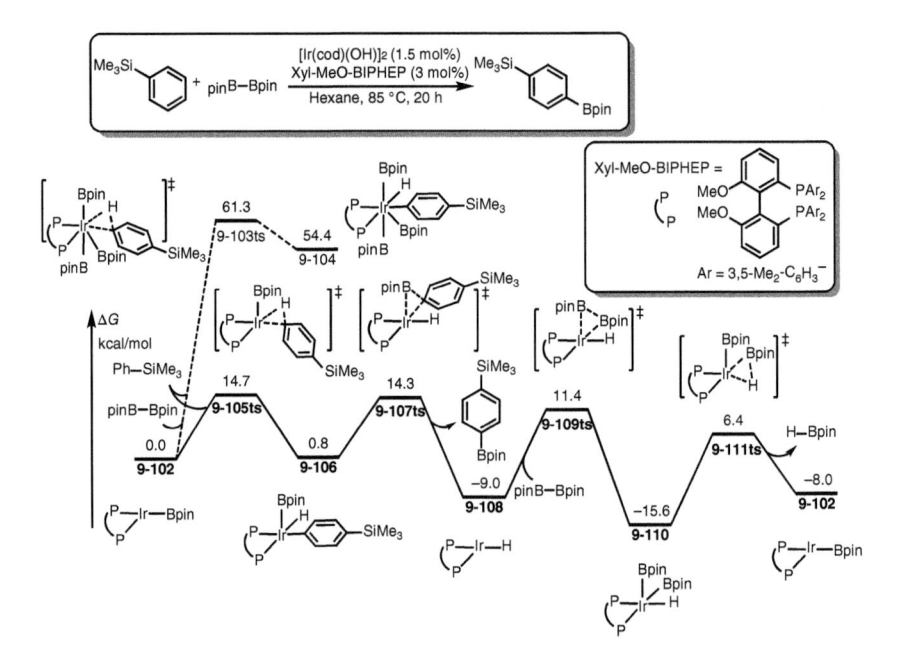

Figure 9.14 The free-energy profiles for the Ir-catalyzed borylation of arenes through an Ir(I)–Ir(III) catalytic cycle. The energies were calculated at N12/6-311++G(d,p) (LANL08(f) for Ir)/SMD(hexane)//B3LYP/6-31G(d) (LANL08(f) for Ir) level of theory.

elimination generates *para*-boralated benzene and forms a hydride Ir(I) intermediate **9-108**. The following oxidative addition with diboron onto Ir(I) occurs via transition state **9-109ts** with an energy barrier of 20.4 kcal/mol to form an Ir(III) intermediate **9-110**. Then, a B—H reductive elimination regenerates Ir(I) active species **9-102**.

9.4 Ir-Catalyzed Aminations

There are two diverse pathways for Ir-catalyzed amination, including transfer hydrogenation of in situ-generated imines and nitrene insertion into C—H bonds. The mechanism of those reactions was studied by DFT calculations [89, 90].

9.4.1 Amination of Alcohols

As shown in Scheme 9.6, in the presence of Cp*Ir(III) catalyst, primary alcohol can react with amine to achieve amination products. The proposed mechanism for this reaction involves two redox neutral steps. The dehydrogenation of primary alcohol to generate hydride Ir(III) complex and corresponding aldehyde. The in situ-generated aldehyde can react with amine to provide imine, which can be hydrogenated by hydride Ir(III) complex to yield amine product.

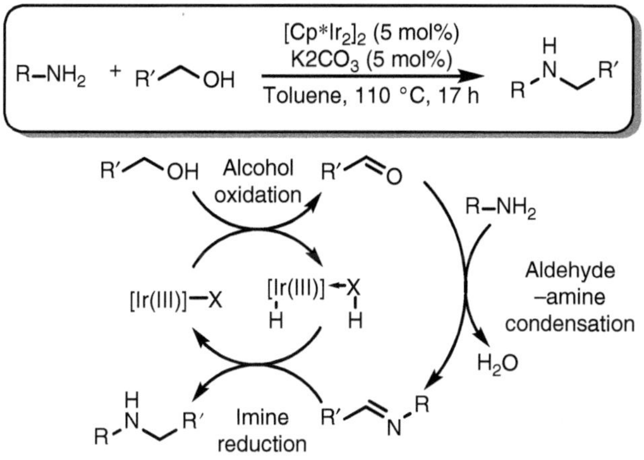

Scheme 9.6 Mechanism of Ir-catalyzed amination of alcohols.

Computational results are shown in Figure 9.15. Carbonate Ir(III) complex **9-112** is set to relative zero in free-energy profiles. The coordination of methanol forms intermediate **9-113** with 4.9 kcal/mol endergonic. Then, a proton transfer from coordinated methanol to carbonate via transition state **9-114ts** affords methoxide Ir(III) intermediate **9-115**. Then, a β-hydride elimination takes place via transition state **9-116ts** to form a formaldehyde-coordinated hydride Ir(III) complex **9-117**. After release of one molecular formaldehyde, hydride Ir(III) species **9-118**

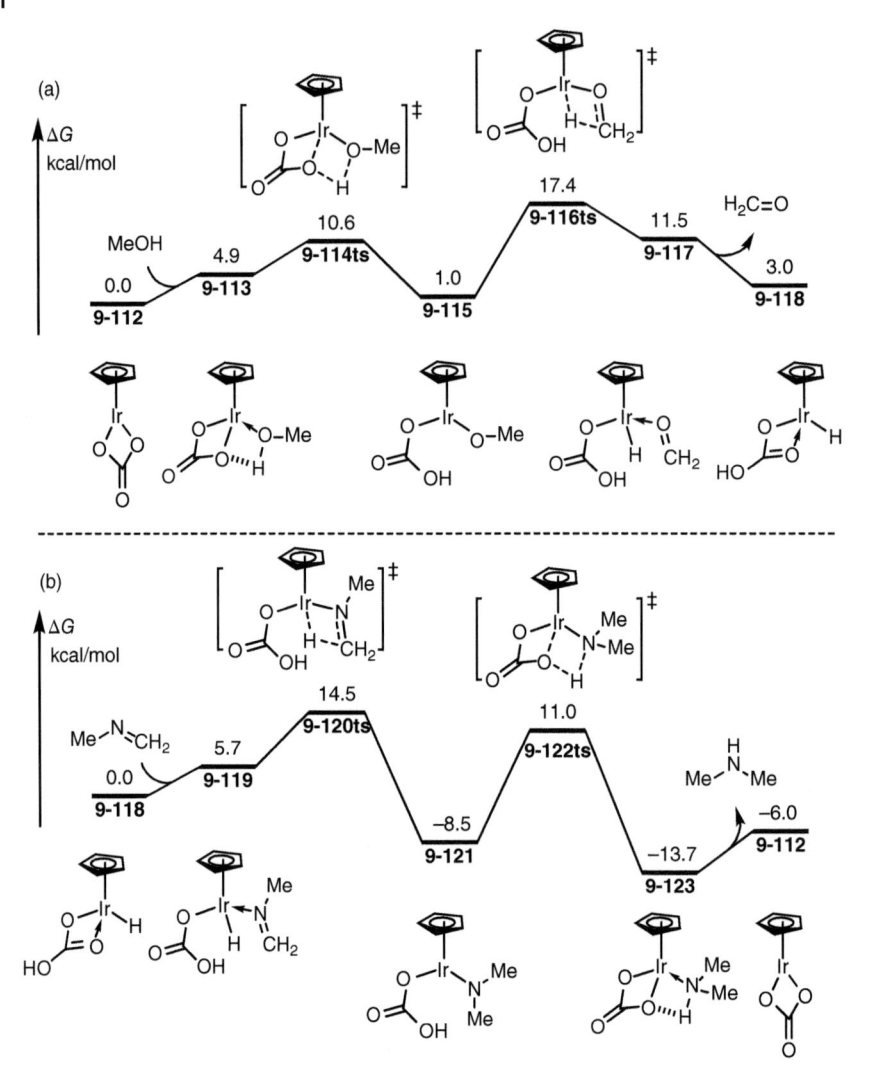

Figure 9.15 The free-energy profiles for the Ir-catalyzed amination of alcohols through (a) alcohol oxidation and (b) imine reduction. The energies were calculated at B3PW91/6-31G(d,p) (SDDALL for Ir) level of theory.

is generated with overall 3.0 kcal/mol endergonic. In an outer-sphere process, the amination of formaldehyde provides imine for further transformations. In another process (Figure 9.15b), the in situ-generated imine coordinates with hydride Ir(III) specie **9-118** forming complex **9-119** with 5.7 kcal/mol endergonic. Then, C=N double bond inserts into Ir—H bond through transition state **9-120ts** to form amino Ir(III) intermediate **9-121**. Protonation by bicarbonate via transition state **9-122ts** can yield amine product and regenerate carbonate Ir(III) active species **9-112**. In the whole catalytic cycle, the oxidative state of Ir keeps at +3.

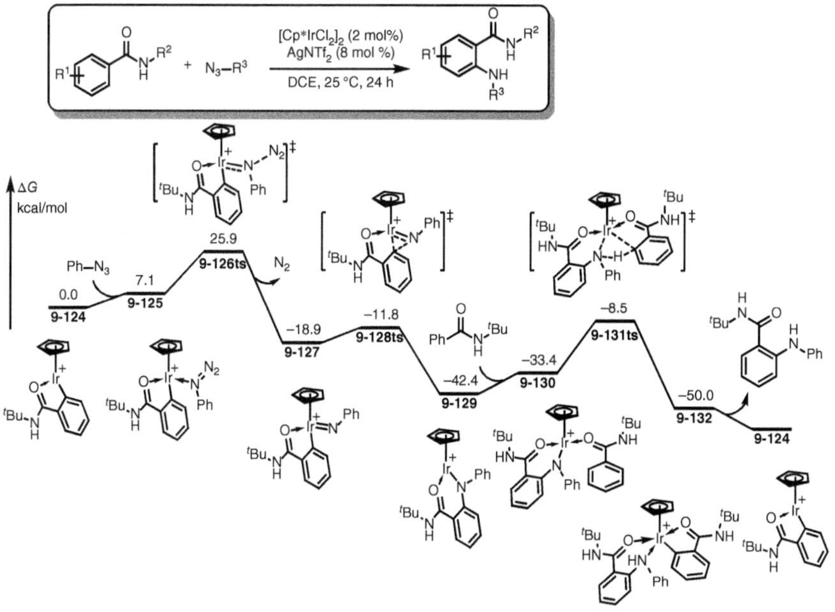

Figure 9.16 The free-energy profiles for the Ir-catalyzed amination of arenes through nitrene insertion. The energies were calculated at M06/6-31G(d,p) (LANL2DZ for Ir)/PCM(DCE) level of theory.

9.4.2 Amination of Arenes

As a group 9 transition metal, Ir can be employed as catalyst for nitrene insertion into C(aryl)—H bond through a C—H activation process [91]. The key step of amination was revealed by DFT calculation [92]. As shown in Figure 9.16, when aryl Ir(III) species **9-124** is formed, the coordination of azide onto Ir forms complex **9-125** with 7.1 kcal/mol endergonic. The denitrogenation can occur via a linear transition state **9-126ts** to afford a Ir(III)–nitrene complex **9-127**. The calculated activation free energy for this step is 29.5 kcal/mol. Then, nitrene inserts into Ir—C(aryl) bond via a three-membered ring-type transition state **9-128ts** to form an amino Ir(III) intermediate **9-129**. The amino group plays as a Brønsted base in the following C—H activation process. When another molecule arene coordinates with Ir in complex **9-130**, a concerted metalation deprotonation takes place via transition state **9-131ts** to form aryl Ir(III) species **9-132**. After release of amination product, active species **9-124** can be regenerated. The calculated activation free energy for the C—H activation step is as high as 33.9 kcal/mol, which is considered to be the rate-determining step for the whole catalytic cycle.

9.5 Ir-Catalyzed C—C Bond Coupling Reactions

As a transition metal, Ir complexes can be used in C—C bond coupling reactions [93–97]. However, there are rare examples for the theoretical investigations of Ir-catalyzed coupling reactions.

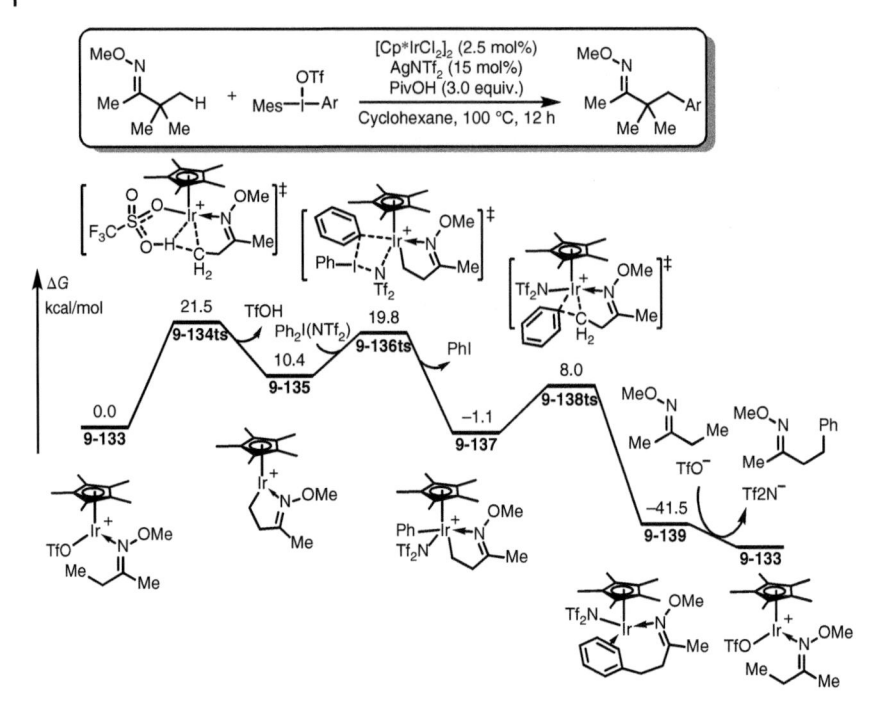

Figure 9.17 The free-energy profiles for the Ir-catalyzed arylation of alkanes. The energies were calculated at M06-D3/6-31G(d) (LANL2DZ for Ir)/PCM(cyclohexane) level of theory.

When a C—C bond is formed from a nucleophile and an electrophile, it is usually named as cross-coupling reaction [98–100]. As an example (Figure 9.17), C(alkyl)—H bond can be considered as nucleophile to react with electrophilic diaryliodonium salts for the construction of new C(alkyl)—C(aryl) bonds in an Ir-catalyzed cross-coupling reaction [101]. DFT calculation found that the reaction starts from a cationic Ir(III) species **9-133**. In the presence of oxime-directing group, the β-C(alkyl)—H activation can occur via a concerted metalation deprotonation-type transition state **9-134ts** to afford an alkyl Ir(III) intermediate **9-135**. The calculated activation free energy of this step is 21.5 kcal/mol. Ir(III) intermediate **9-135** can be oxidized by diaryliodonium salts via an electrophilic phenyl transfer transition state **9-136ts** to form an Ir(V) intermediate **9-137** with an energy barrier of only 9.4 kcal/mol. After a rapid C(aryl)—C(alkyl) reductive elimination, C—C bond coupling product is generated. The active species **9-133** can be regenerated by the following ligand exchange.

In another example of C(aryl)—C(aryl) cross-coupling reaction, aryldiazonium salts play as electrophile to react with arene in the presence of Ir-catalyst [102]. The key step for the generation of new C(aryl)—C(aryl) bond was revealed by DFT calculation. As shown in Figure 9.18, when aryl Ir(III) species **9-140** is formed from concerted metalation–deprotonation process, it can react with aryldiazonium through an aryl transfer transition state **9-142ts**. The calculated energy barrier of this step is 22.6 kcal/mol. After this step, a bisaryl Ir(V) intermediate **9-143** can be formed

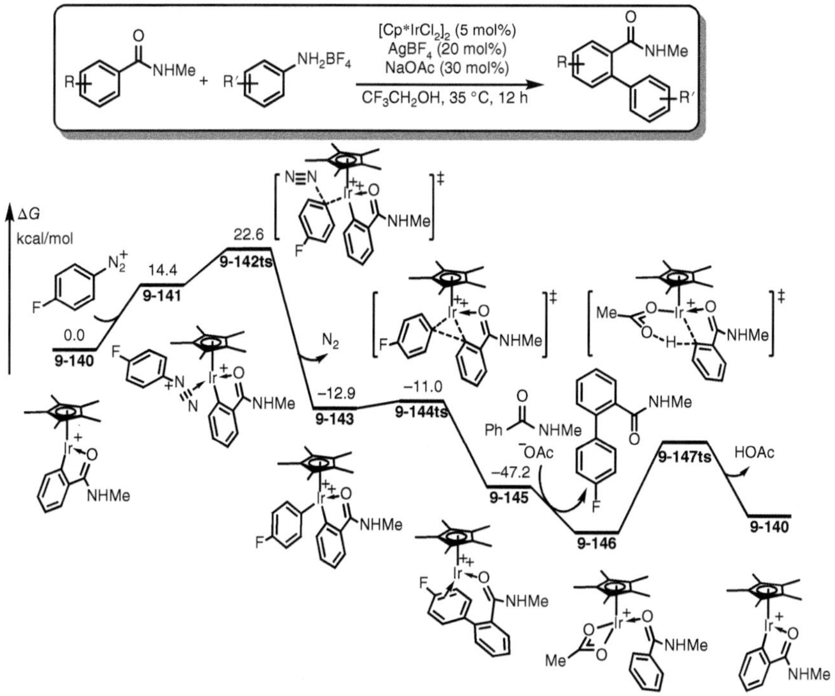

Figure 9.18 The free-energy profiles for the Ir-catalyzed arylation of arenes. The energies were calculated at M06/6-311+G(d,p) (LANL2DZ for Ir)/PCM(2,2,2-trifluoroethanol)//M06/6-31G(d) (LANL2DZ for Ir) level of theory.

with 12.9 kcal/mol exergonic. After a rapid reductive elimination, C(aryl)—C(aryl) bond is formed in Ir(III) complex **9-145**. The ligand exchange with arene reactant and coordination of acetate forms complex **9-146**, which can undergo a concerted metalation–deprotonation type C—H activation to regenerate species **9-140**.

References

1 Church, T.L. and Andersson, P.G. (2008). Iridium catalysts for the asymmetric hydrogenation of olefins with nontraditional functional substituents. *Coordination Chemistry Reviews* 252 (5–7): 513–531.
2 Gooßen, L.J., Huang, L., Arndt, M. et al. (2015). Late transition metal-catalyzed hydroamination and hydroamidation. *Chemical Reviews* 115 (7): 2596–2697.
3 Malacea, R., Poli, R., and Manoury, E. (2010). Asymmetric hydrosilylation, transfer hydrogenation and hydrogenation of ketones catalyzed by iridium complexes. *Coordination Chemistry Reviews* 254 (5–6): 729–752.
4 Ishiyama, T. and Miyaura, N. (2006). Iridium-catalyzed borylation of arenes and heteroarenes via C–H activation. *Pure and Applied Chemistry* 78 (7): 1369–1375.

5 Kawamura, K. and Hartwig, J.F. (2001). Isolated Ir(V) boryl complexes and their reactions with hydrocarbons. *Journal of the American Chemical Society* 123 (34): 8422–8423.

6 Preshlock, S.M., Ghaffari, B., Maligres, P.E. et al. (2013). High-throughput optimization of Ir-catalyzed C–H borylation: a tutorial for practical applications. *Journal of the American Chemical Society* 135 (20): 7572–7582.

7 Liu, Y., Gridnev, I.D., and Zhang, W. (2014). Mechanism of the asymmetric hydrogenation of exocyclic α,β-unsaturated carbonyl compounds with an iridium/BiphPhox catalyst: NMR and DFT studies. *Angewandte Chemie International Edition* 126 (7): 1932–1936.

8 Dietiker, R. and Chen, P. (2004). Gas-phase reactions of the [(PHOX)IrL$_2$]$^+$ ion olefin-hydrogenation catalyst support an Ir^I/Ir^{III} cycle. *Angewandte Chemie International Edition* 43 (41): 5513–5516.

9 Li, S., Zhu, S.F., Xie, J.H. et al. (2010). Enantioselective hydrogenation of α-aryloxy and α-alkoxy α,β-unsaturated carboxylic acids catalyzed by chiral spiro iridium/phosphino–oxazoline complexes. *Journal of the American Chemical Society* 132 (3): 1172–1179.

10 Zhu, L., Qi, X., Li, Y. et al. (2017). Ir(III)/Ir(V) or Ir(I)/Ir(III) catalytic cycle? Steric-effect-controlled mechanism for the para-C–H borylation of arenes. *Organometallics* 36 (11): 2107–2115.

11 Tamura, H., Yamazaki, H., Sato, H. et al. (2003). Iridium-catalyzed borylation of benzene with diboron. Theoretical elucidation of catalytic cycle including unusual iridium(V) intermediate. *Journal of the American Chemical Society* 125 (51): 16114–16126.

12 Mazuela, J., Norrby, P.O., Andersson, P.G. et al. (2011). Pyranoside phosphite-oxazoline ligands for the highly versatile and enantioselective Ir-catalyzed hydrogenation of minimally functionalized olefins. A combined theoretical and experimental study. *Journal of the American Chemical Society* 133 (34): 13634–13645.

13 Cui, C.X., Chen, H., Li, S.J. et al. (2020). Mechanism of Ir-catalyzed hydrogenation: a theoretical view. *Coordination Chemistry Reviews* 412: 213251.

14 Ji, L., Lorbach, A., Edkins, R.M. et al. (2015). Synthesis and photophysics of a 2,7-disubstituted donor–acceptor pyrene derivative: an example of the application of sequential Ir-catalyzed C–H borylation and substitution chemistry. *Journal of Organic Chemistry* 80 (11): 5658–5665.

15 Sevov, C.S., Zhou, J., and Hartwig, J.F. (2014). Iridium-catalyzed, intermolecular hydroamination of unactivated alkenes with indoles. *Journal of the American Chemical Society* 136 (8): 3200–3207.

16 He, H., Liu, W.B., Dai, L.X., and You, S.L. (2009). Ir-catalyzed cross-coupling of styrene derivatives with allylic carbonates: free amine assisted vinyl C–H bond activation. *Journal of the American Chemical Society* 131 (24): 8346–8347.

17 Li, M.L., Yang, S., Su, X.C. et al. (2017). Mechanism studies of Ir-catalyzed asymmetric hydrogenation of unsaturated carboxylic acids. *Journal of the American Chemical Society* 139 (1): 541–547.

18 Liu, J., Krajangsri, S., Singh, T. et al. (2017). Regioselective iridium-catalyzed asymmetric monohydrogenation of 1,4-dienes. *Journal of the American Chemical Society* 139 (41): 14470–14475.

19 Ge, Y., Han, Z., Wang, Z. et al. (2019). Ir-catalyzed double asymmetric hydrogenation of 3,6-dialkylidene-2,5-diketopiperazines for enantioselective synthesis of cyclic dipeptides. *Journal of the American Chemical Society* 141 (22): 8981–8988.

20 Rabten, W., Margarita, C., Eriksson, L. et al. (2018). Ir-catalyzed asymmetric and regioselective hydrogenation of cyclic allylsilanes and generation of quaternary stereocenters via the Hosomi–Sakurai allylation. *Chemistry - A European Journal* 24 (7): 1681–1685.

21 Crabtree, R. (1979). Iridium compounds in catalysis. *Accounts of Chemical Research* 12 (9): 331–337.

22 Helmchen, G. and Pfaltz, A. (2000). Phosphinooxazolines – a new class of versatile, modular P,N-ligands for asymmetric catalysis. *Accounts of Chemical Research* 33 (6): 336–345.

23 Wu, W., Liu, S., Duan, M. et al. (2016). Iridium catalysts with f-amphox ligands: asymmetric hydrogenation of simple ketones. *Organic Letters* 18 (12): 2938–2941.

24 Gu, G., Yang, T., Yu, O. et al. (2017). Enantioselective iridium-catalyzed hydrogenation of α-keto amides to α-hydroxy amides. *Organic Letters* 19 (21): 5920–5923.

25 Petra, D.G.I., Kamer, P.C.J., Spek, A.L. et al. (2000). Aminosulf(ox)ides as ligands for iridium(I)-catalyzed asymmetric transfer hydrogenation. *Journal of Organic Chemistry* 65 (10): 3010–3017.

26 Gruber, S. and Pfaltz, A. (2014). Asymmetric hydrogenation with iridium C,N and N,P ligand complexes: characterization of dihydride intermediates with a coordinated alkene. *Angewandte Chemie International Edition* 53 (7): 1896–1900.

27 O, W.N., Lough, A.J., and Morris, R.H. (2012). Bifunctional mechanism with unconventional intermediates for the hydrogenation of ketones catalyzed by an iridium(III) complex containing an N-heterocyclic carbene with a primary amine donor. *Organometallics* 31 (6): 2152–2165.

28 Soltani, O., Ariger, M.A., Vázquez-Villa, H. et al. (2010). Transfer hydrogenation in water: enantioselective, catalytic reduction of α-cyano and α-nitro substituted acetophenones. *Organic Letters* 12 (13): 2893–2895.

29 Soltani, O., Ariger, M.A., and Carreira, E.M. (2009). Transfer hydrogenation in water: enantioselective, catalytic reduction of (E)-β,β-disubstituted nitroalkenes. *Organic Letters* 11 (18): 4196–4198.

30 Zhu, S.F. and Zhou, Q.L. (2017). Iridium-catalyzed asymmetric hydrogenation of unsaturated carboxylic acids. *Accounts of Chemical Research* 50 (4): 988–1001.

31 Schumacher, A., Bernasconi, M., and Pfaltz, A. (2013). Chiral N-heterocyclic carbene/pyridine ligands for the iridium-catalyzed asymmetric hydrogenation of olefins. *Angewandte Chemie International Edition* 52 (29): 7422–7425.

32 Ponra, S., Yang, J., Kerdphon, S. et al. (2019). Asymmetric synthesis of alkyl fluorides: hydrogenation of fluorinated olefins. *Angewandte Chemie International Edition* 58 (27): 9282–9287.

33 Roseblade, S.J. and Pfaltz, A. (2007). Iridium-catalyzed asymmetric hydrogenation of olefins. *Accounts of Chemical Research* 40 (12): 1402–1411.

34 Yang, X.H., Xie, J.H., Liu, W.P. et al. (2013). Catalytic asymmetric hydrogenation of δ-ketoesters: highly efficient approach to chiral 1,5-diols. *Angewandte Chemie International Edition* 52 (30): 7833–7836.

35 Ge, Y., Han, Z., Wang, Z. et al. (2018). Ir-SpinPHOX catalyzed enantioselective hydrogenation of 3-ylidenephthalides. *Angewandte Chemie International Edition* 57 (40): 13140–13144.

36 Noyori, R., Yamakawa, M., and Hashiguchi, S. (2001). Metal-ligand bifunctional catalysis: a nonclassical mechanism for asymmetric hydrogen transfer between alcohols and carbonyl compounds. *Journal of Organic Chemistry* 66 (24): 7931–7944.

37 Zhang, F.H., Zhang, F.J., Li, M.L. et al. (2020). Enantioselective hydrogenation of dialkyl ketones. *Nature Catalysis* 3 (8): 621–627. https://doi.org/10.1038/s41929-020-0474-5.

38 Glorius, F., Spielkamp, N., Holle, S. et al. (2004). Efficient asymmetric hydrogenation of pyridines. *Angewandte Chemie International Edition* 43 (21): 2850–2852.

39 Zhou, Y.G. (2007). Asymmetric hydrogenation of heteroaromatic compounds. *Accounts of Chemical Research* 40 (12): 1357–1366.

40 Yamaguchi, R., Ikeda, C., Takahashi, Y. et al. (2009). Homogeneous catalytic system for reversible dehydrogenation–hydrogenation reactions of nitrogen heterocycles with reversible interconversion of catalytic species. *Journal of the American Chemical Society* 131 (24): 8410–8412.

41 Li, Z.W., Wang, T.L., He, Y.M. et al. (2008). Air-stable and phosphine-free iridium catalysts for highly enantioselective hydrogenation of quinoline derivatives. *Organic Letters* 10 (22): 5265–5268.

42 Mazet, C., Smidt, S.P., Meuwly, M. et al. (2004). A combined experimental and computational study of dihydrido(phosphinooxazoline)iridium complexes. *Journal of the American Chemical Society* 126 (43): 14176–14181.

43 Sun, Y. and Chen, H. (2016). DFT methods to study the reaction mechanism of iridium-catalyzed hydrogenation of olefins: which functional should be chosen? *ChemPhysChem* 17 (1): 119–127.

44 Fan, Y., Cui, X., Burgess, K. et al. (2004). Electronic effects steer the mechanism of asymmetric hydrogenations of unfunctionalized aryl-substituted alkenes. *Journal of the American Chemical Society* 126 (51): 16688–16689.

45 Yamakawa, M., Yamada, I., and Noyori, R. (2001). CH/π attraction: the origin of enantioselectivity in transfer hydrogenation of aromatic carbonyl compounds catalyzed by chiral η^6-arene-ruthenium(II) complexes. *Angewandte Chemie International Edition* 40 (15): 2818–2821.

46 Wang, M.M., He, L., Liu, Y.M. et al. (2011). Gold supported on mesostructured ceria as an efficient catalyst for the chemoselective hydrogenation of carbonyl compounds in neat water. *Green Chemistry* 13 (3): 602–607.

47 Gnanamgari, D., Moores, A., Rajaseelan, E. et al. (2007). Transfer hydrogenation of imines and alkenes and direct reductive amination of aldehydes catalyzed by triazole-derived iridium(I) carbene complexes. *Organometallics* 26 (5): 1226–1230.

48 Liang, Z., Yang, T., Gu, G. et al. (2018). Scope and mechanism on iridium-f-amphamide catalyzed asymmetric hydrogenation of ketones. *Chinese Journal of Chemistry* 36 (9): 851–856.

49 Zweifel, T., Naubron, J.V., Büttner, T. et al. (2008). Ethanol as hydrogen donor: highly efficient transfer hydrogenations with rhodium(I) amides. *Angewandte Chemie International Edition* 47 (17): 3245–3249.

50 Handgraaf, J.W., Reek, J.N.H., and Meijer, E.J. (2003). Iridium(I) versus ruthenium(II). A computational study of the transition metal catalyzed transfer hydrogenation of ketones. *Organometallics* 22 (15): 3150–3157.

51 Kays (née Coombs), D.L., Day, J.K., Aldridge, S. et al. (2006). Cationic terminal borylene complexes: interconversion of amino and alkoxy borylenes by an unprecedented Meerwein–Ponndorf hydride transfer. *Angewandte Chemie International Edition* 118 (21): 3593–3596.

52 Yu, J., Long, J., Yang, Y. et al. (2017). Iridium-catalyzed asymmetric hydrogenation of ketones with accessible and modular ferrocene-based amino-phosphine acid (f-ampha) ligands. *Organic Letters* 19 (3): 690–693.

53 Dobereiner, G.E., Nova, A., Schley, N.D. et al. (2011). Iridium-catalyzed hydrogenation of N-heterocyclic compounds under mild conditions by an outer-sphere pathway. *Journal of the American Chemical Society* 133 (19): 7547–7562.

54 Hopmann, K.H. and Bayer, A. (2011). On the mechanism of iridium-catalyzed asymmetric hydrogenation of imines and alkenes: a theoretical study. *Organometallics* 30 (9): 2483–2497.

55 Noyori, R. and Ohkuma, T. (2001). Asymmetric catalysis by architectural and functional molecular engineering: practical chemo- and stereoselective hydrogenation of ketones. *Angewandte Chemie International Edition* 40 (1): 40–73.

56 Noyori, R., Kitamura, M., and Ohkuma, T. (2004). Toward efficient asymmetric hydrogenation: architectural and functional engineering of chiral molecular catalysts. *Proceedings of the National Academy of Sciences of the United States of America* 101 (15): 5356–5362.

57 Dub, P.A. and Gordon, J.C. (2017). Metal-ligand bifunctional catalysis: the "Accepted" mechanism, the issue of concertedness, and the function of the ligand in catalytic cycles involving hydrogen atoms. *ACS Catalysis* 7 (10): 6635–6655.

58 Dub, P.A., Henson, N.J., Martin, R.L. et al. (2014). Unravelling the mechanism of the asymmetric hydrogenation of acetophenone by

[RuX$_2$(diphosphine)(1,2-diamine)] catalysts. *Journal of the American Chemical Society* 136 (9): 3505–3521.

59 Dub, P.A., Scott, B.L., and Gordon, J.C. (2017). Why does alkylation of the N–H functionality within M/NH bifunctional Noyori-type catalysts lead to turnover? *Journal of the American Chemical Society* 139 (3): 1245–1260.

60 Schnider, P., Koch, G., Prétôt, R. et al. (1997). Enantioselective hydrogenation of imines with chiral (phosphanodihydrooxazole)iridium catalysts. *Chemistry - A European Journal* 3 (6): 887–892.

61 Baeza, A. and Pfaltz, A. (2010). Iridium-catalyzed asymmetric hydrogenation of imines. *Chemistry - A European Journal* 16 (13): 4003–4009.

62 Kainz, S., Brinkmann, A., Leitner, W. et al. (1999). Iridium-catalyzed enantioselective hydrogenation of imines in supercritical carbon dioxide. *Journal of the American Chemical Society* 121 (27): 6421–6429.

63 Sakamoto, M., Ohki, Y., Kehr, G. et al. (2009). Catalytic hydrogenation of C=O and C=N bonds via heterolysis of H$_2$ mediated by metal–sulfur bonds of rhodium and iridium thiolate complexes. *Journal of Organometallic Chemistry* 694 (17): 2820–2824.

64 Tutkowski, B., Kerdphon, S., Limé, E. et al. (2018). Revisiting the stereodetermining step in enantioselective iridium-catalyzed imine hydrogenation. *ACS Catalysis* 8 (1): 615–623.

65 Salomó, E., Gallen, A., Sciortino, G. et al. (2018). Direct asymmetric hydrogenation of *N*-methyl and *N*-alkyl imines with an Ir(III)H catalyst. *Journal of the American Chemical Society* 140 (49): 16967–16970.

66 Villa-Marcos, B., Li, C., Mulholland, K.R. et al. (2010). Bifunctional catalysis: direct reductive amination of aliphatic ketones with an iridium–phosphate catalyst. *Molecules* 15 (4): 2453–2472.

67 Shirai, S.Y., Nara, H., Kayaki, Y. et al. (2009). Remarkable positive effect of silver salts on asymmetric hydrogenation of acyclic imines with Cp*Ir complexes bearing chiral *N*-sulfonylated diamine ligands. *Organometallics* 28 (3): 802–809.

68 Zhou, H., Li, Z., Wang, Z. et al. (2008). Hydrogenation of quinolines using a recyclable phosphine-free chiral cationic ruthenium catalyst: enhancement of catalyst stability and selectivity in an ionic liquid. *Angewandte Chemie International Edition* 47 (44): 8464–8467.

69 Hesp, K.D., Tobisch, S., and Stradiotto, M. (2010). [Ir(COD)Cl]$_2$ as a catalyst precursor for the intramolecular hydroamination of unactivated alkenes with primary amines and secondary alkyl- or arylamines: a combined catalytic, mechanistic, and computational investigation. *Journal of the American Chemical Society* 132 (1): 413–426.

70 Li, X., Chianese, A.R., Vogel, T. et al. (2005). Intramolecular alkyne hydroalkoxylation and hydroamination catalyzed by iridium hydrides. *Organic Letters* 7 (24): 5437–5440.

71 Hesp, K.D. and Stradiotto, M. (2009). Intramolecular hydroamination of unactivated alkenes with secondary alkyl-and arylamines employing [Ir(COD)Cl]$_2$ as a catalyst precursor. *Organic Letters* 11 (6): 1449–1452.

72 Gray, K., Page, M.J., Wagler, J. et al. (2012). Iridium(III) Cp* complexes for the efficient hydroamination of internal alkynes. *Organometallics* 31 (17): 6270–6277.

73 Xie, X., Zhang, X., Yang, H. et al. (2019). Iridium-catalyzed hydrosilylation of unactivated alkenes: scope and application to late-stage functionalization. *Journal of Organic Chemistry* 84 (2): 1085–1093.

74 Srinivas, V., Nakajima, Y., Sato, K. et al. (2018). Iridium-catalyzed hydrosilylation of sulfur-containing olefins. *Organic Letters* 20 (1): 12–15.

75 Xie, X., Zhang, X., Gao, W. et al. (2019). Iridium-catalyzed Markovnikov hydrosilylation of terminal alkynes achieved by using a trimethylsilyl-protected trihydroxysilane. *Communications Chemistry* 2 (1): 4–11.

76 Song, L.J., Ding, S., Wang, Y. et al. (2016). Ir-catalyzed regio- and stereoselective hydrosilylation of internal thioalkynes: a combined experimental and computational study. *Journal of Organic Chemistry* 81 (15): 6157–6164.

77 Nagamoto, M., Yorimitsu, H., and Nishimura, T. (2018). Iridium-catalyzed hydroarylation of conjugated dienes via π-allyliridium intermediates. *Organic Letters* 20 (3): 828–831.

78 Huang, G. and Liu, P. (2016). Mechanism and origins of ligand-controlled linear versus branched selectivity of iridium-catalyzed hydroarylation of alkenes. *ACS Catalysis* 6 (2): 809–820.

79 Ebe, Y., Onoda, M., Nishimura, T. et al. (2017). Iridium-catalyzed regio- and enantioselective hydroarylation of alkenyl ethers by olefin isomerization. *Angewandte Chemie International Edition* 129 (20): 5699–5703.

80 Romero-Arenas, A., Hornillos, V., Iglesias-Sigüenza, J. et al. (2020). Ir-catalyzed atroposelective desymmetrization of heterobiaryls: hydroarylation of vinyl ethers and bicycloalkenes. *Journal of the American Chemical Society* 142 (5): 2628–2639.

81 Motta, A., Lanza, G., Fragalà, I.L. et al. (2004). Energetics and mechanism of organolanthanide-mediated aminoalkene hydroamination/cyclization. A density functional theory analysis. *Organometallics* 23 (17): 4097–4104.

82 Haibach, M.C., Guan, C., Wang, D.Y. et al. (2013). Olefin hydroaryloxylation catalyzed by pincer–iridium complexes. *Journal of the American Chemical Society* 135 (40): 15062–15070.

83 Lalrempuia, R., Iglesias, M., Polo, V. et al. (2012). Effective fixation of CO_2 by iridium-catalyzed hydrosilylation. *Angewandte Chemie International Edition* 51 (51): 12824–12827.

84 Pérez-Torrente, J.J., Nguyen, D.H., Jiménez, M.V. et al. (2016). Hydrosilylation of terminal alkynes catalyzed by a ONO-pincer iridium(III) hydride compound: mechanistic insights into the hydrosilylation and dehydrogenative silylation catalysis. *Organometallics* 35 (14): 2410–2422.

85 Huang, G., Kalek, M., Liao, R.Z. et al. (2015). Mechanism, reactivity, and selectivity of the iridium-catalyzed C(sp^3)–H borylation of chlorosilanes. *Chemical Science* 6 (3): 1735–1746.

86 Patel, C., Abraham, V., and Sunoj, R.B. (2017). Mechanistic insights and the origin of regioselective borylation in an iridium-catalyzed alkyl C(sp^3)–H bond functionalization. *Organometallics* 36 (1): 151–158.

87 Haines, B.E., Saito, Y., Segawa, Y. et al. (2016). Flexible reaction pocket on bulky diphosphine–Ir complex controls regioselectivity in para-selective C–H borylation of arenes. *ACS Catalysis* 6 (11): 7536–7546.

88 Jover, J. and Maseras, F. (2016). Mechanistic investigation of iridium-catalyzed C–H borylation of methyl benzoate: ligand effects in regioselectivity and activity. *Organometallics* 35 (18): 3221–3226.

89 Balcells, D., Nova, A., Clot, E. et al. (2008). Mechanism of homogeneous iridium-catalyzed alkylation of amines with alcohols from a DFT study. *Organometallics* 27 (11): 2529–2535.

90 Zhao, G.M., Liu, H.L., Huang, X.R. et al. (2015). Mechanistic study on the Cp*iridium-catalyzed N-alkylation of amines with alcohols. *RSC Advances* 5 (29): 22996–23008.

91 Ryu, J., Kwak, J., Shin, K. et al. (2013). Ir(III)-catalyzed mild C–H amidation of arenes and alkenes: an efficient usage of acyl azides as the nitrogen source. *Journal of the American Chemical Society* 135 (34): 12861–12868.

92 Figg, T.M., Park, S., Park, J. et al. (2014). Comparative investigations of cp*-based group 9 metal-catalyzed direct C–H amination of benzamides. *Organometallics* 33 (15): 4076–4085.

93 Sandmeier, T., Goetzke, F.W., Krautwald, S. et al. (2019). Iridium-catalyzed enantioselective allylic substitution with aqueous solutions of nucleophiles. *Journal of the American Chemical Society* 141 (31): 12212–12218.

94 Cheng, Q., Tu, H.F., Zheng, C. et al. (2019). Iridium-catalyzed asymmetric allylic substitution reactions. *Chemical Reviews* 119 (3): 1855–1969.

95 Hamilton, J.Y., Hauser, N., Sarlah, D. et al. (2014). Iridium-catalyzed enantioselective allyl–allylsilane cross-coupling. *Angewandte Chemie International Edition* 126 (40): 10935–10938.

96 Zheng, Y., Yue, B.B., Wei, K. et al. (2018). Iridium-catalyzed enantioselective allyl–allyl cross-coupling of racemic allylic alcohols with allylboronates. *Organic Letters* 20 (24): 8035–8038.

97 Hamilton, J.Y., Sarlah, D., and Carreira, E.M. (2014). Iridium-catalyzed enantioselective allyl–alkene coupling. *Journal of the American Chemical Society* 136 (8): 3006–3009.

98 Xia, Y., Qiu, D., and Wang, J. (2017). Transition-metal-catalyzed cross-couplings through carbene migratory insertion. *Chemical Reviews* 117 (23): 13810–13889.

99 Shang, R. and Liu, L. (2011). Transition metal-catalyzed decarboxylative cross-coupling reactions. *SCIENCE CHINA Chemistry* 54 (11): 1670–1687.

100 Cherney, A.H., Kadunce, N.T., and Reisman, S.E. (2015). Enantioselective and enantiospecific transition-metal-catalyzed cross-coupling reactions of organometallic reagents to construct C–C bonds. *Chemical Reviews* 115 (17): 9587–9652.

101 Gao, P., Guo, W., Xue, J. et al. (2015). Iridium(III)-catalyzed direct arylation of C–H bonds with diaryliodonium salts. *Journal of the American Chemical Society* 137 (38): 12231–12240.

102 Shin, K., Park, S.W., and Chang, S. (2015). Cp*Ir(III)-catalyzed mild and broad C–H arylation of arenes and alkenes with aryldiazonium salts leading to the external oxidant-free approach. *Journal of the American Chemical Society* 137 (26): 8584–8592.

10

Theoretical Study of Fe-Catalysis

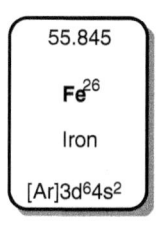

55.845

Fe^{26}

Iron

[Ar]3d⁶4s²

In my opinion, iron is not a fine transition-metal catalyst in homogeneous catalysis. Due to the metallicity of iron, usually, the covalent bond between iron and carbon is relatively unstable compared with other late transition metal–carbon bonds [1–4]. Therefore, there are only a few examples where iron catalysts are applied to the carbon chain growth reactions. However, the reaction mechanism of iron catalysis has not been simplified. The complexity for the iron-catalysis can be attributed to the varieties of oxidative states and electronic configurations [5–10]. In iron chemistry, the oxidative states of iron atom can be from −2 to +6, which can be changed easily. Since the valence shell of iron is often not filled up, the spin configurations of iron are varied. In the same molecular structure of iron complexes, the relative energies of different spin configurations might be close; therefore, the various potential energy surfaces of multiple spin configurations should be considered in mechanism study of iron-catalysis. Unfortunately, describing high-spin multiplicity is a weakness of density functional theory (DFT) calculations, and the calculation results often deviate from the high-accuracy calculations using multireference states [11–18]. Moreover, the high-accuracy algorithm cannot calculate the scale of the existing systems. Effectively, there are still many problems to be solved urgently about the mechanism of iron catalysis [19–25].

In biochemistry, iron is the key atom in heme iron enzymes, which can react with dioxygen to form high-valence iron-oxo intermediates. Subsequent biomimetic studies have designed a series of iron complexes based on heme or nonheme structures, which can also react with dioxygen to generate iron-oxo species. The designed

Computational Methods in Organometallic Catalysis: From Elementary Reactions to Mechanisms,
First Edition. Yu Lan.
© 2021 WILEY-VCH GmbH. Published 2021 by WILEY-VCH GmbH.

iron-oxo species can be used in oxidation reaction to provide an atomic oxygen onto organic compounds [26–31]. The Fe—H bond is important as it can play as a key intermediate in iron-catalyzed hydrogenation and hydrofunctionalization reactions. Moreover, iron reveals radical character, in which radical-type substitution can occur to afford carbon radicals for further transformations [32–35]. This chapter explains these points.

10.1 Fe-Mediated Oxidations

In a manner of speaking, the real mechanism of iron-mediated oxidations is extremely complicated; thus, it can be said that it is only a very simple part of the understanding at present. Generally, a high-valence iron-oxo species plays an important role in oxidation reactions, which can be formed in reactions between the reduced iron complex and dioxygen. Commonly, the binding of dioxygen to iron is very weak. It may even be that in some cases the binding is so weak that the enthalpic binding energy does not fully compensate for the large loss of entropy, and dioxygen will remain unbound. Nevertheless, the coordination of dioxygen onto low-valence iron complexes would lead to the O—O bond activation in further transformations. As shown in Scheme 10.1, after formal absorption of a proton and two electrons, an iron-oxo species can be generated with the increase of two for iron's formal oxidative state [36]. The generated iron-oxo species reveals an active oxygen source in oxidation reactions with C—H bonds, arenes, alcohols [37–39], and heteroatoms [40–42].

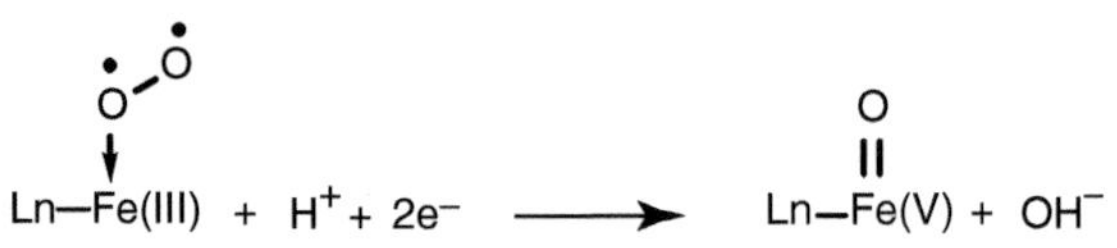

Ln—Fe(III) + H$^+$ + 2e$^-$ ⟶ Ln—Fe(V) + OH$^-$

Scheme 10.1 The generation of iron-oxo species.

10.1.1 Alkane Oxidations

In a general view of iron-mediated alkane oxidation, iron-oxo complexes play the important role of oxidant, which can absorb hydrogen from alkane to afford an alkyl radical. The following alkyl radical substitution with ferric hydroxide leads to the formation of corresponding alcohols [43].

The calculated potential energy surfaces for iron(V)-oxo oxidation of alkane are shown in Figure 10.1 [44]. The iron(V)-oxo complex **10-1** can be quartet or doublet, the energy difference of which is only 0.1 kcal/mol. The hydrogen bond can be formed between reacting alkane and oxygen atom in complex **10-2** with 3.4 kcal/mol exothermic. The hydrogen atom abstraction of camphor can occur via a quartet transition state **10-3ts** with an energy barrier of 21.4 kcal/mol. After this step, an alkyl radical with ferric(IV) hydroxide intermediate **10-4** is generated. Then, the C(alkyl)—O(hydroxide) bond formation can occur via transition state **10-5** with

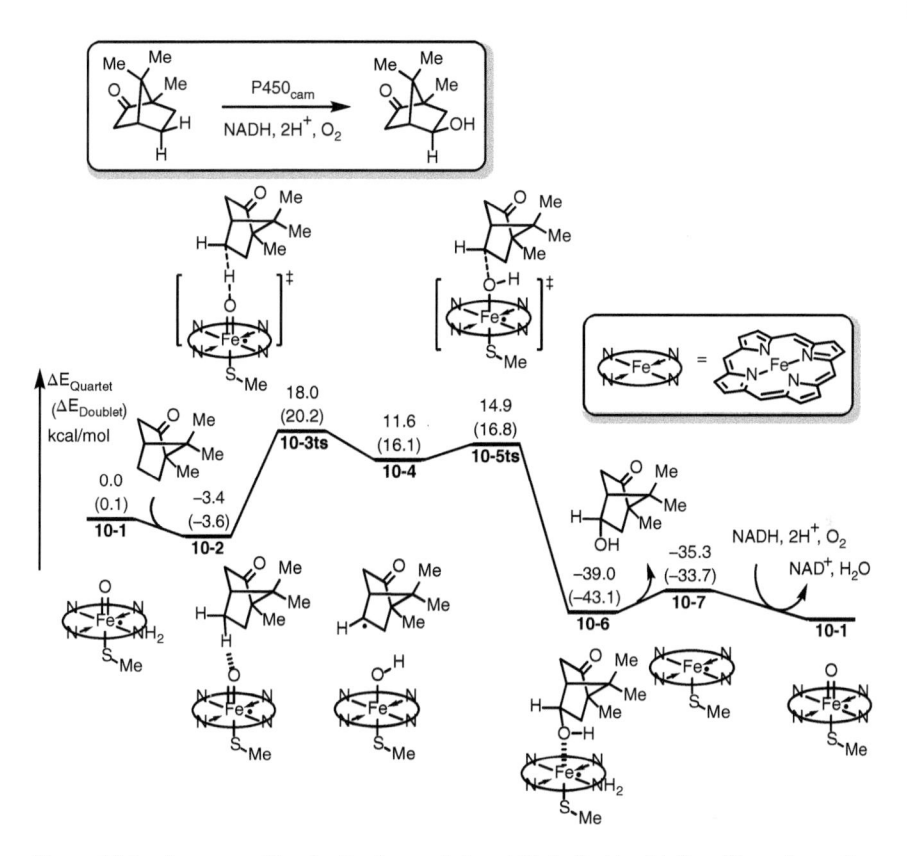

Figure 10.1 Energy profiles for the iron-catalyzed C(alkyl)–H oxidation. The values of energies are calculated at quartet state. The values given in parentheses are relative energy of corresponding doublet states. The energies were calculated at B3LYP/6-311+G(d,p) (LANL2DZ for Fe)//B3LYP/D95 (TZV for Fe) level of theory. Source: Based on Kamachi and Yoshizawa [44].

an energy barrier of only 3.3 kcal/mol to yield corresponding alcohol-coordinated iron(III) complex **10-6**. After release of one molecule alcohol, iron(III) species **10-7** is generated, which can be oxidized by dioxygen to regenerate iron(V)-oxo complex **10-1**.

In a model study of iron(III)-oxo oxidation reaction [31], the sextet state of iron(III)-oxo species **10-8** is 5.8 kcal/mol more stable than its quartet state. When iron(III)-oxo species **10-8** reacts with methane, an agostic complex **10-9** can be formed with 16.4 kcal/mol exothermic in quartet state or 22.8 kcal/mol exothermic in sextet state. A metathesis-type C—H bond cleavage can take place via transition state **10-10ts**. However, quartet state of **10-10ts** is 9.0 kcal/mol more stable than sextet state. Therefore, a spin inversion would occur before C—H bond cleavage. After this process, a methyl ferric(III) hydroxide intermediate **10-11** can be generated. Then, a reductive elimination of methyl and hydroxide takes place via a quartet transition state **10-12ts** to provide a methanol-coordinated iron(I) complex **10-13** (Figure 10.2).

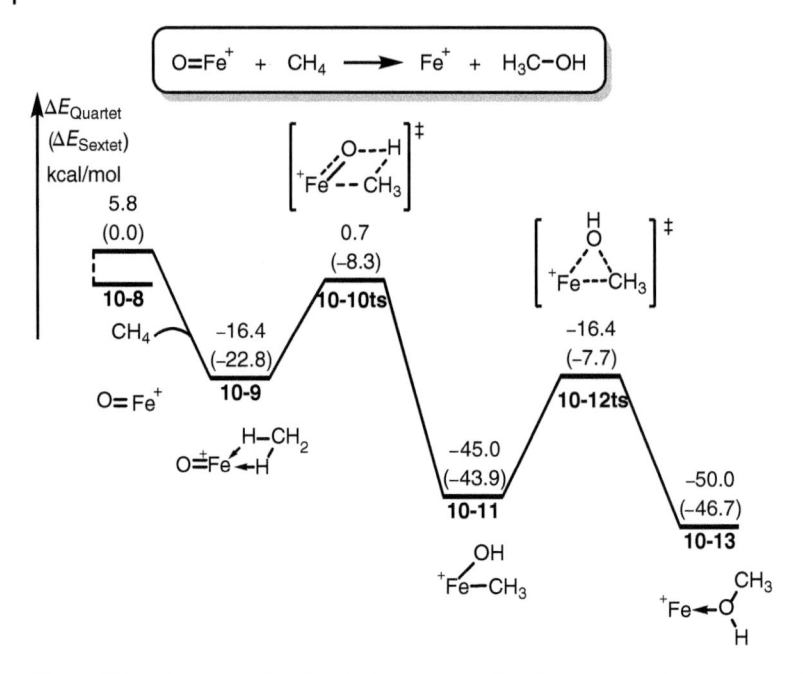

Figure 10.2 Energy profiles for the iron-oxo mediated methane oxidation. The values of energies are calculated at quartet state. The values given in parentheses are relative energy of corresponding sextet states. The energies were calculated at B3LYP/6-311+G(d,p) (LANL2DZ for Fe)//B3LYP/D95 (TZV for Fe) level of theory.

10.1.2 Arene Oxidations

According to alkane oxidations, C—H bond in arenes also can react with iron(IV or V)-oxo complex to form corresponding phenol products [45]. However, the mechanism of arene oxidation is different from alkane oxidation (Scheme 10.2). Generally, oxygen atom in high-valence iron-oxo complex reveals both cationic and radical character, while the reacting arene can play as both nucleophile and radical acceptor. Therefore, two possible pathways were proposed in arene oxidation by iron(IV

Scheme 10.2 Possible reaction modes for the arene oxidation by iron-oxo complex. (a) Radical type oxidation (b) Electrophilic oxidation.

or V)-oxo complex, including radical-type addition followed by 1,2-hydrogen shift [46, 47] and electrophilic addition followed by proton transfer [47].

A comparative study for both radical-type and electrophilic oxidations of benzene by iron(IV)-oxo was carried out to test their feasibilities [47]. As shown in Figure 10.3a, a neutral (**10-14**) or cationic (**10-15**) iron(IV)-oxo complex was chosen as model reactant to simulate enzyme environment. When an iron(IV)-oxo complex **10-14** reacts with benzene (Figure 10.3b), a radical-type attack takes place first to form a new C(arene)—O(oxo) bond via transition state **10-16ts** with an energy barrier of 16.2 kcal/mol to form an aryl radical **10-17**. In this intermediate, spin density is shared by arene ring. Then a 1,2-hydrogen shift via transition state **10-18ts** occurs with an energy barrier of only 5.6 kcal/mol to generate keto-intermediate, which can easily isomerize to phenol. Alternatively, a two-electron transfer process can be expressed as Figure 10.3c. A cationic iron(IV)-oxo complex **10-15** can react with benzene through an electrophilic attack via transition state **10-20ts** to form a cationic arene intermediate **10-21**. Then proton shift also can form the same keto-intermediate. The calculated activation free energy for the electrophilic attack is 10.0 kcal/mol via transition state **10-20ts**, which is 6.2 kcal/mol lower than that of corresponding radical attack via transition state **10-16ts**. Although the radical-type attack cannot be simply excluded because of the different relative zero in free-energy profiles, the computational results might reveal that electrophilic attack might be a favorable process in this oxidation.

DFT calculations revealed the full reaction pathway for the iron-mediated oxidative ortho-hydroxylation of benzoic acid with H_2O_2 [48]. As shown in Figure 10.4, the reaction starts from a benzoic acid–coordinated sextet ferric(III)-hydroperoxo complex **10-24** with a hydrogen bond between benzoic acid and proximal oxygen atom. The proton transfer leads to the O—O bond cleavage through a doublet transition state **10-25ts** with an energy barrier of 16.7 kcal/mol. Then, a doublet iron(V)-oxo benzoate complex **10-26** is formed. The chelation of carboxyl assists the intramolecular radical addition onto arene moiety via doublet transition state **10-27ts**. Then, a sextet iron(IV) species **10-28** is formed. Subsequently, a 1,2-hydrogen shift from carbon to oxygen takes place via transition state **10-29ts** leading to the formation of salicylic acid–coordinated sextet iron(III) species **10-30**.

10.1.3 Alkene Oxidations

A number of nonheme iron complexes have been identified that catalyze the epoxidation [50, 51] and dihydroxylation [52] of olefins with H_2O_2 as oxidant. DFT calculations revealed the mechanism of alkene oxidations. As shown in Figure 10.5 [49], a nonheme cationic ferric(III)-hydroperoxo complex can be coordinated by water molecule to form complex **10-31**. A proton transfer from water to proximal oxygen atom leads to the O—O bond cleavage through transition state **10-32ts** with an energy barrier of 19.2 kcal/mol to afford a hydroxy iron(V)-oxo complex **10-33**. An intermolecular radical addition by oxygen atom of hydroxy group via transition state **10-34ts** can provide dihydroxylation product. The calculated energy barrier of this process is 6.2 kcal/mol.

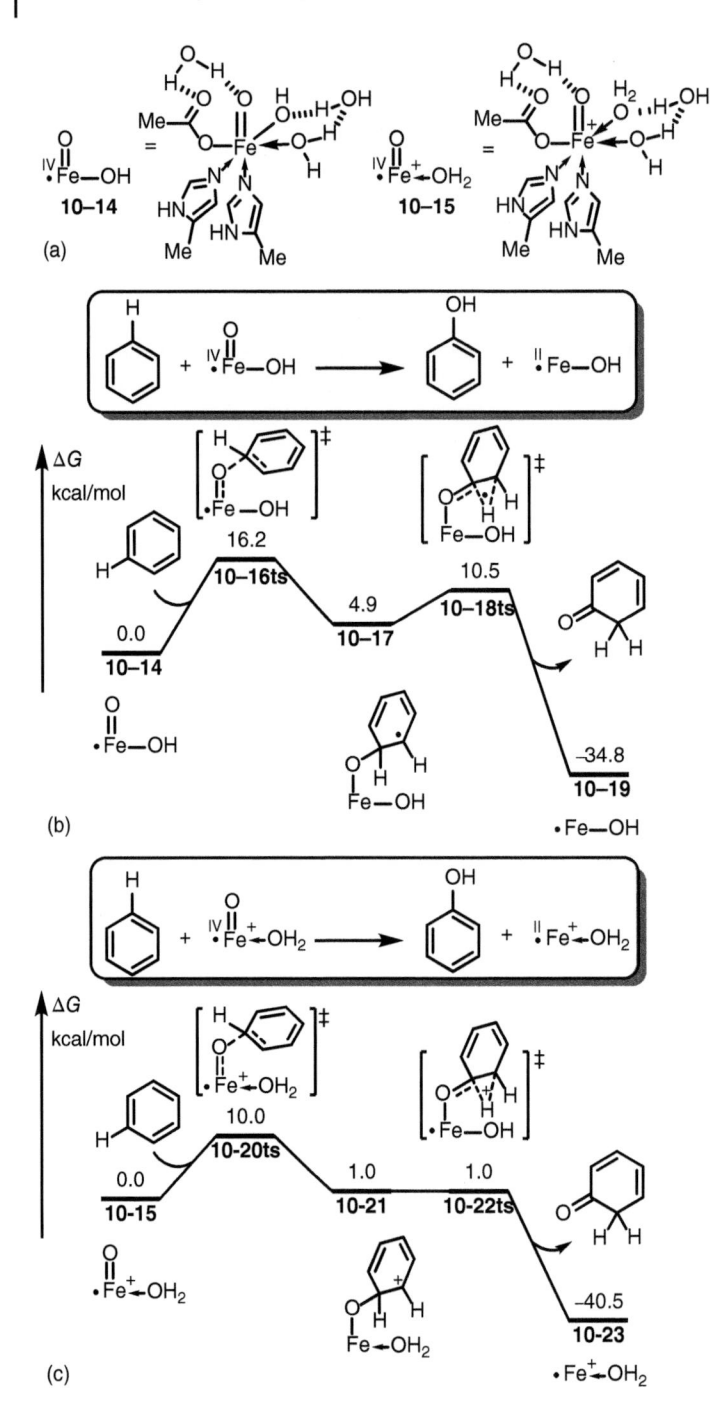

Figure 10.3 (a) A comparative study of the iron-oxo oxidation of benzene. (b) Free-energy profiles of radical-type oxidation at quartet. (c) Free-energy profiles of electrophilic oxidation at quartet. The energies were calculated at B3LYP/lacv3p**/PCM(protein environment, $\varepsilon = 4.0$)//B3LYP/lacvp level of theory.

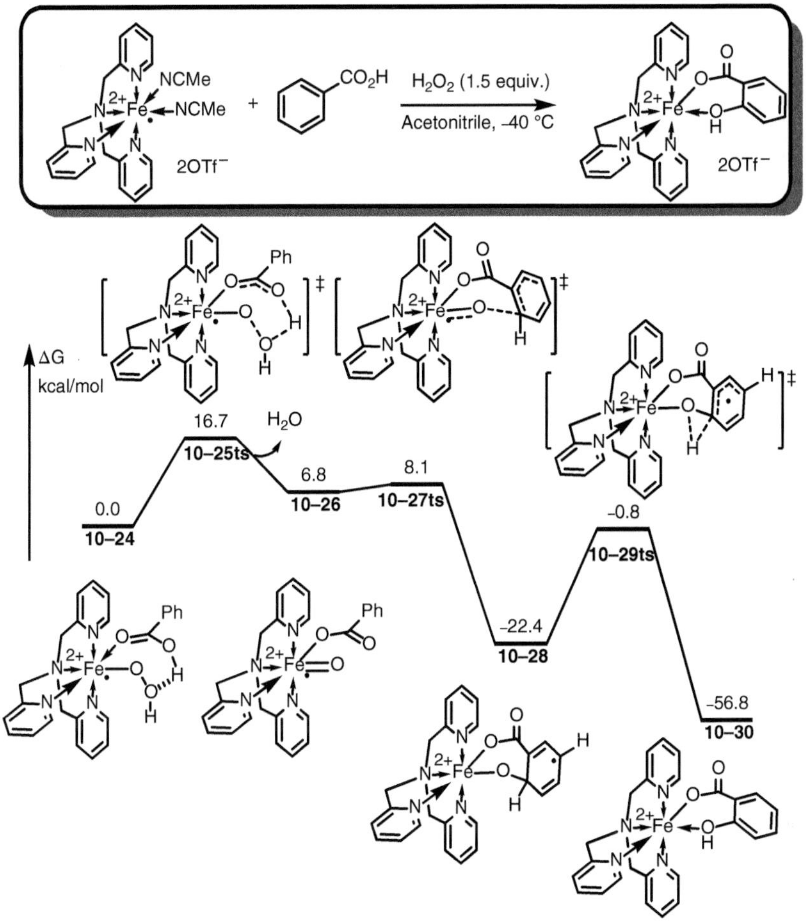

Figure 10.4 Free-energy profiles for the iron-mediated oxidative ortho-hydroxylation of benzoic acid with H_2O_2. The energies were calculated at B3LYP-D3/TZVP/PCM (acetonitrile)//B3LYP-D3/6-31G(d) (LANL2DZ for Fe) level of theory.

10.1.4 Oxidative Catechol Ring Cleavage

Usually, catechols contain a stable six-membered carbon ring. In bioorganic chemistry, protocatechuate 3,4-dioxygenase, an iron(III)-dependent enzyme, can be used in the selective oxidative ring scission of pyrocatechols to yield *cis,cis*-muconic acid [53–55]. The mechanism of this reaction was revealed by DFT calculations [56]. The pyrocatechol can react with iron to form a pyrocatecholate iron(III) complex **10-37**. The coordination of dioxygen generates a peroxo iron(IV) complex **10-38**. Then, the terminal oxygen atom attacks onto pyrocatecholate moiety through transition state **10-39ts** to afford a peroxo-bridged iron(IV) complex **10-40**. The preoxo moiety can be protonated by hydroxyl group in coordinated tyrosine via transition state **10-42ts** to form an iron(IV) complex **10-43**. An acyl migration via transition state **10-44ts** leads to a heterolytic cleavage of the O—O bond. Then an anhydride intermediate

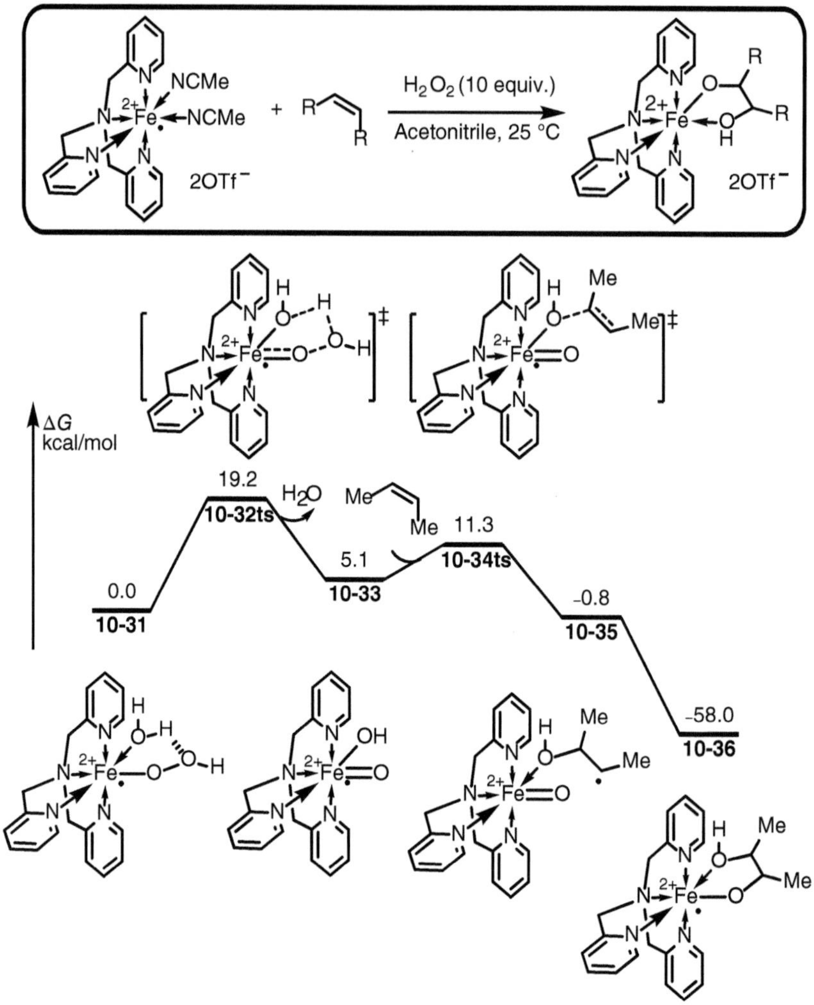

Figure 10.5 Free-energy profiles for the iron-mediated oxidation of alkenes with H_2O_2. The energies were calculated at B3LYP/6-31G(d,p) (LANL2DZ for Fe) level of theory. Source: Based on Oldenburg et al. [49].

10-45 is formed. The carbonyl insertion into Fe—O(hydroxy) bond via transition state **10-46ts** generates intermediate **10-47**. After an intramolecular proton transfer, muconic acid is yielded by ligand exchange (Figure 10.6).

10.2 Fe-Mediated Hydrogenations

Low-valence iron complex can be used to activate dihydrogen molecule to afford hydride iron complex, which can react with unsaturated molecules to achieve hydrogenation. In this chemistry, the activation of dihydrogen is the key step, which can

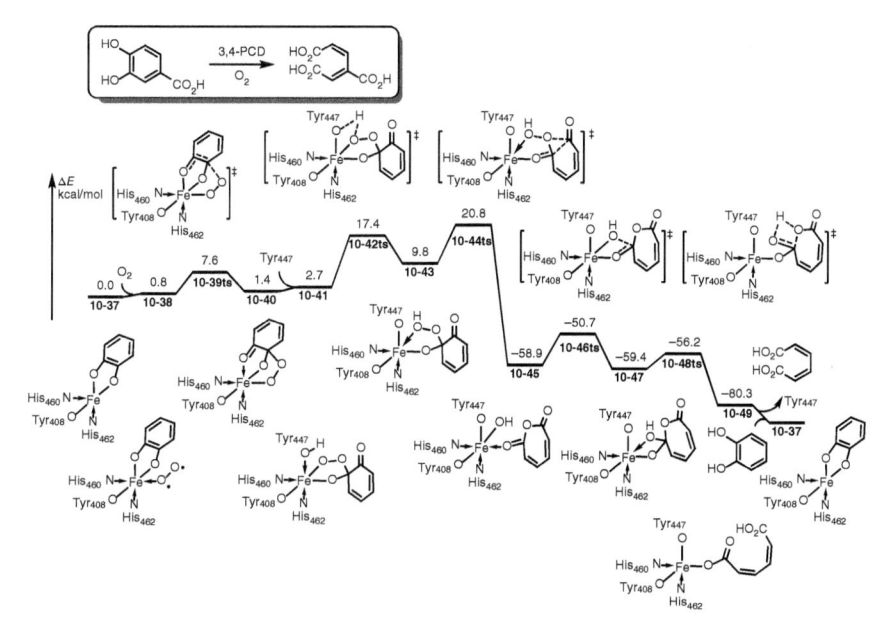

Figure 10.6 Energy profiles for the iron-catalyzed oxidative extradiol oxidation of alkenes with H$_2$O$_2$. The energies were calculated at B3LYP/lacv3p**/PCM(water)//B3LYP/lalcvp level of theory.

undergo both direct oxidative addition and base-assisted processes [48]. The detailed possibilities are discussed in this section.

10.2.1 Hydrogenation of Alkenes

In the presence of Fe(0)CO$_3$ catalyst, the direct dihydrogenation of alkenes can occur to generate the corresponding alkanes [57–61]. The DFT-calculated catalytic pathway for this reaction is shown in Figure 10.7 [48]. The reaction starts from a dihydrogen molecule–coordinated iron(0) complex **10-50**. The oxidative addition of dihydrogen occurs via transition state **10-51ts** to afford a dihydride iron(II) intermediate **10-52** reversibly. Then, the coordinated alkene inserts into Fe—H bond via a four-membered ring-type transition state **10-53ts** from an agostic intermediate **10-54**. The coordination of another molecule alkene leads to the C(alkyl)—H(hydride) reductive elimination via transition state **10-56ts**. After this process, alkane product can be released with the regeneration of iron(0) species **10-50**. The rate-determining step is considered to be the reductive elimination with an energy barrier of 17.5 kcal/mol relating to the second insertion intermediate **10-58**.

When a phosphine-nitrogen-phosphine (PNP)-type pincer ligand is used in iron-catalyzed dihydrogenation reaction of styrenes, a ligand-assisted stepwise bifunctional pathway was proposed and proved by DFT calculations [57]. As shown in Figure 10.8, the reaction starts from an amine-coordinated ferrous species **10-59**. An intermolecular hydride transfer takes place via transition state **10-60ts** with an

428 | *10 Theoretical Study of Fe-Catalysis*

Figure 10.7 Energy profiles for the iron-catalyzed hydrogenation of alkenes. The energies were calculated at B3LYP/6-311+G(2d,p) (LANL2DZ for Fe) level of theory. Source: Based on Asatryan and Ruckenstein [48].

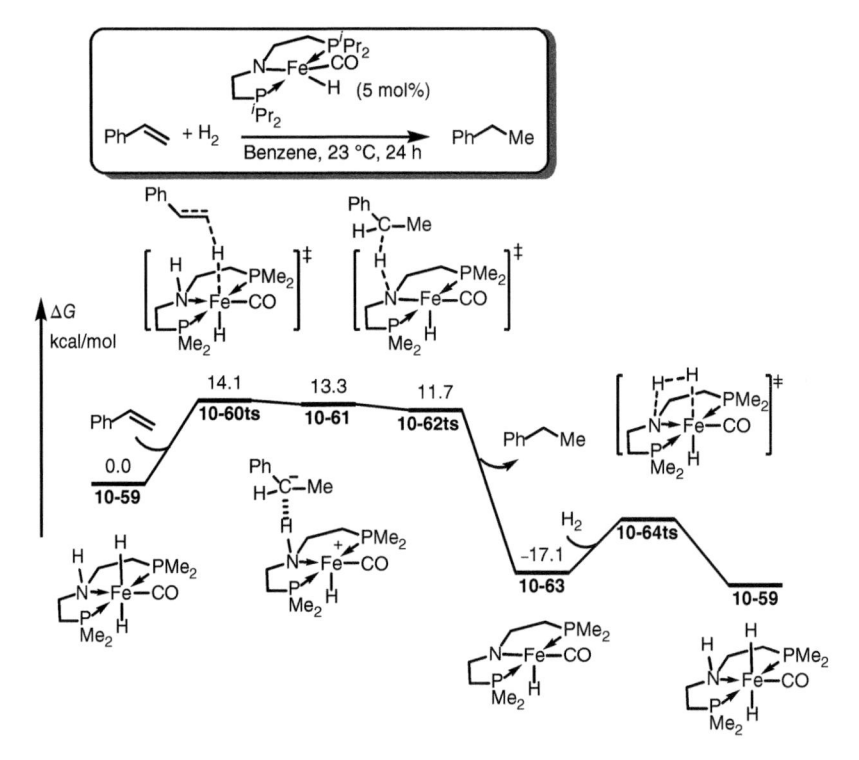

Figure 10.8 Free-energy profiles for the ligand-assisted iron-catalyzed hydrogenation of alkenes. The energies were calculated at B3PW91-D3/def2-QZVPP/SMD(benzene)//B3LYP/def2-SVP level of theory.

energy barrier of 14.1 kcal/mol to form a cationic ferrous intermediate **10-61** with a benzylic anionic counterion. Then, an intermolecular proton transfer from amine ligand to benzylic anion occurs via transition state **10-62ts** to yield phenylethane product and give an amino ferrous species **10-63**, which can react with dihydrogen to regenerate hydride ferrous species **10-59**.

10.2.2 Hydrogenation of Carbonyls

In iron-catalyzed dihydrogenation of carbonyl compounds, a two-step pathway was proposed involving hydride transfer and base-assisted heterolysis of dihydrogen [63–67]. As shown in Figure 10.9 [62], the reaction starts from a hydride ferrous species **10-65**. An intermolecular hydride transfer with acetophenone occurs via a linear transition state **10-66ts** with an energy barrier of 18.1 kcal/mol to form a cationic iron(II) intermediate **10-67** with hydrogen bond of phenylethanolate. The coordination of dihydrogen molecule forms intermediate **10-68**. With the activation of iorn(II), a heterolysis of dihydrogen by phenylethanolate takes place via transition state **10-69ts** to from a hydrogen bond intermediate **10-70**, in which

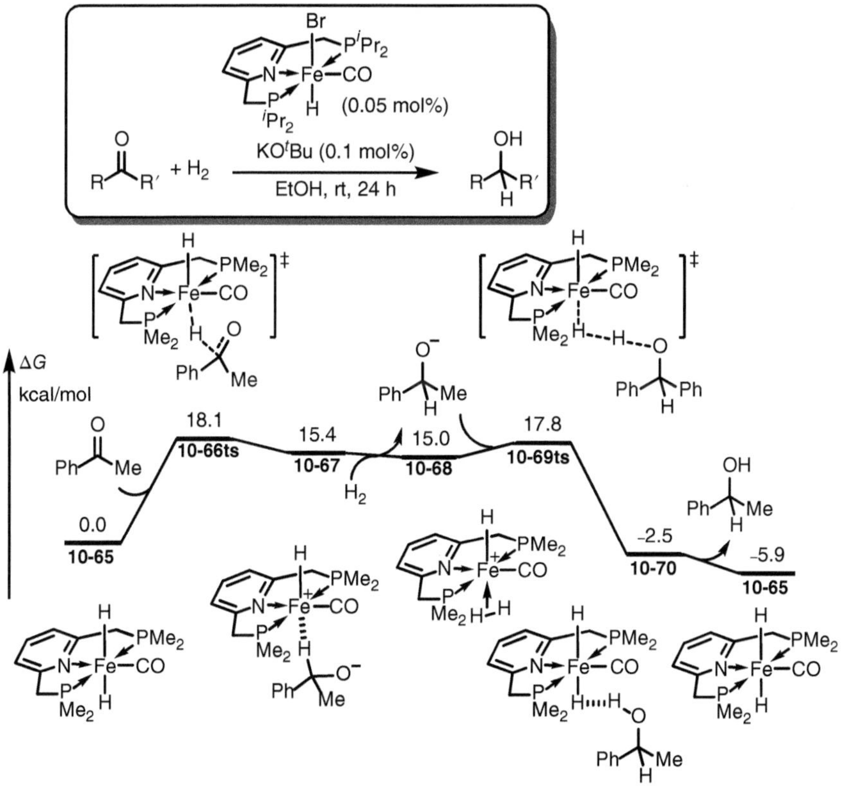

Figure 10.9 Free-energy profiles for the iron-catalyzed hydrogenation of ketones. The energies were calculated at ωB97X-D/6-31++G(d,p)/SMD(ethanol) level of theory. Source: Based on Yang [62].

430 | *10 Theoretical Study of Fe-Catalysis*

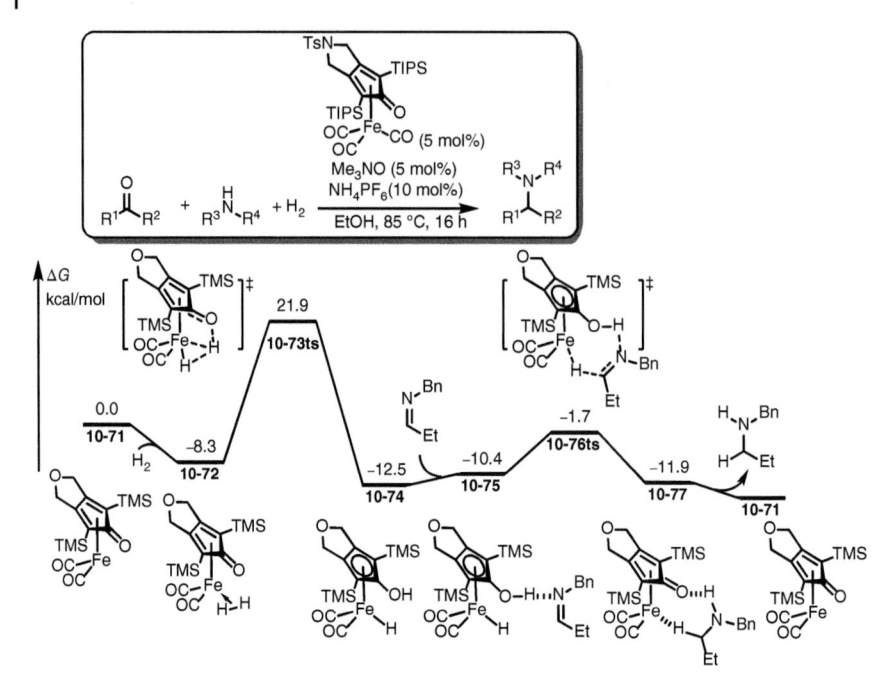

Figure 10.10 Free-energy profiles for the iron-catalyzed hydrogenation of imines. The energies were calculated at M06/TZVP (SDD for Fe)/PCM(ethanol) level of theory. Source: Moulin et al. [68]; Hopmann [69].

two hydrogen atoms formally represent hydride and proton characters. After release of phenylethanol product, active hydride ferrous species **10-65** is regenerated.

10.2.3 Hydrogenation of Imines

When a Knölker-type iron(0) complex is used as catalyst for the dihydrogenation of imines, a ligand-assisted concerted proton/hydride transfer pathway was proposed. As shown in Figure 10.10 [68, 69], when dihydrogen molecule coordinates with Knölker-type iron(0) complex **10-71** in intermediate **10-72**, a ligand-assisted H—H bond cleavage takes place via transition state **10-73ts** with an energy barrier of 30.2 kcal/mol to afford a hydride ferrocene intermediate **10-74**. A hydrogen bond can be formed between hydroxyl group of ligand and reacting imine in intermediate **10-75**. Then, a concerted hydride transfer from iron to imine occurs via transition state **10-76ts** with simultaneous proton transfer from hydroxyl to nitrogen. After release amine product, Knölker-type iron(0) complex **10-71** can be regenerated.

10.2.4 Hydrogenation of Carbon Dioxide

Carbon dioxide can be transferred to formic acid by transition metal-catalyzed hydrogenations in the presence of extra base [71–80]. DFT calculations were carried out to predict this reaction, which has been reported later. As shown in Figure 10.11

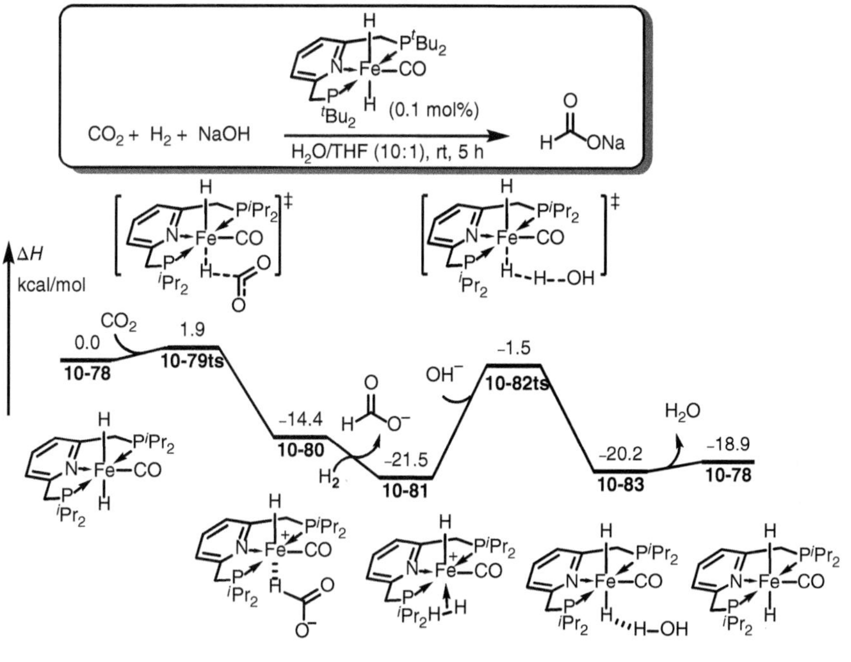

Figure 10.11 Relative enthalpy profiles for the iron-catalyzed hydrogenation of carbon dioxide. The energies were calculated at ωB97X/6-31++G(d,p)/IEF-PCM(water) level of theory. Source: Based on Yang [70].

[70], when a hydride ferrous species **10-78** is formed, an intermolecular hydride transfer with carbon dioxide occurs via a linear transition state **10-79ts** with an energy barrier of only 1.9 kcal/mol. A ferrous intermediate **10-80** is formed by 14.4 kcal/mol exergonic involving a hydrogen bond with formate anion. The coordination of dihydrogen molecule onto iron forms cationic complex **10-81** by further 7.1 kcal/mol exergonic. Then, a hydroxide-promoted deprotonation takes place via transition state **10-82ts** leading to the heterolysis of H—H bond to regenerate active hydride ferrous species **10-78** with release of water.

10.3 Fe-Mediated Hydrofunctionalizations

Due to the notable stability of hydride iron species, iron complexes can be used as catalyst for the hydrofunctionalizations of unsaturated bonds, including hydrosilylations [81, 82], hydroaminations, and hydrophosphinations [83].

10.3.1 Hydrosilylation of Ketones

In the presence of iron catalyst, ketones can react with silane to form siloxanes, which can be further hydrolyzed to generate corresponding alcohols. A unique peripheral catalytic cycle was proposed and proved by DFT calculations [84]. As

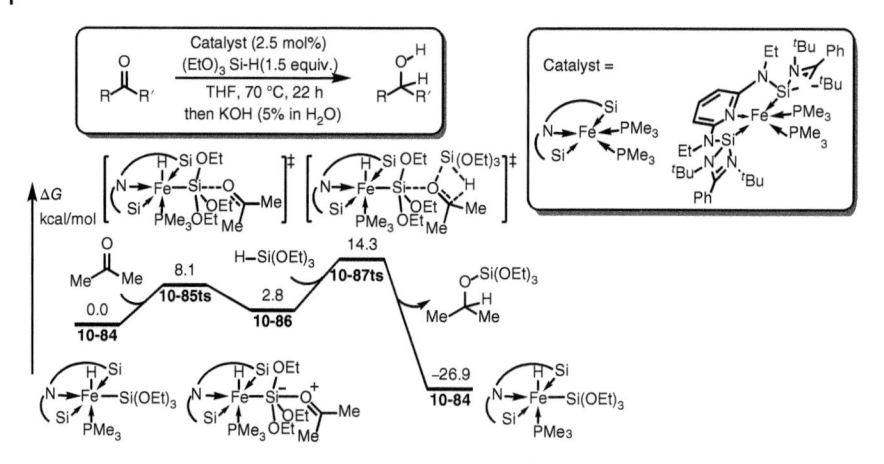

Figure 10.12 Free-energy profiles for the outer-sphere catalytic cycle of iron-catalyzed hydrosilylation of ketones. The energies were calculated at ωB97X-D/6-31G(d) (cc-pVTZ for Fe) level of theory.

shown in Figure 10.12, the reaction starts from a silyl ferrous species **10-84**. A nucleophilic attack of ketone's oxygen atom onto silyl group occurs via transition state **10-85ts** with an energy barrier of 8.1 kcal/mol to form a hypervalence silicon intermediate **10-86**. With the activation of silicon atom, an outer-sphere σ-bond metathesis of ketone with an extra silane takes place via a four-membered ring-type transition state **10-87ts** with an energy barrier of 11.5 kcal/mol. The release of siloxane product 80a regenerates silyl ferrous species **10-84**.

As shown in Figure 10.13, an inner-sphere catalytic cycle was also studied by DFT calculation in another iron-catalyzed hydrosilylation reaction [85]. The coordinated carbonyl in complex **10-89** can insert into Fe—H bond via a four-membered ring-type transition state **10-90ts** to form an alkoxyl iron(II) intermediate **10-91** with an energy barrier of 7.7 kcal/mol. Then, a stepwise σ-bond metathesis of Fe—O bond with H—Si bond occurs via a four-membered ring-type transition state **10-92ts**. The calculated activation free energy for this step is 12.0 kcal/mol, which can be considered as rate-determining step. After this step, a hetero agostic complex **10-93** is formed. The dissociation of siloxane product via transition state **10-94ts** regenerates active hydride ferrous species **10-88**.

10.3.2 Hydroamination of Allenes

Knölker-type iron(0) complex can be used to catalyze an intramolecular hydroamination of allenes [80]. The mechanism of this reaction was revealed by DFT calculation (Figure 10.14). The coordination onto iron(0) leads to the activation of allene in complex **10-95**. Therefore, an intramolecular nucleophilic attack of amino group via transition state **10-96ts** forms a zwitterionic intermediate **10-97**. Then, an intramolecular proton transfer by ligand provides a neutral vinyl ferrous species **10-99**. A ligand-promoted proton transfer through protonation–deprotonation process generates another vinyl ferrous species **10-103**. Then, the protonasis of

10.3 Fe-Mediated Hydrofunctionalizations | 433

Figure 10.13 Free-energy profiles for the inner-sphere catalytic cycle of the iron-catalyzed hydrosilylation of ketones. The energies were calculated at TPSSh/def2-QZVPP/COSMO(toluene)//BP86/def2-TZVP level of theory.

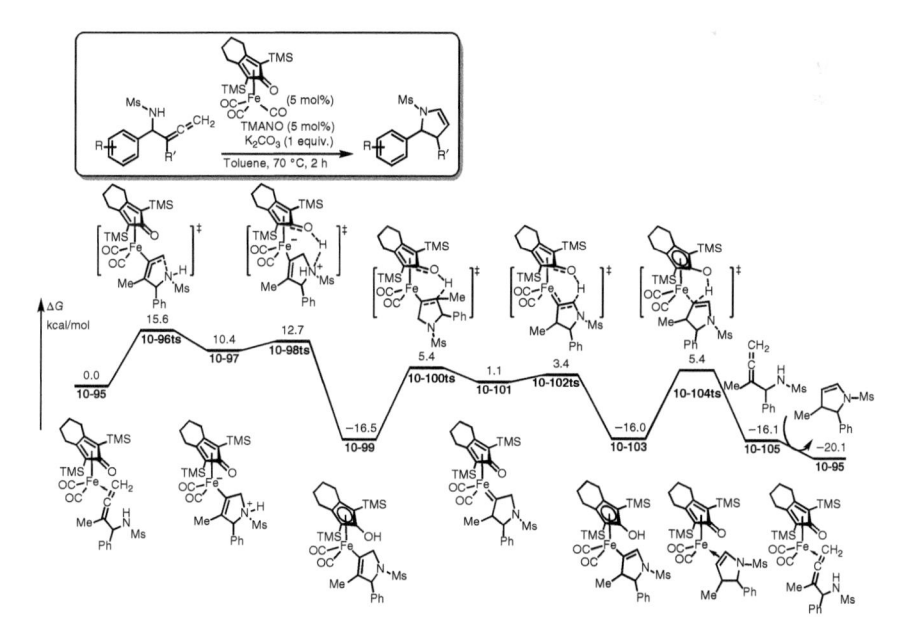

Figure 10.14 Free-energy profiles for the iron-catalyzed intramolecular hydroamination of allenes. The energies were calculated at B3LYP-D3BJ/6-311+G(2d,2p)/SMD(toluene)//B3LYP-D3BJ/6-31G(d,p) (LANL2DZ for Fe) level of theory.

Fe—C(vinyl) bond by the hydroxyl group on ligand via transition state **10-104ts** yields pyrroline product and regenerates Knölker-type iron(0) complex **10-95**.

10.4 Fe-Mediated Dehydrogenations

The notable stability of hydride iron can cause the dehydrogenation of organic compounds for the generation of unsaturated molecules. In this process, two hydrogen atoms transfer from organic molecule to iron in the form of proton and hydride correspondingly [86, 87]. In the following step, hydride iron can be protonated to release gaseous dihydrogen.

In the presence of ferrous catalyst, the dehydrogenation of alcohols can undergo a stepwise ligand-assisted process, in which the ligand plays as a proton acceptor and iron(II) center receives hydride. Interestingly, when methanol is used, the reaction is hard to stop at the formation of formaldehyde, but get carbon dioxide and gaseous dihydrogen directly (Scheme 10.3).

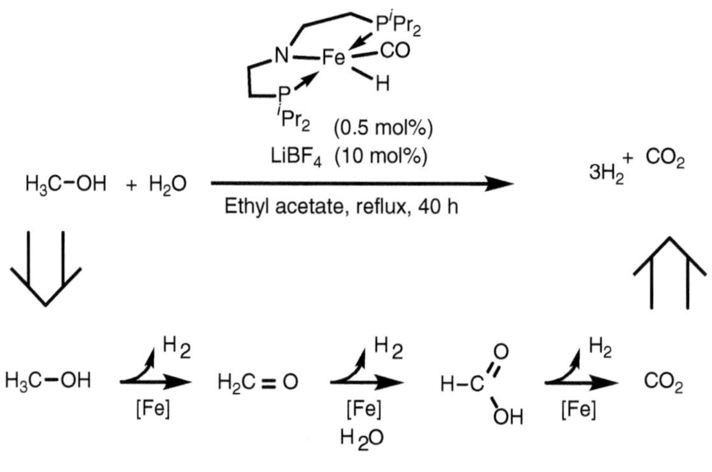

Scheme 10.3 Dehydrogenation of alcohols.

10.4.1 Dehydrogenation of Alcohols

In the dehydrogenation of methanol, a PNP-type pincer ligand is employed to assist deprotonation by using amino moiety [88]. As shown in Figure 10.15, when an amino ferrous complex **10-106** is used as catalyst, the reacting methanol can form a hydrogen bond with nitrogen atom of ligand in complex **10-107**. A hydride transfer from methanol to iron occurs via transition state **10-108ts** with an energy barrier of 15.1 kcal/mol. Synchronously, proton transfer from hydroxyl to amino also occurs. The release of formaldehyde leads to the formation of a hydride ferrous intermediate **10-109**. Then, a methanol-assisted protonation of hydride takes place via transition state **10-110ts** to afford the same intermediate **10-107** with dissociation of dihydrogen molecule. The dissociation of methanol regenerates amino

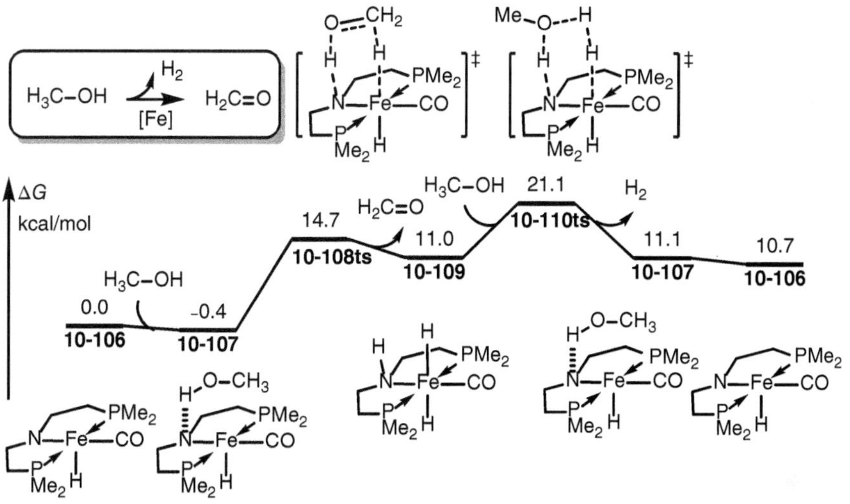

Figure 10.15 Free-energy profiles for the iron-catalyzed dehydrogenation of methanol. The energies were calculated at B3LYP/def2-SVP level of theory.

ferrous species **10-106**. The whole process is 10.7 kcal/mol endergonic indicating a thermodynamically unfavorable process. Therefore, this process can only be carried out by subsequent processes.

10.4.2 Dehydrogenation of Formaldehyde

When formaldehyde is formed, it can react with water to form methanediol reversibly. As shown in Figure 10.16 [88], the dehydrogenation of methanediol is a stepwise process. When a hydrogen bond is formed between methanediol and amino ligand in complex **10-111**, a proton transfer occurs first to form a cationic ferrous intermediate **10-113**. Then, a hydride transfer from hydroxymethanolate to iron occurs via a six-membered ring-type transition state **10-114ts** with an energy barrier of only 3.4 kcal/mol. After this step, a formic acid is generated in intermediate **10-115**. The generated formic acid can be used as Brønsted acid to assist the protonation of hydride via transition state **10-116ts**. The whole process is 37.0 kcal/mol exergonic, which can promote the previous deprotonation of methanol.

10.4.3 Dehydrogenation of Formic Acid

Further deprotonation of formic acid can provide carbon dioxide in the presence of iron catalyst [89–93]. As shown in Figure 10.17 [89], when sodium is used as counterion, an isomerization occurs firstly to form a sodium formate in **10-118**. Subsequently, a hydride transfer takes place via a linear transition state **10-119ts** with an energy barrier of 20.7 kcal/mol. The release of carbon dioxide generates complex **10-120** by appending methanol. Then, a methanol-assisted protonation of hydride occurs via transition state **10-110ts** to regenerate amino ferrous

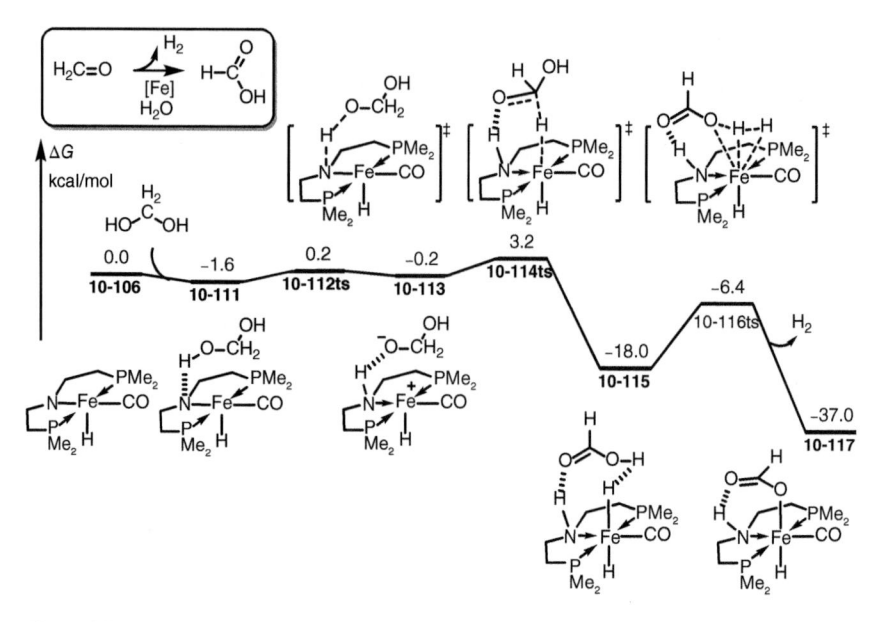

Figure 10.16 Free-energy profiles for the iron-catalyzed dehydrogenation of methanediol. The energies were calculated at B3LYP/def2-SVP level of theory. Source: Based on Bielinski et al. [88].

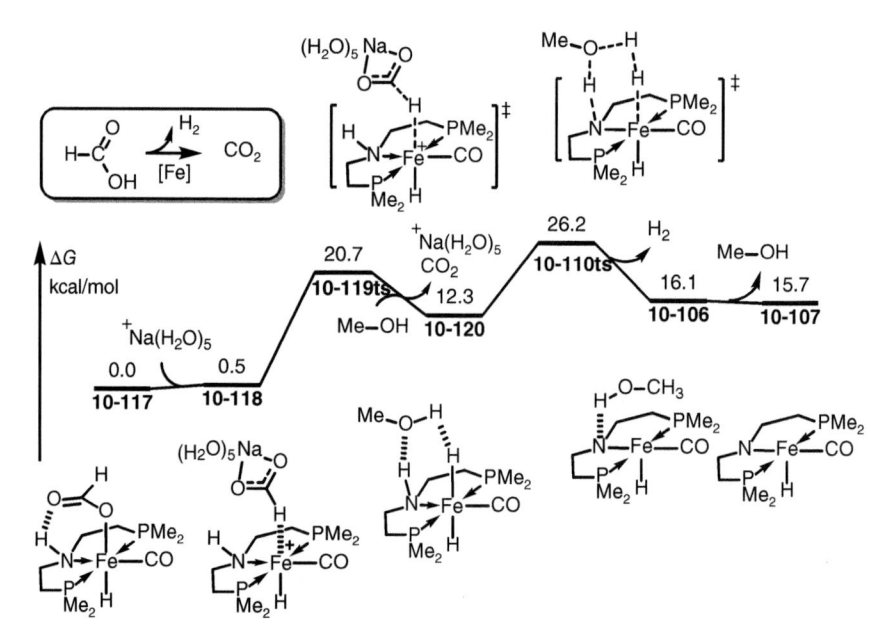

Figure 10.17 Free-energy profiles for the iron-catalyzed dehydrogenation of formic acid. The energies were calculated at B3LYP/def2-SVP level of theory. Source: Based on Zell and Milstein [89].

species **10-106**. The dehydrogenation of formic acid-involved complex **10-107** is 15.7 kcal/mol endergonic. Therefore, the driving force for the whole catalytic cycle of dehydrogenation of methanediol given in Figures 10.16 and 10.17 is the first dehydrogenation step. The overall process of the carbon dioxide formation from methanediol is 21.3 kcal/mol exergonic.

10.4.4 Dehydrogenation of Ammonia-Borane

Interestingly, the same PNP-type pincer ligand-coordinated iron(II) species can be used as catalyst in the dehydrogenation of ammonia-borane Lewis acid–base complex to achieve amino borane polymer [95]. As shown in Figure 10.18 [94], the reaction starts from a hydride ferrous species **10-121**. ammonia-borane plays as a Brønsted acid, which can protonate ferrous hydride via a seven-membered ring-type transition state **10-122ts** to form a dihydrogen-coordinated cationic ferrous intermediate **10-123**. After release of dihydrogen via transition state **10-124ts**, an agostic amino borane–ferrous complex **10-125** is formed. The dissociation of amino borane leads to the regeneration of hydride ferrous species **10-121** via a

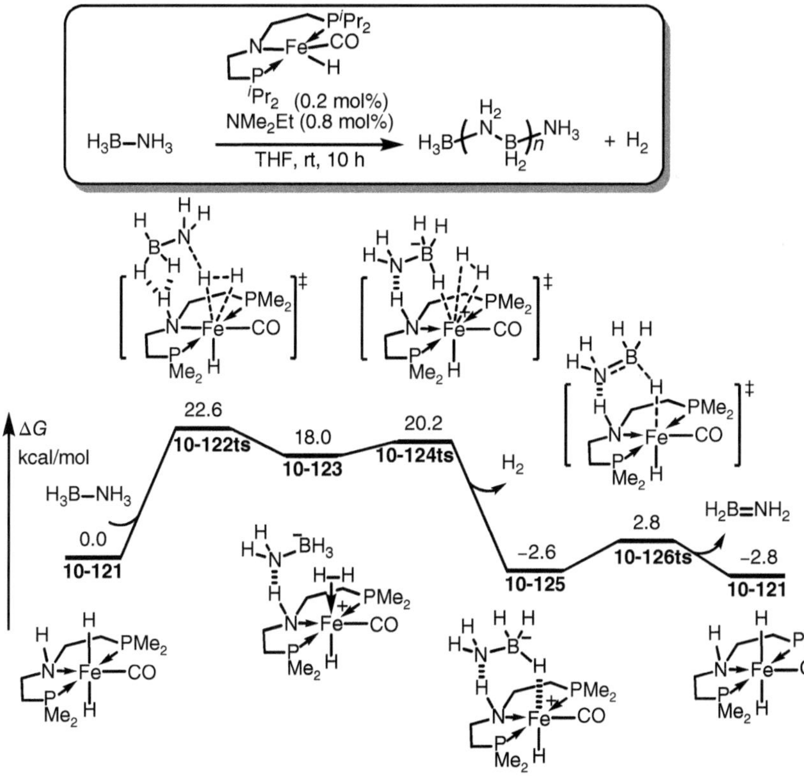

Figure 10.18 Free-energy profiles for the iron-catalyzed dehydrogenation of ammonia-borane. The energies were calculated at B3LYP-D3BJ/def2TZVPP/SMD(THF)// B3LYP-D3BJ/def2-SVP level of theory. Source: Based on Glüer et al. [94].

low-barrier transition state **10-126ts**. The whole process is 2.8 kcal/mol exergonic. The polymerization of amino borane would provide further driving force for this reaction.

10.5 Fe-Catalyzed Coupling Reactions

Early transition metal-catalyzed cross-coupling reactions have been the focus recently, which can be used to construct new C—C covalent bonds. The substitution of early transition metals can remarkably reduce costs and environmental pollution for the cross-coupling reactions. In this chemistry, iron plays good catalytic ability in cross-coupling reactions [1, 96–106]. Although iron-catalyzed cross-coupling reactions can formally produce the same products as post-transition metals, the reaction mechanisms are often different compared with the similar transformations. Therefore, DFT calculations were used to study the mechanism of iron catalysis in cross-coupling reactions.

10.5.1 C—C Cross-Couplings with Aryl Halide

In Pd- or Ni-chemistry, the cross-coupling of aryl halide and alkyl Grignard reagents named Kumada coupling is a powerful way for the synthesis of alkyl arenes. Interestingly, iron also can be used as catalyst in this reaction [107, 108]. The mechanism of iron-catalyzed Kumada-type coupling reaction was revealed by DFT calculations [109]. As shown in Figure 10.19, the reaction starts from an unordinary Mg—Fe complex **10-127**, in which the iron atom expresses a formal −2 oxidative state. The coordination of aryl chloride forms complex **10-128** with 5.6 kcal/mol exergonic. Then, a metathesis-type oxidation occurs via a four-membered ring-type transition state **10-129ts**, in which the Mg—Cl and Fe—C(aryl) bonds are formed synchronously. After this step, an aryl iron(0) intermediate **10-130** is generated with the coordination of Mg salt. Then, intermediate **10-130** can interact with Grignard reagent through ligand exchange to form complex **10-131** with release of Mg salt. The following transmetallation can bear an alkyl migration process via a three-membered ring-type transition state **10-132ts** to form alkyl aryl iron(0) intermediate **10-133**. Finally, reductive elimination via transition state **10-134ts** yields alkylated arene product and regenerates Mg–Fe(−2) species **10-127**.

10.5.2 C—N Cross-Couplings with Aryl Halide

In an iron-catalyzed amination of aryl halide, an Fe(II)–Fe(IV) catalytic cycle was proposed by DFT calculation [110]. As shown in Figure 10.20, the reaction starts from a dimeric ferrous complex **10-135**, which can dissociate to monomeric diamino ferrous species **10-136** with 5.4 kcal/mol endergonic. The oxidative addition of phenyl bromide occurs via a three-membered ring-type transition state **10-137ts** with an overall activation free energy of 26.7 kcal/mol. After this step, a phenyl iron(IV) intermediate **10-138** is generated. Then, a reductive elimination of

10.5 Fe-Catalyzed Coupling Reactions | 439

Figure 10.19 Free-energy profiles for the iron-catalyzed Kumada-type coupling. The energies were calculated at B3LYP/6-311++G(d,p) (SDD for Fe)/C-PCM(THF)//B3LYP/6-311++G(d,p) (SDD for Fe) level of theory.

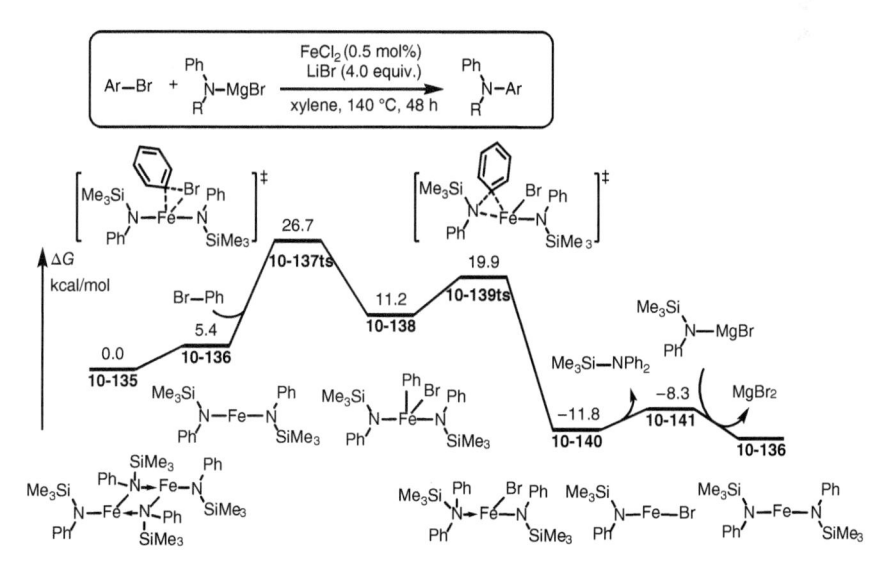

Figure 10.20 Free-energy profiles for the iron-catalyzed amination of aryl bromide. The energies were calculated at B3LYP/6-31G(d) level of theory.

440 | *10 Theoretical Study of Fe-Catalysis*

Figure 10.21 Free-energy profiles for the iron-catalyzed arylation of alkyl chloride through a radical-type pathway. The energies were calculated at B3LYP-D3BJ/BS2/IEF-PCM(THF)//B3LYP-D3BJ/BS1 level of theory. BS1 is the combination of 6-31G(d) (C and H), 6-31+G(d,p) (O, P, Cl, and Mg), and SDD (Fe and Br) basis sets. BS2 is the combination of cc-PVTZ (H, C, O, Mg, P, and Cl) and SDD (Fe and Br) basis sets. Source: Based on Sharma et al. [111].

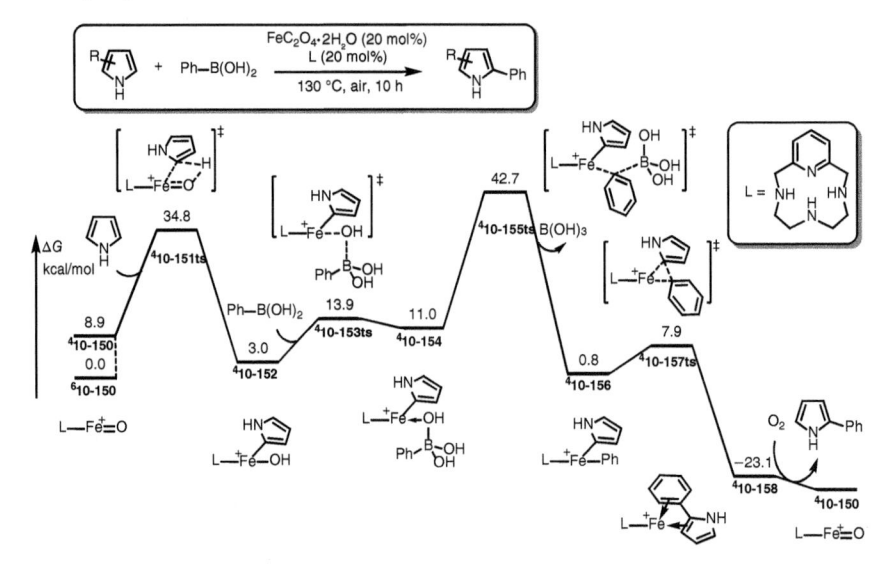

Figure 10.22 Free-energy profiles for the iron-catalyzed oxidative coupling of arenes and aryl boronic acids. The energies were calculated at B3LYP/6-311+G(d,p)/PCM(pyrrole)//B3LYP/6-31G(d) level of theory. Source: Based on Dong et al. [114].

phenyl and amino groups takes place via transition state **10-139ts** to form new C—N bond in ferrous complex **10-140**. Then, the release of amino benzene gives ferrous bromide **10-141**, which can undergo a transmetallation with amino magnesium to regenerate amino ferrous **10-136**.

10.5.3 C—C Cross-Couplings with Alkyl Halide

Different from cross-coupling with aryl halide, the iron-catalyzed cross-couplings with alkyl halide could undergo a radical-type pathway [112, 113]. As shown in Figure 10.21 [111], in an iron-catalyzed cross-coupling of alkyl chloride and aryl Grignard reagents, an Fe(I)–Fe(II)–Fe(III) catalytic cycle was proposed by DFT calculations. The reaction starts from a quartet iron(I) species ^{4}**10-142**. A radical-type substitution with aryl chloride occurs via a linear transition state ^{4}**10-143ts** with an energy barrier of only 1.5 kcal/mol to afford a quintet ferrous dichloride intermediate ^{5}**10-144** by release of an alkyl radical. A transmetallation with aryl Grignard reagent takes place via transition state ^{5}**10-145ts** and forms aryl iron(II) intermediate ^{5}**10-146**. The coordination of alkyl radical onto intermediate ^{5}**10-146** leads to further oxidation, which generates a quartet ferric species ^{4}**10-148**. Then, a reductive elimination via transition state ^{4}**10-149ts** yields aryl alkane product and regenerates quartet iron(I) species ^{4}**10-142**.

10.5.4 Iron-Mediated Oxidative Coupling

Unlike cross-coupling, new covalent bond can be formed between two nucleophiles in the presence of oxidant. Interestingly, ferric compound can be used as both catalyst and oxidant in oxidative coupling of arenes and aryl boronic acids [115]. As shown in Figure 10.22 [114], the reaction starts from a quintet cationic ferric-oxo compound ^{4}**10-150**, whose relative free energy is 8.9 kcal/mol higher than its sextet state. When arene coordinates with ferric, a C—H activation by oxo occurs via a metathesis transition state ^{4}**10-151ts** with an energy barrier of 25.9 kcal/mol to afford aryl ferric–hydroxy intermediate ^{4}**10-152**. Then, a stepwise transmetallation with aryl boronic acid via transition states ^{4}**10-153ts** and ^{4}**10-155ts** provides diaryl ferric intermediate ^{4}**10-156**. The calculated overall activation free energy for this step is 42.7 kcal/mol. Finally, a rapid reductive elimination results in biaryl-coordinated iron(I) species ^{4}**10-158**. Active species ^{4}**10-150** could be regenerated by the oxidation with air.

References

1 Toriyama, F., Cornella, J., Wimmer, L. et al. (2016). Redox-active esters in Fe-catalyzed C–C coupling. *Journal of the American Chemical Society* 138 (35): 11132–11135.

2 Lo, J.C., Kim, D., Pan, C.M. et al. (2017). Fe-catalyzed C—C bond construction from olefins via radicals. *Journal of the American Chemical Society* 139 (6): 2484–2503.

3 Morris, R.H. (2015). Exploiting metal-ligand bifunctional reactions in the design of iron asymmetric hydrogenation catalysts. *Accounts of Chemical Research* 48 (5): 1494–1502.
4 Gorgas, N. and Kirchner, K. (2018). Isoelectronic manganese and iron hydrogenation/dehydrogenation catalysts: similarities and divergences. *Accounts of Chemical Research* 51 (6): 1558–1569.
5 Machala, L., Prochazka, V., Miglierini, M. et al. (2015). Direct evidence of Fe(V) and Fe(IV) intermediates during reduction of Fe(VI) to Fe(III): a nuclear forward scattering of synchrotron radiation approach. *Physical Chemistry Chemical Physics* 17 (34): 21787–21790.
6 Ansari, A. and Rajaraman, G. (2014). Ortho-hydroxylation of aromatic acids by a non-heme Fe(V)=O species: how important is the ligand design? *Physical Chemistry Chemical Physics* 16 (28): 14601–14613.
7 Sahu, S., Zhang, B., Pollock, C.J. et al. (2016). Aromatic C–F hydroxylation by nonheme iron(IV)–oxo complexes: structural, spectroscopic, and mechanistic investigations. *Journal of the American Chemical Society* 138 (39): 12791–12802.
8 Pandey, B., Ansari, A., Vyas, N. et al. (2015). Structures, bonding and reactivity of iron and manganese high-valent metal-oxo complexes: a computational investigation. *Journal of Chemical Sciences* 127 (2): 343–352.
9 Ghafoor, S., Mansha, A., and de Visser, S.P. (2019). Selective hydrogen atom abstraction from dihydroflavonol by a nonheme iron center is the key step in the enzymatic flavonol synthesis and avoids byproducts. *Journal of the American Chemical Society* 141 (51): 20278–20292.
10 Zhang, J., Wei, W.J., Lu, X. et al. (2017). Nonredox metal ions promoted olefin epoxidation by iron(II) complexes with H_2O_2: DFT calculations reveal multiple channels for oxygen transfer. *Inorganic Chemistry* 56 (24): 15138–15149.
11 Verma, P., Varga, Z., Klein, J. et al. (2017). Assessment of electronic structure methods for the determination of the ground spin states of Fe(II), Fe(III) and Fe(IV) complexes. *Physical Chemistry Chemical Physics* 19 (20): 13049–13069.
12 Patra, R. and Maldivi, P. (2016). DFT analysis of the electronic structure of Fe(IV) species active in nitrene transfer catalysis: influence of the coordination sphere. *Journal of Molecular Modeling* 22 (11): 278.
13 Ma, Z., Ukaji, K., Nakatani, N. et al. (2019). Substitution effects on olefin epoxidation catalyzed by oxoiron(IV) porphyrin π-cation radical complexes: a DFT study. *Journal of Computational Chemistry* 40 (19): 1780–1788.
14 Citek, C., Oyala, P.H., and Peters, J.C. (2019). Mononuclear Fe(I) and Fe(II) acetylene adducts and their reductive protonation to terminal Fe(IV) and Fe(V) carbynes. *Journal of the American Chemical Society* 141 (38): 15211–15221.
15 Sirirak, J., Sertphon, D., Phonsri, W. et al. (2017). Comparison of density functionals for the study of the high spin low spin gap in Fe(III) spin crossover complexes. *International Journal of Quantum Chemistry* 117 (9): 25362.
16 Ortuno, M.A., Dereli, B., Chiaie, K.R.D. et al. (2018). The role of alkoxide initiator, spin state, and oxidation state in ring-opening polymerization of ε-caprolactone catalyzed by iron bis(imino)pyridine complexes. *Inorganic Chemistry* 57 (4): 2064–2071.

17 Zima, A.M., Lyakin, O.Y., Bryliakov, K.P. et al. (2019). High-spin and low-spin perferryl intermediates in Fe(PDP)-catalyzed epoxidations. *ChemCatChem* 11 (21): 5345–5352.

18 Torrent-Sucarrat, M., Arrastia, I., Arrieta, A. et al. (2018). Stereoselectivity, different oxidation states, and multiple spin states in the cyclopropanation of olefins catalyzed by Fe–porphyrin complexes. *ACS Catalysis* 8 (12): 11140–11153.

19 Sirois, J.J., Davis, R., and DeBoef, B. (2014). Iron-catalyzed arylation of heterocycles via directed C—H bond activation. *Organic Letters* 16 (3): 868–871.

20 Norinder, J., Matsumoto, A., Yoshikai, N. et al. (2008). Iron-catalyzed direct arylation through directed C—H bond activation. *Journal of the American Chemical Society* 130 (18): 5858–5859.

21 Ilies, L., Asako, S., and Nakamura, E. (2011). Iron-catalyzed stereospecific activation of olefinic C—H bonds with Grignard reagent for synthesis of substituted olefins. *Journal of the American Chemical Society* 133 (20): 7672–7675.

22 Fruchey, E.R., Monks, B.M., and Cook, S.P. (2014). A unified strategy for iron-catalyzed ortho-alkylation of carboxamides. *Journal of the American Chemical Society* 136 (38): 13130–13133.

23 Serrano-Plana, J., Acuna-Pares, F., Dantignana, V. et al. (2018). Acid-triggered O—O bond heterolysis of a nonheme Fe(III) (OOH) species for the stereospecific hydroxylation of strong C—H bonds. *Chemistry – A European Journal* 24 (20): 5331–5340.

24 Yoshikai, N., Matsumoto, A., Norinder, J. et al. (2009). Iron-catalyzed chemoselective ortho arylation of aryl imines by directed C—H bond activation. *Angewandte Chemie International Edition* 48 (16): 2925–2928.

25 Cera, G., Haven, T., and Ackermann, L. (2016). Expedient iron-catalyzed C–H allylation/alkylation by triazole assistance with ample scope. *Angewandte Chemie International Edition* 55 (4): 1484–1488.

26 Wang, G., Chen, M., and Zhou, M. (2004). Matrix isolation infrared spectroscopic and theoretical studies on the reactions of manganese and iron monoxides with methane. *Journal of Physical Chemistry A* 108 (51): 11273–11278.

27 Zhao, L., Wang, Y., Guo, W. et al. (2008). Theoretical investigation of the Fe^+-catalyzed oxidation of acetylene by N_2O. *Journal of Physical Chemistry A* 112 (25): 5676–5683.

28 Ard, S.G., Melko, J.J., Ushakov, V.G. et al. (2014). Activation of methane by FeO^+: determining reaction pathways through temperature-dependent kinetics and statistical modeling. *Journal of Physical Chemistry A* 118 (11): 2029–2039.

29 Altinay, G., Citir, M., and Metz, R.B. (2010). Vibrational spectroscopy of intermediates in methane-to-methanol conversion by FeO^+. *Journal of Physical Chemistry A* 114 (15): 5104–5112.

30 Wang, X., Li, S., and Jiang, Y. (2004). Mechanism of H_2O_2 dismutation catalyzed by a new catalase mimic (a non-heme dibenzotetraaza[14]annulene-Fe(III) complex): a density functional theory investigation. *Inorganic Chemistry* 43 (20): 6479–6489.

31 Yoshizawa, K. (2006). Nonradical mechanism for methane hydroxylation by iron-oxo complexes. *Accounts of Chemical Research* 39 (6): 375–382.

32 Li, Z.-Z., Yu, J., Wang, L.-N. et al. (2018). Cascade radical cyclization/cross-coupling of halobenzamides by synergistic Cu/Fe catalysis: an access to 7-tert-alkylated isoquinolinediones. *Tetrahedron* 74 (45): 6558–6568.

33 Deng, Z., Chen, C., and Cui, S. (2016). Fe(III)-mediated isomerization of α,α-diarylallylic alcohols to ketones via radical 1,2-aryl migration. *RSC Advances* 6 (96): 93753–93755.

34 Hwang, J.Y., Baek, J.H., Shin, T.I. et al. (2016). Single-electron-transfer strategy for reductive radical cyclization: $Fe(CO)_5$ and phenanthroline system. *Organic Letters* 18 (19): 4900–4903.

35 Tao, L., Pattenaude, S.A., Joshi, S. et al. (2020). Radical SAM enzyme HydE generates adenosylated Fe(I) intermediates en route to the [FeFe]-hydrogenase catalytic H-cluster. *Journal of the American Chemical Society* 142 (24): 10841–10848.

36 Schoneboom, J.C., Cohen, S., Lin, H. et al. (2004). Quantum mechanical/molecular mechanical investigation of the mechanism of C–H hydroxylation of camphor by cytochrome P450cam: theory supports a two-state rebound mechanism. *Journal of the American Chemical Society* 126 (12): 4017–4034.

37 Tanaka, S., Kon, Y., Nakashima, T. et al. (2014). Chemoselective hydrogen peroxide oxidation of allylic and benzylic alcohols under mild reaction conditions catalyzed by simple iron–picolinate complexes. *RSC Advances* 4 (71): 37674.

38 Milaczewska, A., Broclawik, E., and Borowski, T. (2013). On the catalytic mechanism of (S)-2-hydroxypropylphosphonic acid epoxidase (HppE): a hybrid DFT study. *Chemistry – A European Journal* 19 (2): 771–781.

39 Tanaka, S., Kon, Y., Ogawa, A. et al. (2016). Mixed picolinate and quinaldinate iron(III) complexes for the catalytic oxidation of alcohols with hydrogen peroxide. *ChemCatChem* 8 (18): 2930–2938.

40 Wang, C. and Chen, H. (2017). Convergent theoretical prediction of reactive oxidant structures in diiron arylamine oxygenases AurF and CmlI: peroxo or hydroperoxo? *Journal of the American Chemical Society* 139 (37): 13038–13046.

41 Chung, L.W., Li, X., Sugimoto, H. et al. (2008). Density functional theory study on a missing piece in understanding of heme chemistry: the reaction mechanism for indoleamine 2,3-dioxygenase and tryptophan 2,3-dioxygenase. *Journal of the American Chemical Society* 130 (37): 12299–12309.

42 Poater, A., Ragone, F., Correa, A. et al. (2011). Comparison of different ruthenium–alkylidene bonds in the activation step with N-heterocyclic carbene Ru-catalysts for olefins metathesis. *Dalton Transactions* 40 (42): 11066–11069.

43 Verma, P., Vogiatzis, K.D., Planas, N. et al. (2015). Mechanism of oxidation of ethane to ethanol at iron(IV)-oxo sites in magnesium-diluted Fe_2(dobdc). *Journal of the American Chemical Society* 137 (17): 5770–5781.

44 Kamachi, T. and Yoshizawa, K. (2003). A theoretical study on the mechanism of camphor hydroxylation by compound I of cytochrome P450. *Journal of the American Chemical Society* 125 (15): 4652–4661.

45 de Visser, S.P., Oh, K., Han, A.R. et al. (2007). Combined experimental and theoretical study on aromatic hydroxylation by mononuclear nonheme iron(IV)-oxo complexes. *Inorganic Chemistry* 46 (11): 4632–4641.

46 de Visser, S.P. (2006). Substitution of hydrogen by deuterium changes the regioselectivity of ethylbenzene hydroxylation by an oxo-iron-porphyrin catalyst. *Chemistry – A European Journal* 12 (31): 8168–8177.

47 Bassan, A., Blomberg, M.R., and Siegbahn, P.E. (2003). Mechanism of aromatic hydroxylation by an activated FeIV=O core in tetrahydrobiopterin-dependent hydroxylases. *Chemistry – A European Journal* 9 (17): 4055–4067.

48 Asatryan, R. and Ruckenstein, E. (2013). Mechanism of iron carbonyl-catalyzed hydrogenation of ethylene. 1. Theoretical exploration of molecular pathways. *Journal of Physical Chemistry A* 117 (42): 10912–10932.

49 Oldenburg, P.D., Mas-Ballesté, R., and Que, L. (2008). Bio-inspired iron-catalyzed olefin oxidations. In: *Mechanisms in Homogeneous and Heterogeneous Epoxidation Catalysis*, 451–469, Elsevier Science: Amsterdam, Netherlands.

50 Kumar, D., de Visser, S.P., and Shaik, S. (2005). Multistate reactivity in styrene epoxidation by compound I of cytochrome P450: mechanisms of products and side products formation. *Chemistry – A European Journal* 11 (9): 2825–2835.

51 Bassan, A., Blomberg, M.R., Siegbahn, P.E. et al. (2005). Two faces of a biomimetic non-heme HO–Fe(V)=O oxidant: olefin epoxidation versus cis-dihydroxylation. *Angewandte Chemie International Edition* 44 (19): 2939–2941.

52 Chow, T.W.-S., Wong, E.L.-M., Guo, Z. et al. (2010). Cis-dihydroxylation of alkenes with oxone catalyzed by iron complexes of a macrocyclic tetraaza ligand and reaction mechanism by ESI–MS spectrometry and DFT calculations. *Journal of the American Chemical Society* 132 (38): 13229–13239.

53 Siegbahn, P.E. and Haeffner, F. (2004). Mechanism for catechol ring-cleavage by non-heme iron extradiol dioxygenases. *Journal of the American Chemical Society* 126 (29): 8919–8932.

54 Deeth, R.J. and Bugg, T.D. (2003). A density functional investigation of the extradiol cleavage mechanism in non-heme iron catechol dioxygenases. *Journal of Biological Inorganic Chemistry* 8 (4): 409–418.

55 Jastrzebski, R., Quesne, M.G., Weckhuysen, B.M. et al. (2014). Experimental and computational evidence for the mechanism of intradiol catechol dioxygenation by non-heme iron(III) complexes. *Chemistry – A European Journal* 20 (48): 15686–15691.

56 Borowski, T. and Siegbahn, P.E. (2006). Mechanism for catechol ring cleavage by non-heme iron intradiol dioxygenases: a hybrid DFT study. *Journal of the American Chemical Society* 128 (39): 12941–12953.

57 Ren, Q., Wu, N., Cai, Y. et al. (2016). DFT study of the mechanisms of iron-catalyzed regioselective synthesis of α-aryl carboxylic acids from styrene derivatives and CO_2. *Organometallics* 35 (23): 3932–3938.

58 Li, L., Lei, M., and Sakaki, S. (2017). DFT mechanistic study on alkene hydrogenation catalysis of iron metallaboratrane: characteristic features of iron species. *Organometallics* 36 (18): 3530–3538.

59 Schroeder, M.A. and Wrighton, M.S. (1976). Pentacarbonyliron(0) photocatalyzed hydrogenation and isomerization of olefins. *Journal of the American Chemical Society* 98 (2): 551–558.

60 Espinal-Viguri, M., Neale, S.E., Coles, N.T. et al. (2018). Room temperature iron-catalyzed transfer hydrogenation and regioselective deuteration of carbon–carbon double bonds. *Journal of the American Chemical Society* 141 (1): 572–582.

61 Xu, R., Chakraborty, S., Bellows, S.M. et al. (2016). Iron-catalyzed homogeneous hydrogenation of alkenes under mild conditions by a stepwise, bifunctional mechanism. *ACS Catalysis* 6 (3): 2127–2135.

62 Yang, X. (2011). Unexpected direct reduction mechanism for hydrogenation of ketones catalyzed by iron PNP pincer complexes. *Inorganic Chemistry* 50 (24): 12836–12843.

63 Prokopchuk, D.E. and Morris, R.H. (2012). Inner-sphere activation, outer-sphere catalysis: theoretical study on the mechanism of transfer hydrogenation of ketones using iron(II) PNNP eneamido complexes. *Organometallics* 31 (21): 7375–7385.

64 Lu, X., Zhang, Y., Yun, P. et al. (2013). The mechanism for the hydrogenation of ketones catalyzed by Knolker's iron-catalyst. *Organic & Biomolecular Chemistry* 11 (32): 5264–5277.

65 Zhang, H., Chen, D., Zhang, Y. et al. (2010). On the mechanism of carbonyl hydrogenation catalyzed by iron catalyst. *Dalton Transactions* 39 (8): 1972–1978.

66 Langer, R., Iron, M.A., Konstantinovski, L. et al. (2012). Iron borohydride pincer complexes for the efficient hydrogenation of ketones under mild, base-free conditions: synthesis and mechanistic insight. *Chemistry – A European Journal* 18 (23): 7196–7209.

67 Langer, R., Leitus, G., Ben-David, Y. et al. (2011). Efficient hydrogenation of ketones catalyzed by an iron pincer complex. *Angewandte Chemie International Edition* 50 (9): 2120–2124.

68 Moulin, S., Dentel, H., Pagnoux-Ozherelyeva, A. et al. (2013). Bifunctional (cyclopentadienone)iron–tricarbonyl complexes: synthesis, computational studies and application in reductive amination. *Chemistry – A European Journal* 19 (52): 17881–17890.

69 Hopmann, K.H. (2015). Iron/bronsted acid catalyzed asymmetric hydrogenation: mechanism and selectivity-determining interactions. *Chemistry – A European Journal* 21 (28): 10020–10030.

70 Yang, X. (2011). Hydrogenation of carbon dioxide catalyzed by PNP pincer iridium, iron, and cobalt complexes: a computational design of base metal catalysts. *ACS Catalysis* 1 (8): 849–854.

71 Hull, J.F., Himeda, Y., Wang, W.H. et al. (2012). Reversible hydrogen storage using CO_2 and a proton-switchable iridium catalyst in aqueous media under mild temperatures and pressures. *Nature Chemistry* 4 (5): 383–388.

72 Tanaka, R., Yamashita, M., and Nozaki, K. (2009). Catalytic hydrogenation of carbon dioxide using Ir(III)–pincer complexes. *Journal of the American Chemical Society* 131 (40): 14168–14169.

73 Jessop, P.G., Hsiao, Y., Ikariya, T. et al. (1996). Homogeneous catalysis in supercritical fluids: hydrogenation of supercritical carbon dioxide to formic acid, alkyl formates, and formamides. *Journal of the American Chemical Society* 118 (2): 344–355.

74 Gaillard, S. and Renaud, J.L. (2008). Iron-catalyzed hydrogenation, hydride transfer, and hydrosilylation: an alternative to precious-metal complexes? *ChemSusChem* 1 (6): 505–509.

75 Yang, X. (2015). Bio-inspired computational design of iron catalysts for the hydrogenation of carbon dioxide. *Chemical Communications* 51 (66): 13098–13101.

76 Ge, H., Chen, X., and Yang, X. (2016). A mechanistic study and computational prediction of iron, cobalt and manganese cyclopentadienone complexes for hydrogenation of carbon dioxide. *Chemical Communications* 52 (84): 12422–12425.

77 Correa, A., García Mancheño, O., and Bolm, C. (2008). Iron-catalysed carbon–heteroatom and heteroatom–heteroatom bond forming processes. *Chemical Society Reviews* 37 (6): 1108.

78 Langer, R., Diskin-Posner, Y., Leitus, G. et al. (2011). Low-pressure hydrogenation of carbon dioxide catalyzed by an iron pincer complex exhibiting noble metal activity. *Angewandte Chemie International Edition* 50 (42): 9948–9952.

79 Federsel, C., Boddien, A., Jackstell, R. et al. (2010). A well-defined iron catalyst for the reduction of bicarbonates and carbon dioxide to formates, alkyl formates, and formamides. *Angewandte Chemie International Edition* 49 (50): 9777–9780.

80 GuÐmundsson, A., Gustafson, K.P.J., Mai, B.K. et al. (2019). Diastereoselective synthesis of N-protected 2,3-dihydropyrroles via iron-catalyzed cycloisomerization of α-allenic sulfonamides. *ACS Catalysis* 9 (3): 1733–1737.

81 Marciniec, B., Kownacka, A., Kownacki, I. et al. (2015). Hydrosilylation vs. dehydrogenative silylation of styrene catalysed by iron(0) carbonyl complexes with multivinylsilicon ligands – mechanistic implications. *Journal of Organometallic Chemistry* 791: 58–65.

82 Guo, C.-H., Liu, X., Jia, J. et al. (2015). Computational insights into the mechanism of iron carbonyl-catalyzed ethylene hydrosilylation or dehydrogenative silylation. *Computational and Theoretical Chemistry* 1069: 66–76.

83 Liu, M., Sun, C., Hang, F. et al. (2014). Theoretical mechanism for selective catalysis of double hydrophosphination of terminal arylacetylenes by an iron complex. *Dalton Transactions* 43 (12): 4813–4821.

84 Metsanen, T.T., Gallego, D., Szilvasi, T. et al. (2015). Peripheral mechanism of a carbonyl hydrosilylation catalysed by an SiNSi iron pincer complex. *Chemical Science* 6 (12): 7143–7149.

85 Bleith, T. and Gade, L.H. (2016). Mechanism of the iron(II)-catalyzed hydrosilylation of ketones: activation of iron carboxylate precatalysts and reaction pathways of the active catalyst. *Journal of the American Chemical Society* 138 (14): 4972–4983.

86 Cheng, L., Wang, J., Wang, M. et al. (2010). Theoretical studies on the reaction mechanism of alcohol oxidation by high-valent iron-oxo complex of non-heme ligand. *Physical Chemistry Chemical Physics* 12 (16): 4092–4103.

87 Chakraborty, S., Lagaditis, P.O., Förster, M. et al. (2014). Well-defined iron catalysts for the acceptorless reversible dehydrogenation–hydrogenation of alcohols and ketones. *ACS Catalysis* 4 (11): 3994–4003.

88 Bielinski, E.A., Förster, M., Zhang, Y. et al. (2015). Base-free methanol dehydrogenation using a pincer-supported iron compound and Lewis acid co-catalyst. *ACS Catalysis* 5 (4): 2404–2415.

89 Zell, T. and Milstein, D. (2015). Hydrogenation and dehydrogenation iron pincer catalysts capable of metal-ligand cooperation by aromatization/dearomatization. *Accounts of Chemical Research* 48 (7): 1979–1994.

90 Chen, Y., Zhang, F., Xu, C. et al. (2012). Theoretical investigation of water gas shift reaction catalyzed by iron group carbonyl complexes M(CO)$_5$ (M = Fe, Ru, Os). *Journal of Physical Chemistry A* 116 (10): 2529–2535.

91 Yang, X. (2013). Mechanistic insights into iron catalyzed dehydrogenation of formic acid: β-hydride elimination vs. direct hydride transfer. *Dalton Transactions* 42 (33): 11987–11991.

92 Zell, T., Butschke, B., Ben-David, Y. et al. (2013). Efficient hydrogen liberation from formic acid catalyzed by a well-defined iron pincer complex under mild conditions. *Chemistry – A European Journal* 19 (25): 8068–8072.

93 Sanchez-de-Armas, R., Xue, L., and Ahlquist, M.S. (2013). One site is enough: a theoretical investigation of iron-catalyzed dehydrogenation of formic acid. *Chemistry – A European Journal* 19 (36): 11869–11873.

94 Glüer, A., Förster, M., Celinski, V.R. et al. (2015). Highly active iron catalyst for ammonia borane dehydrocoupling at room temperature. *ACS Catalysis* 5 (12): 7214–7217.

95 Zhang, Y., Zhang, Y., Qi, Z.-H. et al. (2016). Ammonia-borane dehydrogenation catalyzed by iron pincer complexes: a concerted metal-ligand cooperation mechanism. *International Journal of Hydrogen Energy* 41 (39): 17208–17215.

96 Nakamura, M., Matsuo, K., Ito, S. et al. (2004). Iron-catalyzed cross-coupling of primary and secondary alkyl halides with aryl grignard reagents. *Journal of the American Chemical Society* 126 (12): 3686–3687.

97 Hatakeyama, T., Hashimoto, T., Kondo, Y. et al. (2010). Iron-catalyzed Suzuki–Miyaura coupling of alkyl halides. *Journal of the American Chemical Society* 132 (31): 10674–10676.

98 Adams, C.J., Bedford, R.B., Carter, E. et al. (2012). Iron(I) in Negishi cross-coupling reactions. *Journal of the American Chemical Society* 134 (25): 10333–10336.

99 Adak, L., Kawamura, S., Toma, G. et al. (2017). Synthesis of aryl C-glycosides via iron-catalyzed cross coupling of halosugars: stereoselective anomeric arylation of glycosyl radicals. *Journal of the American Chemical Society* 139 (31): 10693–10701.

100 Nakagawa, N., Hatakeyama, T., and Nakamura, M. (2015). Iron-catalyzed Suzuki–Miyaura coupling reaction of unactivated alkyl halides with lithium alkynylborates. *Chemistry Letters* 44 (4): 486–488.

101 Kawamura, S. and Nakamura, M. (2013). Ligand-controlled iron-catalyzed cross coupling of benzylic chlorides with aryl Grignard reagents. *Chemistry Letters* 42 (2): 183–185.

102 Hatakeyama, T., Fujiwara, Y.-i., Okada, Y. et al. (2011). Kumada–Tamao–Corriu coupling of alkyl halides catalyzed by an iron–bisphosphine complex. *Chemistry Letters* 40 (9): 1030–1032.

103 Hatakeyama, T., Kondo, Y., Fujiwara, Y. et al. (2009). Iron-catalysed fluoroaromatic coupling reactions under catalytic modulation with 1,2-bis(diphenylphosphino)benzene. *Chemical Communications* (10): 1216–1218.

104 Bedford, R.B., Huwe, M., and Wilkinson, M.C. (2009). Iron-catalysed Negishi coupling of benzyl halides and phosphates. *Chemical Communications* (5): 600–602.

105 Bedford, R.B., Carter, E., Cogswell, P.M. et al. (2013). Simplifying iron-phosphine catalysts for cross-coupling reactions. *Angewandte Chemie International Edition* 52 (4): 1285–1288.

106 Sun, C.-L., Krause, H., and Fürstner, A. (2014). A practical procedure for iron-catalyzed cross-coupling reactions of sterically hindered aryl-Grignard reagents with primary alkyl halides. *Advanced Synthesis and Catalysis* 356 (6): 1281–1291.

107 Furstner, A., Leitner, A., Mendez, M. et al. (2002). Iron-catalyzed cross-coupling reactions. *Journal of the American Chemical Society* 124 (46): 13856–13863.

108 Liu, L., Lee, W., Yuan, M. et al. (2018). Mechanisms of bisphosphine iron-catalyzed C(sp^2)–C(sp^3) cross-coupling reactions: inner-sphere or outer-sphere arylation? *Comments on Inorganic Chemistry* 38 (6): 210–237.

109 Ren, Q., Guan, S., Jiang, F. et al. (2013). Density functional theory study of the mechanisms of iron-catalyzed cross-coupling reactions of alkyl Grignard reagents. *Journal of Physical Chemistry A* 117 (4): 756–764.

110 Hatakeyama, T., Imayoshi, R., Yoshimoto, Y. et al. (2012). Iron-catalyzed aromatic amination for nonsymmetrical triarylamine synthesis. *Journal of the American Chemical Society* 134 (50): 20262–20265.

111 Sharma, A.K., Sameera, W.M.C., Jin, M. et al. (2017). DFT and AFIR study on the mechanism and the origin of enantioselectivity in iron-catalyzed cross-coupling reactions. *Journal of the American Chemical Society* 139 (45): 16117–16125.

112 Lee, W., Zhou, J., and Gutierrez, O. (2017). Mechanism of Nakamura's bisphosphine-iron-catalyzed asymmetric C(sp^2)–C(sp^3) cross-coupling reaction: the role of spin in controlling arylation pathways. *Journal of the American Chemical Society* 139 (45): 16126–16133.

113 Jin, M., Adak, L., and Nakamura, M. (2015). Iron-catalyzed enantioselective cross-coupling reactions of α-chloroesters with aryl Grignard reagents. *Journal of the American Chemical Society* 137 (22): 7128–7134.

114 Dong, L., Wen, J., Qin, S. et al. (2012). Iron-catalyzed direct Suzuki–Miyaura reaction: theoretical and experimental studies on the mechanism and the regioselectivity. *ACS Catalysis* 2 (8): 1829–1837.

115 Wen, J., Qin, S., Ma, L.F. et al. (2010). Iron-mediated direct Suzuki–Miyaura reaction: a new method for the ortho-arylation of pyrrole and pyridine. *Organic Letters* 12 (12): 2694–2697.

11

Theoretical Study of Ru-Catalysis

101.07
Ru⁴⁴
Ruthenium
[Kr]4d⁷5s¹

As a fifth periodic late transition metal, Ru plays an important role in organometallic catalysis [1–3]. In Ru catalysis, the most common oxidative state of Ru is +2; moreover, 0 and +4 of oxidative states can also appear in Ru-catalysis [4–9]. Compared with cognate element iron, the single-electron transfer redox is rare for Ru. As an alternative, double-electron transfer is the major way in redox processes of Ru-catalysis [3].

Analogous to Rh [10–13] and Pd [14–18], Ru—C bond reveals certain stability in organometallic catalysis; therefore, when a Ru—C bond is formed, it can be further transferred for the construction of new C—C and C—hetero bonds. Insertion of unsaturated molecules is the common way for this transformation [19–23]. Moreover, Ru—H bond is usually proposed in Ru-catalysis; therefore, Ru can be used as catalyst in the hydrogenations and hydrofunctionalizations of unsaturated molecules [24–26]. Ru-carbene complex plays a unique reactivity, which can react with C=C double bond in olefin metathesis reactions. In this process, the formation of a ruthacycle followed by its decomposition can provide a new alkene and another Ru-carbene complex [27–29].

In this chapter, we would first discuss the mechanism of Ru-catalyzed C—H bond activation and functionalization reactions. Then, the mechanistic study Ru-catalyzed hydrogenation, hydrofunctionalization, and dehydrogenation would be exhibited. Ru-catalyzed olefin metathesis also would be considered in this chapter.

11.1 Ru-Mediated C—H Bond Activation

As one of the most important areas of catalysis, Ru-catalyzed C—H bond activation and functionalization are widely used in modern synthetic applications based on the pioneering work of the groups of Kakiuchi and coworkers [30, 31], Inoue and coworkers [32, 33], Ackermann and coworkers [34–36], Bruneau and coworkers [37, 38]. Following the pioneering work of Lewis in the formation of C—C bond catalyzed by Ru complexes containing ortho-metalated triphenyl phosphite [39], the group of Murai et al. [40] was the first to report Ru(0)-complex-catalyzed coupling of aromatic ketones with olefins in 1993. Many new reactions have been discovered based on initial Ru(0) insertion into the C—H bond and further reactions of the generated C—Ru—H species, especially through unsaturated substrate insertion processes [41–45]. Subsequently, the groups of Oi, Inoue and coworkers [32, 33], Ackermann and coworkers [34, 35], Bruneau and coworkers [38] developed a series of Ru(II)-catalyzed C—H bond functionalization reactions. The success of Ru(II) catalysis is probably because of the easy synthesis of cyclometalated species via C—H bond cleavage, their compatibility with currently used oxidants, and the stability of some of them in both air and water [46–48].

11.1.1 Mechanism of the Ru-Mediated C—H Bond Cleavage

C—H bond cleavage is the key step in Ru-catalyzed hydrocarbon functionalization reactions. The mechanism of this step has been well established by a series of density functional theory (DFT) investigations. In Ru-mediated C—H bond activation, the possible mechanisms include concerted metalation-deprotonation (CMD), base-assisted internal electrophilic-type substitution (BIES), σ-complex-assisted metathesis (σ-CAM), and C—H bond oxidative addition.

CMD is one of the most popular mechanisms proposed for transition-metal-involved C—H activations, in which a C—H bond cleaves by base-assisted deprotonation and a carbon-metal bond simultaneously forms [36, 49]. The CMD process can be facilitated by introducing a weak-coordination-directing group, although some C—H bonds in active arenes can be cleaved without a directing group.

Weak-coordination groups, such as carbonyl, amide, and amino groups, and some heterocycles, are often used as monodentate-directing groups in the transition-metal-involved CMD process [32, 51–53]. Coordination of the directing group partly counteracts the intermolecular entropy loss in the C—H cleavage step, which is beneficial for this step. Roithova and coworkers reported a mechanistic study of Ru-catalyzed C—H activation of phenylpyridines (Scheme 11.1) [50]. By DFT with the B3LYP functional, the calculated activation free energy of C—H bond cleavage with the assistant of acetate as a base is only 15.3 kcal/mol via cationic CMD-type transition state **11-1ts**. In that transition state, the N—Ru bond length is 2.23 Å, which indicates significant coordination of the pyridyl group.

Phosphinite can also act as a base in Ru-mediated C—H activation (Scheme 11.2) [54]. The five-membered-ring CMD-type transition state **11-2ts** can be found by theoretical calculation. In this transition state, the oxygen atom in phosphinite acts as

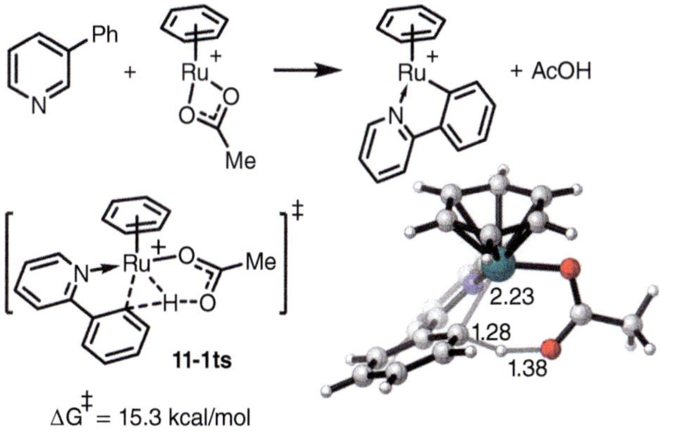

Scheme 11.1 A model reaction of Ru-mediated CMD-type C—H activation. The energies were calculated at B3LYP-D2/6-31G(d) (SDD for Ru) level of theory. Source: Based on Gray et al. [50].

a base to remove the proton from phenylpyridine. Simultaneously, the phosphine atom in phosphinite coordinates with Ru. The calculated activation free energy of this step is 24.7 kcal/mol, which is slightly higher than that of the corresponding step with acetate as a base.

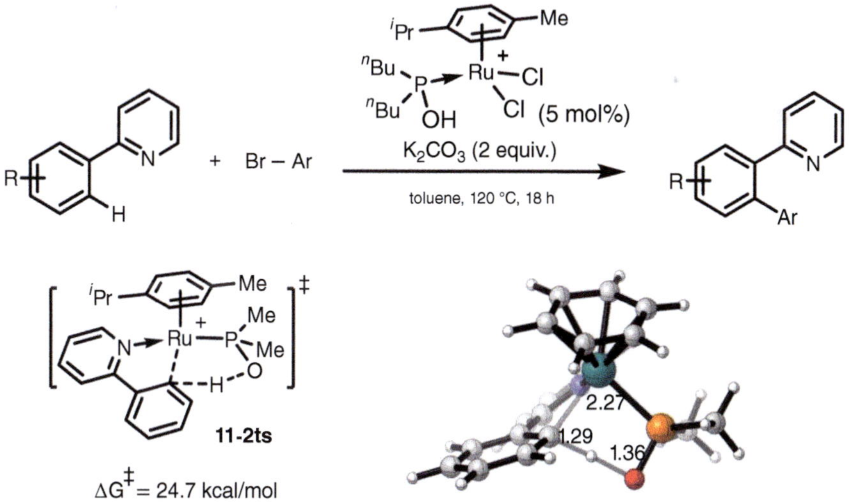

Scheme 11.2 Phosphite Ru(II)-mediated C—H activation of arenes. The energies were calculated at M06-L/6-31+G(d,p) (SDD for Ru)/SMD(toluene) level of theory. Source: Based on Zell et al. [54].

A C—H activation via BIES pathway formally seems same to those via CMD pathway, both on the reaction condition and the transition state geometrical structure. Base-assisted C—H bond cleavage also takes place via six-membered cyclic transition state when a carboxylate is used in the presence of cationic Ru(II) complex, where electron-rich arenes reacted preferentially, which disagrees with CMD mechanism. Different from CMD mechanism, in the BIES mechanism, an electrophilic metal rather than a Brønsted base activates the C—H bond by generating a positively charged hydrogen, which can be treated as an internal electrophilic substitution step. In this step, the lone pair on an M—X ligand (X = O) forms an X—H bond, while the orbital, making up the M—X bond, turns into a coordinating lone pair [55]. That means, a more positive carbon center is beneficial to the BIES step. That is in accordance with the better reactivity of electron-rich arenes.

Ackermann and coworkers reported a Ru-catalyzed acetamide-directed C—H bond alkenylation (Scheme 11.3) [56]. The C—H bond cleavage could take place via transition state **11-3ts**, in which the proton migrates from the phenyl ring to the oxygen of the coordinated acetate in the presence of cationic Ru(II). The calculated activation free energy of this process is 19.5 kcal/mol, which is lower than that of typical CMD process.

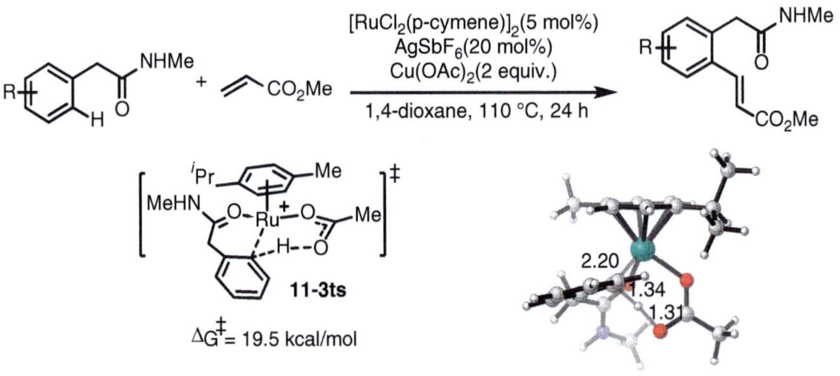

Scheme 11.3 Ru-mediated C—H bond activation of benzamide via BIES-type transition state. The energies were calculated at COSMO-B3LYP-D3BJ/def2-TZVP//TPSS-D3BJ/def2-TZVP level of theory. Source: Modified from Bu et al. [56].

Ru-mediated C—H bond activation can occur via the σ-CAM pathway in the presence of a σ-bond between Ru and the bonding base, in which cleavage of the C—H and Ru–base σ-bonds and formation of the C—Ru and H–base σ-bonds simultaneously occur [57]. In Ru-mediated C—H bond activation with the assistance of acetate, the σ-CAM pathway is often an alternative to the corresponding CMD pathway. When bicarbonate is used as the base (Scheme 11.4) [38], Maseras and Dixneuf found that the activation free energy of σ-CAM-type C—H cleavage via anionic transition state **11-4ts** is only 13.9 kcal/mol, which is favorable compared with the corresponding oxidative addition pathway, although the CMD process was not considered in their theoretical studies.

Prabhu and coworkers reported the competition between the σ-CAM and CMD pathways in Ru-mediated C—H activation of acetophenone [58]. As shown in Scheme 11.5, when acetate is used as the base, the activation free energy of CMD-type C—H bond cleavage via transition state **11-5ts** is only 17.4 kcal/mol.

11.1 Ru-Mediated C–H Bond Activation

Scheme 11.4 Ru-mediated C–H arylation with aryl bromide through σ-CAM-type C–H activation in the presence of bicarbonate. The energies were calculated at B3LYP/6-31+G(d,p) (LANL2DZ for Cl and Ru) level of theory. Source: Modified Ozdemir et al. [38].

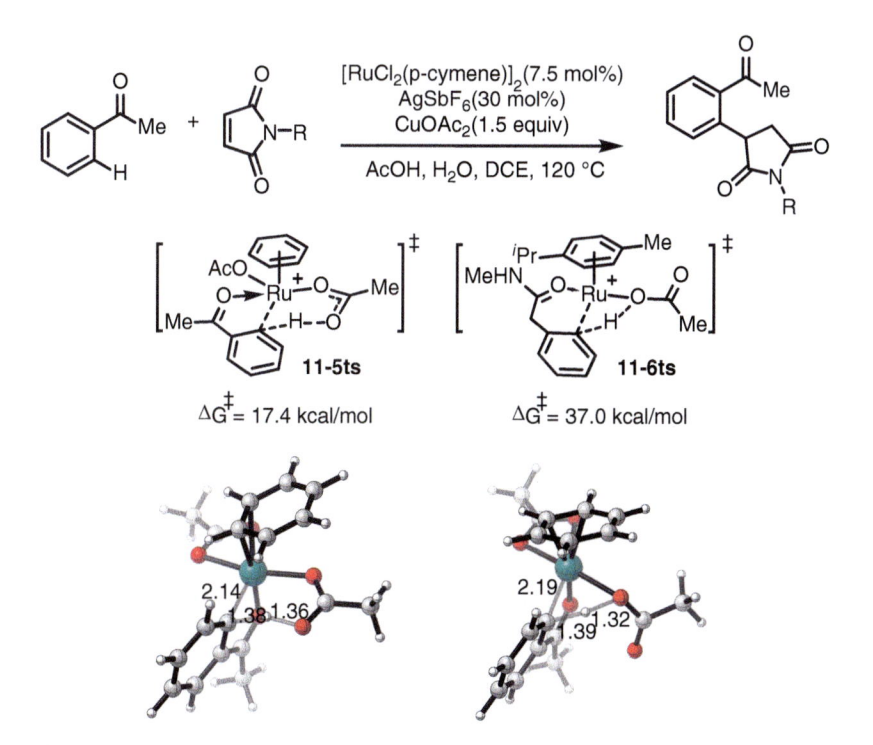

11-4ts
$\Delta G^\ddagger = 13.9$ kcal/mol

11-5ts
$\Delta G^\ddagger = 17.4$ kcal/mol

11-6ts
$\Delta G^\ddagger = 37.0$ kcal/mol

Scheme 11.5 Comparison of CMD- and σ-CAM-type C–H activations in Ru-mediated hydroarylation of alkenes. The energies were calculated at M06/6-31G(d) (LANL2DZ for Ru)/PCM(DCE)//B3LYP/6-31G(d) (LANL2DZ for Ru)/PCM(DCE) level of theory.

However, they also located the four-membered cyclic σ-CAM- type transition state **11-6ts**, whose relative free energy is 19.6 kcal/mol higher than that of **11-5ts**.

The oxidative addition process in Ru-mediated C—H activation usually proceeds via a three-membered cyclic transition state, in which the C—H bond is broken and each of the two atoms is added to Ru to increase the formal oxidation state of Ru by two [60]. Lin and coworkers reported Ru-mediated C—H bond activation by an oxidative addition pathway starting from a Ru(0) species (Scheme 11.6) [59]. Oxidative addition of the arene C—H bond to Ru(0) occurs via transition state **11-7ts**. The calculated activation free energy is only 0.4 kcal/mol from an agostic Ru(0) species.

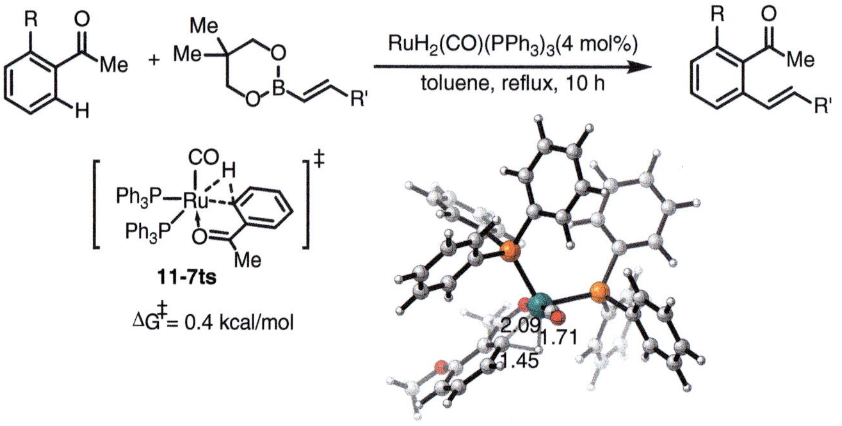

Scheme 11.6 Ru-mediated C—H bond activation/vinylation through oxidative addition with Ru(0) species. The energies were calculated at PBE/6-31G(d,p) (LANL2DZ for Ru and P)/PCM(toluene) level of theory. Source: Based on Wang et al. [59].

Clot and coworkers also reported a theoretical study of oxidation addition with the C—H bond, as shown in Scheme 11.7 [61]. The calculated energy barrier of the oxidation addition step is 1.8 kcal/mol. Both of these two cases show that oxidative addition of the C—H bond to Ru(0) species is a relatively facile process.

11.1.2 Ru-Catalyzed C—H Bond Arylation

Construction of C—C bonds is one of the most important areas in both industrial and academic fields [62]. In the last few decades, the transition-metal-catalyzed cross-coupling reaction has provided a powerful tool to construct new C—C bonds [63]. In this area, a nucleophile, which is a molecule with formal lone-pair electrons, can donate two electrons to its reaction partner, which is called an electrophile, to form a new C—C σ-bond. Alternatively, two nucleophiles can couple during organometallic catalysis in the presence of an oxidant, which is called oxidative coupling [64–73]. In this chemistry, the nucleophile is an electron-rich molecule that contains a lone pair of electrons or a polarized bond. Based on this idea, the C—H bonds of hydrocarbons can be considered to be nucleophiles because the electronegativity of carbon is higher than that of hydrogen. Therefore, Ru-catalyzed

Scheme 11.7 Ru-mediated C−H bond activation/alkanation through oxidative addition with Ru(0) species. The energies were calculated at B3PW91/6-31G(d,p) (SDD for Ru and P) level of theory. Source: Based on Helmstedt and Clot [61].

C—H bond arylation can be categorized as redox-neutral cross-coupling reactions with electrophiles and oxidative-coupling reactions with other nucleophiles in the presence of exogenous oxidants (Scheme 11.8).

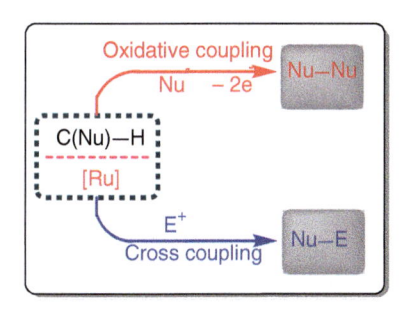

Scheme 11.8 Ru-catalyzed C−H bond functionalization with electrophiles or nucleophiles.

In Ru-catalyzed cross-coupling reactions, aryl halides are often used as electrophiles, which are endogenous oxidants to maintain redox neutrality [34, 36, 38, 76, 77]. The general catalytic cycle is summarized in Scheme 11.9 [51, 74, 75], which is analogous to the corresponding process in Rh-catalysis [78–83]. C—H cleavage occurs from Ru(II) species with arene assisted by a directing group to form aryl Ru(II) intermediate. Subsequently, oxidative addition with an aryl halide generates Ru(IV) intermediate. Reductive elimination then generates a new C—C bond. Finally, ligand exchange releases the cross-coupling product and regenerates active species.

Chatani's group reported Ru-catalyzed ortho-arylation of aromatic amides with a bidentate-directing group by cleavage of a C—H bond (Figure 11.1), which shows high functional group compatibility [74]. A DFT study was performed to investigate the mechanism of this reaction. The calculated potential energy surfaces for Ru-catalyzed ortho-arylation of aromatic amides are shown in Figure 11.1. The

458 | *11 Theoretical Study of Ru-Catalysis*

Scheme 11.9 General mechanism of Ru-catalyzed C–H bond arylation with an electrophilic carbon atom. Sources: Aihara and Chatani [74]; Biafora et al. [75].

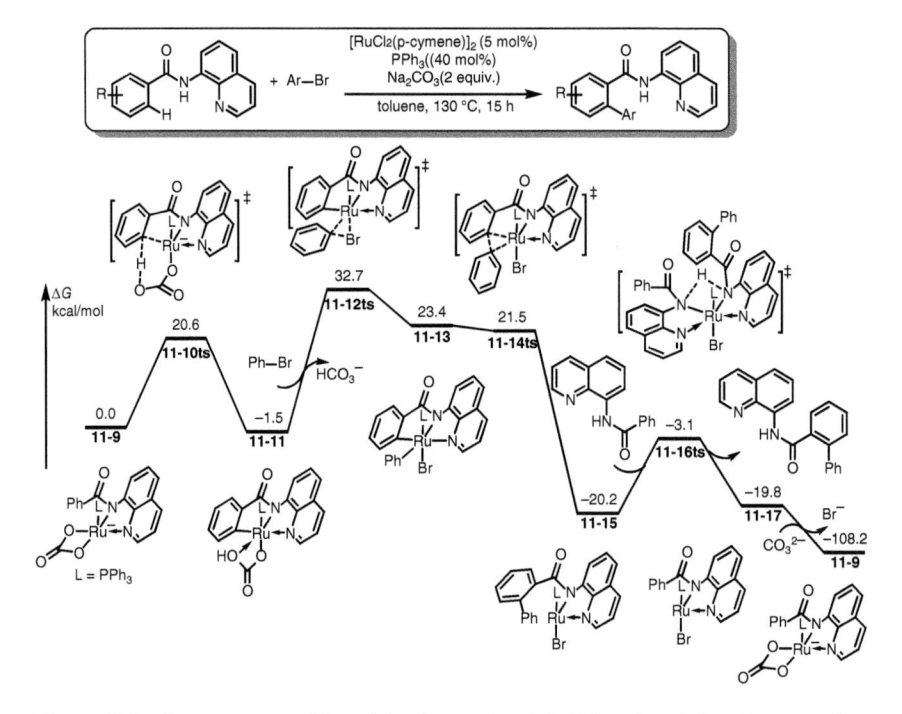

Figure 11.1 Free-energy profiles of the Ru-catalyzed C–H bond arylation. The energies were calculated at M11-L/6-311+G(d,p) (SDD for Ru)/SMD(toluene)//B3LYP/6-31G(d) (SDD for Ru) level of theory. Source: Based on Aihara and Chatani [74].

catalytic cycle starts from the Ru(II) carbonate active catalyst **11-9**. C—H cleavage occurs via transition state **11-10ts** to reversibly generate aryl Ru(II) intermediate **11-11** with an energy barrier of 20.6 kcal/mol. Aryl bromide is loaded by ligand exchange. Oxidative addition occurs via transition state **11-12ts** with an overall activation free energy of 34.2 kcal/mol to form Ru(IV) intermediate **11-13**, which is considered to be the rate-determining step of the catalytic cycle. Reductive elimination rapidly occurs via transition state **11-14ts** to exergonically form the new C—C bond in intermediate **11-15**. The active catalyst **11-9** can be regenerated by proton transfer via transition state **11-16ts** and ligand exchange.

The oxidative coupling reaction between two nucleophiles is an alternative way to construct new C—C bonds [84–86]. Recently, aryl metal reagents, such as aryl boranes [31, 52, 87–89] and aryl silanes [90–92], have been used as the nucleophile to react with the C—H bond in arenes catalyzed by Ru in the presence of exogenous oxidants. Mechanistically, oxidative coupling of the arene and aryl metal complex could occur by two types of reaction pathways, which is determined by the mode of the Ru-mediated C—H activation step.

As shown in Scheme 11.10, Ru-catalyzed C—H arylation with nucleophiles can start from Ru(0) species–mediated C—H cleavage through an oxidative addition process to form Ru(II) hydride [30, 93]. The hydride can be absorbed by the exogenous ketone oxidant. Transmetallation with an aryl metal reagent can generate diaryl Ru(II) complex. Subsequent reductive elimination gives biaryl product and regenerates Ru(0) active catalyst.

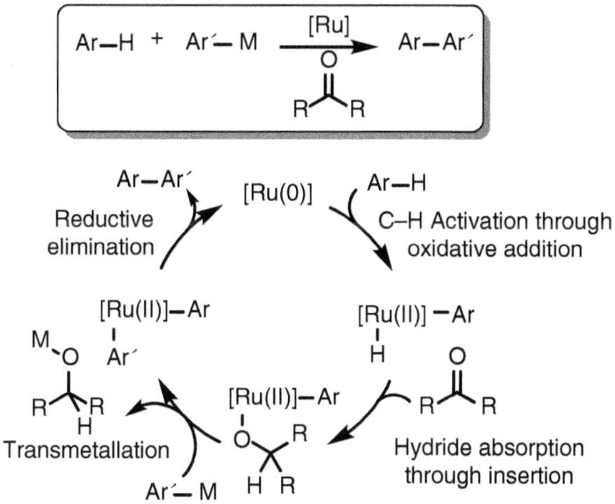

Scheme 11.10 General mechanism of Ru-catalyzed C—H bond activation with nucleophiles starting from a Ru(0) species.

As shown in Figure 11.2, a DFT calculation performed by Lin's group revealed the mechanism of Ru-catalyzed oxidative coupling of arenes with arylboronates [59]. In their theoretical study, agostic Ru(0) species **11-18** was chosen as the relative zero in the free-energy profiles. Oxidative addition of the C—H bond to the Ru atom

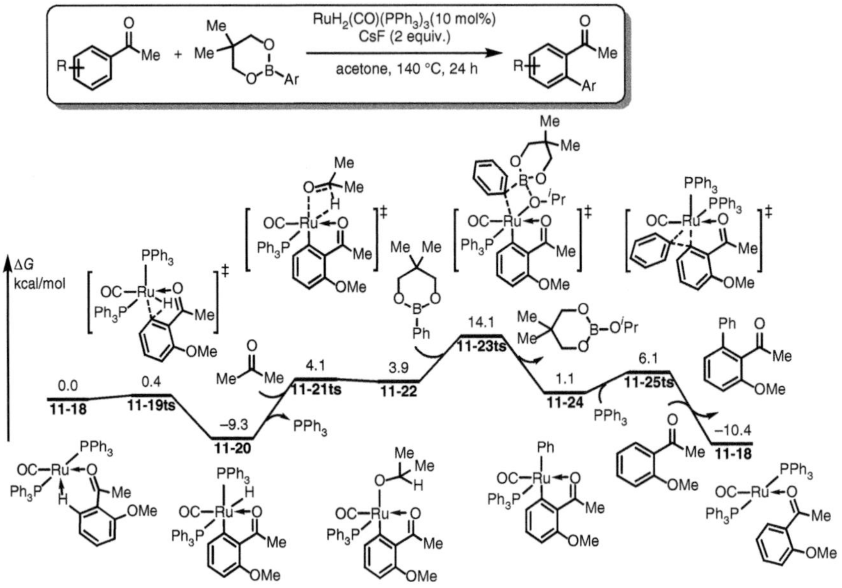

Figure 11.2 Free-energy profiles of Ru-catalyzed C—H bond functionalization of aryl ethers with organoboronates. The energies were calculated at PBE/6-31G(d) (LANL2DZ for Ru) level of theory.

occurs via transition state **11-19ts** with an energy barrier of only 0.4 kcal/mol to exergonically generate Ru(II) hydride intermediate **11-20**. Ligand exchange with acetone leads to carbonyl insertion into the Ru—H bond via transition state **11-21ts** with an energy barrier of only 4.1 kcal/mol to form isopropoxy Ru(II) intermediate **11-22**. Transmetallation then occurs via transition state **11-23ts** with coming phenyl borane to form phenyl Ru(II) intermediate **11-24** with an overall activation free energy of 23.4 kcal/mol, which is considered to be the rate-determining step. Reductive elimination rapidly occurs via transition state **11-25ts**. The active catalyst **11-18** can be regenerated by ligand exchange with the reactant.

11.1.3 Ru-Catalyzed *ortho*-Alkylation of Arenes

Olefins can also be used as the alkyl source in Ru-catalyzed ortho-alkylation reactions. As shown in Scheme 11.11, in the presence of Ru(II) carboxylate species, directed CMD-type C—H cleavage occurs to form aryl Ru(II) intermediate [94–98]. Olefin insertion then generates alkyl Ru(II), which can be protonated by carboxylic acid to regenerate active catalyst and release the alkylation product. The whole catalytic cycle takes place without redox processes.

DFT calculations revealed the mechanism of Ru(II)-catalyzed ortho-alkylation of arenes [58]. As shown in Figure 11.3, CMD-type C—H cleavage occurs from ruthenium dibenzoate species **11-26** via transition state **11-27ts** with a free-energy barrier of 19.3 kcal/mol to from Ru-involved hetero-five-membered cyclic intermediate **11-28**. Olefin insertion then occurs via transition state **11-30ts** with a free-energy

Scheme 11.11 General mechanism of Ru-catalyzed alkylation of arenes with olefins.

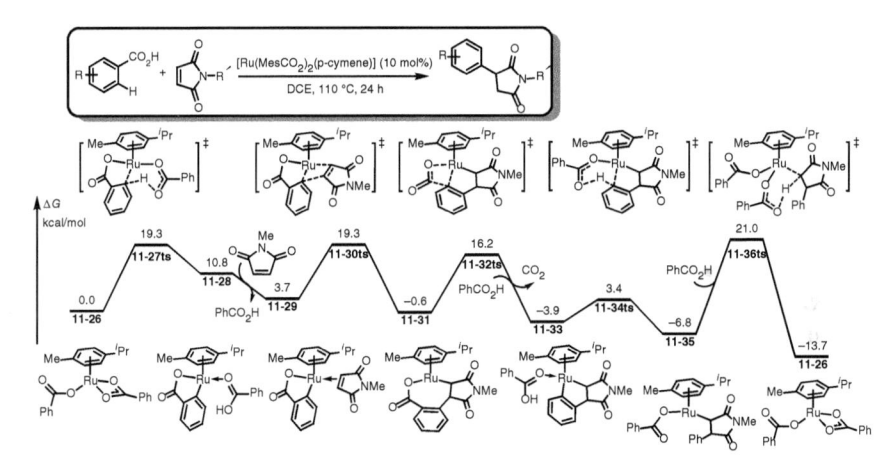

Figure 11.3 Free-energy profiles of Ru-catalyzed decarboxylative ortho-alkylation. The energies were calculated at PBE0-D3BJ/def2-QZVP*/SMD(DCE)//PBE0/def2-SVP level of theory. Source: Based on Bettadapur et al. [58].

barrier of 15.6 kcal/mol to generate Ru-involved hetero-seven-membered cyclic intermediate **11-31**. Intramolecular decarboxylation occurs via transition state **11-32ts** to form arylalkyl Ru(II) intermediate **11-33**. Two tandem protonation processes by benzoic acid give the decarboxylative alkylation product and regenerate active catalyst **11-26** to complete the catalytic cycle. The rate-determining step is the second protonation of the alkylruthenium intermediate **11-35** via transition state **11-36ts**.

11.1.4 Ru-Catalyzed *ortho*-Alkenylation of Arenes

Ru-catalyzed ortho-alkenylation of arenes can be performed to construct styrene derivatives, and the mechanism is analogous to the corresponding arylation process.

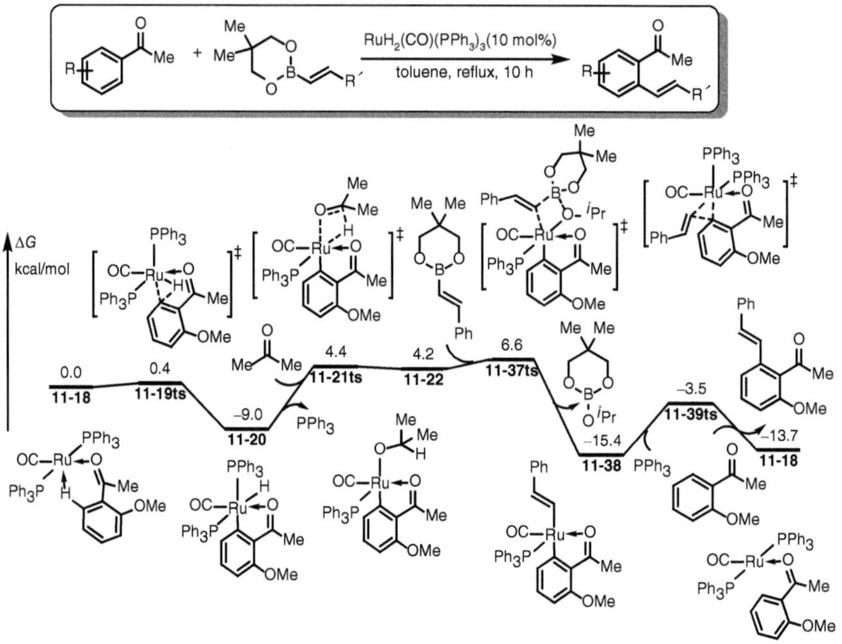

Figure 11.4 Free-energy profiles of Ru-catalyzed oxidative *ortho*-alkenylation with vinyl boronate. The energies were calculated at PBE/6-31G(d) (LANL2DZ for Ru) level of theory.

The activated C—H bond in arenes is considered to be a nucleophile, which can react with an electrophile, such as vinyl acetate, to give a cross-coupling product in a redox-neutral process. Alternatively, another nucleophile could be used in the presence of an outer-sphere oxidant to give the oxidative coupling product.

In the presence of an outer-sphere oxidant, oxidative coupling of the C—H bond in arenes with another nucleophile can give *ortho*-alkenylated arenes using a Ru catalyst. Generally, the other nucleophile is an organometallic compound, such as vinylboronic acid [99]. The C—H bond in simple olefins can also be considered as the nucleophile. Therefore, oxidative coupling can also occur between arenes and olefins.

Lin's group performed a DFT study to reveal the intrinsic mechanism of Ru-catalyzed oxidative C—H alkenylation with vinyl boronate [59]. As shown in Figure 11.4, the reaction starts from a Ru(0) species **11-18** containing an agostic interaction. The oxidative addition of C—H bond onto Rh(0) followed by a carbonyl insertion results in aryl Ru(II) intermediate **11-22**. Then, a transmetallation with coming vinyl boronate gives a vinyl Ru(II) intermediate **11-38**. After coordination of the phosphine ligand, reductive elimination then gives styrene product and regenerates Ru(0) species **11-18**.

In a more atom-economic strategy, oxidative coupling of two C—H bonds can provide styrene derivatives without extra generation of metal salts [101–107]. Frost and coworkers performed theoretical and experimental studies of oxidative coupling of oxooxazolidinyl-directed arenes and acrylates. The calculated free-energy

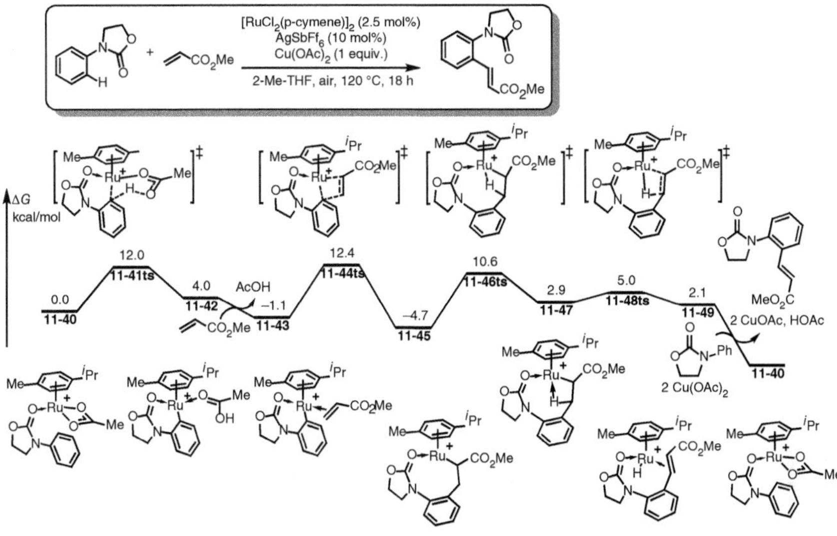

Figure 11.5 Free-energy profiles of Ru-catalyzed oxidative ortho-alkenylation of arenes by using olefins. The energies were calculated at BP86-D3BJ/6-311++G(d,p) (cc-pVTZ for Ru)/IEF-PCM(2-Me-THF)//BP86/6-31G(d,p) (SDD for Ru) level of theory. Source: Based on Leitch et al. [100].

profiles for this reaction are shown in Figure 11.5 [100]. With the coordination of the oxooxazolidinyl-directing group, CMD-type C—H cleavage occurs via transition state **11-41ts** with a free-energy barrier of 12.0 kcal/mol to give ruthenacycle **11-42**. Coordination of acrylate forms intermediate **11-43**, which is exergonic by 5.1 kcal/mol. Olefin insertion occurs via transition state **11-44ts** to reversibly form alkyl Ru intermediate **11-45**. Isomerization of intermediate **11-45** via transition state **11-46ts** gives agostic complex **11-47** with a free-energy barrier of 15.3 kcal/mol, which is considered to be the rate-limiting step. β-hydride elimination then rapidly occurs via transition state **11-48ts** to form Ru hydride species **11-49**. However, the relative free energy of **11-49** is 2.1 kcal/mol higher than that of initial complex **11-40**; therefore, the driving force of the catalytic cycle is considered to be the oxidation step regenerating Ru(II) species **11-40** with a Cu(II) salt.

As a π-acidic transition metal, Ru can react with acetylene to give a vinylideneruthenium intermediate by 1,2-group transfer. Lynam and coworkers reported vinylation of an unsaturated pyridine using [CpRu(Py)$_2$(PPh$_3$)]PF$_6$ as a catalyst [108]. A combination of experiments and calculations showed the fundamental processes that control both C—H activation of the pyridine and subsequent C—C bond formation. The calculated free-energy profiles of the dominant reaction pathway are shown in Figure 11.6. The catalytic cycle starts with cationic Ru(II) species **11-50**, which coordinates to acetylene to form intermediate **11-51**. 1,2-Hydrogen transfer then occurs via transition state **11-52ts** with an activation free energy of 17.0 kcal/mol to give vinylideneruthenium species **11-53**. Outer-sphere nucleophilic attack by pyridine generates vinyl Ru(II) intermediate **11-54**, which is endergonic by 16.7 kcal/mol. However, the transition state of this process was not located.

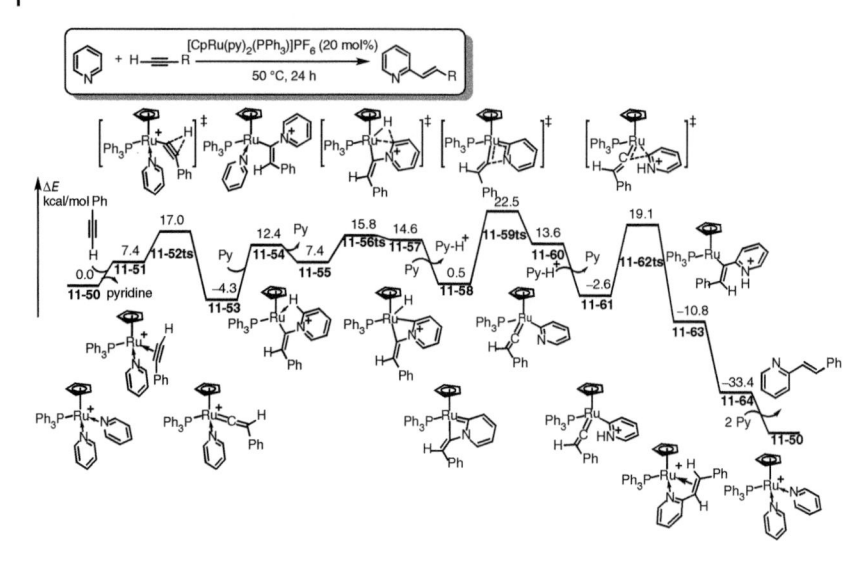

Figure 11.6 Potential energy profiles of Ru-catalyzed pyridine C—H alkenylation with alkyne through acetylene activation. The energies were calculated at PBE0/def2-TZVPP/COSMO(pyridine)//BP86/SV(P) level of theory.

Ortho-C—H bond cleavage occurs via transition state **11-56ts** with a free-energy barrier of 20.1 kcal/mol to give cationic Ru(IV) hydride intermediate **11-57**. Reductive deprotonation of the Ru(IV)—H bond then gives neutral four-membered-ring ruthenacycle **11-58**. Subsequent metathesis with N—C bond cleavage gives pyridyl Ru vinylidene **11-60** with a free-energy barrier of 26.8 kcal/mol, which is considered to be the rate-determining step of the catalytic cycle. The pyridyl moiety in intermediate **11-60** can be protonated to give pyridinium intermediate **11-61**. Migratory insertion of the pyridiniumyl group into the Ru carbenoid moiety then occurs via transition state **11-62ts** with a free-energy barrier of 23.4 kcal/mol to give vinyl Ru(II) intermediate **11-63**. Proton transfer from the pyridinium moiety to the vinyl group then gives the vinyl pyridine product. Active species **11-50** can be regenerated by the ligand exchange with pyridine.

11.2 Ru-Catalyzed Hydrogenations

The hydrogenation of unsaturated molecules can be considered as the addition of a nucleophilic hydride and an electrophilic proton, which would provide corresponding saturated molecules [109, 110]. In Ru-catalysis, the formation of Ru—H bond can be considered as nucleophile for the reaction with unsaturated molecules. The proton source is provided by active hydrogen in ligand or the solvent molecules [26, 111].

11.2.1 Hydrogenation of Alkenes

In the presence of Ru(II) catalyst, the hydrogenation of alkenes can generate the corresponding alkanes by using 30 bar gaseous dihydrogen. The mechanism of this

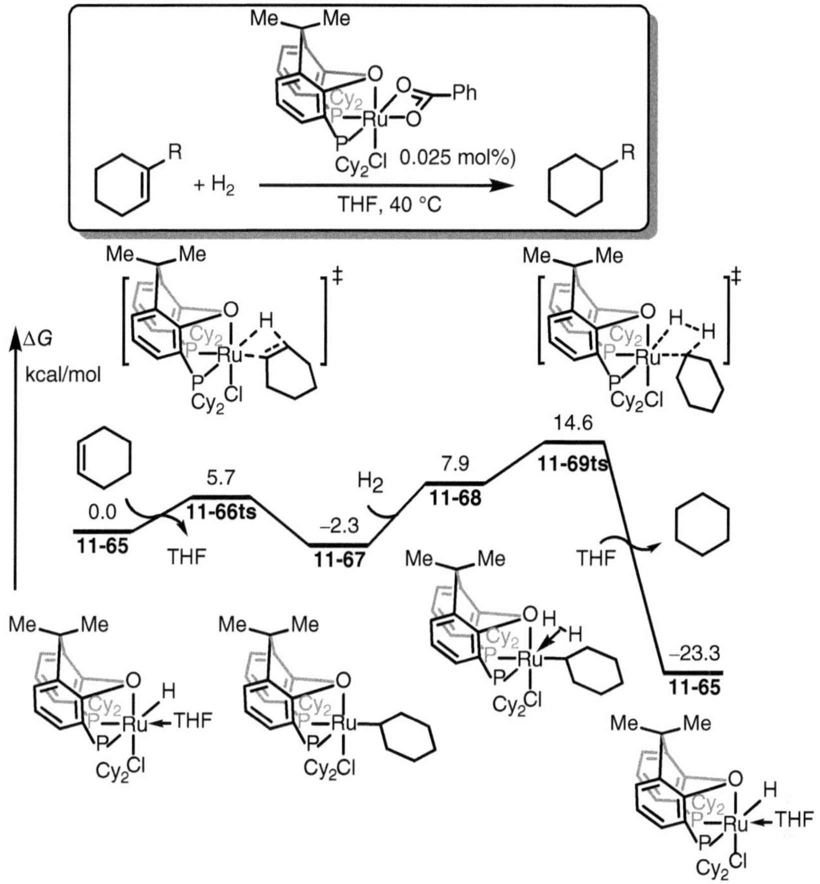

Figure 11.7 Free-energy profiles of Ru-catalyzed hydrogenation of alkenes. The energies were calculated at M06-L/def2-SVP level of theory. Source: Based on Rohmann et al. [112].

reaction was considered by DFT calculations [112]. As shown in Figure 11.7, the reaction starts from a solvent-coordinated hydride Ru(II) species **11-65**. After a ligand exchange with cyclohexene, the C=C double bond inserts into Ru—H bond via transition state **11-66ts** with an energy barrier of 5.7 kcal/mol to afford an alkyl Ru(II) intermediate **11-67**. The coordination of dihydrogen molecular generates complex **11-68**. Then, a σ-bond metathesis between Ru—C(alkyl) bond and H—H bond occurs via a four-membered ring-type transition state **11-69ts** resulting in cyclohexane product with the regeneration of hydride Ru(II) active species **11-65**. DFT calculations found that the whole catalytic cycle is nonredox.

11.2.2 Hydrogenation of Carbonyls

In Ru-catalyzed dehydrogenation of aldehydes, carbonyl group of aldehydes can be considered as a polarized C=O double bond, in which carbon atom reveals electrophilicity, while oxygen atom reveals nucleophilicity. Therefore, in

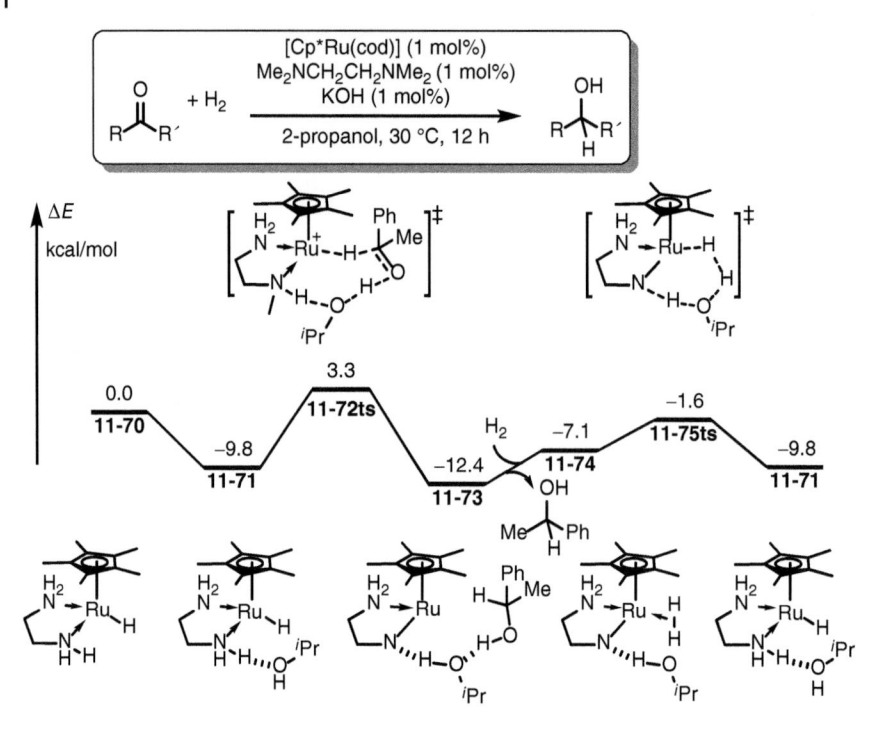

Figure 11.8 Potential energy profiles of Ru-catalyzed hydrogenation of ketones. The energies were calculated at B3LYP/6-31+G(d,p) (LANL2DZ for Ru)/PCM(ethanol)//B3LYP/6-31+G(d,p) (LANL2DZ for Ru) level of theory. Source: Based on Zhang et al. [113].

dehydrogenation, C—H bond can be formed from hydride, and O—H bond is generated by protonation. In this reaction, ligand with active hydrogen can be used as proton source. Interestingly, Fang and coworkers proposed an alcohol hydrogen bond bridge-assisted proton-transfer mechanism [113]. As shown in Figure 11.8, the reaction starts from hydride Ru(II) complex **11-70**, in which a hydrogen bond is formed between N—H in ligand and isopropanol. When acetophenone substrate joins in, another hydrogen bond can be formed between carbonyl and isopropanol. In the presence of hydrogen bond bridge, a hydride transfer from Ru to carbonyl takes place via an eight-membered ring-type transition state **11-72ts** with an energy barrier of 13.1 kcal/mol. Simultaneous proton transfer yields phenylethanol product and generates an amino Ru(II) intermediate **11-73**. The coordination of dihydrogen onto Ru affords complex **11-74**. Then, a hydroxyl-assisted heterolytic cleavage of H—H bond occurs via a six-membered ring-type transition state **11-75ts** to regenerate hydride Ru(II) active species **11-71**. The whole catalytic cycle is nonredox.

11.2.3 Hydrogenation of Esters

In the presence of pincer ligand–coordinated Ru(II) catalyst **11-76**, the hydrogenation of esters can provide corresponding alcohols under a high temperature.

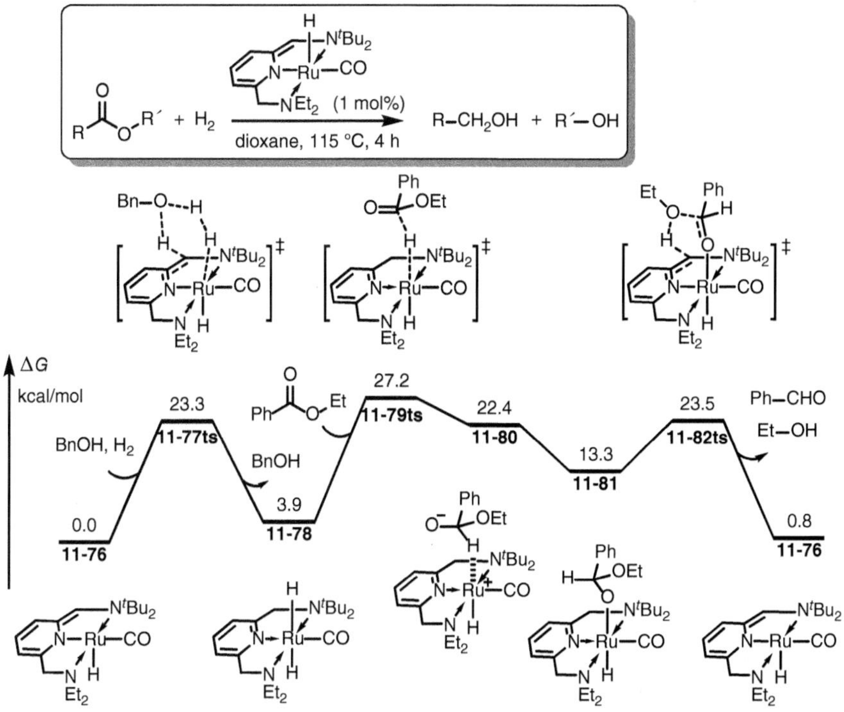

Figure 11.9 Free-energy profiles of Ru-catalyzed hydrogenation of esters. The energies were calculated at B3LYP/6-311++G(d,p) (LANL2DZ(f) for Ru, LANL2DZ(p) for P)/SMD (1,4-dioxane)//B3LYP/6-31G(d,p) (LANL2DZ(f) for Ru, LANL2DZ(d) for P) level of theory.

Interestingly, the pincer ligand involves a methylene pyridinide moiety, which can be dihydrogenated to achieve aromaticity in corresponding pyridyl group. In DFT studies for this reaction, a Milstein-proposed hydrogenation mechanism was considered [114, 115]. As shown in Figure 11.9, an alcohol-assisted heterolytic cleavage of H—H bond via transition state **11-77ts** can provide a hydride Ru(II) species **11-78**. An intermolecular hydride transfer from ruthenium to the coming ester takes place via transition state **11-79ts** resulting in a zwitterionic agostic intermediate **11-80**. An exergonic isomerization of this intermediate can form an alcoholate Rh(II) intermediate **11-81**, which can undergo a protonation by picolylic hydrogen via transition state **11-82ts** with an energy barrier of 10.2 kcal/mol. The protonation also leads to the cleavage of C—O bond in glycolate, which yields corresponding ethanol and benzaldehyde products. In another catalytic cycle, benzaldehyde can be further hydrogenated to achieve benzylalcohol product.

11.2.4 Hydrodefluorination of Fluoroarenes

Aromatic fluorocarbons are an important structural unit and component of many pharmaceutical and agrochemical molecules [116–119]. One hypothetical approach to the selective preparation of di- or trifluorophenyl substituents would be via a metal-catalyzed hydrodefluorination reaction of a pentafluorophenyl ring. As an

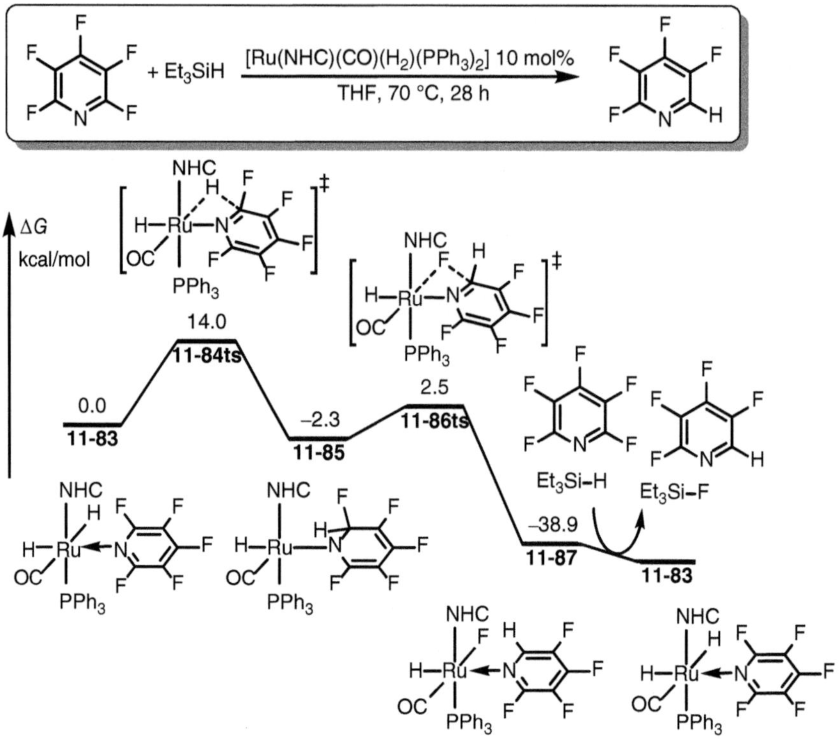

Figure 11.10 Free-energy profiles of Ru-catalyzed hydrodefluorination of fluoroarenes. The energies were calculated at BP86-D3/6-31G(d,p) (SDD for Ru, SDD(p) for P)/PCM(THF)// BP86/6-31G(d,p) (SDD for Ru, SDD(p) for P) level of theory.

example, hydride Ru(II) species [Ru(NHC)(CO)(H$_2$)(PPh$_3$)$_2$] can be used as catalyst in hydrodefluorination of full-fluoro-arenes [120]. In this reaction, silane with a Si—H bond was used as hydrogen source, which also can be used to absorb fluoride to form a Si—F bond. The mechanism of this reaction was revealed by DFT calculations. As shown in Figure 11.10, pentafluoropyridine was chosen as substrate, which can coordinate with hydride Ru(II) to form complex **11-83**. The insertion of C=N bond of pyridine into Ru—H bond occurs via a four-membered ring-type transition state **11-84ts** with an energy barrier of 14.0 kcal/mol to afford a dearomatized amino Ru(II) intermediate **11-85**. Then, a rapid β-fluoro elimination takes place rapidly via transition state **11-86ts** to form a tetrafluoropyridine-coordinated Ru(II) fluoride species **11-87**. This transformation can be considered as aromatic nucleophilic substitution, which involves a nucleophilic addition by hydride and an elimination of fluoride. The rate-determining step of this process is the nucleophilic addition.

11.3 Ru-Catalyzed Hydrofunctionalizations

Similar to other late transition metals, Ru-complexes can be used as catalyst in hydrofunctionalizations [121–125]. Hydrofunctionalization can be considered as

11.3 Ru-Catalyzed Hydrofunctionalizations

the insertion of unsaturated molecule into a polarized H—X covalent bond. The electronegativity of carbon, oxygen, and nitrogen is larger than that of hydrogen; therefore, the hydrogen atom can be considered as electrophilic proton in hydroacylations, hydrocarboxylations, and hydroamidations, while the other atoms would be nucleophile. Alternatively, the electronegativity of silicon and boron atoms is lower than that of hydrogen proving that the formal hydride provides nucleophile and the hetero atoms act as electrophile in hydrosilylations and hydroborations. However, the above-discussed mechanisms are only the common way for hydrofunctionalization. The detailed reaction pathway could be even different from the proposed ones.

11.3.1 Hydroacylations

Hydride Ru(II) complexes can be used as catalyst in hydroacylation reactions of olefins by using aldehyde [126–128]. A common catalytic cycle (Scheme 11.12) can be considered as a nonredox process involving an olefin insertion into Ru—H bond to form an alkyl Ru(II) intermediate, a carbonyl insertion into Ru—C bond, and a β-hydride elimination.

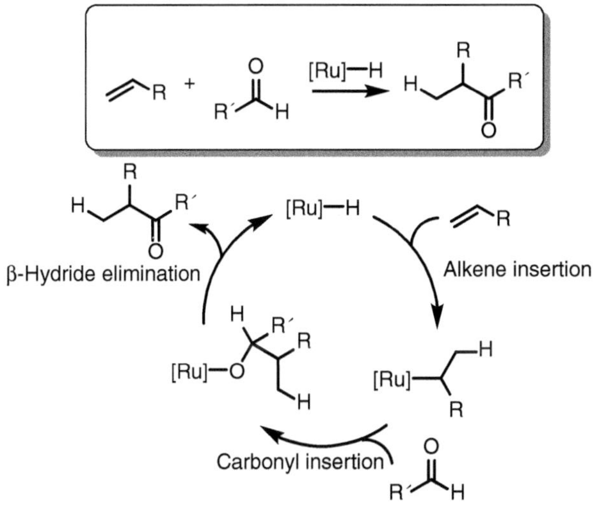

Scheme 11.12 General mechanism of Ru-hydride catalyzed hydroacylation of olefins.

The mechanism of hydride Ru(II) catalyzed 1,2-hydroacylation of conjugative olefin by using aldehyde was revealed by DFT calculations (Figure 11.11) [126]. The reaction starts from an olefin-coordinated hydride Ru(II) species **11-88**. A 1,2-insertion of olefin into Ru—H bond occurs via a four-membered ring-type transition state **11-89ts** to form an allylic Ru(II) species **11-90**. Subsequently, the coordinated aldehyde inserts into C(allyl)—Ru bond via a six-membered ring-type transition state **11-91ts** to form a benzyl alcoholate Ru(II) intermediate **11-92**. The sequential β-hydride elimination via transition state **11-93ts** generates

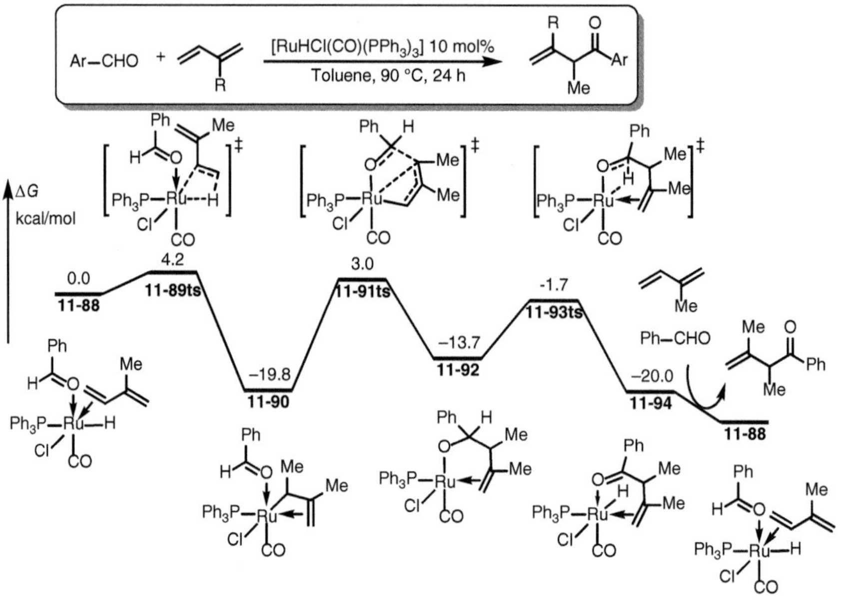

470 | *11 Theoretical Study of Ru-Catalysis*

Figure 11.11 Free-energy profiles of Ru-catalyzed hydrodefluorination of fluoroarenes. The energies were calculated at BP86-D3/6-31G(d,p) (SDD for Ru, SDD(p) for P)/PCM(THF) //BP86/6-31G(d,p) (SDD for Ru, SDD(p) for P) level of theory. Source: Based on Meng et al. [126].

product-coordinated hydride Ru(II) complex **11-94**. The ligand exchange by olefin and aldehyde reactants regenerates active species **11-88**. In the whole catalytic cycle, the oxidative state of Ru remains at +2.

11.3.2 Hydrocarboxylations

In the presence of Ru(II) complex, the hydrocarboxylation of terminal alkynes can provide both Markovnikov-type adduct **11-95** and anti-Markovnikov product **11-96**. Experimentally, the regioselectivity is controlled by the solvent. When dichloromethane (DCM) is used as solvent, Markovnikov-type **11-95** was observed as major product. In tetrahydrofuran (THF) solvent, anti-Markovnikov product **11-96** is the major one (Scheme 11.13) [129].

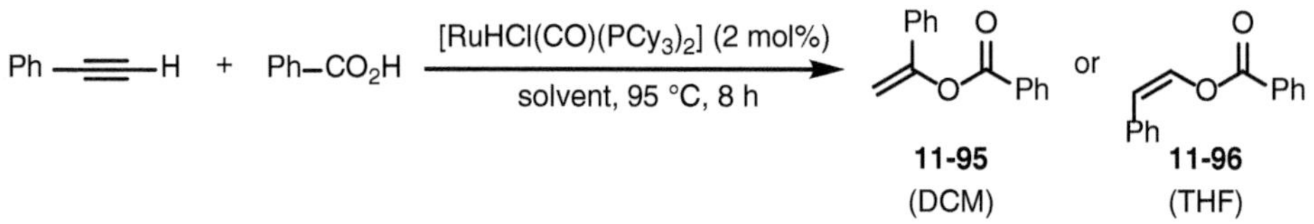

Scheme 11.13 Ru-catalyzed hydrocarboxylation of terminal alkynes.

DFT study was focused on the mechanism and regioselectivity of the hydrocarboxylation of terminal alkynes [130]. The selected calculated free-energy profiles

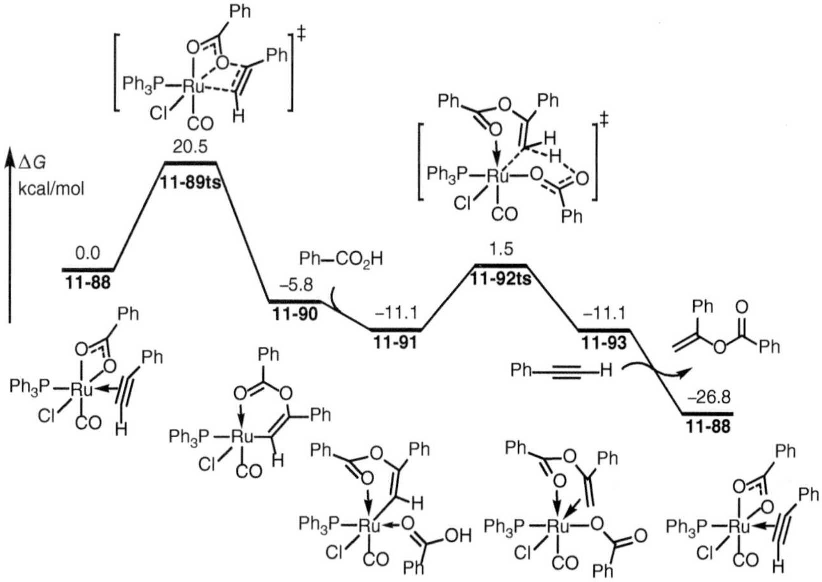

Figure 11.12 Free-energy profiles of Ru-catalyzed hydrocarboxylations of phenylacetylene in DCM. The energies were calculated at M06-L-D3/6-311+G(d,p) (LANL2TZ(f) for Ru)/SMD(DCM)//M06-L/6-31G(d) (LANL2DZ for Ru) level of theory.

for the catalytic cycle in DCM leading to Markovnikov product are given in Figure 11.12. The reaction starts from a phenylacetylene-coordinated Ru(II) species **11-88**. The coordinated alkyne can insert into Ru—O(carboxyl) bond via transition state **11-89ts** with an energy barrier of 20.5 kcal/mol to form a terminal vinyl Ru(II) intermediate **11-90**. The coordination of benzoic acid onto Ru forms complex **11-91**. Then, a protonation of vinyl group occurs via transition state **11-92ts** to form an olefin-coordinated Ru(II) intermediate **11-93**. The ligand exchange with phenylacetylene reactant regenerates the active species **11-88** with release of 1-phenylvinyl benzoate product.

Another case was given in Figure 11.13. When THF is used as solvent, it can coordinate with Ru first to form complex **11-94** with 0.5 kcal/mol further exergonic. Then, the deprotonation of terminal alkyne by benzoate takes place via a CMD-type transition state **11-95ts** with an energy barrier of only 4.2 kcal/mol to afford a phenylethynyl Ru(II) species **11-96**. Then, a protonation by in-situ-generated benzoic acid onto the internal carbon of phenylethynyl group takes place via transition state **11-97ts** to afford a Ru(II) vinylidene complex **11-98**. The vinylidene inserts into Ru—O(benzoate) bond via transition state **11-99ts** with an energy barrier of 16.6 kcal/mol to from a vinyl Ru(II) intermediate **11-100**. Then, the protonation by benzoic acid can provide anti-Markovnikov product **11-96**.

11.3.3 Hydroborations

Interestingly, hydroboration of alkynes could achieve trans-selectivity in the presence of Ru-catalyst [131, 132]. The mechanism of this reaction was studied by DFT

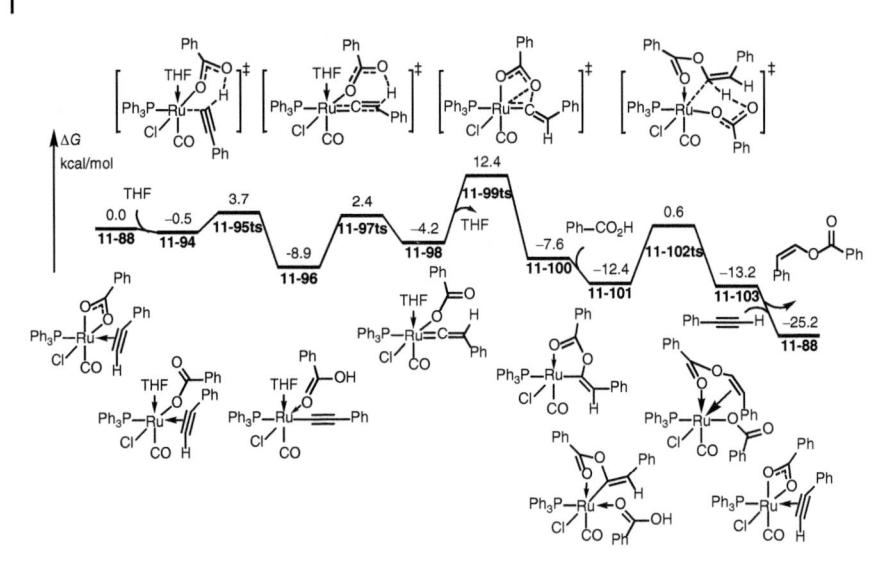

Figure 11.13 Free-energy profiles of Ru-catalyzed hydrocarboxylations of phenylacetylene in THF. The energies were calculated at M06-L-D3/6-311+G(d,p) (LANL2TZ(f) for Ru)/SMD(THF)//M06-L/6-31G(d) (LANL2DZ for Ru) level of theory.

calculation. As shown in Figure 11.14, the reaction starts from a solvent coordinated cationic Ru(II) species **11-104**. The coordination of alkyne and borane forms complex **11-105** with 18.4 kcal/mol endergonic. Then, a Ru-assisted σ-complex metathesis leads to a hydrogen transfer via transition state **11-106ts** with an overall activation free energy of 26.2 kcal/mol to form a metallacyclopropene intermediate **11-107**. Then, a reductive elimination generates borylated product via transition state **11-108ts**, which can be considered as a formal carbene migratory insertion into Ru—B bond. Therefore, the trans-selectivity is achieved.

11.4 Ru-Mediated Dehydrogenations

Homogeneous catalytic dehydrogenation of alcohols is an efficient way of hydrogen production [133–138]. An ideal way to use methanol as a hydrogen carrier is the conversion of a methanol–water mixture into CO_2 and H_2, which means the entire hydrogen content (12 wt%) is used (Scheme 11.14). DFT calculation was employed to reveal the mechanism of the stepwise dehydrogenations [135, 139, 140].

In this section, Beller and coworkers reported pincer bis[2-(diisopropylphosphino) -ethyl]amine (HPNP) ligand-coordinated Ru(II)-hydride complex **11-110** catalyzed full dehydrogenation of methanol was chosen as a model reaction for the investigation of corresponding deprotonation processes of alcohol, formaldehyde, and formic acid [135]. Therefore, the hydrogen desorption of **11-112** is a common process for those three reactions. The calculated free-energy profiles for methanol-assisted hydrogen desorption of **11-112** are given in Figure 11.15 [140]. When Ru(II)-hydride complex **11-111** is formed, the hydrogen bond of coming methanol with the

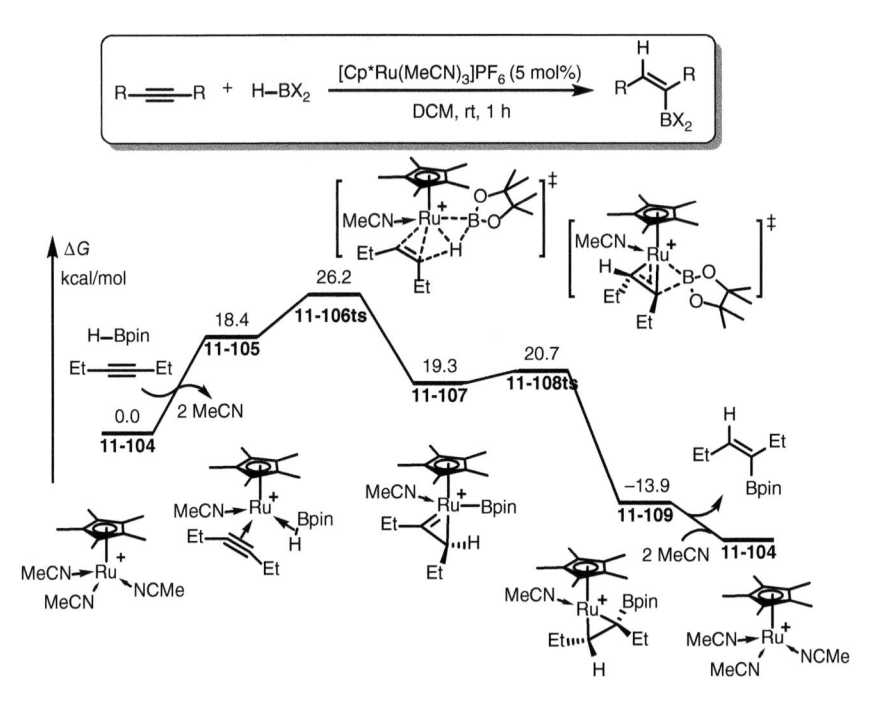

Figure 11.14 Free-energy profiles of Ru-catalyzed trans-hydroboration of acetylenes. The energies were calculated at M06/6-311++G(3df,2pd) (def2-TZVP for Ru)/SMD(DCM)// B3LYP/6-31G(d) (LANL2DZ for Ru) level of theory.

amino group can result in protonation of hydride via a six-membered ring-type transition state **11-113ts** with an energy barrier of 21.8 kcal/mol to afford a dihydrogen-coordinated cationic Ru(II) intermediate **11-114**. Then, a rapid deprotonation of amine moiety in ligand by methanolate via transition state **11-115ts** releases a methanol and a dihydrogen with the generation of complex **11-112**. The overall process is 11.9 kcal/mol endergonic. With complex **11-112** in hand, further dehydrogenation processes are discussed.

11.4.1 Dehydrogenation of Alcohols

As shown in Figure 11.16 [140], the generated amino Ru(II) species **11-112** can react with methanol by deprotonation via transition state **11-116ts** to form a methoxide cationic Ru(II) complex **11-117**. Then, a hydride transfer occurs via a six-membered ring-type transition state **11-118ts** to form Ru(II)-hydride complex **11-111**. This process is only 1.2 kcal/mol exergonic, which cannot cancel the endergonic of hydrogen desorption. Passably, the generated formaldehyde can react with hydroxide to afford hydroxymethanolate anion for further dehydrogenations.

11.4.2 Dehydrogenation of Formaldehyde

A nucleophilic addition of formaldehyde by hydroxide can provide hydroxymethanolate anion, which is an active species for dehydrogenation [140]. As

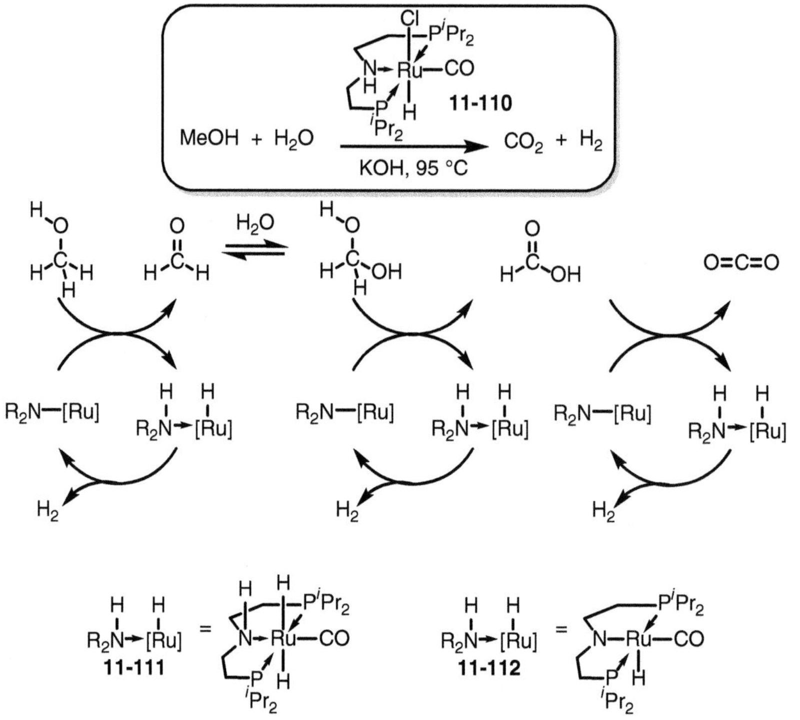

Scheme 11.14 Dehydrogenation of methanol in the presence of pincer-PNP coordinated Ru(II)-hydride catalyst.

shown in Figure 11.17, tThe hydration of amino Ru(II) active species **11-112** via transition state **11-119ts** forms a Ru(II) hydroxide **11-120**, which can react with hydroxymethanolate to form a hydride-coordinated Ru(II) intermediate **11-121** by ligand exchange. Then, a hydride transfer takes place via transition state **11-122ts** to yield formic acid. This process is 5.3 kcal/mol exergonic, which cannot cancel the endergonic of hydrogen desorption from **11-111**.

11.4.3 Dehydrogenation of Formic Acid

As shown in Figure 11.18, amino Ru(II) species **11-112** can be protonated by formic acid to afford a formate-Ru(II) intermediate **11-123** in a barrierless process [141–143]. The generated formate-Ru(II) intermediate **11-123** can isomerize to an agostic intermediate **11-124** reversibly. Then, an intermolecular hydride transfer from formate to Ru(II) cation occurs via transition state **11-125ts** with an overall activation free energy of 23.0 kcal/mol. After release of carbon dioxide, hydride Ru(II) complex **11-111** is generated. DFT calculations found that all the three processes of alcohol, formaldehyde, and formic acid dehydrogenations are endergonic. Therefore, the generation of gaseous dihydrogen and its diffusion would provide the driving force for this reaction.

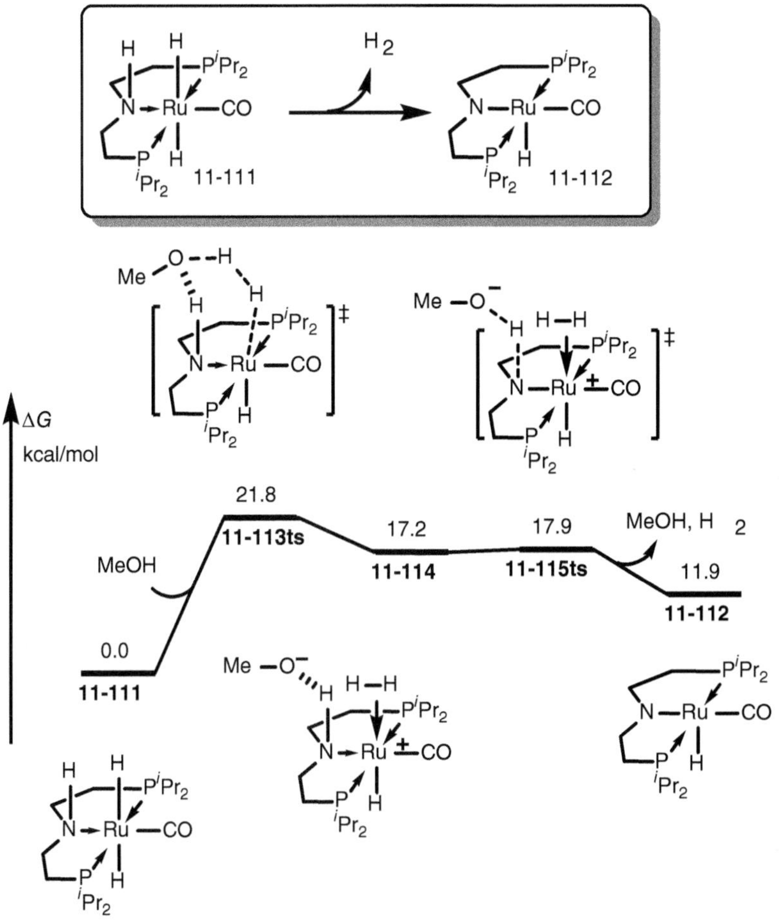

Figure 11.15 Free-energy profiles of methanol-assisted hydrogen desorption in Ru-catalyzed dehydrogenation of methanol. The energies were calculated at M06/6-311++G(d,p) (SDD for Ru)/SMD(methanol) level of theory. Source: Based on Yang [140].

11.5 Ru-Catalyzed Cycloadditions

Ru-complexes show efficiency in promoting the activations of unsaturated molecules, including alkenes, alkynes, carbon monoxide, 1,3-dipoles, and nitriles, which can be used in the cycloadditions of those molecules. Mechanistically, unsaturated C—C bonds can be activated by oxidative cyclizations or insertions. Ru also acts as Lewis acid in the reaction of heteroatoms [144–150].

11.5.1 Ru-Mediated (2+2+2) Cycloadditions

CpRu(II) species are often used in (2+2+2) cycloadditions of diynes with alkynes, olefins, allylic ethers, cyanides, and nitrile oxides [151–153]. In those reactions, the key step is the oxidative coupling of two alkyne ligands to

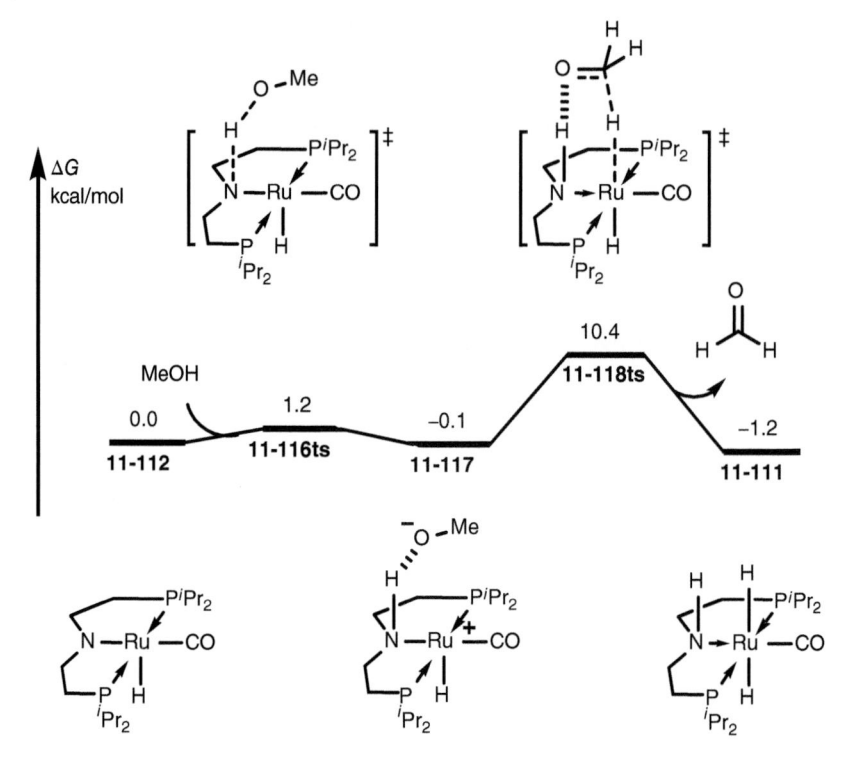

Figure 11.16 Free-energy profiles for the Ru-mediated dehydrogenation of methanol. The energies were calculated at M06/6-311++G(d,p) (SDD for Ru)/SMD(methanol) level of theory. Source: Based on Yang [140].

give a metallacyclopentatriene intermediate (Scheme 11.15), which exhibits metal-involved aromaticity.

Scheme 11.15 The oxidative cycloaddition of diyne onto Ru.

As shown in Figure 11.19, the mechanism of CpRu(II)-catalyzed alkyne trimerization to construct benzene was studied by DFT calculation [151]. The coordination of two acetylene molecules forms complex **11-126**. Then, an oxidative cyclization via transition state **11-127ts** generates a ruthenacyclopentatriene intermediate **11-128** with 38.7 kcal/mol exergonic. The calculated activation free energy for this step is only 13.2 kcal/mol. The geometry information of intermediate **11-128** is shown in the same figure. The averageness of C—C and C—Ru bonds in the five-membered ring revealed a Möbius aromaticity. The coordination of another

11.5 Ru-Catalyzed Cycloadditions | 477

Figure 11.17 Free-energy profiles for the Ru-mediated dehydrogenation of formaldehyde. The energies were calculated at M06/6-311++G(d,p) (SDD for Ru)/SMD(methanol) level of theory.

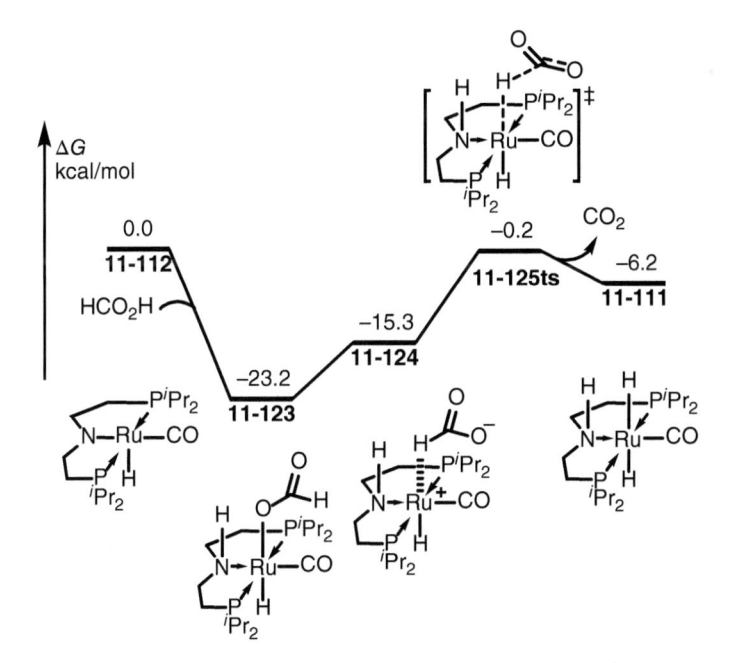

Figure 11.18 Free-energy profiles for the Ru-mediated dehydrogenation of formic acid. The energies were calculated at M06/6-311++G(d,p) (SDD for Ru)/SMD(methanol) level of theory.

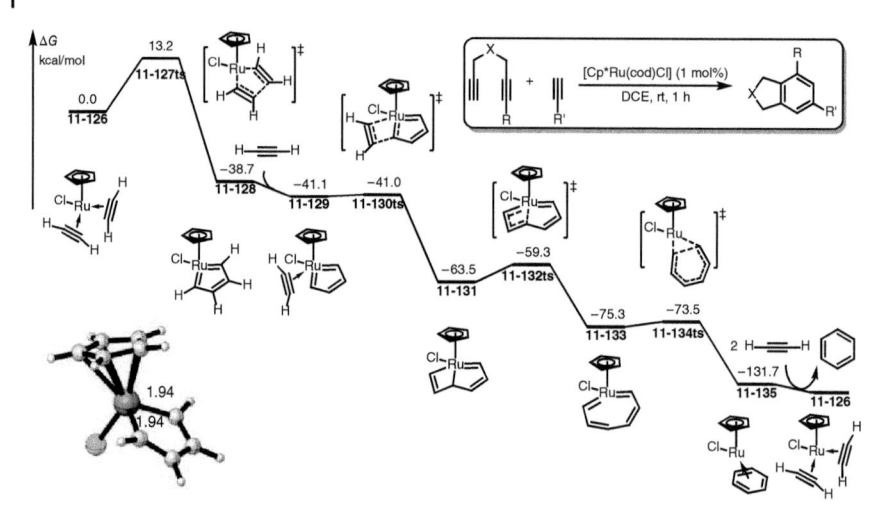

Figure 11.19 Free-energy profiles for the Ru-catalyzed cyclotrimerization of alkynes. The energies were calculated at B3LYP/6-31G(d,p) (SDD for Ru) level of theory.

molecule acetylene forms complex **11-129** with 2.4 kcal/mol exergonic. Subsequently, a π-bond metathesis generates a ruthenabicyclo[3.2.0]heptatriene complex **11-131**, which can rapidly isomerize to an eight-membered ring complex **11-133**. The reductive elimination results in benzene product–coordinated Ru(II) complex **11-135**, and the active species **11-126** can be regenerated by ligand exchange with acetylenes.

11.5.2 Ru-Mediated Pauson–Khand Type (2+2+1) Cycloadditions

In a typical Co-mediated Pauson–Khand reaction with alkene, alkyne, and carbon monoxide, the initiation step is the oxidative cyclization of alkene and alkyne with metal to form a metallacyclopentene intermediate [154–158]. However, in corresponding Ru(0) mediate Pauson–Khand type cycloadditions, it is an unfavorable process detected by DFT calculation. An alternative pathway was proposed by Wu [159, 160]. As shown in Figure 11.20, the coordination of enyne onto Ru(CO)$_5$ forms complex **11-137** with 13.7 kcal/mol endergonic. Then, an oxidative cyclization by the coordinated alkyne and carbon monoxide occurs via a four-membered ring-type transition state **11-138ts** to afford a ruthenacyclobutenone intermediate **11-139**. The calculated activation free energy for this process is 25.7 kcal/mol, which is 13.5 kcal/mol lower than the oxidative cyclization with alkyne and alkene via transition state **11-144ts**. Then, the coordinated alkene moiety inserts into Ru—C(vinyl) bond via transition state **11-140ts**. The reductive elimination yields cyclopentenone product.

11.5.3 Ru-Mediated Click Reactions

In Ru-mediated click (3+2) cyclization of azide and alkyne, Ru(II) species can be used to activate alkyne in a stepwise process. Interestingly, Ru also can play as a

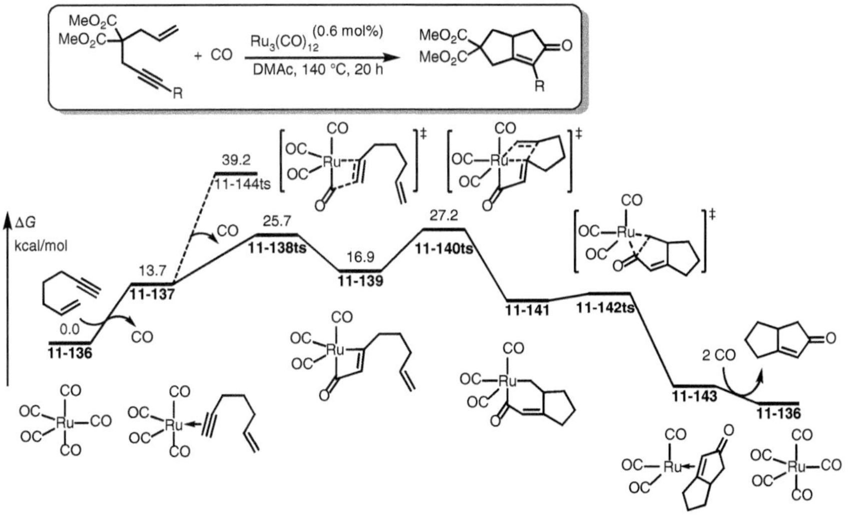

Figure 11.20 Free-energy profiles for the Ru-catalyzed intramolecular Pauson–Khand reaction. The energies were calculated at BP86/6-311+G(d) (SDD for Ru)/IEF-PCM(DMAc)// BP86/6-31G(d) (SDD for Ru) level of theory.

counterion to react with terminal alkynes. The generated alkynyl Ru species can react with azide, which also provides the same triazole products. Both of the two mechanisms have been revealed by DFT calculations [161–167].

The calculated free-energy profiles for the ruthenacycle pathway are shown in Figure 11.21 [162], which provide 1,5-substituted triazoles. The coordination of alkyne and azide onto CpRu(II) can form complex **11-145**, where alkyne is activated by Ru. Then, an oxidative cyclization occurs via transition state **11-146ts** with an energy barrier of 4.1 kcal/mol to form a ruthenatriazabicyclo[3.1.0]hexadiene intermediate **11-147**. Then, the reductive elimination via transition state **11-148ts** generated triazole-coordinated Ru(II) complex **11-149** with an activation free energy of 15.4 kcal/mol. The ligand exchange with alkyne and azide can regenerate active species **11-145**.

In another theoretical study of counterion pathway leading to 1,4-substituted triazoles (Figure 11.22) [164], when phosphine is used as ligand, the (3+2) cyclization of azide and alkyne starts from an azide-coordinated alkynyl Ru(II) species **11-150**. The oxidative coupling via transition state **11-151ts** leads to the formation of a six-membered Ru(IV) vinylidene intermediate **11-152**. Then, an intramolecular reductive insertion of vinylidene moiety into Ru—N bond via transition state **11-153ts** forms a triazolyl Ru(II) species **11-154**. The σ-bond metathesis with terminal alkyne via a four-membered ring-type transition state **11-156ts** releases triazole product and regenerates active species **11-150** after a coordination of azide. The calculated activation free energy for this step is 32.1 kcal/mol, which is considered as the rate-determining step.

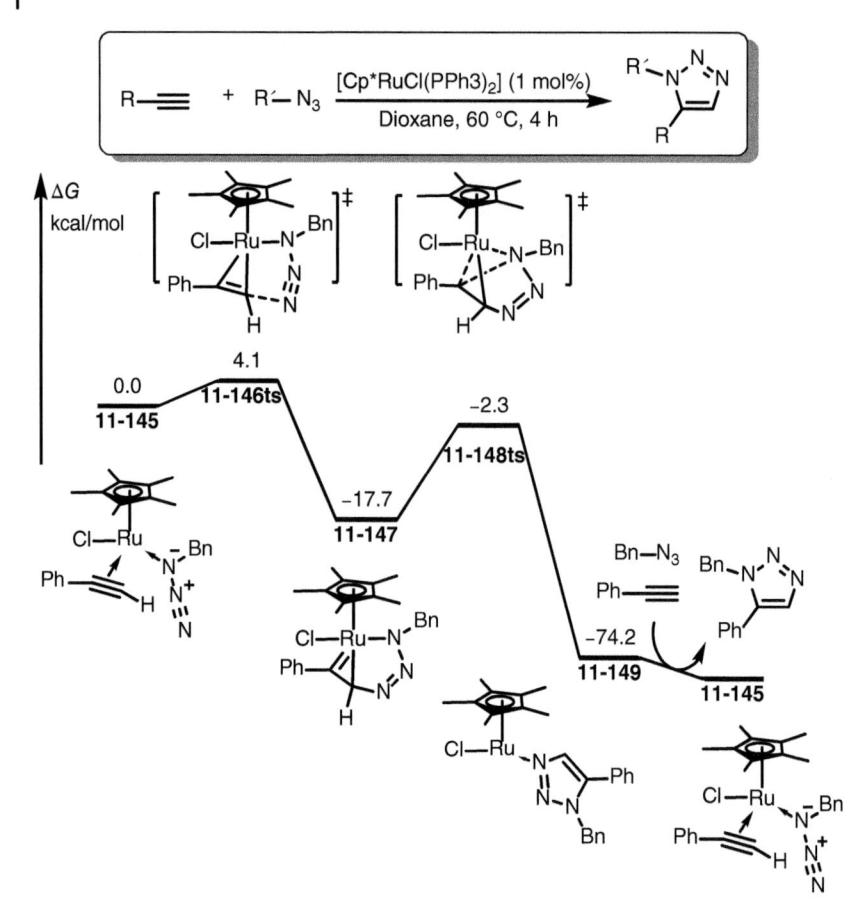

Figure 11.21 Free-energy profiles for the Ru-catalyzed click reaction through ruthenacycle pathway. The energies were calculated at B3LYP/6-31G(d) (LANL2DZ for Ru)/PCM(THF)//B3LYP/6-31G(d) (LANL2DZ for Ru) level of theory. Source: Based on Boz and Tüzün [160].

11.6 Ru-Mediated Metathesis

The field of Ru-mediated olefin metathesis has grown tremendously since the first ruthenium Grubbs-type carbene precatalyst was reported over 20 years ago [168, 169]. A series of theoretical studies focus on the mechanism of Ru-mediated olefin metathesis, which have provided various predictions of reactivity and selectivity [170–174]. Some typical examples for the explorations have been summarized in this section to reveal the mechanism of olefin metathesis. Generally, a well-accepted metal-involved olefin metathesis involves a concerted [2+2] cycloaddition of metal-carbene complex and alkene to afford a metallacyclobutane ring and cycloreversions to form another metal-carbene complex and corresponding alkene (Scheme 11.16).

Scheme 11.16 Ru-mediated olefin metathesis.

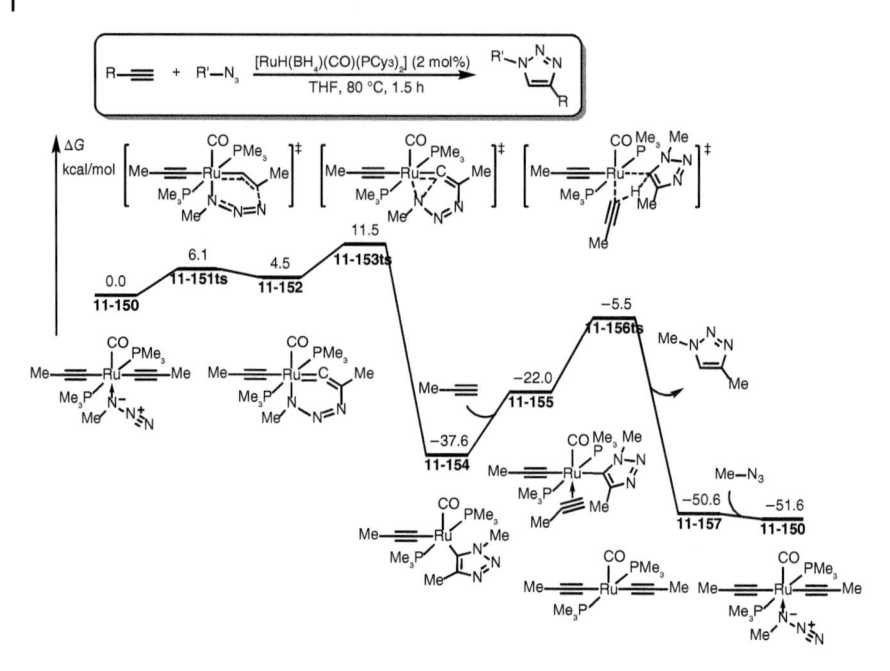

Figure 11.22 Free-energy profiles for the Ru-catalyzed click reaction through counterion pathway. The energies were calculated at B3LYP-D3/6-31G(d) (LANL2DZ for Ru and P)//B3LYP/6-31G(d) (LANL2DZ for Ru and P) level of theory. Source: Based on Liu et al. [164].

11.6.1 Ru-Mediated Intermolecular Olefin Metathesis

Generally, the mechanism of Ru-mediated intermolecular olefin metathesis is composed of the generation of active species and the catalytic cycle. A DFT study on the mechanism of Grubbs II–type catalyst-mediated intermolecular olefin metathesis was employed as an example to reveal the reaction mechanism [175]. The calculated free-energy profiles for the generation of active species are shown in Figure 11.23. The reaction starts from a Grubbs II catalyst **11-158**. A ligand exchange with octene generates complex **11-159** with 8.8 kcal/mol endergonic. Then, a concerted [2+2] cycloaddition takes place via transition state **11-160ts** with an energy barrier of 7.1 kcal/mol to form a ruthenacyclobutane intermediate **11-161**. Then, a cycloreversion via transition state **11-162ts** generates a styrene-coordinated Ru(II)-carbene complex **11-163**. After ligand exchange with reacting 1-octene, a Ru(II)-carbene complex **11-164** is formed by 8.1 kcal/mol overall endergonic, which is the active species in the following catalytic cycle.

In catalytic cycle from octene-coordinated Ru-carbene complex **11-164** (Figure 11.24), a π-bond metathesis occurs through [2+2] cycloaddition transition state **11-165ts** to form a ruthenacyclobutane intermediate **11-166**. The cycloreversion via transition state **11-167ts** yields (E)-tetradec-7-ene product and gives a methylene-Ru complex **11-169**. Another olefin metathesis with octene can regenerate active species **11-164** by release an ethylene. Undoubtedly, the whole catalytic cycle for the self-metathesis of octene is endergonic. The driving force for this reaction can be considered as the irreversible release of gaseous ethylene.

Figure 11.23 Free-energy profiles for the generation of active catalyst in a Grubbs II-type Ru-catalyzed 1-octene self-metathesis. The energies were calculated at PW91/DNP level of theory. DNP; double numeric polarized.

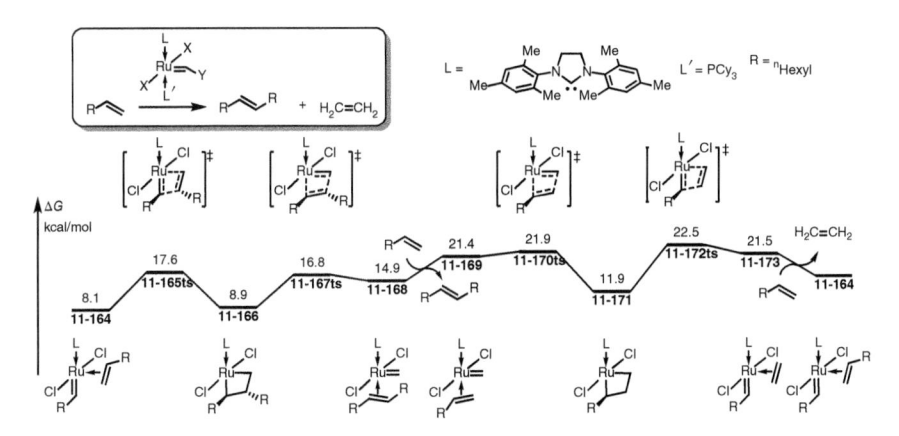

Figure 11.24 Free-energy profiles for the catalytic cycle of Grubbs II-type Ru-catalyzed 1-octene self-metathesis. The energies were calculated at PW91/DNP level of theory.

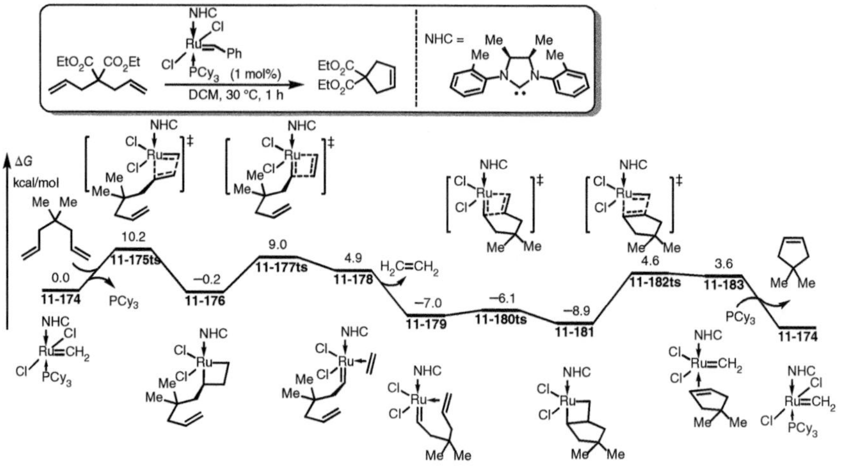

Figure 11.25 Free-energy profiles for the catalytic cycle of Ru-catalyzed intramolecular olefin metathesis. The energies were calculated at BP86/SCP (SDD for Ru)/PCM(DCM)//BP86/SCP (SDD for Ru) level of theory.

11.6.2 Ru-Mediated Intramolecular Diene Metathesis

When a 1,6-diene is used as substrate for Ru-mediated olefin metathesis, cyclopentene derivative can be given in this reaction [176, 177]. The DFT-calculated free-energy profiles are shown in Figure 11.25. The coordination of diene in methylene-Ru(II) complex **11-174** leads to a [2+2] cycloaddition via transition state **11-175ts** to generate a ruthenacyclobutane intermediate **11-176**. The following cycloreversion via transition state **11-177ts** leads to the release of ethylene with the formation of Ru(II)-carbene complex **11-179**. Then, an intramolecular [2+2] cycloaddition can occur via transition state **11-180ts** to form another ruthenacyclobutane intermediate **11-181**. The decomposition of the four-membered ring via transition state **11-182ts** regenerates methylene-Ru(II) species **11-174** by release of cyclopentene product.

11.6.3 Ru-Mediated Alkyne Metathesis

The Ru-mediated metathesis between alkene and alkyne can provide corresponding 1,3-diene product [178, 179]. The DFT-calculated free-energy profiles for the whole catalytic cycle is shown in Figure 11.26. The reaction starts from a phosphine-coordinated Ru(II)-carbene species **11-184**, which can react with acetylene through ligand exchange to form complex **11-185** with 12.8 kcal/mol endergonic. Then, a π-bond metathesis takes place via a four-membered ring-type transition state **11-186ts**. Unlike olefin metathesis, ruthenacyclobutene was not observed in this step. However, only a ring-opened Ru-vinylcarbene complex **11-187** was found. In the following step, the coordination of ethylene followed by a regular [2+2] cycloaddition occurs via transition state **11-189ts** to afford a ruthenacyclobutane species **11-190**. The decomposition of the four-membered ring

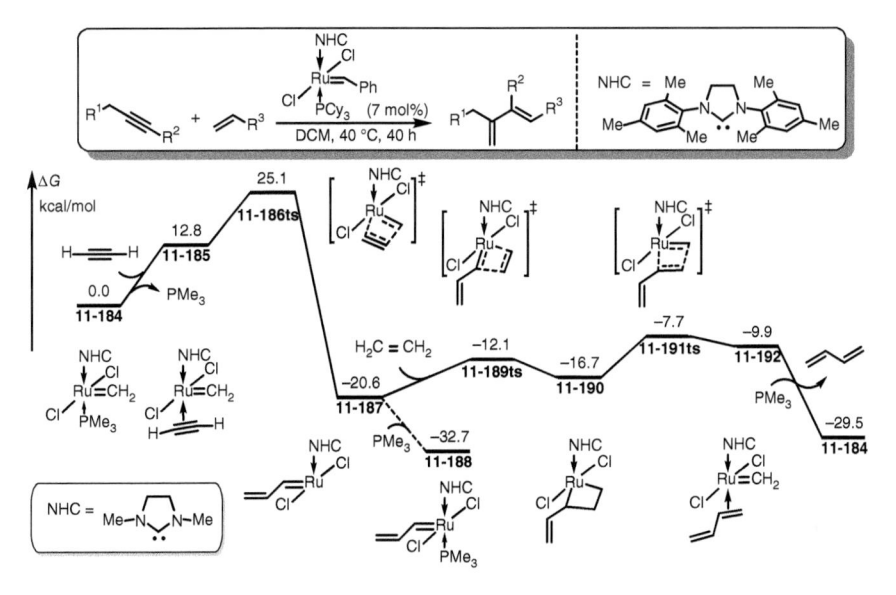

Figure 11.26 Free-energy profiles for the catalytic cycle of Ru-catalyzed intermolecular alkyne-alkene metathesis. The energies were calculated at B3LYP/LACV3P**+//B3LYP/LACVP* level of theory.

via transition state **11-191ts** yields diene product and regenerates Ru(II)-carbene species **11-184**.

References

1 Dragutan, V., Dragutan, I., Delaude, L. et al. (2007). NHC–Ru complexes – friendly catalytic tools for manifold chemical transformations. *Coordination Chemistry Reviews* 251 (5–6): 765–794.
2 Francos, J., García-Garrido, S.E., García-Álvarez, J. et al. (2017). Water-tolerant bis(allyl)-ruthenium(IV) catalysts: an account of their applications. *Inorganica Chimica Acta* 455: 398–414.
3 Naota, T., Takaya, H., and Murahashi, S.I. (1998). Ruthenium-catalyzed reactions for organic synthesis. *Chemical Reviews* 98 (7): 2599–2660.
4 Anaby, A., Schelwies, M., Schwaben, J. et al. (2018). Study of precatalyst degradation leading to the discovery of a new Ru0 precatalyst for hydrogenation and dehydrogenation. *Organometallics* 37 (13): 2193–2201.
5 Bruneau, C. and Achard, M. (2012). Allylic ruthenium(IV) complexes in catalysis. *Coordination Chemistry Reviews* 256 (5–8): 525–536.
6 Cesari, C., Mazzoni, R., Matteucci, E. et al. (2019). Hydrogen transfer activation via stabilization of coordinatively vacant sites: tuning long-range π-system electronic interaction between Ru(0) and NHC pendants. *Organometallics* 38 (5): 1041–1051.

7 Ito, T., Takahashi, K., and Iwasawa, N. (2018). Reactivity of a Ruthenium(0) complex bearing a tetradentate phosphine ligand: applications to catalytic acrylate salt synthesis from ethylene and CO_2. *Organometallics* 38 (2): 205–209.

8 Otsuka, T., Ishii, A., Dub, P.A. et al. (2013). Practical selective hydrogenation of alpha-fluorinated esters with bifunctional pincer-type ruthenium(II) catalysts leading to fluorinated alcohols or fluoral hemiacetals. *Journal of the American Chemical Society* 135 (26): 9600–9603.

9 Qiu, Y., Tian, C., Massignan, L. et al. (2018). Electrooxidative Ruthenium-catalyzed C–H/O–H annulation by weak O-coordination. *Angewandte Chemie International Edition* 57 (20): 5818–5822.

10 Colby, D.A., Bergman, R.G., and Ellman, J.A. (2010). Rhodium-catalyzed C—C bond formation via heteroatom-directed C—H bond activation. *Chemical Reviews* 110 (2): 624–655.

11 Li, Y., Liu, S., Qi, Z. et al. (2015). The mechanism of N—O bond cleavage in Rhodium-catalyzed C—H bond functionalization of quinoline N-oxides with alkynes: a computational study. *Chemistry* 21 (28): 10131–10137.

12 Lin, Y., Zhu, L., Lan, Y. et al. (2015). Development of a Rhodium(II)-catalyzed chemoselective C(sp(3))-H oxygenation. *Chemistry* 21 (42): 14937–14942.

13 Qin, X., Li, X., Huang, Q. et al. (2015). Rhodium(III)-catalyzed ortho C–H heteroarylation of (hetero)aromatic carboxylic acids: a rapid and concise access to pi-conjugated poly-heterocycles. *Angewandte Chemie International Edition* 54 (24): 7167–7170.

14 Arroniz, C., Denis, J.G., Ironmonger, A. et al. (2014). An organic cation as a silver(i) analogue for the arylation of sp2 and sp3 C—H bonds with iodoarenes. *Chemical Science* 5 (9): 3509–3514.

15 Arroniz, C., Ironmonger, A., Rassias, G. et al. (2013). Direct ortho-arylation of ortho-substituted benzoic acids: overriding Pd-catalyzed protodecarboxylation. *Organic Letters* 15 (4): 910–913.

16 Chiong, H.A., Pham, Q.N., and Daugulis, O. (2007). Two methods for direct ortho-arylation of benzoic acids. *Journal of the American Chemical Society* 129 (32): 9879–9884.

17 Engle, K.M., Mei, T.-S., Wasa, M. et al. (2011). Weak coordination as a powerful means for developing broadly useful C–H functionalization reactions. *Accounts of Chemical Research* 45 (6): 788–802.

18 Giri, R., Maugel, N., Li, J.J. et al. (2007). Palladium-catalyzed methylation and arylation of sp2 and sp3 C—H bonds in simple carboxylic acids. *Journal of the American Chemical Society* 129 (12): 3510–3511.

19 Inoue, H., Phan Thi Thanh, N., Fujisawa, I. et al. (2020). Synthesis of forms of a chiral Ruthenium complex containing a Ru—Colefin(sp(2)) bond and their application to catalytic asymmetric cyclopropanation reactions. *Organic Letters* 22 (4): 1475–1479.

20 Li, J., Warratz, S., Zell, D. et al. (2015). N-Acyl amino acid ligands for Ruthenium(II)-catalyzed meta-C–H *tert*-alkylation with removable auxiliaries. *Journal of the American Chemical Society* 137 (43): 13894–13901.

21 Li, Y.M., Li, X., Peng, F.Z. et al. (2011). Catalytic asymmetric construction of spirocyclopentaneoxindoles by a combined Ru-catalyzed cross-metathesis/double Michael addition sequence. *Organic Letters* 13 (23): 6200–6203.
22 Louillat, M.L. and Patureau, F.W. (2013). Toward polynuclear Ru–Cu catalytic dehydrogenative C—N bond formation, on the reactivity of carbazoles. *Organic Letters* 15 (1): 164–167.
23 Watanabe, M., Ikagawa, A., Wang, H. et al. (2004). Catalytic enantioselective Michael addition of 1,3-dicarbonyl compounds to nitroalkenes catalyzed by well-defined chiral ru amido complexes. *Journal of the American Chemical Society* 126 (36): 11148–11149.
24 Glaser, P.B. and Tilley, T.D. (2003). Catalytic hydrosilylation of alkenes by a ruthenium silylene complex. Evidence for a new hydrosilylation mechanism. *Journal of the American Chemical Society* 125 (45): 13640–13641.
25 Rankin, M.A., MacLean, D.F., Schatte, G. et al. (2007). Silylene extrusion from organosilanes via double geminal Si—H bond activation by a Cp*Ru(kappa2-P,N)+ complex: observation of a key stoichiometric step in the glaser-tilley alkene hydrosilylation mechanism. *Journal of the American Chemical Society* 129 (51): 15855–15864.
26 Zaranek, M., Marciniec, B., and Pawluć, P. (2016). Ruthenium-catalysed hydrosilylation of carbon–carbon multiple bonds. *Organic Chemistry Frontiers* 3 (10): 1337–1344.
27 Hong, S.H., Wenzel, A.G., Salguero, T.T. et al. (2007). Decomposition of ruthenium olefin metathesis catalysts. *Journal of the American Chemical Society* 129 (25): 7961–7968.
28 Liberman-Martin, A.L. and Grubbs, R.H. (2017). Ruthenium olefin metathesis catalysts featuring a labile carbodicarbene ligand. *Organometallics* 36 (21): 4091–4094.
29 Sanford, M.S., Ulman, M., and Grubbs, R.H. (2001). New insights into the mechanism of ruthenium-catalyzed olefin metathesis reactions. *Journal of the American Chemical Society* 123 (4): 749–750.
30 Kakiuchi, F., Kan, S., Igi, K. et al. (2003). A ruthenium-catalyzed reaction of aromatic ketones with arylboronates: a new method for the arylation of aromatic compounds via C—H bond cleavage. *Journal of the American Chemical Society* 125 (7): 1698–1699.
31 Ueno, S., Mizushima, E., Chatani, N. et al. (2006). Direct observation of the oxidative addition of the aryl carbon–oxygen bond to a ruthenium complex and consideration of the relative reactivity between aryl carbon–oxygen and aryl carbon–hydrogen bonds. *Journal of the American Chemical Society* 128 (51): 16516–16517.
32 Oi, S., Fukita, S., Hirata, N. et al. (2001). Ruthenium complex-catalyzed direct ortho arylation and alkenylation of 2-arylpyridines with organic halides. *Organic Letters* 3 (16): 2579–2581.

33 Oi, S., Ogino, Y., Fukita, S. et al. (2002). Ruthenium complex catalyzed direct ortho arylation and alkenylation of aromatic imines with organic halides. *Organic Letters* 4 (10): 1783–1785.

34 Ackermann, L. (2005). Phosphine oxides as preligands in ruthenium-catalyzed arylations via C—H bond functionalization using aryl chlorides. *Organic Letters* 7 (14): 3123–3125.

35 Ackermann, L., Althammer, A., and Born, R. (2006). Catalytic arylation reactions by C—H bond activation with aryl tosylates. *Angewandte Chemie International Edition* 45 (16): 2619–2622.

36 Ackermann, L., Vicente, R., and Althammer, A. (2008). Assisted ruthenium-catalyzed C—H bond activation: carboxylic acids as cocatalysts for generally applicable direct arylations in apolar solvents. *Organic Letters* 10 (11): 2299–2302.

37 Ferrer Flegeau, E., Bruneau, C., Dixneuf, P.H. et al. (2011). Autocatalysis for C—H bond activation by ruthenium(II) complexes in catalytic arylation of functional arenes. *Journal of the American Chemical Society* 133 (26): 10161–10170.

38 Ozdemir, I., Demir, S., Cetinkaya, B. et al. (2008). Direct arylation of arene C—H bonds by cooperative action of NH carbene–ruthenium(II) catalyst and carbonate via proton abstraction mechanism. *Journal of the American Chemical Society* 130 (4): 1156–1157.

39 Lewis, L.N. and Smith, J.F. (1986). Catalytic carbon–carbon bond formation via ortho-metalated complexes. *Journal of the American Chemical Society* 108 (10): 2728–2735.

40 Murai, S., Kakiuchi, F., Sekine, S. et al. (1993). Efficient catalytic addition of aromatic carbon–hydrogen bonds to olefins. *Nature* 366 (6455): 529–531.

41 Yan, S.-S., Zhu, L., Ye, J.-H. et al. (2018). Ruthenium-catalyzed umpolung carboxylation of hydrazones with CO_2. *Chemical Science* 9 (21): 4873–4878.

42 Guan, W., Sakaki, S., Kurahashi, T. et al. (2013). Theoretical mechanistic study of Novel Ni(0)-catalyzed [6 − 2 + 2] cycloaddition reactions of isatoic anhydrides with alkynes: origin of facile decarboxylation. *Organometallics* 32 (24): 7564–7574.

43 Kakiuchi, F. and Chatani, N. (2003). Catalytic methods for C—H bond functionalization: application in organic synthesis. *Advanced Synthesis & Catalysis* 345 (910): 1077–1101.

44 Kakiuchi, F. and Kochi, T. (2008). Transition-metal-catalyzed carbon–carbon bond formation via carbon–hydrogen bond cleavage. *Synthesis* 2008 (19): 3013–3039.

45 Kakiuchi, F. and Murai, S. (2002). Catalytic C–H/olefin coupling. *Accounts of Chemical Research* 35 (10): 826–834.

46 Boutadla, Y., Al-Duaij, O., Davies, D.L. et al. (2009). Mechanistic study of acetate-assisted C–H activation of 2-substituted pyridines with [MCl2Cp*]2(M = Rh, Ir) and [RuCl2(p-cymene)]2. *Organometallics* 28 (2): 433–440.

47 Djukic, J.-P., Sortais, J.-B., Barloy, L. et al. (2009). Cycloruthenated compounds – synthesis and applications. *European Journal of Inorganic Chemistry* 2009 (7): 817–853.

48 Dupont, J., Consorti, C.S., and Spencer, J. (2005). The potential of palladacycles: more than just precatalysts. *Chemical Reviews* 105 (6): 2527–2571.

49 Gorelsky, S.I., Lapointe, D., and Fagnou, K. (2008). Analysis of the concerted metalation-deprotonation mechanism in palladium-catalyzed direct arylation across a broad range of aromatic substrates. *Journal of the American Chemical Society* 130 (33): 10848–10849.

50 Gray, A., Tsybizova, A., and Roithova, J. (2015). Carboxylate-assisted C–H activation of phenylpyridines with copper, palladium and ruthenium: a mass spectrometry and DFT study. *Chemical Science* 6 (10): 5544–5553.

51 Al Mamari, H.H., Diers, E., and Ackermann, L. (2014). Triazole-assisted ruthenium-catalyzed C–H arylation of aromatic amides. *Chemistry – A European Journal* 20 (31): 9739–9743.

52 Hiroshima, S., Matsumura, D., Kochi, T. et al. (2010). Control of product selectivity by a styrene additive in ruthenium-catalyzed C–H arylation. *Organic Letters* 12 (22): 5318–5321.

53 Oi, S., Aizawa, E., Ogino, Y. et al. (2005). Ortho-selective direct cross-coupling reaction of 2-aryloxazolines and 2-arylimidazolines with aryl and alkenyl halides catalyzed by ruthenium complexes. *The Journal of Organic Chemistry* 70 (8): 3113–3119.

54 Zell, D., Warratz, S., Gelman, D. et al. (2016). Single-component phosphinous acid Ruthenium(II) catalysts for versatile C–H activation by metal-ligand cooperation. *Chemistry – A European Journal* 22 (4): 1248–1252.

55 Oxgaard, J., Tenn, W.J., Nielsen, R.J. et al. (2007). Mechanistic analysis of iridium heteroatom C–H activation: evidence for an internal electrophilic substitution mechanism. *Organometallics* 26 (7): 1565–1567.

56 Bu, Q., Rogge, T., Kotek, V. et al. (2018). Distal weak coordination of acetamides in Ruthenium(II)-catalyzed C–H activation processes. *Angewandte Chemie International Edition* 57 (3): 765–768.

57 Siegbahn, P.E. and Blomberg, M.R. (2009). A combined picture from theory and experiments on water oxidation, oxygen reduction and proton pumping. *Dalton Transactions* (30): 5832–5840.

58 Bettadapur, K.R., Lanke, V., and Prabhu, K.R. (2015). Ru (II)-catalyzed C–H activation: ketone-directed novel 1,4-addition of ortho C—H bond to maleimides. *Organic Letters* 17 (19): 4658–4661.

59 Wang, Z., Zhou, Y., Lam, W.H. et al. (2017). DFT studies of Ru-catalyzed C—O versus C—H bond functionalization of aryl ethers with organoboronates. *Organometallics* 36 (12): 2354–2363.

60 Lee, S.H., Gorelsky, S.I., and Nikonov, G.I. (2013). Catalytic H/D exchange of unactivated aliphatic C—H bonds. *Organometallics* 32 (21): 6599–6604.

61 Helmstedt, U. and Clot, E. (2012). Hydride ligands make the difference: density functional study of the mechanism of the Murai reaction catalyzed by

[Ru(H)2(H2)2(PR3)2] (R=cyclohexyl). *Chemistry – A European Journal* 18 (36): 11449–11458.

62 Yamaguchi, J., Yamaguchi, A.D., and Itami, K. (2012). C—H bond functionalization: emerging synthetic tools for natural products and pharmaceuticals. *Angewandte Chemie International Edition* 51 (36): 8960–9009.

63 Cheung, M.S., Sheong, F.K., Marder, T.B. et al. (2015). Computational insight into nickel-catalyzed carbon–carbon versus carbon–boron coupling reactions of primary, secondary, and tertiary alkyl bromides. *Chemistry – A European Journal* 21 (20): 7480–7488.

64 da Silva, M.J. and Gusevskaya, E.V. (2001). Palladium-catalyzed oxidation of monoterpenes: novel tandem oxidative coupling–oxidation of camphene by dioxygen. *Journal of Molecular Catalysis A: Chemical* 176 (1–2): 23–27.

65 He, C., Guo, S., Ke, J. et al. (2012). Silver-mediated oxidative C–H/C–H functionalization: a strategy to construct polysubstituted furans. *Journal of the American Chemical Society* 134 (13): 5766–5769.

66 Itahara, T., Hashimoto, M., and Yumisashi, H. (1984). Oxidative dimerization of thiophenes and furans bearing electron-withdrawing substituents by palladium acetate. *Synthesis* 1984 (03): 255–256.

67 Ke, J., He, C., Liu, H. et al. (2013). Oxidative cross-coupling/cyclization to build polysubstituted pyrroles from terminal alkynes and beta-enamino esters. *Chemical Communications* 49 (68): 7549–7551.

68 Liu, C., Yuan, J., Gao, M. et al. (2015). Oxidative coupling between two hydrocarbons: an update of recent C–H functionalizations. *Chemical Reviews* 115 (22): 12138–12204.

69 Liu, C., Zhang, H., Shi, W. et al. (2011). Bond formations between two nucleophiles: transition metal catalyzed oxidative cross-coupling reactions. *Chemical Reviews* 111 (3): 1780–1824.

70 Liu, D., Liu, C., Li, H. et al. (2014). Copper-catalysed oxidative C–H/C–H coupling between olefins and simple ethers. *Chemical Communications* 50 (27): 3623–3626.

71 Shi, R., Lu, L., Zhang, H. et al. (2013). Palladium/copper-catalyzed oxidative C–H alkenylation/N-dealkylative carbonylation of tertiary anilines. *Angewandte Chemie International Edition* 52 (40): 10582–10585.

72 Shi, W., Liu, C., and Lei, A. (2011). Transition-metal catalyzed oxidative cross-coupling reactions to form C—C bonds involving organometallic reagents as nucleophiles. *Chemical Society Reviews* 40 (5): 2761–2776.

73 Zhang, L., Zhu, L., Zhang, Y. et al. (2018). Experimental and theoretical studies on Ru(II)-catalyzed oxidative C–H/C–H coupling of phenols with aromatic amides using air as oxidant: scope, synthetic applications, and mechanistic insights. *ACS Catalysis* 8 (9): 8324–8335.

74 Aihara, Y. and Chatani, N. (2013). Ruthenium-catalyzed direct arylation of C—H bonds in aromatic amides containing a bidentate directing group: significant electronic effects on arylation. *Chemical Science* 4 (2): 664–670.

75 Biafora, A., Krause, T., Hackenberger, D. et al. (2016). ortho-C–H arylation of benzoic acids with aryl bromides and chlorides catalyzed by ruthenium. *Angewandte Chemie International Edition* 55 (47): 14752–14755.

76 Simonetti, M., Cannas, D.M., Panigrahi, A. et al. (2017). Ruthenium-catalyzed C–H arylation of benzoic acids and indole carboxylic acids with aryl halides. *Chemistry – A European Journal* 23 (3): 549–553.

77 Simonetti, M., Perry, G.J., Cambeiro, X.C. et al. (2016). Ru-catalyzed C–H arylation of fluoroarenes with aryl halides. *Journal of the American Chemical Society* 138 (10): 3596–3606.

78 Gao, B., Liu, S., Lan, Y. et al. (2016). Rhodium-catalyzed cyclocarbonylation of ketimines via C—H bond activation. *Organometallics* 35 (10): 1480–1487.

79 Liu, S., Qi, X., Qu, L.-B. et al. (2018). C—H bond cleavage occurring on a Rh(V) intermediate: a theoretical study of Rh-catalyzed arene azidation. *Catalysis Science & Technology* 8 (6): 1645–1651.

80 Luo, Y., Liu, S., Xu, D. et al. (2018). Counterion effect and directing group effect in Rh-mediated C—H bond activation processes: a theoretical study. *Journal of Organometallic Chemistry* 864: 148–153.

81 Qi, X., Li, Y., Bai, R. et al. (2017). Mechanism of rhodium-catalyzed C–H functionalization: advances in theoretical investigation. *Accounts of Chemical Research* 50 (11): 2799–2808.

82 Tan, G., Zhu, L., Liao, X. et al. (2017). Rhodium/copper cocatalyzed highly trans-selective 1,2-diheteroarylation of alkynes with azoles via C–H addition/oxidative cross-coupling: a combined experimental and theoretical study. *Journal of the American Chemical Society* 139 (44): 15724–15737.

83 Zhang, T., Qi, X., Liu, S. et al. (2017). Computational investigation of the role played by Rhodium(V) in the Rhodium(III)-catalyzed ortho-bromination of arenes. *Chemistry* 23 (11): 2690–2699.

84 Tang, S., Wu, X., Liao, W. et al. (2014). Synergistic Pd/enamine catalysis: a strategy for the C–H/C–H oxidative coupling of allylarenes with unactivated ketones. *Organic Letters* 16 (13): 3584–3587.

85 Wang, J., Liu, C., Yuan, J. et al. (2013). Copper-catalyzed oxidative coupling of alkenes with aldehydes: direct access to alpha,beta-unsaturated ketones. *Angewandte Chemie International Edition* 52 (8): 2256–2259.

86 Zhang, H., Shi, R., Gan, P. et al. (2012). Palladium-catalyzed oxidative double C–H functionalization/carbonylation for the synthesis of xanthones. *Angewandte Chemie International Edition* 51 (21): 5204–5207.

87 Alberico, D., Scott, M.E., and Lautens, M. (2007). Aryl–aryl bond formation by transition-metal-catalyzed direct arylation. *Chemical Reviews* 107 (1): 174–238.

88 Ogiwara, Y., Miyake, M., Kochi, T. et al. (2016). Syntheses of RuHCl(CO)(PAr3)3 and RuH2(CO)(PAr3)3 containing various triarylphosphines and their use for arylation of sterically congested aromatic C—H bonds. *Organometallics* 36 (1): 159–164.

89 Paymode, D.J. and Ramana, C.V. (2015). Ruthenium(II)-catalyzed C3 arylation of 2-aroylbenzofurans with arylboronic acids/aryltrifluoroborates via carbonyl-directed C—H bond activation. *The Journal of Organic Chemistry* 80 (22): 11551–11558.

90 Nareddy, P., Jordan, F., and Szostak, M. (2017). Highly chemoselective ruthenium(II)-catalyzed direct arylation of cyclic and N,N-dialkyl benzamides with aryl silanes. *Chemical Science* 8 (4): 3204–3210.

91 Nareddy, P., Jordan, F., and Szostak, M. (2017). Ruthenium(II)-catalyzed ortho-C–H arylation of diverse N-heterocycles with aryl silanes by exploiting solvent-controlled N-coordination. *Organic & Biomolecular Chemistry* 15 (22): 4783–4788.

92 Nareddy, P., Jordan, F., and Szostak, M. (2018). Ruthenium(II)-catalyzed direct C–H arylation of indoles with arylsilanes in water. *Organic Letters* 20 (2): 341–344.

93 Kakiuchi, F., Matsuura, Y., Kan, S. et al. (2005). A RuH(2)(CO)(PPh(3))(3)-catalyzed regioselective arylation of aromatic ketones with arylboronates via carbon–hydrogen bond cleavage. *Journal of the American Chemical Society* 127 (16): 5936–5945.

94 Chatani, N., Asaumi, T., Yorimitsu, S. et al. (2001). Ru(3)(CO)(12)-catalyzed coupling reaction of sp(3) C—H bonds adjacent to a nitrogen atom in alkylamines with alkenes. *Journal of the American Chemical Society* 123 (44): 10935–10941.

95 Kulago, A.A., Van Steijvoort, B.F., Mitchell, E.A. et al. (2014). Directed ruthenium-catalyzed C(sp3)–H α-alkylation of cyclic amines using dioxolane-protected alkenones. *Advanced Synthesis & Catalysis* 356 (7): 1610–1618.

96 Rouquet, G. and Chatani, N. (2013). Ruthenium-catalyzed ortho-C—H bond alkylation of aromatic amides with α,β-unsaturated ketones via bidentate-chelation assistance. *Chemical Science* 4 (5): 2201.

97 Schinkel, M., Marek, I., and Ackermann, L. (2013). Carboxylate-assisted ruthenium(II)-catalyzed hydroarylations of unactivated alkenes through C–H cleavage. *Angewandte Chemie International Edition* 52 (14): 3977–3980.

98 Schinkel, M., Wang, L., Bielefeld, K. et al. (2014). Ruthenium(II)-catalyzed C(sp3)-H alpha-alkylation of pyrrolidines. *Organic Letters* 16 (7): 1876–1879.

99 Ueno, S., Chatani, N., and Kakiuchi, F. (2007). Regioselective alkenylation of aromatic ketones with alkenylboronates using a RuH2(CO)(PPh3)3 catalyst via carbon–hydrogen bond cleavage. *The Journal of Organic Chemistry* 72 (9): 3600–3602.

100 Leitch, J.A., Wilson, P.B., McMullin, C.L. et al. (2016). Ruthenium(II)-catalyzed C–H functionalization using the oxazolidinone heterocycle as a weakly coordinating directing group: experimental and computational insights. *ACS Catalysis* 6 (8): 5520–5529.

101 Hashimoto, Y., Ueyama, T., Fukutani, T. et al. (2011). Ruthenium-catalyzed oxidative alkenylation of arenes via regioselective C—H bond cleavage directed by a nitrogen-containing group. *Chemistry Letters* 40 (10): 1165–1166.

102 Kozhushkov, S.I. and Ackermann, L. (2013). Ruthenium-catalyzed direct oxidative alkenylation of arenes through twofold C—H bond functionalization. *Chemical Science* 4 (3): 886–896.

103 Li, B., Ma, J., Wang, N. et al. (2012). Ruthenium-catalyzed oxidative C—H bond olefination of *N*-methoxybenzamides using an oxidizing directing group. *Organic Letters* 14 (3): 736–739.

104 Manikandan, R., Madasamy, P., and Jeganmohan, M. (2015). Ruthenium-catalyzed ortho alkenylation of aromatics with alkenes at room temperature with hydrogen evolution. *ACS Catalysis* 6 (1): 230–234.

105 Matsuura, Y., Tamura, M., Kochi, T. et al. (2007). The Ru(cod)(cot)-catalyzed alkenylation of aromatic C—H bonds with alkenyl acetates. *Journal of the American Chemical Society* 129 (32): 9858–9859.

106 Ping, Y., Chen, Z., Ding, Q. et al. (2017). Ru-catalyzed ortho-oxidative alkenylation of 2-arylbenzo[d]thiazoles in aqueous solution of anionic surfactant sodium dodecylbenzenesulfonate (SDBS). *Tetrahedron* 73 (5): 594–603.

107 Simon, M.O. and Li, C.J. (2012). Green chemistry oriented organic synthesis in water. *Chemical Society Reviews* 41 (4): 1415–1427.

108 Johnson, D.G., Lynam, J.M., Mistry, N.S. et al. (2013). Ruthenium-mediated C–H functionalization of pyridine: the role of vinylidene and pyridylidene ligands. *Journal of the American Chemical Society* 135 (6): 2222–2234.

109 Bauer, I. and Knolker, H.J. (2015). Iron catalysis in organic synthesis. *Chemical Reviews* 115 (9): 3170–3387.

110 Sun, J. and Deng, L. (2015). Cobalt complex-catalyzed hydrosilylation of alkenes and alkynes. *ACS Catalysis* 6 (1): 290–300.

111 Marciniec, B. (2005). Catalysis by transition metal complexes of alkene silylation–recent progress and mechanistic implications. *Coordination Chemistry Reviews* 249 (21–22): 2374–2390.

112 Rohmann, K., Holscher, M., and Leitner, W. (2016). Can contemporary density functional theory predict energy spans in molecular catalysis accurately enough to be applicable for in silico catalyst design? A computational/experimental case study for the ruthenium-catalyzed hydrogenation of olefins. *Journal of the American Chemical Society* 138 (1): 433–443.

113 Zhang, X., Guo, X., Chen, Y. et al. (2012). Mechanism investigation of ketone hydrogenation catalyzed by ruthenium bifunctional catalysts: insights from a DFT study. *Physical Chemistry Chemical Physics* 14 (17): 6003–6012.

114 Wang, H., Liu, C., and Zhang, D. (2017). Decisive effects of solvent and substituent on the reactivity of Ru-catalyzed hydrogenation of ethyl benzoate to benzyl alcohol and ethanol: a DFT study. *Molecular Catalysis* 440: 120–132.

115 Zhang, J., Leitus, G., Ben-David, Y. et al. (2006). Efficient homogeneous catalytic hydrogenation of esters to alcohols. *Angewandte Chemie International Edition* 45 (7): 1113–1115.

116 Eisenstein, O., Milani, J., and Perutz, R.N. (2017). Selectivity of C–H activation and competition between C—H and C—F bond activation at fluorocarbons. *Chemical Reviews* 117 (13): 8710–8753.

117 Purser, S., Moore, P.R., Swallow, S. et al. (2008). Fluorine in medicinal chemistry. *Chemical Society Reviews* 37 (2): 320–330.

118 Reade, S.P., Mahon, M.F., and Whittlesey, M.K. (2009). Catalytic hydrodefluorination of aromatic fluorocarbons by ruthenium *N*-heterocyclic carbene complexes. *Journal of the American Chemical Society* 131 (5): 1847–1861.

119 Zhou, Y., Wang, J., Gu, Z. et al. (2016). Next generation of fluorine-containing pharmaceuticals, compounds currently in phase II–III clinical trials of major pharmaceutical companies: new structural trends and therapeutic areas. *Chemical Reviews* 116 (2): 422–518.

120 McKay, D., Riddlestone, I.M., Macgregor, S.A. et al. (2015). Mechanistic study of Ru-NHC-catalyzed hydrodefluorination of fluoropyridines: the influence of the NHC on the regioselectivity of C–F activation and chemoselectivity of C—F versus C—H bond cleavage. *ACS Catalysis* 5 (2): 776–787.

121 Geri, J.B. and Szymczak, N.K. (2015). A proton-switchable bifunctional ruthenium complex that catalyzes nitrile hydroboration. *Journal of the American Chemical Society* 137 (40): 12808–12814.

122 Ovchinnikov, M.V., LeBlanc, E., Guzei, I.A. et al. (2001). Hydrofunctionalization of alkenes promoted by diruthenium complexes [[(eta(5)-C(5)H(3))(2)(SiMe(2))(2)]Ru(2)(CO)(3)(eta(2)-CH(2)=CH-R)(mu-H)](+) featuring a kinetically inert proton on a metal–metal bond. *Journal of the American Chemical Society* 123 (46): 11494–11495.

123 Shi, J., Hu, B., Ren, P. et al. (2018). Synthesis and reactivity of metal–ligand cooperative bifunctional ruthenium hydride complexes: active catalysts for β-alkylation of secondary alcohols with primary alcohols. *Organometallics* 37 (16): 2795–2806.

124 Zhang, C., Zhao, J.-P., Hu, B. et al. (2019). Ruthenium-catalyzed β-alkylation of secondary alcohols and α-alkylation of ketones via borrowing hydrogen: dramatic influence of the pendant *N*-heterocycle. *Organometallics* 38 (3): 654–664.

125 Zhang, X., Ji, X., Xie, X. et al. (2018). Construction of highly sterically hindered 1,1-disilylated terminal alkenes. *Chemical Communications* 54 (92): 12958–12961.

126 Meng, Q., Su, P., Wang, F. et al. (2016). Substituent effect and ligand exchange control the reactivity in ruthenium(II)-catalyzed hydroacylation of isoprenes and aldehydes ‖ A DFT study. *Journal of Theoretical and Computational Chemistry* 15 (03): 1650019.

127 Omura, S., Fukuyama, T., Horiguchi, J. et al. (2008). Ruthenium hydride-catalyzed addition of aldehydes to dienes leading to beta,gamma-unsaturated ketones. *Journal of the American Chemical Society* 130 (43): 14094–14095.

128 Wang, F. and Meng, Q. (2017). Mechanism for ruthenium hydride-catalyzed regioselective hydroacylation of enones and aldehydes to give 1,3-diketones: Insights from density functional calculations. *Molecular Catalysis* 433: 55–61.

129 Yi, C.S. and Gao, R. (2009). Scope and mechanistic Investigations on the solvent-controlled regio- and stereoselective formation of enol esters from the ruthenium-catalyzed coupling reaction of terminal alkynes and carboxylic acids. *Organometallics* 28 (22): 6585–6592.

130 Maity, B. and Koley, D. (2018). Solvent-promoted regio- and stereoselectivity in Ru-catalyzed hydrocarboxylation of terminal alkynes: a DFT study. *ChemCatChem* 10 (3): 566–580.

131 Song, L.-J., Wang, T., Zhang, X. et al. (2017). A combined DFT/IM-MS study on the reaction mechanism of cationic Ru(II)-catalyzed hydroboration of alkynes. *ACS Catalysis* 7 (2): 1361–1368.

132 Sundararaju, B. and Furstner, A. (2013). A trans-selective hydroboration of internal alkynes. *Angewandte Chemie International Edition* 52 (52): 14050–14054.

133 Chandra, P., Ghosh, T., Choudhary, N. et al. (2020). Recent advancement in oxidation or acceptorless dehydrogenation of alcohols to valorised products using manganese based catalysts. *Coordination Chemistry Reviews* 411: 213241.

134 Dutta, I., Sarbajna, A., Pandey, P. et al. (2016). Acceptorless dehydrogenation of alcohols on a diruthenium(II,II) platform. *Organometallics* 35 (10): 1505–1513.

135 Nielsen, M., Alberico, E., Baumann, W. et al. (2013). Low-temperature aqueous-phase methanol dehydrogenation to hydrogen and carbon dioxide. *Nature* 495 (7439): 85–89.

136 Samuelsen, S.V., Santilli, C., Ahlquist, M.S.G. et al. (2019). Development and mechanistic investigation of the manganese(iii) salen-catalyzed dehydrogenation of alcohols. *Chemical Science* 10 (4): 1150–1157.

137 Zhong, J.J., To, W.P., Liu, Y. et al. (2019). Efficient acceptorless photo-dehydrogenation of alcohols and *N*-heterocycles with binuclear platinum(ii) diphosphite complexes. *Chemical Science* 10 (18): 4883–4889.

138 Zhu, R., Wang, B., Cui, M. et al. (2016). Chemoselective oxidant-free dehydrogenation of alcohols in lignin using Cp*Ir catalysts. *Green Chemistry* 18 (7): 2029–2036.

139 Alberico, E., Lennox, A.J., Vogt, L.K. et al. (2016). Unravelling the mechanism of basic aqueous methanol dehydrogenation catalyzed by Ru-PNP pincer complexes. *Journal of the American Chemical Society* 138 (45): 14890–14904.

140 Yang, X. (2014). Mechanistic insights into ruthenium-catalyzed production of H_2 and CO_2 from methanol and water: a DFT study. *ACS Catalysis* 4 (4): 1129–1133.

141 Liu, N., Guo, L., Cao, Z. et al. (2016). Mechanisms of the water-gas shift reaction catalyzed by ruthenium carbonyl complexes. *The Journal of Physical Chemistry A* 120 (15): 2408–2419.

142 Mazzone, G., Alberto, M.E., and Sicilia, E. (2014). Theoretical mechanistic study of the formic acid decomposition assisted by a Ru(II)-phosphine catalyst. *Journal of Molecular Modeling* 20 (5): 2250.

143 Zhou, J. (2016). Theoretical investigation on the ruthenium catalyzed dehydrogenation of formic acid and ligand effect. *Applied Catalysis A: General* 515: 101–107.

144 Alshakova, I.D., Gabidullin, B., and Nikonov, G.I. (2018). Ru-catalyzed transfer hydrogenation of nitriles, aromatics, olefins, alkynes and esters. *ChemCatChem* 10 (21): 4860–4869.

145 Fukuyama, T., Higashibeppu, Y., Yamaura, R. et al. (2007). Ru-catalyzed intermolecular [3+2+1] cycloaddition of alpha,beta-unsaturated ketones with silylacetylenes and carbon monoxide leading to alpha-pyrones. *Organic Letters* 9 (4): 587–589.

146 Ho, G.M., Judkele, L., Bruffaerts, J. et al. (2018). Metal-catalyzed remote functionalization of omega-Ene unsaturated ethers: towards functionalized vinyl species. *Angewandte Chemie International Edition* 57 (27): 8012–8016.

147 Kang, S.K., Kim, K.J., Yu, C.M. et al. (2001). Ru-catalyzed cyclocarbonylation of alpha- and beta-allenic sulfonamides: synthesis of gamma- and delta-unsaturated lactams. *Organic Letters* 3 (18): 2851–2853.

148 Ma, X., Li, W., Li, X. et al. (2012). Ru-catalyzed highly chemo- and enantioselective hydrogenation of gamma-halo-gamma,delta-unsaturated-beta-keto esters under neutral conditions. *Chemical Communications* 48 (43): 5352–5354.

149 Ritleng, V., Sirlin, C., and Pfeffer, M. (2002). Ru-, Rh-, and Pd-catalyzed C—C bond formation involving C–H activation and addition on unsaturated substrates: reactions and mechanistic aspects. *Chemical Reviews* 102 (5): 1731–1770.

150 Tsujita, H., Ura, Y., Wada, K. et al. (2005). Synthesis of 2-alkylidenetetrahydrofurans by Ru-catalyzed regio- and stereoselective codimerization of dihydrofurans with alpha,beta-unsaturated esters. *Chemical Communications* 40: 5100–5102.

151 Kirchner, K., Calhorda, M.J., Schmid, R. et al. (2003). Mechanism for the cyclotrimerization of alkynes and related reactions catalyzed by CpRuCl. *Journal of the American Chemical Society* 125 (38): 11721–11729.

152 Varela, J.A. and Saá, C. (2009). CpRuCl- and CpCo-catalyzed or mediated cyclotrimerizations of alkynes and [2+2+2] cycloadditions of alkynes to alkenes: a comparative DFT study. *Journal of Organometallic Chemistry* 694 (2): 143–149.

153 Yamamoto, Y., Ogawa, R., and Itoh, K. (2000). Highly chemo- and regio-selective [2+2+2] cycloaddition of unsymmetrical 1,6-diynes with terminal alkynes catalyzed by Cp*Ru(cod)Cl under mild conditions. *Chemical Communications* (7): 549–550.

154 Blanco-Urgoiti, J., Anorbe, L., Perez-Serrano, L. et al. (2004). The Pauson–Khand reaction, a powerful synthetic tool for the synthesis of complex molecules. *Chemical Society Reviews* 33 (1): 32–42.

155 Brummond, K.M. and Kent, J.L. (2000). Recent advances in the Pauson–Khand reaction and related [2+2+1] cycloadditions. *Tetrahedron* 56 (21): 3263–3283.

156 Ingate, S.T. and Marco-Contellers, J. (1998). The asymmetric Pauson–Khand reaction. A review. *Organic Preparations and Procedures International* 30 (2): 121–143.

References

157 Jeong, N., Hwang, S.H., Lee, Y. et al. (1994). Catalytic version of the Intramolecular Pauson–Khand reaction. *Journal of the American Chemical Society* 116 (7): 3159–3160.

158 Keun Chung, Y. (1999). Transition metal alkyne complexes: the Pauson–Khand reaction. *Coordination Chemistry Reviews* 188 (1): 297–341.

159 Kondo, T., Suzuki, N., Okada, T. et al. (1997). First ruthenium-catalyzed intramolecular Pauson–Khand reaction. *Journal of the American Chemical Society* 119 (26): 6187–6188.

160 Wang, C. and Wu, Y.-D. (2008). Theoretical studies on Ru-catalyzed Pauson–Khand-Type [2+2+1] and related [2+2+1+1] cycloadditions. *Organometallics* 27 (23): 6152–6162.

161 Boren, B.C., Narayan, S., Rasmussen, L.K. et al. (2008). Ruthenium-catalyzed azide-alkyne cycloaddition: scope and mechanism. *Journal of the American Chemical Society* 130 (28): 8923–8930.

162 Boz, E. and Tüzün, N.Ş. (2013). Reaction mechanism of ruthenium-catalyzed azide–alkyne cycloaddition reaction: a DFT study. *Journal of Organometallic Chemistry* 724: 167–176.

163 Lamberti, M., Fortman, G.C., Poater, A. et al. (2012). Coordinatively unsaturated ruthenium complexes As efficient alkyne–azide cycloaddition catalysts. *Organometallics* 31 (2): 756–767.

164 Liu, P.N., Li, J., Su, F.H. et al. (2012). Selective formation of 1,4-disubstituted triazoles from ruthenium-catalyzed cycloaddition of terminal alkynes and organic azides: scope and reaction mechanism. *Organometallics* 31 (13): 4904–4915.

165 Wang, C., Ikhlef, D., Kahlal, S. et al. (2016). Metal-catalyzed azide-alkyne "click" reactions: mechanistic overview and recent trends. *Coordination Chemistry Reviews* 316: 1–20.

166 Wang, T.-H., Wu, F.-L., Chiang, G.-R. et al. (2014). Preparation of ruthenium azido complex containing a Tp ligand and ruthenium-catalyzed cycloaddition of organic azides with alkynes in organic and aqueous media: Experimental and computational studies. *Journal of Organometallic Chemistry* 774: 57–60.

167 Zhang, L., Chen, X., Xue, P. et al. (2005). Ruthenium-catalyzed cycloaddition of alkynes and organic azides. *Journal of the American Chemical Society* 127 (46): 15998–15999.

168 Novak, B.M. and Grubbs, R.H. (1988). Catalytic organometallic chemistry in water: the aqueous ring-opening metathesis polymerization of 7-oxanorbornene derivatives. *Journal of the American Chemical Society* 110 (22): 7542–7543.

169 Ogba, O.M., Warner, N.C., O'Leary, D.J. et al. (2018). Recent advances in ruthenium-based olefin metathesis. *Chemical Society Reviews* 47 (12): 4510–4544.

170 Bernardi, F., Bottoni, A., and Miscione, G.P. (2003). DFT study of the olefin metathesis catalyzed by ruthenium complexes. *Organometallics* 22 (5): 940–947.

171 Fomine, S., Vargas, S.M., and Tlenkopatchev, M.A. (2003). Molecular modeling of ruthenium alkylidene mediated olefin metathesis reactions. DFT study of reaction pathways. *Organometallics* 22 (1): 93–99.

172 Minenkov, Y., Occhipinti, G., and Jensen, V.R. (2013). Complete reaction pathway of ruthenium-catalyzed olefin metathesis of ethyl vinyl ether: kinetics and mechanistic insight from DFT. *Organometallics* 32 (7): 2099–2111.

173 Nelson, J.W., Grundy, L.M., Dang, Y. et al. (2014). Mechanism of Z-selective olefin metathesis catalyzed by a ruthenium monothiolate carbene complex: a DFT study. *Organometallics* 33 (16): 4290–4294.

174 Zhu, B.-L., Pang, X.-Y., and Wang, G.-C. (2016). Reaction mechanism and Z-selectivity for chelated Ru-catalyzed AROCM of endic anhydride and propene: A DFT study. *International Journal of Quantum Chemistry* 116 (1): 35–41.

175 Marx, F.T., Jordaan, J.H., Lachmann, G. et al. (2014). A comparison of low and high activity precatalysts: do the calculated energy barriers during the self-metathesis reaction of 1-octene correlate with the precatalyst metathesis activity? *Journal of Computational Chemistry* 35 (19): 1464–1471.

176 Costabile, C., Mariconda, A., Cavallo, L. et al. (2011). The pivotal role of symmetry in the ruthenium-catalyzed ring-closing metathesis of olefins. *Chemistry* 17 (31): 8618–8629.

177 Solans-Monfort, X., Pleixats, R., and Sodupe, M. (2010). DFT mechanistic study on diene metathesis catalyzed by Ru-based Grubbs–Hoveyda-type carbenes: the key role of pi-electron density delocalization in the Hoveyda ligand. *Chemistry – A European Journal* 16 (24): 7331–7343.

178 Kim, M., Park, S., Maifeld, S.V. et al. (2004). Regio- and stereoselective enyne cross metathesis of silylated internal alkynes. *Journal of the American Chemical Society* 126 (33): 10242–10243.

179 Lippstreu, J.J. and Straub, B.F. (2005). Mechanism of enyne metathesis catalyzed by Grubbs ruthenium-carbene complexes: a DFT study. *Journal of the American Chemical Society* 127 (20): 7444–7457.

12

Theoretical Study of Mn-Catalysis

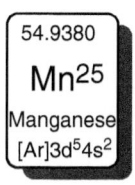

Manganese is the third most abundant transition metal in the earth, which is only lower than iron and titanium. Manganese plays an important role in life as an essential trace element [1–4]. It is found to be the active center of numerous enzymes that can operate a series of redox processes in organisms. Therefore, the manganese chemistry can be considered as starting from bioinspired processes. Usually, manganese complexes play unique activity in homogeneous oxidation and reduction reactions [5–8]. The oxidative state of manganese can be from −1 to +7. Both single-electron- and double-electron-transfer can occur with manganese complexes.

As a typical 3d transition metal, manganese reveals significant metallicity, which can form stable covalent bond with oxygen atom. Therefore, it has been widely used as homogeneous catalyst in inorganic reactions. However, in organometallic transformations, manganese often only plays as an oxidant, which can react with unsaturated bonds or inert C—H bonds. Recently, the generation and transformation of Mn—C bonds were reported, in which manganese can be used as transition metal catalyst for the construction of new C—C bonds [9–16].

Generally, the mechanism of manganese-mediated redox reactions is simple, which only contain a few typical steps. Therefore, theoretical studies of Mn-catalysis are less than the cases with other late transition metals. In this chapter, we focus on the Mn-mediated C—H oxidations and functionalizations. Some other theoretical studies of Mn-catalysis are also given.

Computational Methods in Organometallic Catalysis: From Elementary Reactions to Mechanisms,
First Edition. Yu Lan.
© 2021 WILEY-VCH GmbH. Published 2021 by WILEY-VCH GmbH.

12.1 Mn-Mediated Oxidation of Alkanes

In the presence of Mn complexes, inert C(sp^3)—H bonds can be directly functionalized to achieve halogenations [17, 18], hydroxylations [19, 20], azidations [21, 22], and isocyanations [23]. In these reactions, a common outer-sphere radical-type oxidation mechanism was proposed [24–26]. As shown in Scheme 12.1, the catalytic cycle starts from a high-valent Mn-oxo complex, which can react with C—H bond through a radical substitution process to generate a free alkyl radical and Mn–hydroxide. The counterion exchange introduces nucleophile onto Mn in the generated intermediate. It can further react with alkyl radical to form functionalized product through radical substitution. The oxidative state of Mn is reduced by 2 in those two steps. Then, the high-valent Mn species can be regenerated by the exogenous oxidants. The detailed mechanisms were revealed by the corresponding theoretical calculations.

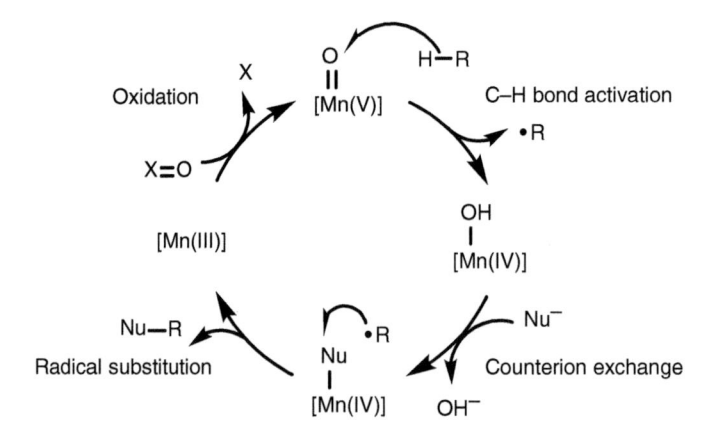

Scheme 12.1 General mechanism for Mn-mediated alkane oxidation.

12.1.1 C—H Hydroxylations

Porphyrin-coordinated Mn(V) species are a common catalyst for the oxidative hydroxylation of C—H bond in alkanes, which often involve a Mn-oxo moiety [27]. DFT calculations found that singlet state of this type Mn(V) species is more stable. However, the reaction would start from a triplet state Mn(V), where the spin density is majorly located on Mn and oxygen atoms (Scheme 12.2) [28].

As shown in Figure 12.1, the radical character of oxygen atom in triplet Mn(V)-oxo complex leads to a radical substitution with alkane via transition state **12-3ts** with an energy barrier of only 2.7 kcal/mol, which undergoes a C—H bond cleavage to afford an alkyl radical and a Mn(IV) hydroxide **12-4**. Then, alkyl radical attaches onto hydroxy group via transition state **12-5ts** to form new C—O bond. After release of corresponding alcohol product, a porphyrin-coordinated Mn(III) species **12-6** is formed.

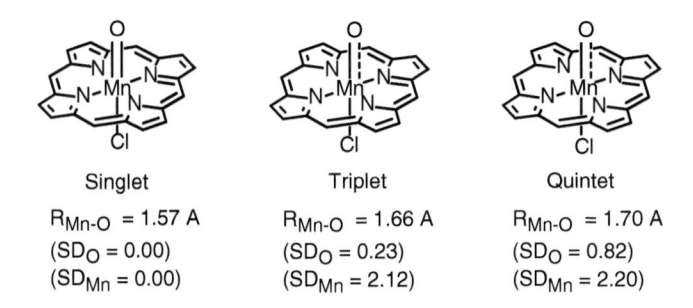

Singlet	Triplet	Quintet
R_{Mn-O} = 1.57 A	R_{Mn-O} = 1.66 A	R_{Mn-O} = 1.70 A
(SD_O = 0.00)	(SD_O = 0.23)	(SD_O = 0.82)
(SD_{Mn} = 0.00)	(SD_{Mn} = 2.12)	(SD_{Mn} = 2.20)

Scheme 12.2 Porphyrin-coordinated Mn(V) species. Corresponding spin densities of oxygen and manganese are given in parentheses. Source: Based on Balcells et al. [28].

12.1.2 C—H Halogenations

By using the same catalyst, the halogenation of C—H bond in alkane can occur in the presence of extra halide [18, 29]. DFT calculations found that, when an alkyl radical is formed, the energy barrier for the following radical attack onto Mn(IV) halide is only 3.1 kcal/mol [30] (Scheme 12.3).

$\Delta G^{\ddagger}$ = 3.1 kcal/mol
12-7ts

Scheme 12.3 Manganese porphyrin–catalyzed C(alkyl)-H fluorination with fluoride ion. The energies were calculated at B3LYP/LACVP** level of theory.

12.1.3 C—H Azidations

Azide can be considered as an alternative of halide, which also can react with Mn(IV) hydroxide to afford a Mn(IV) azide species. DFT calculations found that in a quintet state [22], alkyl radical attacks onto terminal nitrogen atom would occur via transition state **12-8ts**, which provides the lowest activation free energy of only 7.0 kcal/mol (Scheme 12.4).

12.1.4 C—H Isocyanations

When isocyanate is used as nucleophile, it can react with Mn(IV) hydroxide to afford a Mn(IV) isocyanate species. DFT calculations for a model reaction found

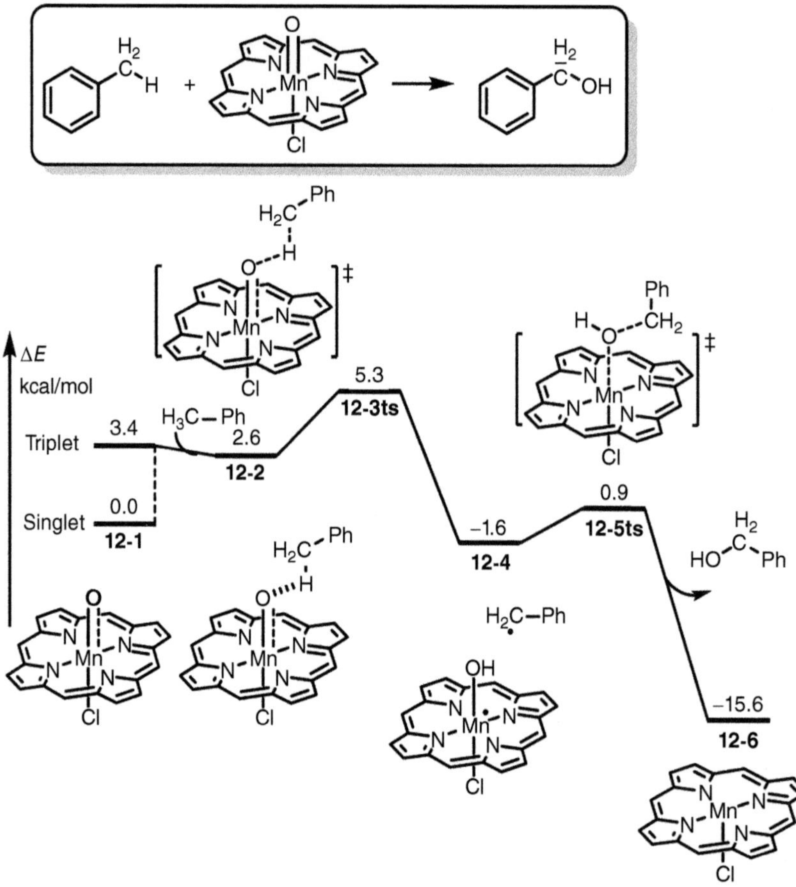

Figure 12.1 Potential energy profiles for the C(alkyl)-H oxidation by using Mn(V)-oxo complex. The energies were calculated at BP86/6-31G(d,p) (SDD for Mn and Cl)//BP86/6-31G (SDD for Mn and Cl) level of theory.

that either nitrogen or oxygen atoms can be attacked by alkyl radical via transition state **12-10ts** or **12-12ts** [31], and the calculated energy barrier for them is close. However, the formation of isocyanatoalkane is much more stable than that of cyanatoalkane. Therefore, isocyanatoalkane is observed experimentally through a thermodynamic process (Figure 12.2).

12.2 Mn-Mediated C—H Activations

Direct C—H activation-functionalization by using second- and third-row transition metal plays an important role in synthetic chemistry. However, the development of corresponding reactions employing Mg has received much less attention despite their abundance in the earth's crust and economic benefits. So far, only sporadic examples of Mn-catalyzed C—H activation reactions have been reported.

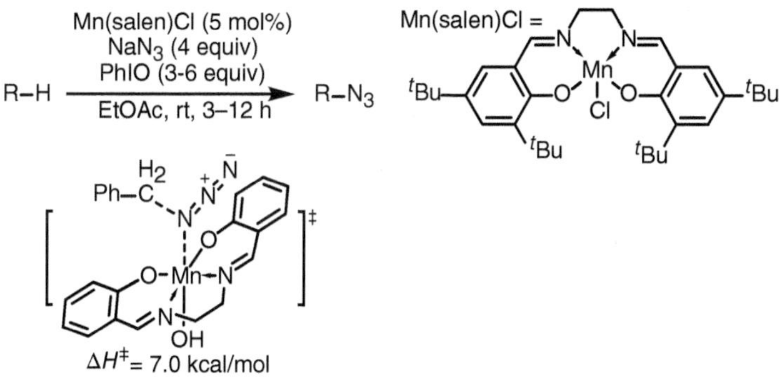

Scheme 12.4 Manganese porphyrin–catalyzed C(alkyl)–H azidation with azide. The energies were calculated at B3LYP/6-311++G(d,p) (SDD for Mn)/CPCM(ethyl acetate)//B3LYP/6-31G(d) (SDD for Mn) level of theory.

Meanwhile, the theoretical studies for this type of reactions are also rare. Generally, Mn often plays metallo character, which can be used as electrophile to activate arene C—H bond through electrophilic deprotonation. In inner-sphere process, C—H bond activation also can undergo σ-complex-assisted metathesis or concerted metalation–deprotonation (Scheme 12.5).

(a) Electrophilic deprotonation

(b) σ-Complex assisted metathesis

(c) Concerted-metalation–deprotonation

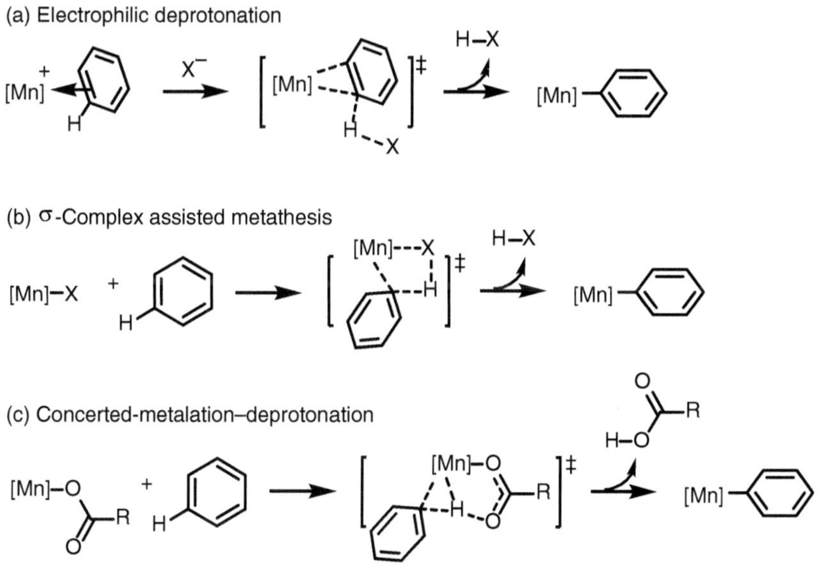

Scheme 12.5 General modes of Mn-mediated C—H activations.

12.2.1 Electrophilic Deprotonation

In 2014, Chen, Wang, and coworkers reported a Mn-catalyzed C—H activation-alkylation, where a 20% amine is necessary. The following DFT calculation revealed the mechanism of this reaction [32]. As shown in Figure 12.3, the catalytic cycle

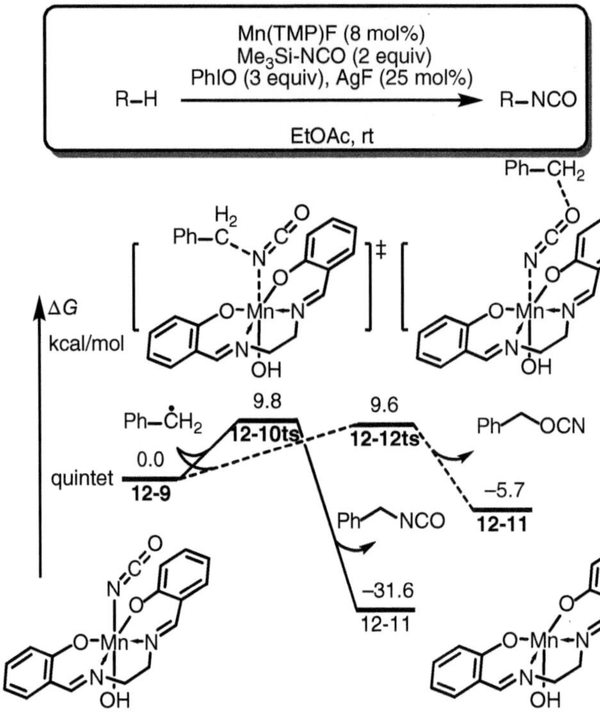

Figure 12.2 Free-energy profiles for the Mn-catalyzed C(alkyl)–H isocyanation. The energies were calculated at B3LYP/6-311++G(d,p) (SDD for Mn)/PCM(ethyl acetate)//B3LYP/6-31G(d) (SDD for Mn) level of theory.

starts from a directing group-coordinated Mn(I) species **12-13**, where reacting arene is activated by electrophilic Mn(I) cation. Therefore, an outer-sphere deprotonation by extra amine takes place via transition state **12-14ts** with an energy barrier of 22.5 kcal/mol to afford a neutral aryl Mn(I) intermediate **12-15**. The coordinated olefin inserts into C(aryl)—Mn bond via transition state **12-16ts** with an energy barrier of 23.4 kcal/mol. The generated enolate Mn(I) intermediate **12-17** can be protonated by ammonium to form a product-coordinated Mn(I) species **12-19**. The ligand exchange with reactant releases alkylated arene product and regenerates active species **12-13**.

12.2.2 σ-Complex-Assisted Metathesis

σ-Complex-assisted metathesis can be an alternative in Mn-mediated C—H bond activation. As shown in Figure 12.4, Chen, Wang, and coworkers reported a Mn-catalyzed vinylation of arene [33]. DFT calculation was employed to reveal the mechanism of this reaction. The calculated catalytic cycle starts from a neutral alkynyl Mn(I) species **12-20**, where metal center is coordinated by the directing group. Then, a σ-complex-assisted metathesis takes place via a four-membered ring-type transition state **12-21ts** with an activation free energy of only 16.9 kcal/mol.

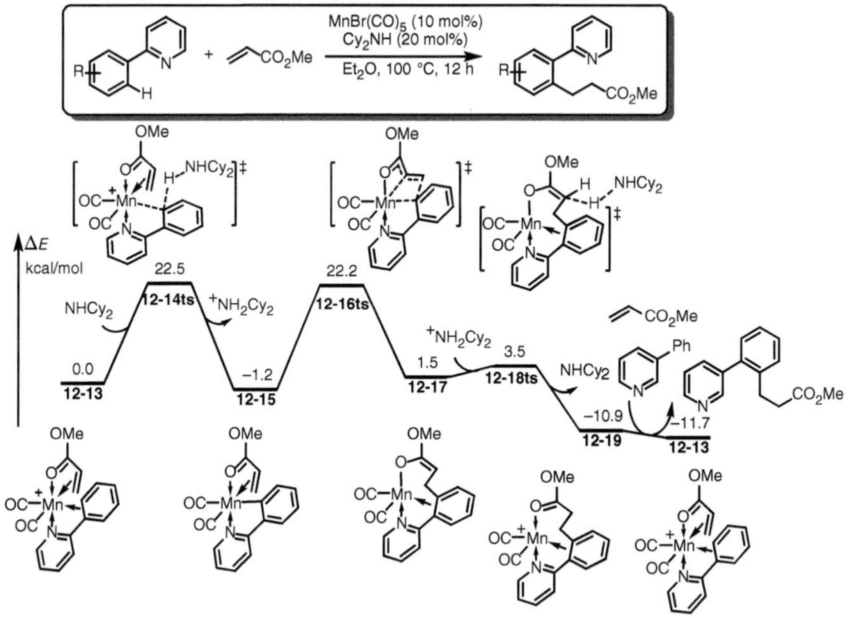

Figure 12.3 Potential energy profiles for the Mn-catalyzed C(aryl)-H alkylation through a key step of electrophilic deprotonation. The energies were calculated at B3LYP/cc-pVTZ/SMD(diethyl ether)//BP86/cc-pVDZ level of theory.

The generated alkyne inserts into C(aryl)—Mn bond via transition state **12-23ts** to form a vinyl Mn(I) intermediate **12-24**. Then, another molecule alkyne coordinates onto Mn in intermediate **12-25**. Another σ-complex-assisted metathesis takes place via a four-membered ring-type transition state **12-26ts**, which yields vinylation product and regenerates alkynyl Mn(I) species **12-20** by the following ligand exchange.

In another case, Shi, Fu, and coworkers reported a theoretical study on the mechanism of Mn-catalyzed dehydrogenative coupling of imines and alkynes [33] (Figure 12.5). The reaction starts from the coordination of imine onto a Mn(I) hydride complex **12-28** to afford complex **12-29**. A σ-bond metathesis of C(aryl)—H and Mn—H bonds occurs via transition state **12-30ts** to form aryl Mn(I) intermediate **12-31** with release of dihydrogen molecule. Then, alkyne inserts into C(aryl)—Mn bond via transition state **12-32ts** to form a vinyl Mn(I) intermediate **12-33**. The calculated overall activation free energy for this step is as high as 37.3 kcal/mol. A [3,3] σ tropic rearrangement via transition state **12-34ts** can afford a six-membered ring in intermediate **12-35**. Then, a β-hydride elimination takes place via transition state **12-36ts,** releases isoquinoline product, and regenerates active species **12-28**.

12.2.3 Concerted Metalation–Deprotonation

In the presence of acetate, a concerted metalation–deprotonation type C—H activation was proposed in Mn-catalyzed vinylation of arenes [34]. In this reaction, imine

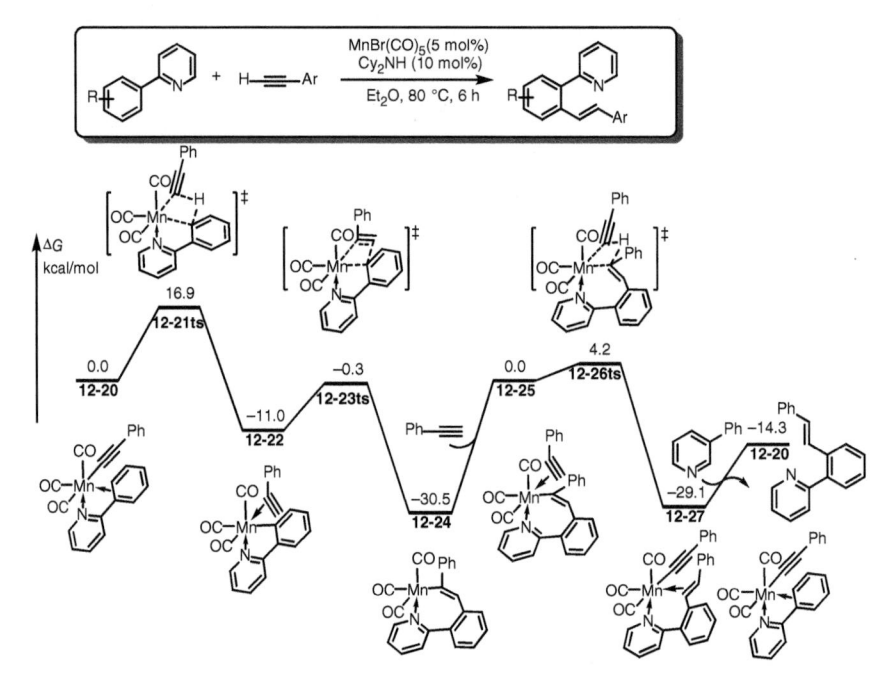

Figure 12.4 Free-energy profiles for the Mn-catalyzed C(aryl)-H alkylation through a key step of σ-complex-assisted metathesis. The energies were calculated at B3LYP/cc-pVTZ/SMD(diethyl ether)//BP86/cc-pVDZ level of theory.

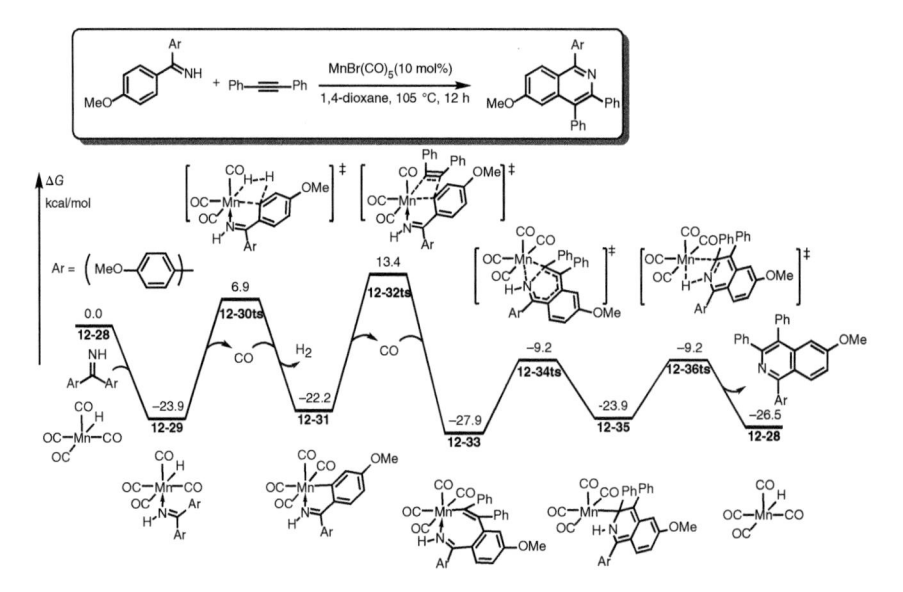

Figure 12.5 Free-energy profiles for the Mn-catalyzed dehydrogenative annulation of aryl imines and alkynes through a key step of σ-complex-assisted metathesis. The energies were calculated at M06-L/6-311+G(d,p) (SDD for Mn)/SMD(dioxane)//M06-L/6-31G(d) (LANL2DZ for Mn) level of theory.

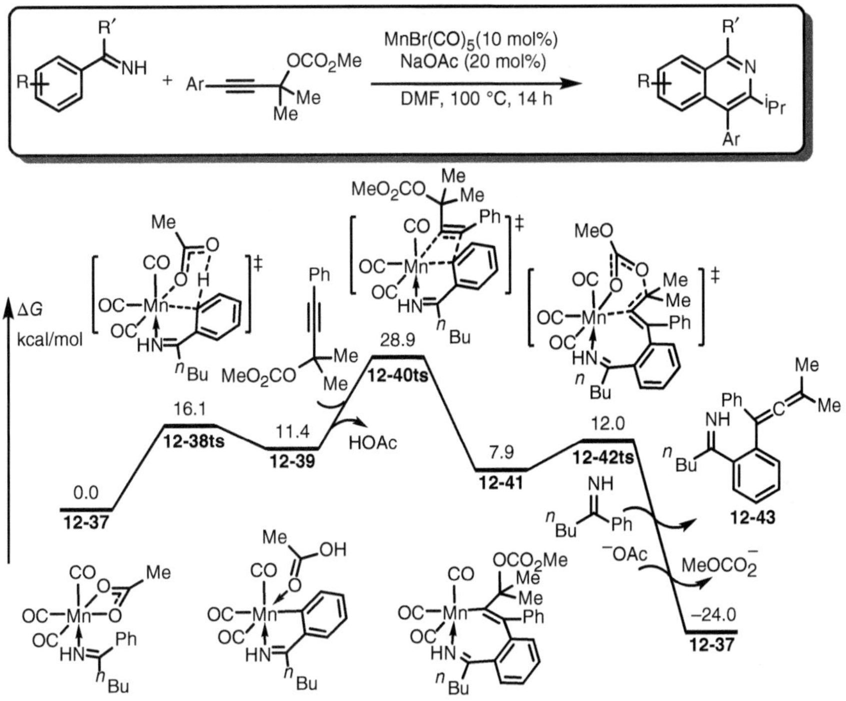

Figure 12.6 Free-energy profiles for the Mn-catalyzed annulation of aryl imines and propargyl esters through a key step of CMD-type C−H activation. The energies were calculated at M11-L/6-311 + G(d,p) (LANL2DZ for Mn)/SMD(DMF)//B3LYP/6-31G(d) (LANL2DZ for Mn) level of theory.

plays as directing group, which can coordinate onto Mn(I) acetate to form intermediate **12-37**. A concerted metalation deprotonation takes place via transition state **12-38ts** with an activation free energy of 16.1 kcal/mol to form aryl Mn(I) intermediate **12-39**. Then, propargyl ester reactant inserts into C(aryl)−Mn bond via transition state **12-40ts**, which is considered as rate-determining step for the whole catalytic cycle. When vinyl Mn(I) intermediate **12-41** is formed, a β-acetate elimination takes place via transition state **12-42ts** to release allenyl arene product **12-43** and regenerate acetate Mn(I) species **12-37** by the coordination of imine reactant. In an outer-sphere process, cyclization of allenylarylimine **12-43** can provide final isoquinoline product (Figure 12.6).

12.3 Mn-Mediated Hydrogenations

As a first-row transition metal, Mn can be used to activate dihydrogen, which is involved in the hydrogenation reactions of carbon dioxide and carbonates.

12.3.1 Hydrogenation of Carbon Dioxide

Generally, the hydrogen source for the hydrogenation of carbon dioxide can be formally considered as a proton and a hydride. In Mn-catalyzed ligand-assisted

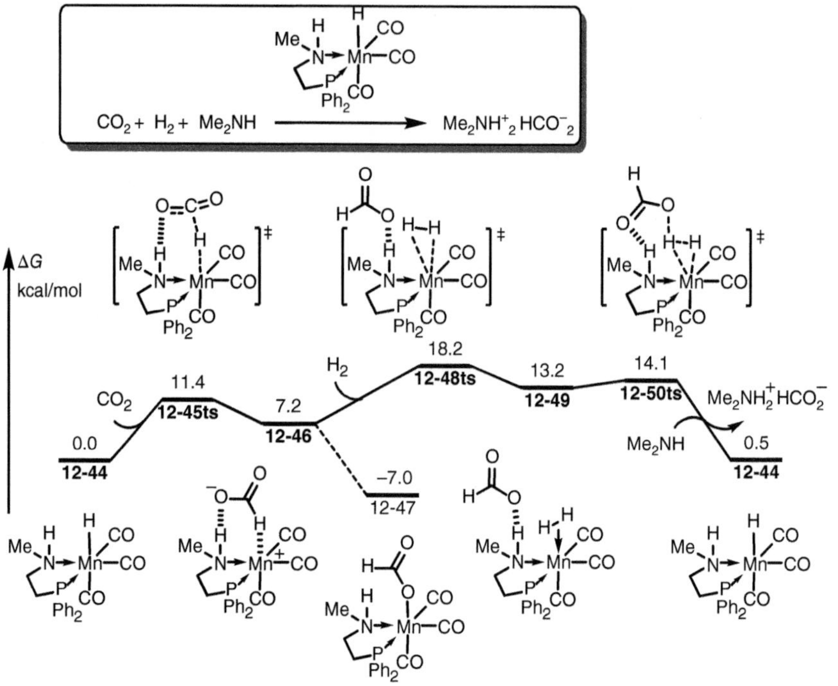

Figure 12.7 Free-energy profiles for the Mn-catalyzed hydrogenation of carbon dioxide. The energies were calculated at B3LYP-D3/6-31++G(d,p) (LANL2DZ for Mn)/CPCM(THF) level of theory. Source: Based on Singh Rawat et al. [42].

hydrogenation, the Mn—H moiety can be used as hydride source, while the proton part is provided from amine ligand [35–41]. Pathak group performed a systematic theoretical study on the mechanism of a model reaction of Mn-catalyzed hydrogenation of carbon dioxide. As shown in Figure 12.7 [42], a bidentate phosphine-amine ligand-coordinated amino Mn(I) hydride **12-44** was used as active catalyst. Carbon dioxide can be activated by hydrogen bond with amine ligand. Therefore, a nucleophilic attack by hydride takes place via transition state **12-45ts** to afford a formate-coordinated cationic Mn(I) species **12-46**, which can reversibly isomerize to a more stable Mn(I)-formate salt **12-47** by 14.2 kcal/mol exergonic. The coordination of dihydrogen via a substitution transition state **12-48ts** forms complex **12-49** with overall 20.2 kcal/mol endergonic. A formate-assisted heterolytic cleavage of H—H bond occurs via an eight-membered ring-type transition state **12-50ts**. After releasing a formic acid molecule, hydride Mn(I) species **12-44** can be regenerated.

12.3.2 Hydrogenation of Carbonates

Carbonates can be considered as the homologs of carbon dioxide, which can react with dihydrogen to afford corresponding reductive products in the presence of Mn-catalyst [43]. As shown in Figure 12.8, the mechanism of Mn-catalyzed

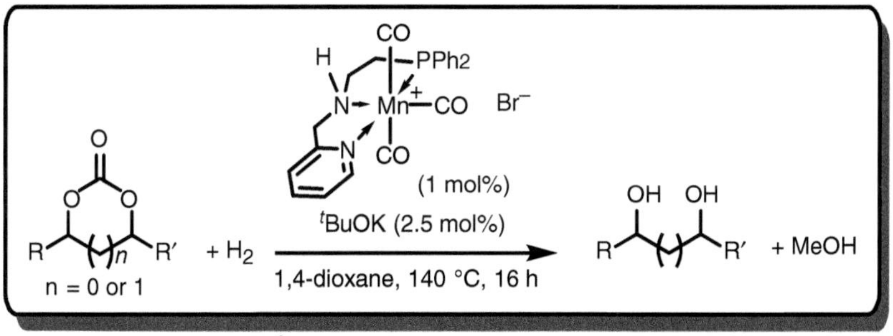

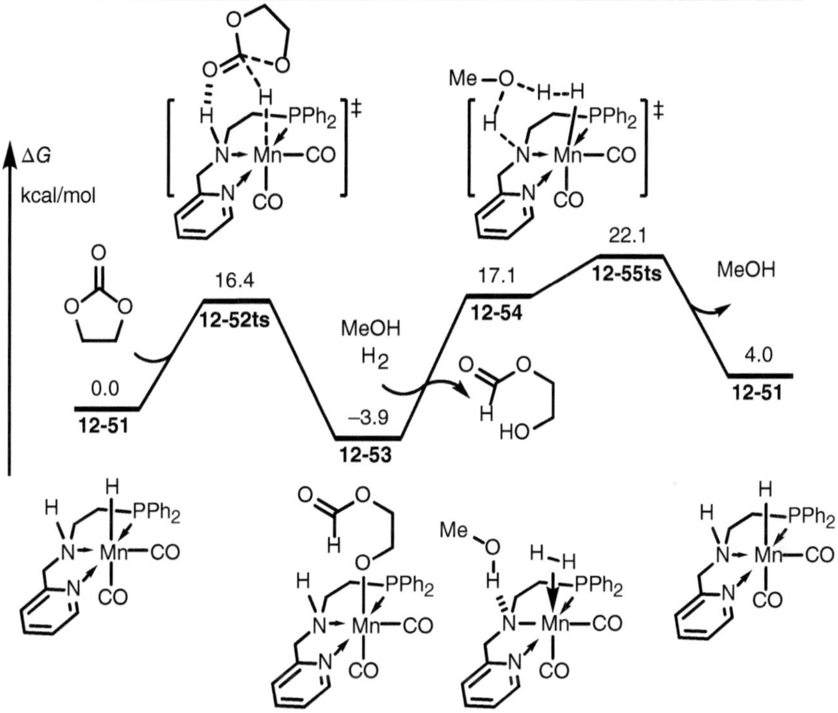

Figure 12.8 Free-energy profiles for the Mn-catalyzed hydrogenation of carbonates. The energies were calculated at M06/TZVP/PCM(dioxane)//ωB97XD/SVP (TZVP for Mn)/PCM(dioxane) level of theory.

hydrogenation of dioxolanone is revealed by DFT calculation at M06 level. When an N—N—P type ligand was used, Mn(I)-hydride species **12-51** was considered as active species in the catalytic cycle. With the activation of hydrogen bond between carbonyl and N—H bond in ligand, a hydride transfer takes place via transition state **12-52ts** with an energy barrier of 16.4 kcal/mol. A simultaneous C—O bond cleavage generates an alkoxy Mn(I) intermediate **12-53**. Then, an alcohol exchange forms methoxy Mn(I) intermediate **12-54**. A methanol-assisted H—H bond cleavage via transition state **12-55ts** regenerates Mn(I)-hydride species **12-51**, which is considered as the rate limit for the whole catalytic cycle. Computational study found that the whole catalytic cycle is 4.0 kcal/mol endothermic. Therefore, high pressure of gaseous dihydrogen is necessary to drive the reaction forward.

12.4 Mn-Mediated Dehydrogenations

Dehydrogenation can be considered as the inverse reaction of hydrogenation, which can release a molecule of dihydrogen and construct new covalent π or σ bonds.

12.4.1 Dehydrogenation of Alcohols

Both the experimental and theoretical studies found that the hydrogenation of carbonyl compounds is reversible [44–48], which is determined by the reaction conditions such as the reacting temperature, solvent, and pressure of substrates. In the presence of Mn-catalyst, the dehydrogenation of alcohols can afford corresponding aldehydes, ketones, and even carboxylic acids. As shown in Figure 12.9 the mechanism for the Mn-catalyzed dehydrogenation of methanol was studied by Jiao group

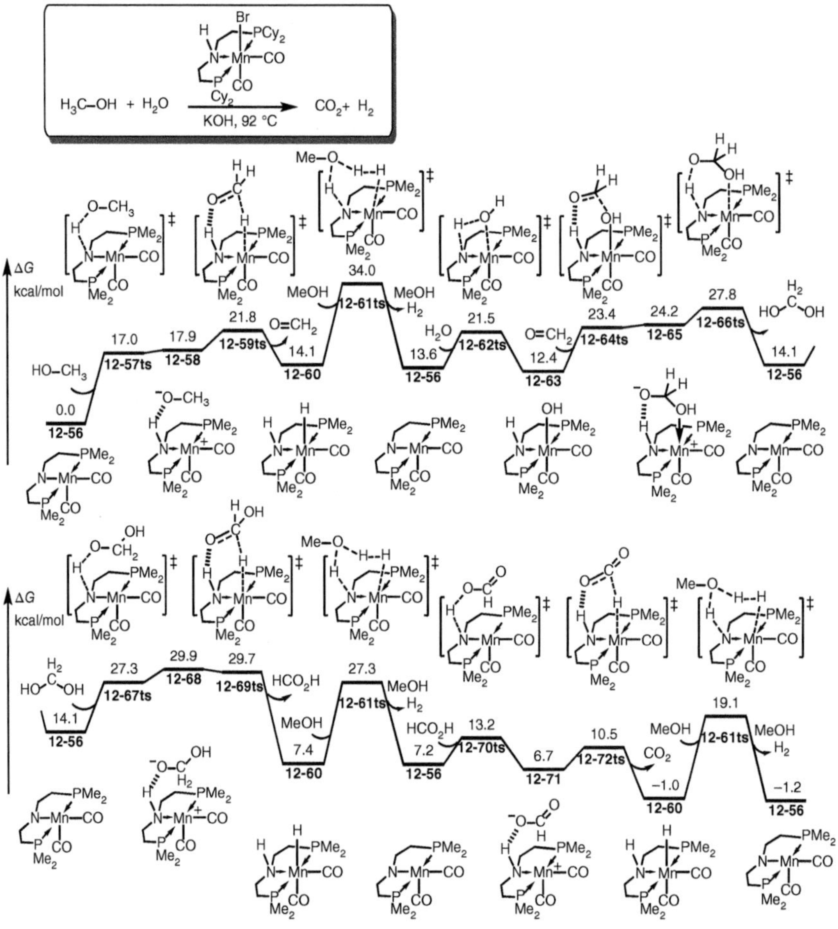

Figure 12.9 Free-energy profiles for the Mn-catalyzed dehydrogenation of methanol. The energies were calculated at B3PW91/TZVP (LANL2DZ for Mn) level of theory.

[49]. The reaction starts from an pincer amino Mn(I) species **12-56**, which can form a hydrogen bond with methanol reactant. After a proton transfer via transition state **12-57ts**, a cationic Mn(I) intermediate **12-58** is formed with a hydrogen-bonding methoxide. Then, hydride transfer takes place via a linear transition state **12-59ts** to afford hydride Mn(I) species **12-60** and release a formaldehyde. The whole process is 14.1 kcal/mol endergonic, which requires an extra driving force for further process. Interestingly, the dehydrogenation of complex **12-60** is only 0.5 kcal/mol exergonic, which cannot be considered as driving force for this process. When the active species **12-56** is regenerated, the coordination of water leads to the protonation of amino group via a four-membered ring-type transition state **12-62ts** to afford a hydroxide Mn(I) intermediate **12-63**. Then, a nucleophilic attack of hydroxide onto formaldehyde via transition state **12-64ts** can form a cationic Mn(I) intermediate **12-65**. Subsequently, proton transfer via transition state **12-66ts** regenerates amino Mn(I) species **12-56** and forms methanediol reversibly. Then, proton transfer between methanediol and **12-56** takes place via transition state **12-67ts** followed by a hydride transfer via transition state **12-69ts** can afford formic acid and dihydrogen, which is 6.7 kcal/mol exergonic. The following dehydrogenation of formic acid can provide further driving force, which releases carbon dioxide and dihydrogen. The whole catalytic cycle is 1.2 kcal/mol exergonic. Indeed, the formation of gaseous dihydrogen and carbon dioxide can provide further driving force for this transformation.

12.4.2 Dehydrogenative Couplings

In the absence of extra oxidant, dehydrogenation of two nucleophiles involving H—X (X = C or N) bonds can achieve corresponding oxidative coupling by release gaseous dihydrogen.

The dehydrogenative coupling of nitriles and alcohols can be used for the α-olefination of nitriles [33, 50, 51]. DFT calculations revealed that the mechanism of this dehydrogenative coupling can be split into two cycles involving dehydrogenation of alcohol to achieve ketone and α-olefination of nitrile by using in-situ-generated aldehyde [52]. The denitrogenation of alcohol was already discussed above. Therefore, we focus on the mechanism of α-olefination process. As shown in Figure 12.10, the reaction starts from a pincer-ligand involved amino Mn(I) species **12-73**. Due to the alkalinity of amino group, the α-deprotonation of nitrile can occur via transition state **12-74ts** to afford a vinylideneamide Mn(I) intermediate **12-75**. When aldehyde is activated by the hydrogen bond with bridged water molecule, a nucleophilic attack by vinylideneamide can occur via a 10-membered ring-type transition state **12-76ts** to provide a new C—C covalent bond in intermediate **12-77**. Then, a water-assisted proton transfer takes place via transition state **12-78ts**, which generates a functionalized vinylideneamide Mn(I) intermediate **12-79**. Then, a water-assisted dehydration via transition state **12-80ts** results acrylonitrile product and regenerates active catalyst **12-73**.

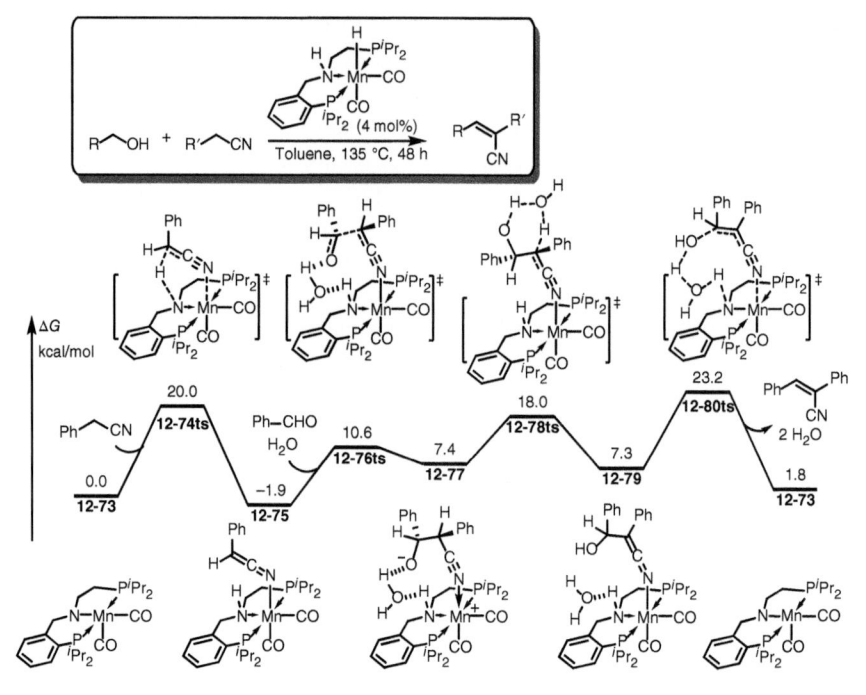

Figure 12.10 Free-energy profiles for the Mn-catalyzed dehydrogenative coupling of nitriles and alcohols. The energies were calculated at M06/6-311++G(d,p) (SDD for Mn)/SMD(toluene)//M06/6-31G(d,p) (SDD for Mn) level of theory.

References

1 Gorgas, N. and Kirchner, K. (2018). Isoelectronic manganese and iron hydrogenation/dehydrogenation catalysts: similarities and divergences. *Accounts of Chemical Research* 51 (6): 1558–1569.
2 Kovacs, J.A. (2015). Tuning the relative stability and reactivity of manganese dioxygen and peroxo intermediates via systematic ligand modification. *Accounts of Chemical Research* 48 (10): 2744–2753.
3 Sun, W. and Sun, Q. (2019). Bioinspired manganese and iron complexes for enantioselective oxidation reactions: ligand design, catalytic activity, and beyond. *Accounts of Chemical Research* 52 (8): 2370–2381.
4 Trovitch, R.J. (2017). The emergence of manganese-based carbonyl hydrosilylation catalysts. *Accounts of Chemical Research* 50 (11): 2842–2852.
5 He, R., Huang, Z.-T., Zheng, Q.-Y. et al. (2014). Manganese-catalyzed dehydrogenative [4+2] annulation of N–H imines and alkynes by C–H/N–H activation. *Angewandte Chemie International Edition* 53 (19): 4950–4953.
6 Kuninobu, Y., Fujii, Y., Matsuki, T. et al. (2009). Rhenium-catalyzed insertion of nonpolar and polar unsaturated molecules into an olefinic C–H bond. *Organic Letters* 11 (12): 2711–2714.

7 Kuninobu, Y., Nishina, Y., Takeuchi, T. et al. (2007). Manganese-catalyzed insertion of aldehydes into a C–H bond. *Angewandte Chemie International Edition* 46 (34): 6518–6520.

8 Zhou, B., Chen, H., and Wang, C. (2013). Mn-catalyzed aromatic C-H alkenylation with terminal alkynes. *Journal of the American Chemical Society* 135 (4): 1264–1267.

9 Cai, S.-H., Ye, L., Wang, D.-X. et al. (2017). Manganese-catalyzed synthesis of monofluoroalkenes via C–H activation and C–F cleavage. *Chemical Communications* 53 (62): 8731–8734.

10 Chen, S.Y., Han, X.L., Wu, J.Q. et al. (2017). Manganese(I)-catalyzed regio- and stereoselective 1,2-diheteroarylation of allenes: combination of C–H activation and Smiles rearrangement. *Angewandte Chemie International Edition* 56 (33): 9939–9943.

11 Lu, Q., Gressies, S., Cembellin, S. et al. (2017). Redox-neutral manganese(I)-catalyzed C–H activation: traceless directing group enabled regioselective annulation. *Angewandte Chemie International Edition* 56 (41): 12778–12782.

12 Wang, C., Wang, A., and Rueping, M. (2017). Manganese-catalyzed C–H functionalizations: hydroarylations and alkenylations involving an unexpected heteroaryl shift. *Angewandte Chemie International Edition* 56 (33): 9935–9938.

13 Yahaya, N.P., Appleby, K.M., Teh, M. et al. (2016). Manganese(I)-catalyzed C–H activation: the key role of a 7-membered manganacycle in H-transfer and reductive elimination. *Angewandte Chemie International Edition* 55 (40): 12455–12459.

14 Yang, X. and Wang, C. (2018). Dichotomy of manganese catalysis via organometallic or radical mechanism: stereodivergent hydrosilylation of alkynes. *Angewandte Chemie International Edition* 57 (4): 923–928.

15 Zhou, B., Ma, P., Chen, H. et al. (2014). Amine-accelerated manganese-catalyzed aromatic C–H conjugate addition to α,β-unsaturated carbonyls. *Chemical Communications* 50 (93): 14558–14561.

16 Zhu, C., Schwarz, J.L., Cembellin, S. et al. (2018). Highly selective manganese(I)/Lewis acid cocatalyzed direct C–H propargylation using bromoallenes. *Angewandte Chemie International Edition* 57 (2): 437–441.

17 Liu, W. and Groves, J.T. (2010). Manganese porphyrins catalyze selective C–H bond halogenations. *Journal of the American Chemical Society* 132 (37): 12847–12849.

18 Liu, W. and Groves, J.T. (2015). Manganese catalyzed C–H halogenation. *Accounts of Chemical Research* 48 (6): 1727–1735.

19 Bahramian, B., Mirkhani, V., Moghadam, M. et al. (2006). Selective alkene epoxidation and alkane hydroxylation with sodium periodate catalyzed by cationic Mn(III)-salen supported on Dowex MSC1. *Applied Catalysis A: General* 301 (2): 169–175.

20 de Freitas Silva, G., da Silva, D.C., Guimarães, A.S. et al. (2007). Cyclohexane hydroxylation by iodosylbenzene and iodobenzene diacetate catalyzed by a new β-octahalogenated Mn–porphyrin complex: the effect of meso-3-pyridyl substituents. *Journal of Molecular Catalysis A: Chemical* 266 (1–2): 274–283.

21 Chen, Y., Tian, T., and Li, Z. (2019). Mn-catalyzed azidation–peroxidation of alkenes. *Organic Chemistry Frontiers* 6 (5): 632–636.

22 Huang, X., Bergsten, T.M., and Groves, J.T. (2015). Manganese-catalyzed late-stage aliphatic C-H azidation. *Journal of the American Chemical Society* 137 (16): 5300–5303.

23 Robins, J. (1965). Structural effects in metal ion catalysis of isocyanate–hydroxyl reactions. *Journal of Applied Polymer Science* 9 (3): 821–838.

24 Friedermann, G.R., Halma, M., de Freitas Castro, K.A.D. et al. (2006). Intermediate species generated from halogenated manganese porphyrins electrochemically and in homogeneous catalysis of alkane oxidation. *Applied Catalysis A: General* 308: 172–181.

25 Modén, B., Oliviero, L., Dakka, J. et al. (2004). Structural and functional characterization of redox Mn and Co sites in AlPO materials and their role in alkane oxidation catalysis. *Journal of Physical Chemistry B* 108 (18): 5552–5563.

26 Slaughter, L.M., Collman, J.P., Eberspacher, T.A. et al. (2004). Radical autoxidation and autogenous O_2 evolution in manganese-porphyrin catalyzed alkane oxidations with chlorite. *Inorganic Chemistry* 43 (17): 5198–5204.

27 Balcells, D., Raynaud, C., Crabtree, R.H. et al. (2009). C-H oxidation by hydroxo manganese(v) porphyrins: a DFT study. *Chemical Communications* 45 (13): 1772–1774.

28 Balcells, D., Raynaud, C., Crabtree, R.H. et al. (2008). The rebound mechanism in catalytic C-H oxidation by MnO(tpp)Cl from DFT studies: electronic nature of the active species. *Chemical Communications* (6): 744–746.

29 Liu, W. and Groves, J.T. (2013). Manganese-catalyzed oxidative benzylic C-H fluorination by fluoride ions. *Angewandte Chemie International Edition* 52 (23): 6024–6027.

30 Gring, M., Kuhnert, M., Langen, T. et al. (2012). Relaxation and prethermalization in an isolated quantum system. *Science* 337 (6100): 1318–1322.

31 Huang, X., Zhuang, T., Kates, P.A. et al. (2017). Alkyl isocyanates via manganese-catalyzed C–H activation for the preparation of substituted ureas. *Journal of the American Chemical Society* 139 (43): 15407–15413.

32 Zhou, B., Ma, P., Chen, H. et al. (2014). Amine-accelerated manganese-catalyzed aromatic C–H conjugate addition to alpha,beta-unsaturated carbonyls. *Chemical Communications* 50 (93): 14558–14561.

33 Yang, Y., Zhang, Q., Shi, J. et al. (2016). Mechanism study of Mn(I) complex-catalyzed imines and alkynes dehydrogenation coupling reaction. *Acta Chimica Sinica* 74 (5): 422.

34 Wang, Z., Zhu, L., Zhong, K. et al. (2018). Mechanistic insights into manganese (I)-catalyzed chemoselective hydroarylations of alkynes: a theoretical study. *ChemCatChem* 10 (22): 5280–5286.

35 Bertini, F., Glatz, M., Gorgas, N. et al. (2017). Carbon dioxide hydrogenation catalysed by well-defined Mn(I) PNP pincer hydride complexes. *Chemical Science* 8 (7): 5024–5029.

36 Biswas, S., Pramanik, A., and Sarkar, P. (2018). Computational design of quaterpyridine-based Fe/Mn–complexes for the direct hydrogenation of CO_2

to HCOOH: a direction for atom-economic approach. *ChemistrySelect* 3 (18): 5185–5193.

37 Chen, X., Ge, H., and Yang, X. (2017). Newly designed manganese and cobalt complexes with pendant amines for the hydrogenation of CO_2 to methanol: a DFT study. *Catalysis Science & Technology* 7 (2): 348–355.

38 Ge, H., Chen, X., and Yang, X. (2016). A mechanistic study and computational prediction of iron, cobalt and manganese cyclopentadienone complexes for hydrogenation of carbon dioxide. *Chemical Communications* 52 (84): 12422–12425.

39 Mandal, S.C., Rawat, K.S., Nandi, S. et al. (2019). Theoretical insights into CO_2 hydrogenation to methanol by a Mn–PNP complex. *Catalysis Science & Technology* 9 (8): 1867–1878.

40 Mandal, S.C., Rawat, K.S., and Pathak, B. (2019). A computational study on ligand assisted vs. ligand participation mechanisms for CO_2 hydrogenation: importance of bifunctional ligand based catalysts. *Physical Chemistry Chemical Physics* 21 (7): 3932–3941.

41 Rawat, K.S. and Pathak, B. (2018). Flexible proton-responsive ligand-based Mn(I) complexes for CO_2 hydrogenation: a DFT study. *Physical Chemistry Chemical Physics* 20 (18): 12535–12542.

42 Singh Rawat, K., Garg, P., Bhauriyal, P. et al. (2019). Metal-ligand bifunctional based Mn-catalysts for CO_2 hydrogenation reaction. *Molecular Catalysis* 468: 109–116.

43 Zubar, V., Lebedev, Y., Azofra, L.M. et al. (2018). Hydrogenation of CO_2-derived carbonates and polycarbonates to methanol and diols by metal-ligand cooperative manganese catalysis. *Angewandte Chemie International Edition* 57 (41): 13439–13443.

44 Andérez-Fernández, M., Vogt, L.K., Fischer, S. et al. (2017). A stable manganese pincer catalyst for the selective dehydrogenation of methanol. *Angewandte Chemie International Edition* 56 (2): 559–562.

45 Chakraborty, S., Gellrich, U., Diskin-Posner, Y. et al. (2017). Manganese-catalyzed N-formylation of amines by methanol liberating H_2: a catalytic and mechanistic study. *Angewandte Chemie International Edition* 56 (15): 4229–4233.

46 Das, K., Mondal, A., Pal, D. et al. (2019). Phosphine-free well-defined Mn(I) complex-catalyzed synthesis of amine, imine, and 2,3-dihydro-1*H*-perimidine via hydrogen autotransfer or acceptorless dehydrogenative coupling of amine and alcohol. *Organometallics* 38 (8): 1815–1825.

47 Nguyen, D.H., Trivelli, X., Capet, F. et al. (2017). Manganese pincer complexes for the base-free, acceptorless dehydrogenative coupling of alcohols to esters: development, scope, and understanding. *ACS Catalysis* 7 (3): 2022–2032.

48 Wei, Z., de Aguirre, A., Junge, K. et al. (2018). Benzyl alcohol dehydrogenative coupling catalyzed by defined Mn and Re PNP pincer complexes – a computational mechanistic study. *European Journal of Inorganic Chemistry* 2018 (42): 4643–4657.

49 Wei, Z., de Aguirre, A., Junge, K. et al. (2018). Exploring the mechanisms of aqueous methanol dehydrogenation catalyzed by defined PNP Mn and Re pincer complexes under base-free as well as strong base conditions. *Catalysis Science & Technology* 8 (14): 3649–3665.

50 Chakraborty, S., Das, U.K., Ben-David, Y. et al. (2017). Manganese catalyzed α-olefination of nitriles by primary alcohols. *Journal of the American Chemical Society* 139 (34): 11710–11713.

51 Luque-Urrutia, J.A., Sola, M., Milstein, D. et al. (2019). Mechanism of the manganese-pincer-catalyzed acceptorless dehydrogenative coupling of nitriles and alcohols. *Journal of the American Chemical Society* 141 (6): 2398–2403.

52 Lu, Y., Zhao, R., Guo, J. et al. (2019). A unified mechanism to account for manganese- or ruthenium-catalyzed nitrile alpha-olefinations by primary or secondary alcohols: a DFT mechanistic study. *Chemistry A European Journal* 25 (15): 3939–3949.

13

Theoretical Study of Cu-Catalysis

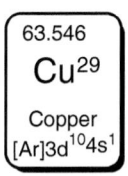

63.546
Cu²⁹
Copper
[Ar]3d¹⁰4s¹

Copper is one of the earliest metals used by human beings. As early as prehistoric times, people began to excavate open-pit copper mines and use the acquired copper to make weapons, utensils, and other utensils. The use of copper had a far-reaching impact on the progress of early human civilization. Copper is a metal existing in the earth's crust and ocean. The content of copper in the earth's crust is about 0.01%. It is a cheap and readily available metal and is widely used in modern industry. In fact, copper was also one of the first transition metals used in synthetic chemistry, which reveals unique reactivity for the construction of new C—C bonds. In the early 1900s, Ullmann and Goldberg gave a good example where copper was used to facilitate nucleophilic substitution of arenes [1–3]. In the original protocol, stoichiometric amounts of copper salts together with high reaction temperatures (≥200 °C) and long reaction times are required. Copper has become an acceptable alternative for noble metal catalysis because of its low cost and easy availability, although the early-reported copper-mediated reactions were often accompanied by harsh reaction conditions. In recent decades, with the development of new ligands and reaction systems, copper catalysis has become more and more mild and selective; therefore, it has been widely used in synthetic chemistry [4–8].

As a late transition metal, copper occurs in a range of oxidation states, including 0, +1, +2, +3, and +4, while the copper ions readily form complexes yielding a variety of coordination compounds. In homogeneous catalysis, Cu(I) and Cu(II) are two most common species, which usually are used as a precatalyst [9–11]. Free Cu(I) iron is unstable in aqueous solution, according to the reported oxidation potentials leading to a disproportionation equilibrium: $2Cu(I) \rightarrow Cu(0) + Cu(II)$ [12, 13]. However, the stability can be regulated by the outstanding coordination of ligands, solvents, and counterions. Cu(III) is often considered as an unstable

Computational Methods in Organometallic Catalysis: From Elementary Reactions to Mechanisms,
First Edition. Yu Lan.
© 2021 WILEY-VCH GmbH. Published 2021 by WILEY-VCH GmbH.

intermediate in Cu-chemistry, which can be generated through the oxidation of Cu(I) by double-electron transfer or Cu(II) by single-electron transfer. The generated Cu(III) intermediate usually reveals strong oxidizability, which can contribute to the coupling of two coordinated nucleophiles [14–17].

The variety of oxidative states for copper leads to the complexity for the mechanism of Cu-catalyzed reactions. In early years, copper was often considered as an alternative of palladium [18, 19]. Therefore, the copper catalytic cycle was constructed after palladium in corresponding catalytic cycle was replaced by copper, which was supposed to undergo a double-electron transfer oxidative addition-reductive elimination process through Cu(III) intermediates [20, 21]. However, more and more examples showed that single-electron transfer would be included in Cu-mediated transformations. Especially with the development of computational organometallic chemistry, the mechanism of copper catalysis is more clearly demonstrated, and single-electron transfer (SET) redox process is more recognized. Mechanistically, the single-electron transfer redox process in Cu-catalysis would inevitably bring about different reactivity from that of classical transition metal catalysis leading to a broader potential application prospects [14, 22–24].

Copper reveals a typical π-acidity, which can be coordinated by acetylenes to achieve triple-bond activation. The combination of π-acidity and redox property leads to a series of functionalization reactions of acetylenes [25, 26]. Moreover, copper also can play as a Lewis acid, which can be used to activate carbonyl group to further increase its electron-withdrawing effect. Therefore, with proper ligands, copper can be used to catalyze asymmetric pericyclic reaction [27, 28].

In this chapter, we begin our discussion from the mechanism of Ullmann and modified Ullmann couplings, which has a history longer than 100 years. As a late transition metal, copper also can be used in C—H activation reactions, which provide a potential nucleophile for further coupling reactions. As a π-acid, acetylenes, carbenes, and nitrenes can coordinate with copper, which leads to a new mechanistic concept in the functionalizations of those species. The various oxidative states of copper lead to redox active in Cu-catalysis; therefore, the mechanism of oxidation and hydrogenation is discussed separately.

13.1 Cu-Mediated Ullmann Condensations

Copper-mediated aromatic nucleophilic substitution (ANS) was first reported by Ullmann and Goldberg, which has a history longer than 100 years. As shown in Scheme 13.1, the original protocol for this reaction required the use of stoichiometric amounts of copper salts together with high reaction temperatures and long reaction time [1–3]. On this basis, a series of copper-based coupling reactions have been developed [29, 30]. When amine, phenol, or thiophenol is used as nucleophile to react with aryl halide, the corresponding aryl -amine, -ether, or -thioether compounds can be constructed, respectively, which is referred to as Ullmann condensation reaction [31–34]. The successful development of improved catalytic

13.1 Cu-Mediated Ullmann Condensations

versions of this grand old chemistry has caused a veritable "renaissance" of what is now known as the "modified Ullmann reaction."

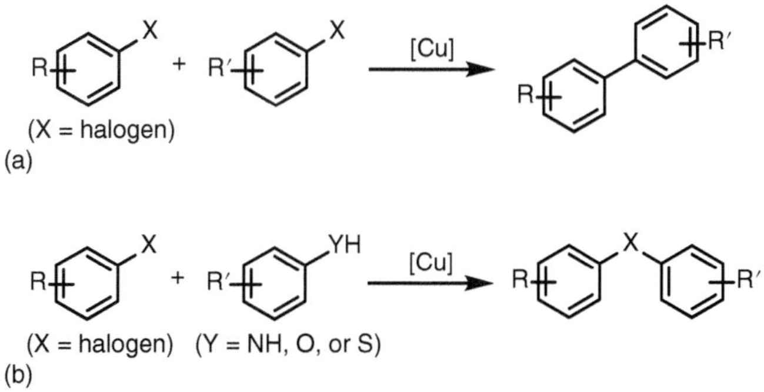

(a) (X = halogen)

(b) (X = halogen) (Y = NH, O, or S)

Scheme 13.1 (a) Ullmann reactions and (b) Ullmann condensations.

Although the reaction system of Ullmann condensations seems simple, the mechanism is hard to explore and understand. Experiments show that the reaction has undergone redox process; nevertheless, the key is that there are many possibilities for the change of the oxidative state of copper in reaction. Notably, almost ranging Cu(I) and Cu(II) salts, and even metallic copper may be used as source of catalyst in Ullmann condensations; therefore, the question of which oxidative state of copper is present in the active catalyst was the first to catch the interest of researchers [35]. On the basis of previous experimental and theoretical observations, four possible mechanisms are proposed for the Ullmann condensations, which are dependent on the cleavage of C(aryl)—halide bond and the formation of C(aryl)—X (X = O, N, or S) bonds.

The four possible mechanisms for the coupling reaction of aryl halides and nucleophiles are summarized in Scheme 13.2 [36, 37]. According to the mechanism of Pd-catalysis of Hartwig–Buchwald amination reaction [38], an oxidative addition-reductive elimination pathway through a Cu(III) intermediate was proposed in Scheme 13.2a, though the existence of copper(III) complexes has been questioned for a long time [39–42]. The reaction starts from a ligand exchange with nucleophile to afford Cu(I) species, which can undergo an oxidative addition by aryl halide to achieve Cu(III). The reductive elimination yields nucleophilic substitution product and regenerates Cu(I) species. In an SET involving pathway (Scheme 13.2b), a halogen atom transfer (HAT) from aryl halide to Cu(I) species leads to the oxidation of copper to afford a Cu(II) intermediate with the release of aryl radical [43–47]. A radical addition with nucleophile results in an anionic radical with the formation of C(aryl)—X (X = O, N, or S) bond, which can undergo a SET with Cu(II) intermediate to complete the catalytic cycle. As shown in Scheme 13.2c, the halide in substrate can be activated by Cu(I) salt, which can undergo a four-membered metathesis transition state with nucleophile to directly achieve nucleophilic substitution through a nonredox process [48–50]. Alternatively (Scheme 13.2d), aryl group also can be activated by Cu(I) salt resulting in an intermolecular nucleophilic substitution [51, 52].

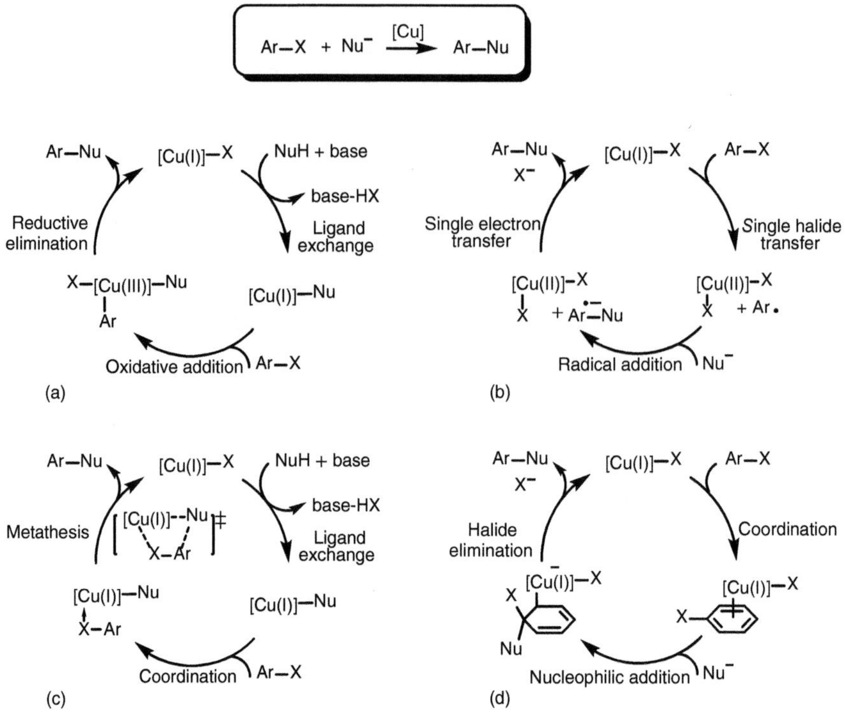

Scheme 13.2 General mechanism of Ullmann condensations: (a) oxidative addition-reductive elimination (OA-RE) pathway, (b) halogen atom transfer-single-electron transfer (HAT-SET) pathway, (c) metathesis pathway, and (d) aromatic nucleophilic substitution (ANS) pathway. Sources: Andrada et al. [36]; Sperotto et al. [37].

Following this idea, a series of theoretical calculations focus on the mechanism study of Cu-mediated Ullmann condensation reactions. In this section, we try to categorize them by the type of nucleophiles.

13.1.1 C—N Bond Couplings

When pyrrolyl was used as nucleophile, Ullmann condensation with iodobenzene in the presence of Cu(II) salt can provide corresponding phenyl pyrrole [53, 54]. A density functional theory (DFT) study was carried out by Norrby group to compare the oxidative addition-reductive elimination pathway and dissociative SET pathway [55]. As shown in Figure 13.1, when pyrrolyl Cu(I) species **13-1** is formed, the oxidative addition of iodobenzene takes place via transition state **13-2ts** with an energy barrier of 16.3 kcal/mol to afford a biaryl Cu(III) intermediate **13-3**. Then, a rapid reductive elimination results in phenyl pyrrole product and generates a Cu(I) species **13-5**. By contrast, radical substitution with iodobenzene would occur via transition state **13-6ts** suffering an energy barrier of 29.2 kcal/mol to provide a phenyl radical and a Cu(II) species **13-7**. Then, another radical substitution via transition state **13-8ts** could afford the same phenyl pyrrole product. Moreover,

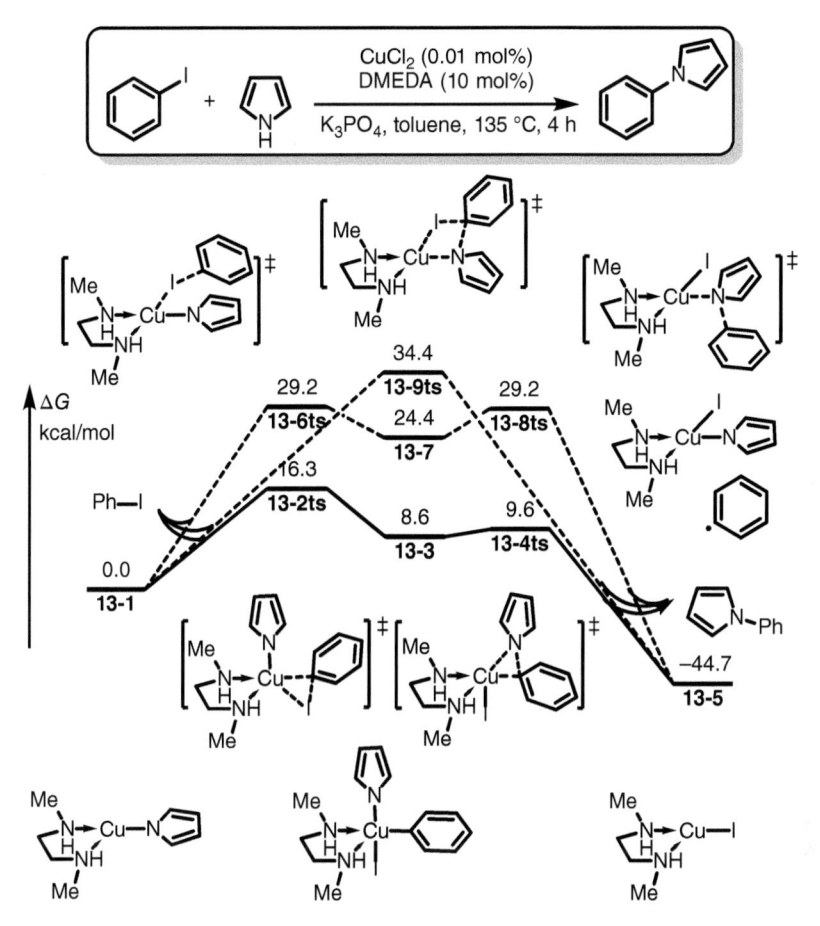

Figure 13.1 Free-energy profiles for the Cu-catalyzed Ullmann condensation of pyrrole and iodobenzene. The energies were calculated at B3LYP-D3/LACVP*/PBF(benzene) //B3LYP-D3/LACVP* level of theory.

the calculated relative free energy for the metathesis-type transition state **13-9ts** is as high as 34.4 kcal/mol. In this work, oxidative addition-reductive elimination pathway was determined to be preferred.

In another case, Ciofini, Grimaud, and Taillefer reported a combination of theoretical and experimental study on the mechanism of Cu-catalyzed Ullmann condensation of diazonium salt and pyrazole [56, 57]. As shown in Figure 13.2, the complexation of diazonium and anionic Cu(I) species **13-10** affords a neutral species **13-11** with 17.6 kcal/mol exergonic. The oxidative addition of diazonium onto Cu(I) takes place via transition state **13-12ts** with an energy barrier of 21.6 kcal/mol resulting in a Cu(III) species **13-13**. The sequential denitrogenation and deprotonation reversibly form a pyrazolyl Cu(III) intermediate **13-16** smoothly. Then, a reductive elimination takes place via transition state **13-17ts** yielding phenyl pyrazole product. The following ligand exchange regenerates the active catalyst Cu(I) species **13-10**.

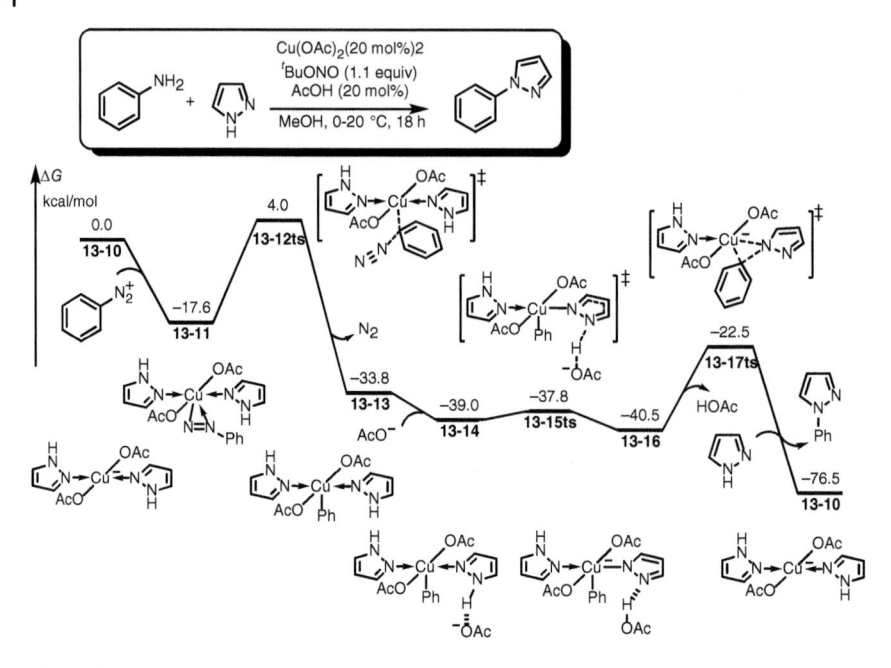

Figure 13.2 Free-energy profiles for the Cu-catalyzed Ullmann condensation of pyrazole and diazonium. The energies were calculated at B3LYP-D3/6-311+G(d,p) (LANL2DZ for Cu)/PCM(methanol) level of theory.

13.1.2 C—O Bond Couplings

Generally, the mechanism of Cu-mediated Ullmann condensation for the C—O bond coupling is considered through an oxidative addition-reductive elimination pathway [58, 59]. As shown in Figure 13.3, phenol was detected as the major product in the absence of Cs_2CO_3 [60]. In DFT-calculated free-energy profiles, the oxidative addition by iodobenzene of hydroxy Cu(I) species **13-18** was considered as the rate-determining step with an energy barrier of 13.3 kcal/mol to form a five-coordinated Cu(III) species **13-20**. After the dissociation of iodide, reductive elimination occurs from a four-coordinated Cu(III) intermediate **13-22** via transition state **13-23ts** to form phenol-coordinated Cu(I) species **13-24**. The ligand exchange with base releases phenol product and regenerates active species **13-18**.

13.1.3 C—F Bond Couplings

When fluoride is used as nucleophile, the fluorination of an electrophilic aryl can be considered as an analogy of Ullmann condensation, which also undergoes an oxidative addition-reductive elimination pathway [61, 62]. As shown in Figure 13.4, Canty and Sanford reported a Cu-catalyzed fluorination of diaryliodonium salts. In the mechanism study, this reaction was considered as starting from an anionic fluoride Cu(I) species **13-25**, which can form a complex **13-26** with coordination of diaryliodonium. DFT calculation found that the oxidative addition can occur via a three-membered ring-type transition state **13-27ts** with an overall activation free

13.1 Cu-Mediated Ullmann Condensations | 523

Figure 13.3 Free-energy profiles for the Cu-catalyzed Ullmann condensation to achieve hydroxylation of iodobenzene. The energies were calculated at PBE0/6-31+G(d) (SDD for Cu)/PCM(DMF) level of theory.

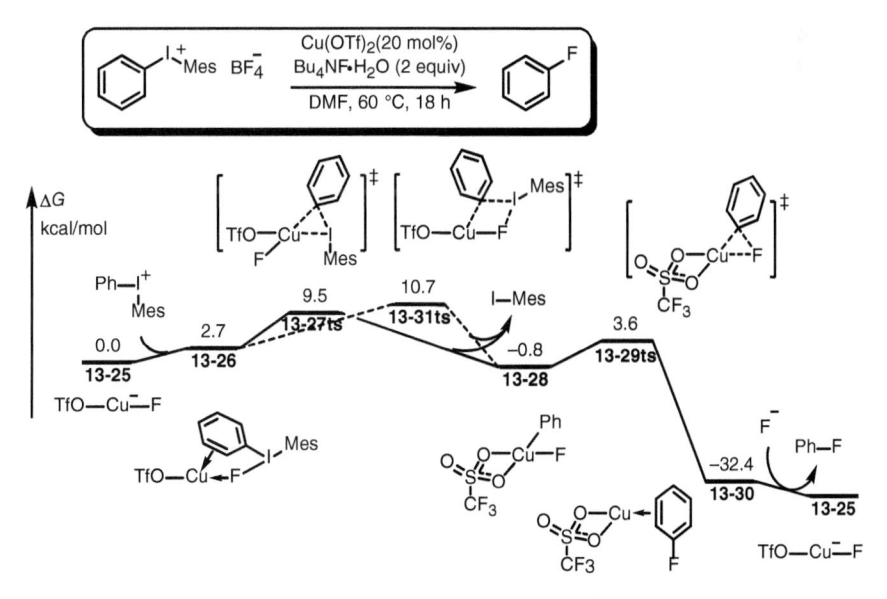

Figure 13.4 Free-energy profiles for the Cu-catalyzed Ullmann condensation to achieve fluorination of diaryliodonium salts. The energies were calculated at BP86/6-311+G(2d,p) (def2-QZVP for Cu and I)/IEF-PCM(DMF) level of theory.

energy of 9.5 kcal/mol to form a Cu(III) intermediate **13-28**. Then, a C(aryl)—F reductive elimination via transition state **13-29ts** results in fluorobenzene. In an alternative pathway, the iodonium moiety can be activated by fluoride, which results in a four-membered ring-type metathesis transition state **13-31ts**. The calculated relative free energy of transition state **13-31ts** is only 1.2 kcal/mol higher than that of **13-27ts**, indicating it is also a possible process.

13.2 Cu-Mediated Trifluoromethylations

According to the strong electron-attracting effect of fluorine atom, trifluoromethyl group can play as a nucleophilic group, which can react with electrophilic carbon to achieve trifluoromethylations through redox-neutral cross-coupling. Alternatively, in the presence of exogenous oxidants, Cu-mediated trifluoromethylations also can occur with another nucleophile through oxidative coupling with removing one pair of electrons. Moreover, trifluoromethyl radical reveals unique stability; therefore, trifluoromethylations can undergo a radical process through a single-electron oxidation of Cu(I) [63–65].

13.2.1 Trifluoromethylations Through Cross-Coupling

Anionic trifluoromethyl group can be considered as a nucleophile to react with aryl halide to achieve trifluoromethylation [66, 67]. Grushin group took a DFT calculation to reveal the mechanism of trifluoromethylation of aryl halide with CuCF$_3$ salt [68]. As shown in Figure 13.5, the oxidative addition of iodobenzene onto Cu(I)CF$_3$ salt takes place via transition state **13-33ts** with an energy barrier of 21.9 kcal/mol to form an aryl Cu(III) intermediate **13-34**. Then, a C(CF$_3$)–C(aryl) reductive elimination occurs via transition state **13-35ts** to form trifluoromethyl benzene product.

13.2.2 Trifluoromethylations Through Oxidative Coupling

When exogenous oxidants were used, nucleophilic trifluoromethyl can react with another nucleophile in the presence of copper catalyst. Qing and coworkers reported a Cu(I)I-catalyzed oxidative coupling of terminal alkyne and trifluoromethyl silane, which can proceed in air to construct trifluoromethyl acetylenes [69–71]. Maseras group reported a mechanistic study on the mechanism of this reaction [72]. As shown in Figure 13.6, the reaction starts from a trifluoromethyl Cu(I) intermediate **13-37**. The coordination of dioxygen molecule onto Cu(I) would form a side-on complex, which can be further assembled to corresponding side-on dimer **13-38**. Climbing over an energy barrier of 4.0 kcal/mol via oxidation transition state **13-39ts**, the O—O covalent bond is broken in a Cu(III) dimer **13-40**. Terminal alkyne can react with Cu(III) dimer **13-40** through metathesis-type C(alkynyl)—H bond cleavage to form an alkynyl Cu(III) intermediate **13-42**. Subsequently, another alkynylation results in decomposition of dimer to an alkynyl Cu(III) species **13-44**.

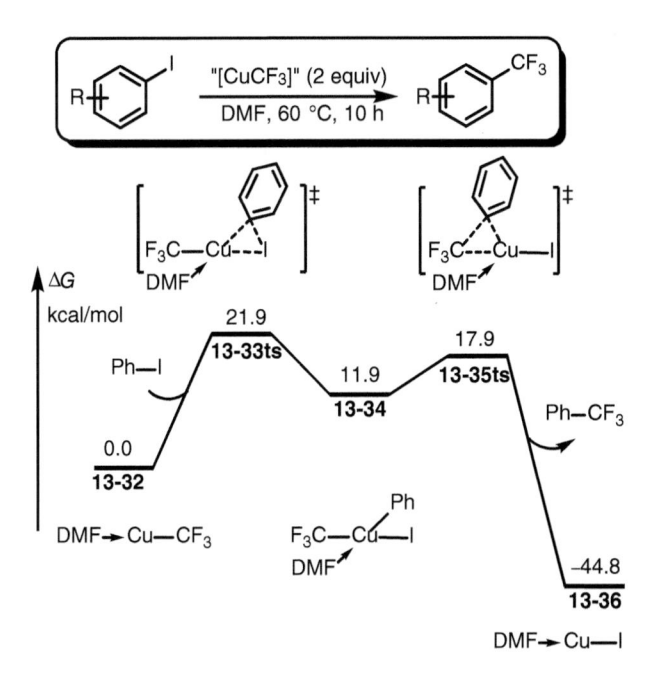

Figure 13.5 Free-energy profiles for the trifluoromethylation of iodobenzene using CuCF$_3$. The energies were calculated at mPW2PLYPD/6-311++G(2d,p) (SDD for Cu and I)/PCM(DMF)//B3LYP-D/6-31+G(d) (SDD for Cu and I)/PCM(DMF) level of theory. Sources: Holub and Kirshenbaum [25]; Kolb et al. [26].

Then, a reductive elimination of C(alkynyl)—C(CF$_3$) takes place via transition state **13-45ts** to yield trifluoromethyl acetylene product and afford Cu(I) species **13-46**. Active species **13-37** can be regenerated through a transmetallation with Me$_3$Si(CF$_3$) reactant assisted by potassium fluoride (KF) salt.

13.2.3 Radical-Type Trifluoromethylations

Togni's reagents, involving a hypervalent iodine, are usually considered as electrophilic trifluoromethyl source, which can react with other nucleophiles to achieve trifluoromethylation [73, 74]. In this chemistry, Cu(I) represents one-electron reductant to react with Togni's reagent through SET resulting in a trifluoromethyl free radical. Therefore, a radical-type mechanism can be proposed for this chemistry [75]. As shown in Scheme 13.3, Cu(I) can react with Togni's reagent through a SET process to afford trifluoromethyl radical and a Cu(II) species. The generated trifluoromethyl radical reacts with nucleophile resulting in an electron-rich radical, which can be oxidized by Cu(II) to yield cross-coupling product.

Following this idea, Li reported a mechanistic study of Cu(I)-catalyzed allylic trifluoromethylation, where C(allyl)—H bond plays as nucleophile to couple with electrophilic Togni's reagent II [76]. As shown in Figure 13.7, Cu(I)Cl salt can react with Togni's reagent II through a radical substitution via a singlet diradical transition state **13-48ts** to afford a mixture **13-49** of trifluoromethyl radical and Cu(II) species. The

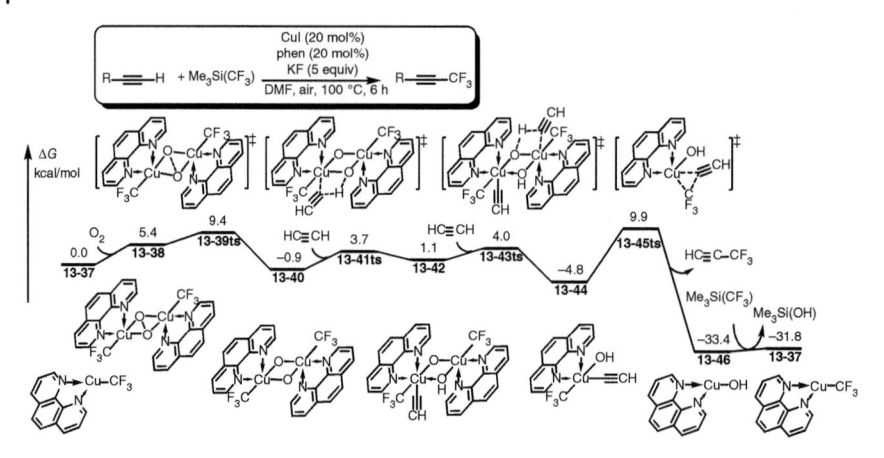

Figure 13.6 Free-energy profiles for the Cu-catalyzed trifluoromethylation of terminal alkynes. The energies were calculated at PBE-D3/6-311+G(d,p) (aug-cc-pVTZ for Cu)/SMD(DMF)//PBE/6-31+G(d) (SDD for Cu)/SMD(DMF) level of theory.

Scheme 13.3 The generation of CF$_3$ radical.

generated trifluoromethyl radical can react with alkene through radical addition via transition state **13-50ts** to form an alkyl radical **13-51**. The alkyl radical coordinates onto Cu(II) species via transition state **13-52ts** resulting in a further oxidation to form Cu(III)-alkyl intermediate **13-53**. Then, a benzoate-assisted reductive deprotonation takes place via transition state **13-54ts** and the trifluoromethylation product is formed with the formation of Cu(I)Cl salt **13-55**.

In another work, Yu, Bai, and Lan studied the mechanism of copper-catalyzed oxytrifluoromethylation of allylamines with CO$_2$ [77]. In this theoretical study (Figure 13.8), aminomethyl indole **13-64** was chosen as the model reactant, which can react with the copper catalyst to afford Cu(I) carboxylate complex **13-56^s** in the presence of CO$_2$. Coordination of Togni's reagent II leads to homolytic cleavage of the I—CF$_3$ bond by spin crossover via MECP **13-58^{MECP}**. Release of the trifluoromethyl radical is accompanied by formation of Cu(II) dicarboxylate complex **13-59^d**. Intermolecular radical addition of the trifluoromethyl radical to the C=C double bond in the indole moiety then occurs via transition state **13-60ts^t** to give a benzylic radical in triplet intermediate **13-61^t**. In this triplet intermediate, the coordinated carboxylate shows some radical character, and it can intramolecularly react with the benzylic radical to form a new C—O bond. C—O bond formation also requires spin crossover via MECP **13-62^{MECP}**. When singlet Cu(I) intermediate **13-63^s** forms, product **13-65** can be released by ligand exchange with reactant **13-64**. Active species **13-56^s** can be regenerated by CO$_2$ insertion.

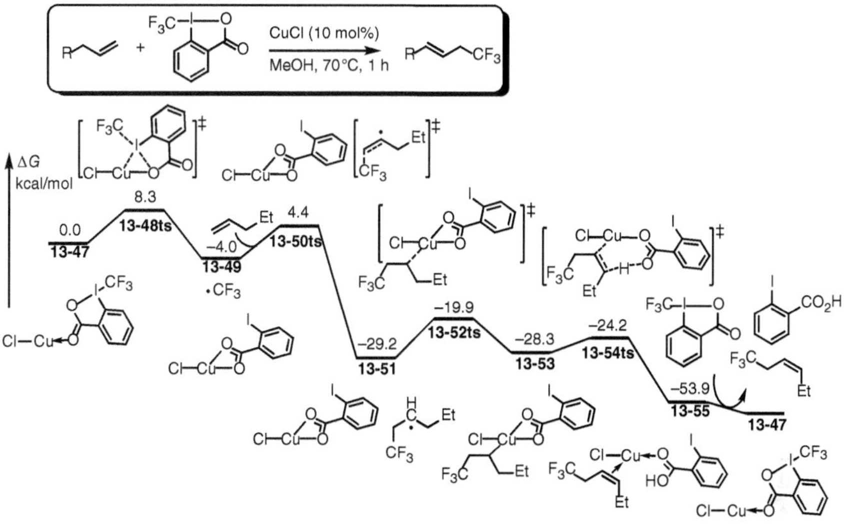

Figure 13.7 Free-energy profiles for the Cu-catalyzed trifluoromethylation of alkenes using Togni's reagent II. The energies were calculated at B3LYP/6-31+G(d,p) (aug-cc-pVTZ for Cu, SDD for I)/SMD(methanol) level of theory.

13.3 Cu-Mediated C—H Activations

During the past decade, copper-mediated C—H activation has attracted more attention as a complementary method; however, various reaction properties of copper catalysis led to the complexity of the Cu-mediated C—H activations [78–80]. As a late transition metal, copper-catalysis can undergo a double-electron transfer process, where copper can have an electrophilic active arene for the deprotonation of nucleophilic attack. Moreover, SETs are often involved in copper-catalysis. Therefore, redox activation of arenes also could lead to sequential C—H cleavage through electrophilic substitution or proton transfer.

A series of theoretical studies focus on the mechanism of copper-mediated C—H activation and functionalizations. Some computational examples are given in this section to elaborate the detailed mechanism for copper-mediated C—H functionalizations.

13.3.1 C—H Arylations

C(aryl)—H can be considered as nucleophile to react with aryl halide to achieve redox-neutral cross-coupling [81]. A deprotonation-metallation mechanism was proposed by Lin group and supported by their DFT calculations [57]. As shown in Figure 13.9, the reaction starts from a methoxy Cu(I) σ-complex **13-66**, which can react with benzoxazole to break the C—H bond via a four-membered ring-type metathesis transition state **13-67ts**. The calculated energy barrier of this step is only 14.8 kcal/mol. The generated aryl Cu(I) intermediate **13-68** undergoes a metathesis

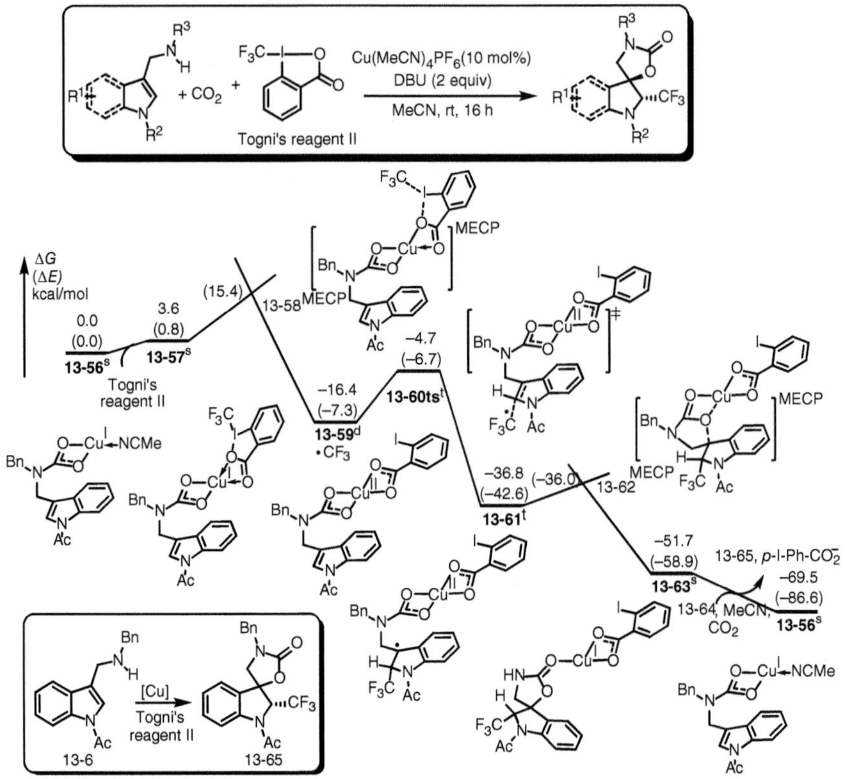

Figure 13.8 Free-energy profiles of copper-catalyzed oxytrifluoromethylation of allylamines. The energies were calculated at the B3LYP/6-311+G(d,p) (SDD for Cu and I)/IEF-PCM(acetonitrile)//B3-LYP/6-31G(d) (SDD for Cu and I) level of theory. The values in parentheses are the relative electronic energies.

with phenyl iodide via transition state **13-69ts** to yield phenyl benzooxazole product. Active species **13-66** can be generated through counterion exchange.

In another example, Li and Wu focused on the mechanism of copper-catalyzed meta-arylation of anilide, which was reported by Gaunt group [82]. In this work, a Heck-like mechanism was proposed to support the unique meta-selectivity [83]. As shown in Figure 13.10, the reaction starts from an amide-coordinated aryl-Cu(III) species **13-71**. The chelated coordination of aryl group onto electron-deficient Cu(III) center takes place via transition **13-72ts** to generate complex **13-73** with 13.3 kcal/mol endergonic. Arene is activated by the coordination onto Cu(III) center, which leads to an insertion-type nucleophilic attack by phenyl group via a four-membered ring-type transition state **13-74ts**. The calculated overall activation free energy of this Heck-like insertion is 19.2 kcal/mol, which indicates the rate-determining step. After this process, a dearomatic alkyl Cu(III) intermediate **13-75** is formed. The following reductive deprotonation takes place by trifluoromethanesulfonate via transition state **13-76ts** to yield meta-arylation product.

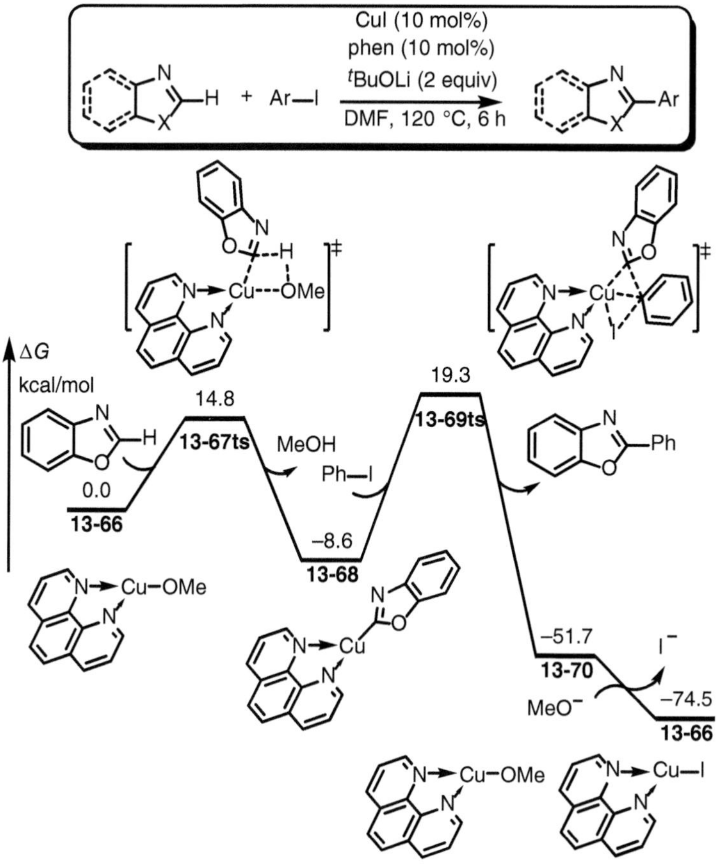

Figure 13.9 Free-energy profiles of copper-catalyzed arylation of heterocycles through C–H activation. The energies were calculated at the B3LYP/6-31G(d,p) (6-311G(d) for Cu, and LANL2DZ for I)/CPCM(DMF)//B3LYP/6-31G(d) (6-311G(d) for Cu, and LANL2DZ for I) level of theory.

Active species **13-71** can be regenerated by the oxidation of Cu(I) species **13-77** by Ph$_2$IOTf reactant.

13.3.2 C–H Aminations

A theoretical study taken by Fu and coworkers revealed the mechanism of intramolecular C—H amination of benzimidamide to construct benzoimidazole ring [84]. As shown in Figure 13.11, concerted metallation-deprotonation mechanism is supported by DFT calculations. The reaction starts from the coordination of benzimidamide onto Cu(OAc)$_2$ to form complex **13-79** with 4.8 kcal/mol exergonic. Then, a concerted ortho-metallation-deprotonation takes place via a six-member ring transition state **13-80ts** with an energy barrier of 22.0 kcal/mol to form an aryl Cu(II) intermediate **13-81**. An oxidative disproportionation with another

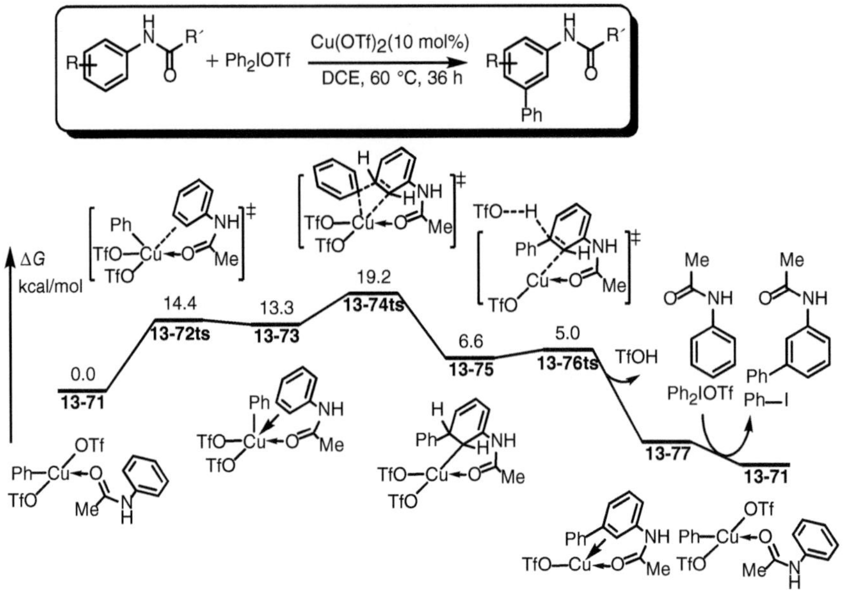

Figure 13.10 Free-energy profiles of copper-catalyzed *meta*-arylation of C(aryl)–H bond. The energies were calculated at the B3LYP/6-31G(d,p) (6-311G(d) for Cu, and 6-31+G(d) for O)/IEF-PCM(DCE) level of theory.

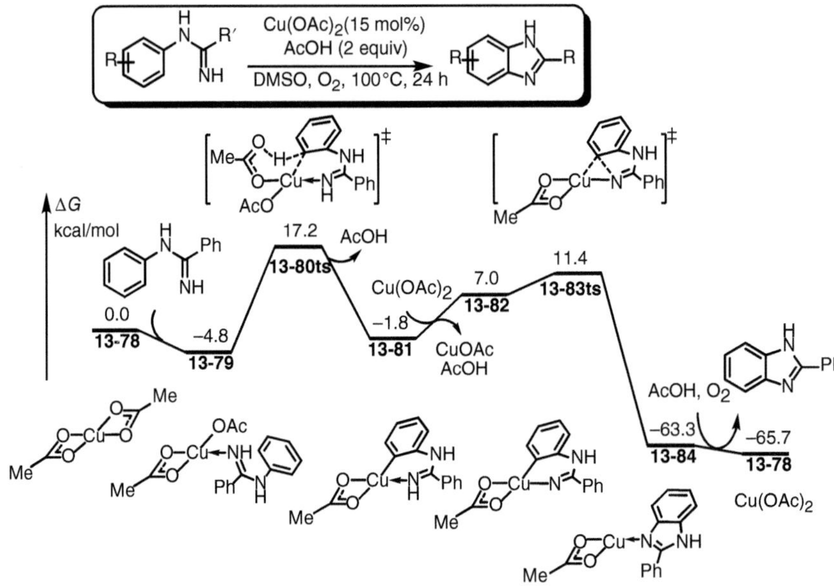

Figure 13.11 Free-energy profiles of copper-catalyzed intramolecular C(aryl)-H amination of benzimidamide to construct benzoimidazole. The energies were calculated at the M06/6-311++G(d,p)/SMD(DMSO)//B3P86/6-31+G(d) level of theory.

Cu(OAc)$_2$ provides Cu(III) intermediate **13-82** with 8.8 kcal/mol endergonic, which can bear a rapid C(aryl)—N reductive elimination via transition state **13-83ts** to yield benzoimidazole product-coordinated Cu(I) complex **13-84**. The oxidation by dioxygen regenerates Cu(II) species **13-78** and releases benzoimidazole product.

13.3.3 C—H Hydroxylation

The hydroxylation of a C(aryl)—H bond can undergo copper-mediated oxidation with hydrogen peroxide. In a combination of theoretical and experimental study on the mechanism of copper-catalyzed benzene oxidation reaction, a radical-type C—H activation was proposed [85]. A high-spin trispyrazolylborate-coordinated

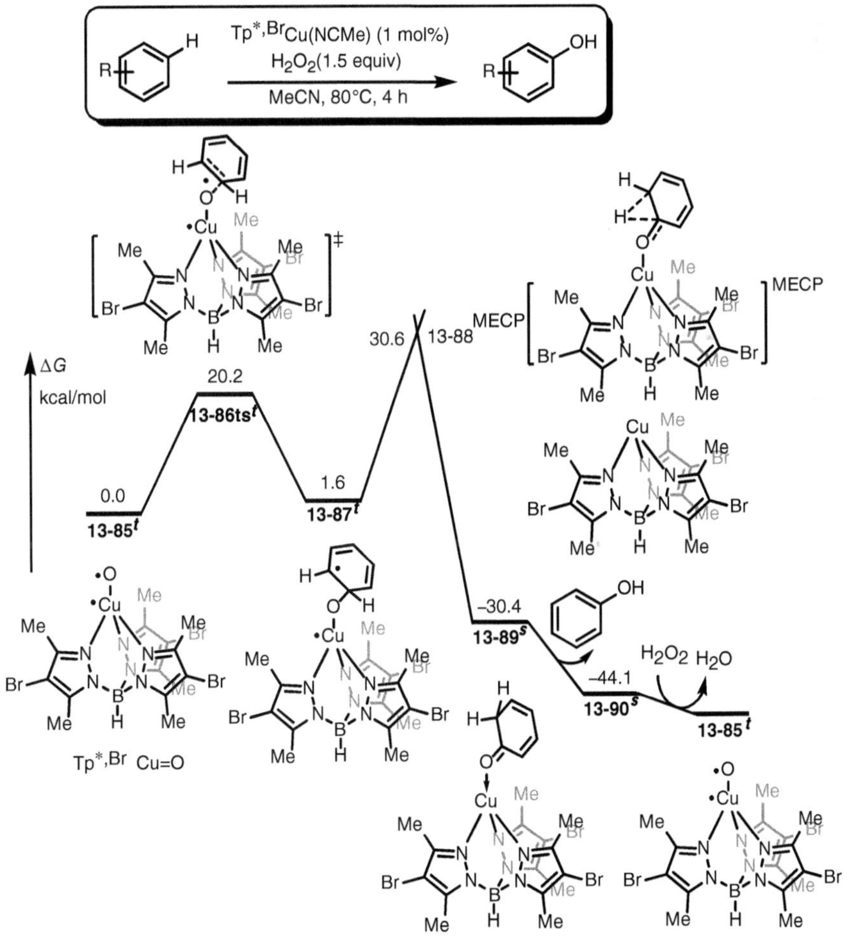

Figure 13.12 Free-energy profiles of copper-catalyzed oxidative hydroxylation of a C(aryl)—H bond. The energies were calculated at the BHandHLYP/6-311+G(d,p) (SDD for Cu)/SMD(acetonitrile)//BHandHLYP/6-31G(d,p) (SDD for Cu) level of theory.

532 | *13 Theoretical Study of Cu-Catalysis*

Cu(II)-oxo radical **13-85^t** was used as catalyst. A radical attack of oxo moiety onto benzene takes place via transition state **13-86ts^t** to give a diradical species **13-87^t**. A 1,2-hydrogen shift occurs through a minimum energy crossing point **13-88^{MECP}** to form a singlet cyclohexadienone-coordinated Cu(I) species **13-89^s** Phenol can be easily achieved through isomerization of cyclohexadienone, while Cu(I) species **13-90^s** can be oxidized by hydrogen peroxide to rerun the catalytic cycle (Figure 13.12).

13.3.4 C−H Etherifications

When Cu(III) species was used to assist C(acyl)−H activation, a simultaneous C−H activation/reductive elimination can occur as the key step [86]. Fang performed a DFT calculation to reveal the mechanism of etherification of N,N-dimethylformamide (DMF), where a simultaneous C−H activation/reductive elimination was proposed [87]. As shown in Figure 13.13, Cu(I) species **13-91** reacts with *tert*-butyl hydroperoxide oxidant through an open-shield triplet process (**13-92ts**) resulting in Cu(II) species **13-93** and a *tert*-butoxy radical. Then, a DMF-assisted oxidation of Cu(II) species **13-93** to Cu(III) species **13-96** occurs through an outer-sphere hydrogen transfer from DMF to *tert*-butoxy radical followed by intermolecular hydrogen transfer from acetoacetic ester ligand to DMF radical via transition state **13-95ts**. Then, a hydroxyl-assisted deprotonation of coordinated DMF takes place via transition state **13-99ts**; however, the generated Cu(III) intermediate was not located, which can further transfer to a Cu(I)-intermediate **13-100** through reductive elimination. After ligand exchange, carbamate product is yielded with the regeneration of Cu(I) species **13-91**.

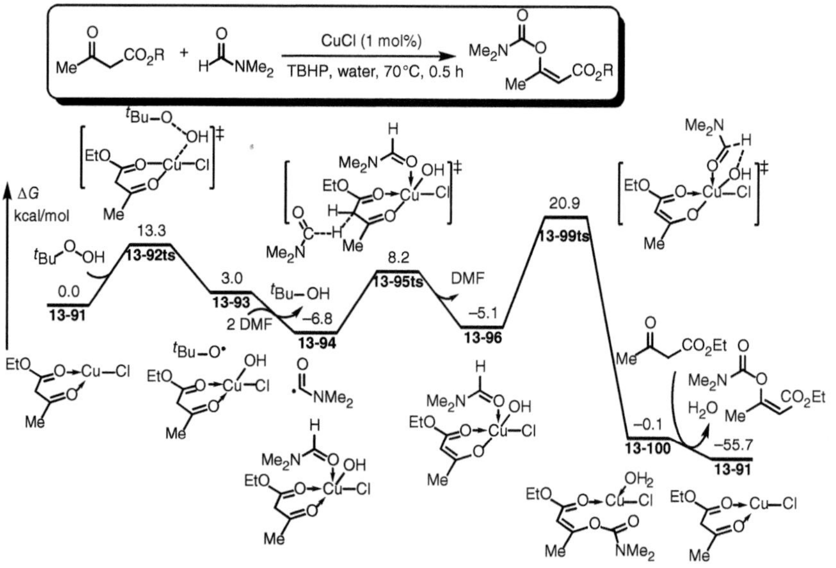

Figure 13.13 Free-energy profiles of copper-catalyzed C-H etherifications of DMF. The energies were calculated at the B3LYP/DZVP/PCM(DMF) level of theory.

13.4 Cu-Mediated Alkyne Activations

Copper as a late group IB transition metal can be used to activate alkynes through various possible pathways depending on the reaction site of alkynes. As shown in Scheme 13.4, the coordination of alkyne onto copper would bring an insertion of alkyne into Cu—nucleophile bond, which can be understood as a typical transition metal character [88]. As a variant of insertion, alkyne-coordinated Cu-species can react with other unsaturated bond through an oxidative cyclization to achieve copper cycle [89]. Copper can play a π-acid, which can be coordinated by the π-orbital of alkyne to significantly reduce the electron density. It would lead to a further intermolecular nucleophilic attack with exogenous nucleophiles [90]. When copper reacts with terminal alkynes in the presence of base, the coordination onto copper increases the acidity of alkyne, which can cause a deprotonation resulting in an alkynyl Cu species. The generated alkynyl Cu species can be used as a nucleophile to react with electrophile or nucleophile to achieve redox-neutral cross-coupling or oxidative coupling [91]. Following this idea, the mechanism of copper-catalyzed alkyne activation reactions is discussed in this section.

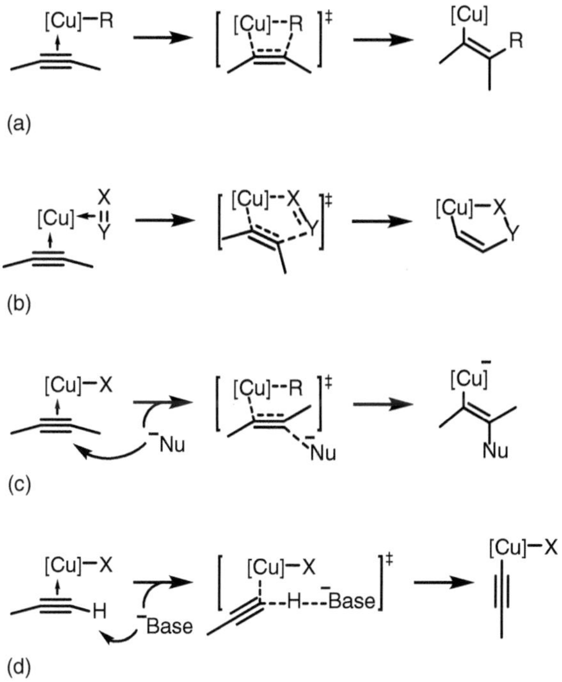

Scheme 13.4 Cu-mediated alkyne activation modes. (a) Alkyne insertion, (b) oxidative cycloaddition, (c) nucleophilic attack, and (d) base-assisted deprotonation.

13.4.1 Azide–Alkyne Cycloadditions

The Cu(I)-catalyzed [3+2] cycloadditions of terminal alkynes and organic azides to give 1,4-disubstituted 1,2,3-triazoles exhibit remarkably broad scope and exquisite

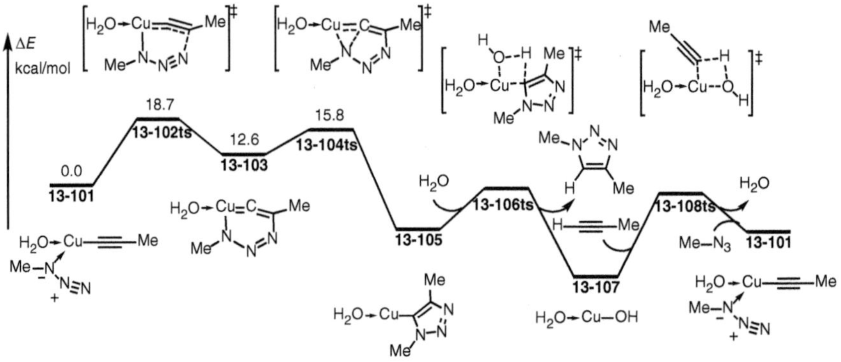

Figure 13.14 Potential energy profiles of Cu-catalyzed [3+2] cycloadditions of terminal alkynes and azides. The energies were calculated at the B3LYP/6-311+G(2d,2p)/COSMO(water)//B3LYP/6-311G(d,p) level of theory.

selectivity (Scheme 13.5) [25, 26]. The best click reaction to date, it quickly found applications in chemistry, biology, and materials science. Thinking up mechanistic schemes that might explain such a robust process is an interesting but daunting challenge. While most "incredible" reactivity findings remain just that, on the rare occasion that important and unprecedented reactivity is proven bona fide, it is likely to represent the signature of a new intermediate and/or a pathway.

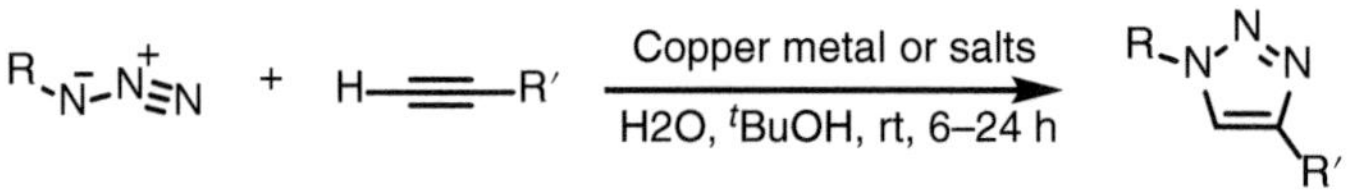

Scheme 13.5 Cu-catalyzed [3+2] cycloadditions of terminal alkynes and organic azides.

As everyone knows, the uncatalyzed [3+2] cyclization of azides and alkynes through concerted pathway would suffer a rather high-energy barrier (about 25–30 kcal/mol), which is hard to occur in room temperature. Therefore, an alkynyl Cu(I) intermediate was proposed and considered as key species in pioneering works of mechanistic study reported by Noodleman, Sharpless, and Fokin [92]. A stepwise pathway is given in Figure 13.14. When alkynyl Cu(I) species **13-107** is formed, it can be coordinated by azide to form complex **13-101**. The terminal nitrogen of the coordinated azide attacks the internal carbon of alkynyl group via transition state **13-102ts** with an energy barrier of 18.7 kcal/mol to afford an unusual six-membered copper(III) metallacycle **13-103**. This step is endothermic by 8.2 kcal/mol, and the calculated barrier is 14.9 kcal/mol, which is considerably lower than that for the uncatalyzed reaction. Then, a reductive migration of nitrogen forms another C—N bond via transition state **13-104ts** resulting in a triazolyl-copper species **13-105**. The Bronsted alkalinity of triazolyl is stronger than that of terminal alkyne, which leads to the regeneration of active species **13-107** through a metathesis process.

In a more detailed DFT study, Tämm, Sikk and co-workers reported free-energy profiles for a binuclear copper-catalyzed azide alkyne [3+2] cycloaddition through

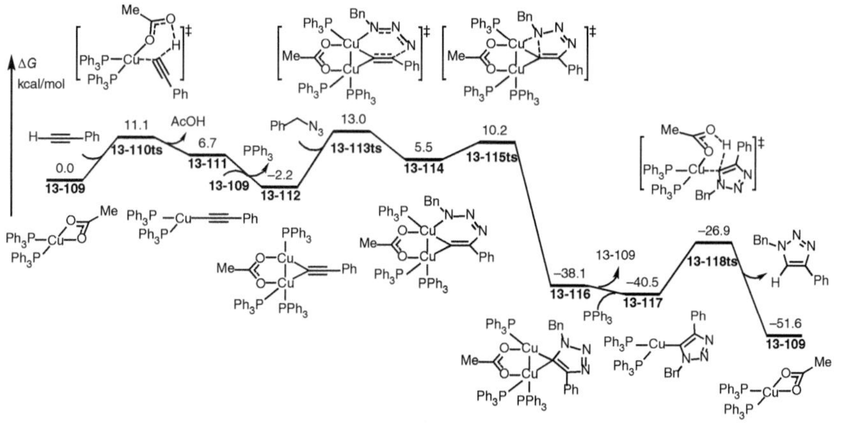

Figure 13.15 Free-energy profiles of Cu-catalyzed [3+2] cycloadditions of terminal alkynes and azides through a dimeric copper-mediated pathway. The energies were calculated at the B3LYP/6-311+G(2d,2p)/COSMO(water)//B3LYP/6-311G(d,p) level of theory.

a dimeric copper-mediated pathway [93]. As shown in Figure 13.15, the reaction starts from a mononuclear Cu(I) acetate complex **13-109**, which can undergo a concerted metallation-deprotonation with terminal acetylene via transition state **13-110ts** to form an alkynyl Cu(I) species **13-111**. Then, it can react with another Cu(I) acetate complex **13-109** to afford binuclear species **13-112** with 8.9 kcal/mol exergonic, where the terminal carbon in alkynyl group plays as bridge between the two copper atoms. The coordination of azide leads to a nucleophilic attack of terminal nitrogen onto alkynyl group via transition state **13-113ts** to form a six-membered cycle **13-114**. A reductive elimination occurs via transition-state **13-115ts** to achieve triazolyl cycle. The dissociation of binuclear copper and protonation yields triazole product and regenerates mononuclear Cu(I) acetate complex **13-109**. The rate-determining step is detected to be the nucleophilic attack step with an observed energy barrier of 15.2 kcal/mol.

Interestingly, tetramer of alkynyl Cu(I) species was reported as real structure, which has been considered as starting material for copper-catalyzed azide alkyne [3+2] cycloadditions [94]. As shown in Figure 13.16, the coordination of azide onto alkynyl Cu(I) tetramer **13-119** forms complex **13-120** with 8.5 kcal/mol endergonic. Subsequently, the cycloaddition occurs through a similar stepwise pathway. The calculated activation free energy for this case is 25.2 kcal/mol, where the key step is nucleophilic attack of azide onto alkynyl via transition state **13-121ts**.

13.4.2 Nucleophilic Attack onto Alkynes

Cu(I) species can be coordinated by π-acidic alkynes to achieve electrophilic π-bond activation, which would lead to a nucleophilic attack to afford trans-vinyl copper. Alternatively, alkyne also can undergo an insertion process to achieve cis-vinyl copper species.

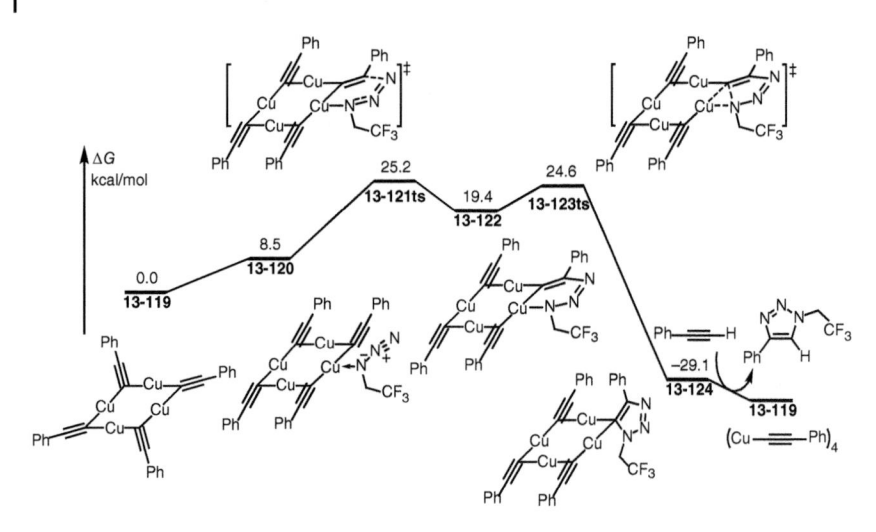

Figure 13.16 Free-energy profiles of Cu-catalyzed [3+2] cycloadditions of terminal alkynes and azides through a tetramer copper-mediated pathway. The energies were calculated at the B3LYP/6-311+G(d,p)/IEF-PCM(acetonitrile)//B3LYP/6-311+G(d,p) level of theory.

As an example, Janesko studied the mechanism of copper-catalyzed cyclization of *ortho*-alkynylbenzamides, which provide Z-enisoindolinones through 5-exo attack [90, 95]. As shown in Figure 13.17, DFT calculations found that when an amido Cu(I) species **13-126** is formed with the coordination of alkynyl group onto Cu(I), an intramolecular nucleophilic 5-exo-attack of amido group onto copper-actived alkynyl takes place via transition state **13-127ts** with an energy barrier of 7.2 kcal/mol, which provides five-membered vinyl copper species **13-128**. An alternative 6-endo nucleophilic attack was also considered, which would occur via transition state **13-129ts**. The calculated energy barrier is 3.3 kcal/mol higher than that of 5-exo-attack.

Alternatively, Gunnoe and Cundari reported an experimental and theoretical study on the mechanism of copper-catalyzed intramolecular hydroalkoxylation of alkynes, which would undergo an insertion process to achieve alkyne functionalizations [96]. In their DFT calculations (Figure 13.18), the coordination of alkynyl group onto copper takes place via transition state **13-132ts** with an energy barrier of 10.9 kcal/mol to afford an alkynyl-coordinated Cu(I) species **13-133**. Then, an intramolecular alkyne insertion occurs via a four-membered ring-type transition state **13-134ts** to provide exo-addition vinyl copper intermediate **13-135**. Ligand exchange with alkynyl alcohol can yield methylene furan product.

13.4.3 Alkynyl Cu Transformations

When a terminal alkyne reacts with copper in the presence of base, a deprotonation-metallation mechanism provides alkynyl copper species, which plays as nucleophile in coupling reactions.

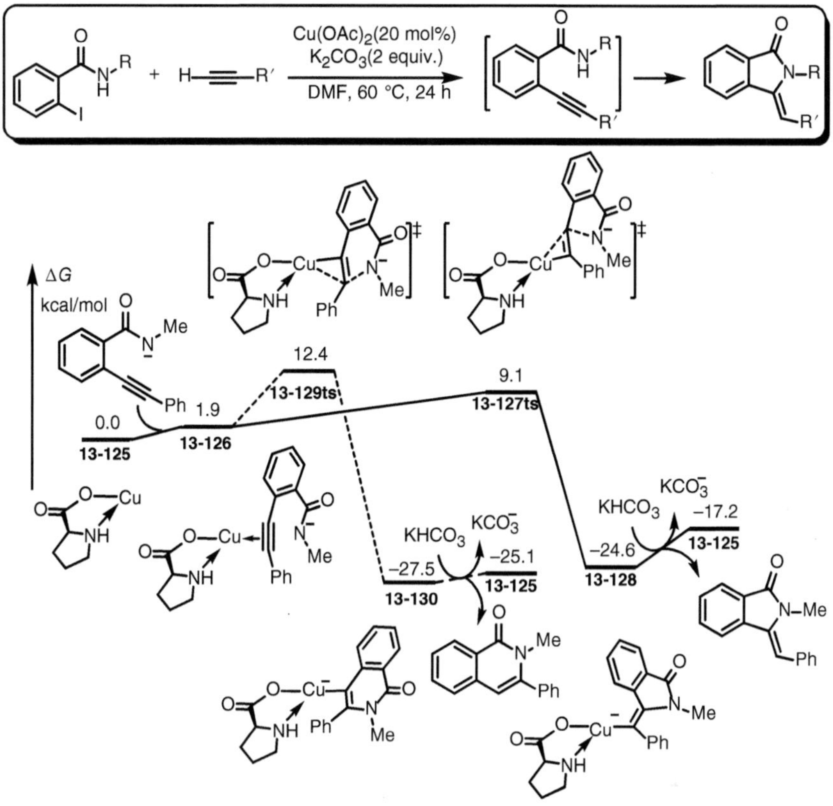

13.4 Cu-Mediated Alkyne Activations | 537

Figure 13.17 Free-energy profiles of the competition of intramolecular 5-exo/6-endo nucleophilic attack of Cu-activated alkyne. The energies were calculated at the B3LYP/6-31+G(d,p) (LANL2DZ for Cu)/CPCM(DMSO) level of theory.

In 2014, Lin reported a mechanistic study of copper-catalyzed cross-coupling of terminal alkynes and allylic chloride (Figure 13.19) [97]. The reaction starts from a Cu(I) chloride species **13-136**, which can react with acetylene in the presence of base through a metallation-deprotonation transition state **13-137ts** to afford alkynyl Cu(I) intermediate **13-138** with 10.0 kcal/mol exergonic. In this alkynyl Cu(I) intermediate, alkynyl group plays a nucleophile, which can react with carbon dioxide through carbonyl insertion via transition state **13-139ts** to generate a carboxylate Cu(I) intermediate **13-140**. Then, an intermolecular nucleophilic substitution with allyl chloride takes place via transition state **13-141ts** with a free-energy barrier of 25.1 kcal/mol and forms the coupling product–coordinated Cu(I) intermediate **13-142**. Ligand exchange with chloride regenerates CuCl species **13-136** and releases coupling product. The whole catalytic cycle is nonredox process, where the oxidative state of Cu kept at +1.

Copper(II) dichloride **13-144**[d] can be rapidly reduced by phenylacetylene to afford dimeric cuprous species **13-145**[s] and the corresponding butadiyne under very

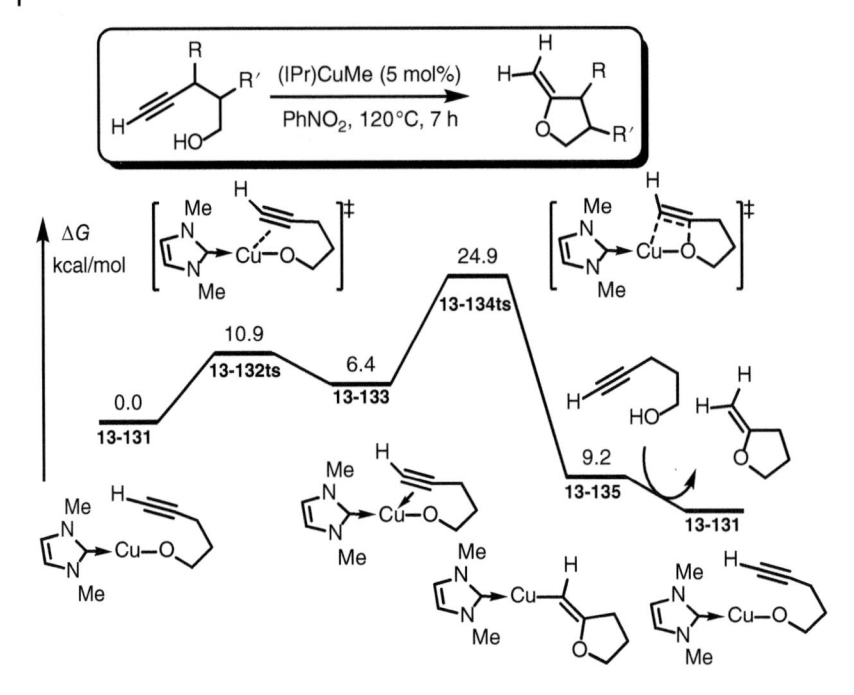

Figure 13.18 Free-energy profiles of the competition of intramolecular exo-hydroalkoxylation of alkyne through insertion pathway. The energies were calculated at the B3LYP/6-311+G(d) (SDD for Cu) level of theory.

mild conditions (−20 °C, 10 minutes) in the presence of catalytic species **13-145^s**. Obviously, a powerful oxidant that can oxidize Cu(II) to Cu(III) does not exist in this system. Therefore, a bimetallic single-electron reductive elimination mechanism has been proposed for this reaction [98, 99]. The DFT-calculated free-energy profile for the key step of this reaction is shown in Figure 13.20. The reaction starts from two molecules of alkynyl Cu(II) species **13-144^d**. Dimerization of **13-144^d** forms triplet complex **13-146^t**, where two η^2-alkynyl groups bridge the copper centers. After release of a chloride ion, cationic dimer **13-147^t** can be generated with another η^2-chloride bridge. The geometry information of **13-147^t** is shown in Figure 13.20, which reveals two asymmetric η^2-alkynyl groups. Inside, the terminal sp hybrid orbital of one alkynyl group coordinates to one copper atom by end-on coordination, while side-on coordination of the alkynyl π-orbital to another copper achieves a η^2-bridge. After spin crossover via MECP **13-148^{MECP}**, triplet complex **13-147^t** can be reversibly isomerized to the corresponding singlet **13-149^s** with an energy barrier of 9.1 kcal/mol. The geometry information of **13-149^s** shows that the two alkynyl groups are symmetric, and each of them shares its terminal sp hybrid orbital with two copper atoms to achieve end-on coordination. DFT calculations revealed that the free-energy barrier of reductive elimination in triplet state is 22.1 kcal/mol via transition state **13-150ts^t**, but the corresponding free-energy barrier in the singlet state is only 4.4 kcal/mol via transition state **13-150ts^s**. Therefore, the whole

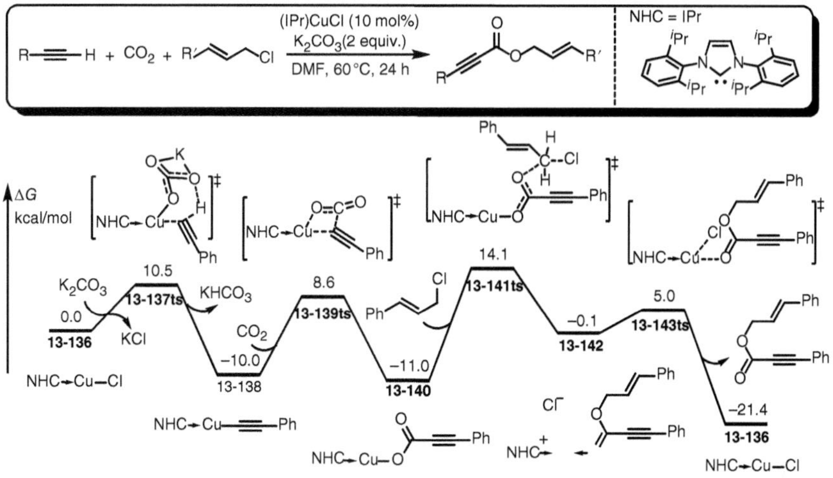

Figure 13.19 Free-energy profiles for the copper-catalyzed cross-coupling of terminal alkynes and allylic chloride with carbon dioxide insertion. The energies were calculated at the B3P86/6-31G(d,p) (6-311+G(d) for Cu, Cl, and O)/PCM(DMF) level of theory. Source: Modified from Yuan and Lin [97].

pathway can be concluded to be a three-step process. Dimerization of alkynyl Cu(II) species **13-144^d** forms triplet dimeric Cu(II) intermediate **13-147^t**, which can undergo spin crossover to afford singlet Cu(II) intermediate **13-149^s**. Rapid reductive elimination then gives the butadiyne product with generation of singlet Cu(I) dimer **13-145^s** by coordination of the leaving chloride ion.

13.5 Cu-Mediated Carbene Transformations

The combination of copper and carbene precursors can afford a relatively stable Fischer-type copper-carbene complex, where the carbon atom of carbene represents strong electrophilicity. Therefore, when copper-carbene complex is formed, it can react with various nucleophiles to achieve the construction of new covalent bonds [100–104]. In this section, the mechanism for the transformations of copper-carbene complexes would be discussed in the light of the employment of nucleophiles.

13.5.1 [2+1] Cycloadditions with Alkenes

When copper-carbene complex is generated, it can react with ethylene to achieve cyclopropanation. Two possible pathways have been proposed for this reaction [105–108]. As shown in Scheme 13.6, when copper-carbene complex is formed, it can react with acetylene through a concerted cycloaddition to form cyclopropane product, where the carbene moiety is considered as an electrophile. Alternatively, a stepwise pathway would undergo an intermolecular metathesis to afford a metallacyclobutane intermediate, which can suffer a reductive elimination to result

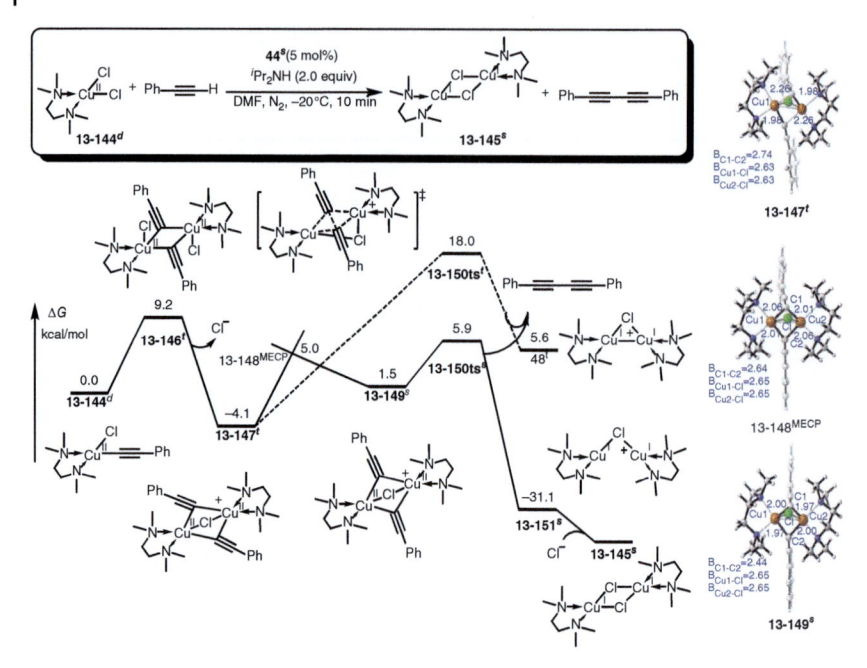

Figure 13.20 Free-energy profiles of Cu(II)-mediated oxidative homo-coupling of phenylacetylene. The energies were calculated at the M11-L/6-311+G(d) (SDD for Cu)/SMD(DMF)//B3LYP/6-31G(d) (SDD for Cu) level of theory. The bond lengths in the structures are given in Å.

in the same cyclopropane product. In this case, metal-carbene complex exhibits metal-carbene double-bond character.

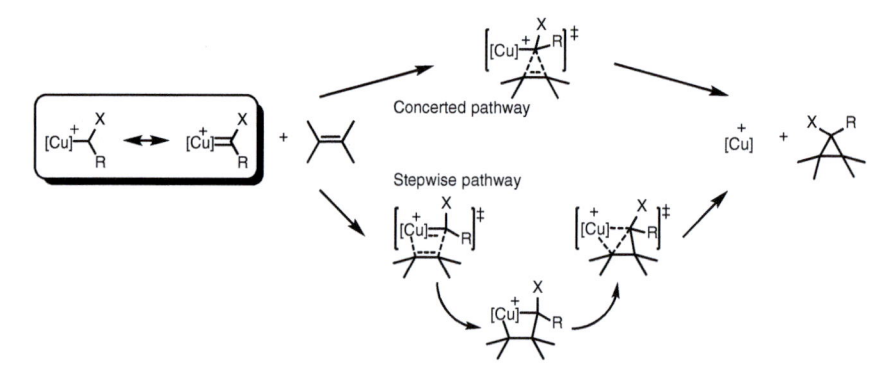

Scheme 13.6 Possible mechanism of Cu-mediated [2+1] cycloadditions of carbenes and alkenes.

DFT calculation was used to reveal the mechanism of copper-mediated cyclopropanation of simple ethylene [109]. As shown in Figure 13.21, alkene-coordinated cationic Cu(I) species **13-152** was chosen as relative zero in calculated free-energy profiles. The ligand exchange loads diazo reactant in complex **13-153** with

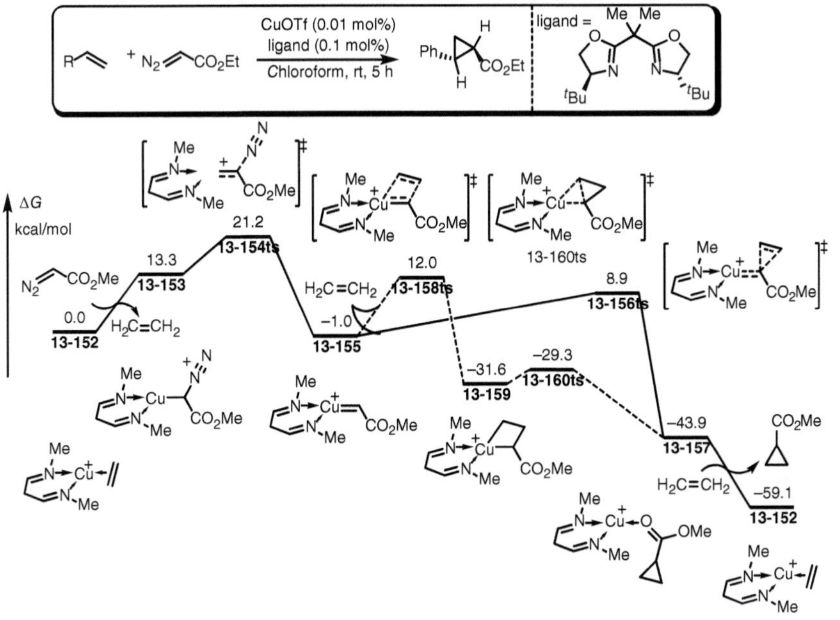

Figure 13.21 Free-energy profiles of Cu-catalyzed [2+1] cycloadditions of carbenes and alkenes. The energies were calculated at the B3LYP/6-311+G(2d,p)/IPCM(DCM)//B3LYP/6-31G(d) level of theory.

13.3 kcal/mol endergonic. The sequential denitrogenation takes place via transition state **13-154ts** with an overall activation free energy of 21.2 kcal/mol and provides copper-carbene complex **13-155**. A concerted cycloaddition can occur via transition state **13-156ts** with a free-energy barrier of 9.9 kcal/mol. Then, cyclopropane product can be yielded with the regeneration of Cu(I) catalyst **13-152**. In an alternative process, π-bond metathesis via transition state **13-158ts** would occur to afford metallacyclobutane intermediate **13-159**, which can undergo a rapid reductive elimination via transition state **13-160ts** to achieve the same product. The calculated activation energy for the metathesis step is 13.0 kcal/mol, which is 3.1 kcal/mol higher than that of concerted cycloaddition process.

13.5.2 Carbene Insertions

In the presence of copper-catalyst, carbene can formally insert into C(aryl)—H bond to construct new C(aryl)—C(alkyl) bond [110–112]. As shown in Figure 13.22, DFT calculation has been performed to reveal the mechanism of carbene insertion reaction. In this reaction, coordinated base in complex **13-161** reacts with N-iminopyridinium first to provide pyridyl lithium intermediate **13-163**, which can undergo a transmetallation with copper to form pyridyl Cu(I) species **13-165**. The denitrogenation of coordinated diazo takes place via transition state **13-166ts** with an energy barrier of 19.2 kcal/mol to form a copper-carbene complex **13-167**. Then, a carbene insertion occurs via a three-membered ring-type transition state **13-168ts**

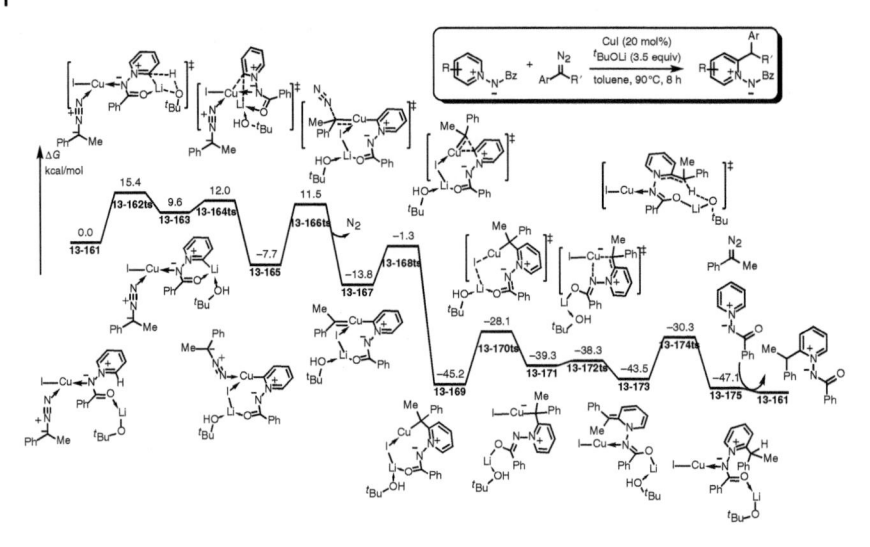

Figure 13.22 Free-energy profiles of Cu-catalyzed carbene insertion into C(aryl)−H bond. The energies were calculated at the B3LYP/6-311+G(d,p) (SDD for Cu and I)/SMD(toluene) level of theory.

to afford an alkyl copper intermediate **13-169**. Then, sequential rearrangements via transition states **13-170ts** and **13-172ts** result in a methylene pyridine intermediate **13-173**, which can be protonated by alcohol to yield corresponding product.

13.5.3 Rearrangement of Carbenes

Interestingly, alkyl copper-carbene species can undergo an intramolecular 1,2-hydrogen shift to provide corresponding alkene. Jiao reported a cascade reaction for the directly synthesis of alkenyl nitriles from simple alkynes by using azidosilane in the presence of copper-catalyst and oxygen [113]. The reaction would occur through an α-diazonitrile intermediate. DFT calculation was used to reveal the mechanism of transformation from α-diazonitrile to acrylonitrile. As shown in Figure 13.23, the coordination of α-diazonitrile onto cationic copper species **13-176** forms intermediate **13-177**. A sequential denitrogenation takes place via transition state **13-178ts** to afford copper-carbene complex **13-179**. A 1,2-hydrogen shift occurs via transition state **13-180ts** with an energy barrier of only 3.3 kcal/mol to achieve the isomerization of carbene and yield acrylonitrile product.

13.6 Cu-Mediated Nitrene Transformations

Although nitrene can be considered as the isoelectronic species of carbene, the high-spin state of copper-nitrene complexes is usually more stable than the singlet one that is different from copper-carbene complex [114–116]. Therefore, the spin states should be considered in theoretical studies on the mechanism of copper-nitrene transformations.

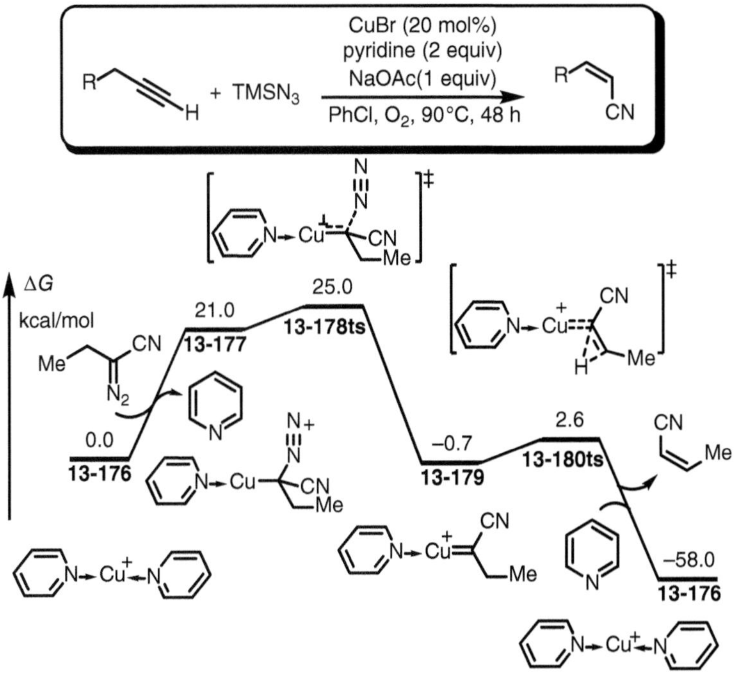

Figure 13.23 Free-energy profiles of Cu-mediated transformation from α-diazonitrile to acrylonitrile through carbene rearrangement. The energies were calculated at the B3LYP/6-311+G(d,p) (SDD for Cu)/CPCM(PhCl)//B3LYP/6-31G(d) (SDD for Cu) level of theory.

13.6.1 [2+1] Cycloadditions with Alkenes

As shown in Figure 13.24, nitrene precursor can react with styrene to synthesize aziridine derivatives in the presence of copper-catalyst [117–119]. DFT calculation found that the triplet copper-nitrene complex **13-181^s** is 4.7 kcal/mol more stable than the corresponding singlet one **13-181^t** [120]. It can react with styrene through a radical-type attack via transition state **13-182ts^t** with an energy barrier of 12.3 kcal/mol to afford a diradical intermediate **13-183^t**. Then, a spin cross-over takes place to afford singlet intermediate **13-183^s** with 2.8 kcal/mol exergonic. Then, the second C—N bond is formed rapidly in aziridine product. In an alternative singlet transformation, the first C—N bond formation could occur via transition state **13-182ts^s**. The calculated relative free energy for the singlet C—N bond formation transition state **13-182ts^s** is 5.1 kcal/mol higher than that of triplet one.

13.6.2 Amination of Nitrenes

Copper-nitrene complex can undergo an intramolecular nucleophilic attack by nitrogen to achieve annulation in the synthesis of heterocycles [122]. In 2014, Li and Yan reported a DFT study on the mechanism of a transformation from azide to indazole (Figure 13.25) [121]. In this work, a binuclear Cu(I) species **13-186**

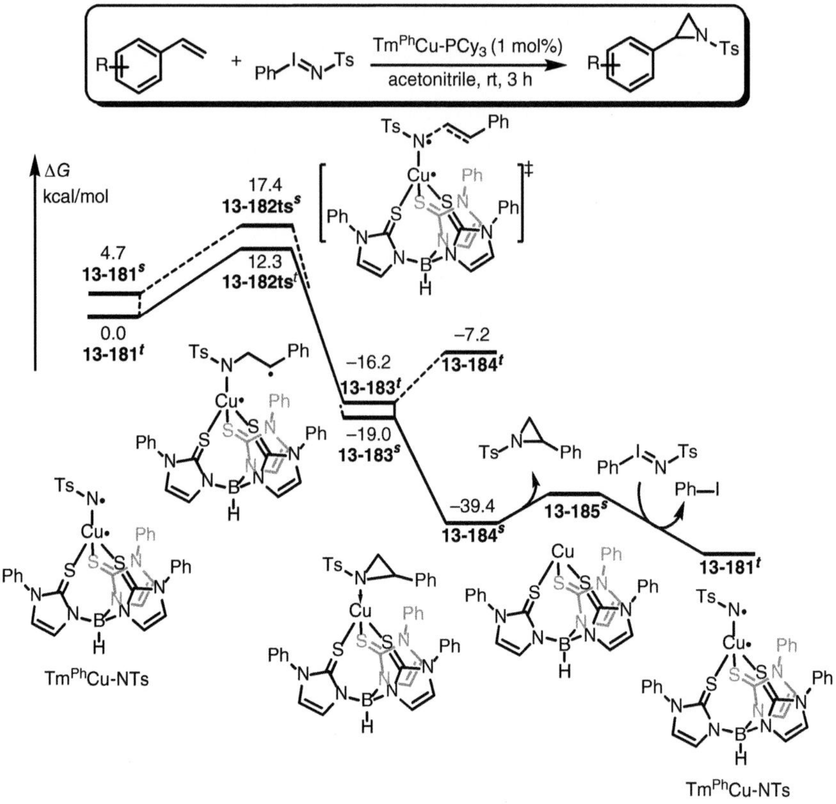

Figure 13.24 Free-energy profiles of Cu-catalyzed [2+1] cycloadditions of nitrenes and styrene. The energies were calculated at the M06/6-311+G(d) (SDD for Cu)/PCM(acetonitrile)//M06-L/6-31G(d) (SDD for Cu) level of theory.

was considered as active species, which can be coordinated by o-azidophenyl methanimine to form a bidentate complex **13-187**. Then, a copper-assisted N—N bond cleavage takes place via transition state **13-188ts** to provide a copper-nitrene complex **13-189**. The calculated overall activation free energy of denitrogenation is 19.6 kcal/mol, which is considered as the rate-determining step. Interestingly, a resonant structure of copper-nitrene **13-189** can be described as a dearomatized imino copper because of the presence of *ortho*-unsaturated bond, which can be considered as the driving force for the formation of copper-nitrene species. Accordingly, a reductive elimination constructs new N—N bond via transition state **13-190ts** to result in a 2*H*-indazole-coordinated copper complex **13-191**.

13.6.3 Nitrene Insertions

In the presence of copper catalyst, the in-situ-generated nitrene can insert into C(aryl)—H bond to achieve arene amination [123]. The mechanism for this type of reactions has been discussed by DFT calculation [124]. As shown in Figure 13.26,

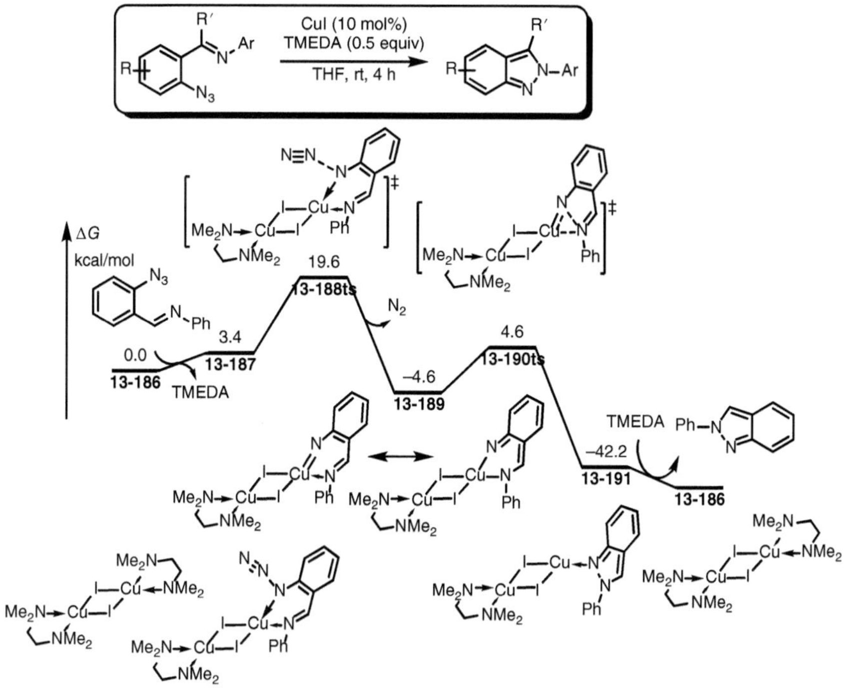

Figure 13.25 Free-energy profiles of Cu-mediated intramolecular amination of nitrene. The energies were calculated at the BPW91/6-311++G(d,p) (LANL2DZ(f) for Cu and LANL2DZ(d) for I)/SMD(THF)//BPW91/6-31G(d) (LANL2DZ(f) for Cu and LANL2DZ(d) for I) level of theory. Source: Modified from Li et al. [121].

the triplet copper-nitrene complex **13-192^t** is 12.4 kcal/mol more stable than singlet one. An intramolecular radical-type addition of nitrene with benzene takes place via transition state **13-193ts^t** to afford a diradical intermediate **13-194^t**, where the spin density is shared by arene and copper. The oxidative state of copper can be considered as +2 in intermediate **13-194^t**. After a spin cross-over, another C—N bond is formed to achieve a singlet aziridine-coordinated Cu(I) complex **13-195^s**. A C—N bond cleavage takes place via transition state **13-196ts^s** to form intermediate **13-197^s**. Then, a hydrogen shift occurs via transition state **13-198ts^s** to achieve the formal insertion of nitrene into C(aryl)—H bond.

13.7 Cu-Catalyzed Hydrofunctionalizations

In copper chemistry, a hydride copper often plays as an important species in catalytic cycle, which can achieve hydrofunctionalizations [125–130]. In this type of reactions, hydride often acts nucleophile, while the combination of an electrophile would often lead to redox-neutral process. The Cu-catalyzed hydrofunctionalizations also can be considered as an insertion of unsaturated bonds into

13 Theoretical Study of Cu-Catalysis

H-heteroatom covenant bonds. The general mechanism of this type of reaction is shown in Scheme 13.7. The reaction starts from a cuprous-hydride species. Then, an insertion of unsaturated bond into Cu—H bond generates new C—H bond. A σ-bond metathesis or transmetallation can yield hydrofunctionalization product and regenerate cuprous-hydride species.

Scheme 13.7 General mechanism of cuprous-hydride-catalyzed hydrofunctionalizations.

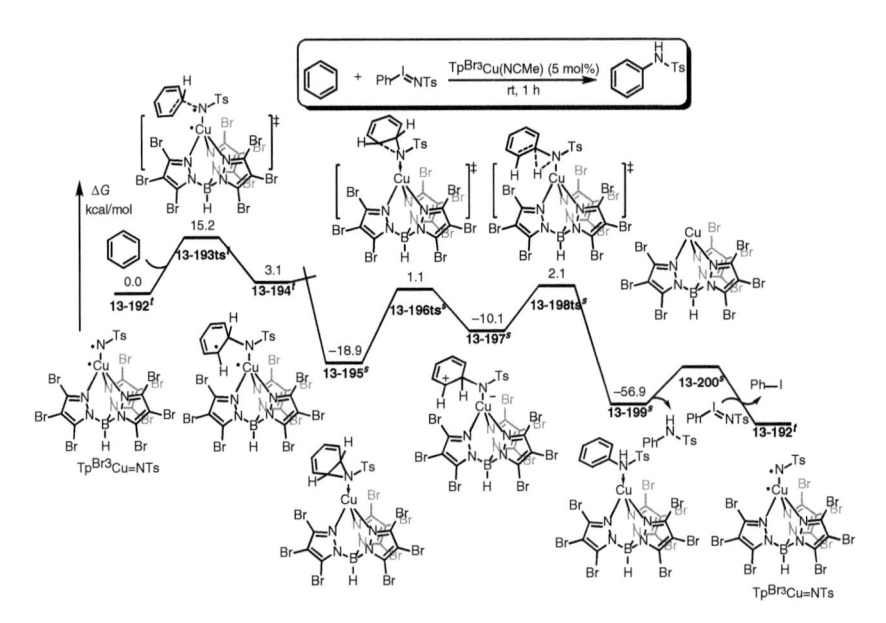

Figure 13.26 Free-energy profiles of Cu-catalyzed nitrene insertion into C(aryl)—H bond. The energies were calculated at the M06/6-311++G(d,p) (LANL2DZ for Cu)/SMD(benzene) //M06/6-31G(d) (LANL2DZ for Cu and Br) level of theory.

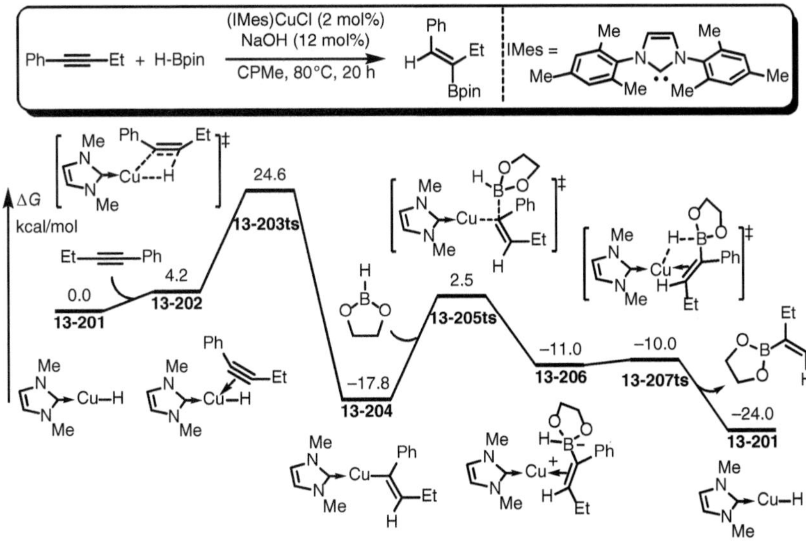

Figure 13.27 Free-energy profiles of Cu-catalyzed hydroborylation of alkynes. The energies were calculated at the B3LYP-D3/6-311+G(2d,2p)/PCM(CPME)//B3LYP/6-31G(d) (6-311G(d) for Cu) level of theory.

13.7.1 Hydroborylations

In the presence of Cu(I) catalyst, alkynes can insert into H—B bond to achieve hydroborylations. For this reaction, hydride-Cu(I) was considered as the key species in the catalytic cycle [131, 132]. As shown in Figure 13.27, the coordination of alkyne onto hydride-Cu(I) species **13-201** generates complex **13-202** with endergonic 4.2 kcal/mol. Subsequently, alkyne inserts into H—Cu bond via transition state **13-203ts** with an overall activation free energy of 24.6 kcal/mol to afford a vinyl Cu(I) intermediate **13-204**. In this intermediate, vinyl group reveals nucleophilicity, which can react with H-Bpin via transition state **13-205ts** to form a dipolar intermediate **13-206**. Then, a rapid hydride transfer from borane to copper takes place via transition state **13-207ts** to achieve the hydroborylation and regenerate hydride-Cu(I) species.

13.7.2 Hydrosilylation

The copper-catalyzed hydrosilylation of ketones can be considered as the insertion of carbonyl into H—Si bond, where hydride plays as nucleophile bonding with carbon, while silyl acts an electrophile to bond with oxygen (Figure 13.28) [133–135]. DFT calculations showed that the carbonyl insertion into Cu—H bond takes place via transition state **13-209ts** with an energy barrier of 13.8 kcal/mol. Then, the generated copper alcoholate **13-210** can react with silane through a σ-bond metathesis via a four-membered ring-type transition state **13-211ts** with an energy barrier of 37.1 kcal/mol, which is considered to be the rate-determining step. After σ-bond

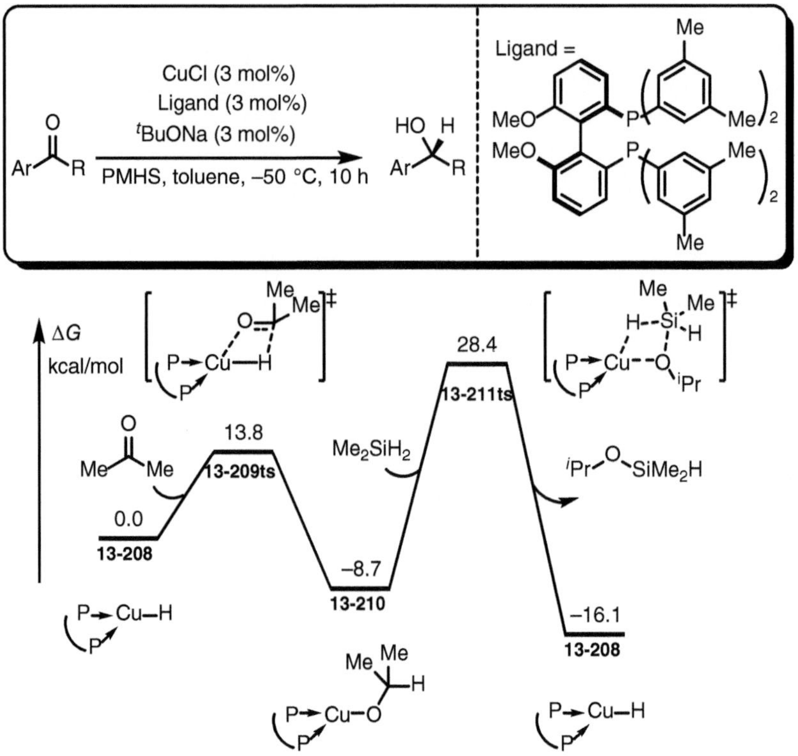

Figure 13.28 Free-energy profiles of Cu-catalyzed hydrosilylation of ketones. The energies were calculated at the B3LYP/6-31G(d,p) (def2-SVP for Cu) level of theory. Sources: Issenhuth et al. [133]; Lipshutz et al. [134]; Zhang et al. [135].

metathesis, silyl ether product is yielded with the regeneration of hydride-copper species **13-208**.

13.7.3 Hydrocarboxylations

Copper-catalyzed hydrocarboxylations can be considered as a cascade insertion of alkyne and carbon dioxide into a Cu—H bond, which is powerful tool for the synthesis of acrylic acid derivatives [136–138]. As shown in Figure 13.29, DFT calculation was employed to reveal the mechanism of copper-catalyzed hydrocarboxylation of alkynes [139]. In this study, hydride-copper **13-212** was set to relative zero in free-energy profiles. The coordination of alkyne onto copper forms complex **13-213** with exergonic 6.1 kcal/mol. Subsequently, alkyne inserts into Cu—H bond via transition state **13-214ts** with an energy barrier of 14.0 kcal/mol to form a vinyl-Cu(I) intermediate **13-215**. Then, carbon dioxide inserts into Cu—C(vinyl) bond via transition state **13-216ts** with an energy barrier of 15.0 kcal/mol to form an acrylate copper intermediate **13-217**. The nucleophilic addition of acrylate onto silane takes place via transition state **13-218ts** to form a hyper valence silicon in dipolar intermediate **13-219**. A sequential hydride transfer occurs via

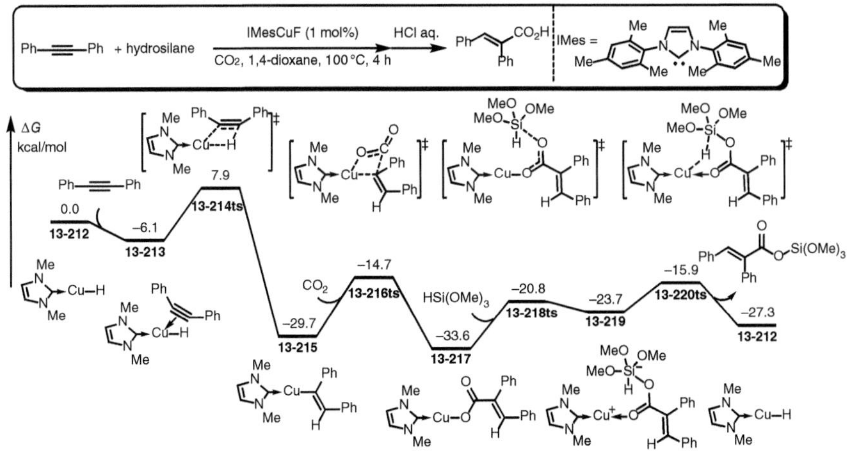

13.8 Cu-Catalyzed Borylations

Figure 13.29 Free-energy profiles of Cu-catalyzed hydrocarboxylation of alkynes. The energies were calculated at the B3LYP/6-311G(d,p)/6-31G/CPCM(dioxane)//B3LYP/6-311G(d,p)/6-31G level of theory.

transition state **13-220ts** to achieve the stepwise σ-bond metathesis with release of hydrocarboxylation product and regeneration of hydride-copper species **13-212**.

13.8 Cu-Catalyzed Borylations

Boryl-copper complexes play an important role in copper-catalyzed borylation reactions, where unsaturated bonds can insert into Cu—B bond to construct new C—B bond [8, 140, 141]. As shown in Scheme 13.8, boryl often acts nucleophile in first step to react with cuprous complex through transmetallation. Then, alkenes or alkynes

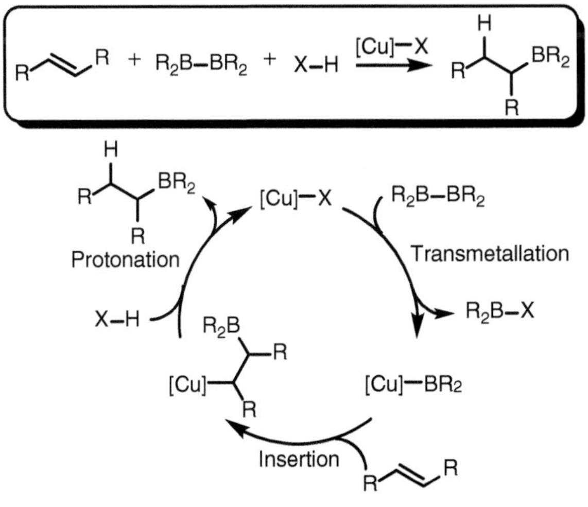

Scheme 13.8 General mechanism of cuprous-catalyzed borylations.

Figure 13.30 Free-energy profiles of Cu-catalyzed alkylborylation of alkenes. The energies were calculated at the M06/6-311+G(2d,p) (SDD for Cu)/SMD(DMA)//B3LYP/6-31G(d) (LANL2DZ for Cu) level of theory.

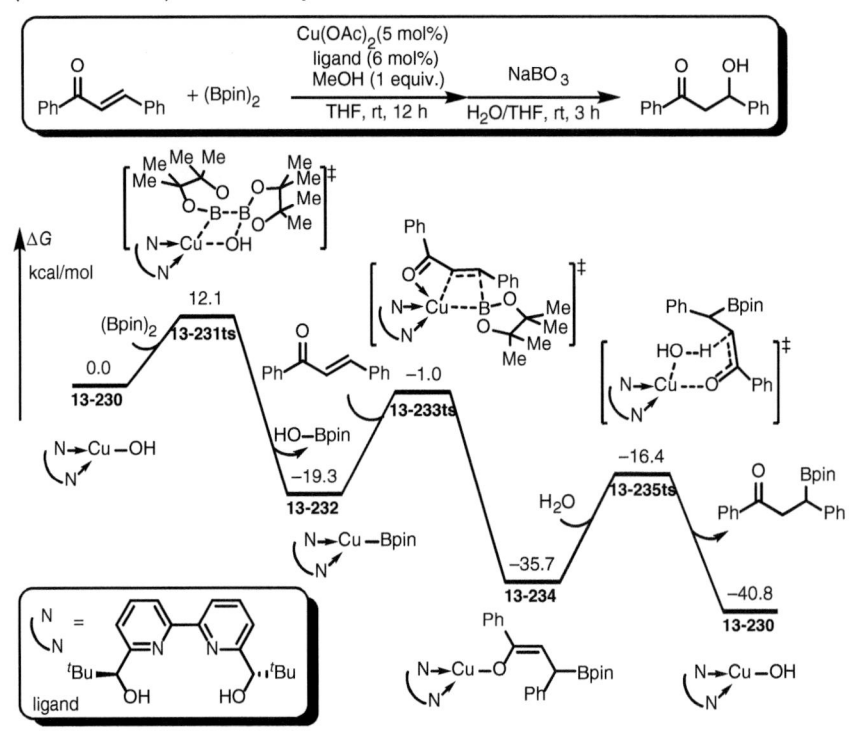

Figure 13.31 Free-energy profiles of Cu-catalyzed hydroborylation of alkenes. The energies were calculated at the M06-L/def2-TZVP (SDD for Cu)/SMD(diethyl ether)//M06-L/6-31G(d) (SDD for Cu)/SMD(diethyl ether) level of theory.

can insert into Cu—B bond to form corresponding alkyl or alkenyl cuprous intermediate, which can react with electrophile to achieve borylations.

13.8.1 Borylation of Alkenes

In copper-catalyzed borylation, boryl group can be considered as nucleophile. When haloalkanes is used as electrophile, alkylborylation can be achieved in the presence of Cu(I) catalyst [142–145]. As shown in Figure 13.30, DFT calculation was used to reveal the mechanism of this reaction [146]. When boryl-copper(I) species **13-221** is used as active catalyst, the coordination of alkene onto copper generates complex **13-222** with exergonic 11.7 kcal/mol. Sequentially, alkene inserts into Cu—B bond via transition state **13-223ts** with an energy barrier of 9.0 kcal/mol to irreversibly form alkyl-cuprous intermediate **13-224**. Then, an oxidative addition of haloalkane onto copper via a three-membered ring-type transition state **13-226ts** takes place with an energy barrier of 29.7 kcal/mol to form a bisaryl-Cu(III) intermediate **13-227**. A rapid reductive elimination occurs via transition state **13-228ts** to yield alkylborylation product and provide cuprous bromide **13-229**. In the presence of methoxide, a σ-bond metathesis with diboryl can regenerate cuprous boryl active species **13-221**.

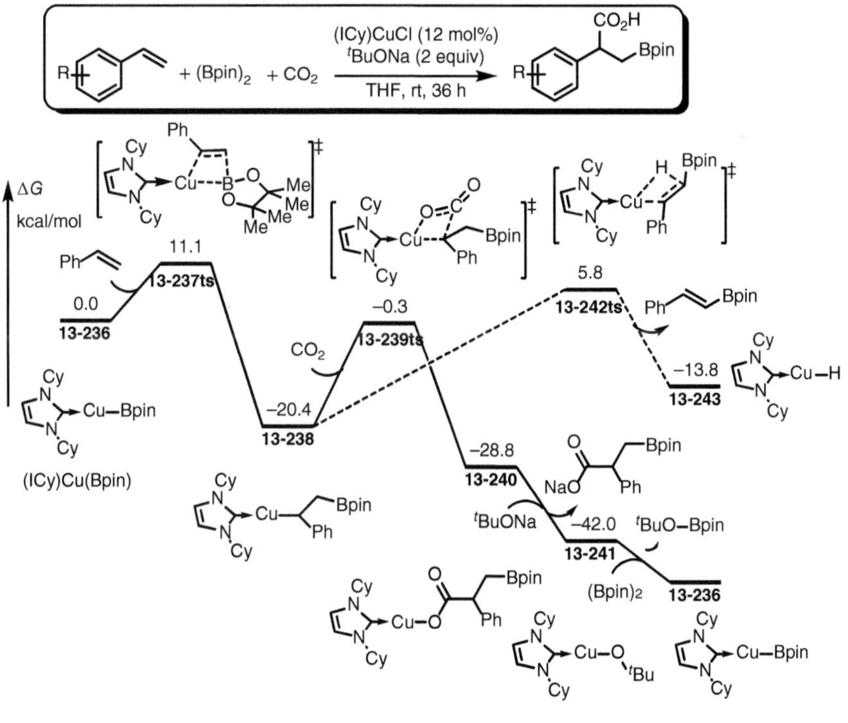

Figure 13.32 Free-energy profiles of Cu-catalyzed boracatboxylation of styrene. The energies were calculated at the M06/6-311++G(d,p) (SDD for Cu)/SMD(THF)//B3LYP/6-31G(d) (SDD for Cu)/SMD(diethyl ether) level of theory. Source: Modified from Lv et al. [148].

13 Theoretical Study of Cu-Catalysis

In another example, water was considered as electrophile. Notably, the generation of boryl-copper was considered by DFT calculations [147]. As shown in Figure 13.31, when a cuprous hydroxide **13-230** is chosen as the initial species in the catalytic cycle, it can react with diboryl via a four-membered ring-type metathesis transition state **13-231ts** with an energy barrier of 12.1 kcal/mol to provide cuprous boryl complex **13-232**. Then, chalcone inserts into Cu—B bond via transition state **13-233ts** in the assistance of carbonyl and an enolate copper-intermediate **13-234** is formed. The sequential protonation by water takes place via transition state **13-235ts** and releases borylation product with the regeneration of cuprous hydroxide **13-230**.

When carbon dioxide is used as reactant, a cascade insertion would achieve boracarboxylation of styrene in the presence of copper catalyst (Figure 13.32) [148]. DFT calculations found that the insertion of styrene into Cu—B bond in complex **13-236** can occur via transition state **13-237ts** with an energy barrier of 11.1 kcal/mol to form an alkyl copper intermediate **13-238**. Then, a carbon dioxide inserts into Cu—C(alkyl) bond via transition state **13-239ts** with an energy barrier of 20.1 kcal/mol to generate cupric carboxylate **13-240**. Transmetallation with sodium alkoxide results in carboxylic sodium as product. DFT calculations also found that the competitive β-hydride elimination via transition state **13-242ts** is kinetically unfavorable, when N-hetero carbene is used as ligand in this reaction.

13.8.2 Borylation of Alkynes

The key step for the borylation of alkynes can be considered as an insertion of alkyne into Cu—B bond [150–153]. The generated vinyl copper can be protonated by alcohol solvent to achieve hydroboration. As shown in Figure 13.33 [149], the

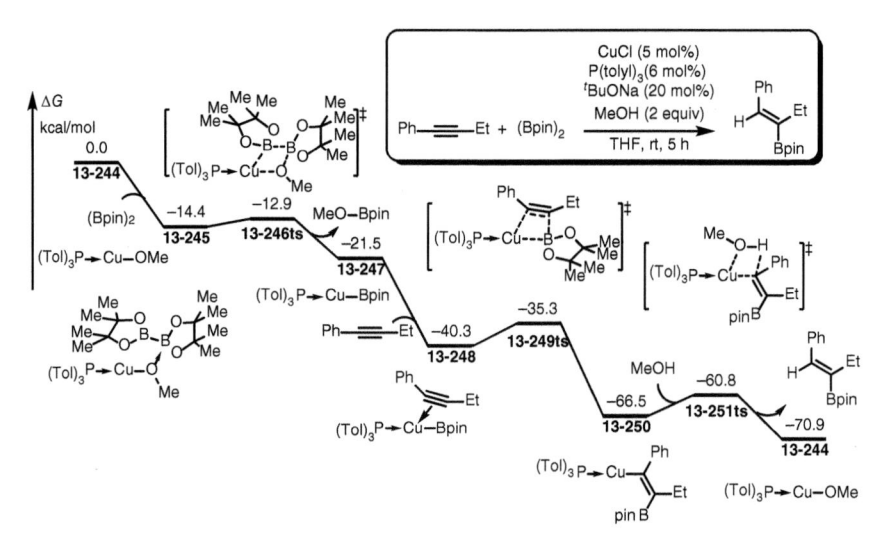

Figure 13.33 Free-energy profiles of Cu-catalyzed hydroborylation of enternal alkynes. The energies were calculated at the M06-2X/6-31G(d) level of theory. Source: Based on Moon et al. [149].

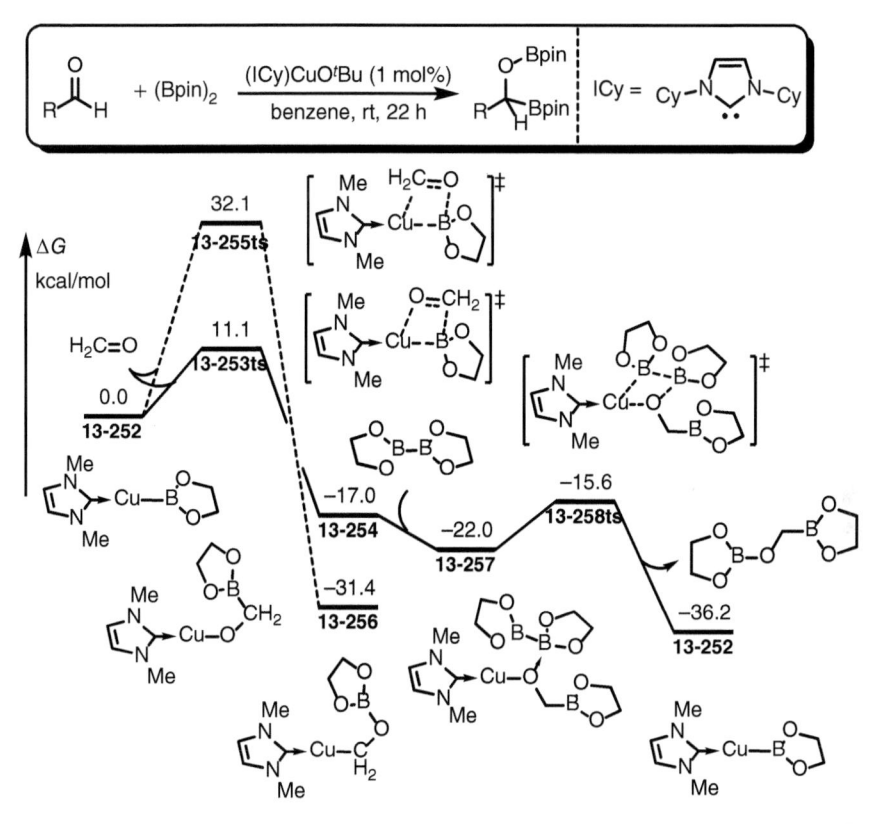

Figure 13.34 Free-energy profiles of Cu-catalyzed diborylation of aldehydes. The energies were calculated at the B3LYP/6-31G (6-311G(d) for Cu)/PCM(benzene)//B3LYP/6-31G (6-311G(d) for Cu) level of theory.

reaction starts from a cuprous methoxide species **13-244**, which can coordinate with diboryl to afford complex **13-245** with exergonic 14.4 kcal/mol. A σ-metathesis takes place via a four-membered ring-type transition state **13-246ts** with an energy barrier of only 1.5 kcal/mol. The generated cuprous boryl intermediate **13-247** can be coordinated by reacting alkyne to afford complex **13-248** with further exergonic 18.8 kcal/mol. The coordinated alkyne inserts into Cu—B bond via transition state **13-249ts** with an energy barrier of only 5.0 kcal/mol, which is significantly lower than that of alkene insertion. The generated vinyl cuprous **13-250** can be protonated by methanol via a four-membered ring-type metathesis transition state **13-251ts**. The release of hydroboration product regenerates cuprous methoxide species **13-244**.

13.8.3 Borylation of Carbonyls

The nucleophilic boryl can react with carbonyl to afford new C—B bond. As an example, Sadighi group reported a copper-catalyzed diborylation of aldehydes [154]. The reaction starts from a copper boryl complex **13-252**. Two types of carbonyl

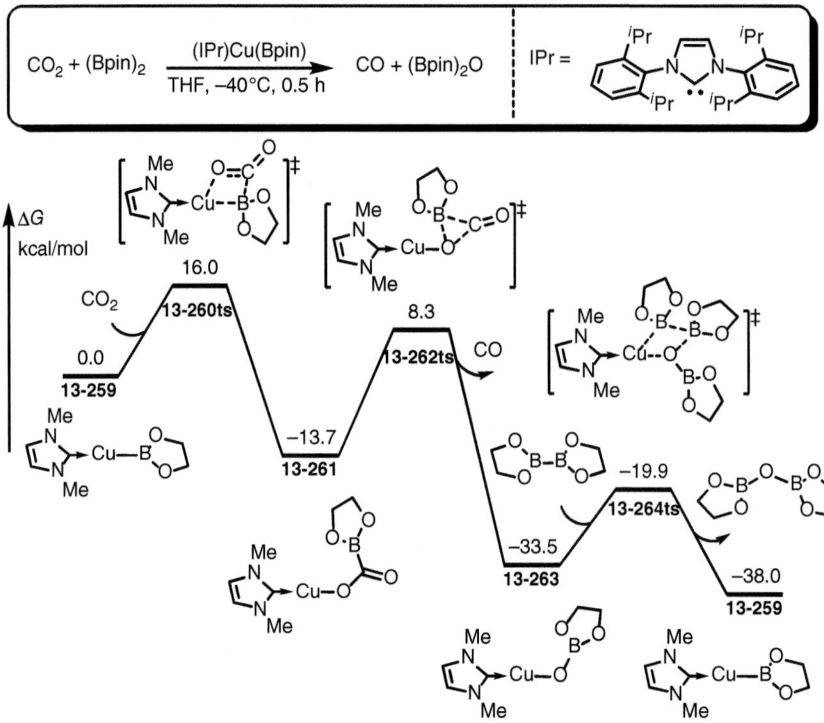

Figure 13.35 Free-energy profiles of Cu-catalyzed reduction of carbon dioxide by using diboryl. The energies were calculated at the B3LYP/6-31G (6-311G(d) for Cu) level of theory.

insertion transition states **13-253ts** and **13-255ts** were found by DFT calculations [155]. The relative free energy of **13-253ts** leading to the formation of C—B bond is 21.0 kcal/mol lower than that of **13-255ts** for the formation of O—B bond. The generated cuprous borylmethanolate **13-254** can react with diboryl through σ-bond metathesis via transition state **13-258ts** to yield diborylation product with the regeneration of cuprous boryl complex **13-252** (Figure 13.34).

Interestingly, diboryl can play as a reductant to reduce carbon dioxide to carbon monoxide in the presence of cuprous catalyst [156]. As shown in Figure 13.35, carbonyl in carbon dioxide inserts into Cu—B bond via a four-membered ring-type transition state **13-260ts** to afford a cuprous borylcarboxylate complex **13-261** [157]. Then, a B—O elimination takes place via a three-membered ring-type transition state **13-262ts** to release carbon monoxide. The generated cuprous borate **13-263** can react with diboryl through σ-bond metathesis (transition state **13-264ts**) to regenerate copper boryl complex **13-259**.

References

1 Ullmann, F. (1903). Ueber eine neue Bildungsweise von Diphenylamin-derivaten. *Berichte der Deutschen Chemischen Gesellschaft* 36 (2): 2382–2384.

2 Ullmann, F. and Bielecki, J. (1901). Ueber Synthesen in der Biphenylreihe. *Berichte der Deutschen Chemischen Gesellschaft* 34 (2): 2174–2185.
3 Ullmann, F. and Sponagel, P. (1905). Ueber die Phenylirung von Phenolen. *Berichte der Deutschen Chemischen Gesellschaft* 38 (2): 2211–2212.
4 Allen, S.E., Walvoord, R.R., Padilla-Salinas, R. et al. (2013). Aerobic copper-catalyzed organic reactions. *Chemical Reviews* 113 (8): 6234–6458.
5 Cheng, L.J. and Mankad, N.P. (2020). C–C and C–X coupling reactions of unactivated alkyl electrophiles using copper catalysis. *Chemical Society Reviews*.
6 Evano, G., Blanchard, N., and Toumi, M. (2008). Copper-mediated coupling reactions and their applications in natural products and designed biomolecules synthesis. *Chemical Reviews* 108 (8): 3054–3131.
7 Gamez, P., Aubel, P.G., Driessen, W.L. et al. (2001). Homogeneous bio-inspired copper-catalyzed oxidation reactions. *Chemical Society Reviews* 30 (6): 376–385.
8 Hemming, D., Fritzemeier, R., Westcott, S.A. et al. (2018). Copper-boryl mediated organic synthesis. *Chemical Society Reviews* 47 (19): 7477–7494.
9 Hathaway, B.J. (1981). Copper. *Coordination Chemistry Reviews* 35: 211–252.
10 Hathaway, B.J. and Billing, D.E. (1970). The electronic properties and stereochemistry of mono-nuclear complexes of the copper(II) ion. *Coordination Chemistry Reviews* 5 (2): 143–207.
11 Krause, N. (ed.) (2002). *Modern Organocopper Chemistry*, 1–44. Germany: Wiley-VCH Verlag GmbH.
12 Jones, G.R., Anastasaki, A., Whitfield, R. et al. (2018). Copper-mediated reversible deactivation radical polymerization in aqueous media. *Angewandte Chemie International Edition* 57 (33): 10468–10482.
13 Kaga, A. and Chiba, S. (2017). Engaging radicals in transition metal-catalyzed cross-coupling with alkyl electrophiles: recent advances. *ACS Catalysis* 7 (7): 4697–4706.
14 Li, J.-M., Wang, Y.-H., Yu, Y. et al. (2017). Copper-catalyzed remote C–H functionalizations of naphthylamides through a coordinating activation strategy and single-electron-transfer (SET) mechanism. *ACS Catalysis* 7 (4): 2661–2667.
15 Li, S.J. and Lan, Y. (2020). Is Cu(iii) a necessary intermediate in Cu-mediated coupling reactions? A mechanistic point of view. *Chemical Communications* 56 (49): 6609–6619.
16 Pampana, V.K.K., Sagadevan, A., Ragupathi, A. et al. (2020). Visible light-promoted copper catalyzed regioselective acetamidation of terminal alkynes by arylamines. *Green Chemistry* 22 (4): 1164–1170.
17 Suess, A.M., Ertem, M.Z., Cramer, C.J. et al. (2013). Divergence between organometallic and single-electron-transfer mechanisms in copper(II)-mediated aerobic C–H oxidation. *Journal of the American Chemical Society* 135 (26): 9797–9804.
18 Beletskaya, I.P. and Cheprakov, A.V. (2012). The complementary competitors: palladium and copper in C–N cross-coupling reactions. *Organometallics* 31 (22): 7753–7808.
19 Hickman, A.J. and Sanford, M.S. (2012). High-valent organometallic copper and palladium in catalysis. *Nature* 484 (7393): 177–185.

20 Beletskaya, I.P. and Cheprakov, A.V. (2004). Copper in cross-coupling reactions. *Coordination Chemistry Reviews* 248 (21–24): 2337–2364.

21 Ma, D. and Cai, Q. (2008). Copper/amino acid catalyzed cross-couplings of aryl and vinyl halides with nucleophiles. *Accounts of Chemical Research* 41 (11): 1450–1460.

22 Goj, L.A., Blue, E.D., Delp, S.A. et al. (2006). Single-electron oxidation of monomeric Copper(I) alkyl complexes: evidence for reductive elimination through bimolecular formation of alkanes. *Organometallics* 25 (17): 4097–4104.

23 Lu, F.D., Liu, D., Zhu, L. et al. (2019). Asymmetric propargylic radical Cyanation enabled by dual organophotoredox and copper catalysis. *Journal of the American Chemical Society* 141 (15): 6167–6172.

24 Zhang, C., Tang, C., and Jiao, N. (2012). Recent advances in copper-catalyzed dehydrogenative functionalization via a single electron transfer (SET) process. *Chemical Society Reviews* 41 (9): 3464–3484.

25 Holub, J.M. and Kirshenbaum, K. (2010). Tricks with clicks: modification of peptidomimetic oligomers via copper-catalyzed azide-alkyne [3 + 2] cycloaddition. *Chemical Society Reviews* 39 (4): 1325–1337.

26 Kolb, H.C., Finn, M.G., and Sharpless, K.B. (2001). Click chemistry: Diverse chemical function from a few good reactions. *Angewandte Chemie International Edition* 40 (11): 2004–2021.

27 Johnson, J.S. and Evans, D.A. (2000). Chiral bis(oxazoline) copper(II) complexes: versatile catalysts for enantioselective cycloaddition, aldol, Michael, and carbonyl ene reactions. *Accounts of Chemical Research* 33 (6): 325–335.

28 Reymond, S. and Cossy, J. (2008). Copper-catalyzed Diels–Alder reactions. *Chemical Reviews* 108 (12): 5359–5406.

29 Fanta, P.E. (1964). The Ullmann synthesis of biaryls, 1945–1963. *Chemical Reviews* 64 (6): 613–632.

30 Hassan, J., Sevignon, M., Gozzi, C. et al. (2002). Aryl-aryl bond formation one century after the discovery of the Ullmann reaction. *Chemical Reviews* 102 (5): 1359–1470.

31 Bhunia, S., Pawar, G.G., Kumar, S.V. et al. (2017). Selected copper-based reactions for C–N, C–O, C–S, and C–C bond formation. *Angewandte Chemie International Edition* 56 (51): 16136–16179.

32 Finet, J.-P., Fedorov, A., Combes, S. et al. (2002). Recent advances in Ullmann reaction: Copper(II) diacetate C at alysed N- ,O- and S- arylation involving polycoordinate hetero atomic derivatives. *Current Organic Chemistry* 6 (7): 597–626.

33 Monnier, F. and Taillefer, M. (2009). Catalytic C–C, C–N, and C–O Ullmann-type coupling reactions. *Angewandte Chemie International Edition* 48 (38): 6954–6971.

34 Sambiagio, C., Marsden, S.P., Blacker, A.J. et al. (2014). Copper catalysed Ullmann type chemistry: from mechanistic aspects to modern development. *Chemical Society Reviews* 43 (10): 3525–3550.

35 Weston, P.E. and Adkins, H. (1928). Catalysis with copper in the Ullmann reaction. *Journal of the American Chemical Society* 50 (3): 859–866.

36 Andrada, D., Soria-Castro, S., Caminos, D. et al. (2017). Understanding the heteroatom effect on the Ullmann copper-catalyzed cross-coupling of X-arylation (X = NH, O, S) mechanism. *Catalysts* 7 (12): 388.

37 Sperotto, E., van Klink, G.P., van Koten, G. et al. (2010). The mechanism of the modified Ullmann reaction. *Dalton Transactions* 39 (43): 10338–10351.

38 Heravi, M.M., Kheilkordi, Z., Zadsirjan, V. et al. (2018). Buchwald–Hartwig reaction: an overview. *Journal of Organometallic Chemistry* 861: 17–104.

39 Bowman, W.R., Heaney, H., and Smith, P.H.G. (1984). Copper(1) catalysed aromatic nucleophilic substitution: a mechanistic and synthetic comparison with the SRN1 reaction. *Tetrahedron Letters* 25 (50): 5821–5824.

40 Cohen, T. and Cristea, I. (1975). Copper(I)-induced reductive dehalogenation, hydrolysis, or coupling of some aryl and vinyl halides at room temperature. *The Journal of Organic Chemistry* 40 (25): 3649–3651.

41 Cohen, T. and Cristea, I. (1976). Kinetics and mechanism of the copper(I)-induced homogeneous Ullmann coupling of o-bromonitrobenzene. *Journal of the American Chemical Society* 98 (3): 748–753.

42 Lockhart, T.P. (1983). Mechanistic investigation of the copper-catalyzed reactions of diphenyliodonium salts. *Journal of the American Chemical Society* 105 (7): 1940–1946.

43 Bowman, W.R., Heaney, H., and Smith, P.H.G. (1982). Intramolecular aromatic substitution (SRN1) reactions; use of entrainment for the preparation of benzothiazoles. *Tetrahedron Letters* 23 (48): 5093–5096.

44 Hey, D.H. and Waters, W.A. (1937). Some organic reactions involving the occurence of free radicals in solution. *Chemical Reviews* 21 (1): 169–208.

45 Jenkins, C.L. and Kochi, J.K. (1972). Homolytic and ionic mechanisms in the ligand-transfer oxidation of alkyl radicals by copper(II) halides and pseudohalides. *Journal of the American Chemical Society* 94 (3): 856–865.

46 Kornblum, N., Michel, R.E., and Kerber, R.C. (1966). Chain reactions in substitution processes which proceed via radical-anion intermediates. *Journal of the American Chemical Society* 88 (23): 5662–5663.

47 Russell, G.A. and Danen, W.C. (1966). Coupling reactions of the 2-nitro-2-propyl anion 1. *Journal of the American Chemical Society* 88 (23): 5663–5665.

48 Bacon, R.G.R. and Hill, H.A.O. (1964). 212. Metal ions and complexes in organic reactions. Part III. Reduction of 1-bromonaphthalene in some reaction systems containing copper(I). *Journal of the Chemical Society (Resumed)*: 1112.

49 Bacon, R.G.R. and Hill, H.A.O. (1964). 211. Metal ions and complexes in organic reactions. Part II. Substitution reactions of aryl halides promoted by cuprous oxide. *Journal of the Chemical Society (Resumed)*: 1108.

50 Bacon, R.G.R. and Hill, H.A.O. (1964). 210. Metal ions and complexes in organic reactions. Part I. Substitution reactions between aryl halides and cuprous salts in organic solvents. *Journal of the Chemical Society (Resumed)*: 1097.

51 Dargel, T.K., Hertwig, R.H., and Koch, W. (1999). How do coinage metal ions bind to benzene? *Molecular Physics* 96 (4): 583–591.

52 Weingarten, H. (1964). Mechanism of the Ullmann condensation. *The Journal of Organic Chemistry* 29 (12): 3624–3626.

53 Larsson, P.F., Bolm, C., and Norrby, P.O. (2010). Kinetic investigation of a ligand-accelerated sub-mol% copper-catalyzed C–N cross-coupling reaction. *Chemistry – A European Journal* 16 (46): 13613–13616.

54 Larsson, P.F., Correa, A., Carril, M. et al. (2009). Copper-catalyzed cross-couplings with part-per-million catalyst loadings. *Angewandte Chemie International Edition* 48 (31): 5691–5693.

55 Larsson, P.-F., Wallentin, C.-J., and Norrby, P.-O. (2014). Mechanistic aspects of submol % copper-catalyzed C–N cross-coupling. *ChemCatChem*: 1277–1282.

56 Fabre, I., Perego, L.A., Bergès, J. et al. (2016). Antagonistic effect of acetates in C–N bond formation with in situ generated diazonium salts: a combined theoretical and experimental study. *European Journal of Organic Chemistry* 2016 (35): 5887–5896.

57 Wang, M., Fan, T., and Lin, Z. (2012). DFT studies on copper-catalyzed arylation of aromatic C–H bonds. *Organometallics* 31 (2): 560–569.

58 Giri, R., Brusoe, A., Troshin, K. et al. (2018). Mechanism of the Ullmann biaryl ether synthesis catalyzed by complexes of anionic ligands: evidence for the reaction of iodoarenes with ligated anionic Cu(I) intermediates. *Journal of the American Chemical Society* 140 (2): 793–806.

59 Lefèvre, G., Franc, G., Tlili, A. et al. (2012). Contribution to the mechanism of copper-catalyzed C–N and C–O bond formation. *Organometallics* 31 (22): 7694–7707.

60 Lefevre, G., Tlili, A., Taillefer, M. et al. (2013). Discriminating role of bases in diketonate copper(I)-catalyzed C–O couplings: phenol versus diarylether. *Dalton Transactions* 42 (15): 5348–5354.

61 Ichiishi, N., Canty, A.J., Yates, B.F. et al. (2013). Cu-catalyzed fluorination of diaryliodonium salts with KF. *Organic Letters* 15 (19): 5134–5137.

62 Ichiishi, N., Canty, A.J., Yates, B.F. et al. (2014). Mechanistic investigations of Cu-catalyzed fluorination of diaryliodonium salts: elaborating the Cu(I)/Cu(III) manifold in copper catalysis. *Organometallics* 33 (19): 5525–5534.

63 Nie, J., Guo, H.C., Cahard, D. et al. (2011). Asymmetric construction of stereogenic carbon centers featuring a trifluoromethyl group from prochiral trifluoromethylated substrates. *Chemical Reviews* 111 (2): 455–529.

64 Tomashenko, O.A. and Grushin, V.V. (2011). Aromatic trifluoromethylation with metal complexes. *Chemical Reviews* 111 (8): 4475–4521.

65 Umemoto, T. (1996). Electrophilic perfluoroalkylating agents. *Chemical Reviews* 96 (5): 1757–1778.

66 Dubinina, G.G., Furutachi, H., and Vicic, D.A. (2008). Active trifluoromethylating agents from well-defined Copper(I)-CF$_3$ complexes. *Journal of the American Chemical Society* 130 (27): 8600–8601.

67 Lishchynskyi, A., Novikov, M.A., Martin, E. et al. (2013). Trifluoromethylation of aryl and heteroaryl halides with fluoroform-derived CuCF$_3$: scope, limitations, and mechanistic features. *The Journal of Organic Chemistry* 78 (22): 11126–11146.

68 Konovalov, A.I., Lishchynskyi, A., and Grushin, V.V. (2014). Mechanism of trifluoromethylation of aryl halides with CuCF$_3$ and the ortho effect. *Journal of the American Chemical Society* 136 (38): 13410–13425.

69 Chu, L. and Qing, F.L. (2010). Copper-mediated aerobic oxidative trifluoromethylation of terminal alkynes with Me$_3$SiCF$_3$. *Journal of the American Chemical Society* 132 (21): 7262–7263.

70 Chu, L. and Qing, F.L. (2010). Copper-mediated oxidative trifluoromethylation of boronic acids. *Organic Letters* 12 (21): 5060–5063.

71 Jiang, X., Chu, L., and Qing, F.L. (2012). Copper-catalyzed oxidative trifluoromethylation of terminal alkynes and aryl boronic acids using (trifluoromethyl)trimethylsilane. *The Journal of Organic Chemistry* 77 (3): 1251–1257.

72 Jover, J. and Maseras, F. (2013). Computational characterization of a mechanism for the copper-catalyzed aerobic oxidative trifluoromethylation of terminal alkynes. *Chemical Communications* 49 (89): 10486–10488.

73 Charpentier, J., Fruh, N., and Togni, A. (2015). Electrophilic trifluoromethylation by use of hypervalent iodine reagents. *Chemical Reviews* 115 (2): 650–682.

74 Eisenberger, P., Gischig, S., and Togni, A. (2006). Novel 10-I-3 hypervalent iodine-based compounds for electrophilic trifluoromethylation. *Chemistry – A European Journal* 12 (9): 2579–2586.

75 Wang, X., Ye, Y., Zhang, S. et al. (2011). Copper-catalyzed C(sp$_3$)–C(sp$_3$) bond formation using a hypervalent iodine reagent: an efficient allylic trifluoromethylation. *Journal of the American Chemical Society* 133 (41): 16410–16413.

76 Ling, L., Liu, K., Li, X. et al. (2015). General reaction mode of hypervalent iodine trifluoromethylation reagent: a density functional theory study. *ACS Catalysis* 5 (4): 2458–2468.

77 Zhu, L., Ye, J.-H., Duan, M. et al. (2018). The mechanism of copper-catalyzed oxytrifluoromethylation of allylamines with CO$_2$: a computational study. *Organic Chemistry Frontiers* 5 (4): 633–639.

78 Guo, X.X., Gu, D.W., Wu, Z. et al. (2015). Copper-catalyzed C–H functionalization reactions: efficient synthesis of heterocycles. *Chemical Reviews* 115 (3): 1622–1651.

79 Li, Z., Bohle, D.S., and Li, C.J. (2006). Cu-catalyzed cross-dehydrogenative coupling: a versatile strategy for C–C bond formations via the oxidative activation of sp(3) C–H bonds. *Proceedings of the National Academy of Sciences of the United States of America* 103 (24): 8928–8933.

80 Wendlandt, A.E., Suess, A.M., and Stahl, S.S. (2011). Copper-catalyzed aerobic oxidative C–H functionalizations: trends and mechanistic insights. *Angewandte Chemie International Edition* 50 (47): 11062–11087.

81 Do, H.Q., Khan, R.M., and Daugulis, O. (2008). A general method for copper-catalyzed arylation of arene C–H bonds. *Journal of the American Chemical Society* 130 (45): 15185–15192.

82 Phipps, R.J. and Gaunt, M.J. (2009). A meta-selective copper-catalyzed C–H bond arylation. *Science* 323 (5921): 1593–1597.

83 Chen, B., Hou, X.-L., Li, Y.-X. et al. (2011). Mechanistic understanding of the unexpected meta selectivity in copper-catalyzed anilide C–H bond arylation. *Journal of the American Chemical Society* 133 (20): 7668–7671.

84 Tang, S., Gong, T., and Fu, Y. (2012). Mechanistic study of copper-catalyzed intramolecular ortho-C–H activation/carbon–nitrogen and carbon–oxygen cyclizations. *SCIENCE CHINA Chemistry* 56 (5): 619–632.

85 Vilella, L., Conde, A., Balcells, D. et al. (2017). A competing, dual mechanism for catalytic direct benzene hydroxylation from combined experimental-DFT studies. *Chemical Science* 8 (12): 8373–8383.

86 Barve, B.D., Wu, Y.-C., El-Shazly, M. et al. (2012). Synthesis of carbamates by direct C–H bond activation of formamides. *European Journal of Organic Chemistry* 2012 (34): 6760–6766.

87 Zhang, L.L., Li, S.J., Zhang, L. et al. (2016). Theoretical studies on CuCl-catalyzed C–H activation/C–O coupling reactions: oxidant and catalyst effects. *Organic & Biomolecular Chemistry* 14 (19): 4426–4435.

88 Yuan, B., He, R., Shen, W. et al. (2017). Effects of base strength on the copper-catalyzed cycloisomerization of propargylic acetates to form indolizines: a DFT study. *Tetrahedron* 73 (42): 6092–6100.

89 Worrell, B.T., Malik, J.A., and Fokin, V.V. (2013). Direct evidence of a dinuclear copper intermediate in Cu(I)-catalyzed azide-alkyne cycloadditions. *Science* 340 (6131): 457–460.

90 Li, L. and Janesko, B.G. (2016). 3-Methyleneisoindolin-1-one assembly via base- and CuI/l-proline-catalyzed domino reaction: mechanism of regioselective anionic cyclization. *The Journal of Organic Chemistry* 81 (22): 10802–10808.

91 Hein, J.E. and Fokin, V.V. (2010). Copper-catalyzed azide-alkyne cycloaddition (CuAAC) and beyond: new reactivity of copper(I) acetylides. *Chemical Society Reviews* 39 (4): 1302–1315.

92 Himo, F., Lovell, T., Hilgraf, R. et al. (2005). Copper(I)-catalyzed synthesis of azoles. DFT study predicts unprecedented reactivity and intermediates. *Journal of the American Chemical Society* 127 (1): 210–216.

93 Kalvet, I., Tammiku-Taul, J., Mäeorg, U. et al. (2016). NMR and DFT study of the Copper(I)-catalyzed cycloaddition reaction: H/D scrambling of alkynes and variable reaction order of the catalyst. *ChemCatChem* 8 (10): 1804–1808.

94 Ozen, C. and Tuzun, N.S. (2012). The mechanism of copper-catalyzed azide-alkyne cycloaddition reaction: a quantum mechanical investigation. *Journal of Molecular Graphics & Modelling* 34: 101–107.

95 Pan, J., Xu, Z., Zeng, R. et al. (2013). Copper(II)-catalyzed tandem synthesis of substituted 3-methyleneisoindolin-1-ones. *Chinese Journal of Chemistry* 31 (8): 1022–1026.

References

96 Pouy, M.J., Delp, S.A., Uddin, J. et al. (2012). Intramolecular hydroalkoxylation and hydroamination of alkynes catalyzed by Cu(I) complexes supported by N–heterocyclic carbene ligands. *ACS Catalysis* 2 (10): 2182–2193.

97 Yuan, R. and Lin, Z. (2014). Mechanism for the carboxylative coupling reaction of a terminal alkyne, CO_2, and an allylic chloride catalyzed by the Cu(I) complex: a DFT study. *ACS Catalysis* 4 (12): 4466–4473.

98 Qi, X., Bai, R., Zhu, L. et al. (2016). Mechanism of synergistic Cu(II)/Cu(I)-mediated alkyne coupling: dinuclear 1,2-reductive elimination after minimum energy crossing point. *The Journal of Organic Chemistry* 81 (4): 1654–1660.

99 Zhang, G., Yi, H., Zhang, G. et al. (2014). Direct observation of reduction of Cu(II) to Cu(I) by terminal alkynes. *Journal of the American Chemical Society* 136 (3): 924–926.

100 Cardin, D.J., Cetinkaya, B., and Lappert, M.F. (1972). Transition metal-carbene complexes. *Chemical Reviews* 72 (5): 545–574.

101 Doyle, M.P. (1986). Catalytic methods for metal carbene transformations. *Chemical Reviews* 86 (5): 919–939.

102 Doyle, M.P. and Forbes, D.C. (1998). Recent advances in asymmetric catalytic metal carbene transformations. *Chemical Reviews* 98 (2): 911–936.

103 Kirmse, W. (2003). Copper carbene complexes: advanced catalysts, new insights. *Angewandte Chemie International Edition* 42 (10): 1088–1093.

104 Meng, Q. and Li, M. (2006). Theoretical insights of copper(I) carbenes. *Journal of Molecular Structure: Theochem* 765 (1-3): 13–20.

105 Evans, D.A., Woerpel, K.A., Hinman, M.M. et al. (1991). Bis(oxazolines) as chiral ligands in metal-catalyzed asymmetric reactions. Catalytic, asymmetric cyclopropanation of olefins. *Journal of the American Chemical Society* 113 (2): 726–728.

106 Özen, C., Konuklar, F.A.S., and Tuzun, N.S. (2009). Mechanistic study on [3+2] cycloaddition and cyclopropanation reactions of 1,3-dioxepine derivatives in the presence of Copper(I) catalyst. *Organometallics* 28 (17): 4964–4973.

107 Straub, B.F., Gruber, I., Rominger, F. et al. (2003). Mechanism of copper(I)-catalyzed cyclopropanation: a DFT study calibrated with copper(I) alkene complexes. *Journal of Organometallic Chemistry* 684 (1-2): 124–143.

108 Zhang, H., Wu, G., Yi, H. et al. (2017). Copper(I)-catalyzed chemoselective coupling of cyclopropanols with diazoesters: ring-opening C–C bond formations. *Angewandte Chemie International Edition* 56 (14): 3945–3950.

109 Fraile, J.M., García, J.I., Martínez-Merino, V. et al. (2001). Theoretical (DFT) insights into the mechanism of copper-catalyzed cyclopropanation reactions. Implications for enantioselective catalysis. *Journal of the American Chemical Society* 123 (31): 7616–7625.

110 Xiao, Q., Ling, L., Ye, F. et al. (2013). Copper-catalyzed direct ortho-alkylation of *N*-iminopyridinium ylides with *N*-tosylhydrazones. *The Journal of Organic Chemistry* 78 (8): 3879–3885.

111 Ye, F., Ma, X., Xiao, Q. et al. (2012). C(sp)–C(sp3) bond formation through Cu-catalyzed cross-coupling of *N*-tosylhydrazones and trialkylsilylethynes. *Journal of the American Chemical Society* 134 (13): 5742–5745.

112 Zhao, X., Wu, G., Zhang, Y. et al. (2011). Copper-catalyzed direct benzylation or allylation of 1,3-azoles with *N*-tosylhydrazones. *Journal of the American Chemical Society* 133 (10): 3296–3299.

113 Huang, X., Li, X., and Jiao, N. (2015). Copper-catalyzed direct transformation of simple alkynes to alkenyl nitriles via aerobic oxidative *N*-incorporation. *Chemical Science* 6 (11): 6355–6360.

114 Badiei, Y.M., Dinescu, A., Dai, X. et al. (2008). Copper–nitrene complexes in catalytic C–H amination. *Angewandte Chemie International Edition* 47 (51): 9961–9964.

115 Che, C.M., Lo, V.K.Y., and Zhou, C.Y. (2014). 7.02 oxidation by metals (nitrene). In: *Comprehensive Organic Synthesis II*, 2e (ed. P. Knochel), 26–85. Amsterdam: Elsevier.

116 Pellissier, H. (2010). Recent developments in asymmetric aziridination. *Tetrahedron* 66 (8): 1509–1555.

117 Amme, M.J., Kazi, A.B., and Cundari, T.R. (2009). Copper-catalyzed phosphinidene transfer to ethylene, acetylene, and carbon monoxide: A computational study. *International Journal of Quantum Chemistry* 110 (9): 1702–1711.

118 Comba, P., Lang, C., Lopez de Laorden, C. et al. (2008). The mechanism of the (bispidine)copper(II)-catalyzed aziridination of styrene: a combined experimental and theoretical study. *Chemistry - A European Journal* 14 (17): 5313–5328.

119 Montero-Campillo, M.M. and Cordeiro, M.N.D.S. (2013). Mechanism of aziridination of styrene catalyzed by copper(I) bis(oxazoline). *International Journal of Quantum Chemistry* 113 (16): 2002–2011.

120 Lam, T.L., Tso, K.C., Cao, B. et al. (2017). Tripodal S-ligand complexes of Copper(I) as catalysts for alkene aziridination, sulfide sulfimidation, and C–H amination. *Inorganic Chemistry* 56 (8): 4253–4257.

121 Li, J., Zhang, Q., Wu, C. et al. (2014). A DFT study of the mechanism of copper-catalyzed synthesis of 2*H*-indazoles from aryl azide. *Dalton Transactions* 43 (1): 55–62.

122 Hu, J., Cheng, Y., Yang, Y. et al. (2011). A general and efficient approach to 2H-indazoles and 1H-pyrazoles through copper-catalyzed intramolecular N–N bond formation under mild conditions. *Chemical Communications* 47 (36): 10133–10135.

123 Díaz-Requejo, M.M., Belderraín, T.R., Nicasio, M.C. et al. (2003). Cyclohexane and benzene amination by catalytic nitrene insertion into C–H bonds with the copper-homoscorpionate catalyst TpBr$_3$Cu(NCMe). *Journal of the American Chemical Society* 125 (40): 12078–12079.

124 Hou, K., Hrovat, D.A., and Bao, X. (2015). Computational exploration of the mechanism of copper-catalyzed aromatic C–H bond amination of benzene via a nitrene insertion approach. *Chemical Communications* 51 (84): 15414–15417.

125 Deutsch, C. and Krause, N. (2008). CuH-catalyzed reactions. *Chemical Reviews* 108 (8): 2916–2927.

126 Díez-González, S. and Nolan, S.P. (2008). Copper, silver, and gold complexes in hydrosilylation reactions. *Accounts of Chemical Research* 41 (2): 349–358.

127 Jordan, A.J., Lalic, G., and Sadighi, J.P. (2016). Coinage metal hydrides: synthesis, characterization, and reactivity. *Chemical Reviews* 116 (15): 8318–8372.

128 Pirnot, M.T., Wang, Y.-M., and Buchwald, S.L. (2016). Copper hydride catalyzed hydroamination of alkenes and alkynes. *Angewandte Chemie International Edition* 55 (1): 48–57.

129 Rendler, S. and Oestreich, M. (2007). Polishing a diamond in the rough: "Cu–H" catalysis with silanes. *Angewandte Chemie International Edition* 46 (4): 498–504.

130 Tsuji, Y. and Fujihara, T. (2016). Copper-catalyzed transformations using Cu–H, Cu–B, and Cu–Si as active catalyst species. *The Chemical Record* 16 (5): 2294–2313.

131 Li, Y. (2017). Copper(I)-catalyzed reaction of unsymmetrical alkyne with HB(pin): a density functional theory study. *Journal of Physical Organic Chemistry* 30 (6): e3630.

132 Schmid, S.C., Van Hoveln, R., Rigoli, J.W. et al. (2015). Development of N–heterocyclic carbene–copper complexes for 1,3-halogen migration. *Organometallics* 34 (16): 4164–4173.

133 Issenhuth, J.-T., Notter, F.-P., Dagorne, S. et al. (2010). Mechanistic studies on the copper-catalyzed hydrosilylation of ketones. *European Journal of Inorganic Chemistry* 2010 (4): 529–541.

134 Lipshutz, B.H., Noson, K., and Chrisman, W. (2001). Ligand-accelerated, copper-catalyzed asymmetric hydrosilylations of aryl ketones. *Journal of the American Chemical Society* 123 (51): 12917–12918.

135 Zhang, W., Li, W., and Qin, S. (2012). Origins of enantioselectivity in the chiral diphosphine-ligated CuH-catalyzed asymmetric hydrosilylation of ketones. *Organic & Biomolecular Chemistry* 10 (3): 597–604.

136 Han, S.-M., Wang, X., Miao, S. et al. (2017). Mechanistic study on ligand-controlled copper-catalyzed regiodivergent silacarboxylation of allenes with carbon dioxide and silylborane. *RSC Advances* 7 (46): 29035–29041.

137 Yi, Z., Yong-Jun, L.I.U., and Shu-Ping, Z. (2015). Reaction mechanism and regioselectivity of Cu(I)-catalyzed hydrocarboxylation of 1-phenyl-propyne with carbon dioxide. *Acta Physico-Chimica Sinica* 31 (2): 237–244.

138 Zhao, Y., Liu, Y., Bi, S. et al. (2013). Theoretical study on copper-catalyzed reaction of hydrosilane, alkyne and carbon dioxide: a hydrocarboxylation or a hydrosilylation process? *Journal of Organometallic Chemistry* 745–746: 166–172.

139 Fan, T., Sheong, F.K., and Lin, Z. (2013). DFT studies on copper-catalyzed hydrocarboxylation of alkynes using CO_2 and hydrosilanes. *Organometallics* 32 (18): 5224–5230.

140 Cid, J., Gulyás, H., Carbó, J.J. et al. (2012). Trivalent boron nucleophile as a new tool in organic synthesis: reactivity and asymmetric induction. *Chemical Society Reviews* 41 (9): 3558–3570.

141 Neeve, E.C., Geier, S.J., Mkhalid, I.A.I. et al. (2016). Diboron(4) compounds: from structural curiosity to synthetic workhorse. *Chemical Reviews* 116 (16): 9091–9161.

142 Butcher, T.W., McClain, E.J., Hamilton, T.G. et al. (2016). Regioselective copper-catalyzed boracarboxylation of vinyl arenes. *Organic Letters* 18 (24): 6428–6431.

143 Dang, L., Zhao, H., Lin, Z. et al. (2007). DFT studies of alkene insertions into Cu–B bonds in Copper(I) boryl complexes. *Organometallics* 26 (11): 2824–2832.

144 Dang, L., Zhao, H., Lin, Z. et al. (2008). Understanding the higher reactivity of B2cat2 versus B2pin2 in Copper(I)-catalyzed alkene diboration reactions. *Organometallics* 27 (6): 1178–1186.

145 Su, W., Gong, T.J., Lu, X. et al. (2015). Ligand-controlled regiodivergent copper-catalyzed alkylboration of alkenes. *Angewandte Chemie International Edition* 54 (44): 12957–12961.

146 Xu, Z.Y., Jiang, Y.Y., Su, W. et al. (2016). Mechanism of ligand-controlled regioselectivity-switchable copper-catalyzed alkylboration of alkenes. *Chemistry - A European Journal* 22 (41): 14611–14617.

147 Isegawa, M., Sameera, W.M.C., Sharma, A.K. et al. (2017). Copper-catalyzed enantioselective boron conjugate addition: DFT and AFIR study on different selectivities of Cu(I) and Cu(II) catalysts. *ACS Catalysis* 7 (8): 5370–5380.

148 Lv, X., Wu, Y.-B., and Lu, G. (2017). Computational exploration of ligand effects in copper-catalyzed boracarboxylation of styrene with CO_2. *Catalysis Science & Technology* 7 (21): 5049–5054.

149 Moon, J.H., Jung, H.-Y., Lee, Y.J. et al. (2015). Origin of regioselectivity in the copper-catalyzed borylation reactions of internal aryl alkynes with bis(pinacolato)diboron. *Organometallics* 34 (11): 2151–2159.

150 Chae, Y.M., Bae, J.S., Moon, J.H. et al. (2014). Copper-catalyzed monoborylation of silylalkynes; regio- and stereoselective synthesis of (Z)-β-(borylvinyl)silanes. *Advanced Synthesis & Catalysis* 356 (4): 843–849.

151 Kim, H.R. and Yun, J. (2011). Highly regio- and stereoselective synthesis of alkenylboronic esters by copper-catalyzed boron additions to disubstituted alkynes. *Chemical Communications* 47 (10): 2943–2945.

152 Moure, A.L., Gomez Arrayas, R., Cardenas, D.J. et al. (2012). Regiocontrolled Cu(I)-catalyzed borylation of propargylic-functionalized internal alkynes. *Journal of the American Chemical Society* 134 (17): 7219–7222.

153 Zhu, G., Kong, W., Feng, H. et al. (2014). Synthesis of (Z)-1-thio- and (Z)-2-thio-1-alkenyl boronates via copper-catalyzed regiodivergent hydroboration of thioacetylenes: an experimental and theoretical study. *The Journal of Organic Chemistry* 79 (4): 1786–1795.

154 Laitar, D.S., Tsui, E.Y., and Sadighi, J.P. (2006). Catalytic diboration of aldehydes via insertion into the copper–boron bond. *Journal of the American Chemical Society* 128 (34): 11036–11037.

155 Zhao, H., Dang, L., Marder, T.B. et al. (2008). DFT Studies on the mechanism of the diboration of aldehydes catalyzed by Copper(I) boryl complexes. *Journal of the American Chemical Society* 130 (16): 5586–5594.

156 Laitar, D.S., Müller, P., and Sadighi, J.P. (2005). Efficient homogeneous catalysis in the reduction of CO_2 to CO. *Journal of the American Chemical Society* 127 (49): 17196–17197.

157 Zhao, H., Lin, Z., and Marder, T.B. (2006). Density functional theory studies on the mechanism of the reduction of CO_2 to CO catalyzed by Copper(I) boryl complexes. *Journal of the American Chemical Society* 128 (49): 15637–15643.

14

Theoretical Study of Ag-Catalysis

107.8682

Ag⁴⁷

Silver

[Kr]4d¹⁰5s¹

Silver catalysis has become an attractive option for synthetic transformations in organic chemistry because of its favorable features, which has high versatility, providing access to a wide variety of reactions with a broad range of conditions, mostly C—C multiple bond activation and functionalization [1–3]. As a typical ds-block element, the atomic radius of silver is only slightly larger than that of copper. The stronger controllability of valence electron leads to silver becoming more inactive than copper. Therefore, silver salts often act as oxidant in other transition metal–catalyzed oxidative coupling reactions, which can be easily reduced to elemental silver.

Different from the changeable oxidative state of copper, the common oxidative state of silver in its salts is +1. Therefore, the redox process is usually not involved in silver catalysis, though the +2 oxidative state of silver only occurs in some radical-involving processes. In silver salts, the filled-up d orbital of silver can donate back electron to the ligand for the formation of back-donation π-bond, which leads to the fact that the silver salt can be used to activate unsaturated molecules for the further transformations. Because the coordination environment of silver is relatively simple as well as the oxidation state of silver is invariable, the side reactions of silver-catalyzed unsaturated bond activation are often less and the products are single, compared with other post transition metal catalysts [4, 5]. In this chemistry, carbenoids [6, 7], nitrenoids [8, 9], silylenoids [10, 11], alkynes [12, 13], and alkenes [14, 15] can be activated for the further transformations, whose mechanisms would be discussed in this chapter [16–20].

Computational Methods in Organometallic Catalysis: From Elementary Reactions to Mechanisms,
First Edition. Yu Lan.
© 2021 WILEY-VCH GmbH. Published 2021 by WILEY-VCH GmbH.

14.1 Ag-Mediated Carbene Complex Transformations

Silver is favorable for the formation of Fischer-type metallacarbenes bearing a neutral divalent carbon ligand, in which carbene is in the singlet state and carbene carbon has an sp²-hybridized lone pair (σ electrons) and an unoccupied p orbital. The σ lone pair of the carbene is donated to the vacant orbital of silver, and simultaneously π back-donation occurs from the silver d orbital to the unoccupied carbene p orbital, which collectively constitutes the silver–carbon bond in this Fischer-type metallacarbene complex [21, 22] (Scheme 14.1).

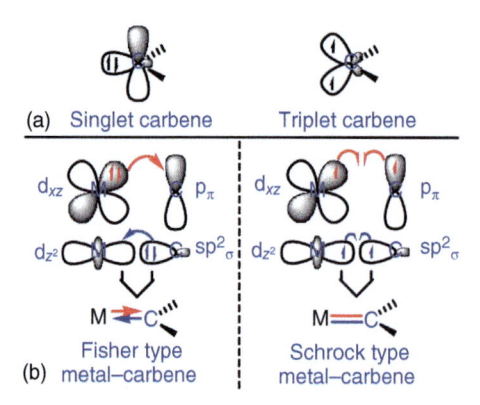

Scheme 14.1 (a) Electron configuration of singlet and triplet carbene; (b) metal–carbon bonds in Fischer- and Schrock-type metal–carbene complexes.

As shown in Scheme 14.2, the carbene center in Fischer-type metallacarbene complexes is electrophilic in character, especially when electron-withdrawing groups are appended to carbene to further decrease its electron density, and is susceptible to nucleophilic attack [23–25]. Therefore, many moderate or inactive nucleophiles, such as Y—H (Y=C, Si, O, S, N, etc.), halogens, oxygen, and olefins, are used in the subsequent carbene transformation reactions, including Y—H insertion, ylide formation, cyclopropanation, and cyclopropenation [26–34].

In silver chemistry, Ag–carbene complexes in which the oxidation state of silver is +1 could be generated by the dissociation of a leaving group. Theoretical calculations have revealed that such a low-oxidation state indicates that Ag–carbene complexes could be considered as Fischer-type metallacarbene complexes, which could further react with nucleophiles to construct new bifunctionalized carbenoid compounds [35–39]. In these types of reactions, how to generate Ag–carbene and its further transformation are important to understand the mechanism of these reactions.

14.1.1 Silver–Carbene Formation

In Ag–carbene chemistry, diazo compounds are popular carbenoid precursors that can react with silver complexes to generate Ag–carbene intermediates [33, 40, 41]. As shown in Scheme 14.3, diazo compounds are often considered to be 1,3-dipolar,

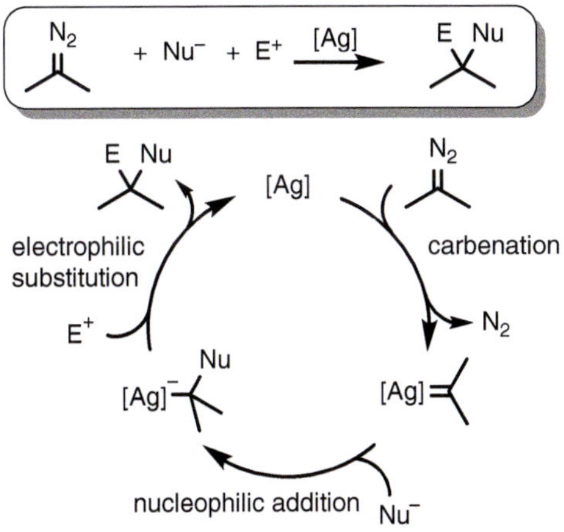

Scheme 14.2 General mechanism of silver-catalyzed carbene transfer reactions.

one of the important resonance structures, which contains a formal N≡N triple bond and a negative charge located on the carbon atom. When a silver catalyst is used, the negative carbon could coordinate with silver followed by the dissociation of N_2 gas, which leads to irreversible reactions [42–46]. The loss of N_2 gas leads to the generation of carbene, which could be stabilized by silver. Moreover, another resonance structure of diazo compounds could be considered to possess a cumulene-type Lewis electron structure in which the formal negative charge is located on the terminal nitrogen atom. In this case, the terminal nitrogen atom could be activated by a silver catalyst to generate a free carbene intermediate by release of N_2 gas. The complexity of diazo compounds leads to a series of reaction pathways in the Ag–carbene generation step. Therefore, some theoretical investigations have focused on this process to understand how Ag–carbene intermediates are generated from diazo compounds [47–51].

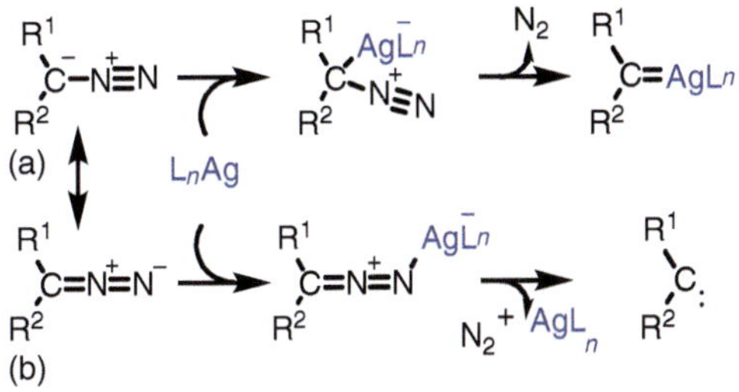

Scheme 14.3 Two activation models of a diazo compound with an Ag species: (a) carbon activation to generate an Ag–carbene complex and (b) nitrogen activation to generate free carbene.

As an example, Li and Gao conducted a theoretical study on the mechanism of Ag-catalyzed intermolecular carbenoid arylation for the synthesis of 3-alkylideneoxindoles from N-aryl-α-diazoamides [51]. In this reaction, the main

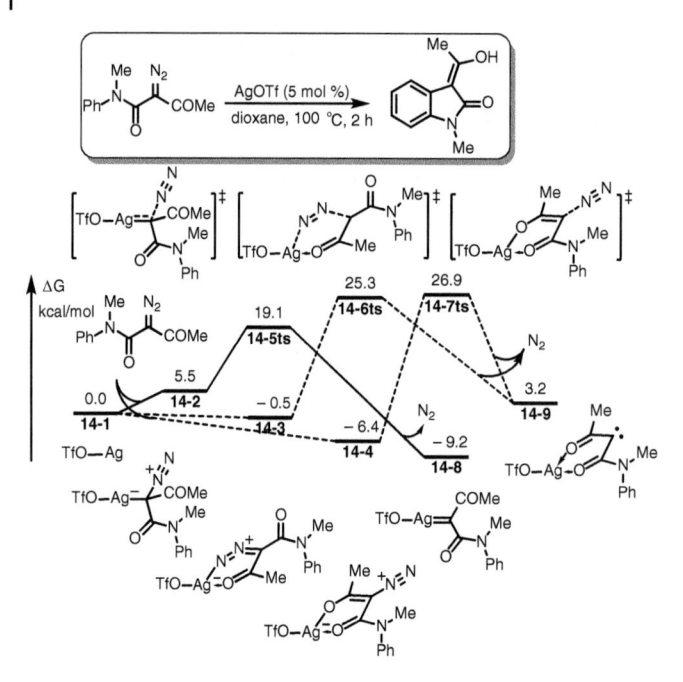

Figure 14.1 Free-energy profiles of possible pathways for carbene formation in Ag-catalyzed intramolecular carbenoid arylation. The energies were calculated at B3LYP/6-31+G(d) (LANL2DZ+p for Ag)/CPCM(THF)//B3LYP/6-31G(d) (LANL2DZ for Ag) level of theory.

step is the formation of the Ag–carbene complex, which was clearly revealed by density functional theory (DFT) calculations. In their study, three possible pathways for carbene formation were considered theoretically. As shown in Figure 14.1, AgOTf **14-1** could coordinate to the electron-rich carbon atom of diazo substrate to form complex **14-2** endothermically. Following the activation by silver, the C—N bond cleavage could take place via transition state **14-5ts** with an activation free energy of 19.1 kcal/mol, which led to the formation of Ag–carbene complex **14-8**.

Analysis of the Lewis electron structure of the diazo compound showed that the negative charge was also shared with the terminal nitrogen atom and the electron-withdrawing groups on carbon; therefore, silver could also coordinate with the terminal nitrogen atom of the diazo group or with a carbonyl group. Li and colleagues found that the coordination of the terminal nitrogen atom with silver to form intermediate **14-3** was exothermic with a free energy of −0.5 kcal/mol. However, according to the Curtin–Hammett principle, an intermediate with lower energy does not necessarily lead to a transition state with a lower activation barrier. Although the free energy of **14-3** is 6.0 kcal/mol lower than that of **14-2**, the C—N bond in diazo is difficult to break because of the shorter C—N bond length than that in **14-2** (1.33 vs 1.37 Å). The free energy of transition state **14-6ts** was 6.2 kcal/mol higher than that of **14-5ts**. Furthermore, the free energy of the free-carbene complex **14-9** generated via this pathway was 12.4 kcal/mol higher than that of

14-8. Therefore, this pathway could be excluded because of its considerably higher activation free energy.

Another activation mode was also considered theoretically in which Ag was used to activate the electron-withdrawing substituent groups of the diazo compound, generating stable intermediate **14-4**. In this intermediate, the C—N bond length in diazo moiety was only 1.32 Å, which reveals that it would be even harder to cleave than that of **14-2**. As expected, the theoretical calculations showed that the relative free energy of transition state **14-7ts** was 26.9 kcal/mol, which was 7.8 kcal/mol higher that of **14-5ts**; therefore, this pathway could also be excluded. Based on the above theoretical calculations, Ag–carbene complex formation could occur through the activation induced by the coordination of the diazo carbon atom with silver to weaken the C—N bond followed by the dissociation of N_2. The driving force for this step is the generation of an Ag–carbene complex and release of N_2 gas.

Similar competition between diazo activation with diazo carbon coordination and electron-withdrawing group coordination in Ag–carbene formation was also considered by Davies's group in their theoretical study on the mechanism of silver-catalyzed carbene insertion into methanol [50] (Figure 14.2). DFT calculations showed that when diazoacetate reacted with AgOTf **14-1**, the coordination of the diazo carbon in complex **14-10** led to the activation of the diazo C—N bond.

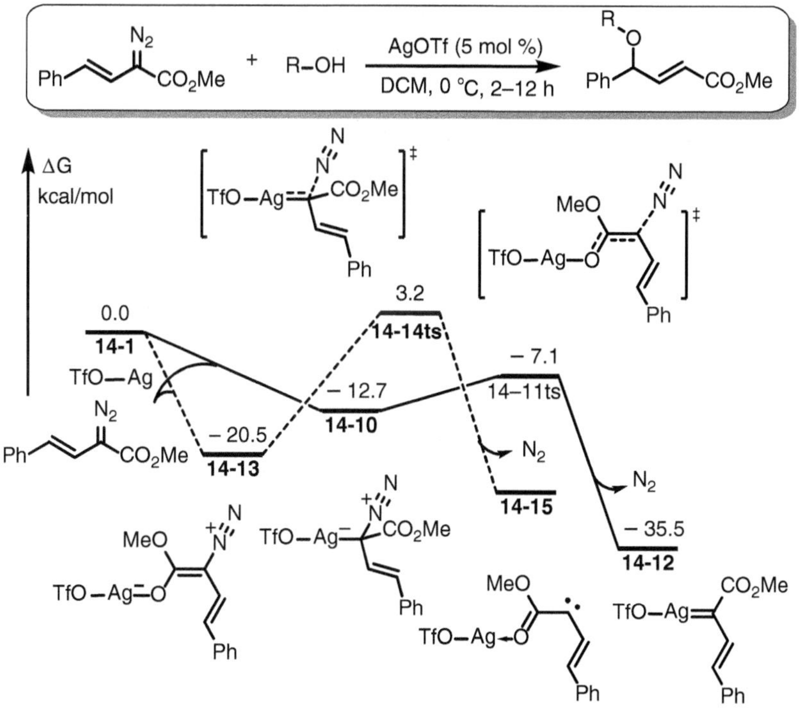

Figure 14.2 Free-energy profiles of carbene formation in Ag-catalyzed carbene hydroalkoxylation. The energies were calculated at B3LYP/6-311+G(2d,2p) (RSC+4f for Ag)/IEF-PCM(DCM)//B3LYP/6-31G(d) (RSC+4f for Ag) level of theory.

The calculated activation barrier via transition state **14-11ts** was only 5.6 kcal/mol and afforded Ag–carbene complex **14-12** irreversibly. An enolate activation model through intermediate **14-13** was also considered theoretically. The calculated activation free energy was 23.7 kcal/mol via transition state **14-14ts**, which indicates an unfavorable process. This result is consistent with Li's theoretical analysis.

14.1.2 Carbene Insertion into C−Cl Bond

In halohydrocarbon derivatives, the electronegativity of the halogen atom is higher than that of the carbon atom; therefore, the halogen possesses weak nucleophilicity. When a halohydrocarbon reacts with an Ag–carbene species, the halogen atom could nucleophilically attack the carbon atom of the carbene moiety, leading to the insertion of carbene into the C—halogen bond. Pérez and coworkers reported combined experimental and theoretical study of carbene insertion into haloalkanes [52] (Figure 14.3). Their DFT calculations showed that Ag–carbene intermediate **14-16** could react with dichloromethane (DCM) to form chloronium ylide intermediate **14-17** exothermically by nucleophilic addition of the chlorine atom. Intermediate **14-17** also exhibits β-oxoalkyl silver character; therefore, it could be isolated as an enolate tautomer **14-19** via transition state **14-18ts** with an energy barrier of 10.2 kcal/mol. Then, a Stevens-type rearrangement of the chloronium ylide moiety could occur via transition state **14-20ts** to yield carbene insertion product–coordinated silver complex **14-21** irreversibly. The calculated

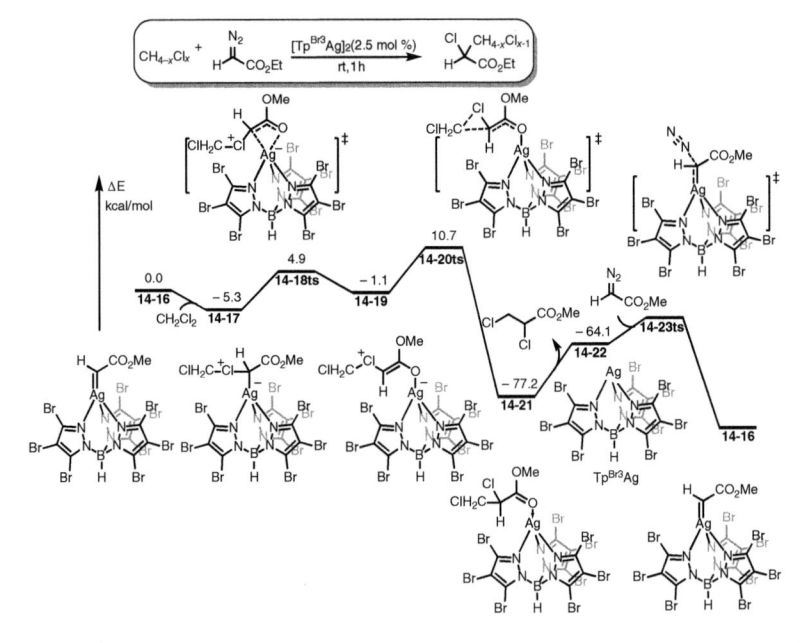

Figure 14.3 Potential energy profiles of Ag-catalyzed carbene insertion into the C−Cl bond of a halohydrocarbon. The energies were calculated at B3LYP/6-31G(d) (LANL2DZ for Ag and Br) level of theory.

overall activation energy of the carbene insertion process was only 16.0 kcal/mol, indicating a rather easy course. The release of product followed by coordination and denitrogenation via transition state **14-23ts** regenerates silver–carbene complex **14-16**.

14.1.3 Carbene Insertion into O–H Bond

Hydroxyl groups possess an oxygen atom that could react nucleophilically with Ag–carbene species to form a new C—O bond. In 2011, Davies and coworker reported a mechanistic study of Ag-catalyzed vinylcarbene insertion into O—H bonds [50]. As shown in Scheme 14.4, Ag–vinylcarbene species **14-12** could react with methanol through two possible pathways. The nucleophilic attack of the carbene carbon could occur via transition state **14-26ts** to form alkyl silver intermediate **14-27**. Alternatively, the nucleophilic attack also could occur at the vinyl site of **14-12** to form vinyl silver intermediate **14-29** via transition state **14-28ts**. The calculated activation barrier of the former process was 5.0 kcal/mol and that of the latter was 2.3 kcal/mol. The computational results revealed that the nucleophilic attack at the vinyl site was the kinetically favorable process, which is in reasonable agreement with experimental observations. The higher energy of transition state **14-26ts** relative to that of **14-28ts** could be attributed to the strain originating from the strong hydrogen-bonding interaction between the attacking methanol molecule and triflate counterion leading to the distortion of the linear-type coordination structure of Ag.

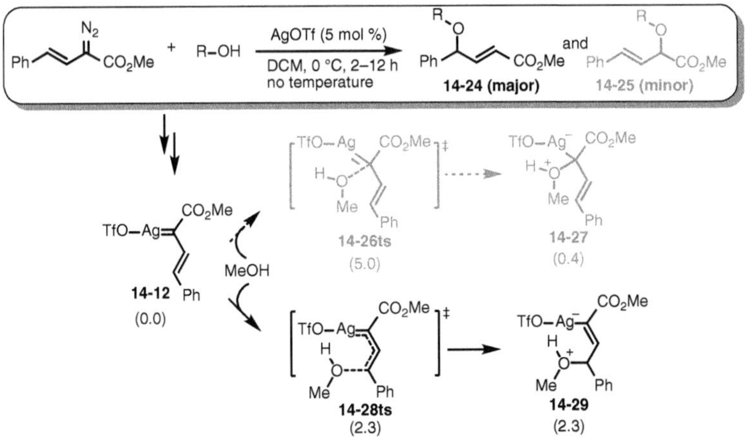

Scheme 14.4 Regioselectivity of the nucleophilic addition of a hydroxyl group to an Ag–vinylcarbene complex. The energies were calculated at B3LYP/6-311+G(2d,2p) (RSC+4f for Ag)/IEF-PCM(DCM)//B3LYP/6-31G(d) (RSC+4f for Ag) level of theory.

14.1.4 Nucleophilic Attack by Carbonyl Groups

Carbonyl oxygen atoms are nucleophilic. DFT calculations have been used to study Ag–carbene transformation when carbonyl oxygen is used as the nucleophile. As

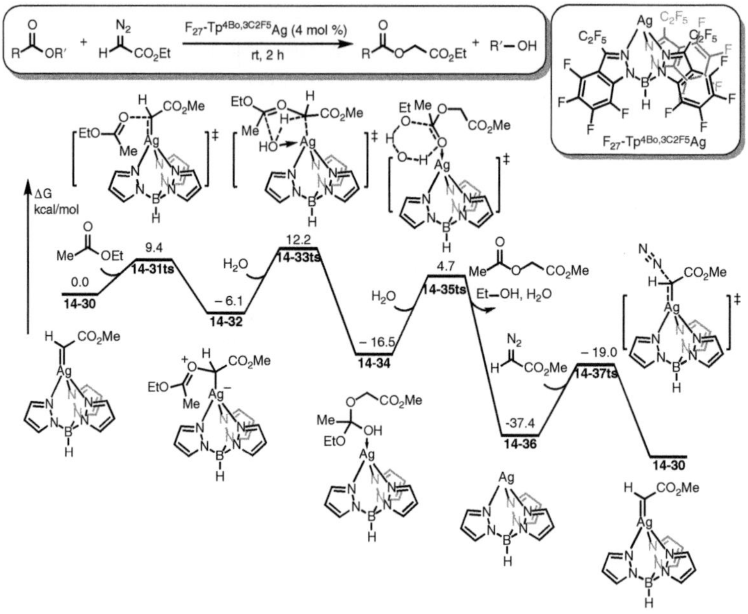

Figure 14.4 Free-energy profiles for the nucleophilic addition of the carbonyl oxygen atom in an ester to an Ag–carbene complex. The energies were calculated at B97-D/6-31+G(d) (LANL2DZ for Ag)/SMD(methyl acetate) level of theory.

illustrated in Figure 14.4, Pérez and coworkers performed a mechanistic study of a silver-catalyzed coupling reaction between esters and ethyl diazoacetate [49]. Their computational results showed that Ag–carbene complex **14-30** underwent nucleophilic attack by the carbonyl oxygen via transition state **14-31ts** with an activation energy of 9.4 kcal/mol to generate oxonium ylide intermediate **14-32** exothermically. Subsequently, the coming water molecule can nucleophilically attack the oxonium moiety via transition state **14-33ts** with a free-energy barrier of 18.3 kcal/mol. Simultaneously, a proton transfer from coming water to C(alkyl)—Ag can generate a dialkyl orthoacetate–coordinated silver complex **14-34**. Then, a water-assisted decomposition of orthoacetate takes place via a six-membered ring-type transition state **14-35ts** to yield corresponding ester and alcohol products with the formation of silver complex **14-36**. The following coordination and denitrogenation via transition state **14-37ts** regenerates silver–carbene complex **14-30**. The calculated activation free energy of denitrogenation is 18.4 kcal/mol, which is almost as high as that of nucleophilic attack by water via transition state **14-33ts**, which indicates that both of these processes could be the rate-determining step.

14.1.5 Carbene Insertion into C—H Bond

A nucleophile is an electron-rich molecule that contains a lone pair of electrons or polarized bond. Therefore, the C—H bonds of hydrocarbons can be considered nucleophilic because the electronegativity of carbon is higher than that of hydrogen.

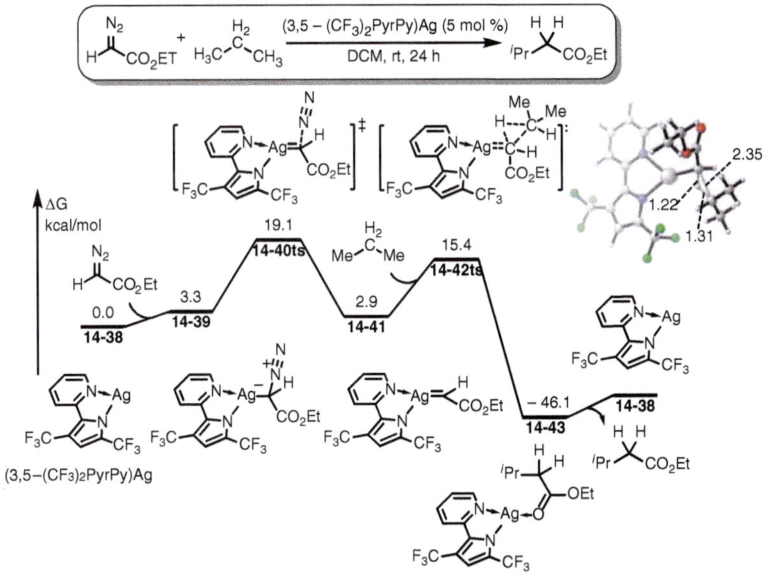

14.1 Ag-Mediated Carbene Complex Transformations | 575

Figure 14.5 Free-energy profiles of Ag-catalyzed carbene insertion into the C—H bonds of alkanes. The energies were calculated at B3LYP/cc-pVTZ(-f) (LACV3P for Ag)//PCM(DCM)//B3LYP/6-31G(d,p) (LACVP for Ag) level of theory.

A series of DFT investigations have indicated that transition metal–supported carbenes can undergo nucleophilic attack by weak nucleophiles, such as C—H bonds of hydrocarbons. The general mechanism for this carbene transformation can be explained as the attack of a poor nucleophile (the C—H bond) by a strong electrophile (the carbene carbon), which causes cleavage of a C—H bond accompanied with formation of carbene–carbon and carbene–hydrogen bonds in a concerted manner.

As shown in Figure 14.5, Mindiola and coworkers studied the mechanism of carbene insertion into the C—H bonds of alkanes [48]. When Ag complex **14-38** was considered as active species in catalytic cycle, the coordination of diazo compound onto Ag forms complex **14-39** with endergonic 3.3 kcal/mol. The denitrogenation takes place via transition state **14-40ts** with an activation free energy of 19.1 kcal/mol to generate silver-carbene complex **14-41**. Carbene insertion could take place via transition state **14-42ts**, whose computed geometry is shown in Figure 14.5. The lengths of the forming C—C and C—H bonds in **14-42ts** were 2.35 and 1.22 Å, respectively, and the length of the breaking C—C bond was 1.31 Å, which indicates that the insertion process is concerted. The computed free-energy barrier for the reaction with the secondary C—H in propane was 12.5 kcal/mol. Active species **14-38** can be regenerated after carbene insertion by release of product.

Similar to those in alkanes, C(aryl)—H bonds in electron-rich arenes also display nucleophilicity in Ag-mediated carbene insertion reactions. Yang and coworkers reported an Ag-catalyzed synthesis of alkylideneoxindoles from diazo compound via intramolecular carbene insertion into an arene sp² C—H bond [45] (Figure 14.6).

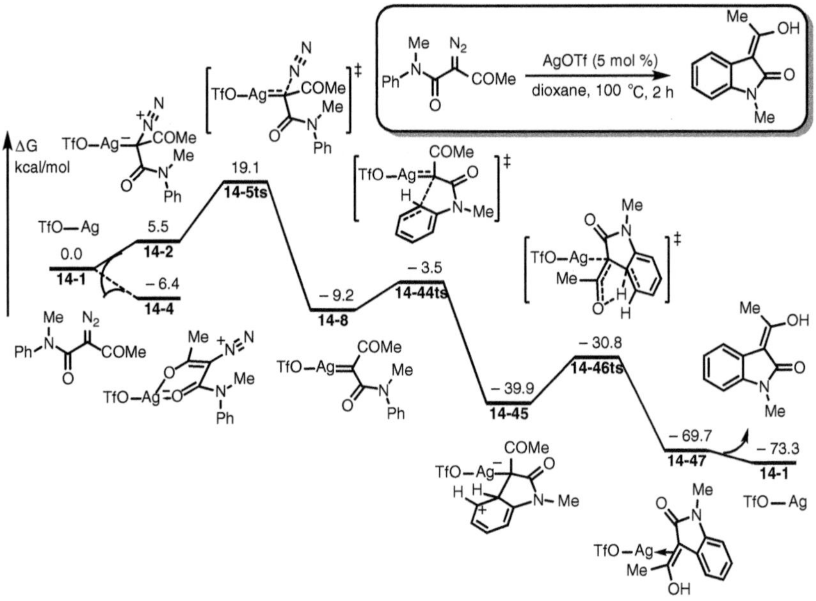

Figure 14.6 Free-energy profiles of possible pathways for carbene insertion into C(aryl)—H bond. The energies were calculated at B3LYP/6-31+G(d) (LANL2DZ+p for Ag)/CPCM(THF)//B3LYP/6-31G(d) (LANL2DZ for Ag) level of theory.

In the subsequent DFT studies by Li's group, an aromatic electrophilic substitution mechanism was proposed for this reaction. As shown in Figure 14.6, after Ag–carbene species **14-8** has formed, an intramolecular electrophilic addition onto an aryl with an electron-donating group could take place via transition state **14-44ts** with a free-energy barrier of only 5.7 kcal/mol to form zwitterionic intermediate **14-45** irreversibly. In this intermediate, the cationic carbon could be stabilized by an electron-donating amido group. The following rapid proton transfer occurs via transition state **14-46ts** to generate the alkylideneoxindole product. The computational results indicated that aromatic electrophilic substitution was a stepwise process, which is different from the concerted process for carbene insertion into an alkane C—H bond.

14.2 Ag-Mediated Nitrene Transformations

Nitrene (Scheme 14.5a) is a neutral monovalent species with six valence electrons that is the isoelectronic equivalent of carbene, sharing some similar characteristics. However, nitrogen is more electronegative than carbon, which leads to essential differences when it reacts with a transition metal to form a metal–nitrene complex. A carbene can react with silver to form a Fischer-type Ag–carbene complex in which the carbene moiety uses one lone-pair electron to coordinate with silver and has a formal positive charge on the carbon atom of the carbene moiety [53–56]. Conversely, if a Fischer-type Ag–nitrene complex is formed, the nitrogen atom is more electronegative than carbon. Thus, the donation of a lone-pair electron from the nitrogen atom of nitrene is unfavorable compared with that from a carbon of carbene. Meanwhile, the formal positive charge located on nitrogen atom in nitrene moiety is also invalid. Therefore, the higher electronegativity of nitrogen compared with that of carbon leads to Ag–nitrene complexes possessing a more stable triplet electronic structure compared with the corresponding singlet state, which maintains the electronic neutrality of the nitrogen atom.

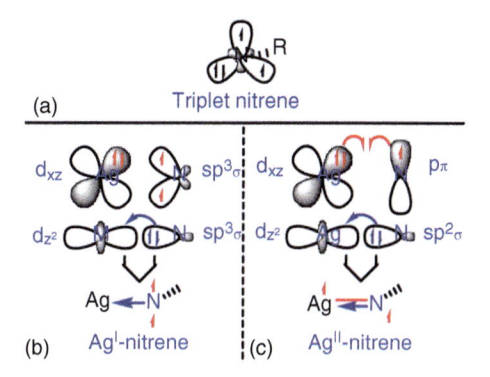

Scheme 14.5 (a) A triplet nitrene. (b) Triplet Ag(I)–nitrene complex with two unpaired electrons on the nitrogen atom. (c) Triplet Ag(II)–nitrene complex with one unpaired electron on silver and one on nitrogen.

In pioneering theoretical studies of Ag–nitrene complexes, Pérez and coworkers proposed that the triplet state of an Ag–nitrene complex has the two unpaired electrons mainly localized on the nitrogen atom of arene and the oxidative state of silver is +1 based on the results of their DFT calculations [57] (Scheme 14.5b). Schomaker and coworkers proposed that the lowest-energy state of the Ag–nitrene intermediate was also a triplet, in which an Ag(II) ion is ferromagnetically coupled to a doublet nitrene anion radical [58] (Scheme 14.5c). The various types of electronic structures of Ag–nitrene species lead to greater complexity in Ag–nitrene chemistry than in Ag–carbene chemistry. Ag-mediated nitrene transfer reactions also represent an effective approach to introduce new C—N bonds into molecules because of the complexity of Ag–nitrene chemistry [59–71].

14.2.1 Silver–Nitrene Complex Formation

Nitrenoids, including chloramine-T (TsN=(Cl)Na), bromamine-T (TsN=(Br)Na), organic azides (RN$_3$), and iminoiodinanes (PhI=NR), are often used as nitrene precursors to afford metal–nitrene species [8, 9, 72]. In particular, iminoiodinanes, which could be formed as isolated materials or in situ by oxidizing sulfonamides or other amines with PhI(OAc)$_2$ or PhIO, are usually reacted with Ag(I) species to form the corresponding Ag–nitrene complexes [73, 74].

The mechanism for the generation of triplet Ag–nitrene complexes was clarified by Pérez's group [57]. As shown in Figure 14.7, their DFT calculations revealed that

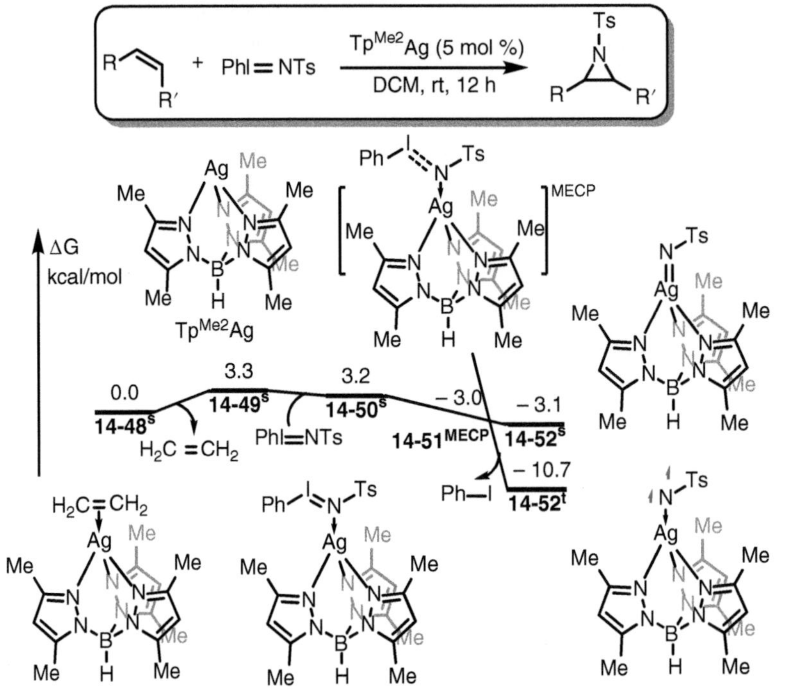

Figure 14.7 Free-energy profiles for the formation of a triplet Ag–nitrene complex in an Ag-catalyzed olefin aziridination reaction. The energies were calculated at M06/6-311++G(d,p) (SDD for Ag and I)//M06/6-31G(d) (SDD for Ag and I) level of theory.

the coordination of iminoiodinane with a silver species formed intermediate **14-50^s** reversibly. The cleavage of the N—I bond leading to the release of iodobenzene could then occur via a minimum energy cross-point **14-51^{MECP}** to yield triplet Ag–nitrene species **14-52^t**. The relative free energy of **14-51^{MECP}** was 6.2 kcal/mol lower than that of the singlet intermediate **14-50^s**; however, this calculated relative free energy is usually inaccurate because of the free-energy correction for a nonstationary point. Therefore, the generation of Ag–nitrene species is a rather rapid process. Interestingly, the relative free energy of triplet Ag–nitrene complex **14-52^t** was 7.6 kcal/mol more stable than that of the closed-shell singlet state **14-52^s**, as stated above. The calculated spin densities of nitrogen and silver atoms in triplet Ag–nitrene complex **14-52^t** were 1.32 and 0.30, respectively, which revealed that the two unpaired electrons were mainly localized on the nitrogen atom, and the oxidative state of silver was +1.

In another example, Pérez and coworkers reported a combined theoretical and experimental study on the mechanism of Ag-catalyzed nitrene transfer [75] (Figure 14.8). In this work, a triplet Ag–nitrene complex was also formed from the reaction of an Ag catalyst with iminoiodinane. Their theoretical study revealed that the formation of Ag–iminoiodinane complex Tp*,^{Br}Ag(PhI=NTs) **14-55^s** was the rate-determining step, whose relative free energy was 23.4 kcal/mol higher than that of the dimeric Ag catalyst **14-53^s**. After this complex had formed, N—I bond

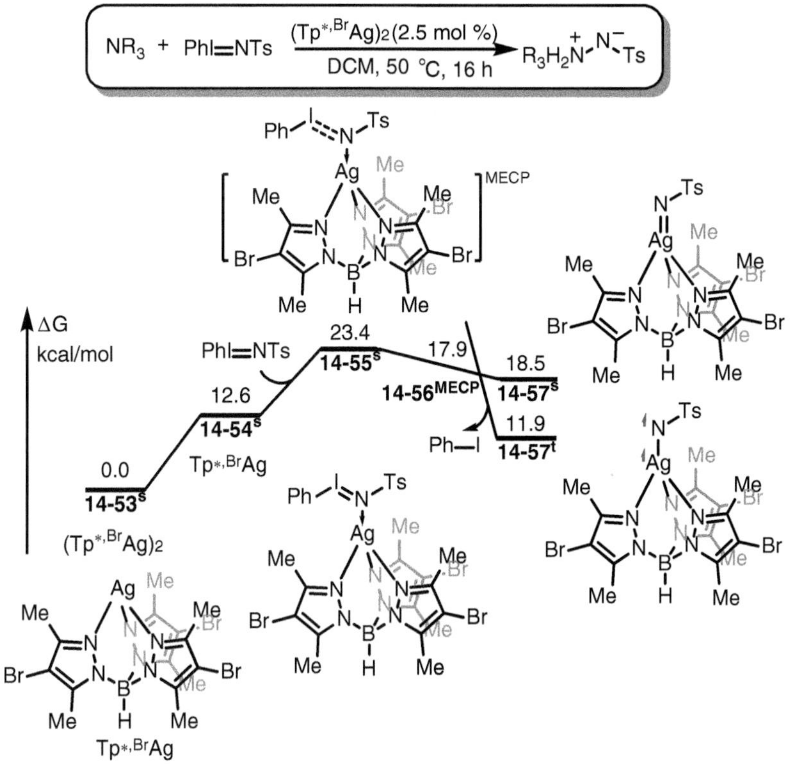

Figure 14.8 Free-energy profiles for Ag–nitrene complex formation in N-amidation of tertiary amines. The energies were calculated at M06/6-311++G(d,p) (SDD for Ag, Br, and I)/SMD(DCM)//M06/6-31G(d) (SDD for Ag, Br, and I)/SMD(DCM) level of theory.

cleavage could take place via low-energy **14-56**^{MECP} to form triplet Ag–nitrene complex **14-57**^{t}. Alternatively, singlet Ag–nitrene complex **14-57**^{s} could also be generated from Ag–iminoiodinane complex **14-55**^{s}, the relative free energy of which was 6.6 kcal/mol higher than that of triplet complex **14-57**^{t}. Interestingly, the calculated spin-density distributions of the silver and nitrogen atoms in triplet complex **14-57**^{t} were 1.34 and 0.28, respectively, which are consistent with previous observations. However, in their discussion, the silver and nitrogen atoms were considered to possess an unpaired electron each, and the proposed oxidation state of Ag was +2.

In Ag-catalyzed nitrene transfer reactions, the triplet Ag–nitrene complex could be generated from the singlet nitrene precursor through a singlet-to-triplet spin crossover characterized by a minimum energy cross-point (MECP), because the triplet complex is more stable than the corresponding singlet one. The spin density of the nitrogen atom in a triplet Ag–nitrene complex is higher than 1.0, indicating that the two unpaired electrons majorly localize on the nitrogen atom. Therefore, the nitrogen atom could be considered to be electrophilic because the number of valence electrons of nitrogen is lower than that needed to obey the octet rule.

14.2.2 Nucleophilic Attack by Unsaturated Bonds

Because of the electrophilicity of the nitrogen atom in an Ag–nitrene complex, Ag-catalyzed nitrene transfer reactions provide an effective route to construct hetero C—N bonds, which is often realized through nucleophilic attack of the generated Ag–nitrene complex. Previous experimental studies showed that various nucleophiles, including π bonds, C—H bonds, and amines, could be used in corresponding reactions, such as aziridination of olefins, C—H amidation, and N amidation [59–71]. Theoretical studies have been performed to reveal the detailed reaction pathway and gain mechanistic insight into these nitrene transformations [57, 58, 75, 76].

A metal–nitrene complex with an electrophilic nitrogen atom can preferentially react with olefins to form the corresponding aziridines, which are strained-ring compounds. Usually, both concerted and stepwise pathways have been predicted for the cycloaddition of nitrene and olefin in previous experimental and theoretical studies [56, 77] (Scheme 14.6). Generally, a closed-shell reaction between singlet nitrene and an olefin is considered to possess a concerted mechanism, in which the two C—N bonds are formed simultaneously. Meanwhile, an open-shell reaction between triplet nitrene and an olefin could follow a stepwise pathway in which a radical addition takes place first to construct one C—N bond and form a diradical intermediate. As mentioned in Section 14.2.1, previous theoretical calculations revealed that the triplet Ag–nitrene complex is more stable than the singlet one; therefore, the stepwise reaction pathway is usually considered in Ag–nitrene transformations.

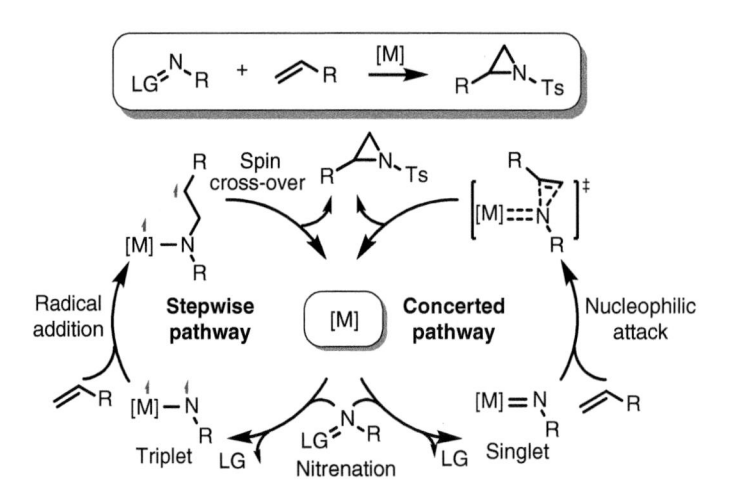

Scheme 14.6 Two possible pathways for transition metal-supported olefin aziridination with nitrene: a stepwise pathway through radical addition with an olefin (left) and a concerted pathway through nucleophilic addition of nitrene (right).

As shown in Figure 14.9, Pérez and coworkers performed a comprehensive computational study on Ag-catalyzed nitrene transformation to explore olefin aziridination using hydrotris(pyrazolyl)borate (Tpx) as a ligand [57]. In their calculated triplet

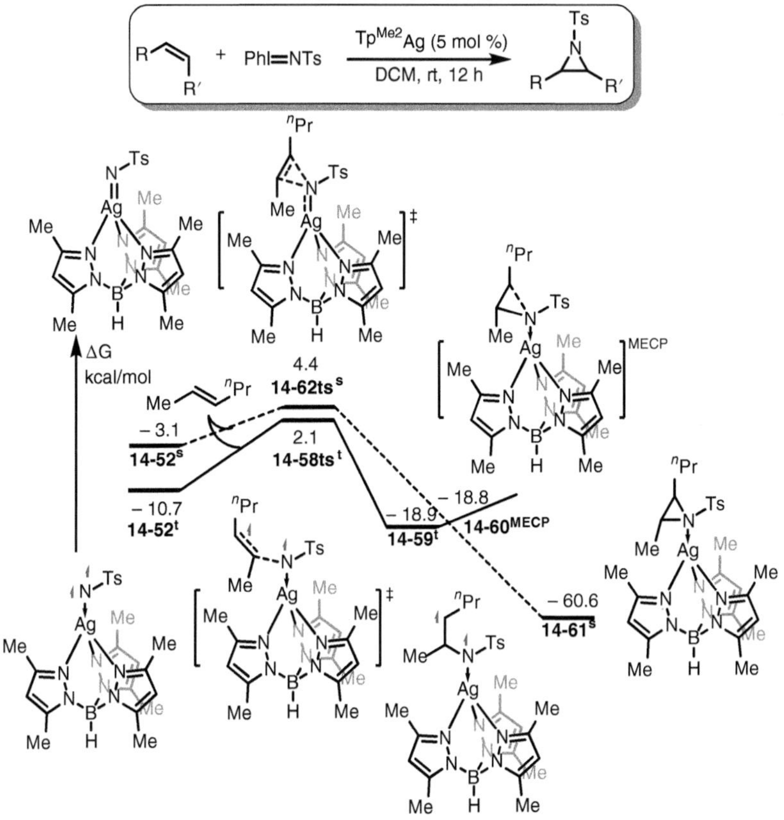

14.2 Ag-Mediated Nitrene Transformations | 581

Figure 14.9 Free-energy profiles for the Ag-mediated aziridination of alkenes. The energies were calculated at M06/6-311++G(d,p) (SDD for Ag and I)//M06/6-31G(d) (SDD for Ag and I) level of theory.

free-energy surface, the reaction starts with the formation of triplet silver nitrene intermediate **14-52^t**, where the radical character is mostly localized on the nitrogen atom of the nitrene moiety. Nitrene radical addition with an olefin could take place via transition state **14-58ts^t** with a free-energy barrier of 12.8 kcal/mol to form the first C—N bond. In the triplet free-energy surface, intermediate **14-59^t** could be generated over transition state **14-58ts^t**, which obviously displays diradical character. In intermediate **14-59^t**, one unpaired electron is located on the carbon atom coming from the olefin and the other one is shared by silver and nitrogen atoms. Moreover, these researchers also calculated the corresponding singlet free-energy surface for this reaction. As illustrated in Figure 14.9, the relative free energy of singlet Ag–nitrene complex **14-52^s** was 7.6 kcal/mol higher than that of the triplet one. They also identified singlet transition state **14-62ts^s**, the relative free energy of which was 2.3 kcal/mol higher than that of the singlet one. The authors considered that going through the singlet transition state **14-62ts^s**, two N—C bonds could be formed synchronously to produce aziridine **14-61^s**, which was 41.7 kcal/mol more stable than triplet intermediate **14-59^t**. The authors also located a cross-point of these two free-energy surfaces (**14-60**[MECP]), the relative free energy of which was 0.1 kcal/mol higher than that of **14-59**[t]. Therefore, the detailed mechanism of this reaction could be described as radical addition of a triplet Ag–nitrene complex with an olefin to form the first N—C bond via transition state **14-58ts**[t] followed by a spin crossover via **14-60**[MECP] to form another N—C bond and give the final product aziridine.

14.2.3 Nucleophilic Attack by Amines

Nucleophiles, including amines, pyridines, and other aromatic N-heterocycles, can attack metal–nitrene complexes to yield aminimides (R_3N^+–N^-R), which have a wide range of applications because of their zwitterionic nature. DFT calculations were used to study this Ag-catalyzed nitrene amidation reaction of a tertiary amine with PhI=NTs as a nitrene source [75]. As shown in Figure 14.10, the triplet Ag–nitrene complex **14-57**[t] was 6.6 kcal/mol more stable than the singlet one **14-57**[s]. Nucleophilic attack of the amine led to triplet intermediate **14-64**[t] by an endergonic process with a free energy of 4.0 kcal/mol, i.e. is without an energy barrier. Pérez's group found that spin crossover could occur before the formation of intermediate **14-64**[t] via **14-63**[MECP] to form singlet species **14-66**[s], which had an N—N bond length of 1.45 Å, i.e. a single bond. The dissociation of the aminimide led to the regeneration of the silver catalyst. The authors also studied the singlet free-energy surface of this reaction. In singlet species **14-57**[s], the singlet Fischer-type Ag–nitrene complex could be nucleophilically attacked by the amine via transition

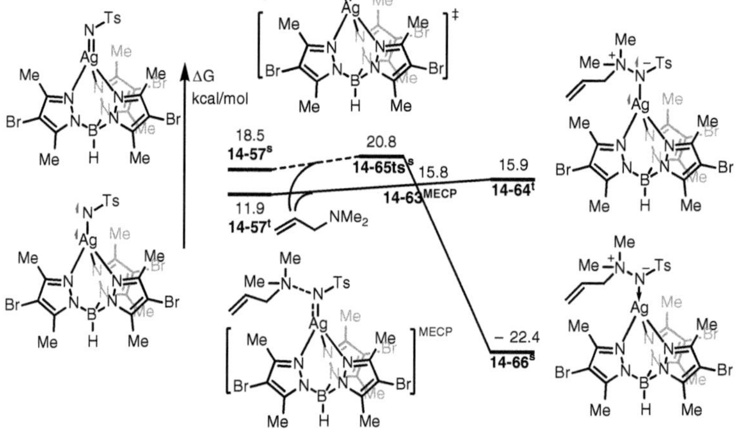

Figure 14.10 Free-energy profiles for the nucleophilic addition of an amine to an Ag–nitrene complex to generate an aminimide species. The energies were calculated at M06/6-311++G(d,p) (SDD for Ag, Br, and I)/SMD(DCM)//M06/6-31G(d) (SDD for Ag, Br, and I)/SMD(DCM) level of theory.

state **14-65ts^s** with a free-energy barrier of only 2.3 kcal/mol. However, the higher relative free energy of the singlet Ag–nitrene complex compared with that of the triplet complex makes the singlet pathway unfavorable.

14.3 Ag-Mediated Silylene Transformations

Silylenes (R^{1}R^{2}Si:) are active intermediates containing a neutral divalent silicon atom and are considered heavier analogues of carbene. As discussed above, carbenes are observed in either the triplet or singlet state depending on the nature of the substituents due to the lower-energy gap between 2s and 2p orbitals of carbon atom. By contrast, silylenes typically have a singlet ground state because the large-energy gap between the 3s and 3p orbitals of silicon induces a substantial singlet–triplet gap [78–81]. Similar to the bonding mode of a metal–carbene complex, the silylene moiety in a Fischer-type metal–silylene complex behaves as an sp σ donor and p π acceptor. The π bonding in a metal–silylene complex is weaker than that in a metal–carbene one because of the lower electronegativity and the longer bond length between silicon and a metal than those of carbon. As a result, the metal–silicon double bond is highly polar and can rotate almost freely. The silicon center in a metal–silylene species is electrophilic in character just like the carbon atom in a Fischer-type metal–carbene complex. Thus, it is also susceptible to nucleophilic attack by nucleophiles such as alkenes and alkynes, providing the corresponding silacyclopropanation and silacyclopropenation products, respectively [82–94].

Silylene intermediates can be generated through thermolysis of various cyclic silanes, photolysis of cyclic silanes or trisilanes, and the reduction of dihalosilanes by strong reductants such as lithium or potassium [95–100]. The generated reactive free silylene intermediates can participate in a range of silylene transfer reactions, including insertion and addition. In addition to interest in the chemistry of silylenes, the involvement of transition metal catalysts has also attracted attention; great effort has been expended in developing their synthesis and transformation. As shown in Scheme 14.7, Woerpel and coworkers discovered that silver complexes exhibit interesting activity to mediate silylene transfer reactions. Silver–silylene complexes can be obtained by extrusion from a sacrificial silacyclopropane under mild reaction conditions [101–106]. The generated electrophilic silver–silylene complexes can react with a variety of functional groups to effectively afford three- or five-membered heterocycles [82–94].

Salvatella's group reported a theoretical study on the mechanism of Ag-catalyzed silylene transfer with olefins to afford siliranes [107] (Figure 14.11). In their theoretical calculations, a cationic silver species **14-67** reacted with silirane to yield an Ag–silylene complex **14-69** with the release of an olefin. The calculated free-energy barrier was 2.2 kcal/mol via transition state **14-68ts**, which indicates a rather rapid process. The strain release of silirane leads to the exergonic formation of an Ag–silylene complex with a free energy of 2.1 kcal/mol. The intramolecular steric repulsion between a *tert*-butyl group and the flagpole hydrogen atoms of

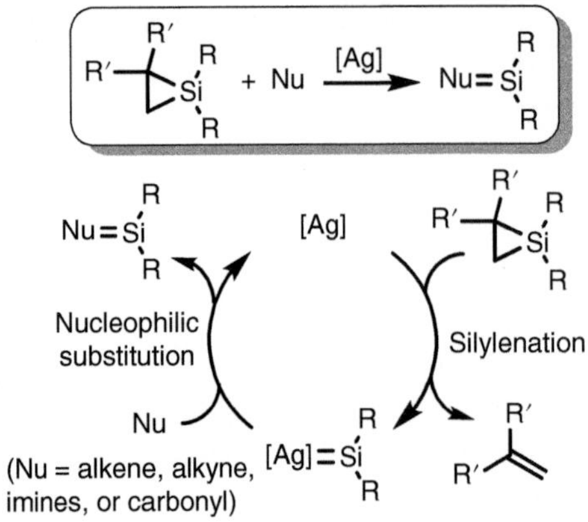

Scheme 14.7 Mechanism of Ag-catalyzed silylene transfer reactions.

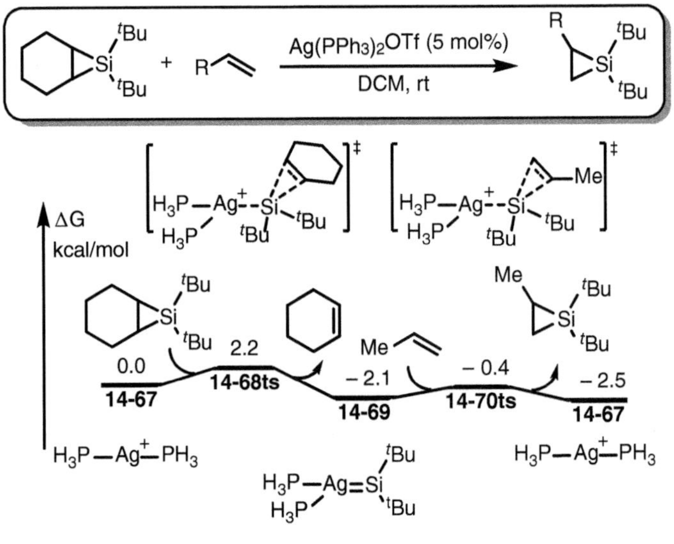

Figure 14.11 Free-energy profiles for the Ag-catalyzed silylene transfer reaction. The energies were calculated at B3LYP/6-31G(d) (SDD for Ag) level of theory.

the boat-shaped six-membered ring in cyclohexene silirane further increased the reactivity of this process. Interestingly, the silylene plane was oriented perpendicular to bisphosphine silver to avoid the steric repulsion between the *tert*-butyl group and phosphine ligand, which was facilitated by the free rotation of the Si—Ag bond.

After Ag–silylene complex **14-69** is generated, silylene transformation could take place with a less steric propene to generate a three-membered silirane through a nucleophilic substitution process via transition state **14-70ts**. The silylene transformation can easily be performed to afford silirane because it has a free-energy barrier of only 1.7 kcal/mol. The driving force of this whole process is considered to be the intramolecular steric repulsion between a *tert*-butyl group and flagpole hydrogen atoms of the boat-shaped six-membered ring in cyclohexene silirane.

14.4 Ag-Mediated Alkyne Activations

As a ds-block late transition metal, silver can be easily coordinated by a π-acidic ligand alkyne to afford silver–alkyne complex, where the unsaturated bond in alkyne can be activated by silver to react with nucleophiles for the synthesis of vinyl compounds. Interestingly, in the presence of Brønsted base, the coordinated terminal alkyne can be deprotonated to achieve alkynyl silver complex, which can be used as nucleophile in further transformations [13, 108] (Scheme 14.8).

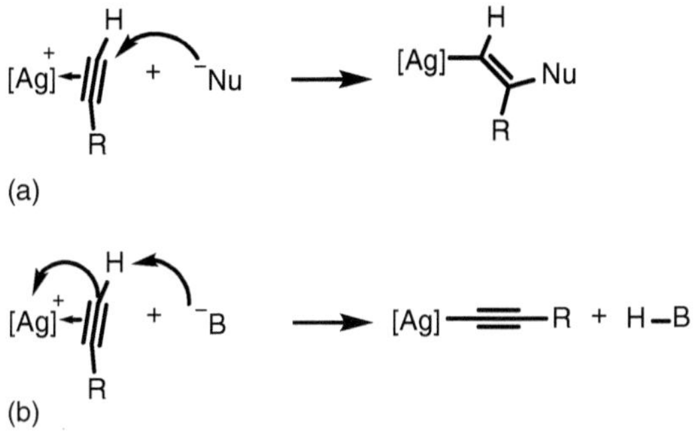

Scheme 14.8 Ag-meditated alkyne activation modes: (a) π-activation and (b) C–H activation.

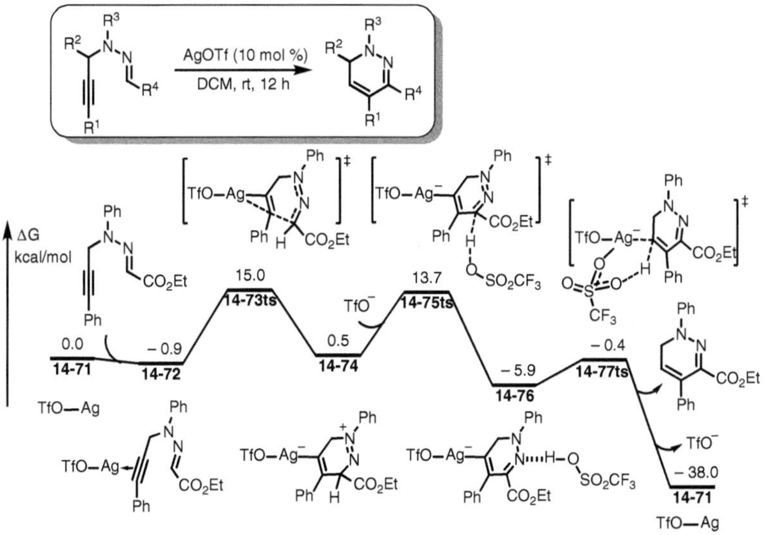

Figure 14.12 Free-energy profiles for the Ag-catalyzed annulation of propargyl hydrazones. The energies were calculated at M06/6-311++G(d,p) (LANL2DZ for Ag)/PCM(DCM)//B3LYP/6-31G(d) (LANL2DZ for Ag) level of theory.

14.4.1 π-Activation of Alkynes

In the presence of Ag(I) salt, alkynes can be active as an electrophile, which can react with an even weak nucleophile to form a vinyl–silver complex for further transformations.

As an example, carbon atom in hydrazones can play as a nucleophile, which can react with silver-activated alkynes through an intramolecular nucleophilic attack for the construction of pyridazine derivatives [109] (Figure 14.12). In the DFT discussion of reaction mechanism, the coordination of N-propargyl hydrazine onto silver(I) salt **14-71** forms complex **14-72** with exergonic 0.9 kcal/mol. The intramolecular nucleophilic attack of carbon atom in hydrazine takes place via transition state **14-73ts** with an energy barrier of 15.9 kcal/mol to afford a six-membered vinyl silver intermediate **14-74**. Then, an OTf-assisted 1,4-proton shift takes place through a stepwise process, which first undergoes a deprotonation via transition state **14-75ts** to form a silver species **14-76**. Then, protonation by TfOH via transition state **14-77ts** results in pyridazine product and regenerates silver salt **14-71**.

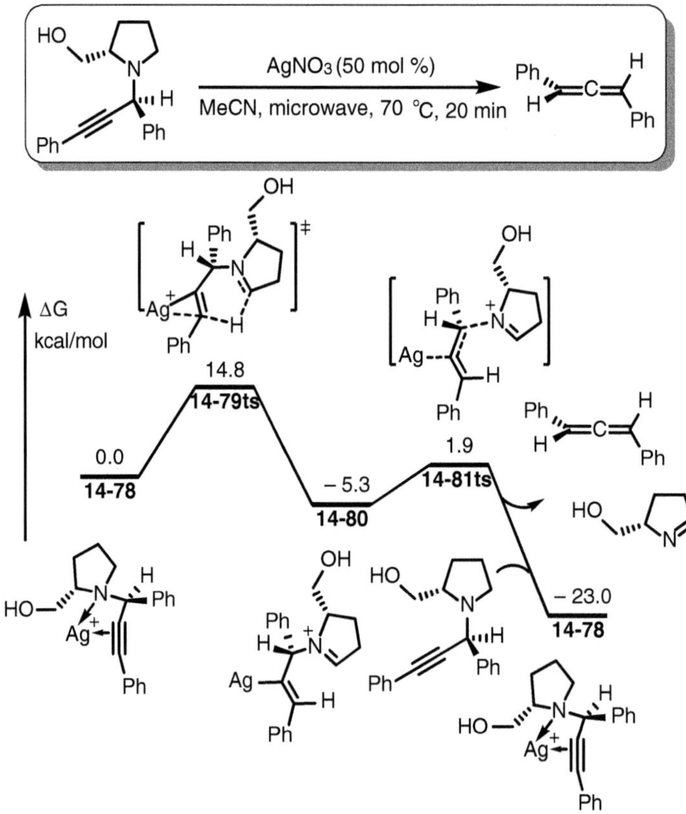

Figure 14.13 Free-energy profiles for the Ag-catalyzed decomposition of propargylamines through a key step of nucleophilic attack onto alkyne by hydrogen. The energies were calculated at B3LYP/6-311++G(d,p) (LANL2DZ for Ag)/PCM(acetonitrile)//B3LYP/6-31G(d,p) (LANL2DZ for Ag) level of theory.

Even a very weak nucleophile, such as the α-hydrogen of amino group, can nucleophilically attach onto silver–alkyne complex to result in corresponding vinyl silver species. As shown in Figure 14.13, in the presence of silver salt, propargylamine can directly transfer to allene with a yield of up to 99% in microwave condition [110]. DFT calculations found that, when propargylamine coordinates onto cationic silver in complex **14-78**, a 1,5-hydrogen shift takes place via a six-membered ring-type transition state **14-79ts** with an energy barrier of only 14.8 kcal/mol to result in a vinyl silver intermediate **14-80**. In this step, amino group increases the nucleophilicity of α-hydrogen, which can be considered as a hydride transfer to generate an iminium moiety. Subsequently, the C—N bond cleavage via transition state **14-81ts** releases a neutral imine and an allene. The active species **14-78** can be regenerated by the coordination of propargylamine.

14.4.2 C—H Activation of Alkynes

In the presence of Brønsted base, the coordinated terminal alkyne can be deprotonated to result in a silver–alkynyl complex, which can be used as nucleophile to react with carbon dioxide for the synthesis of propionate salts [111]. As shown in Figure 14.14, when phosphine ligand is used in a theoretical prediction, silver carbonate can form a hydrogen bond complex **14-83** with terminal alkyne by endergonic 5.9 kcal/mol. Then, a proton transfer to carbonate takes place via transition state **14-84ts** to result in alkynyl silver complex **14-85** with an overall activation free energy of 13.4 kcal/mol. Then, an intermolecular nucleophilic attack onto carbon dioxide

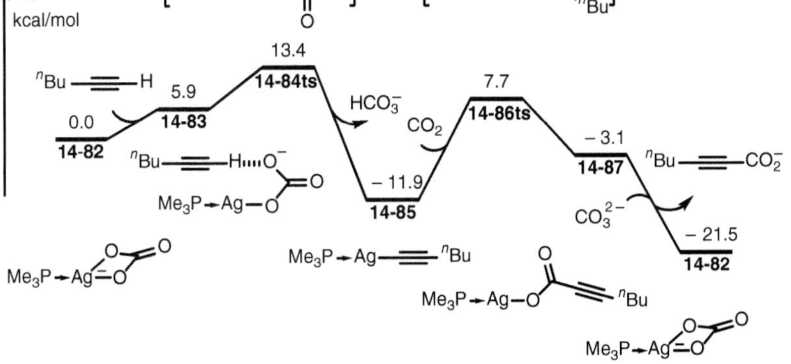

Figure 14.14 Free-energy profiles for the Ag-catalyzed carboxylation of terminal alkynes. The energies were calculated at B97-D/6-311++G(d,p) (aug-CC-pVTZ for Ag)/SMD(DMSO)//B97-D/6-31+G(d) (SDD for Ag)/SMD(DMSO) level of theory.

occurs via transition state **14-86ts** with an energy barrier of 19.6 kcal/mol. In this step, silver also shifts to the oxygen atom with the generation of propionate–silver complex **14-87**. The ligand exchange with carbonate regenerates active species **14-82** and results in propionate salts.

References

1 Alvarez-Corral, M., Munoz-Dorado, M., and Rodriguez-Garcia, I. (2008). Silver-mediated synthesis of heterocycles. *Chemical Reviews* 108 (8): 3174–3198.
2 Pellissier, H. (2016). Enantioselective silver-catalyzed transformations. *Chemical Reviews* 116 (23): 14868–14917.
3 Zhang, J., Shan, C., Zhang, T. et al. (2019). Computational advances aiding mechanistic understanding of silver-catalyzed carbene/nitrene/silylene transfer reactions. *Coordination Chemistry Reviews* 382: 69–84.
4 Liu, Z., Sivaguru, P., Zanoni, G. et al. (2018). Catalyst-dependent chemoselective formal insertion of diazo compounds into C—C or C—H bonds of 1,3-dicarbonyl compounds. *Angewandte Chemie International Edition* 57 (29): 8927–8931.
5 Zheng, Q.Z. and Jiao, N. (2016). Ag-catalyzed C—H/C—C bond functionalization. *Chemical Society Reviews* 45 (16): 4590–4627.
6 Liu, Z., Li, Q., Liao, P. et al. (2017). Silver-catalyzed [2+1] cyclopropenation of alkynes with unstable diazoalkanes: *N*-nosylhydrazones as room-temperature decomposable diazo surrogates. *Chemistry – A European Journal* 23 (20): 4756–4760.
7 Liu, Z., Liu, B., Zhao, X.-F. et al. (2017). Silver-catalyzed cross-olefination of donor and acceptor diazo compounds: use of *N*-nosylhydrazones as diazo surrogate. *European Journal of Organic Chemistry* 2017 (4): 928–932.
8 Ajay Kumar, K., Lokanatha Rai, K.M., and Umesha, K.B. (2001). A new approach for the transformation of alkenes to pyrrolines via aziridine intermediates. *Tetrahedron* 57 (32): 6993–6996.
9 Ishii, T., Aizawa, N., Kanehama, R. et al. (2002). Cocrystallites consisting of metal macrocycles with fullerenes. *Coordination Chemistry Reviews* 226 (1–2): 113–124.
10 Ager, B.J., Bourque, L.E., Buchner, K.M. et al. (2010). Silylene transfer to allylic sulfides: formation of substituted silacyclobutanes. *Journal of Organic Chemistry* 75 (16): 5729–5732.
11 Calad, S.A. and Woerpel, K.A. (2005). Silylene transfer to carbonyl compounds and subsequent Ireland–Claisen rearrangements to control formation of quaternary carbon stereocenters. *Journal of the American Chemical Society* 127 (7): 2046–2047.
12 Liu, Z., Liao, P., and Bi, X. (2015). General silver-catalyzed hydroazidation of terminal alkynes by combining TMS-N$_3$ and H$_2$O: synthesis of vinyl azides. *Organic Letters* 16 (14): 3668–3671.

13 Yuan, H., Xiao, P., Zheng, Y. et al. (2017). DFT studies on the mechanism of Ag$_2$CO$_3$-catalyzed hydroazidation of unactivated terminal alkynes with TMS-N$_3$: an insight into the silver(I) activation mode. *Journal of Computational Chemistry* 38 (27): 2289–2297.

14 Li, Y.M., Sun, M., Wang, H.L. et al. (2013). Direct annulations toward phosphorylated oxindoles: silver-catalyzed carbon-phosphorus functionalization of alkenes. *Angewandte Chemie International Edition* 52 (14): 3972–3976.

15 Li, Z., Song, L., and Li, C. (2013). Silver-catalyzed radical aminofluorination of unactivated alkenes in aqueous media. *Journal of the American Chemical Society* 135 (12): 4640–4643.

16 Cui, Y. and He, C. (2003). Efficient aziridination of olefins catalyzed by a unique disilver(I) compound. *Journal of the American Chemical Society* 125 (52): 16202–16203.

17 Huang, J., Li, L., Chen, H. et al. (2017). Silver-catalyzed geminal aminofluorination of diazoketones with anilines and N-fluorobenzenesulphonimide. *Organic Chemistry Frontiers* 4 (4): 529–533.

18 Li, Z., Capretto, D.A., Rahaman, R. et al. (2007). Silver-catalyzed intermolecular amination of C-H groups. *Angewandte Chemie International Edition* 119 (27): 5276–5278.

19 Liu, Z., Zhang, X., Virelli, M. et al. (2018). Silver-catalyzed regio- and stereoselective formal carbene insertion into unstrained C-Cσ-bonds of 1,3-dicarbonyls. *iScience* 8: 54–60.

20 Luo, H., Wu, G., Zhang, Y. et al. (2015). Silver(I)-catalyzed N-trifluoroethylation of anilines and O-trifluoroethylation of amides with 2,2,2-trifluorodiazoethane. *Angewandte Chemie International Edition* 127 (48): 14711–14715.

21 de Frémont, P., Marion, N., and Nolan, S.P. (2009). Carbenes: synthesis, properties, and organometallic chemistry. *Coordination Chemistry Reviews* 253 (7–8): 862–892.

22 Taylor, T.E. and Hall, M.B. (1984). Theoretical comparison between nucleophilic and electrophilic transition metal carbenes using generalized molecular orbital and configuration interaction methods. *Journal of the American Chemical Society* 106 (6): 1576–1584.

23 Liu, F., Zhu, L., Zhang, T. et al. (2019). Nucleophilicity versus Brønsted basicity controlled chemoselectivity: mechanistic insight into silver- or scandium-catalyzed diazo functionalization. *ACS Catalysis* 10 (2): 1256–1263.

24 Schoeller, W.W., Eisner, D., Grigoleit, S. et al. (2000). On the transition metal complexation (Fischer-type) of phosphanylcarbenes. *Journal of the American Chemical Society* 122 (41): 10115–10120.

25 Schrock, R.R. (2002). High oxidation state multiple metal-carbon bonds. *Chemical Reviews* 102 (1): 145–179.

26 Bachmann, S., Fielenbach, D., and Jorgensen, K.A. (2004). Cu(I)-carbenoid- and Ag(I)-Lewis acid-catalyzed asymmetric intermolecular insertion of α-diazo compounds into N-H bonds. *Organic & Biomolecular Chemistry* 2 (20): 3044–3049.

27 Davies, P.W., Albrecht, S.J., and Assanelli, G. (2009). Silver-catalysed Doyle-Kirmse reaction of allyl and propargyl sulfides. *Organic & Biomolecular Chemistry* 7 (7): 1276–1279.

28 Dias, H.V., Browning, R.G., Polach, S.A. et al. (2003). Activation of alkyl halides via a silver-catalyzed carbene insertion process. *Journal of the American Chemical Society* 125 (31): 9270–9271.

29 Dias, H.V. and Polach, S.A. (2000). An isolable, oxygen-coordinated silver(I) complex of dimethyl diazomalonate: synthesis and characterization of [HB(3,5-[CF$_3$]$_2$Pz)$_3$]Ag[OC(OCH$_3$)]$_2$CN$_2$ (where Pz = pyrazolyl). *Inorganic Chemistry* 39 (21): 4676–4677.

30 Dias, H.V.R., Browning, R.G., Richey, S.A. et al. (2004). Silver(I) scorpionate mediated insertion of carbenes into aliphatic C−H bonds. *Organometallics* 23 (6): 1200–1202.

31 Sudrik, S.G., Sharma, J., Chavan, V.B. et al. (2006). Wolff rearrangement of α-diazoketones using in situ generated silver nanoclusters as electron mediators. *Organic Letters* 8 (6): 1089–1092.

32 Sun, J. and Kozmin, S.A. (2006). Silver-catalyzed hydroamination of siloxy alkynes. *Angewandte Chemie International Edition* 45 (30): 4991–4993.

33 Thompson, J.L. and Davies, H.M. (2007). Enhancement of cyclopropanation chemistry in the silver-catalyzed reactions of aryldiazoacetates. *Journal of the American Chemical Society* 129 (19): 6090–6091.

34 Urbano, J., Belderraín, T.R., Nicasio, M.C. et al. (2005). Functionalization of primary carbon−hydrogen bonds of alkanes by carbene insertion with a silver-based catalyst. *Organometallics* 24 (7): 1528–1532.

35 Grasse, P.B., Brauer, B.E., Zupancic, J.J. et al. (1983). Chemical and physical properties of fluorenylidene: equilibration of the singlet and triplet carbenes. *Journal of the American Chemical Society* 105 (23): 6833–6845.

36 Melaimi, M., Soleilhavoup, M., and Bertrand, G. (2010). Stable cyclic carbenes and related species beyond diaminocarbenes. *Angewandte Chemie International Edition* 49 (47): 8810–8849.

37 Munz, D. (2018). Pushing electrons – which carbene ligand for which application? *Organometallics* 37 (3): 275–289.

38 Nemirowski, A. and Schreiner, P.R. (2007). Electronic stabilization of ground state triplet carbenes. *Journal of Organic Chemistry* 72 (25): 9533–9540.

39 Skell, P.S. and Woodworth, R.C. (1956). Structure of carbene, CH$_2$. *Journal of the American Chemical Society* 78 (17): 4496–4497.

40 Sudrik, S.G., Chaki, N.K., Chavan, V.B. et al. (2006). Silver nanocluster redox-couple-promoted nonclassical electron transfer: an efficient electrochemical Wolff rearrangement of α-diazoketones. *Chemistry - A European Journal* 12 (3): 859–864.

41 Sudrik, S.G., Maddanimath, T., Chaki, N.K. et al. (2003). Evidence for the involvement of silver nanoclusters during the Wolff rearrangement of α-diazoketones. *Organic Letters* 5 (13): 2355–2358.

42 Dias, H.V. and Lovely, C.J. (2008). Carbonyl and olefin adducts of coinage metals supported by poly(pyrazolyl)borate and poly(pyrazolyl)alkane ligands and silver mediated atom transfer reactions. *Chemical Reviews* 108 (8): 3223–3238.

43 Diaz-Requejo, M.M., Belderrain, T.R., Nicasio, M.C. et al. (2006). The carbene insertion methodology for the catalytic functionalization of unreactive hydrocarbons: no classical C–H activation, but efficient C–H functionalization. *Dalton Transactions* 35 (47): 5559–5566.

44 Diaz-Requejo, M.M. and Perez, P.J. (2008). Coinage metal catalyzed C-H bond functionalization of hydrocarbons. *Chemical Reviews* 108 (8): 3379–3394.

45 Wang, H.L., Li, Z., Wang, G.W. et al. (2011). Silver catalyzed intramolecular cyclization for synthesis of 3-alkylideneoxindoles via C-H functionalization. *Chemical Communications* 47 (40): 11336–11338.

46 Yue, Y., Wang, Y., and Hu, W. (2007). Regioselectivity in Lewis acids catalyzed X–H (O, S, N) insertions of methyl styryldiazoacetate with benzyl alcohol, benzyl thiol, and aniline. *Tetrahedron Letters* 48 (23): 3975–3977.

47 Braga, A.A.C., Caballero, A., Urbano, J. et al. (2011). Mechanism of side reactions in alkane C-H bond functionalization by diazo compounds catalyzed by Ag and cu homoscorpionate complexes – a DFT study. *ChemCatChem* 3 (10): 1646–1652.

48 Flores, J.A., Komine, N., Pal, K. et al. (2012). Silver(I)-catalyzed insertion of carbene into alkane C–H bonds and the origin of the special challenge of methane activation using DFT as a mechanistic probe. *ACS Catalysis* 2 (10): 2066–2078.

49 Gava, R., Fuentes, M.Á., Besora, M. et al. (2014). Silver-catalyzed functionalization of esters by carbene transfer: the role of ylide zwitterionic intermediates. *ChemCatChem* 6 (8): 2206–2210.

50 Hansen, J.H. and Davies, H.M.L. (2011). Vinylogous reactivity of silver(I) vinylcarbenoids. *Chemical Science* 2 (3): 457–461.

51 Li, Z. and Gao, H.X. (2012). Theoretical study on the mechanism of Ag-catalyzed synthesis of 3-alkylideneoxindoles from N-aryl-alpha-diazoamides: a Lewis acid or Ag-carbene pathway? *Organic & Biomolecular Chemistry* 10 (31): 6294–6298.

52 Vetter, A.J., Rieth, R.D., Brennessel, W.W. et al. (2009). Selective C-H activation of haloalkanes using a rhodiumtrispyrazolylborate complex. *Journal of the American Chemical Society* 131 (30): 10742–10752.

53 Cenini, S. and La Monica, G. (1976). Organic azides and isocyanates as sources of nitrene species in organometallic chemistry. *Inorganica Chimica Acta* 18: 279–293.

54 Kvaskoff, D., Bednarek, P., George, L. et al. (2006). Nitrenes, diradicals, and ylides. Ring expansion and ring opening in 2-quinazolylnitrenes. *Journal of Organic Chemistry* 71 (11): 4049–4058.

55 Muller, P. and Fruit, C. (2003). Enantioselective catalytic aziridinations and asymmetric nitrene insertions into C-H bonds. *Chemical Reviews* 103 (8): 2905–2920.

56 Scamp, R.J., Rigoli, J.W., and Schomaker, J.M. (2014). Chemoselective silver-catalyzed nitrene insertion reactions. *Pure and Applied Chemistry* 86 (3): 381–393.

57 Maestre, L., Sameera, W.M., Diaz-Requejo, M.M. et al. (2013). A general mechanism for the copper- and silver-catalyzed olefin aziridination reactions: concomitant involvement of the singlet and triplet pathways. *Journal of the American Chemical Society* 135 (4): 1338–1348.

58 Dolan, N.S., Scamp, R.J., Yang, T. et al. (2016). Catalyst-controlled and tunable, chemoselective silver-catalyzed intermolecular nitrene transfer: experimental and computational studies. *Journal of the American Chemical Society* 138 (44): 14658–14667.

59 Alderson, J.M., Phelps, A.M., Scamp, R.J. et al. (2014). Ligand-controlled, tunable silver-catalyzed C–H amination. *Journal of the American Chemical Society* 136 (48): 16720–16723.

60 Arenas, I., Fuentes, M.A., Alvarez, E. et al. (2014). Syntheses of a novel fluorinated trisphosphinoborate ligand and its copper and silver complexes. Catalytic activity toward nitrene transfer reactions. *Inorganic Chemistry* 53 (8): 3991–3999.

61 Beltrán, Á., Lescot, C., Mar Díaz-Requejo, M. et al. (2013). Catalytic C–H amination of alkanes with sulfonimidamides: silver(I)-scorpionates vs. dirhodium(II) carboxylates. *Tetrahedron* 69 (22): 4488–4492.

62 Besora, M., Braga, A.A.C., Sameera, W.M.C. et al. (2015). A computational view on the reactions of hydrocarbons with coinage metal complexes. *Journal of Organometallic Chemistry* 784: 2–12.

63 Díaz-Requejo, M.M. and Pérez, P.J. (2005). Copper, silver and gold-based catalysts for carbene addition or insertion reactions. *Journal of Organometallic Chemistry* 690 (24–25): 5441–5450.

64 Flores, J.A., Pal, K., Carroll, M.E. et al. (2014). Mechanistic understanding of a silver pyridylpyrrolide as a catalyst for 3 + 2 cyclization of a nitrile with diazo ester. *Organometallics* 33 (7): 1544–1552.

65 Grant, L.N., Carroll, M.E., Carroll, P.J. et al. (2016). An unusual cobalt azide adduct that produces a nitrene species for carbon-hydrogen insertion chemistry. *Inorganic Chemistry* 55 (16): 7997–8002.

66 Huang, M., Corbin, J.R., Dolan, N.S. et al. (2017). Synthesis, characterization, and variable-temperature NMR studies of silver(I) complexes for selective nitrene transfer. *Inorganic Chemistry* 56 (11): 6725–6733.

67 Jain, S.S., Anet, F.A., Stahle, C.J. et al. (2004). Enzymatic behavior by intercalating molecules in a template-directed ligation reaction. *Angewandte Chemie International Edition* 43 (15): 2004–2008.

68 Ju, M., Weatherly, C.D., Guzei, I.A. et al. (2017). Chemo- and enantioselective intramolecular silver-catalyzed aziridinations. *Angewandte Chemie International Edition* 56 (33): 9944–9948.

69 Martinez-Garcia, H., Morales, D., Perez, J. et al. (2010). 1,3,5-Tris(thiocyanatomethyl)mesitylene as a ligand. Pseudooctahedral molybdenum, manganese, and rhenium carbonyl complexes and copper and silver dimers. Copper-catalyzed carbene- and nitrene-transfer reactions. *Inorganic Chemistry* 49 (15): 6974–6985.

70 Perez, P.J., White, P.S., Brookhart, M. et al. (1994). Nitrene transfer to trimethylphosphine from cationic tungsten tosylnitrene complexes [Tp'(CO)$_2$W(NTs)][X]. *Inorganic Chemistry* 33 (26): 6050–6056.

71 Rigoli, J.W., Weatherly, C.D., Vo, B.T. et al. (2013). Chemoselective allene aziridination via Ag(I) catalysis. *Organic Letters* 15 (2): 290–293.

72 Levason, W. and Spicer, M.D. (1987). The chemistry of copper and silver in their higher oxidation states. *Coordination Chemistry Reviews* 76: 45–120.

73 Evans, D.A., Faul, M.M., and Bilodeau, M.T. (1991). Copper-catalyzed aziridination of olefins by (N-[p-toluenesulfonyl]imino)phenyliodinane. *Journal of Organic Chemistry* 56 (24): 6744–6746.

74 Heuss, B.D., Mayer, M.F., Dennis, S. et al. (2003). Iron mediated nitrenoid transfer: [(η^5-C$_5$H$_5$)Fe(CO)$_2$(THF)]$^+$[BF$_4$]$^-$ catalyzed aziridination of olefins. *Inorganica Chimica Acta* 342: 301–304.

75 Maestre, L., Dorel, R., Pablo, O. et al. (2017). Functional-group-tolerant, silver-catalyzed N-N bond formation by nitrene transfer to amines. *Journal of the American Chemical Society* 139 (6): 2216–2223.

76 Llaveria, J., Beltran, A., Sameera, W.M. et al. (2014). Chemo-, regio-, and stereoselective silver-catalyzed aziridination of dienes: scope, mechanistic studies, and ring-opening reactions. *Journal of the American Chemical Society* 136 (14): 5342–5350.

77 Alderson, J.M., Corbin, J.R., and Schomaker, J.M. (2017). Tunable, chemo- and site-selective nitrene transfer reactions through the rational design of silver(I) catalysts. *Accounts of Chemical Research* 50 (9): 2147–2158.

78 Cirakovic, J., Driver, T.G., and Woerpel, K.A. (2002). Metal-catalyzed silacyclopropanation of mono- and disubstituted alkenes. *Journal of the American Chemical Society* 124 (32): 9370–9371.

79 Driver, T.G. and Woerpel, K.A. (2004). Mechanism of silver-mediated di-tert-butylsilylene transfer from a silacyclopropane to an alkene. *Journal of the American Chemical Society* 126 (32): 9993–10002.

80 Hengge, E. and Weinberger, M. (1993). Mechanistische aspekte zur übergangsmetall-katalysierten dehydrierenden polymerisation von disilanen und höheren silanen. *Journal of Organometallic Chemistry* 443 (2): 167–173.

81 Waterman, R., Hayes, P.G., and Tilley, T.D. (2007). Synthetic development and chemical reactivity of transition-metal silylene complexes. *Accounts of Chemical Research* 40 (8): 712–719.

82 Bourque, L.E., Cleary, P.A., and Woerpel, K.A. (2007). Metal-catalyzed silylene insertions of allylic ethers: stereoselective formation of chiral allylic silanes. *Journal of the American Chemical Society* 129 (42): 12602–12603.

83 Bourque, L.E., Haile, P.A., and Woerpel, K.A. (2009). Silylene-mediated ring contraction of homoallylic ethers to form allylic silanes. *Journal of Organic Chemistry* 74 (18): 7180–7182.

84 Bourque, L.E. and Woerpel, K.A. (2008). Intermolecular silacarbonyl ylide cycloadditions: a direct pathway to oxasilacyclopentenes. *Organic Letters* 10 (22): 5257–5260.

85 Brook, A.G., Azarian, D., Baumegger, A. et al. (1993). Ring insertion reactions of silaaziridines with aldehydes and isocyanates. *Organometallics* 12 (2): 529–534.

86 Cirakovic, J., Driver, T.G., and Woerpel, K.A. (2004). Metal-catalyzed di-tert-butylsilylene transfer: synthesis and reactivity of silacyclopropanes. *Journal of Organic Chemistry* 69 (12): 4007–4012.

87 Clark, T.B. and Woerpel, K.A. (2004). Formation and utility of oxasilacyclopentenes derived from functionalized alkynes. *Journal of the American Chemical Society* 126 (31): 9522–9523.

88 Clark, T.B. and Woerpel, K.A. (2005). Silver-catalyzed silacyclopropenation of 1-heteroatom-substituted alkynes and subsequent rearrangement reactions. *Organometallics* 24 (25): 6212–6219.

89 Clark, T.B. and Woerpel, K.A. (2006). Formation and reactivity of silacyclopropenes derived from siloxyalkynes: stereoselective formation of 1,2,4-triols. *Organic Letters* 8 (18): 4109–4112.

90 Ottosson, H. and Steel, P.G. (2006). Silylenes, silenes, and disilenes: novel silicon-based reagents for organic synthesis? *Chemistry - A European Journal* 12 (6): 1576–1585.

91 Parvin, N., Dasgupta, R., Pal, S. et al. (2017). Strikingly diverse reactivity of structurally identical silylene and stannylene. *Dalton Transactions* 46 (20): 6528–6532.

92 Prevost, M. and Woerpel, K.A. (2009). Insertions of silylenes into vinyl epoxides: diastereoselective synthesis of functionalized, optically active trans-dioxasilacyclooctenes. *Journal of the American Chemical Society* 131 (40): 14182–14183.

93 Rotsides, C.Z., Hu, C., and Woerpel, K.A. (2013). Diastereoselective synthesis of eight-membered-ring allenes from propargylic epoxides and aldehydes by silylene insertion into carbon-oxygen bonds. *Angewandte Chemie International Edition* 52 (49): 13033–13036.

94 Ventocilla, C.C. and Woerpel, K.A. (2011). Silylene-mediated polarity reversal of dienoates: additions of dienoates to aldehydes at the δ-position to form trans-dioxasilacyclononenes. *Journal of the American Chemical Society* 133 (3): 406–408.

95 Denk, M., Lennon, R., Hayashi, R. et al. (1994). Synthesis and structure of a stable silylene. *Journal of the American Chemical Society* 116 (6): 2691–2692.

96 Franz, A.K. and Woerpel, K.A. (2000). Development of reactions of silacyclopropanes as new methods for stereoselective organic synthesis. *Accounts of Chemical Research* 33 (11): 813–820.

97 Grumbine, S.K. and Tilley, T.D. (1994). Transition-metal-mediated redistribution at silicon: a bimolecular mechanism involving silylene complexes. *Journal of the American Chemical Society* 116 (15): 6951–6952.

98 Moiseev, A.G. and Leigh, W.J. (2006). Diphenylsilylene. *Journal of the American Chemical Society* 128 (45): 14442–14443.

99 Schäfer, A., Weidenbruch, M., Peters, K. et al. (1984). Hexa-*tert*-butylcyclotrisilan, ein gespanntes Molekül mit ungewöhnlich langen Si-Si- und Si-C-Bindungen. *Angewandte Chemie International Edition* 96 (4): 311–312.

100 Straus, D.A., Grumbine, S.D., and Tilley, T.D. (1990). Base-free silylene complexes [(η^5-C$_5$Me$_5$)(PMe$_3$)$_2$Ru:Si(SR)$_2$]BPh$_4$ (R = et, p-MeC$_6$H$_4$). *Journal of the American Chemical Society* 112 (21): 7801–7802.

101 Calad, S.A. and Woerpel, K.A. (2007). Formation of chiral quaternary carbon stereocenters using silylene transfer reactions: enantioselective synthesis of (+)-5-epi-acetomycin. *Organic Letters* 9 (6): 1037–1040.

102 Cleary, P.A. and Woerpel, K.A. (2005). Metal-catalyzed rearrangement of homoallylic ethers to silylmethyl allylic silanes in the presence of a di-*tert*-butylsilylene source. *Organic Letters* 7 (24): 5531–5533.

103 Driver, T.G. and Woerpel, K.A. (2003). Mechanism of di-*tert*-butylsilylene transfer from a silacyclopropane to an alkene. *Journal of the American Chemical Society* 125 (35): 10659–10663.

104 Greene, M.A., Prevost, M., Tolopilo, J. et al. (2012). Diastereoselective synthesis of seven-membered-ring trans-alkenes from dienes and aldehydes by silylene transfer. *Journal of the American Chemical Society* 134 (30): 12482–12484.

105 Howard, B.E. and Woerpel, K.A. (2007). Synthesis of tertiary α-hydroxy acids by silylene transfer to α-keto esters. *Organic Letters* 9 (22): 4651–4653.

106 Nevarez, Z. and Woerpel, K.A. (2007). Metal-catalyzed silylene transfer to imines: synthesis and reactivity of silaaziridines. *Organic Letters* 9 (19): 3773–3776.

107 Mayoral, J.A., Rodríguez-Rodríguez, S., and Salvatella, L. (2010). A theoretical insight into the mechanism of the silver-catalysed transsiliranation reaction. *European Journal of Organic Chemistry* 2010 (7): 1231–1234.

108 Ruan, C., Yang, L., Yuan, Y. et al. (2015). The reaction mechanism of incorporation of carbon dioxide into *o*-alkynylaniline derivatives catalyzed by silver salt. *Computational and Theoretical Chemistry* 1058: 34–40.

109 Yuan, B., He, R., Shen, W. et al. (2017). Influence of base strength on the proton-transfer reaction by density functional theory. *European Journal of Organic Chemistry* 2017 (27): 3947–3956.

110 Liao, C., Li, B., Wang, J. et al. (2012). Mechanism of silver(I)-catalyzed enantioselective synthesis of axially chiral allenes based on propargylamines. *Chinese Journal of Chemistry* 30 (4): 951–958.

111 Jover, J. and Maseras, F. (2014). Computational characterization of the mechanism for coinage-metal-catalyzed carboxylation of terminal alkynes. *Journal of Organic Chemistry* 79 (24): 11981–11987.

15

Theoretical Study of Au-Catalysis

Due to its stable chemical properties, high ductility, and rarity, gold is not only a special currency for reserve and investment, but also an important material for jewelry, electronics, modern communication, aerospace, and other sectors [1–3]. Not only that, gold also provides a productive alternative choice for both of homogeneous and heterogeneous transition metal catalysis [4–11]. As a late-transition metal, gold is more abundant than platinum, palladium, rhodium, and some other noble metals that are used even in large-scale processes. Therefore, gold-catalyzed transformations have been developed with remarkable frequency during the past several decades [12–18].

Generally, series of gold salts play catalytic activity, where the oxidative state of gold can be +1 or +3 [19–24]. In gold chemistry, nonredox processes are well accepted because the oxidation of Au(I) to Au(III) is difficult. The majority of the nonredox processes are based on the propensity of Au(I) to act as an especially soft and carbophilic Lewis acid to activate unsaturated C—C bonds [25–27]. The back-donation bond between C—C π-bonds and the coordinated Au(I) increases the electrophilicity of the π-bonds, which would lead to a sequential nucleophilic attack by frequently used nucleophiles, thus allowing the formation of corresponding C—C, C—O, C—S, and C—N bonds [28–32]. In particular, gold exhibits strong alkynophilic but not as oxophilic as most other Lewis acids, which brings a true advantage to the chemists in the sense that oxygen, water, and alcohols are often well tolerated in sharp contrast to most air- and moisture-sensitive Lewis acids [33–37]. Following this idea, a general catalytic cycle of Au(I)–catalyzed nonredox unsaturated C—C bonds functionalizations can be proposed in Scheme 15.1 [26, 38–41]. The coordination of unsaturated bond forms gold complex, which can undergo a nucleophilic attack to afford an alkyl or vinyl gold species. Then, further transformations, such as protonation, can yield corresponding adduct and regenerate gold catalyst.

Computational Methods in Organometallic Catalysis: From Elementary Reactions to Mechanisms,
First Edition. Yu Lan.
© 2021 WILEY-VCH GmbH. Published 2021 by WILEY-VCH GmbH.

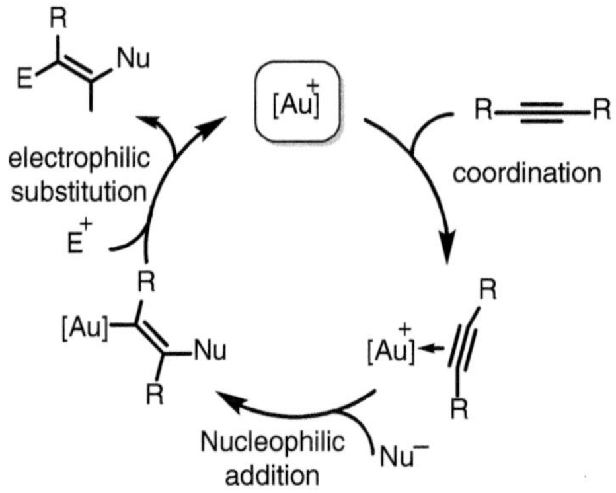

Scheme 15.1 General mechanism of gold-catalyzed unsaturated bond functionalizations. Source: Furstner 2009 [26]; Wang et al. [38]; Liu and Hammond [39]; Jimenez-Nunez and Echavarren [40]; Ghosh et al. [41].

In Au(I) chemistry, chemists are often interested in the catalytic capability and regioselectivity for the activations of unsaturated bonds, which can be easily explored by the theoretical calculations [42–47]. In this chapter, the author groups the reactions by the type of reacting C—C unsaturated bonds, which would cause slight difference in the reaction mechanisms. As the author knows, redox processes through Au(I)–Au(III) transformations are also important in gold chemistry; however, there are really rare reports that focus on this chemistry [48–53].

15.1 Au-Mediated Alkyne Activations

Undoubtedly, Au(I) complexes assume one of the most effective auxiliaries for the electrophilic activation of alkynes in homogeneous catalysis, which have been well documented for the construction of new carbon—carbon and carbon—heteroatom bonds. The Au(I) complex demonstrates strong alkynophilicity, which can be attributed to the relativistic effects. Structurally, Au(I) predominantly forms linear two-coordinate complexes, although higher coordination numbers are also possible [54–59]. Usually, Au(I) can be coordinated by alkynes to afford a η^2–Au complex, which can be attacked by the nucleophile to afford a *trans*-vinyl gold species as the key intermediate (Scheme 15.2) [60–65]. In this section, reactions are grouped by the use of nucleophiles.

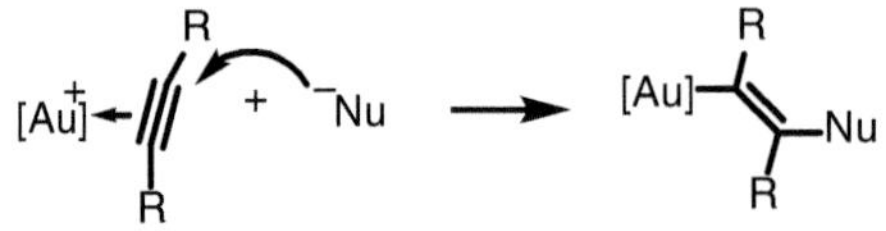

Scheme 15.2 Trans-nucleophilic attack of gold–alkyne complex. Source: Zhdanko and Maier [60]; Zhang et al. [61]; Ma et al. [62]; Luo et al. [63]; Bruneau [64]; Akana et al. [65].

15.1.1 Isomerization of Alkynes

In the beginning of this section, the Brøsted base–assisted isomerization of alkynes should be discussed, which plays an important role for the gold-assisted alkyne activations. As we know, alkynes, allenes, and dienes have the same unsaturation of two, which can be transferred to each other in equilibrium reactions. As shown in Figure 15.1, a theoretical study on mechanism of the alkyne isomerization to conjugative diene via allene intermediate was selected as example to reveal this process, where a basic amino group was installed on the phosphine ligand and revealed a key rule for this reaction [66]. In theoretical calculations, the reaction starts from an alkyne-coordinated Au(I) species **15-1**, where gold increases the acidity of the propargyl hydrogen. A base-assisted deprotonation of this position takes place via transition state **15-2ts** with an energy barrier of 22.4 kcal/mol to afford an allenyl gold intermediate **15-3**. Then, the protonation of vinyl carbon occurs via transition state **15-4ts** leading to the formation of allene-coordinated gold species **15-5**, whose relative free energy is 0.8 kcal/mol higher than that of alkyne coordinated one. Gold atom can shift from on C=C double bond of allene to another smoothly via transition state **15-6ts** with an energy barrier of only 5.4 kcal/mol. Then, the second base-assisted deprotonation of allylic hydrogen takes place via transition state **15-8ts** to form a vinyl gold intermediate **15-9**. Subsequently, the protonation of vinyl carbon occurs via transition state **15-10ts** to result in diene-coordinated gold species **15-11**. The relative free energy of **15-11** is 5.7 kcal/mol lower than that of alkyne coordinated one, which indicates that diene is more stable than the corresponding alkyne. Computational study found that the suitable base can assist the isomerization between each other of corresponding alkynes, allenes, and dienes.

15.1.2 Nucleophilic Attack by Oxygen-Involved Nucleophiles

Alcohols should be one of the most conceivable nucleophiles, which involves a nucleophilic oxygen atom [67, 68]. However, the nucleophilicity of alcohols is often weak, which is hard to employ in nucleophilic attack reactions. Therefore, a weak Brøsted base–assisted nucleophilic attack was proposed by density functional theory (DFT) calculation for the gold-catalyzed hydroalkoxylation of alkynes. As shown in Figure 15.2, the reaction was considered starting from an Au(OTs) species **15-12**, which can undergo a ligand substitution by the reacting alkyne to afford an alkyne-coordinated gold intermediate **15-13** [69]. In this intermediate, methanol reactant can form a hydrogen bond with toluenesulfonate, while toluenesulfonate is stabilized by the ion pair with cationic gold. The hydrogen bond significantly increases the nucleophilicity of methanol, which leads to a nucleophilic antiattack on coordinated alkyne via transition state **15-14ts** with an energy barrier of only 15.6 kcal/mol. Notably, an explicit solvent model was used in DFT calculation, which involved an extra methanol bridge to link reacting methanol and toluene sulfonate. Then, the generated vinyl gold **15-15** can be protonated by the methanolated proton via transition state **15-16ts** to yield methoxyethene product.

The oxygen atom in carbonyl plays as a weak nucleophile, when an intramolecular nucleophilic annulation occurs with alkynyl group in the presence of gold catalyst

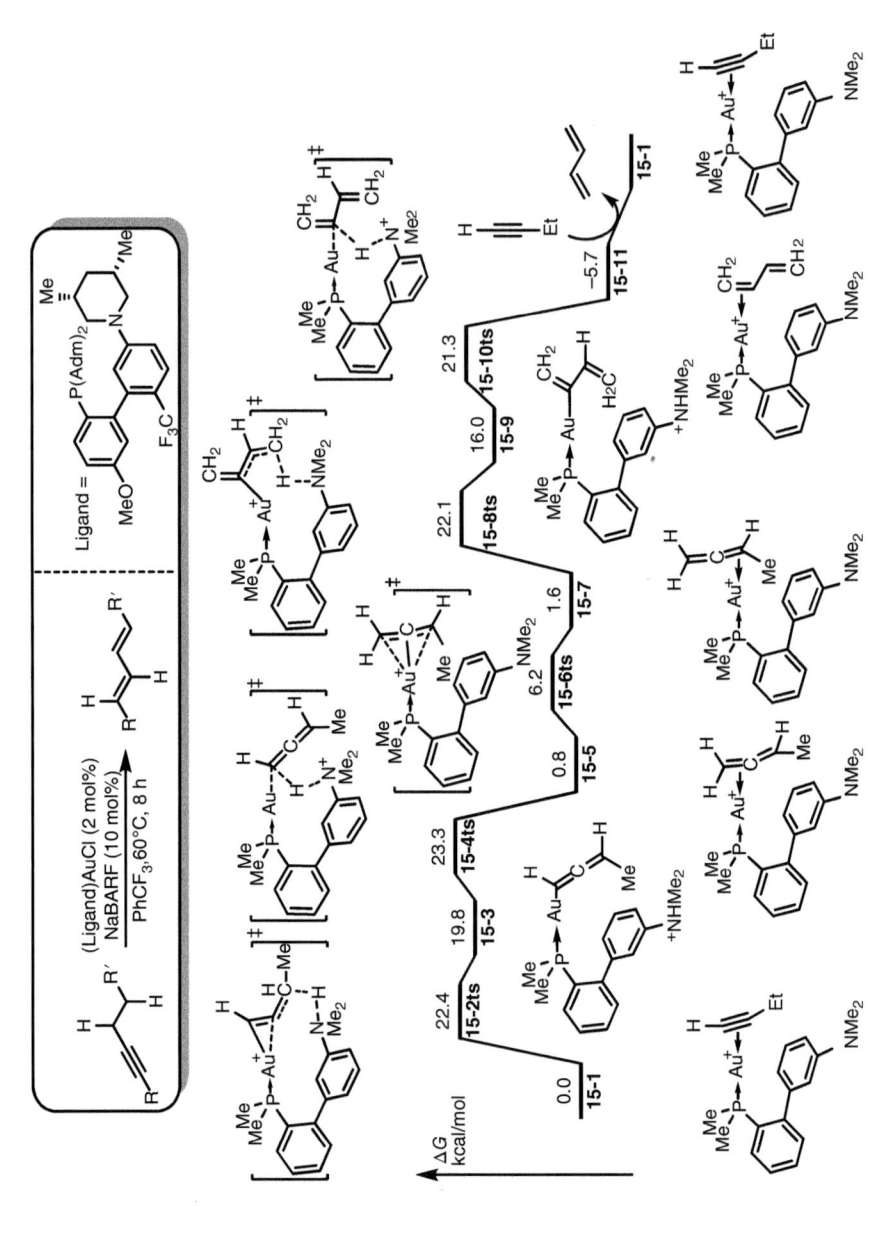

Figure 15.1 The free-energy profiles for the gold-catalyzed isomerization of alkyne to diene. The energies were calculated at B3LYP/6-31G(d,p) (LANL2DZ for Au)/PCM(DCM)//B3LYP/6-31G(d,p) (LANL2DZ for Au) level of theory.

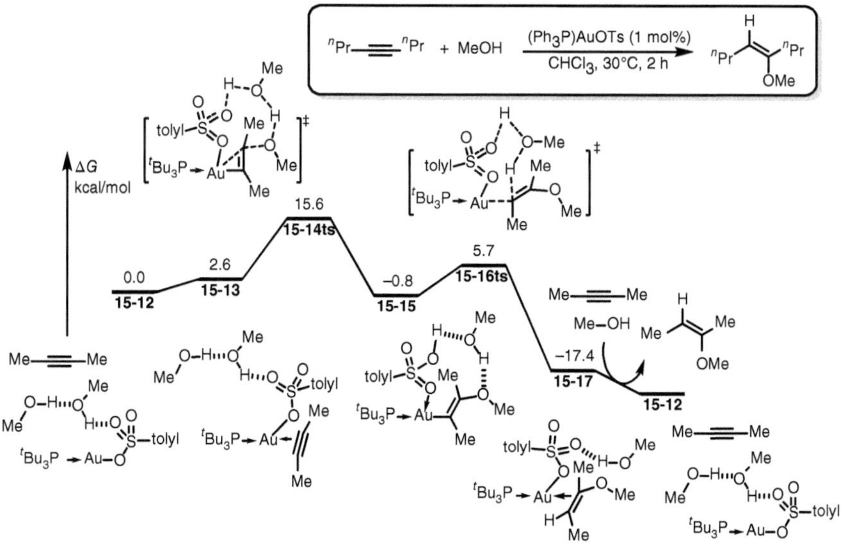

Figure 15.2 The free-energy profiles for the gold-catalyzed hydroalkoxylation of alkynes. The energies were calculated at B2PLYP/def2-TZVP/SMD(chloroform)//BP86/def2-TZVP level of theory.

[70–72]. Yamamoto group reported a 4+2 annulation of alkynyl benzaldehyde and alkynes [73]. The mechanism of this reaction was revealed by DFT calculations. As shown in Figure 15.3, the reaction starts from an alkyne-coordinated Au(I) species **15-18**, which can undergo an intramolecular nucleophilic attack by the ortho-carbonyl group via transition state **15-19ts** to afford a six-membered ring intermediate **15-20** with exergonic 9.6 kcal/mol. Intermediate **15-20** can be considered as resonance structures of aromatic chromenylium and dipolar carbonyl ylide. Then, a hetero-Diels–Alder [4+2] cycloaddition takes place with coming alkyne via transition state **15-21ts** with an energy barrier 27.8 kcal/mol to form a [2,2,2]-bicyclic intermediate **15-22**. Then, a vinyl shift (via transition state **15-23ts**) followed by a C—O bond cleavage (via transition state **15-25ts**) results in the final 4+2 annulation product to construct naphthaldehyde.

The oxygen atom in amine-N-oxide can be considered as an internal oxidant to provide oxygen in a redox-neutral process, which would undergo an intramolecular oxygen transfer through nucleophilic attack by the oxygen atom [74–76]. As shown in Figure 15.4, DFT study is performed to reveal the mechanism of gold-catalyzed rearrangements of acetylenic amine-N-oxide [77]. The reaction starts from an amine-N-oxide coordinated Au(I) species **15-27**, which can undergo an intramolecular syn-nucleophilic attack of oxygen onto alkyne moiety via transition state **15-28ts** with an energy barrier of 24.9 kcal/mol. This process also can be considered as the insertion of alkyne into Au—O bond via a four-membered ring-type transition state. Then, the proton transfer from aminomethyl group to vinyl group occurs accompanied by the cleavage of N—O bond via transition state **15-30ts** to form an iminium intermediate **15-31**. The intramolecular nucleophilic attack by alkyl onto

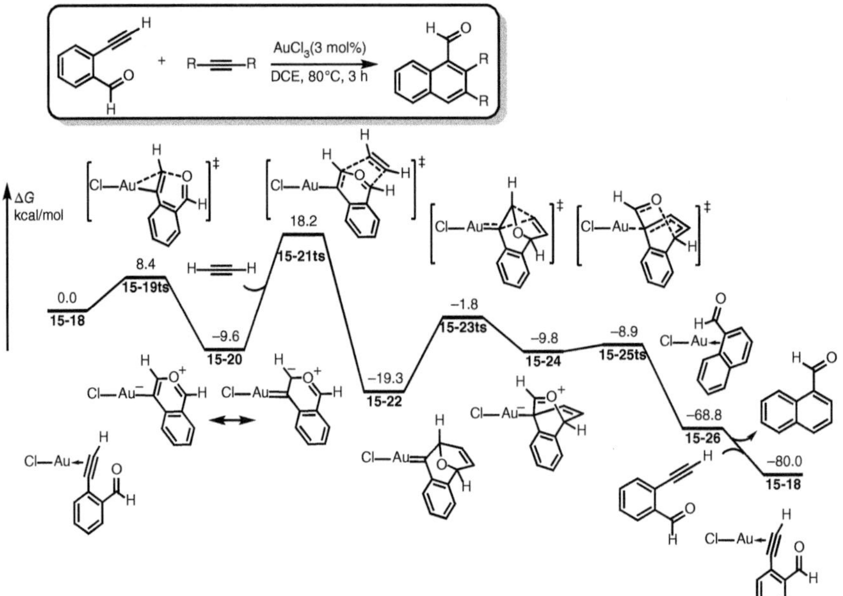

Figure 15.3 The free-energy profiles for the gold-catalyzed 4+2 annulation of alkynyl benzaldehyde and alkynes. The energies were calculated at B3LYP/6-311G(d,p) (LANL2TZ for Au)//B3LYP/6-31G(d) (LANL2DZ for Au) level of theory.

iminium via transition state **15-32ts** achieves the annulation to yield azepanone product.

In an intermolecular nucleophilic attack example, oxyarylation of alkynes was achieved in the presence of Au(I) catalyst [78]. As shown in Figure 15.5, the coordination of alkyne onto cationic gold forms complex **15-33**. Then, an intermolecular anti-nucleophilic attack by sulfoxide takes place via transition state **15-34ts** with an energy barrier of only 4.3 kcal/mol to form a vinyloxysulfonium intermediate **15-35**. Subsequently, a [3,3′]-sigmatropic rearrangement occurs via transition state **15-36ts** to afford a cationic intermediate **15-37**, where the S—O bond is broken. Then, a 1,2-proton shift results in the oxyarylation product with the regeneration of gold catalyst **15-33**.

15.1.3 Nucleophilic Attack by Nitrogen-Involved Nucleophiles

Amines are common nucleophile containing nitrogen atom, which can attack gold-activated alkynes to afford enamine species [79–82]. As shown in Figure 15.6, the gold-catalyzed hydroamination of alkynes can be used to synthesize corresponding imine molecules [83]. In a DFT study for the mechanism of this reaction, an acetonitrile-coordinated cationic Au(I) species **15-40** was chosen as relative zero, which can undergo an amine-assisted substitution with reacting alkyne to afford intermediate **15-41** with endergonic 5.7 kcal/mol. Then, an intermolecular nucleophilic attack by amine takes place via transition state **15-42ts** to result in a vinyl

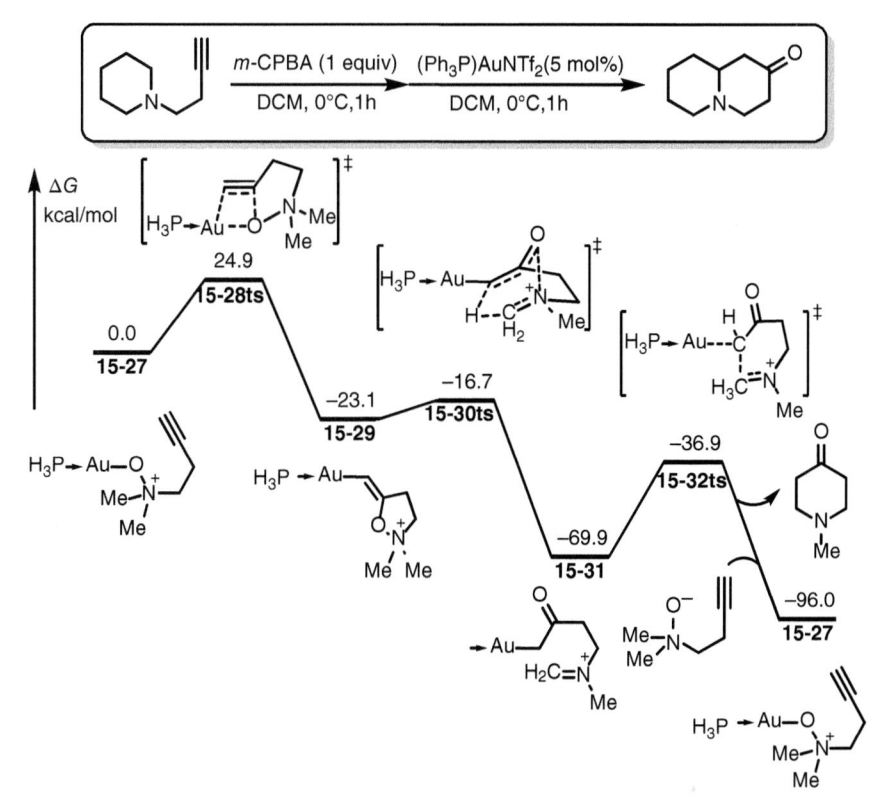

Figure 15.4 The free-energy profiles for the gold-catalyzed rearrangements of acetylenic amine-*N*-oxide. The energies were calculated at B3LYP/6-31G(d) (LANL2DZ for Au)/CPCM(DCM) level of theory.

gold intermediate **15-43**, where the positive charge is located on the ammonium moiety. A 1,3-proton transfer from ammonium to vinyl can be processed under the assistance of extra amine, which experiences a stepwise deprotonation–protonation process via transition states **15-44ts** and **15-46ts**, respectively. The enamine product can be released by the ligand exchange with acetonitrile molecule. The deprotonation step (**15-44ts**) was considered as the rate-determining step for the whole catalytic cycle.

In an intramolecular case, Bi reported a theoretical study for the mechanism of gold-catalyzed cyclization of alkynyl-*N*-propargylanilines, which can result in an indole derivative [84]. As shown in Figure 15.7, the reaction starts from an alkyne-coordinated cationic gold species **15-48**, which can undergo an intramolecular nucleophilic attack by amino group via transition state **15-49ts** to afford indolyl gold intermediate **15-50**. The calculated activation energy for this step is only 5.5 kcal/mol. The generated propargyl ammonium can undergo a Stevens-type [3,3]-sigmatropic rearrangement via transition state **15-51ts** to afford an allenyl indole intermediate. The allenyl group can be active by gold in complex **15-53**, which can undergo a nucleophilic attack by 2-position of indole via transition state

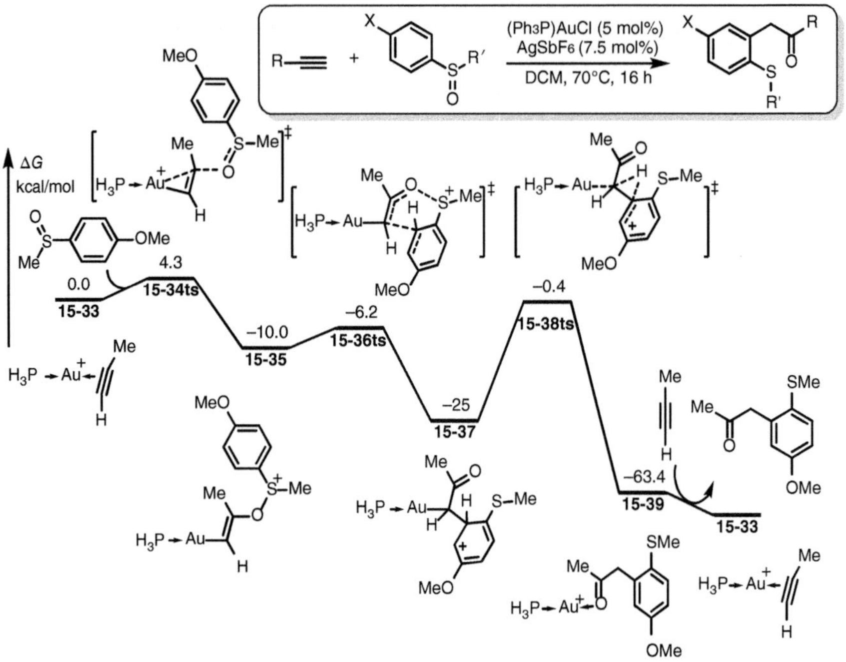

Figure 15.5 The free-energy profiles for the gold-catalyzed oxyarylation of alkynes. The energies were calculated at B3LYP/6-311++G(d,p) (SDD for Au)/CPCM(DCM)//B3LYP/6-31G(d) (SDD for Au) level of theory.

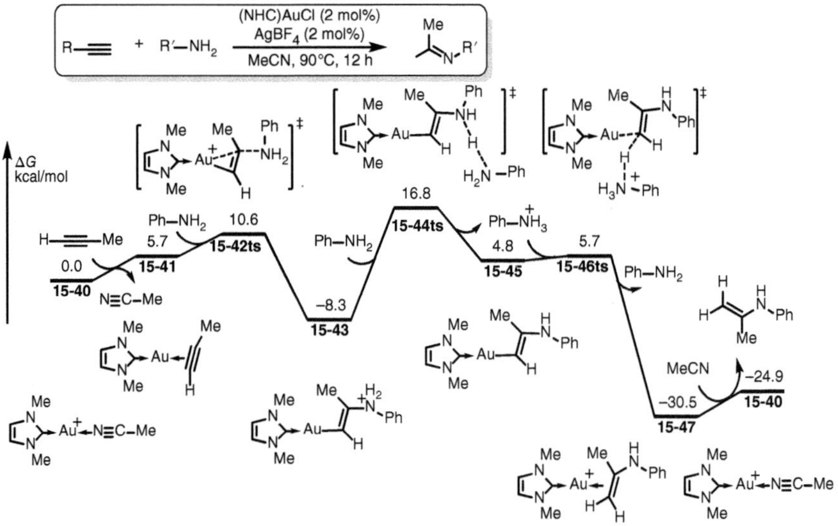

Figure 15.6 The free-energy profiles for the gold-catalyzed intermolecular hydroamination of alkynes. The energies were calculated at B3LYP/TZVP (SDD for Au)/PCM(acetonitrile)//B3LYP/6-31G(d) (SDD for Au) level of theory.

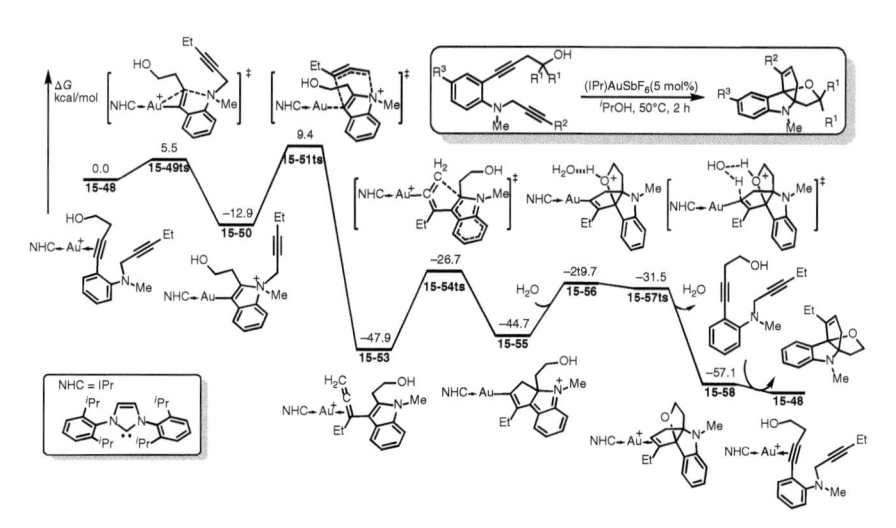

Figure 15.7 The free-energy profiles for the gold-catalyzed cyclization of alkynyl-N-propargylanilines. The energies were calculated at M06/6-311G(d,p) (SDD for Au)/PCM(acetonitrile)//M06/6-31G(d,p) (LANL2DZ-f for Au) level of theory.

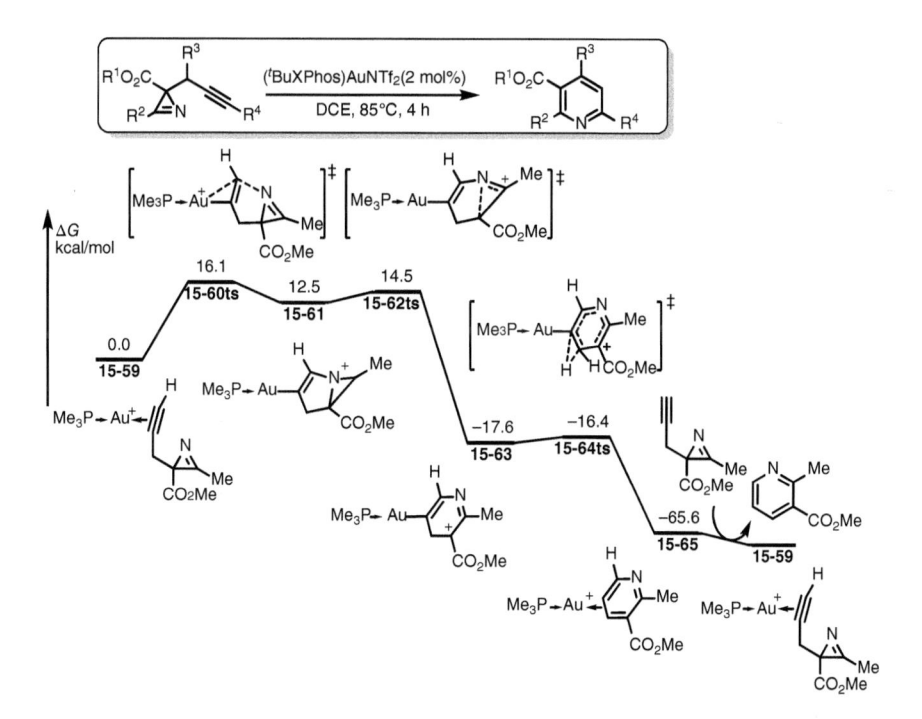

Figure 15.8 The free-energy profiles for the gold-catalyzed isomerization of propargyl azirines. The energies were calculated at B2PLYP/6-311+G(d,p) (def2-TZVP for Au)/IEF-PCM(DCM)//PBE0/6-311+G(d,p) (def2-TZVP for Au)/IEF-PCM(DCM) level of theory.

15-54ts to form another five-membered ring intermediate **15-55**. Then, the C—O bond formation affords an oxonium intermediate **15-56** in a barrierless process, which can undergo a water-assisted proton transfer to yield the final product.

The intramolecular nucleophilic attack by the nitrogen atom in azirines onto gold-activated alkynes achieves the formation of pyridines in the following processes [85]. As shown in Figure 15.8, when the alkyne moiety coordinates onto cationic gold in complex **15-59**, an intramolecular nucleophilic attack takes place via transition state **15-60ts** with an energy barrier of 16.1 kcal/mol to afford a bicyclic intermediate **15-61**. Sequentially, C—N bond cleavage occurs via transition state **15-62ts** to afford a six-membered ring-type intermediate **15-63**, which can undergo a rapid 1,2-proton shift via transition state **15-64ts** to yield the final pyridine product.

15.1.4 Nucleophilic Attack by Arenes

In some cases, electron-rich arenes can act as nucleophile in gold-mediated alkyne activations, which results in the functionalization of arenes [86–90]. As shown in Figure 15.9, Zhang reported a mechanistic study for the mechanism of gold-catalyzed intramolecular hydroarylation of alkynes [91]. When alkyne coordinates onto gold in species **15-66**, an intramolecular nucleophilic attack by the electron-rich arene can occur via transition state **15-67ts** with an energy barrier of 19.5 kcal/mol to result in a Friedel–Crafts-type intermediate **15-68**. Then, a sulfonate-assisted 1,3-proton transfer takes place through a stepwise process via deprotonative transition state **15-70ts** and protonative transition state **15-72ts** sequentially to yield the dihydroquinoline product.

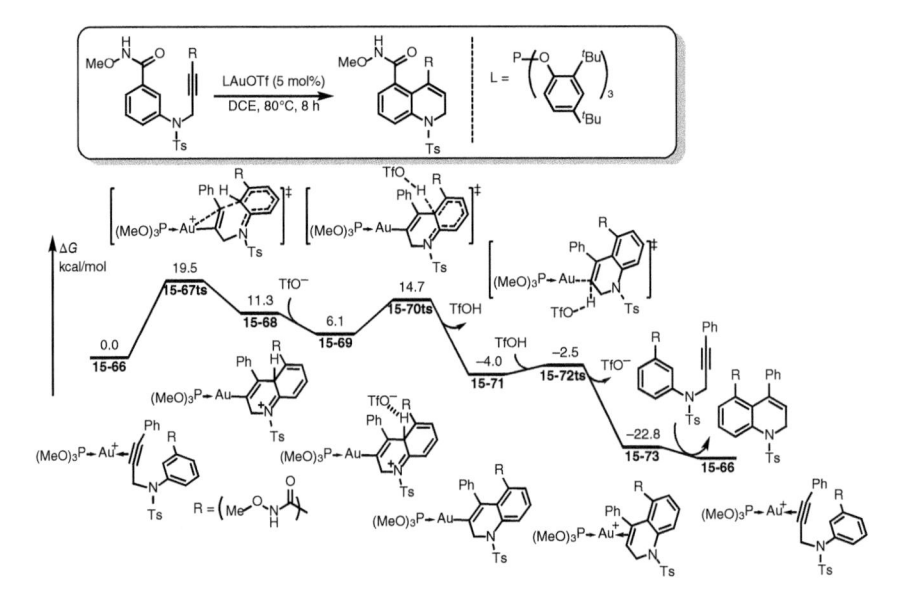

Figure 15.9 The free-energy profiles for the gold-catalyzed intramolecular hydroarylation of alkynes. The energies were calculated at M06/6-311+G(d,p) (LANL2DZ-f for Au)/PCM(DCE)//M06/6-31G(d,p) (LANL2DZ-f for Au) level of theory.

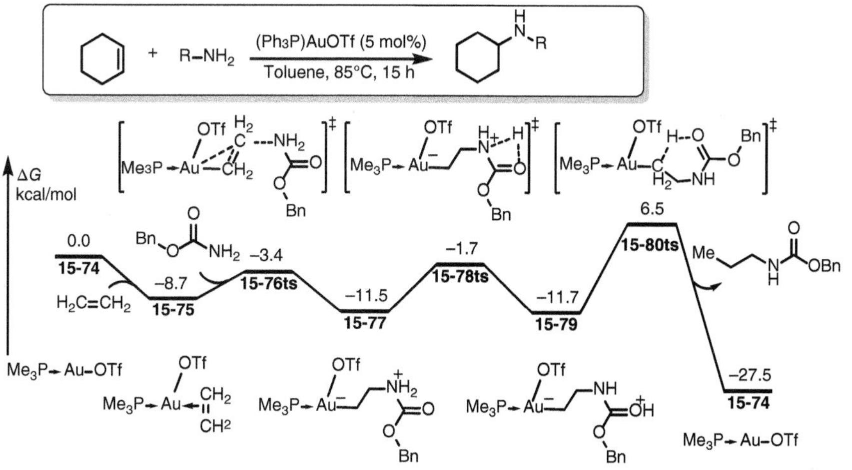

Figure 15.10 The free-energy profiles for the gold-catalyzed intermolecular hydroamination of alkenes. The energies were calculated at M06/6-311++G(d,p) (SDD for Au)/PCM(toluene)//B3LYP/6-31G(d) (SDD for Au) level of theory.

15.2 Au-mediated Alkene Activations

The coordination of alkene onto Au(I) could lead to the activation of alkene to bear a nucleophilic attack; however, the protonation of generated alkyl gold species would undergo a proton transfer to an sp^3 hybrid carbon, which would be considered as rate limit for this transformation.

15.2.1 Nucleophilic Addition of Alkenes

As shown in Figure 15.10, the mechanism for the gold-catalyzed hydroamination of alkenes was revealed by DFT calculations [92]. The coordination of alkene onto gold salt forms complex **15-75** with exergonic 8.7 kcal/mol. Then, an intermolecular nucleophilic attack of amine occurs via transition state **15-76ts** with an energy barrier of 5.3 kcal/mol to afford an ammonium intermediate **15-77**. The deprotonation of ammonium by acyl group takes place via transition state **15-78ts** to afford an oxonium intermediate **15-79** reversibly. Subsequently, the protonation of C(alkyl)—Au bond can occur via transition state **15-80ts** with an energy barrier of 18.2 kcal/mol, which is considered as the rate-determining step for the whole catalytic cycle.

In another case, after the nucleophilic attack of gold active alkene, the generated alkyl gold can play as a nucleophile for further transformations [93]. The mechanism of gold-catalyzed intramolecular cyclopropanation of alkenes and sulfonium ylides was revealed by DFT calculations [94]. As shown in Figure 15.11, when alkene coordinates onto cationic gold in complex **15-81**, an intramolecular nucleophilic attack of sulfonium ylide takes place via transition state **15-82ts** with an energy barrier of 12.3 kcal/mol to afford a sulfonium intermediate **15-83**. The sulfur atom can be considered as a good leaving group, which leads to a sequential intramolecular nucleophilic substitution by the alkyl gold moiety via transition state **15-84ts**. After

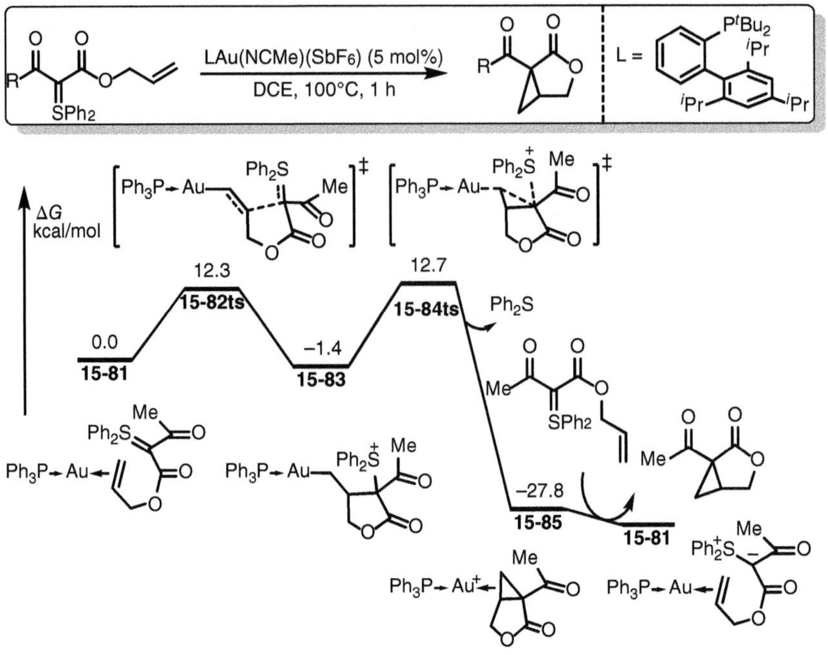

Figure 15.11 The free-energy profiles for the gold-catalyzed intermolecular intramolecular cyclopropanation of alkenes and sulfonium ylides. The energies were calculated at PBE0/6-311++G(d,p) (SDD-f for Au)/PCM(DCE)//PBE0/6-31G(d,p) (SDD-f for Au) level of theory.

this step, cyclopropane species is released as the product to accomplish the catalytic cycle.

15.2.2 Allylic Substitutions

As shown in Scheme 15.3, when a leaving group is installed at the allylic position, after the nucleophilic addition onto gold-activated alkene, the leaving of allylic substituent group can result in a new C=C double bond to achieve allylic substitution.

In a simple example, DFT calculation was employed to study the mechanism of gold-catalyzed sigmatropic rearrangement of allylic acetates [95]. A stepwise process was given in this reaction, which involves a nucleophilic addition of acetate onto gold-activated alkene via transition state **15-87ts** to result in a six-membered 1,3-acetoxonium intermediate **15-88**. Then, a C—O bond cleavage occurs via transition state **15-89ts** to result in the rearrangement product (Figure 15.12).

Interestingly, the gold-catalyzed etherification of allylic alcohols was considered to be processed through stepwise pathway [96]. As shown in Figure 15.13, the coordinated allylic alcohol also can form a hydrogen bond with the reacting ethanol in complex **15-91**. A nucleophilic attack by ethanol via transition state **15-92ts** forms a hydrogen-bonded six-membered ring-type intermediate **15-93** with endergonic 10.2 kcal/mol. Then, a C—O bond cleavage via transition state **15-94ts** results in the etherification product-coordinated gold complex **15-95**.

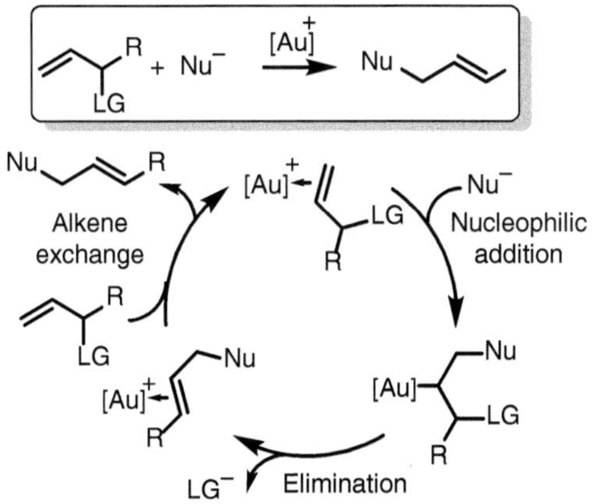

Scheme 15.3 General mechanism of gold-catalyzed allylic substitutions.

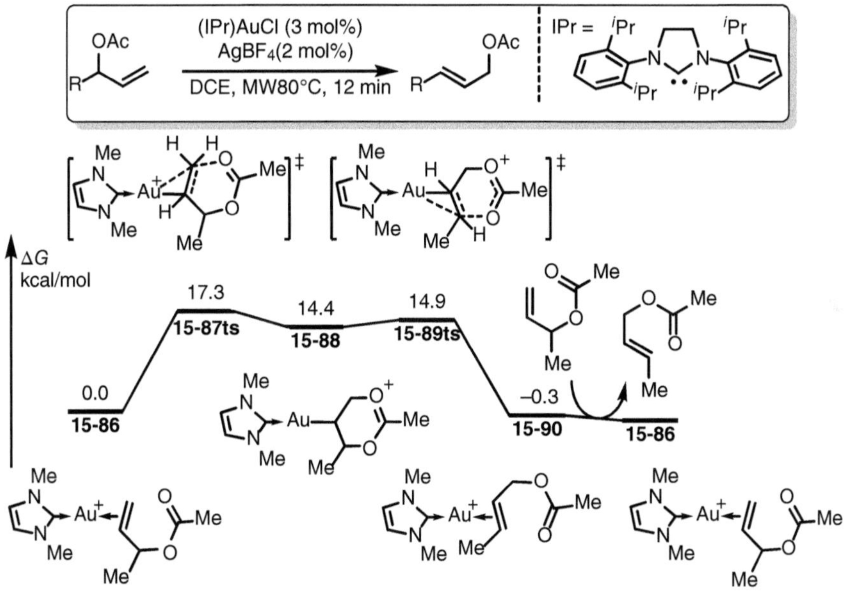

Figure 15.12 The free-energy profiles for the gold-catalyzed sigmatropic rearrangement of allylic acetates. The energies were calculated at B3LYP/6-31+G(d,p) (SDD for Au)/PCM(DCE)//B3LYP/6-31+G(d,p) (SDD for Au) level of theory.

15.3 Au-mediated Allene Activations

Allenes contain two cumulative C=C double bonds, which often reveal considerable reactivity in addition reactions [97–101]. The coordination of allene onto gold leads to the activation of allenes, which can react with the coming nucleophiles to achieve the formation of new covalent bonds [29, 102–107]. Generally, the mechanism of gold-mediated allene activation is similar as the corresponding one of alkyne

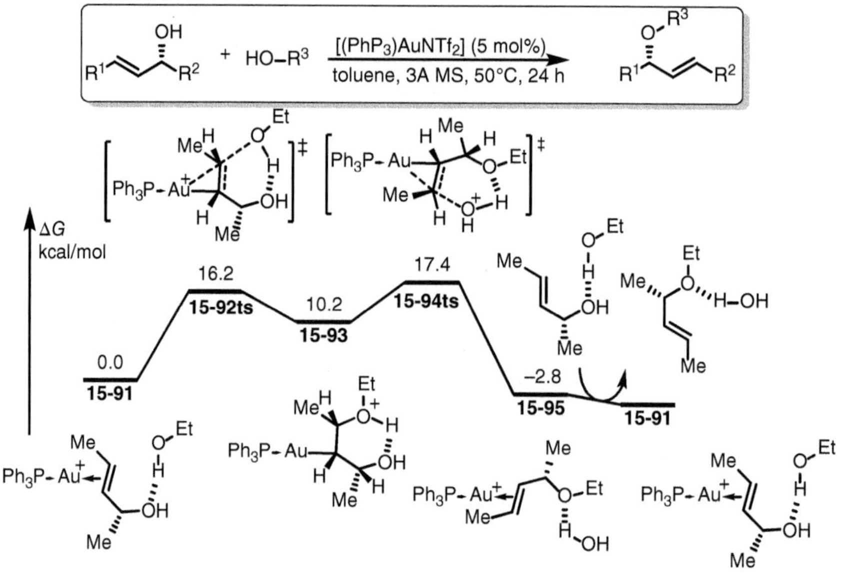

Figure 15.13 The free-energy profiles for the gold-catalyzed etherification of allylic alcohols. The energies were calculated at BP86-D3/6-31G(d,p) (SDD for Au and P)/PCM(ethanol)//BP86/6-31G(d,p) (SDD for Au and P) level of theory.

activation, which involved an anti-nucleophilic attack and protonation of the generated vinyl gold complex. Some gold-mediated allene activation reactions would be discussed in this section.

15.3.1 Hydroamination of Allenes

Different from using alkynes, the gold-catalyzed hydroamination of allenes can be used to synthesize allyl amines [108, 109]. DFT calculations found that the mechanism of gold-mediated allene activation also has peculiarity [110]. As shown in Figure 15.14, when allene coordinates onto cationic gold in complex **15-96**, it would first undergo an isomerization via transition state **15-97ts** to afford an allylic carbon cation intermediate **15-98**, where the C—Au bond is formed with the center carbon in allene moiety. Then, a nucleophilic attack can occur via transition state **15-99ts** with a lower barrier of only 1.3 kcal/mol to result in an allyl ammonium intermediate **15-100**. Further proton transfer can achieve hydroamination.

15.3.2 Hydroalkoxylation of Allenes

In the presence of gold catalyst, the hydroalkoxylation of allenes can afford corresponding allylic ethers [102, 111, 112]. The mechanism of this reaction was given by DFT calculations [113]. As shown in Figure 15.15, when allene coordinates onto gold in species **15-103**, the nucleophilic attack by methanol takes place via transition state **15-104ts** to afford an oxonium intermediate **15-105**, which can be stabilized

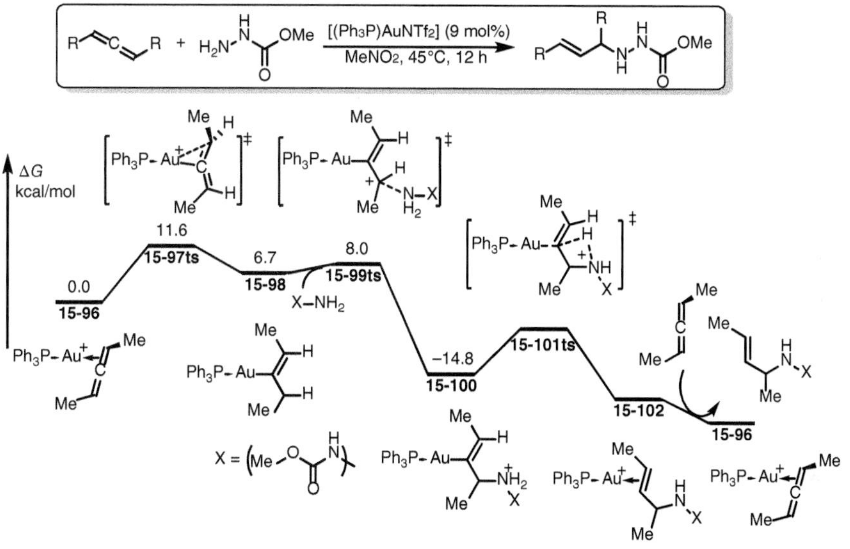

15.3 Au-mediated Allene Activations | 611

Figure 15.14 The free-energy profiles for the gold-catalyzed hydroamination of allenes. The energies were calculated at M06 level of theory.

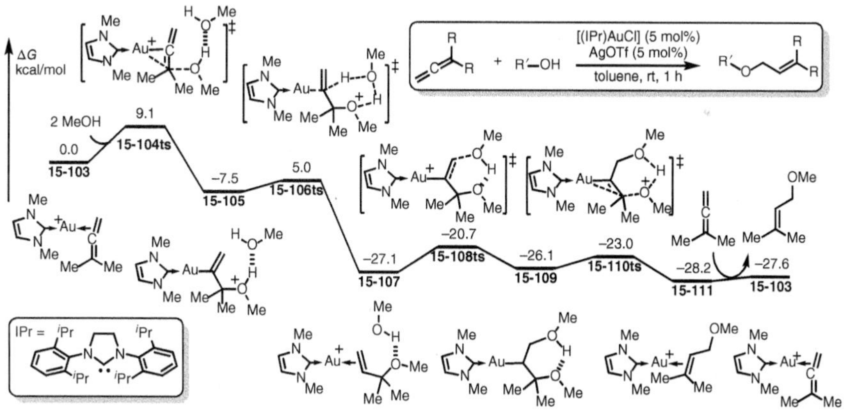

Figure 15.15 The potential energy profiles for the gold-catalyzed hydroalkoxylation of allenes. The energies were calculated at PBE0/6-31G(d,p) (SDD for Au)/PCM(toluene) //PBE0/6-31G(d,p) (SDD for Au) level of theory.

by the hydrogen bond with another methanol molecule. Then, a methanol-assisted 1,3-hydrogen transfer from oxonium to vinyl carbon can occur via a six-membered ring-type transition state **15-106ts** with an energy barrier of only 2.5 kcal/mol. After this step, allylic ether is yielded with the regeneration of cationic gold catalyst. Interestingly, the generated branched allylic ether can isomerize to its thermodynamically stable linear isomer in the presence of gold catalyst through a stepwise allylic substitution via transition states **15-108ts** and **15-110ts**, which is consistent with experimental observations.

Figure 15.16 The free-energy profiles for the gold-catalyzed cycloisomerization of allenyl ketones. The energies were calculated at B3LYP/6-31G(d) (LANL2DZ for Au and Br)/CPCM(toluene)//B3LYP/6-31G(d) (LANL2DZ for Au and Br) level of theory.

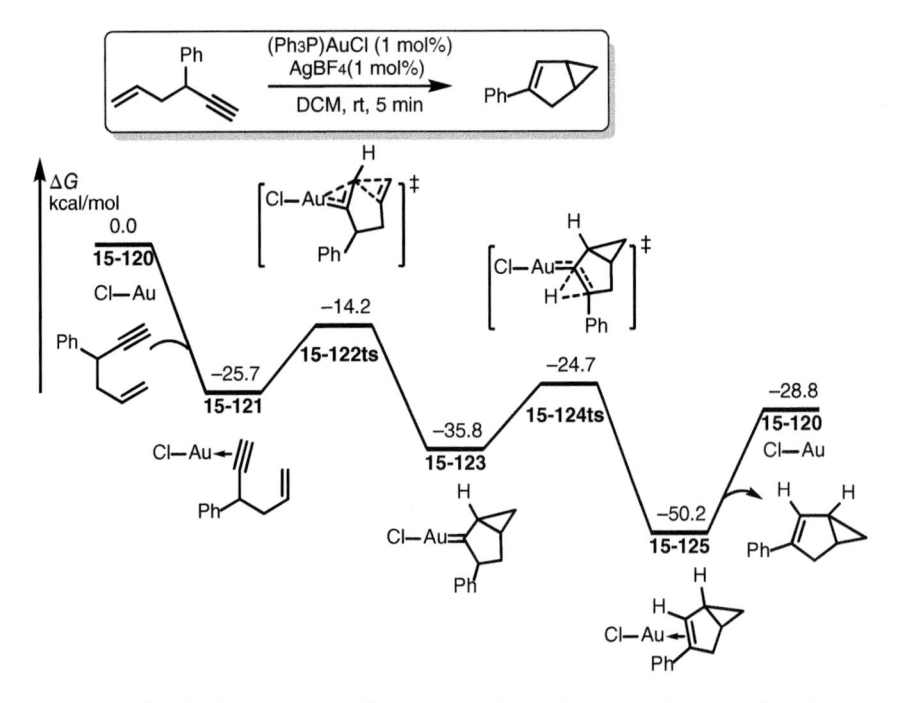

Figure 15.17 The free-energy profiles for the gold-catalyzed cycloisomerization of 1,5-enynes. The energies were calculated at B3LYP/6-31G(d,p) (LANL2DZ for Au)/PCM(DCE)//B3LYP/6-31G(d,p) (LANL2DZ for Au) level of theory.

15.3.3 Cycloisomerization of Allenyl Ketones

In the presence of gold catalyst, the cycloisomerization of allenyl ketones can afford corresponding furan derivatives [114, 115]. As shown in Figure 15.16, the coordination of allenyl ketone onto neutral AuCl complex **15-112** affords intermediate **15-113** with endergonic 20.8 kcal/mol. Then, an intramolecular nucleophilic attack by the oxygen atom in carbonyl occurs via transition state **15-114ts**. The calculated overall activation free energy is 31.3 kcal/mol, which can be attributed to the endergonic coordination of allene onto neutral gold. The generated five-membered ring intermediate **15-115** can be considered as a resonant between zwitterionic gold anion and neutral gold carbene. Then, a stepwise 1,3-proton transfer occurs in the assistance of chloride via transition state **15-116ts** and **15-118ts** to result in the final furan product [116].

15.4 Au-mediated Enyne Transformations

Cationic Au(I) species have emerged as a powerful catalyst for electrophilic activation of alkynes toward a variety of nucleophiles. In a common process, the coordination of alkyne onto gold leads to the activation of unsaturated C—C

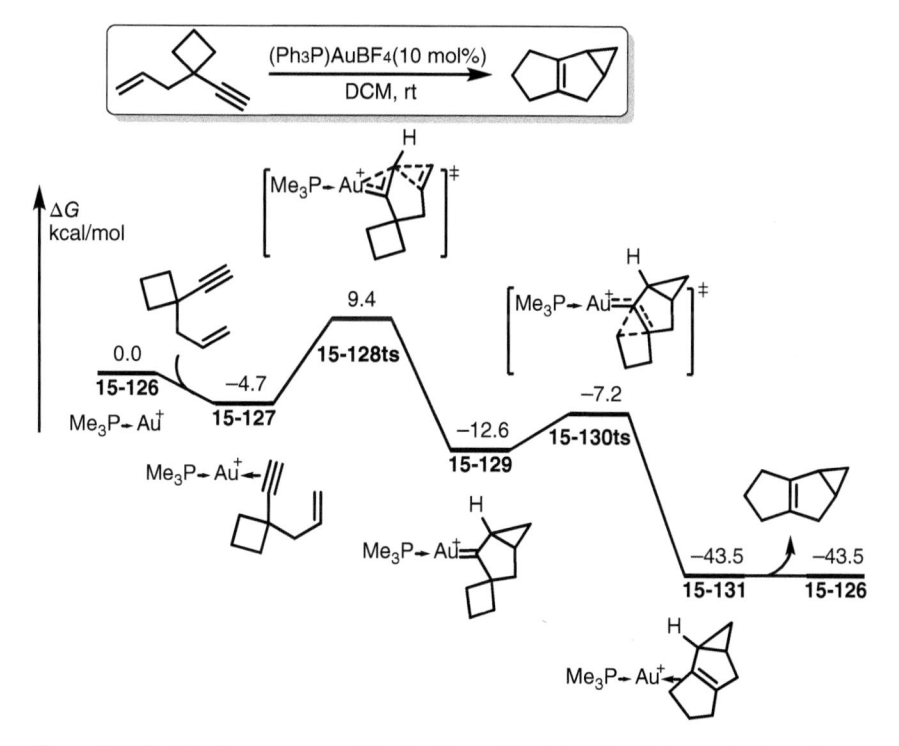

Figure 15.18 The free-energy profiles for the gold-catalyzed cycloisomerization of 1,5-enynes with ring-expansion. The energies were calculated at B3LYP/6-31G(d) (LANL2DZ for Au and P) level of theory.

triple bond, which can undergo a nucleophilic attack to achieve *trans*-vinyl gold complexes. When enynes are used in this reaction, the alkene moiety could act as nucleophile to attack the gold-activated alkyne moiety, which would result in cycloisomerizations for the construction of ring molecules [40, 117–119]. In particular, 1,5-enynes and 1,6-enynes are often used in this reaction.

15.4.1 1,5-Enyne Cycloisomerizations

In the cycloisomerizations of 1,5-enynes, the strain of the forming ring restricts the nucleophilic attack of alkene onto gold-activated alkyne through 5-endo position, which would result in a five-membered ring-type intermediate [120–122].

As shown in Figure 15.17, in the presence of gold catalyst, the cycloisomerizations of 1,5-enynes can result in a bicyclo[3.1.0]hexene in room temperature [123]. DFT calculations found that the coordination of enyne onto AuCl salt **15-120** gives complex **15-121** with exergonic 25.7 kcal/mol. Then, an intramolecular anti-nucleophilic attack of alkene moiety onto gold-activated alkyne takes place via transition state **15-122ts** with an energy barrier of only 11.5 kcal/mol. After the nucleophilic attack, a gold–carbene complex **15-123** is formed, which can undergo a 1,2-hydrogen shift via transition state **15-124ts** to afford bicyclo[3.1.0]hexene-coordinated gold complex **15-125**. Finally, the active catalyst **15-121** can be regenerated by ligand exchange with enyne reactant.

In another case, the strain release would result in 1,2-alkyl shift in a similar process [124]. As shown in Figure 15.18, when a cyclobutyl-substituted

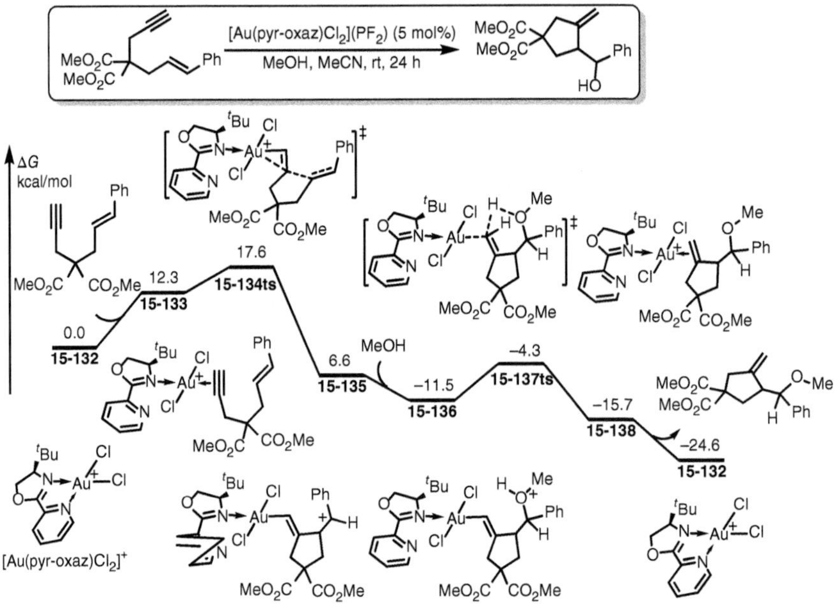

Figure 15.19 The free-energy profiles for the gold-catalyzed cycloisomerization of 1,6-enynes through 5-exo nucleophilic attack. The energies were calculated at ωB97X-D/def2-TZVPP/SMD(acetonitrile)//ωB97X-D/def2-SVP/SMD(acetonitrile) level of theory.

1,5-enyne is used as substrate, the cycloisomerization can result in a ring-expanded [3.3.0]bicyclic product. In the mechanism study, the coordination of 1,5-enyne onto cationic gold species **15-126** generates complex **15-127** with 4.7 kcal/mol exergonic. Then, an intramolecular nucleophilic attack by the alkene moiety onto gold-activated alkyne takes place via transition state **15-128ts** with an energy barrier of 14.1 kcal/mol to afford a gold–carbene intermediate **15-129**. Then, a 1,2-alkyl shift occurs via transition state **15-130ts** to result in an alkene-coordinated gold complex **15-131**. The calculated activation free energy for the 1,2-alkyl shift is only 5.4 kcal/mol, which can be attributed to the strain release of cyclobutyl substituent.

15.4.2 1,6-Enyne Cycloisomerizations

When 1,6-enyne is used as substrate, both 5-exo and 6-endo attacks are possible in gold-catalyzed cycloisomerizations [125–127].

As shown in Figure 15.19, in the presence of Au(III) complex, the cycloisomerizations of 1,6-enyne can give 5-exo product. In mechanism study, the coordination of enyne onto cationic Au(III) complex **15-132** results in complex **15-133** with 12.3 kcal/mol endergonic because of the release of pyridine

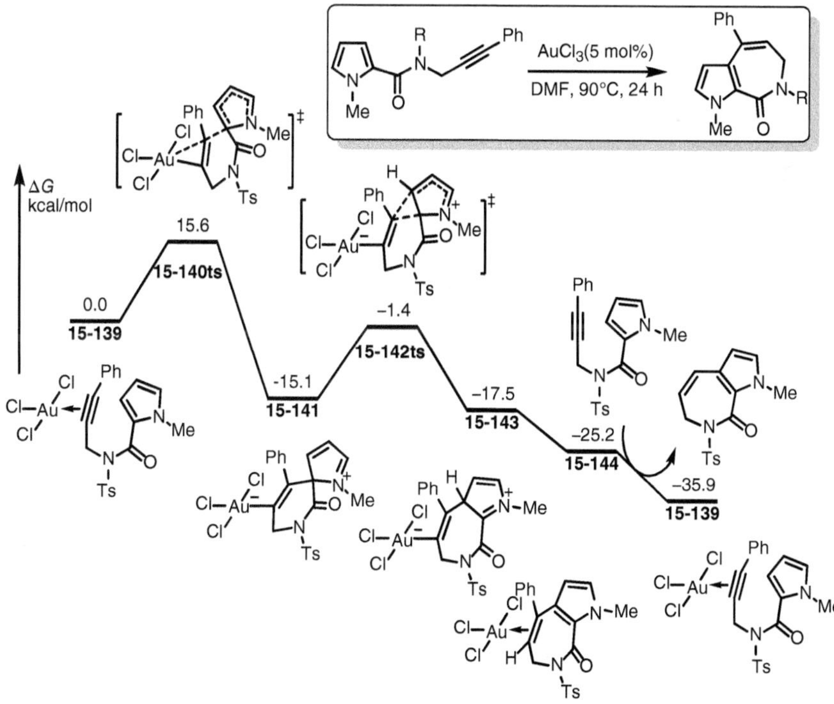

Figure 15.20 The free-energy profiles for the gold-catalyzed cycloisomerization of 1,6-enynes through 6-endo nucleophilic attack. The energies were calculated at B3LYP/6-31G(d) (SDD for Au, S, and Cl)/CPCM(DMF)//B3LYP/6-31G(d) (SDD for Au, S, and Cl) level of theory.

ligand. Then, a facile cyclization occurs via transition state **15-134ts** with overall barrier of 17.6 kcal/mol. After this step, a vinyl-gold complex with a carbocation **15-135** is formed instead of the formation of three-membered ring. Then, a methanol molecule reacts with carbocation to afford an oxonium intermediate **15-136**, which can undergo a 1,5-proton shift via transition state **15-137ts** to result in 5-exo alkene-coordinated gold complex **15-138**. Active catalyst **15-132** can be regenerated by the release of 5-exo alkene product [128].

When pyrrolyl is used instead of regular C=C double bond, a 6-endo nucleophilic attack would result through a stepwise process [129]. As shown in Figure 15.20, when alkyne coordinated onto Au(III) salt in complex **15-139**, a 6-endo nucleophilic takes place via transition state **15-140ts** with a free-energy barrier of 15.6 kcal/mol to result in a spiral intermediate **15-141**. Then, a 1,2-vinyl shift from 2-position of pyrrolyl to 3-position takes place via a three-membered ring-type transition state **15-142ts** to generate a bicyclic intermediate **15-143**. Then, a 1,2-proton shift provides a fused pyrrole-coordinated Au(III) complex **15-144**. The ligand exchange with enyne reactant yields final product and regenerates Au(III) active species **15-139**.

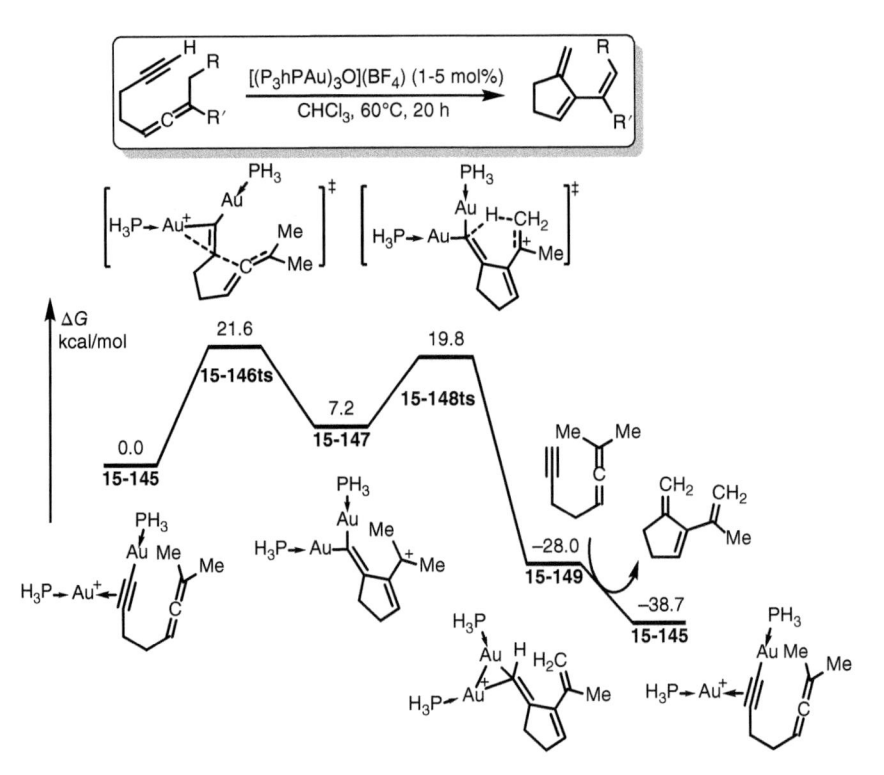

Figure 15.21 The free-energy profiles for the gold-catalyzed cycloisomerization of allenynes through 5-exo attack. The energies were calculated at HF/6-31+G(d,p) (LANL2DZ for Au)/PCM(chloroform)//B3LYP/6-31+G(d) (LANL2DZ for Au) level of theory.

15.4.3 Allenyne Cycloisomerizations

Gold-catalyzed cycloisomerizations of allenynes can be considered as a result of Alder-ene reaction; however, the reaction temperature would be much lower than that of uncatalyzed Alder-ene reaction [130–132]. As shown in Figure 15.21, the reaction starts from an alkynyl–Au(I)-coordinated Au(I) complex **15-145**. Then, an intramolecular 5-exo nucleophilic attack of the middle carbon of allene moiety onto gold-activated C—C triple bond via transition state **15-146ts** with an energy barrier of 21.6 kcal/mol generates a five-membered ring-type allylic carbocation **15-147**. Then, a 1,5-proton shift takes place via a six-membered ring-type transition state **15-148ts** resulting in a vinyl digold complex **15-149**. The ligand exchange with allenyne yields the conjugative triene product and regenerates the active digold complex **15-145** [133].

15.4.4 Conjugative Enyne Cycloisomerizations

Interestingly, when a conjugative enyne is used as substrate for gold-catalysis, the key step of cycloisomerization would undergo a Nazarov electrocyclic reaction [134]. As shown in Figure 15.22, in the presence of gold-catalyst, the cycloisomerization of conjugative enyne can afford cyclopentadiene product under room temperature. The mechanism of this reaction was revealed by DFT calculations [135]. When the propargylic acetoxyl enyne coordinates onto cationic gold in complex **15-150**, a reversible 1,3-acetoxyl migration takes place via a [3,3]-sigmatropic rearrangement transition state **15-151ts** to generate a six-membered ring intermediate **15-152**. Then, a C—O bond cleavage occurs via transition state **15-153ts** and affords a

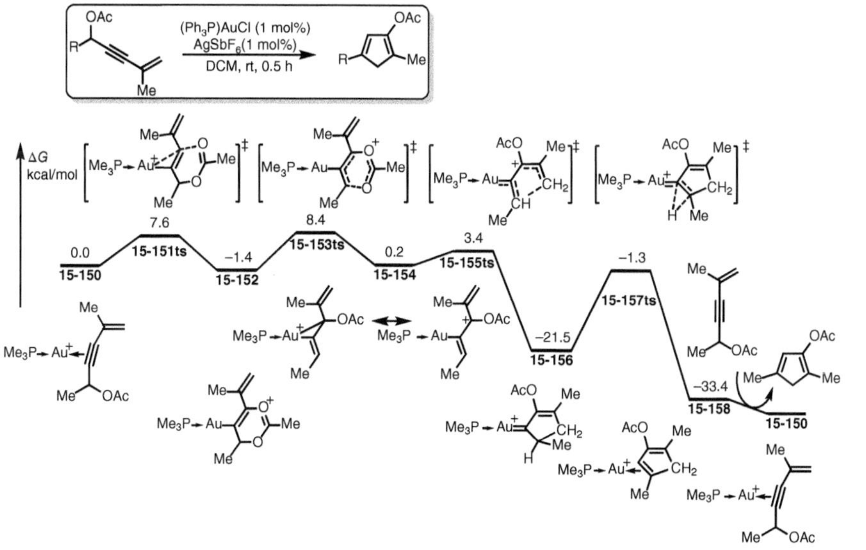

Figure 15.22 The free-energy profiles for the gold-catalyzed cycloisomerization of conjugative enynes. The energies were calculated at B3LYP/6-31G(d) (SDD for Au)/CPCM(DCM)//B3LYP/6-31G(d) (SDD for Au) level of theory.

vinylallene-coordinated gold complex **15-154**. This complex also can be considered as a resonance structure that involves an allylic carbocation with a vinyl gold because of the electron donation of acetoxyl group installing on the carbocation. According to this structure, a Nazarov electrocyclic rearrangement takes place via transition state **15-155ts** with an energy barrier of only 3.4 kcal/mol to irreversibly generate a gold–carbene complex **15-156**. A water-assisted 1,2-proton shift takes place via transition state **15-157ts** resulting in a cyclopentadiene-coordinated gold complex **15-158**.

References

1 Holliday, R. and Goodman, P. (2002). Going for gold [gold in electronics industry]. *IEE Review* 48 (3): 15–19.
2 Cooke, T. (1986). The demand for gold by the electronics industry in Europe. *Gold Bulletin* 19 (3): 87–89.
3 Fourie, J.T. (1982). Gold in electron microscopy. *Gold Bulletin* 15 (1): 2–6.
4 Zhao, J. and Jin, R. (2018). Heterogeneous catalysis by gold and gold-based bimetal nanoclusters. *Nano Today* 18: 86–102.
5 Zhang, X., Llabrés i Xamena, F.X., and Corma, A. (2009). Gold(III) – metal organic framework bridges the gap between homogeneous and heterogeneous gold catalysts. *Journal of Catalysis* 265 (2): 155–160.
6 Wang, Y.M., Lackner, A.D., and Toste, F.D. (2014). Development of catalysts and ligands for enantioselective gold catalysis. *Accounts of Chemical Research* 47 (3): 889–901.
7 Takei, T., Akita, T., Nakamura, I. et al. (2012). Heterogeneous catalysis by gold. *Advances in Catalysis* 55: 1–126.
8 Li, Y., Li, W., and Zhang, J. (2017). Gold-catalyzed enantioselective annulations. *Chemistry – A European Journal* 23 (3): 467–512.
9 Guan, Y. and Hensen, E.J.M. (2009). Ethanol dehydrogenation by gold catalysts: the effect of the gold particle size and the presence of oxygen. *Applied Catalysis A: General* 361 (1–2): 49–56.
10 Cheong, W.C., Yang, W., Zhang, J. et al. (2019). Isolated iron single-atomic site-catalyzed chemoselective transfer hydrogenation of nitroarenes to arylamines. *ACS Applied Materials & Interfaces* 11 (37): 33819–33824.
11 Boronat, M., Concepcion, P., Corma, A. et al. (2007). A molecular mechanism for the chemoselective hydrogenation of substituted nitroaromatics with nanoparticles of gold on TiO_2 catalysts: a cooperative effect between gold and the support. *Journal of the American Chemical Society* 129 (51): 16230–16237.
12 Zhang, L. (2014). A non-diazo approach to alpha-oxo gold carbenes via gold-catalyzed alkyne oxidation. *Accounts of Chemical Research* 47 (3): 877–888.
13 Zhang, D.H., Tang, X.Y., and Shi, M. (2014). Gold-catalyzed tandem reactions of methylenecyclopropanes and vinylidenecyclopropanes. *Accounts of Chemical Research* 47 (3): 913–924.

14 Obradors, C. and Echavarren, A.M. (2014). Intriguing mechanistic labyrinths in gold(I) catalysis. *Chemical Communications* 50 (1): 16–28.

15 Obradors, C. and Echavarren, A.M. (2014). Gold-catalyzed rearrangements and beyond. *Accounts of Chemical Research* 47 (3): 902–912.

16 Li, Z., Brouwer, C., and He, C. (2008). Gold-catalyzed organic transformations. *Chemical Reviews* 108 (8): 3239–3265.

17 Hashmi, A.S. (2007). Gold-catalyzed organic reactions. *Chemical Reviews* 107 (7): 3180–3211.

18 Fensterbank, L. and Malacria, M. (2014). Molecular complexity from polyunsaturated substrates: the gold catalysis approach. *Accounts of Chemical Research* 47 (3): 953–965.

19 Zhou, C.Y., Chan, P.W., and Che, C.M. (2006). Gold(III) porphyrin-catalyzed cycloisomerization of allenones. *Organic Letters* 8 (2): 325–328.

20 Zhang, X. and Corma, A. (2008). Supported gold(III) catalysts for highly efficient three-component coupling reactions. *Angewandte Chemie International Edition* 120 (23): 4430–4433.

21 Muzart, J. (2008). Gold-catalysed reactions of alcohols: isomerisation, inter- and intramolecular reactions leading to C—C and C—heteroatom bonds. *Tetrahedron* 64 (25): 5815–5849.

22 Melhado, A.D., Amarante, G.W., Wang, Z.J. et al. (2011). Gold(I)-catalyzed diastereo- and enantioselective 1,3-dipolar cycloaddition and Mannich reactions of azlactones. *Journal of the American Chemical Society* 133 (10): 3517–3527.

23 Georgy, M., Boucard, V., and Campagne, J.M. (2005). Gold(III)-catalyzed nucleophilic substitution of propargylic alcohols. *Journal of the American Chemical Society* 127 (41): 14180–14181.

24 Corma, A., Gonzalez-Arellano, C., Iglesias, M. et al. (2007). Gold nanoparticles and gold(III) complexes as general and selective hydrosilylation catalysts. *Angewandte Chemie International Edition* 46 (41): 7820–7822.

25 Furstner, A. and Davies, P.W. (2007). Catalytic carbophilic activation: catalysis by platinum and gold π acids. *Angewandte Chemie International Edition* 46 (19): 3410–3449.

26 Furstner, A. (2009). Gold and platinum catalysis – a convenient tool for generating molecular complexity. *Chemical Society Reviews* 38 (11): 3208–3221.

27 Frenking, G. and Frohlich, N. (2000). The nature of the bonding in transition-metal compounds. *Chemical Reviews* 100 (2): 717–774.

28 Rudolph, M. and Hashmi, A.S. (2011). Heterocycles from gold catalysis. *Chemical Communications* 47 (23): 6536–6544.

29 Krause, N. and Winter, C. (2011). Gold-catalyzed nucleophilic cyclization of functionalized allenes: a powerful access to carbo- and heterocycles. *Chemical Reviews* 111 (3): 1994–2009.

30 Cossy, J. (2010). Efficient cyclization routes to substituted heterocyclic compounds mediated by transition-metal catalysts. *Pure and Applied Chemistry* 82 (7): 1365–1373.

31 Corma, A., Leyva-Perez, A., and Sabater, M.J. (2011). Gold-catalyzed carbon–heteroatom bond-forming reactions. *Chemical Reviews* 111 (3): 1657–1712.

32 Bandini, M. (2011). Gold-catalyzed decorations of arenes and heteroarenes with C—C multiple bonds. *Chemical Society Reviews* 40 (3): 1358–1367.

33 Xie, L., Wu, Y., Yi, W. et al. (2013). Gold-catalyzed hydration of haloalkynes to α-halomethyl ketones. *The Journal of Organic Chemistry* 78 (18): 9190–9195.

34 Shen, H.C. (2008). Recent advances in syntheses of heterocycles and carbocycles via homogeneous gold catalysis. Part 1: Heteroatom addition and hydroarylation reactions of alkynes, allenes, and alkenes. *Tetrahedron* 64 (18): 3885–3903.

35 Marion, N., Ramon, R.S., and Nolan, S.P. (2009). [(NHC)Au(I)]-catalyzed acid-free alkyne hydration at part-per-million catalyst loadings. *Journal of the American Chemical Society* 131 (2): 448–449.

36 Hashmi, A.S. (2010). Homogeneous gold catalysis beyond assumptions and proposals – characterized intermediates. *Angewandte Chemie International Edition* 49 (31): 5232–5241.

37 Enomoto, T., Girard, A.L., Yasui, Y. et al. (2009). Gold(I)-catalyzed tandem reactions initiated by hydroamination of alkynyl carbamates: application to the synthesis of nitidine. *The Journal of Organic Chemistry* 74 (23): 9158–9164.

38 Wang, W., Hammond, G.B., and Xu, B. (2012). Ligand effects and ligand design in homogeneous gold(I) catalysis. *Journal of the American Chemical Society* 134 (12): 5697–5705.

39 Liu, L.-P. and Hammond, G.B. (2012). Recent advances in the isolation and reactivity of organogold complexes. *Chemical Society Reviews* 41 (8): 3129.

40 Jimenez-Nunez, E. and Echavarren, A.M. (2008). Gold-catalyzed cycloisomerizations of enynes: a mechanistic perspective. *Chemical Reviews* 108 (8): 3326–3350.

41 Ghosh, N., Nayak, S., and Sahoo, A.K. (2011). Gold-catalyzed regioselective hydration of propargyl acetates assisted by a neighboring carbonyl group: access to α-acyloxy methyl ketones and synthesis of (±)-actinopolymorphol B. *The Journal of Organic Chemistry* 76 (2): 500–511.

42 Wegener, M., Huber, F., Bolli, C. et al. (2015). Silver-free activation of ligated gold(I) chlorides: the use of [Me3NB12Cl11]- as a weakly coordinating anion in homogeneous gold catalysis. *Chemistry – A European Journal* 21 (3): 1328–1336.

43 Sanguramath, R.A., Hooper, T.N., Butts, C.P. et al. (2011). The interaction of gold(I) cations with 1,3-dienes. *Angewandte Chemie International Edition* 50 (33): 7592–7595.

44 Johnson, A., Laguna, A., and Gimeno, M.C. (2014). Axially chiral allenyl gold complexes. *Journal of the American Chemical Society* 136 (37): 12812–12815.

45 Dorel, R. and Echavarren, A.M. (2015). Gold(I)-catalyzed activation of alkynes for the construction of molecular complexity. *Chemical Reviews* 115 (17): 9028–9072.

46 Dias, H.V. and Wu, J. (2007). Thermally stable gold(I) ethylene adducts: [(HB{3,5-(CF$_3$)$_2$Pz}$_3$)Au(CH$_2$=CH$_2$)] and [(HB{3-(CF$_3$),5-(Ph)Pz}$_3$)Au(CH$_2$=CH$_2$)]. *Angewandte Chemie International Edition* 46 (41): 7814–7816.

47 Brown, T.J., Sugie, A., Dickens, M.G. et al. (2010). Solid-state and dynamic solution behavior of a cationic, two-coordinate gold(I) π-allene complex. *Organometallics* 29 (19): 4207–4209.

48 Levin, M.D. and Toste, F.D. (2014). Gold-catalyzed allylation of aryl boronic acids: accessing cross-coupling reactivity with gold. *Angewandte Chemie International Edition* 53 (24): 6211–6215.

49 Lauterbach, T., Livendahl, M., Rosellon, A. et al. (2010). Unlikeliness of Pd-free gold(I)-catalyzed Sonogashira coupling reactions. *Organic Letters* 12 (13): 3006–3009.

50 Joost, M., Zeineddine, A., Estevez, L. et al. (2014). Facile oxidative addition of aryl iodides to gold(I) by ligand design: bending turns on reactivity. *Journal of the American Chemical Society* 136 (42): 14654–14657.

51 Hopkinson, M.N., Gee, A.D., and Gouverneur, V. (2011). Au(I)/Au(III) catalysis: an alternative approach for C–C oxidative coupling. *Chemistry – A European Journal* 17 (30): 8248–8262.

52 Ball, L.T., Lloyd-Jones, G.C., and Russell, C.A. (2014). Gold-catalyzed oxidative coupling of arylsilanes and arenes: origin of selectivity and improved precatalyst. *Journal of the American Chemical Society* 136 (1): 254–264.

53 Ball, L.T., Lloyd-Jones, G.C., and Russell, C.A. (2012). Gold-catalyzed direct arylation. *Science* 337 (6102): 1644–1648.

54 Gimeno, M.C. and Laguna, A. (1997). Three- and four-coordinate gold(I) complexes. *Chemical Reviews* 97 (3): 511–522.

55 Carvajal, M.A., Novoa, J.J., and Alvarez, S. (2004). Choice of coordination number in d10 complexes of group 11 metals. *Journal of the American Chemical Society* 126 (5): 1465–1477.

56 Bott, R.C., Healy, P.C., and Smith, G. (2007). Structural studies of two-coordinate complexes of tris(2-methoxylphenyl)phosphine and tris(4-methoxyphenyl)phosphine with gold(I) halides. *Polyhedron* 26 (12): 2803–2809.

57 Balch, A.L. and Fung, E.Y. (1990). Two- and four-coordinate gold(I) complexes of tris(2-(diphenylphosphino)ethyl)phosphine. *Inorganic Chemistry* 29 (23): 4764–4768.

58 Baker, M.V., Barnard, P.J., Berners-Price, S.J. et al. (2006). Cationic, linear Au(I) N-heterocyclic carbene complexes: synthesis, structure and anti-mitochondrial activity. *Dalton Transactions* (30): 3708–3715.

59 Baker, M.V., Barnard, P.J., Berners-Price, S.J. et al. (2005). Synthesis and structural characterisation of linear Au(I) N-heterocyclic carbene complexes: new analogues of the Au(I) phosphine drug Auranofin. *Journal of Organometallic Chemistry* 690 (24–25): 5625–5635.

60 Zhdanko, A. and Maier, M.E. (2014). Explanation of counterion effects in gold(I)-catalyzed hydroalkoxylation of alkynes. *ACS Catalysis* 4 (8): 2770–2775.

61 Zhang, L., Sun, J., and Kozmin, S.A. (2006). Gold and platinum catalysis of enyne cycloisomerization. *Advanced Synthesis & Catalysis* 348 (16–17): 2271–2296.

62 Ma, S., Yu, S., and Gu, Z. (2005). Gold-catalyzed cyclization of enynes. *Angewandte Chemie International Edition* 45 (2): 200–203.

63 Luo, K., Cao, T., Jiang, H. et al. (2017). Gold-catalyzed ring expansion of enyne-lactone: generation and transformation of 2-oxoninonium. *Organic Letters* 19 (21): 5856–5859.

64 Bruneau, C. (2005). Electrophilic activation and cycloisomerization of enynes: a new route to functional cyclopropanes. *Angewandte Chemie International Edition* 44 (16): 2328–2334.

65 Akana, J.A., Bhattacharyya, K.X., Muller, P. et al. (2007). Reversible C—F bond formation and the Au-catalyzed hydrofluorination of alkynes. *Journal of the American Chemical Society* 129 (25): 7736–7737.

66 Basak, A., Chakrabarty, K., Ghosh, A. et al. (2016). Theoretical study on the isomerization of propargyl derivative to conjugated diene under Au(I)-catalyzed reaction: a DFT study. *Computational and Theoretical Chemistry* 1083: 38–45.

67 Cala, L., Mendoza, A., Fananas, F.J. et al. (2013). A catalytic multicomponent coupling reaction for the enantioselective synthesis of spiroacetals. *Chemical Communications* 49 (26): 2715–2717.

68 Belting, V. and Krause, N. (2006). Gold-catalyzed tandem cycloisomerization-hydroalkoxylation of homopropargylic alcohols. *Organic Letters* 8 (20): 4489–4492.

69 D'Amore, L., Ciancaleoni, G., Belpassi, L. et al. Unraveling the anion/ligand interplay in the reaction mechanism of gold(I)-catalyzed alkoxylation of alkynes. *Organometallics* 36: 2364–2376.

70 Zhu, S., Huang, X., Zhao, T.Q. et al. (2015). Metal-catalyzed formation of 1,3-cyclohexadienes: a catalyst-dependent reaction. *Organic & Biomolecular Chemistry* 13 (4): 1225–1233.

71 Liu, L.P., Malhotra, D., Paton, R.S. et al. (2010). The [4+2], not [2+2], mechanism occurs in the gold-catalyzed intramolecular oxygen transfer reaction of 2-alkynyl-1,5-diketones. *Angewandte Chemie International Edition* 49 (48): 9132–9135.

72 Das, A., Chang, H.K., Yang, C.H. et al. (2008). Platinum- and gold-catalyzed hydrative carbocyclization of oxo diynes for one-pot synthesis of benzopyrones and bicyclic spiro ketones. *Organic Letters* 10 (18): 4061–4064.

73 Straub, B.F. (2004). Gold(I) or gold(III) as active species in AuCl(3)-catalyzed cyclization/cycloaddition reactions? A DFT study. *Chemical Communications* 40 (15): 1726–1728.

74 Yeom, H.S. and Shin, S. (2014). Catalytic access to alpha-oxo gold carbenes by N—O bond oxidants. *Accounts of Chemical Research* 47 (3): 966–977.

75 Xiao, J. and Li, X. (2011). Gold alpha-oxo carbenoids in catalysis: catalytic oxygen-atom transfer to alkynes. *Angewandte Chemie International Edition* 50 (32): 7226–7236.

76 Iuliis, M.Z.D., Watson, I.D.G., Yudin, A.K. et al. (2009). A DFT investigation into the origin of regioselectivity in palladium-catalyzed allylic amination. *Canadian Journal of Chemistry* 87 (1): 54–62.

77 Noey, E.L., Luo, Y., Zhang, L. et al. (2012). Mechanism of gold(I)-catalyzed rearrangements of acetylenic amine-N-oxides: computational investigations lead to a new mechanism confirmed by experiment. *Journal of the American Chemical Society* 134 (2): 1078–1084.

78 Cuenca, A.B., Montserrat, S., Hossain, K.M. et al. (2009). Gold(I)-catalyzed intermolecular oxyarylation of alkynes: unexpected regiochemistry in the alkylation of arenes. *Organic Letters* 11 (21): 4906–4909.

79 Kramer, S., Dooleweerdt, K., Lindhardt, A.T. et al. (2009). Highly regioselective Au(I)-catalyzed hydroamination of ynamides and propiolic acid derivatives with anilines. *Organic Letters* 11 (18): 4208–4211.

80 Fustero, S., Ibáñez, I., Barrio, P. et al. (2013). Gold-catalyzed intramolecular hydroamination of o-alkynylbenzyl carbamates: a route to chiral fluorinated isoindoline and isoquinoline derivatives. *Organic Letters* 15 (4): 832–835.

81 Xia, Y., Dudnik, A.S., Li, Y. et al. (2010). On the validity of Au-vinylidenes in the gold-catalyzed 1,2-migratory cycloisomerization of skipped propargylpyridines. *Organic Letters* 12 (23): 5538–5541.

82 Prechter, A., Henrion, G., Faudot dit Bel, P. et al. (2014). Gold-catalyzed synthesis of functionalized pyridines by using 2H-azirines as synthetic equivalents of alkenyl nitrenes. *Angewandte Chemie International Edition* 53 (19): 4959–4963.

83 Katari, M., Rao, M.N., Rajaraman, G. et al. (2012). Computational insight into a gold(I) N-heterocyclic carbene mediated alkyne hydroamination reaction. *Inorganic Chemistry* 51 (10): 5593–5604.

84 Duan, Y., Liu, Y., Bi, S. et al. (2016). Theoretical study of gold-catalyzed cyclization of 2-alkynyl-N-propargylanilines and rationalization of kinetic experimental phenomena. *The Journal of Organic Chemistry* 81 (19): 9381–9388.

85 Jin, L., Wu, Y., and Zhao, X. (2015). Theoretical insight into the mechanism of gold(I)-catalyzed rearrangement of 2-propargyl 2H-azirines to pyridines. *The Journal of Organic Chemistry* 80 (7): 3547–3555.

86 Shu, C., Li, L., Chen, C.B. et al. (2014). Gold-catalyzed 6-exo-dig cycloisomerization: a versatile approach to functionalized phenanthrenes. *Chemistry – An Asian Journal* 9 (6): 1525–1529.

87 Pflasterer, D., Rettenmeier, E., Schneider, S. et al. (2014). Highly efficient gold-catalyzed synthesis of dibenzocycloheptatrienes. *Chemistry – A European Journal* 20 (22): 6752–6755.

88 Ferrer, C. and Echavarren, A.M. (2006). Gold-catalyzed intramolecular reaction of indoles with alkynes: facile formation of eight-membered rings and an unexpected allenylation. *Angewandte Chemie International Edition* 45 (7): 1105–1109.

89 Ferrer, C., Amijs, C.H., and Echavarren, A.M. (2007). Intra- and intermolecular reactions of indoles with alkynes catalyzed by gold. *Chemistry – A European Journal* 13 (5): 1358–1373.

90 Cervi, A., Aillard, P., Hazeri, N. et al. (2013). Total syntheses of the coumarin-containing natural products pimpinellin and fraxetin using Au(I)-catalyzed intramolecular hydroarylation (IMHA) chemistry. *The Journal of Organic Chemistry* 78 (19): 9876–9882.

91 Yang, Y., Liu, Y., Lv, P. et al. (2018). Theoretical insight into the mechanism and origin of ligand-controlled regioselectivity in homogenous gold-catalyzed intramolecular hydroarylation of alkynes. *The Journal of Organic Chemistry* 83 (5): 2763–2772.

92 Kovács, G., Lledós, A., and Ujaque, G. (2010). Mechanistic comparison of acid- and gold(I)-catalyzed nucleophilic addition reactions to olefins. *Organometallics* 29 (22): 5919–5926.

93 Klimczyk, S., Huang, X., Kahlig, H. et al. (2015). Stereoselective gold(I) domino catalysis of allylic isomerization and olefin cyclopropanation: mechanistic studies. *The Journal of Organic Chemistry* 80 (11): 5719–5729.

94 Huang, X., Klimczyk, S., Veiros, L.F. et al. (2013). Stereoselective intramolecular cyclopropanation through catalytic olefin activation. *Chemical Science* 4 (3): 1105.

95 Gourlaouen, C., Marion, N., Nolan, S.P. et al. (2009). Mechanism of the [(NHC)Au(I)]-catalyzed rearrangement of allylic acetates. A DFT study. *Organic Letters* 11 (1): 81–84.

96 Barker, G., Johnson, D.G., Young, P.C. et al. (2015). Chirality transfer in gold(I)-catalysed direct allylic etherifications of unactivated alcohols: experimental and computational study. *Chemistry – A European Journal* 21 (39): 13748–13757.

97 Yu, S. and Ma, S. (2012). Allenes in catalytic asymmetric synthesis and natural product syntheses. *Angewandte Chemie International Edition* 51 (13): 3074–3112.

98 Sydnes, L.K. (2003). Allenes from cyclopropanes and their use in organic synthesis-recent developments. *Chemical Reviews* 103 (4): 1133–1150.

99 Ma, S. (2009). Electrophilic addition and cyclization reactions of allenes. *Accounts of Chemical Research* 42 (10): 1679–1688.

100 Ma, S. (2005). Some typical advances in the synthetic applications of allenes. *Chemical Reviews* 105 (7): 2829–2872.

101 Bates, R.W. and Satcharoen, V. (2002). Nucleophilic transition metal based cyclization of allenes. *Chemical Society Reviews* 31 (1): 12–21.

102 Zhang, Z. and Widenhoefer, R.A. (2007). Gold(I)-catalyzed intramolecular enantioselective hydroalkoxylation of allenes. *Angewandte Chemie International Edition* 46 (1–2): 283–285.

103 Yang, W. and Hashmi, A.S. (2014). Mechanistic insights into the gold chemistry of allenes. *Chemical Society Reviews* 43 (9): 2941–2955.

104 Widenhoefer, R.A. (2008). Recent developments in enantioselective gold(I) catalysis. *Chemistry – A European Journal* 14 (18): 5382–5391.

105 Shu, X.Z., Nguyen, S.C., He, Y. et al. (2015). Silica-supported cationic gold(I) complexes as heterogeneous catalysts for regio- and enantioselective lactonization reactions. *Journal of the American Chemical Society* 137 (22): 7083–7086.

106 LaLonde, R.L., Sherry, B.D., Kang, E.J. et al. (2007). Gold(I)-catalyzed enantioselective intramolecular hydroamination of allenes. *Journal of the American Chemical Society* 129 (9): 2452–2453.

107 Brown, T.J., Weber, D., Gagne, M.R. et al. (2012). Mechanistic analysis of gold(I)-catalyzed intramolecular allene hydroalkoxylation reveals an off-cycle bis(gold) vinyl species and reversible C—O bond formation. *Journal of the American Chemical Society* 134 (22): 9134–9137.

108 Zhang, Z., Bender, C.F., and Widenhoefer, R.A. (2007). Gold(I)-catalyzed dynamic kinetic enantioselective intramolecular hydroamination of allenes. *Journal of the American Chemical Society* 129 (46): 14148–14149.

109 Zhang, Z., Bender, C.F., and Widenhoefer, R.A. (2007). Gold(I)-catalyzed enantioselective hydroamination of N-allenyl carbamates. *Organic Letters* 9 (15): 2887–2889.

110 Wang, Z.J., Benitez, D., Tkatchouk, E. et al. (2010). Mechanistic study of gold(I)-catalyzed intermolecular hydroamination of allenes. *Journal of the American Chemical Society* 132 (37): 13064–13071.

111 Zhang, Z., Liu, C., Kinder, R.E. et al. (2006). Highly active Au(I) catalyst for the intramolecular exo-hydrofunctionalization of allenes with carbon, nitrogen, and oxygen nucleophiles. *Journal of the American Chemical Society* 128 (28): 9066–9073.

112 Hamilton, G.L., Kang, E.J., Mba, M. et al. (2007). A powerful chiral counterion strategy for asymmetric transition metal catalysis. *Science* 317 (5837): 496–499.

113 Paton, R.S. and Maseras, F. (2009). Gold(I)-catalyzed intermolecular hydroalkoxylation of allenes: a DFT study. *Organic Letters* 11 (11): 2237–2240.

114 Dudnik, A.S., Xia, Y., Li, Y. et al. (2010). Computation-guided development of Au-catalyzed cycloisomerizations proceeding via 1,2-Si or 1,2-H migrations: regiodivergent synthesis of silylfurans. *Journal of the American Chemical Society* 132 (22): 7645–7655.

115 Dudnik, A.S. and Gevorgyan, V. (2007). Metal-catalyzed [1,2]-alkyl shift in allenyl ketones: synthesis of multisubstituted furans. *Angewandte Chemie International Edition* 46 (27): 5195–5197.

116 Xia, Y., Dudnik, A.S., Gevorgyan, V. et al. (2008). Mechanistic insights into the gold-catalyzed cycloisomerization of bromoallenyl ketones: ligand-controlled regioselectivity. *Journal of the American Chemical Society* 130 (22): 6940–6941.

117 Ghosh, A., Basak, A., Chakrabarty, K. et al. (2018). Au-catalyzed hexannulation and Pt-catalyzed pentannulation of propargylic ester bearing a 2-alkynyl-phenyl substituent: a comparative DFT study. *ACS Omega* 3 (1): 1159–1169.

118 Fan, T., Chen, X., Sun, J. et al. (2012). DFT studies on gold-catalyzed cycloisomerization of 1,5-enynes. *Organometallics* 31 (11): 4221–4227.

119 Borsini, E., Broggini, G., Fasana, A. et al. (2011). Access to pyrrolo–pyridines by gold-catalyzed hydroarylation of pyrroles tethered to terminal alkynes. *Beilstein Journal of Organic Chemistry* 7: 1468–1474.

120 Mamane, V., Gress, T., Krause, H. et al. (2004). Platinum- and gold-catalyzed cycloisomerization reactions of hydroxylated enynes. *Journal of the American Chemical Society* 126 (28): 8654–8655.

121 Luzung, M.R., Markham, J.P., and Toste, F.D. (2004). Catalytic isomerization of 1,5-enynes to bicyclo[3.1.0]hexenes. *Journal of the American Chemical Society* 126 (35): 10858–10859.

122 Gagosz, F. (2005). Unusual gold(I)-catalyzed isomerization of 3-hydroxylated 1,5-enynes: highly substrate-dependent reaction manifolds. *Organic Letters* 7 (19): 4129–4132.

123 Liu, Y., Zhang, D., Zhou, J. et al. (2010). Theoretical elucidation of Au(I)-catalyzed cycloisomerizations of cycloalkyl-substituted 1,5-enynes: 1,2-alkyl shift versus C—H bond insertion products. *The Journal of Physical Chemistry A* 114 (20): 6164–6170.

124 Liu, Y., Zhang, D., and Bi, S. (2010). Theoretical investigation on the isomerization reaction of 4-phenyl-hexa-1,5-enyne catalyzed by homogeneous Au catalysts. *The Journal of Physical Chemistry A* 114 (49): 12893–12899.

125 Nieto-Oberhuber, C., Munoz, M.P., Lopez, S. et al. (2006). Gold(I)-catalyzed cyclizations of 1,6-enynes: alkoxycyclizations and exo/endo skeletal rearrangements. *Chemistry – A European Journal* 12 (6): 1677–1693.

126 Nieto-Oberhuber, C., Munoz, M.P., Bunuel, E. et al. (2004). Cationic gold(I) complexes: highly alkynophilic catalysts for the exo- and endo-cyclization of enynes. *Angewandte Chemie International Edition* 43 (18): 2402–2406.

127 Mezailles, N., Ricard, L., and Gagosz, F. (2005). Phosphine gold(I) bis-(trifluoromethanesulfonyl)imidate complexes as new highly efficient and air-stable catalysts for the cycloisomerization of enynes. *Organic Letters* 7 (19): 4133–4136.

128 Reiersolmoen, A.C., Csokas, D., Papai, I. et al. (2019). Mechanism of Au(III)-mediated alkoxycyclization of a 1,6-enyne. *Journal of the American Chemical Society* 141 (45): 18221–18229.

129 Lin, Y.L.Z. (2015). Gold(III)-catalyzed intramolecular cyclization of α-pyrroles to pyrrolopyridinones and pyrroloazepinones: a DFT study. *Organometallics* 34: 3538–3545.

130 Zriba, R., Gandon, V., Aubert, C. et al. (2008). Alkyne versus allene activation in platinum- and gold-catalyzed cycloisomerization of hydroxylated 1,5-allenynes. *Chemistry – A European Journal* 14 (5): 1482–1491.

131 Yang, C.Y., Lin, G.Y., Liao, H.Y. et al. (2008). Gold-catalyzed hydrative carbocyclization of 1,5- and 1,7-allenynes mediated by π-allene complex: mechanistic evidence supported by the chirality transfer of allenyne substrates. *The Journal of Organic Chemistry* 73 (13): 4907–4914.

132 Lemiere, G., Gandon, V., Agenet, N. et al. (2006). Gold(I)- and gold(III)-catalyzed cycloisomerization of allenynes: a remarkable halide effect. *Angewandte Chemie International Edition* 45 (45): 7596–7599.

133 Cheong, P.H., Morganelli, P., Luzung, M.R. et al. (2008). Gold-catalyzed cycloisomerization of 1,5-allenynes via dual activation of an ene reaction. *Journal of the American Chemical Society* 130 (13): 4517–4526.

134 Zhang, L. and Wang, S. (2006). Efficient synthesis of cyclopentenones from enynyl acetates via tandem Au(I)-catalyzed 3,3-rearrangement and the Nazarov reaction. *Journal of the American Chemical Society* 128 (5): 1442–1443.

135 Li, Y., Kirillov, A.M. et al. (2017). Effect of substituent on the mechanism and chemoselectivity of the gold(I)-catalyzed propargyl ester tandem cyclization. *Organometallics* 36 (6): 1164–1172.

Index

a

acetate Mn(I) species 507
acetylene-coordinated hydride Ir(I)
    species 400
acetylene-coordinated silyl Ir(III) complex
    401
acetylene insertion 203
acetylene moiety 260
acetylenes 400
acetyl-protected aminoethyl quinoline
    additive 217
π-acidic catalyst Pt(II) species 264
π-acidic compounds 297
acidic hydrogen 128
activation free energy 66, 83, 101, 104,
    128, 134, 137, 141, 201, 203, 207
π-activation of alkynes 586
active digold complex 617
active hydride ferrous species 432
acyl aryl reductive elimination 214
3-acyloxy-1,4-enynes 368
Ag-catalyzed nitrene transfer 578–580
Ag-catalyzed vinylcarbene insertion 573
Ag-mediated alkyne activations
    C—H activation of alkynes 587–588
    π-activation of alkynes 586–587
    vinyl compounds 585
Ag-mediated carbene complex
    transformation
    carbene insertion into C—Cl bond
        572–573
    carbene insertion into C—H bond
        574–576

carbene insertion into O—H bond 573
Fischer-type metallacarbene complex
    568
general mechanism 568
nucleophilic attack by carbonyl groups
    573–574
silver–carbene formation 568–572
Ag-mediated nitrene transfer reactions
    577
Ag-mediated nitrene transformations
    nitrene 576–577
    nucleophilic attack by amines
        582–583
    nucleophilic attack by unsaturated
        bonds 580–582
    silver–nitrene complex formation
        577–579
Ag-mediated silylene transformations
    583–585
agostic amino borane-ferrous complex
    437
agostic Ru(0) species 459
alchohol dehydrogenation 434, 510
alcoholate Rh(II) intermediate 467
alcohol-coordinated iron(III) complex
    421
alcohol dehydrogenations 473
alcohols 405
Alder-ene reaction 617
alkane-coordinated Ir(III) complex
    390
alkane oxidation 420
alkanes 402

Computational Methods in Organometallic Catalysis: From Elementary Reactions to Mechanisms,
First Edition. Yu Lan.
© 2021 WILEY-VCH GmbH. Published 2021 by WILEY-VCH GmbH.

Index

alkene-coordinated cationic Cu(I) species   540
alkene-coordianted cationic Rh(I) species   347
alkene coordinated Rh(I) species   347
alkene diboration   347
alkene hydrogenations   464
alkene insertion   294
alkene oxidations   423
alkenes   388, 427
alkenes, Cu-catalyzed borylations   551
alkenyl-Rh(III) complex   338
alkoxy/amino Ir(III) complex   391
alkoxyl iorn(II) intermediate   432
alkoxyl Rh(III) intermediate   352
alkoxy Mn(I) intermediate   509
alkyl arenes   438
alkyl aryl iron(0) intermediate   438
alkylated bipyridine-coordinated Rh(I) intermediate   334
alkylated enamine-coordinated Rh(I) complex   335
α-alkylated ketone product   335
alkyl bromide   135
alkyl copper-carbene species   542
alkyl-cuprous intermediate   551
α-alkyl elimination   349
alkylideneoxindole product   576
3-alkylideneoxindoles   569, 575
alkyl Ir(III) complex   398
alkyl Ir(V) complex   389, 391
alkyl Ir(III) intermediate   389
alkyl Ir(V) intermediate   388
alkyl Ir(V) trihydride intermediate   389
alkyl metal compounds   6
alkyl Rh(III) intermediate   334, 335, 338, 347
alkylruthenium intermediate   461
alkyne-coordinated Au(I) species   599, 601
alkyne-coordinated Rh(I) species   351
alkyne hydroamination   349
alkyne metathesis   484
alkynes, Cu-catalyzed borylations   552
alkynes, isomerization of   599

alkynyl-Au(I) coordinated Au(I) complex   617
alkynyl benzaldehyde   601
alkynyl-coordinated Cu(I) species   536
alkynyl Cu(I) species   535
alkynyl Cu(II) species   538
alkynyl Cu transformations   536
alkynyl-N-propargylanilines   603, 605
alkynyl nucleophile   80
allene-coordinated hydride-Rh(III)   351
allene-coordinated Rh(I) intermediate   350
allene intermediate   599
allene moiety   70
allenes, hydroalkoxylation of   610
allenes, hydroamination of   610
allenyl arene product   507
allenylarylimine   507
allenyl gold intermediate   599
allenyl indole intermediate   603
allenyl ketones, cycloisomerization of   613
allenyne cycloisomerizations   617
allyl ammonium intermediate   610
allylic carbon cation intermediate   610
allylic ether   611
allylic hydrogen   599
allylic ions   363
allylic Pd complex   234
   formation, from allene insertion   238–239
   formation, from allylic C—H activation   237–238
   formation, from allylic nucleophilic substitution   236–237
   formation, from allylic oxidative addition   235–236
   formation, from nucleophilic attack onto allene   237
allylic Ru(II) species   469
allylic Rh(III) intermediate   365
allylic substitutions   608
amide-coordinated aryl-Cu(III) species   528

Index | 631

amide-directed C—H bond activation   332
amination of nitrenes   543
amine-coordinated ferrous species   427
amine coordinated Ir(I) complex   397
amine-N-oxide coordinated Au(I) species   601
aminimides   582
aminocarbonylation   230
amino ferrous   441
amino ferrous complex   434
amino ferrous species   429, 435, 437
amino Ir(III) intermediate   406
aminomethyl indole   526
amino Ru(II) species   474
ammonia-borane Lewis acid-base complex   437
ammonium iridate species   397
amphiphilic palladium acetate   212
Amsterdam Density Functional (ADF)   39
animo olefin coordinated IrI species   397
anionic fluoride Cu(I) species   522
anionic sulfonate   197
anionic trifluoromethyl group   524
π antibond   5
anti-Markovnikov product   470
arene oxidation   422
arenes   403, 407
arenes/indole derivatives   293
arenes, nucleophilic attack by   606
aromatic chromenylium   601
aromatic fluorocarbons   467
aromatic ketones   452
aromatic N-heterocycles   582
arylalkyl Ru(II) intermediate   461
arylation of carbenes   357
arylation-product-coordinated Rh(I) complex   332
aryl boranes   459
aryl bromide   459
aryl Cu(I) intermediate   527
aryldiazonium salts   408
β-aryl elimination   88, 343

aryl halides   457
aryl iron(0) intermediate   438
aryl-Rh(III) intermediate   332, 337
aryl Rh-nitrene complex   361
aryl sulfamates   186
atom-economical reactions   204
atomic orbitals   21
Au-catalysis   258
Au(I) catalyzed non-redox unsaturated C—C bonds functionalization   597
Au-mediated alkene activations
  allylic substitutions   608–609
  nucleophilic addition of alkenes   607–608
Au-mediated alkyne activations
  isomerization of alkynes   599
  nucleophilic attack by arenes   606
  nucleophilic attack by nitrogen-involved nucleophiles   602–606
  nucleophilic attack by oxygen involved nucleophiles   599–602
  trans-nucleophilic attack   598
Au-mediated allene activations
  cycloisomerization of allenyl ketones   613
  hydroalkoxylation of allenes   610–612
  hydroamination of allenes   610
Au-mediated enyne transformations
  allenyne cycloisomerizations   617
  conjugative enyne cycloisomerizations   617–618
  1,5-enyne cycloisomerization   614–615
  1,6-enyne cycloisomerizations   615–616
azaindoline   334
7-azaindoline cocatalyst   335
azaindoline-coordinated Rh(I) complex   334
azepanone product   602
azide–alkyne cycloadditions   533
azide-coordianted alkynyl Ru(II) species   479
aziridination of nitrenes   361
aziridination of olefins   580

b

back-donation bonds 53
back-donation π-bond formation 567
base-assisted C—H bond cleavage 453
base-assisted internal electrophilic-type substitution (BIES) 452
benzaldehyde 467
benzamides 332
benzamide substrate 296
benzoate-assisted reductive deprotonation 526
benzoic acid-coordinated sextet ferric(III)–hydroperoxo complex 423
benzyl alcoholate Ru(II) intermediate 469
benzylalcohol product 467
B(boryl)–H(hydride) elimination 402
biaryl coordinated iron(I) species 441
biaryl Cu(III) intermediate 520
biaryl moiety 186
[2,2,2]-bicyclic intermediate 601
bicyclo[3.1.0]hexene 614
bidentate olefin ligands 330
bidentate phosphine 134
bidentate phosphine-amine ligand coordinated amino Mn(I) hydride 508
bidentate phosphine ligand 129, 146, 147
bifunctionalized carbenoid compound 568
bimetallic reductive elimination 75–77
bimetallic single-electron reductive elimination mechanism 538
binuclear copper catalyzed azide alkyne [3+2] cycloaddition 534
binuclear Cu(I) species 543
bisaryl-Cu(III) intermediate 551
bisaryl Ir(V) intermediate 408
bisboryl Rh(III) intermediate 345
bisphosphine-coordinated allylic Pd(II) complex 74
bisphosphinesilver 584
bis-silylation 202
bissilylethane derivatives 202
π-bonding electrons 4
Born–Oppenheimer approximation 20, 22
boryl aryl hydride Ir(III) intermediate 403
boryl-copper complexes 549
boryliminoacyl intermediate 342
boryl Ir species 387
borylisocyanide-coordinated phenyl Rh(I) complex 342
boryl Rh(I) species 342, 343
branched allylic ether 611
bromamine-T 577
Brønsted acid 437
Brønsted base-assisted isomerization of alkynes 599
butylbenzene-coordinated Ir(III) complex 391

c

carbene 82
carbene insertion into C—Cl bond 572–573
carbene insertion into C—H bonds 261–264, 354–357, 574–576
carbene insertion into O—H bond 573
carbene insertions 541–542
carbenes rearrangement 542
carbenoids 332
carbonate hydrogenations 508
carbonate Ir(III) complex 405
carbon-cyano bond actviation 342
carboncycles 362
carbon dioxide 430
carbon dioxide hydrogenation 507
β-carbon elimination 89, 343
carbon-halogen bonds 5, 184
carbon monoxide 81, 229
carbon radical attacks 75
carbonyl compounds 5, 80, 391
carbonyl-coordinated aryl Rh(I) intermediate 343
carbonyl-coordinated five-membered rhoda(III)cycle 349

Index | 633

carbonyl hydrogenation 465
carbonyl insertion 214
carbonyl oxygen atoms 573
carbonyls 429
carbonyls, Cu-catalyzed borylations 553
carboxylate-assisted C—H bond activation 128
carboxylic anhydrides 184
C2-arylated indole intermediate 332
catechols 425
cationic Au(I) species 613
cationic Cp*Rh(III) complex 337
cationic ferrous intermediate 429
cationic iorn(II) intermediate 429
cationic iron(IV)–oxo complex 423
cationic oxidant 63
cationic quinolinium intermediate 396
cationic Rh(I) catalyzed dihydrogenation of alkenes 347
cationic Rh(I) species 352
cationic Ru(IV) hydride intermediate 464
C(aryl)—B reductive elimination 403
C(vinyl)—C(vinyl) bond 183
C(aryl)-C(alkyl) bond formation 332
C—C/C—hetero bonds 329
C(aryl)—C(aryl) cross-coupling reaction 408
C—C cross-couplings with alkyl halide 441
C—C cross-couplings with aryl halide 438
C(aryl)—Cl bond 184
cesium pivalate 330
C—F bond couplings 522
C(aryl)—F reductive elimination 524
β-C(alkyl)—H activation 408
C—H activation of alkynes 587–588
C(aryl)—halide bond 519
C—H amidations 580
C—H aminations 529
C—H arylations 527
C—H azidations 501
C—H bond arylation, Ru-catalyzed 456
C—H bond cleavage 330

C—H bond oxidative addition 452
chelated Rh(I)–NHC complex 334
chelation-assisted meta-C(aryl)—H activation 214–216
chelation-free C(sp^2)—H activation 208–211
chelation-free C(sp^3)—H activation 206–207
C—H etherifications 532
C—H halogenations 501
C—H hydroxylations 500, 531
C—H isocyanations 501
chloramine-T 577
chloroaryl triflate 184
chloronium ylide intermediate 572
chloronium ylide moiety 572
C—H metalation 223
C(alkyl)—H reductive elimination 398
click (3+2) cyclization 478
C—N bond activation 344
C—N bond couplings 520
C—N cross-couplings with aryl halide 438
cobalt (Co) 289
C(arene)—O(oxo) bond 423
C—O bond activation 345
C—O bond coupling 522
C(alkyl)—O(hydroxide) bond formation 420
Co-catalysis, theoretical study of
  co-catalyzed hydroformylation 307
    direct hydroformylation, by H$_2$ and CO 308
    transfer hydroformylation 309–310
  co-catalyzed hydrogenation 301
    of carbon dioxide 301–304
    hydrogenation, of alkenes 304–306
    hydrogenation, of alkynes 306–307
  co-mediated carbene activation 310
    arylation, of carbene 310–312
    carboxylation, of carbene 312–313
  co-mediated carbene activation
    arylation, of carbene 312
Co-mediated C—H bond activation 289, 290

Co-catalysis, theoretical study of (contd.)
  hydroarylation, of alkenes 291–293
  hydroarylation, of alkynes 294–296
  hydroarylation, of allenes 293–294
  hydroarylation, of nitrenoid 296–297
  oxidative C—H alkoxylation 297
  Co-mediated cycloadditions 297–298
  co-catalyzed [2 + 2] cyclizations 300–301
  co-catalyzed [2 + 2 + 2] cyclizations 299–300
  co-catalyzed [4 + 2] cyclizations 299
  Pauson-Khand reaction 298–299
  co-mediated nitrene activation 313, 314
    amination, of C—H bonds 315
    amination, of isonitriles 314–316
    aziridination, of olefins 314
co-catalyzed [2 + 2 + 2] cyclizations 299–301
co-catalyzed [4 + 2] cyclizations 299–300
complete basis set (CBS) methods 23
concerted metalation-deprotonation (CMD) 206, 258, 290, 452, 505
concerted oxidative addition 60
concerted reductive elimination 71–73
conductor-like PCM (C—PCM) 33
Conia-ene type addition 367
co-nitrene complex 313, 314
conjugative diene 599
conjugative enyne cycloisomerizations 617–618
conjugative insertion 83–85
conjugative olefin 469
conjugative triene product 617
conjugative unsaturated bonds 83
conventional organocatalysis 6
coordinative chelation-assisted C(sp^3)—H activation 216, 217
copper 517
copper-based coupling reactions 518
copper catalytic cycle 518

copper-catalyzed benzene oxidation reaction 531
copper(II) dichloride 537
copper-mediated cyclopropanation 540
correlation-consistent basis sets 31, 32
coupled cluster (CC) methods 39
coupled cluster theory 39
coupling 183
covalent chelation-assisted ortho-C(aryl)—H activation 212, 213
Cp*Rh(III)–catalyzed C—H bond alkylations 335
CpRh(III) species 361
CpRu(II) catalyzed alkyne trimerization 476
[CpRu(Py)$_2$(PPh$_3$)]PF$_6$ 463
CpRu(II) species 475
C(aryl)—Rh bond 332
C—Rh bond transformation 330
cross-coupling reactions 4
Cu(I) catalyzed allylic trifluoromethylation 525
Cu-catalyzed borylations
  alkenes 551–552
  alkynes 552–553
  carbonyls 553–554
Cu(I)–catalyzed [3+2] cycloadditions 533
Cu-catalyzed hydrofunctionalizations
  hydroborylations 547
  hydrocarboxylations 548–549
  hydrosilylation 547–548
Cu(II) dicarboxylate complex 526
Cu-mediated alkyne activations
  alkynyl Cu transformations 536–539
  azide–alkyne cycloadditions 533–535
  nucleophilic attack onto alkynes 535–536
  schematic of 533
Cu-mediated carbene transformations
  carbene insertions 541–542
  [2+1] cycloadditions with alkenes 539–541
  rearrangement of carbenes 542
Cu-mediated C—H activations

C—H aminations 529
C—H arylations 527
C—H etherifications 532
C—H hydroxylation 531
Cu-mediated nitrene transformations
  amination of nitrenes 543–544
  [2+1] cycloadditions with alkenes 543
  nitrene insertions 544–545
Cu-mediated trifluoromethylations
  radical type trifluoromethylations 525–527
  through cross-coupling 524
  through oxidative coupling 524–525
Cu-mediated Ullmann condensation
  C—F bond couplings 522–524
  C—N bond couplings 520–522
  C—O bond coupling 522
  coupling reaction of aryl halides and nucleophiles 519
  single electron transfer involving pathway 519
cumulene-type Lewis electron structure 569
cupric carboxylate 552
cuprous borylcarboxylate complex 554
cuprous boryl complex 552, 554
cuprous borylmethanolate 554
cuprous methoxide species 553
Curtin–Hammett principle 570
cyclepentadiene derivatives 330
(2+2+2) cycloadditions 475–478
(3+2) cycloadditions 365–367
(5+2) cycloadditions 367–369
(5+2+1) cycloadditions 369
[2+1] cycloadditions with alkenes 539–541, 543
cyclobutanol 343
cyclobutanolate Rh(I) species 343
cyclobutenones 204
cyclobutyl substituted 1,5-enyne 614–615
cycloheptatriene product 369
cycloisomerization of allenyl ketones 613
cyclometalation intermediate 212

cyclopentadiene 617
cyclopentadienyl (Cp) ligands 330
cyclopentenone derivatives 349
cyclopentenone product 478
cyclopropanation 266–268
cyclopropanation of carbenes 358, 359
cyclopropenes 273

d

dearomatic alkyl Cu(III) intermediate 528
dearomatized amino Ru(II) intermediate 468
decarbonylation 224
decarbonylation reaction 90
decarboxylation transition state 145
dehydrogenative coupling 511
density functional theory (DFT) 23, 24, 259, 570
  Jacob's ladder, of density functionals 25
  correction, of dispersion 27–29
  fifth rung 26
  fourth rung 26
  second rung 25–26
  third rung 26
deprotonation-metallation mechanism 527
dialkylbiarylphosphine ligand 186
dialkyl orthoacetate 574
diaryl ferric intermediate 441
diaryl-Rh(III) complex 332
diaryl-Rh(III) intermediate 332
diaryl Ru(II) complex 459
diazo compound 310
diazo compounds 310
α-diazonitrile intermediate 542
diazonium salt 521
diboryl alkane 349
dichloromethane (DCM) 572
dielectric formulation (D-PCM) 33
1,6-diene 484
diene insertion 167
diethylene-coordinated Pd(0) species 232

dihalosilanes   583
dihydride iron(II) intermediate   427
dihydride Rh(III) intermediate   347, 352
dihydride Rh(III) species   347, 353
dihydrogen-coordinated cationic ferrous intermediate   437
dihydrogen-coordinated cationic Ir(III) complex   396
dihydrogen-coordinated Ir(III) dihydride catalyst   389
dihydrogen molecule-coordinated iron(0) complex   427
dihydroquinoline product   606
dimeric ferrous complex   438
dimeric Rh(II)   330
N,N-dimethylformamide (DMF)   532
diphenylethanimine   395
(E)-1,2-diphenyl-1-propene   389
diphosphine ligand   304
diphosphine ligand Xyl-MeO–BIPHEP   403
dipolar carbonyl ylide   601
direct hydrogenation   391
directing group-coordinated Mn(I) species   504
dirhodium catalyst   361
dirhodium tetracarboxylate complexes   356
dispersion effect   35
dissociative mechanism   57
distortion-interaction analysis   184
1,4-disubstituted 1,2,3-triazoles   533
d orbital   5, 41, 52, 53, 82
doublet iron(V)–oxo benzoate complex   423
double-zeta (DZ) basis   30
Drude oscillators   27

e

effective core potential (ECP)   32
eight-membered Rhoda(III)cycle   368
electron-deficient olefins   335
electronegativity, of carbon atom   206
electron-electron repulsion   21
electron pair filling one sp^2 hybrid orbital   5
electron-poor phosphine ligand   330
electron-rich arenes   606
electrophile   456
electrophilic carbon subtracts   181
electrophilic deprotonation   503
electrophilic proton   397
electrophilic silver–silylene complexes   583
α-elimination   90, 91
β-elimination   87–89, 91
enamine type 1,4-dihydrogenated intermediate   396
enamine type 1,4-dihydroquinoline   396
enantioselective direct hydrogenation reactions of ketones   391
6-endo nucleophilic attack   616
ene-vinylcyclopropane   369
enolate-Rh(III) complex   335
enolate tautomer   572
entropy effect   41
1,5-enyne cycloisomerizations   614
1,6-enyne cycloisomerizations   615
enynol   365
equivalent formate Rh(III) hydride complex   353
ester hydrogenation   466
ethanol   467
etherification product coordinated gold complex   608
ethynylbenzene reactant   262
ethynyl-benzylalcohol   261, 262
exact solvation effect   41
exo-addition vinyl copper intermediate   536
5-exo alkene   616

f

Fe-catalyzed coupling reactions
  C—C cross-couplings with alkyl halide   441
  C—C cross-couplings with aryl halide   438

C—N cross-couplings with aryl halide 438–441
iron-mediated oxidative coupling 441
Fe-mediated dehydrogenations
  alchohols 434–435
  ammonia-borane dehydrogenation 437–438
  formaldehyde 435
  formic acid 435–437
Fe-mediated hydrofunctionalizations
  hydroamination of allenes 432–434
  hydrosilylation of ketones 431–432
Fe-mediated hydrogenations
  alkenes 427–429
  carbon dioxide 430–431
  carbonyls 429–430
  imines 430
Fe-mediated oxidations
  alkane oxidation 420–422
  alkene oxidations 423–425
  arene oxidation 422–423
  iron-oxo species 420
  oxidative catechol ring cleavage 425–426
ferric(IV) hydroxide intermediate 420
Fischer-type metallacarbene complex 568
Fischer type Rh-carbene complex 354, 358, 359
five-coordinated Cu(III) species 522
five-membered rhoda(III)cycle 344
five-membered rhodacycle complex 339
five-membered-ring CMD-type transition state 452
five-membered ring type allylic carbocation 617
formaldehyde-coordinated hydride Ir(III) complex 405
formaldehyde dehydrogenation 435, 473
formate-coordinated cationic Mn(I) species 508
formate-Ru(II) intermediate 474
formic acid dehydrogenation 435, 474
four-coordinate anionic Rh(I) complex 331

four-coordinated Cu(III) intermediate 522
four-membered cyclic σ-CAM-type transition state 456
four-membered rhodacycle 359
free electron gas 24
free radical 67
frontier molecular orbitals (FMOs) 60
fused pyrrole coordinated Au(III) complex 616

g

gaseous dihydrogen 102, 140, 157, 302, 353, 388, 391, 396
Gaussian-n methods 23
Gaussian-type orbitals (GTOs) 29
generalized gradient approximation (GGA) 24
Gibbs free energy 55
gold-assisted alkyne activations 599
Grubbs II type catalyst-mediated intermolecular olefin metathesis 482

h

Hamiltonian operator 19
Hartree–Fock (HF) theory 20–22, 39
Hartwig–Buchwald amination reaction 519
Heck–Mizokori reaction 183
heme iron enzymes 419
hetero agostic complex 432
β-heteroatom 88
heteroatoms 265
hetero-Diels-Alder [4+2] cycloaddition 601
hexenylidene Pt intermediate 267
high-spin triplet Rh(III)–nitrene complex 360
high spin trispyrazolylborate-coordinated Cu(II)–oxo radical 532
high valence transition metal 5
H-Ir covalent bond 397
Hiyama coupling 183, 190, 192, 193
η^6-oligoacene coordinated $Cr(CO)_3$ 54

η¹-oligoacene coordinated intermediate 54
η³-oligoacene coordinated transition state 54
homogeneous catalytic dehydrogenation of alcohols 472
hydride–alkyl reductive elimination 389
hydride-coordinated Ru(II) intermediate 474
hydride copper 545, 548
β-hydride elimination 210, 294
β-hydride elimination yields 227
hydride Ir(III) complex 397
hydride Rh(III) intermediate 351
hydride Ru(II) species 467
hydroacetoxylation of alkynes 351
hydroacylation of ketones 353
hydroacylations 469
hydroalkoxylation of allenes 610
hydroamination of allenes 610
hydroamination reaction 270
hydroborations 471
hydroborylations 547
hydrocarboxylations 470, 548
hydrocylation of alkynes 349
hydrodefluorination of fluoroarenes 467
hydrogenation of alkenes 346
hydrogenation of carbon dioxide 353
hydrogenation of ketones 352
hydrogen-bonded six-membered ring type intermediate 608
β-hydrogen elimination 78, 88, 268
hydrosilylation 547
hydrosilylation of ketones 431
hydrothiolation of alkynes 351
hydrotris(pyrazolyl)borate (Tpx) 580
hydroxymethanolate anion 473
hypervalence silicon intermediate 432

i

imines 5, 393, 430
imine type 3,4-dihydroquinoline 396
iminium intermediate 601
iminoiodinanes 577
imino pyrrolone 213
indolyl gold intermediate 603
inert $C(sp^3)$—H bonds 500
inner-sphere oxidative addition 235
insertion reaction 78
in-situ generated benzoic acid 471
integral equation formalism PCM (IEF-PCM) 33
intermolecular hydrogen transfer 156
intermolecular olefin metathesis 482
intramolecular C—C bond formation 267
intramolecular [2+2] cycloaddition 484
intramolecular diene metathesis 484
intramolecular nucleophilic 5-exo-attack 536
intramolecular phenylsulfane oxidative addition 198
intramolecular radical type reduction 75
intrinsic reaction coordinate (IRC) 260
Ir-catalyzed aminations
   alcohols 405–407
   arenes 407
Ir-catalyzed borylations
   alkanes 402–403
   arenes 403–405
Ir-catalyzed C—C bond coupling reactions 407
Ir-catalyzed hydroamination 397
Ir-catalyzed hydroarylations 397–399
Ir-catalyzed hydrogenations
   alkenes 388–391
   carbonyl compounds 391–393
   imines 393–396
   quinolines 396–397
Ir-catalyzed hydrosilylation 399–401
Ir-C(aryl) bond 398
Ir(PHOX) complex 389
Ir(III) dihydride complex 389
Ir-mediated dihydrogenation 387
Ir(III)–nitrene complex 407
iron-catalyzed Kumada type coupling reaction 438
iron-mediated oxidative coupling 441
iron-mediated oxidative *ortho*-hydroxylation of benzoic acid 423

iron(V)–oxo complex   420
iron-oxo species   420
Ir-(PHOX)–mediated imine hydrogenation   394
Ir(III) trihydride complex   396
isocyanatoalkane   502
isocyanide   231
isocyanide insertion   213
isodensity PCM (I-PCM)   33
isomerization of alkynes   599
isonicotinamide   213
isonitrile   213
isopropoxy Ru(II) intermediate   460
isoquinoline product   507

k

ketone α-alkylation reaction   334
kinetic isotope effect (KIE)   332
Knölker's type iron(0) complex   430, 432, 434
Kumada coupling   7, 438

l

β-lactams   344
Lewis acid   597
  additives   142
  property, of Pt   257
ligand-free reaction condition   232
linear combination of atomic orbitals (LCAO) theory   21
lithium silylphenyl methanolate intermediate   192
local-density approximation (LDA)   24
local spin density approximation (LSDA)   24

m

Mäbius aromaticity   476
manganese   499
Markovnikov type adduct   470
maximum electron correlation energy   31
metalation-deprotonation mechanism   529
metal-(η²-silane) interactions   53

metal-involved intramolecular Alder-ene type reaction   268
metallacarbene complexes   82
metallacyclobutane intermediate   102, 103, 539, 541
metallacyclopentatriene intermediate   476
metal-organic bond   84
metal-organic cooperative catalysis   334
metal–silicon double bond   583
metathesis–insertion type Ir(III)oendash Ir(V) catalytic cycle B   390
metathesis type C—H bond cleavage   421
methanediol   435, 511
methanol-assisted H—H bond cleavage   509
methanol assisted 1,3-hydrogen transfer   611
methanol-coordinated iron(I) complex   421
methoxide cationic Ru(II) complex   473
methoxy Cu(I) σ-complex   527
methoxyethanolate   392
methoxy Mn(I) intermediate   509
methylene pyridine intermediate   542
methylene pyridinide moiety   467
methylene-Ru(II) species   484
methyl ferric(III) hydroxide intermediate   421
8-methylquinolines   337
Mg-Fe(-2) species   438
minima-energy cross-point (MECP)   68, 75
Mizoroki–Heck reaction   79, 87
Mn(IV) azide species   501
Mn-catalyzed C—H activation-alkylation   503
Mn-catalyzed ligand-assisted hydrogenation   508
Mn(I)–hydride species   509
Mn(IV) hydroxide   500
Mn-mediated C—H activations
  concerted metalation-deprotonation   505–507
  electrophilic deprotonation   503–504

Mn-mediated C—H activations (contd.)
  σ-complex assisted metathesis 504–505
Mn-mediated dehydrogenation
  alchohol dehydrogenation 510–511
  dehydrogenative coupling 511–512
Mn-mediated hydrogenations
  carbonates 508–509
  carbon dioxide 507–508
Mn-mediated oxidation of alkanes
  C—H azidations 501
  C—H halogenations 501
  C—H hydroxylations 500–501
  C—H isocyanations 501–502
  general mechanism 500
modified Chalk-Harrod pathway 398
modified Ullmann reaction 519
molecular mechanics (MM) force fields 27
molecular vinylcyclopropane substrate 167
Møller–Plesset perturbation theory 22, 23, 39
monodentate phosphine 147
monomeric diamino ferrous species 438
mononuclear Cu(I) acetate complex 535
muconic acid 426

n

N amidation 580
naphthaldehyde 601
N-aryl-α-diazoamides 569
N-bromosuccinimide (NBS) 62, 340
Negishi coupling 7, 186, 188, 189
neutral aryl Mn(I) intermediate 504
neutral/cationic iron(IV)–oxo complex 423
neutral four-membered-ring ruthenacycle 464
neutral gold carbene 613
neutral Ir(III) hydride complex 394
neutral Ir(III) trihydride complex 396
neutral vinyl ferrous species 432
N—H bond cleavage/formation 392
NHC–oxazole ligands 390
N-hetero carbene IMes[Me] 345

N-heterocyclic carbene 148, 186
Ni(0)–carbene catalyzed intramolecular (5+2) cycloadditions of enynes 161
Ni-catalysis, theoretical study of
  Ni-mediated C—C bond cleavage
    C=C double bond activation 152–153
    C—C single bond activation 151–152
  Ni-mediated C—halogen bond cleavage 133
    by β-halide elimination 137–139
    concerted oxidative addition of C—Halogen bond 133–135, 137
    nucleophilic substitution 139
    radical type substitution, of C—halogen bond 135–137
  Ni-mediated C—H bond activation 128
    Ni-mediated aldehyde C—H activation 132–133
    Ni-mediated arene C—H activation 128–132
  Ni-mediated C—N bond cleavage 148–150
  Ni-mediated C—O bond activation 140
    ester C—O bond activation 142–148
    ether C—O bond activation 140–143
  Ni-mediated unsaturated bond activation 153
    electrophilic addition 156–158
    nucleophilic addition 159, 160
    oxidative cyclization 153–155
    unsaturated compounds insertion 157–159
Ni-catalyzed Negishi type cross-coupling reactions 135, 136, 138
Ni-catalyzed olefin hydroarylation 129
nickel catalyst 143
nickel hydride 159
Ni-mediated α-aryl elimination 91

Ni-mediated concerted metalation-deprotonation type C—H bond activation 128
Ni-mediated cyclization 160
  Ni-mediated cycloadditions 161–164
  Ni-mediated ring extensions 166–168
  Ni-mediated ring substitutions 163–166
Ni-mediated ring substitutions 164
nine-membered Rhoda(III)cycle 369
N-iodosuccinimide (NIS) 340
nitrene 83, 576
nitrene insertion into C—H bonds 360
nitrene insertions 544
nitrenoids 577
nitrogen-involved nucleophiles 602
nonheme cationic ferric(III)–hydroperoxo complex 423
nonheme iron complexes 423
non-redox catalytic cycles 271
non-redox mechanism 304
norbornadiene derivatives 330
N-oxide-directed CMD-type C—H bond cleavage 335
N-oxide-directed electrophilic deprotonation 10
N-oxide moiety 335
Noyori-type catalysis 392
N-phenoxyacetamides 339
N-propargyl hydrazine 586
nuclear-nuclear repulsion energy 20
nucleophiles 4
nucleophilic addition of alkenes 607–608
nucleophilic attack by amines 582–583
nucleophilic attack by arenes 606–607
nucleophilic attack by carbonyl groups 573
nucleophilic attack by nitrogen-involved nucleophiles 602
nucleophilic attack by oxygen involved nucleophiles 599
nucleophilic attack by unsaturated bonds 580
nucleophilic attack onto alkynes 535
nucleophilic carbon 196
nucleophilic hydride 397
nucleophilic substitution 57

O

o-alkynylbenzacetal derivatives 265
O–atom transfer 335
o-azidophenyl methanimine 544
octene-coordinated Ru-carbene complex 482
olefin aziridination 580
olefin-coordinated hydride Ru(II) species 469
olefin-coordinated Ru(II) intermediate 471
olefin coordinated silyl Ir(III) complex 401
olefin insertion 291
olefin metathesis 102
orbital interaction 184
organic azides 313, 339, 577
organometallic chemistry
  computational methods in
    accuracy, of DFT methods 40–41
    basis set 29, 30, 36–37
    computational programs 37–40
    correlation-consistent basis sets 31–32
    density function 34–36
    DFT methods 23–29, 34
    diffuse functions 31
    entropy effect 41
    exact solvation effect 41
    excited state 41
    history of, quantum chemistry 19–21
    mechanism 37
    polarization functions 31
    pople's basis set 30
    post-HF methods 21–23
    pseudo potential basis sets 32
    reaction mechanism 41–42
  coupling reactions 4
  electronegativity, of carbon 4
  history 3, 6–8
  mechanism of, transition metal catalysis 8–12
  synthetic chemistry 3

organometallic chemistry, elementary
reactions 51, 52
   coordination bond 52–55
   dissociation 55–57
   ligand exchange 57–59
   oxidative addition 59, 60
      concerted 60–63
      oxidative cyclization 68–70
      radical type addition 67, 69
      substitution-type oxidative addition
         62–64, 66–68
organosilicon compounds 190
ortho-alkenylation of arenes 461
ortho-alkenylations 293
ortho-alkylation of arenes 460
ortho-alkynylbenzamides 536
ortho-aminated products 339
ortho-phenylpyridine 292
outer-sphere σ-bond metathesis 432
oxazinedione 165
oxazinoquinolinium-coordinated Rh(I)
   complex 335
oxazolidinone-coordianted Cu(I) complex
   75
oxidative addition, of C—H bond 259
oxidative catechol ring cleavage 425
oxidative cyclization 68
oxidative Heck-type coupling 335
oxime esters 329
oxime ethers 329
oxiranyl propargylic esters 265
β-oxoalkyl silver character 572
oxonium intermediate 606, 610
oxonium ylide intermediate 574
oxooxazolidinyl directed arenes and
   acrylates 462
oxygen involved nucleophiles 599

p

para-boralated benzene 405
Pauson–Khand reactions 227, 298
Pauson–Khand type (2+2+1)
   cycloadditions 367, 478
(Xantphos)Pd(CH$_2$NBn$_2$)$^+$ 10
Pd-catalysis, theoretical study

Pd-catalyzed cross-coupling reactions
   182, 183
   Heck-Mizokori reaction 192–196
   Hiyama coupling 190–193
   Negishi coupling 186–189
   Stille coupling 189–191
   Suzuki-Miyaura cross-coupling
      reaction 183–186
Pd-mediated activation, of unsaturated
   molecules 224
   alkene activation 225, 227
   alkyne activation 225–227
   carbene activation 231–233
   CO activation 229–231
   enyne activation 226–228
   imine activation 229
   isocyanide activation 231
Pd-mediated C—H activation reactions
   204, 206
   C—H bond activation, electrophilic
      deprotonation 219–221
   C—H bond activation, oxidative
      addition 223–224
   C—H bond activation, σ-complex-
      assisted metathesis 221–223
   chelation-assisted *meta*-C(aryl)—H
      activation 214–216
   chelation-free C(sp^2)—H activation
      208–211
   chelation-free C(sp^3)—H activation
      206–208
   coordinative chelation-assisted
      C(sp^3)—H activation 216, 217
   coordinative chelation-assisted
      ortho-C(aryl)—H activation
      210–212
   covalent chelation-assisted
      C(sp^3)—H activation 216–219
   covalent chelation-assisted
      ortho-C(aryl)—H activation
      212–214
Pd-mediated C—hetero bond formation
   196
   C—B bond formation 196–197
   C—I bond formation 200–201

C—S bond formation   197–200
C—Si bond formation   201–204
Pd-catalyzed C—H activation, of methane   207
Pd-catalyzed cross-coupling reactions   60
Pd-catalyzed intramolecular carbene insertion   223
Pd-catalyzed Negishi coupling   187
Pd-catalyzed olefin hydroarylation reaction   225
Pd-C bond   181
pentafluoropyridine   468
Perdew–Burke–Ernzerh (PBE)   26
pericyclic process   51
peroxo-bridged iron(IV) complex   425
peroxo iron(IV) complex   425
phenol-coordinated Cu(I) species   522
phenyl acetate   143
phenylacetylene-coordinated Ru(II) species   471
phenyl benzooxazole product   528
phenylboronate   148
phenylboronic acid   144
phenyl bromide   210
phenylethane product   429
phenylethynyl Ru(II) species   471
phenyl iron(IV) intermediate   438
2-phenylpyridine   310, 339, 452
phenyl pyrrole product   520
phenylpyrrolidine   357
phenyl-Rh(III) intermediate   331
phenyl Ru(II) intermediate   460
phenyl(trifluoromethyl)sulfane   135
1-phenylvinyl benzoate product   471
phenyl yrazole product   521
phosphate-assisted intermolecular C—H bond activation   145
phosphine-coordinated Ir-catalyzed tetrahydrogenation   396
phosphine-coordinated Ru(II)–carbene species   484
phosphine ligand   105, 201
phosphinite   452
phosphinooxazoline ligand   389
pincer amino Mn(I) species   511
pincer bis[2-(diisopropylphosphino)-ethyl]amine (HPNP) ligand coordinated Ru(II)–hydride complex   472
pincer ligand coordinated Ru(II) catalyst   466
PNP type pincer ligand   427
PNP type pincer ligand-coordinated iron(II) species   437
porphyrin-coordinated Mn(III) species   500
porphyrin-coordinated Mn(V) species   500, 501
post-HF linear variational method   22
primary alkyl iodide   63
product-coordinated Mn(I) species   504
propargylamine   587
propargyl ester reactant   507
propargylic acetoxyl enyne   617
propenaminium   236
protocatechuate 3,4-dioxygenase   425
proton accepter   221, 434
proton concerted electron transfer (PCET)   290
pseudo potential basis sets   32
Pt-catalysis, theoretical study
   Pt-catalyzed alkene activation   270
      cyclopropenes, isomerization of   273–274
      hydroamination, of alkenes   270–273
   Pt-catalyzed alkyne activation, mechanism   264
      cyclopropanation   266–267
      nucleophilic additions   264–266
      oxidative cycloaddition   268–270
   Pt-catalyzed C—H activation, mechanism of   258
      carbene insertion, into C—H bonds   261–264
      electrophilic dehydrogenation   259, 260
      oxidative addition, of C—H bond   259

Index

Pt-catalyzed cycloisomerization, of enynes   266
Pt-mediated alkene functionalizations   270
Pt-mediated C—H activations   258
pyrazolyl Cu(III) intermediate   521
pyridinone   165
pyridotriazole   359
pyridyl Ru vinylidene   464
pyrocatechol   425
pyrrolyl   520

q

quantum chemical computation   10
quantum chemistry composite methods   22
quinoline-N-oxide   329, 335
quinolines   396

r

radical-substitution-type reductive elimination   74–76
radical type substitution, of C—halogen bond   135–138
radical type trifluoromethylations   525–527
rearrangement of carbenes   542, 543
redox-neutral cross-coupling   4, 329, 330, 457, 524, 527, 533
redox-neutral process reaction   232, 233
reductive elimination   70, 198
   α-elimination   90, 91
   β-elimination   86–91
   bimetallic reductive elimination   75–78
   concerted reductive elimination   71–73
   eliminative reduction   77–78
   insertion   78
     conjugative insertion   83–85
     1,1-insertion   80–84
     1,2-insertion   79–81
     outer-sphere insertion   84–86
   radical-substitution-type reductive elimination   74–76
   substitution-type reductive elimination   73–75
reductive elimination yields   135, 139, 140, 150, 161, 166, 167, 183, 188, 196, 219, 222, 224, 228, 232, 300, 342, 345, 351, 478, 519
regioselectivity   140, 145–147, 184, 185, 214, 216, 334, 470, 573, 598
relative enthalpy   53, 54, 316, 431
relative free energy   55, 62, 83, 140, 144–146, 185–186, 214, 220, 261, 311, 343, 349, 369, 397, 456, 463, 521, 524, 543, 554, 571, 578–579, 581–583, 599
Rh(acac)$_3$ 332
Rh-assisted intramolecular cope-type 3,3-sigmatropic migration   367
Rh-catalyzed addition reactions of carbonyl compounds
   hydroacylation of ketones   353–355
   hydrogenation of carbon dioxide   353–354
   hydrogenation of ketones   352–853
Rh-catalyzed alkene functionalizations
   alkene diboration   347–349
   hydrogenation of alkenes   346–347
Rh-catalyzed alkenylation of C—H bond   335–338
Rh-catalyzed alkyne functionalizations
   alkyne hydroamination   349–351
   hydroacetoxylation of alkynes   351
   hydrocylation of alkynes   349–350
   hydrothiolation of alkynes   351
Rh-catalyzed amination of C—H bond   338–340
Rh-catalyzed arylation of C—H bond   330–332
Rh-catalyzed carbene transformations
   arylation of carbenes   357–358
   carbene insertion into C—H bonds   354–357
   cyclopropanation of carbenes   358–359
   cyclopropenation of carbenes   359
Rh(I)–catalyzed carbonyl insertion reaction   344

Index

Rh-catalyzed C—C bond activations and transformations
  β-carbon elimination 343
  carbon-cyano bond actviation 342–343
  strain-driven oxidative addition 341
Rh-catalyzed C—H bond alkylation 332, 335
Rh-catalyzed cycloadditions
  (3+2) cycloadditions 365–367
  (5+2) cycloadditions 367–369
  (5+2+1) cycloadditions 369
  Pauson–Khand type (2+2+1) cycloadditions 367
Rh-catalyzed halogenation of C—H bond 340–341
Rh(III)–catalyzed intermolecular arene C—H bond amination 339
Rh-catalyzed nitrene transformations
  aziridination of nitrenes 361–362
  nitrene insertion into C—H bonds 360–361
Rh(III)–catalyzed ortho-bromination/iodination of arenes 340
Rh(III)–catalyzed oxidative C—H/C—H cross-coupling 332, 333
Rh(III)–catalyzed oxidative Heck-type coupling reaction 335, 337
Rh-catalyzed redox-neutral cross-coupling reaction 329
Rh-C(alkenyl) bond 368
Rh-C(vinyl) bond 11, 349
[Rh(coe)$_2$Cl]$_2$-catalyzed C2-selective arylation of indoles 330
Rh(III)–hydride intermediate 337
Rh(III)–hydride species 334
Rh-mediated C-hetero bond activations
  C—N bond activation 344–345
  C—O bond activation 345–346
Rh(V)-nitrene complex 340
Rh-nitrene complexes 338, 339
rhoda(III)cycle 342, 344, 349, 354, 363, 367–369
ring-expanded [3.3.0]bicyclic product 615

ring-opened Ru-vinylcarbene complex 484
[Ru(NHC)(CO)(H$_2$)(PPh$_3$)$_2$] 468
Ru-carbene complex 104, 451, 482
Ru(II)–carbene complex 482, 484
Ru carbenoid moiety 464
Ru(II) carboxylate species 460
Ru-catalyzed acetamide-directed C—H bond alkenylation 454
Ru-catalyzed cycloadditions
  click (3+2) cyclization 478
  (2+2+2) cycloadditions 475–478
  Pauson–Khand type (2+2+1) cycloadditions 478–479
Ru-catalyzed hydrofunctionalization
  hydroacylations 469–470
  hydroborations 471–472
  hydrocarboxylation 470–471
Ru-catalyzed hydrogenations
  alkenes 464–465
  carbonyls 465–466
  esters 466–467
  hydrodefluorination of fluoroarenes 467–468
Ru(II) hydride intermediate 460
Ru-involved hetero-seven-membered cyclic intermediate 461
Ru-mediated C—H bond activation
  C—H bond arylation 456–460
  mechanism 452–456
  ortho-alkenylation of arenes 461–464
  ortho-alkylation of arenes 460
Ru-mediated dehydrogenations
  alcohols 473, 476
  formaldehyde 473, 477
  formic acid 474, 477
Ru-mediated metathesis
  alkyne metathesis 484–485
  intermolecular olefin metathesis 482–484
  intramolecular diene metathesis 484
ruthenabicyclo[3.2.0]heptatriene complex 478
ruthenacyclobutane intermediate 482, 484

Index

ruthenacyclobutenone intermediate 478
ruthenacyclopentatriene intermediate 476
ruthenatriazabicyclo[3.1.0]hexadiene intermediate 479
ruthenium dibenzoate species 460
Ru(II) vinylidene complex 471

S

σ* antibonding orbital of Si—H 53
σ-bonding orbital 53
σ-bond metathesis 223
σ-CAM-type C—H cleavage 454
Schrödinger equation 19, 22
σ-complex-assisted metathesis (σ-CAM) 452, 504
σ covalent bond 86
selectfluor 63
self-consistent IPCM (SCI-PCM) 33
seven-membered Rhoda(III)cycle 342, 369
sextet iron(IV) species 423
silacyclobutanes 204, 205
silacyclopropanation 583
silacyclopenation 583
siliranes 583
silver–carbene formation 568–572
silver catalysis 567
silver–nitrene complex formation 577–579
silver–silylene complexes 583
silyl 99, 157, 203, 204, 273, 397, 400, 401, 404, 432, 547, 548
silylated cyclopropene coordination 273
silyl-benzene 403
silylenes 583
silyl ferrous species 432
silyl Ir(III) complex 401
silyltriflate-coordinated hydride Ir(III) complex 400
single-electron transfer (SET) 289, 290, 297, 313, 451, 518–520, 525
single electron transfer involving pathway 519

single-electron transfer redox process 518
singlet cyclohexadienone coordinated Cu(I) species 532
singlet Fischer-type Ag-nitrene complex 582
singlet Rh-nitrene complex 362
six-membered 1,3-acetoxonium intermediate 608
six-membered amino-Rh(III) species 340
six-membered rhoda(III)cycle 349
six-membered Ru(IV) vinylidene intermediate 479
six-membered vinyl silver intermediate 586
Slater-type orbital (STO) 21, 29, 30
small hindrance bipyridine 403
small to medium-sized cyclic organic compounds 362
S_N2 type substitution transition state 64
soft and carbophilic Lewis acid 597
solvent coordinated cationic Ru(II) species 472
solvent-coordinated hydride Ru(II) species 465
solvent effect 33–34, 39
split-valence basis sets 30, 31
stepwise 1,3-proton transfer 613
Stevens type [3,3]-sigmatropic rearrangement 603
Stille coupling 7, 189–191, 235
Stille–Kosugi–Migita coupling 183
Stille type cross-coupling reaction 235
strain-driven oxidative addition 341–342
styrene 131, 159, 195, 292, 314, 358, 398, 427, 461, 462, 482, 543, 544, 551, 552
styrene-coordinated Ru(II)–carbene complex 482
1,4-substituted triazoles 479
1,5-substituted triazoles 479
substitution-type oxidative addition 62–68, 76

substitution-type reductive elimination 73–76
substrate-coordinated cationic Cp*Rh(III) acetate complex   335
sulfonate-assisted 1,3-proton transfer   606
sulfoxide   129, 194, 197–199, 347, 602
sulfoximide Cu(II) complex   75
Suzuki coupling   7
Suzuki–Miyaura coupling   183–186
Suzuki–Miyaura cross-coupling reaction   72, 183, 184, 186
Suzuki–Miyaura type cross-coupling product   148

t

terminal alkynes   261, 294, 470, 471, 479, 524, 526, 533–537, 539, 585, 587
terminal vinyl Ru(II) intermediate   471
tert-butoxide lithium   201
tetracarboxylate ligand   330
(E)-tetradec-7-ene product   482
tetradentate liganted Rh(III) catalyst   360
tetrafluoropyridine-coordinated Ru(II) fluoride species   468
tetrahydroquinolines   396
thermodynamically stable linear isomer   611
thiolate Rh(I) species   351
thiovinyl hydride Rh(III) species   351
Togni's reagent   67, 68, 525–527
toluenesulfonate   599
total free energy   53, 193
tractional Noyori-type hydrogenation   392
transfer hydrogenation   388, 391, 393, 405
transition metal catalysis   5–13
transition-metal-catalyzed hydroformylation, of olefins   272
transmetallation process   92, 96, 99, 144, 181
  concerted ring type   92–98
  electrophilic substitution   98–100
  metathesis   100–106

alkyne metathesis   106
σ-bond metathesis   100–105
stepwise transmetallation   99, 101
trans to cis isomerization   58
trans-vinyl gold complexes   614
triazole-coordinated Ru(II) complex   479
triazolyl-copper species   534
triazolyl Ru(II) species   479
tricyclohexylphosphine (PCy$_3$)   184
trifluoromethanolate alkyl Pd(IV) complex   78
trifluoromethylations through cross-coupling   524–525
trifluoromethylations through oxidative coupling   524, 526
trifluoromethyl Cu(I) intermediate   524
trifluoromethyl silane   524
trigonal bipyramidal geometry   54
triplet Ag-nitrene complexes   577
triplet copper-nitrene complex   543, 545
triplet dimeric Cu(II) intermediate   539
triplet Mn(V)–oxo complex   500
triplet nitrene   577, 580
triple zeta (TZ)   30
truncated Ir(I)–(PHOX) catalyst   389

u

unordinary Mg-Fe complex   438
α,β-unsaturated carbonyl compounds   292

v

van der Waals density functionals (vdW-DFs)   27
vinyl acetate   462
vinyl compounds   585
vinyl cyclopropane-coordinated Rh(I) species   365
vinyl cyclopropanes   167, 363, 365
vinylcyclopropanes   368
vinyl digold complex   617
vinyl gold intermediate   599
vinylideneamide Mn(I) intermediate   511
vinylidene Pt species   262

vinylideneruthenium intermediate 463
vinylideneruthenium species 463
vinyl Ir(I) complex 401
vinyl Ir(III) complex 401
vinyl Mn(I) intermediate 507
vinyloxysulfonium intermediate 602
vinyl Rh(III)–hydride complex 334
vinyl Ru(II) intermediate 462, 463

w
water-assisted 1,2-proton shift 618
water molecule 149, 225, 303, 423, 511, 574

weak Brøsted base assisted nucleophilic attack 599

z
Z-configuration ethylenes 400
Z-enisoindolinones 536
zwitterionic agostic intermediate 467
zwitterionic gold anion 613
zwitterionic iminium rhodate intermediate 357
zwitterionic intermediate 432, 576